DYNAMICS OF MERCURY POLLUTION ON REGIONAL AND GLOBAL SCALES:
Atmospheric Processes and Human Exposures Around the World

DYNAMICS OF MERCURY POLLUTION ON REGIONAL AND GLOBAL SCALES:
Atmospheric Processes and Human Exposures Around the World

Edited by

Nicola Pirrone
CNR-Institute for Atmospheric Pollution, Rende, Italy

and

Kathryn R. Mahaffey
U. S. Environmental Protection Agency, Washington, DC, USA

 Springer

Library of Congress Control Number: 2005043442

ISBN-10: 0-387-24493-X   e-ISBN-10: 0-387-24494-8
ISBN-13: 978-0-387-24493-8

Printed on acid-free paper.

Printed in the United States of America.

9 8 7 6 5 4 3 2

springer.com

TABLE OF CONTENTS

PART-IV: HUMAN EXPOSURE

PREFACE

This book brings together authors with expertise in a wide range of fields to provide an up to the minute overview of the most important problems relating to mercury in the environment. The book reflects growing concern over the likelihood of harmful effects to human health and sensitive ecosystems posed by mercury in the light of increasing fossil fuel combustion, mercury use in a range of manufactured goods, and the lack of emission control policies. Concern has been expressed at local, national, and international levels; in the last three years both the European Commission and UNEP have published reports on mercury in the environment. Growing concern has led to an increased effort to understand the fate of mercury in the environment, including primary production and trade in mercury, emissions from manufacturing and power generation, natural emissions and re-emission, atmospheric transport and transformation, deposition patterns, uptake by biota, and eventual health impacts on living organisms. The increasing specialisation, and amount of research in the numerous scientific fields associated with the study of the fate of mercury in the environment, make the publication of this book both necessary and timely. For experts, and non-experts, who require both the broader picture, as well as an awareness of the latest progress in the fields relevant to mercury research, this book provides the most comprehensive and up to date overview available.

The book has five sections. The first section covers *'International and Regional Perspectives'*, looking firstly at mercury production, trade and use over the last ten years and future trends. Current and projected industrial emissions of mercury, and their assessment methods are presented. The last part of this section describes the European Union and the United States Environmental Protection Agency perspectives and activities regarding the limitation of risk to human health.

The second section addresses *'Monitoring and Analytical Methods'*, that is the determination of the concentration of mercury and mercury compounds in the atmosphere, and in aquatic and terrestrial ecosystems, including biological samples. The problems involved and precautions required when measuring trace concentrations in complex media are described.

The third section deals with the *'Chemical and Physical Processes'* which determine the partitioning of mercury between environmental interfaces, air-sea, water-sediment, air-soil, soil-plant, canopy-air, and the role these have in the global biogeochemical mercury cycle. Recent advances in the understanding of mercury kinetics in the atmosphere are described.

The fourth section includes six studies on 'Human *Exposure*', reviewing studies reports on different populations from around the world, as well as methods for assessing mercury exposure. The major risks associated with high and extreme cases of mercury exposure are described.

The last section provides seven *'Regional Case Studies'*, for regions where substantial research on the fate of mercury in the environment has been carried out including, the Mediterranean Basin, North and North-Eastern Europe, the Polar regions, the Great Lakes and the North-Eastern U.S., Florida, Asia (particularly China), and the area covered by EMEP, the UNECE-Long Range Transport Convention reference area for Europe.

Nicola Pirrone
CNR-Institute for Atmospheric Pollution, Division of Rende, 87036 Rende, Italy

Kathryn R. Mahaffey
Office of Science Coordination and Policy, United States Environmental Protection Agency, Washington, D.C. 20640, USA

Chapter-1

WHERE WE STAND ON MERCURY POLLUTION AND ITS HEALTH EFFECTS ON REGIONAL AND GLOBAL SCALES

Nicola Pirrone[1] and Kathryn R. Mahaffey[2] *

[1] *CNR-Institute for Atmospheric Pollution, Division of Rende, 87036 Rende, Italy*
[2] *Office of Science Coordination and Policy, United States Environmental Protection Agency, Washington, D.C. 20640,* * see disclaimer below.

INTRODUCTION

It is widely accepted in the scientific community that mercury (Hg) contamination of ecosystems and subsequent human exposure remains a serious environmental hazard. The ability of Hg to distribute globally via the atmosphere has received increasing attention in recent years and has emphasized the need for a global perspective in both research, monitoring and policy making. The aim of this chapter is to provide an overview of our understanding of the mercury pollution problem in relation to both its global cycle and its negative effects on human health.

MERCURY IN THE GLOBAL ENVIRONMENT

It is well known that mercury is released to the global environment from a multitude of natural and anthropogenic sources. Once released to soil, water and atmospheric ecosystems it is re-distributed in the environment through a complex combination of chemical, physical and biological processes that can

---

* *Disclaimer:* The statements in this publication are the professional views and opinions of the author and should not be interpreted to be the policies of the United States Environmental Protection Agency.

act with different time scales. Recent estimates indicate that natural sources (volcanoes, surface waters, soil and vegetation) contribute with 2700 tonnes of mercury released annually to the global atmosphere, whereas the contribution from major industrial sources account for 2250 tonnes per year (Pirrone et al., 1996; Pirrone et al., 2001; Pacyna et al., 2003). Mercury emissions in Europe and North America contribute less than 25% to the global atmospheric emissions, where Asia account for about 40% of global total. The majority of the emissions originate from combustion of fossil fuels, particularly in the Asian countries (i.e., China, India). Combustion of coal is and will remain in the near future as the main source of energy in these countries. The emissions from stationary combustion of fossil fuels (especially coal) and incineration of waste materials accounts for approximately 70% of the total quantified atmospheric emissions from significant anthropogenic sources. As combustion of fossil fuels is increasing in order to meet the growing energy demands of both developing and developed nations, mercury emissions can be expected to increase accordingly in the absence of the deployment of control technologies or the use of alternative energy sources.

Once released to the atmosphere, mercury and its compounds can be transported over long distances before being removed by particle dry deposition and wet scavenging by precipitation (i.e., Pirrone et al., 2000; Pirrone et al., 2003a; 2003b; Hedgecock and Pirrone, 2001; 2004). The temporal and spatial scales of mercury transport in the atmosphere and its transfer to aquatic and terrestrial receptors depends primarily on the chemical and physical forms of mercury which drive their interactions with other atmospheric contaminants and with surface marine waters as well. Gaseous elemental mercury (Hg^0) is relatively inert to chemical reactions with other atmospheric constituents, and is only sparingly soluble in pure water. Therefore, once released to the atmosphere, mercury can be dispersed and transported for long distances over hemispheric and global scales before being deposited to terrestrial and aquatic receptors. The concentration of Hg^0 in ambient air is mainly determined by the background concentration of around 1.5-1.8 ng m^{-3} in the Northern Hemisphere and 0.9 – 1.5 ng m^{-3} in the Southern Hemisphere (see Table I). Oxidised mercury (Hg(II)) and mercury bound to particulate matter (Hg(p)) are typically present in concentrations less than 1 % of the Hg^0 (Table I).

Studies carried out in the last decade have shown that mercury is transported and deposited to very remote locations such as the Arctic as well as the Antarctica (i.e., Schroeder et al., 1998; Ebinghaus et al., 2002; Lindberg et al., 2002; Sprovieri et al., 2002). The mechanism that primarily influence the transfer of mercury from the atmosphere to snow and ice pack is known as "Mercury Depletion Event (MDE)", this event or mechanism takes place (high deposition rate of mercury to the surface) primarily during the first few months of the Polar sunrise. The mercury depletion happens at the same time as the surface-level ozone depletion (a separate phenomenon from the better known ozone depletion in the stratosphere). The net atmospheric input to Polar ecosystems resulting from this phenomena is not known in detail. Re-emissions of mercury occur from the snow surface and during snowmelt, but the depletion events may still result in significant input to the aquatic environment. In case this phenomenon shows up to be resulting in higher yearly mercury deposition rates in the Polar regions than in other regions of the world, this could mean that the Polar regions serve as "mercury cold traps" collecting an un-proportionally high part of the global mercury emissions. This would fit well with the observed high mercury concentrations in the Arctic aquatic environment. Mercury depletion has now been observed in Alert, Canada (Schroeder et al., 1998; Lu et al., 2001), in Barrow, Alaska, USA (Lindberg et al., 2002), Svalbard (Berg et al., 2003; Sprovieri et al., 2005, see also Chapter 28), in Greenland (Skov, 2002) as well as in the Antarctic (Ebinghaus et al., 2002), and can thus be described as a generally occurring polar phenomena which may influence the total input to Polar ecosystems.

Atmospheric deposition to marine waters is primarily driven by particle dry deposition and wet scavenging by precipitation mechanisms. Generally, the relative contribution of wet deposition accounts for about two thirds of the overall mercury budget entering to the marine system compared to particle dry deposition. However, in warm and dry region (i.e., Mediterranean) dry deposition was found to account for nearly 50% of the total flux (Pirrone et al., 2003a). Gas exchange of gaseous mercury between the top water microlayer and the atmosphere is considered the major mechanisms driving gaseous mercury from the seawater to the air (e.g., Pirrone et al., 2001b; Pirrone et al., 2003a).

Once released to marine waters, it undergoes a number of chemical and physical transformations (i.e., Mason et al., 2001). Hg^0 is found in the mixed layer and in deeper waters of the ocean with concentrations generally ranging from 0.01 to 0.5 pM (e.g., Horvat et al., 2003). Gas exchange via Hg reduction and volatilization is the major loss term for marine Hg. Due to the low solubility of Hg^0 in water, almost all the aqueous mercury is present as Hg(II) in the inorganic form and organic methylmercury. Mercury levels in fish constitute a long-standing health hazard and this environmental problem relates predominantly to the conversion of inorganic Hg to neurotoxic monomethylmercury (MMHg) and dimethylmercury (DMHg) (e.g., IARC, 1994).

Table 1. Typical concentrations of mercury species in the Planetary Boundary Layer (PBL).

Species	Concentration	Location	References
Hg^0 ($ng\ m^{-3}$)	0.5 – 1.2	Atlantic air, southern hemisphere	UNEP, 2002
	1.1 – 1.8	Atlantic air, continental background, northern hemisphere	Wängberg et al., 2001 EC, 2001
	0.8 – 2.2	Mediterranean air	Sprovieri et al., 2003
	1.5 – 15	Continental air, urbanized, industrial	Pirrone et al., 2001; 2003a
	0.1 – 1.4	Arctic	Sprovieri et al., 2000
	0.1 – 1.1	Antarctica	Sprovieri et al., 2002
	1.7 – 4.1	United States	Ebinghaus et al., 2002
			Keeler et al., 1995 Landis et al., 2002
Hg(II) ($pg\ m^{-3}$)	< 30	Background air	Sprovieri et al., 2003
	up to 40	marine and continental	Pirrone et al., 2001; 2003a
	5 – > 50	near sources	Wängberg et al., 2003
	up to 200	Antarctica and Arctic	Sprovieri et al., 2002
Hg(p) ($ng\ m^{-3}$)	0.1 – 5	Background air	Sprovieri et al., 2003
	0.1 – 25	Marine (Mediterranean) air	Pirrone et al., 2001; 2003a
	5 - >50	Continental background, higher near sources.	Wängberg et al., 2003
	up to 100	Antarctica and Arctic	Sprovieri et al., 2002
CH_3HgX ($pg\ m^{-3}$)	0.1 – 10	Background air	Lee et al., 2003
$(CH_3)_2Hg$ ($pg\ m^{-3}$)	< 5	Background air	Lee et al., 2003
	-30	Marine polar air	
Hg(II) in precip. ($ng\ L^{-1}$)	1 – 20	Background / marine locations	Wängberg et al., 2001 Keeler et al., 1995

Anthropogenic activities presumably increased the surface water marine Hg concentration by a factor three, an increase which resulted amongst others in elevated Hg concentrations in marine fishes (e.g., Amyot et al., 1997; Horvat et al., 2001). It is currently thought that most of the methylated Hg found in the water column and the biota of the marine waters is generated by *in-situ* production, though the reaction mechanisms are not yet clearly understood (e.g., Mason et al., 2002; Hintelman et al., 1997).

Once entered to terrestrial ecosystems, mercury is accumulated in forest soils (Steinnes et al., 1993) from where it is only slowly transported to surface and deep waters. In aquatic ecosystems, a fraction of the mercury directly deposited and transported from surrounding catchments is transformed into methylmercury compounds which are readily taken up and bioaccumulated in aquatic food-chains.

Industrial discharges of mercury directly to water systems will have the same effect. Accumulation of mercury in forest soils may also lead to adverse effects on soil micro-organisms, which has a potential impact on mineralisation processes (Pirrone et al., 2001 and ref. herein).

Mercury in the Technosphere

Mercury is a natural component of the Earth, with an average abundance of approximately 0.05 $\mu g\ g^{-1}$ in the Earth's crust, with significant local variations. Mercury ores that are mined generally contain about 1% mercury, although the strata mined in Spain typically contain up to 12-14% mercury. While about 25 principal mercury minerals are known, virtually the only deposits that have been harvested for the extraction of mercury are cinnabar. Mercury is also present at very low levels throughout the biosphere. Its absorption by plants may account for the presence of mercury within fossil fuels like coal, oil and gas, since these fuels are conventionally thought to be formed from geologic transformation of organic residues. As described in detail by Maxon (this volume – Chapter 2) the mercury available on the world market is supplied from a number of different sources, including:

- Mine production of primary mercury either as the main product of the mining activity, or as by-product of mining or refining of other metals (such as zinc, gold, silver) or minerals;
- Recovered primary mercury from refining of natural gas (actually a by-product, when marketed, however, is not marketed in all countries);
- Reprocessing or secondary mining of historic mine tailings containing mercury;
- Recycled mercury recovered f rom s pent p roducts a nd w aste f rom industrial production processes. Large amounts (reservoirs) of mercury are "stored" in society within products still in use and "on the users shelves";

- Mercury from government reserve stocks or inventories;
- Private stocks (such as mercury in use in chlor-alkali and other industries), some of which may later be returned to the market.

Since the industrial revolution, due to its unique physico-chemical properties (i.e., high specific gravity, low electrical resistance, constant volume of expansion), mercury has been employed in a wide variety of applications (i.e., manufacturing, dentistry, metallurgy). As a result of its use the amount of mercury mobilised and released into the atmosphere has increased compared to the pre-industrial levels. In the past, a number of organic mercury compounds were used quite widely, for example in pesticides (extensive use in seed dressing among others) and biocides in some paints, pharmaceuticals and cosmetics. While many of these uses have diminished in some parts of the world, organic mercury compounds are still used for several purposes. Some examples are the use of seed dressing with mercury compounds in some countries, use of dimethylmercury in small amounts as a reference standard for some chemical tests, and thimerosal (which contains ethylmercury) used as a preservative in some vaccines and other medical and cosmetic products since the 1930's. As the awareness of mercury's potential adverse effects to health and the environment has been rising, the number of applications (for inorganic and organic mercury) as well as the volume of mercury used have been reduced significantly in many of the industrialised countries, particularly during the last two decades. Therefore as metal, mercury uses (just to cite few applications and uses) are (UNEP, 2003; see also Chapter-2 herein):

- for extraction of gold and silver
- as a catalyst for chlor-alkali production
- in manometers for measuring and controlling pressure
- in thermometers
- in electrical and electronic switches
- in fluorescent lamps
- in dental amalgam fillings

As chemical compounds (among others):

- in batteries (as a dioxide)
- biocides in paper industry, paints and on seed grain
- as antiseptics in pharmaceuticals
- laboratory analyses reactants
- catalysts
- pigments and dyes (may be historical)

- detergents (may be historical)
- explosives (may be historical)

HEALTH EFFECTS

Both humans and wildlife are adversely affected by multiple chemical forms or chemical species of mercury, although specific changes within the organ system predominantly affected differs with the chemical form of mercury. For example, renal or kidney dysfunction accompanies exposure to inorganic mercury, but the nervous system is adversely affected by all three major forms of mercury found in the environment: mercury vapor, inorganic mercury, and methylmercury. It is important, however, to recognize that the specific types of neurological damage produced following mercury exposures differ with the chemical form of mercury. For all three forms the severity of the damage varies with the intensity and duration of exposure (i.e., the dose).

Adverse human health effects range from those detectable only with specialized testing protocols and sophisticated instruments to gross, clinically evident abnormalities, as well as death. It is unclear at this time the extent to which neurological damage produced by concurrent exposure to multiple forms of mercury produces additive or cumulative neurological damage. Concurrent exposures to both mercury vapor, inorganic mercury, and methylmercury have been identified in people living in artisanal mining areas with long-term environmental contamination secondary to mercury in mining wastes. Within these regions bioaccumulation of methylmercury by the aquatic food chain causing elevated methylmercury accumulation among fish-consuming workers and their families who live in these geographic areas has been found.

The effects of mercury on organ systems in addition to the nervous system include the cardiac, immune, and endocrine functions. Although described in the medical literature, these adverse effects have not yet been incorporated into risk assessments used by countries and world public health organizations in setting regulatory standards or policies aimed to protect public health.

Vulnerability to effects of methylmercury in particular depends on age, in addition to dose and duration of exposure. Specifically methylmercury adversely affects the developing fetal brain at far lower exposures than adversely affect the adult's nervous system. This was first observed in Minamata, Japan during the major outbreak in the 1960s where women who themselves were minimally symptomatic gave birth to infants with substantial neurological problems (Harada, 1977). This difference reflects methylmercury's interference with fetal brain development.

Neurological development during fetal life must progress in an exquisitely programmed series of steps that must occur in a timed sequence for normal neurological outcomes. A number of mechanisms through which methylmercury impairs *in utero* development have been identified (Rice and Barone, 2000). It is not entirely clear which of these is the "most critical", but it is clear that there are many opportunities for methylmercury to impair neurological development.

METHYLMERCURY

What makes methylmercury important to wildlife and human health is that it bioaccumulates in the aquatic food chain. Some wildlife are obligate piscivores consuming only fish and shellfish. Examples, include other fish, birds, and mammals. Methylmercury (released from other organomercurials) which had been added to seed grains in the 1950s and 1960s as a preservative resulted in death of birds in Europe and the United States (US EPA, 1997). Methylmercury is now understood not simply to kill birds at high doses and produce overt symptoms at lower doses, but also to prevent reproduction in wild birds including the common loon (Barr, 1986) and common tern (Fimreite, 1974) and cause neurological damage (Henny et al., 2002).

Fish which are generally thought of as a source of methylmercury to piscivores, including humans. However, as additional toxicology information has been obtained in the past decade, fish are no longer simply regarded as a source of methylmercury, but are themselves adversely impacted by methylmercury exposure as shown by reduced growth in walleye (Freidmann et al., 1996) and reduced reproduction of fish spescies including the fathead minnow (Hammerschmidt et al., 2002) through alteration of reproductive endocrinology (Drevnick and Sandheinrich, 2003). Because effects of methylmercury on wildlife reproduction and health are complex and publication of significant key studies occur at a rapidly accelerating pace, no attempt has been made to include these in this volume despite their importance.

Because methylmercury exposure is so closely linked to consumption of fish and shellfish, nutritional considerations are a major issue, particularly in geographic regions with few choices in available food resources (Mahaffey, 2004). Fish and shellfish supply protein, omega-3 fatty acids, vitamins and minerals (IOM/NAS, 2002). Omega-3 fatty acids, in particular, are critical to normal development of the fetal nervous system (IOM, 2002). A complex epidemiological situation is emerging in which the same variable (i.e., fish and shellfish consumption) is associated with both beneficial (e.g., omega-3 fatty acids) and adverse (e.g., methylmercury) effects on neurological development. Although affected by both of these constituents of fish, different domains of neurological function are affected by these chemicals. Recognizing that fish provide important nutrients, actions to control pollution that preserve fish and shellfish resources for both wildlife and people are essential.

Although environmental releases of inorganic mercury and mercury vapor raise great concern for human health and wildlife because these are methylated and bioaccumulate in the aquatic food web, humans also are directly exposed to additional forms of mercury. Multiple uses of mercury in products that may be sold to the general population, such as cosmetics and both regulated and unregulated "medical" remedies, can result in exposures to both inorganic mercury and organo-mercurials. Occupational exposures to mercury vapor and inorganic mercury through industry and mining (particularly Artisanal gold mining) dramatically increase the risk of mercury toxicity for part of the population. Combined with methylmercury exposure the risk of mercury toxicity is further increased.

RISK ASSESSMENTS

Most risk assessments for methylmercury are based on damage to the fetal nervous system as the most sensitive health endpoint (Table 2). Many government regulations and public health decisions rely on these risk assessments. The World Health Organization's assessment in 1990 indicated that there was a 5% risk of damage to fetal neurological development when maternal mercury exposures resulted in maternal hair mercury concentrations exceeding 10 ppm (WHO, 1990). Subsequent to this assessment, a series of epidemiological studies have been carried out using both longitudinal and cross-sectional approaches. Most of these investigations are still active and continue to yield new data. There has been a clear trend in the past decade to adoption of more public health protective standards for methylmercury.

Comparison of risk assessments for methylmercury developed during the past decade emphasizes differences in the accepted margins between exposures that produce recognized adverse effects and those judged to be an accepted level of exposure. These differences, frequently referred to as "uncertainty factors" are intended to protect members of the population by allowing for variability and uncertainty in toxicodynamics and toxicokinetics of methylmercury. Uncertainty factors broadly reflect two areas: variability between individuals and/or groups, and effects or differences that simply are not recognized at the time the assessment is made.

Dealing first with variability described as differences in toxicodynamics and toxicokinetics. Generally person-to-person variability in toxicodynamics is under-described and risk assessments often need to rely on default values which are typically not data-derived for the specific assessment (Reference Dose/Reference Concentration Technical Panel, 2002). Toxicokinetic factors are more frequently data derived (Reference Dose/Reference Concentration Technical Panel, 2002). Typically the data-derived component of toxicokinetic factors substantiates the range of person-to-person variability, or when fetal risk is the health end-point of concern maternal/fetal pair-maternal/fetal pair variability. Occasionally an area of variability may be known qualitatively and only as data are assessed through more advanced statistical procedures can the magnitude of the variability be better described. An example of this is the concentration of methylmercury across the placenta from maternal blood to cord blood.

Uncertainty factors are present to reflect effects that are only partially understood and/or differences that there are not yet data sufficient to provide quantitative estimates of variability. Examples for methylmercury include the possible effects of methylmercury on coronary heart disease (Salonen et al., 1995; Guallar et al., 2002; Yoshizawa et al., 2002), as well as methylmercury's effects on the endocrine and immune systems. Over time, as evidence for the effect of a chemical on an organ system accumulates, such data may change the basis of risk assessments. An example was seen for inorganic lead between the 1970s and the 1980s. During the 1970s almost public health screening programs and risk recommendations for health intervention to protect children against lead poisoning were based on changes in the hematopoietic pathway, specifically increases in free erythrocyte protoporphyrin (Centers for Disease Control, 1978). Free erythrocyte protoporphyrin increased exponentially with increasing blood lead concentration with an apparent threshold effect at a blood lead concentration between about 15 and 18 µg Pb/dL whole blood (Piomelli et al., 1982). This strategy was used in public health screening programs for children at a time when the neurobehavioral effects of lead were thought to occur if blood lead concentrations exceeded 30 µg/dL (Centers for Disease Control, 1978). After approximately the mid-1980s as the effects of early childhood lead exposure on intellectual development associated with blood lead exposures near 10 µg Pb/dL whole blood became clear, risk assessments shifted in two ways. The assessments were based on inorganic lead's impact on intelligence in young children rather than on impaired hematopoiesis. The second change was that rather being concerned about exposures producing blood lead concentrations in the range of 25 µg/dL (associated with hematopoietic changes), exposures producing blood lead concentrations of ~ 10 µg/dL became of concern because of neuro-behavioral effects (United States Centers for Disease Control, 1991).

What will the future holds for risk assessments of methylmercury? Inclusion of cardiac effects and/or adult neurotoxicity as sensitive health endpoints would greatly modify the size of the population of immediate concern. It is also possible that as complex, highly adaptable organ systems - of particular interest, mercury's effects on the immune system and on the endocrine system - are more throughly evaluated, these may respond adversely to methylmercury at exposure levels even lower than those currently of concern as adversely affecting fetal neuro-development.

It is abundantly clear that fish, shellfish, and other constituents of the aquatic food web are extraordinarily important food sources of high quality protein, omega-3 fatty acids, vitamins, and minerals. Maintaining low methylmercury concentrations in food sources that supply these nutrients is needed for the well being of all. Continued contamination of these aquatic food sources with methylmercury will further diminish the food supply of this planet.

GAPS IN OUR UNDERSTANDING

Although our understanding of the global atmospheric cycle including its interfaces with land, water and vegetation has improved greatly in recent decades, we are not yet at a scientific level where we can explain observations of Hg levels in different ecosystems globally or precisely predict the benefit of different scenarios of emission reduction.

In assessing the relative contribution of different patterns/mechanisms affecting the cycle of mercury within and between different ecosystems and its impact on ecosystems and human health, a number of questions, though significant improvement have been made in recent years, still remain to be answered, these questions are briefly reported below.

With reference to the *Retention of Mercury in the Ecosystems*:

- How much of atmospherically-deposited Hg is returned in ecosystems in short-term and in long-term?
- Can we better predict rates of volatilisation of deposited Hg?
- Can we better understand the difference between levels of deposition and re-emission for different ecosystem types?
- Is there any development of watershed budgeting methods for Hg including significant but poorly understood influences such as forest fires?

Table 2. Exposure Limits for Methylmercury.

	Date	Recommended Limits	Critical Effects and Target Group
US FDA	1970s	Acceptable Daily Intake = 0.4 µg/kgbw/day	Paresthesia in adults. 200 µg Hg/L whole blood. 50 ppm Hg in hair.
Health Directorate Canada	1990	0.47 µg/kgbw/day.	General population. Same as US FDA 1970.
World Health Organization	1990	0.48 µg/kgbw/day Maternal hair mercury levels in the 10 ppm - 20 ppm range	Paresthesias in adults. Same as US FDA 1970. 5% risk of neurological deficits in the child following fetal exposure secondary to maternal ingestion of methyl-mercury sufficient to produce maternal hair mercury levels in the 10 ppm - 20 ppm range.
US-EPA	2000	Reference Dose = 0.1 µg/kgbw/day	Maternal/fetal pair. BMDL of 11 ppm in hair. UF of 10. Fetal/cord blood [Hg] of 58 µg/L. Delays and deficits in neuropsychological development and neuromotor function following *in utero* methylmercury exposure.
US Agency for Toxic Substances and Disease Registry	1999	Minimal Risk Level = 0.3 µg/kgbw/day	Maternal/fetal pair. Delays ad deficits in neuropsychological development and neuromotor function following *in utero* methylmercury exposure.
Health Canada - Health Protection Branch	1998	Provisional Tolerable Daily Intake = 0.2 µg/kgbw/day	Maternal/fetal pair. BMDL of 11 ppm in hair. UF of 5.
Kommission "Human-Biomonitoring" des Umwelbundesa mtes (Germany)	1999	Recommended limit values for inorganic and organic mercury for general populations, occupationally exposed groups, and sensitive subpopulations. HBM I of 5 µg/L for organic mercury among women of reproductive age: corresponds to maternal hair mercury concentration of 1.5 µg/g using a 1:300 conversion	Fetal nervous system. HMBI of 5 ug/L in blood or hair of 1.5 ppm mercury. HMB II of 15 µg Hg/L blood or ~ 4 to 5 ppm Hg in hair. HMB I: Women whose blood mercury exceeds these levels are advised to restrict fish consumption and/or restrict the use of methylmercury-containing pharmaceuticals. HMB II: Additional interventions recommended.
Joint Expert Committee on Food Additives	2003	1.6 µg/kgbw/day Provisional Tolerable Weekly Intake (PTWI). JECFA Committee utilized a mean maternal hair:blood ratio of 250 and a factor of 2 for likely inter-individual variability. For inter-individual pharmacokinetic variability, a UF of 3.2 was used in converting maternal blood concentration to a steady-state dietary intake.	PTWI considered sufficient to protect the developing fetus. Committee calculated a composite hair mercury from Faroes and Seychelles of 14 mg/kg maternal hair to be without appreciable adverse effects in the offspring. Total UF of 6.4 (2 x 3.2).

With reference to the *Ecosystem Sensitivity*:

- How can we predict/understand the wide variability among lakes/rivers in biotic Hg concentrations?
- Have the effects of watershed manipulation (i.e., fishery, agriculture) on fish Hg levels been adequately understood?

With reference to the *Ecosystem Toxicity*:

- What are the key receptors?
- What environmentally concentrations are of key importance to be monitored?
- What impact do elevated fish Hg concentrations have on fish and on their predators?
- What appropriate Hg threshold values to protect soil micro-biota under different ecological conditions?

With reference to the *Ecosystem Response Time*:

- How much time is needed for environmental concentrations to respond to changes in atmospheric Hg depositions?

With reference to the *Human Health*:

- What are the toxic effects of different levels and combined species of inorganic and organic Hg?
- Are there known mixture effects of mercury exposure and exposure to other neurotoxicants commonly found in fish and shell fish?
- What are the long-term effects of low dose exposure at critical life stages in addition to the recognised neurotoxic effects of mercury and methylmercury exposure?

To answer these questions, there is a need to fill existing gaps in our understanding of different chemical and physical mechanisms involved in the dynamics of mercury within and between atmospheric, marine and terrestrial ecosystems. The following may represent the most significant questions in relation to atmospheric and marine processes:

- what are the variations in the regional and global mercury cycle between atmospheric, marine and terrestrial ecosystems over time that can occur with changes in emissions of mercury and other atmospheric contaminants (e.g., NO_x, SO_2) as well as with climate change. The effects driven by climate change on the global mercury cycle has not received a great attention, though on short- and long-term, it is believed to represent the major driving mechanism that may influence the re-distribution of mercury on global and regional scales. The effects of climate change can be classified as primary and secondary effects. Primary effects account for an increase in air and sea temperatures, wind speeds and variation in precipitation patterns, whereas secondary effects are related to an increase in O_3 concentration and aerosol loading, to a decrease of sea ice cover in the Arctic and changes in plant growth regimes. All these primary and secondary effects may act with difference time scales and influence the atmospheric residence time of mercury and ultimately its dynamics from local to regional and global scale;

- recent research suggests that through consideration of the r ole o f halogen and OH radical chemistry involving Hg compounds in the marine boundary layer (MBL) better deposition estimates of Hg (and its compounds) could be obtained;

- gaseous Hg exchange at the air-water interface is primarily driven by chemistry in the lower layer of the atmosphere, chemical and biological processes in the marine system and water wave dynamics; the combination of these three mechanisms and their relative magnitude are still unclear;

- in order to develop global assessment models for mercury, there is a need to promote a global mercury monitoring network aimed to assess long-term changes in mercury concentrations in the atmosphere, marine and freshwater reservoirs with reference to primary ecological and public health indicators;

- although stocks of different Hg compounds in the marine system are relatively well quantified, translocations of Hg from one compartment to the other remain largely unknown. In addition the role of sediments and micro-organisms in the biogeochemical cycling of Hg is not yet completely understood;

- qualitative as well as quantitative information about complexing ligands for Hg that act as carriers from one compartment to another (water to plankton, plankton to higher trophic level) as well as from one ecosystem to another is scarce and requires a further investigation;

- bacterially mediated production of organomercury compounds is recognised as an important control function of the Hg introduction in the food chain. Preliminary studies have also shown that demethylation may also occur in seawater (due to photodegradation) or in sediments (due to bacterial activity), simultaneously with the methylation process. A better understanding of these factors/mechanisms will certainly help improve our capabilities in modeling the fate of mercury in the marine system.

INTERNATIONAL ACTIONS ON MERCURY

Past a nd o n-going i nitiatives a imed t o r educe t he i mpact o f m ercury pollution on the environment and human health, including waste management practices, have been taking place at national and international levels (see UNEP, 2003; EC, 2003). Detailed information on regional and global agreements, instruments, organisations and programmes tackling aspects of the mercury problem is provided in detail elsewhere (UNEP, 2003; EC, 2004; Chapters 4, 5, 6 of this volume), therefore, only a brief overview of the main initiatives is given here.

- *The 1998 Protocol on Heavy Metals under the UNECE Convention on Long-Range Transboundary Air Pollution (LRTAP).* Provisions of the protocol require parties to reduce total annual emissions of mercury into the atmosphere, secure application of the best available techniques for stationary sources, and consider applying additional product controls. The protocol entered into force on 29 December 2003.
- *The OSPAR Convention for the Protection of the Marine Environment of the North-East Atlantic.* The Convention's objective of preventing and eliminating pollution is reflected in a strategy on hazardous substances, agreed in 1998. This has the ultimate aim of achieving concentrations in the marine environment near background values for naturally occurring substances (such as mercury) and close to zero for man-made synthetic substances, with every endeavour to be made to move towards the target of cessation of discharges, emissions and losses of hazardous substances by 2020.
- *The Helsinki Convention on the Protection of the Marine Environment of the Baltic Sea Area.* The Convention aims to prevent and eliminate pollution in order to promote the ecological restoration of the Baltic Sea Area and the preservation of its ecological balance. Its objective is to prevent pollution by continuously reducing discharges, emissions and losses of hazardous substances towards the target of their cessation by 2020. The ultimate aim is to achieve concentrations in the environment near background values for naturally occurring substances and close to zero for man-made synthetic substances.
- *The UNEP Mediterranean Action Plan (MAP).* MAP is an effort involving 21 countries bordering the Mediterranean Sea, as well as the

EU. There are three protocols which control pollution to the sea, including the input of hazardous substances.

- *The Basel Convention on the Control of Transboundary Movements of Hazardous Wastes and their Disposal.* The Convention strictly regulates the transboundary movements of hazardous wastes and establishes obligations for parties to ensure such wastes are managed and disposed of in an environmentally sound manner. Any waste containing or contaminated by mercury or its compounds is considered hazardous waste and is covered by the provisions of the Convention. Hazardous wastes may not be exported from the EU or OECD for disposal, recovery or recycling in other countries.

- *The Rotterdam Convention on the Prior Informed Consent Procedure for Certain Hazardous Chemicals and Pesticides in International Trade.* The Convention establishes the principle that export of specified chemicals and pesticides can only take place with the prior informed consent of the importing party. At present, mercury compounds used as pesticides are covered by the PIC procedure, but mercury and its compounds intended for industrial use are not.

- *The Arctic Council Action Plan to Eliminate Pollution of the Arctic (ACAP).* The Arctic Council is a high-level intergovernmental forum that provides a mechanism to address the common concerns and challenges faced by the Arctic governments and peoples. Planned activities include identification and quantification of major point sources, with the aim of implementing concrete emission reduction pilot projects.

- *The Nordic Environmental Action Programme 2001-2004.* This programme establishes environmental priorities within the framework of Nordic cooperation in the fields of nature and the environment. It follows up on commitments in a Nordic sustainable development strategy, which has as one of its objectives the discontinuation within 25 years of discharges of chemicals posing a threat to health and the environment.

- *International action relating to artisanal gold mining.* A number of international bodies have worked on this issue, including the International Labour Organisation, the World Bank, and the United Nations Industrial Development Organisation.

- *UNEP Mercury Programme.* As widely referred to in this paper considerable work has been undertaken under the auspices of UNEP Chemicals in the context of the Global Mercury Assessment.

The brief overview reported above, shows that a considerable range of measures have been implemented at the national and regional levels to deal with mercury and mercury compounds. Through such measures, a number of countries have achieved substantial reductions in emissions and releases of mercury from products and industrial processes. In addition, a number of coordinated regional approaches, both binding and non-binding have supported national measures and contributed to additional reductions beyond national borders.

Despite these successful national and regional initiatives (see UNEP, 2003 for details), some countries consider that they might not be sufficient to ensure adequate protection of human health and the environment from the adverse effects of mercury, and are calling for the consideration of coordinated initiatives at the international level. If it is found that there are global problems related to mercury that should be addressed, it might be essential to the effectiveness of any reduction measures for the substantive commitments to be discussed and agreed at the international level. Any specific regional or national considerations may be addressed taking into account common but differentiated responsibilities within the commitments agreed to.

REFERENCES

Agency for Toxic Substances and Disease Registry (US). Toxicological profile for mercury. Atlanta: Centers for Disease Control and Prevention. (http://www.atsdr.cdc. gov/ toxprofiles/ tp46.html)

Amyot, M., Gill, G.A., Morel, F.M.M. Production and loss of dissolved gaseous mercury in coastal sea water. *Env. Sci. Technol.,* 31, 3606 – 3611, 1997.

Barr, J.F. Population dynamics of the common loon (*Garvia immer*) associated with mercury-contaminated waters in north-wester Ontario. Occasional Paper 56. Canadian Wildlife Service, Ottawa, Ontario. Canada.

Berg, T., Sekkesæter, S., Steinnes, E., Valdal, A-K., Wibetoe, G. Arctic springtime depletion of mercury in the European Arctic as observed at Svalbard. . *Sci. Tot. Env.*, 304, 43-51, 2003.

Drevnick, P.E., Sanheinrich M.R. Effects of dietary methylmercury on reproductive endocrinology of fathead minnows. *Env. Sci Technol.*; 37, 4390-96, 2003.

Ebinghaus, R., Kock, H.H, Temme, C., Einax, J.W., Lowe, A.G., Richter, A., Burrows, J.P., Schroeder, W.H. Antarctic springtime depletion of atmospheric mercury. *Env. Sci. Technol.,* 36, 1238-1244, 2002.

European Commission (EC) Consultation Document on Development of an EU Mercury Strategy, EC-DG Environment (ENV.G., ENV.G.2), draft released on 15 March 2004.

Fimreite, N. Mercury contamination of aquatic birds in northwestern Ontario. *J. Wildlife Management*, 38, 120-131, 1974.

Food and Agriculture Organization of the United Nations. World Health Organization. Joint FAO/WHO Expert Committee on Food Additives. Summary and Conclusions of the Sixth-First meeting. 10-19 June 2003. <ftp://ftp.fao.org/es/esn/jecfa/ jecfa/jecfa61sc.pdf>

Friedmann, A.S., Watzin, M.C., Brinck-Johnsen, T., Leiter, J.C. Low levels of dietary methylmercury inhibit growth and gonadal development in juvenile walleye (*Stizostedion vitreum*). *Aquatic Toxicology*, 35, 265-278, 1996.

Guallar, E., Sanz-Gallardo, M.I., van't Veer, P., Bode, P., Aro, A., Gomez-Aracena, J., Kark , J.D., Riemersma, R.A., Matin-Moreno, J.M., Kok, F.J. Heavy Metals and Myocardial Infarction Study Group. Mercury fish soils, and the risk of myocardial infarction. *New England J. Medicine*, 347, 1747-54, 2002.

Hammerschmidt, C.R., Sandheinrich, M.B., Wiener J.G., Rada R.G. Effects of dietary methylmercury on reproduction of fathead minnows. *Env. Sci Technol.* 36, 877-83, 2002.

Harada Y. Congenital Minamata Disease. In: Tsubaki T., Irukayama K. Editors. Minamata Disease. Elsevier, Amsterdam, 209-39, 1977.

Health Protection Branch, Bureau of Chemical Safety (Canada). Review of the Tolerable Daily Intake for Methylmercury. Ottawa: Health Canada: 1998 Apr 27.

Hedgecock, I., Pirrone, N., Sprovieri, F., Pesenti, E. Reactive Gaseous Mercury in the Marine Boundary Layer: Modeling and Experimental Evidence of its Formation in the Mediterranean. *Atmos. Environ.*, 37, S1, 41-50, 2003.

Hedgecock, I.M., Pirrone, N. Chasing Quicksilver: Modeling the Atmospheric Lifetime of $Hg^0{}_{(g)}$ in the Marine Boundary Layer at Various Latitudes. *Env. Sci. Technol.*, 38, 69-76. 2004.

Henny C.J., Hill, E.F., Hoffman, D.J., Spalding, M.G., Grove, R.A. Nineteenth century mercury: hazard to wading birds and cormorants of the Carson River, Nevada. *Ecotoxicology*, 11, 213-31, 2002.

Hintelmann, H.; Falter, R.; Ilgen, G., Evans, R.D. Determination of artifactual formation of monomethylmercury in environmental samples using stable Hg(II) isotopes with ICP/MS detection: calculation of contents applying species specific isotope addition. *Fresenius J. Anal. Chem., 358*, 363-370, 1997.

Horvat, M., Kotnik, J., Fajon, V., Logar, M., Zvonaric, T., Pirrone, N. Speciation of Mercury in Waters of the Mediterranean Sea. In: *Mat. Geoenv.*, Hines, M., Horvat, M., Faganeli, J. (Editors). Proceedings of the *International Workshop on Mercury in the Northern Adriatic Sea*, May 13-15, 2001, Portoroz, Slovenia, Vol. 48, 241-252, 2001.

Horvat, M., Kotnik, J., Fajon, V., Logar, M., Zvonaric, T., Pirrone, N. Speciation of Mercury in Surface and Deep Seawater in the Mediterranean Sea. *Atmos. Environ.*, 37, S1, 93-108, 2003.

IARC Evolution of Carcinogenic Risk to Humans. Vol. 58. Mercury and Mercury Compounds. Lyon, France, 1994.

IOM (Institute of Medicine). National Academy of Sciences. Dietary fats, total fats, and fatty acids. In: dietary Reference Intakes for Energy, Carbohydrates, Fiber, Fatty Acids, Cholesterol, Protein, and Amino Acids. Part 1. Summary and Chapters 1 through 9. Food and Nutrition Board. Panel on Macronutrients. National Academy Press. Washington, D.C. pp. 8-1-9-97 (prepublication copy; unedited proofs) (Chapter 8), 2002.

Keeler, G.J., Glinsorn, G., Pirrone, N. Particulate Mercury in the Atmosphere: Its Significance, Transport, Transformation and Sources. *Water, Air Soil Pollut.*, 80, 159-168, 1995.

Kommission "Human-Biomonitoring" des Umweltbundesamtes. Stoffmonographie Quecksilver-Referenz-und-Human-Biomonitoring-Werte (HBM). Berlin. Komission "Human-Biomonitoring" des Umweltbundesamtes, 1999.

Landis, M.S., Stevens, R.K., Shaedlich, F., Prestbo, D.E.M. Development and characterization of an annular denuder methodology for the measurement of divalent inorganic reactive gaseous mercury in ambient air. *Env. Sci. Technol.*, 36, 3000-3009, 2002.

Lee, Y.H., Wängberg, I., Munthe, J. Sampling and analysis of gas-phase methylmercury in ambient air. *Sci. Tot. Env.* 304,107-113, 2003.

Lindberg, S.E., Brooks, S., Lin, C.J., Scott, K.J., Landis, M.S., Stevens, R. K., Goodsite, M., Richter, A. Dynamic oxidation of gaseous mercury in the Arctic troposphere at polar sunrise. *Env. Sci. Technol.*, 36, 1245-1256, 2002b.

Lu, J. Y., Schroeder, W. H., Barrie, L. A., Steffen, A., Welch, H. E., Martin, K., Lockhart, L., Hunt, R. V., Boila, G., R ichter, A ., M agnification o f a tmospheric m ercury deposition to polar regions in springtime: the link to tropospheric ozone depletion chemistry. *Geophy. Res. Lett.* 28, 3219-3222, 2001.

Mahaffey, KR. Fish and shellfish as dietary sources of methylmercury and the ω-3 fatty acids, eicosahexaenoic acid and docosahexaenoic acid: risks and benefits. *Environ. Res.* 95, 414-428, 2004.

Mason, R.P., Lawson, N.M., Sheu, G.R. Mercury in the Atlantic Ocean: factors controlling air-sea exchange of mercury and its distribution in the upper waters. *Deep-Sea Research II*, 48, 2829-2853, 2001.

Mason, R.P.; Sheu, G.-R. *Global Biogeochem. Cycles*, 16, 1093, doi:10.10229/2001 GB001440, 2002.

Munthe, J., Wangberg, I., Pirrone, N., Iverfeld, A., Ferrara, R., Ebinghaus, R., Feng., R., Gerdfelt, K., Keeler, G.J., Lanzillotta, E., Lindberg, S.E., Lu, J., Mamane, Y., Prestbo, E., Schmolke, S., Schroder, W.H., Sommar, J., Sprovieri, F., Stevens, R.K., Stratton, W., Tuncel, G., Urba, A. Intercomparison of Methods for Sampling and Analysis of Atmospheric Mercury Species. *Atmos. Environ.*, 35, 3007-3017, 2001.

Piomelli S, Seaman, C, Zullow, D, Curran, A, Davidow, B. Threshold for lead damage to heme synthesis in urban children. Proc Natl Acad Sci USA. 79, 3335-39, 1982.

Pirrone, N., Keeler, G.J., Nriagu, J.O. Regional Differences in Worldwide Emissions of Mercury to the Atmosphere. *Atmos. Env.*, 30, 2981-2987, 1996.

Pirrone, N., Hedgecock, I., Forlano, L. The Role of the Ambient Aerosol in the Atmospheric Processing of Semi-Volatile Contaminants: A Parameterised Numerical Model (GASPAR). *J. Geophys. Res.*, 105, D8, 9773-9790, 2000.

Pirrone, N., Munthe, J., Barregård, L., Ehrlich, H.C., Petersen, G., Fernandez, R., Hansen, J.C., Grandjean, P., Horvat, M., Steinnes, E., Ahrens, R., Pacyna, J.M., Borowiak, A., Boffetta, P., Wichmann-Fiebig, M. EU Ambient Air Pollution by Mercury (Hg) - Position Paper. Office for Official Publications of the European Communities, 2001.

Pirrone, N., Costa, P., Pacyna, J.M., Ferrara, R. Atmospheric Mercury Emissions from Anthropogenic and Natural Sources in the Mediterranean Region. *Atmos. Environ.*, 35, 2997-3006, 2001a.

Pirrone, N., Pacyna, J.M., Barth, H. Atmospheric Mercury Research in Europe. Special Issue of Atmospheric Environment, volume 35 (17), Elsevier Science, Amsterdam, Netherlands, 2001b.

Pirrone, N., Ferrara, R., Hedgecock, I.M., Kallos. G., Mamane, Y., Munthe, J., Pacyna, J.M., Pytharoulis, I., Sprovieri, F., Voudouri, A., Wangberg, I. Dynamic Processes of Atmospheric Mercury Over the Mediterranean Region. *Atmos. Environ.*, 37, S1, 21-40, 2003a.

Pirrone, N., Pacyna, J.M., Munthe, J., Barth, H. Dynamic Processes of Mercury and Other Atmospheric Contaminants in the Marine Boundary Layer of European Seas. Special Issue of Atmospheric Environment, volume 37 (S1), Elsevier Science, Amsterdam, Netherlands, 2003b.

Reference Dose/Reference Concentration (RfD/RfC) Technical Panel. Risk Assessment Forum. A Review of the Reference Dose and Reference Concentration Processes. EPA/630/P-02/002F. December 2002. Final Report. http://cfpub2.epa.gov/ncea/raf/

Rice D, Barone, S. Jr. Critical periods of vulnerability for the developing nervous system: evidence from humans and animal models. *Environ. Health Perspect*, 108 (Suppl 3), 511-33, 2000.

Salonen, J.T., Seppanen, K., Nyyssonen, K., Korpela, H., Kauhanen, J., Kantol, M., Tuomilehto, J., Esterbauer H., Tatzber F., Salonen, R. Intake of mercury from fish, lipid peroxidation, and the risk of myocardial infarction and coronary, cardiovascular, and any death in eastern Finnish men. *Circulation*, 91, 645-55, 1995.

Schroeder, W. H., Anlauf, K.G., Barrie, L.A., Lu, J.Y., Steffen, A. Arctic springtime depletion of mercury. *Nature* 394, 331-332, 1998.

Skov, H. Personal communication, NERI, Denmark, 2002.

Sprovieri, F., Pirrone, N. A Preliminary Assessment of Mercury Levels in the Antarctic and Arctic Troposphere. *J. Aerosol. Sci.*, 31, 757-758, 2000.

Sprovieri, F., Pirrone, N., Hedgecock, I. M., Landis, M., Stevens, B, K. Intensive atmospheric mercury measurements at Terra Nova Bay in Antarctica during November and December 2000. *J. Geophys. Res.* 107, D23, 4722-4729, 2002.

Sprovieri, F., Pirrone, N., Gardfeldt, K., Sommar, J. Atmospheric Mercury Speciation in the Marine Boundary Layer along 6000 km Cruise path over the Mediterranean Sea. *Atmos. Environ.*, 37/S1, 63-72, 2003.

Sprovieri, F., Pirrone, N., Landis, M., Stevens, B, K. Mercury depletion events in the Arctic during the Intensive Spring 2003 campaign. *Environ. Sci. Technol.* (submitted), 2005.

Steinnes, E., Andersson, Jakobsen. E.M. Atmospheric deposition of mercury in Norway. In Allan and Nriagu (eds.) Proceedings of the International Conference of Heavy Metals in the Environment, Toronto, September 1993, 70-73, 1993.

UNEP Global Mercury Assessment (GMS) report. United Nations Environment Programme,

Geneva, Switzerland, 2002.

United States Centers for Disease Control Preventing lead poisoning in young children. J Pediatr, 93, 709-720, 1978.

United States Centers for Disease Control. Preventing Lead Poisoning in Young Children: A Statement by the Centers for Disease Control. Report No. 99-2230. Atlanta, GA: CDC U.S. Department of Health and Human Services, 1991.

United States Environmental Protection Agency. Report to Congress on Mercury. Volume VI. An Ecological Assessment for Anthropogenic Mercury Emissions in the United States. Pages 2.28-2.34. EPA-452/R-97-008. Washington D.C. USA.

Wängberg, I., Munthe, J., Pirrone, N., Iverfeldt, Å., Bahlman, E., Costa, P., Ebinghaus, R., Feng, X., Ferrara, R., Gårdfeldt, K., Kock, H., Lanzillotta, E., Mamane, Y., Mas, F., Melamed, E., Osnat, Y., Prestbo E., Sommar, J., Schmolke, S., Spain, G., Sprovieri, F., Tuncel, G. Atmospheric Mercury Distributions in North Europe and in the Mediterranean Region. *Atmos. Environ.*, 35, 3019-3025, 2001.

World Health Organization. Environmental health criteria. 101: methylmercury. Geneva (WHO): 1990.

Yoshizawa, K., Rimm, E.B., Morris, J.S., Spate, V.L., Hsieh, C.C., Spiegelman, D., Stampfer, M.J., Willett, W.C. Mercury and the risk of coronary heart disease in men. *New England J. Medicine*, 347, 1755-60, 2002.

PART-I:
INTERNATIONAL AND REGIONAL PERSPECTIVES

Chapter-2

GLOBAL MERCURY PRODUCTION, USE AND TRADE

Peter A. Maxson

Concorde East/West Sprl, 10 Av. René Gobert, 1180 Brussels, Belgium

INTRODUCTION

The intent of this chapter is to provide an overview of the global economics of mercury - supply and demand, mercury markets and pricing, and in more detail, global trade in mercury, with an occasional focus on the particularities of the European Union and the US. The chapter is based extensively on recent work of the author for the European Commission (Maxson, 2004), and includes a number of updates of data and other relevant information built upon that foundation. Where not otherwise noted, that reference may be consulted for further detail.

The connection between the economic perspective offered in this chapter, and the rest of the book, lies in three main areas:

1. The economics of mercury need to be a necessary component to any policy-making exercise, and especially to any international initiative or strategy.
2. This analysis contributes a fresh perspective to setting priorities when some groups – and rightly so - are discussing reducing human exposures by micrograms, while the final destination of hundreds of tonnes of traded mercury annually is not known with any certainty.
3. An eventual linking of the physical movements of traded mercury, as described here, with emission and transport models would be highly valuable both to scientists and to policy-makers.

If this chapter succeeds in demonstrating the value of an economic perspective to the international policy debate, the result will be sufficient reward for the author's efforts.

MERCURY SUPPLY

The global supply of mercury comes mainly from prime mined (virgin) mercury, secondary mercury recovered as a by-product of mining other ores, secondary mercury coming from recycling or waste processing, residual mercury recovered from decommissioned chlor-alkali facilities, and other mercury occasionally released from government or industry inventories or stockpiles.

The supply of prime mercury to the world market is dominated by three key nations that mine mercury for export (Spain, Kyrgyzstan and Algeria), and China, which has long supplied its domestic market, and supplemented that supply with substantial imports of mercury as well during the last 10-15 years. China has made some noise recently about closing its larger mercury mines, apparently in response to increased international scrutiny, although there is no sign as yet that these operations have stopped. Spain stopped its underground mercury mining operations in 2002, but continues to produce mercury from a massive stockpile of previously mined cinnabar, and "reserves the right" to resume mining as necessary to meet its customers' ongoing demands for some 1000 tonnes of elemental mercury per year. World production of prime mercury in 2003 amounted to over 1600 tonnes, of which 745 tonnes were produced by Spain (MAYASA, 2004), 234 tonnes by Algeria [1], 397 tonnes by Kyrgyzstan [2], and probably well over 200 tonnes by China [3].

The production of secondary "by-product" mercury, i.e., mercury that is recovered from mining or processing activities where the primary mineral is gold, silver, copper, zinc, etc., amounted to at least 1000 tonnes in 2003 [4].

Recycled and recovered mercury from products and wastes are quite difficult to estimate worldwide. However, based on information from the major recyclers in Europe and the US, total recycled/recovered mercury (not including mercury recovered from decommissioned chlor-alkali facilities) was probably in the range of 600 tonnes in 2003 [5].

Residual mercury recovered from decommissioned chlor-alkali facilities amounted to 227 tonnes in Europe in 2003 [6]; details from other parts of the world were unavailable. Considering that most of the European mercury cell chlor-alkali plants will be closed or converted by 2020, in line with an industry commitment [7], and that these plants hold some 12,000 tonnes of commodity-grade mercury [8,9], 2003 was a below-average year for this source of mercury in the EU.

On the other hand, the members of the industry association Euro Chlor and the Spanish mercury mining and trading company, MAYASA, have signed an agreement whereby the chlor-alkali industry has agreed to sell its residual mercury to "an established mercury producer so as to displace new production of the equivalent quantity of virgin mercury." Thus, even during years when quantities of residual mercury from chlor-alkali plants are greater than they were in 2003, total world production of mercury would, in principle, not be affected by such an increase. While the Euro Chlor/MAYASA agreement clearly represents an effort by the signatories to responsibly address the potential problem of surplus mercury on the world market, some hold the view that there are not yet adequate controls on where the Euro Chlor mercury might eventually be sold or how it might be used.

Due to a range of factors - government subsidies to mercury production by mines, sales of mercury holdings from the US and the USSR governments, significant sales of mercury inventories from closing chlor-alkali plants, significant and increasing recovery of secondary mercury as a by-product of other mining operations, increasing mercury recycling activities, and greatly shrunken demand for mercury within industrialized economies - there has been an abundance of mercury available to the market during recent decades, forcing mercury prices down and holding them at a historically low level. This has led to the closure of all private mines, and curtailment of production at any but the lowest-cost remaining government sponsored (or owned) mines. According to Hylander (2003) and Maxson (2004), the global commodity mercury supply from 1991-2000 has averaged about 4000 tonnes per year, and based on the various sources described above, in 2003 is estimated in the vicinity of 3600 tonnes.

MERCURY DEMAND

Demand for mercury has long been widespread, although the global mercury commodity market is small in both tonnage and value of sales.

Even though mercury may routinely be traded several times before final "consumption," the available statistics demonstrate that yearly trades of

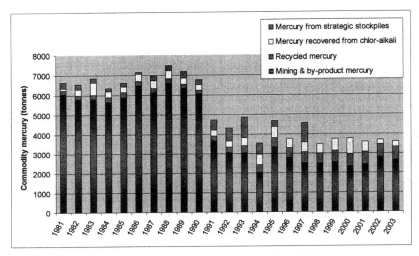

Figure 1. Global Mercury Supply 1981-2003.

Sources: Hylander (2003), Maxson (2004) and author calculations for 1998-2003 based on data concerning MAYASA production, chlor-alkali residual Hg, secondary sources, etc.

mercury and its compounds probably do not exceed €20 million, or $US 25 million. Most transactions are among private parties and not publicly reported. While continuing its long-term decline in most of the OECD countries, there is evidence that demand for mercury remains relatively robust in many developing economies, yet there is little public data pertaining to its end use in many nations.

Mercury is consumed in a broad range of products and processes around the world. The major categories of mercury demand in OECD countries include:

- chlor-alkali production
- dental amalgams
- fever and other thermometers
- other measuring and control equipment
- mercuric oxide and other batteries
- neon, compact fluorescent, HID and other energy-efficient lamps
- electrical switches, contacts and relays
- laboratory and educational uses
- other industrial processes requiring catalysts, etc.

- pharmaceutical processes, products and preservatives
- other product uses, such as cosmetics, fungicides, toys, etc.

Additional categories of mercury demand more prevalent in, but not exclusive to, less developed countries include:

- artisanal gold mining
- cosmetics
- cultural uses and traditional medicine
- paints and pesticides/agricultural chemicals.

While the last 15-20 years have shown a significant reduction of mercury use in the OECD countries, mercury demand in many developing countries, especially South and East Asia (in the case of mercury use in products and artisanal gold mining), and Central and South America (in the case of artisanal gold mining) has been robust. The main factors behind the decrease in mercury demand in the OECD are the substantial reduction or substitution of mercury content in regulated products and processes (paints, batteries, pesticides, chlor-alkali, etc.), and a general shift of mercury product

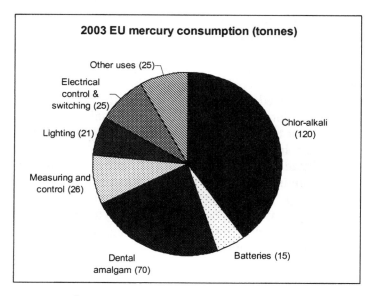

Figure 2. 2003 EU mercury consumption (tonnes).

Sources: Maxson (2004) Nordic Council (2004).

manufacturing operations from OECD countries to third countries (thermometers, batteries, etc.).

A breakdown of mercury demand among different categories of use within the EU is presented below. Other than the chlor-alkali industry, where the mercury cell process is more widely used than in other parts of the OECD, this distribution of mercury uses is reasonably representative of the rest of the OECD.

With regard to the largest category shown, mercury demand by chlor-alkali plants in the EU-15 region is estimated at 120 tonnes for 2003 [10]. It should also be noted that the chlor-alkali plants located in the 10 newest EU Member States may consume another 100 tonnes of mercury or more [11], although formal reporting of these data is not yet routine for all of these plants. While the addition of the 10 new Member States to the EU will change this graph to show a higher mercury demand in all categories – although no increase in demand will be as substantial as that for the chlor-alkali industry – in general mercury use in all categories may be expected to decrease gradually, with the exception of mercury in dental amalgams and lighting.

The characteristics of mercury demand for the EU should be compared with the breakdown for global demand below. While global chlor-alkali demand for mercury remains large, it is probably still exceeded by mercury demand in batteries and small-scale gold mining, which are far less significant in the EU.

With regard to specific mercury demand categories, on average, the non-OECD chlor-alkali facilities consume considerably more mercury than those in the OECD, as suggested in the table below for the year 2000. Reported chlor-alkali consumption of mercury by individual plants may vary significantly from year to year, but the total global consumption for 2003 is roughly the same as below.

Recent detailed studies of mercury demand around the world for artisanal (small-scale) gold mining give an estimate of 800-1000 tonnes of mercury per year (Veiga, 2004). This demand may be drastically reduced in various ways, but the barriers are formidable with regard to the education and awareness-raising of frequently transient miners engaged in an activity that is formally illegal in many countries.

Table 1. Global mercury cell chlorine production capacity and mercury consumption (2000).

Region	Mercury cell chlorine production capacity ('000 tonnes)	Mercury consumption (tonnes)
Western Europe	6,592	95
United States	1,409	72
Rest of world	4,200	630
Totals	12,201	797

Sources: Euro Chlor (2001a, 2001b, 2002a); Chlorine Institute (2002); UNEP Chemicals (2002a).

Previous estimates (Maxson, 2004) based on trade data, of global mercury demand for batteries as high as 1000 tonnes have found support in information gathered recently in China, presented in further detail in Chapter-27. In general, however, this use for mercury is believed to be declining worldwide.

Finally, a demand category not often mentioned is acetylene-based production of vinyl chloride monomer (VCM, a raw material for PVC production) via mercury catalysts. This is based on preliminary information from Russia (ACAP, 2004) and China (Feng, 2004) and the China Plastics Industry Yearbook (2002-2003) statistic of 58 Chinese factories using the acetylene process in 2002, with a capacity of 2.42 million tonnes of PVC per annum. Based on a range of factors that need to be further analyzed, these Chinese plants could be consuming up to 300 tonnes of mercury per year, not to mention some additional plants in Russia and India using a similar process, etc.

For purposes of this chapter, and pending further information and analysis, the author's highly uncertain but surely conservative working estimate of mercury consumption by all of these plants is at least 150 tonnes.

At the same time, however, it must be stressed that until further information is received these facilities (similar to those implicated in the Minamata disaster) deserve to be the focus of immediate scrutiny because of their production process that generates large quantities of methylmercury, and their typically high mercury losses.

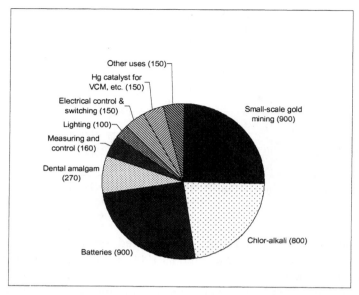

Figure 3. 2003 global mercury consumption (tonnes).

Sources: Maxson (2004) and revisions, Nordic Council (2004).

Largely as a result of increasing awareness and regulation, the global demand for mercury has declined from more than nine thousand tonnes annual average in the 1960s, to over eight thousand tonnes in the 1970s, and just under seven thousand tonnes in the 1980s. The average annual global demand for mercury continued declining to around four thousand tonnes in the 1990s [12] and, as described above, is currently below that level, of which less than 10 percent is consumed in the EU, and less than 10 percent in North America.

Global mercury demand broken down by geographical region is estimated in the table below.

MERCURY PRICES AND MARKETS

As evident in the figure below, mercury prices have been on a downhill slide for most of the past 40 years.

Table 2. Global mercury demand by region (2003).

Region	Mercury demand (tonnes)
European Union (15)	302
North America	308
Other OECD	100
Central & Eastern Europe/CIS	500
Arab States	100
East Asia and Pacific	1350
Latin America & Caribbean	370
South Asia	450
Sub-Saharan Africa	100
TOTAL	3580

Sources: Maxson (2004) and revisions, Nordic Council (2004).

During the last 10 years they have stabilized at about their lowest levels ever – in the range of €4-5 per kg of mercury – before spiking up considerably in the middle of 2004 . Adjusting for inflation, mercury at €5 per kg is worth less than five percent of its peak price during the 1960s.

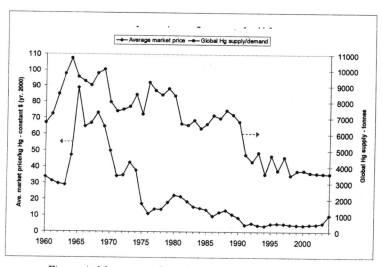

Figure 4. Mercury market price vs global mercury supply.

Sources and notes: Maxson (2004) Market prices for mercury are an average of London prices(Metallgesellschaft, 1939-98), US prices (US Bureau of Mines, USGS 2002) and "world" prices from Roskill Information Services (Roskill, 2000). They have been converted to constant $US of 2000 for purposes of better long-term comparison. Significant data and methodological input provided by L. Hylander, Uppsala University.

This price level reflects a chronic oversupply driven, increasingly, by the regulatory pressures on industry, e.g., to reduce emissions, to organize separate collection of mercury products, and to deal with the increasing restrictions and costs of mercury waste disposal by sending the wastes to recyclers.

It may be argued that low mercury prices actually shelter existing uses, discouraging certain product and process innovations that would otherwise take place to reduce mercury demand if mercury were a more expensive commodity. For example, chlor-alkali factories, and even small-scale gold miners are under no significant economic pressure to reduce consumption of mercury, since the cost of mercury is such a low percentage of the value of their output. Therefore, there is little at present besides regulation or awareness-raising - both highly variable from one country or region to another - to discourage the use of mercury in products and processes. Meanwhile, economic theory suggests that low prices are slowing mercury substitution in products/processes and, combined with often half-hearted enforcement of existing regulations, and in the absence of a coherent international strategy to address the problem, may even be inviting new and additional uses of mercury.

It has been observed that the market price for commodity mercury has increased from about $US 150 per flask in January 2001 to $US 205 per flask in January 2004. Not coincidentally, this is identical – within a few percentage points - to the appreciation of the value of the euro (and other major currencies) relative to the $US during the same period (see following figure). Therefore, for mercury brokers around the world who account for transactions and keep their books in currencies other than dollars, world Hg prices have not – until about mid-2004 – measurably increased since the early 1990s.

As mentioned, the small market for commodity mercury is characterized by a limited number of virgin mercury producers, and a larger number of secondary mercury producers. These actors are complemented by another relatively small group of mercury traders and brokers, mostly located (in addition to the main mining sites) in the Netherlands, the UK, Germany, the US and Hong Kong. All of these "market-makers" buy and sell mercury, timing their trades to influence market movements and profit from price fluctuations.

MAYASA, the Spanish mercury mining and trading company, purchased most or all of the USSR stockpile in the 1990s, for example. In recent years, as noted previously, MAYASA has also purchased residual mercury inventories from Western European chlor-alkali plants as they close or convert to a mercury-free process.

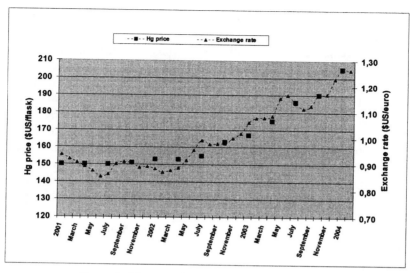

Figure 5. Mercury price vs. Exchange rate correlation.

Note: Mercury prices from London Metals Exchange.

Through a variety of public policies, governments may influence the mercury market; they may influence the available mercury supply, price, and even the number of customers on a global basis. Regulatory measures influence mercury movements and markets by encouraging educational programs, collection and recycling, substitutes for mercury products, etc. At the same time, it may be argued that regulatory programs keep mercury prices low by putting an effective negative value on mercury products and wastes, so that recyclers are sometimes even paid to accept them. In such circumstances, recyclers can process, recover and resell the mercury at a very low price while still making a profit. One can only wonder whether, in such a regulatory environment, the mercury market can really be described as a free market. At the same time, one should perhaps also question, in a world where free market mechanisms are generally held out as desirable, whether encouraging a free market in a toxic substance is in fact in the best interest of society.

MERCURY TRADE

Overview. An analysis of COMTRADE[13] mercury trade statistics confirms that regionally, North America and Europe have dominated mercury markets in the past, but in recent years they have been overtaken by East Asia - especially China; and South Asia – particularly India and Pakistan. However, the EU continues to play a predominant role in the global trade of commodity grade mercury. The European Union imported nearly 400 tonnes of elemental mercury from non-EU countries in 2000, and exported over 1400 tonnes, while global movements of elemental mercury appeared to be well over 8000 tonnes – much of it changing hands repeatedly. In fact, on average, some two to three tonnes of elemental mercury appear in international trade statistics for each tonne of mercury consumed during the same year.

Trade statistics concerning mercury are far from perfect, but they are able to reveal surprises, such as the COMTRADE evidence that, despite restrictions in OECD countries, there appears to be a very active trade in mercuric oxide batteries, especially through China, but also through the EU and the US. Tracing the flows of mercury through the economy demonstrates how fluid and global mercury trade really is. Mercury could be recovered from a Western European mercury cell chlor-alkali plant, sold to the Spanish mercury mining and trading company, shipped from Spain to Germany for further conversion into mercuric oxide, sold to mainland China for the manufacture of button-cell batteries, and the batteries exported to Hong Kong for incorporation into mass-produced watches for export to the European Union and the US.

Purpose of trade flow analysis. The mercury supply and demand factors discussed previously have led to diverse and ever-changing trade flows, not to mention considerable challenges for those researchers trying to get a handle on them. As a check on overall levels of mercury supply and demand cited above, and to bring some transparency to mercury flows around the world, this section describes world trade flows of commodity mercury and mercuric oxide batteries. It takes a "snapshot" picture of world trade in the year 2000, which should not be assumed to mirror trade flows in other years (no two years are alike in the mercury business), but certainly provides a rough idea of typical trading partners, volumes and general market structure. This analysis also permits a better informed assessment of the eventual impact of the supplies of mercury that could be brought onto world markets as chlor-alkali plants continue to close and/or convert to other processes, and as secondary mercury production and recycling increase.

Limitations of trade data. There are several factors that render mercury trade data less transparent than many other trade data. The international commodity characteristics of mercury mean that it is frequently sold for speculative reasons rather than to satisfy immediate demand, resulting in the same mercury often being traded several times. Also, the regional locations of mercury product manufacturers are often not the same as the regions of final product consumption – or at least not in the same volumes. Therefore, one cannot assume that a shipment of mercury sent to India, for example, will end its life-cycle in India. One needs to further understand how the mercury is used in India, whether it goes into an industrial process, and whether or where mercury products are eventually used or exported. Nevertheless, the geographic locations of major mercury dealers are generally evident from the trade data unless, as occasionally happens, for example, elemental mercury is converted to a compound such as mercuric chloride in order to disguise the movement of commodity mercury.

Trade data, whether from Eurostat, the International Trade Commission, the UN COMTRADE database or others, are open to criticism by analysts, and are well known to be incomplete and occasionally inaccurate. However, for the most part, these international agencies merely organize and publish the data submitted to them by national agencies. For a relatively small overall market such as that for mercury, any inaccuracies or omissions take on an even greater importance. Furthermore, the more interest researchers demonstrate in mercury data, the less willing some providers of the data seem to become, as the US Geological Survey (Reese, 1991-99) has discovered over many years of collecting mercury data from industry sources. Nevertheless, a careful analysis of the data can produce some very useful findings and raise some interesting questions, as will be seen below.

Source of mercury trade statistics. For the purposes of this analysis, COMTRADE statistics have been used. They are collected and presented under the responsibility of the United Nations Department of Economic and Social Affairs – Statistics Division. Statistics for trade in commodity ("raw") mercury were downloaded for the year 2000 - selected largely because most information and data that will ultimately be gathered for 2000 has by now been assembled. In order to reduce the database to a more manageable size, the database was requested to overlook any trade movements valued at less than $US 10,000, which means that movements of several tonnes, of which there are surely many, have not been included in this analysis.

EU trade in raw mercury. The raw mercury trade data for imports and exports among the EU-15 are summarized in the figure below, which also shows the larger exports from the EU (thick arrows) but not imports. As often observed with trade data, the data submitted by country A for mercury imports from trade partners B and C were sometimes not consistent with the figures submitted by countries B and C as mercury exports to trade partner A. Such discrepancies have been "reconciled" by including all reported trade movements, whether they were reported by one country as exports, or by a trade partner as imports.

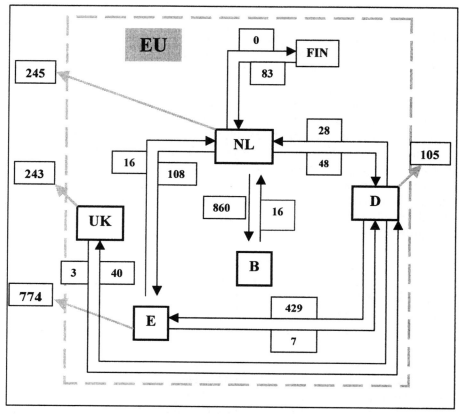

Figure 6. EU Trade in raw mercury, 2000 (metric tonnes).

Source: Maxson (2004) E = Spain, UK = United Kingdom, B = Belgium, NL = Netherlands, D = Germany, FIN = Finland.

Among other intra-EU movements, Germany recorded a shipment of 429 tonnes to Spain, which is a clear indication that the origin of the mercury was a chlor-alkali plant, of which Germany closed several in 1999 and 2000. The quantities of mercury passing through the Netherlands and Spain invited further analysis of the data in order to better clarify the ultimate destinations of this mercury. Such an analysis produced several interesting findings that are not evident from this figure:

1. In 2000 the Netherlands shipped 245 tonnes of commodity mercury to at least 18 countries outside the EU – about half of that amount to

2. countries in the Latin America/Caribbean region. Much of this mercury traded below the spot market price, implying relatively low
3. purity. While it is impossible to know from the trade statistics whether some of that mercury went to small-scale gold mining, it is clear that the quantities in question were well in excess of the countries' typical needs in other industrial sectors.
4. In 2000 Spain shipped 774 tonnes of mercury to at least 20 countries outside the EU – about two-thirds of it to the East Asia/Pacific region. Virtually all of this mercury was also low value, including 50 tonnes exported to Latin America.
5. In 2000 Germany shipped 105 tonnes of mercury to at least 10 countries outside the EU – most of this mercury appears in the statistics at a somewhat higher value, reflecting a higher purity and/or more specialized applications.
6. In 2000 the UK shipped over 220 tonnes of mercury to three countries in South Asia which, along with the nearly 200 tonnes shipped by Spain, made South Asia one of the key destinations for mercury that year. In this case, the mercury shipped from the UK was not formally recorded in the COMTRADE statistics as a UK export, probably because that data was not submitted by UK authorities to COMTRADE; however, it was formally recorded by each trade partner as having been imported from the UK.

In the analysis of this and other trade data, several points should be kept in mind:

- Trade data show only discrete freight movements – the origin and destination of a shipment, and the quantity (and sometimes the value) transported. There is no indication whether the source of the shipment is the real origin of the material, or whether the destination of the shipment is the final destination.
- The trade statistics for commodity mercury do not include trade in mercury compounds (discussed further below), which would substantially increase the mercury flows described here.
- Likewise, if one were to assume a certain number of unreported trade movements, or recognize the internal trade within large countries such as China, Spain, the US, etc., that does not appear in these statistics, or even the instances of shipments under $US 10,000 that have been excluded from this analysis, one would have an even more substantial trade picture.

Global trade in raw mercury. The same analytical approach has been applied using data for inter-regional global trade in raw mercury. This trade is summarized in the following figure. One should ignore the data on the diagonal (shaded in grey), that appear to represent trade within a region, because these data are not comparable. For example, the figure of 1930 tonnes for EU-to-EU represents reported trade between EU countries, while the figure of 15.6 tonnes for North America-to-North America represents only the reported trade between the US and Canada, and therefore overlooks all trade between states (within the US) and provinces (within Canada). For the same reason, these data are not included in the inter-regional or global trade totals – the last column and bottom row of this figure

Table 3. Reconciled global raw mercury movements for 2000 (tonnes) for reported transactions greater than or equal to $US 10,000.

Hg Transferred from:	European Union (15)	North America	Other OECD	Central & Eastern Europe and CIS	Arab States	East Asia and Pacific	Latin America & Caribbean	South Asia	Sub-Saharan	Not specified	Total regional transfers
European Union (15)	1930.0	43.1	74.7	27.2	30.9	529.8	164.0	463.2	24.1	49.3	1406
North America	67.2	15.6	16.9	1.2	1.9	17.4	964.6	97.7		1.5	1168
Other OECD	29.4	62.9				7.1		18.0		33.0	150
Central & Eastern Europe and CIS	27.0	20.6	7.0	144.4		501.9	61.7	15.9	3.6	1.5	639
Arab States	142.2		4.1			17.3		10.0			174
East Asia and Pacific	4.7					20.6		5.2			10
Latin America & Carribbean		21.2					60.7				21
South Asia	89.4	171.0	85.0			10.0		4.1			355
Sub-Saharan Africa								17.0		1.4	18
Not specified	17.2		4.0	15.2	1.2	16.8	6.7	1.1	5.9		68
Total global transfers	377	319	192	44	34	1100	1197	628	34	87	4011

Batteries transferred to:

Source: COMTRADE statistics, United Nations Department of Economic and Social Affairs – Statistics Division, as interpreted and presented by the author.)

The three regions that imported the most raw mercury in 2000, as seen in the summary figure below (showing only the larger trade flows), were Latin America/Caribbean with 1197 tonnes (see comment below), East Asia with 1100 tonnes, and South Asia with 628 tonnes. Of those amounts, the EU supplied about half of the mercury needs of East Asia, and virtually all of the mercury needs of South Asia.

Furthermore, as mentioned previously, Spain, the Netherlands and the UK (followed some distance behind by Germany) were the main suppliers from within the EU, especially for lower-priced, lower-quality mercury.

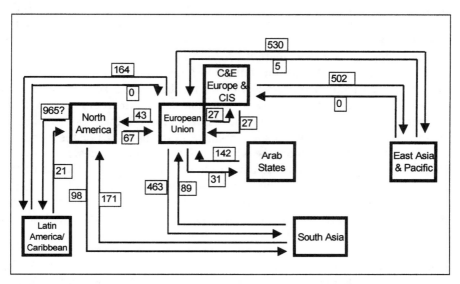

Figure 7. Major EU raw mercury trade with the rest of the world in 2000 (metric tonnes).

Source. Maxson (2004).

Further analysis of the Latin American/Caribbean transfers of mercury showed that 965 tonnes of the total 1197 tonnes was reported by Mexico to have been imported from the US in 2000. While this figure appears aberrant, and was not reported by the US as an export, 2001 imports of mercury reported by Mexico as coming from the US were even greater (1340 tonnes) – clearly more than a coincidence and therefore deserving of further investigation.

Trade in mercuric oxide batteries. Due to various reports that the battery sector may remain a significant consumer of mercury at the global level, the trade data for mercuric oxide batteries (containing typically 30 percent by weight of mercury) were also scrutinized. Specifically, the relevant COMTRADE data cover world trade (2000) in mercuric oxide primary cell batteries (volume less than 300 cc), including both HS1996 code 850630, and HS1992 code 850612. In order to limit the quantity of data, a search was carried out only for reported trade movements of a value equal to or greater than $US 50,000.

EU trade in mercuric oxide batteries. The analysis of EU trade in primary cell (non-rechargeable) mercuric oxide batteries revealed over 500 tonnes of batteries transferred among EU countries in 2000, representing potentially well over 100 tonnes of mercury. Belgium and Spain were the primary net exporters to other Member States in 2000. While this does not necessarily represent demand for raw mercury in addition to what has already been presented, it seems a disturbing statistic in a use category where the battery industry categorically dismisses any concerns about mercury in batteries due to the fact that the major players have phased out mercury use in their European and North American production facilities.[14] It therefore remains to be confirmed whether there is some mistake in the statistics or their coding, and if not, to determine precisely what this battery trade represents, to what extent it may reflect military demand (which is unlikely, since most of the trade is high volume, low value per item), whether these may be batteries originating outside the EU, whether the demand is persistent, etc. In any case, it is not an issue that has been highlighted in recent years, and deserves further investigation.

Global trade in mercuric oxide batteries. Global trade statistics for mercuric oxide batteries were also analyzed for the year 2000. These statistics (summarized in the figure below) report a remarkable 2358 tonnes of mercuric oxide batteries imported by the EU, and 1118 tonnes of mercuric oxide batteries imported by North America (406 tonnes coming from the EU). With regard to exports, the statistics indicate that in 2000 North America and East Asia exported, respectively, 2037 and 2718 tonnes of mercuric oxide batteries, the EU exported 986 tonnes and other OECD countries exported 1623 tonnes. This also raises questions. If North America no longer produces these batteries in significant numbers, how did it apparently export nearly 900 tonnes more than it imported in 2000? [15] And if the EU no longer permits the marketing or use of these batteries, why did it apparently import nearly 1400 tonnes more than it exported in 2000?

East and South Asia have continued to consume large quantities of mercury in battery production. Sznopek & Goonan 2000 have confirmed that the Peoples' Republic of China had legislation on the books to eliminate mercuric oxide battery production from 2002. Meanwhile, Xinbin Feng calculated (see chapter 27) that Chinese consumption of mercury in batteries exceeded 800 tonnes in 1999, and seemed to be increasing. A full explanation of ongoing mercury use in batteries remains elusive, but is likely attributable to remaining inventories being sold off, possible mislabeling of customs documents, substantial ongoing production in some countries, and extensive global trade in batteries of all kinds.

Since the major international battery producers are no longer using mercury in batteries produced in the EU or the US, it is surprising that they would not protest loudly about such low-cost (and illegal, as mentioned above) competition from overseas producers. Until further investigation provides some answers, these statistics would suggest that a large amount of the mercury going to the East Asia/Pacific region may still go into the production of batteries.

Table 4. Reconciled global trade (tonnes) in mercuric oxide primary cell batteries, 2000, for which the volume <300 cc (for reported trade movements valued at equal to or greater than $US 50,000)

Batteries transferred from:	Batteries transferred to:										Total regional transfers
	European Union (15)	North America	Other OECD	Central & Eastern Europe and CIS	Arab States	East Asia and Pacific	Latin America & Carribbean	South Asia	Sub-Saharan	Not specified	
European Union (15)	567	406	13	300	147	38			50	32	986
North America	1713	30		9	7	76	191			41	2037
Other OECD	24	120	19	330	13	1084	38			14	1623
Central & Eastern Europe and CIS	44	59		2	1	302	189				595
Arab States				33							33
East Asia and Pacific	402	254	333	48	255	2385	980		366	80	2718
Latin America & Carribbean		260					122				260
South Asia	24						172				196
Sub-Saharan Africa									32	22	22
Not specified	151	19	87	1	46	15	200		132		651
Total global transfers	2358	1118	433	721	469	1515	1770	0	548	189	9121

Source: COMTRADE statistics, United Nations Department of Economic and Social Affairs – Statistics Division, as interpreted and presented by the author.

It is expected, of course, that production of mercuric oxide batteries will gradually decrease in the coming years, and be eventually replaced by less hazardous alternatives. But the level of production in 2000 appeared to be so massive, and the quantities of mercury involved so great, it is difficult to imagine a serious phase-out of mercury-containing batteries worldwide in the absence of substantial international pressure.

Trade in mercury compounds. Mercury compounds are still commonly used in many countries in cosmetics, batteries, pharmaceuticals, paints and biocides, according to CADTSC (2001), as well as in various chemical processes. The compounds in most frequent use include mercury oxide, mercury chloride, ethyl mercury and phenylmercuric acetate, all of which appear in commerce under a range of trade names. According to USEPA (1999) the only major mercury compounds still imported by the US for use in products are organo-mercury compounds. In a recent year, U.S. imports of organo-mercury compounds were said to be 37 tonnes. This author has not analyzed the statistics for global trade in these compounds, but believes these statistics would also raise a number of interesting questions.

Conclusions and observations concerning mercury trade flows. Analysis of trade flows over the years shows clearly that mercury is traded as a commodity - to be routinely bought and sold according to market opportunities rather than purchased to respond directly to a specific demand. This greatly complicates the understanding of, and eventual political influence over, the international mercury market.

Raw or commodity grade mercury is extensively traded around the world, at the rate of some two to three times the annual consumption. This could be taken as an indicator of reasonable market efficiency – matching sellers with buyers worldwide.

The world market for commodity-grade mercury is small, with relatively few key players, leaving it potentially susceptible to manipulation. On the other hand, the influence of those individuals may be weakening somewhat as by-product and other secondary mercury sources increase, and a more diverse group of secondary suppliers and recyclers appears, who don't depend on mercury production costs to stay in business.

Despite reduced OECD region demand for mercury over the years, the EU (and to a lesser extent, the US) is integrally involved in global mercury markets, supplying a substantial part of global demand through a combination of production from the mercury mining company in Spain, and residual mercury recovered from decommissioned chlor-alkali facilities in various EU countries. Overall, the EU provides 25-30 percent of the global mercury supply, it is a partner in over 50 percent of global mercury trade, and it consumes nearly 10 percent of global demand.

The EU remains a major mercury supplier mainly through the large quantities of mercury (about 1000 tonnes annually) routinely sold through the Spanish mining and marketing company MAYASA. Whether or not MAYASA receives mercury from chlor-alkali plants as they periodically close and/or convert to alternative production processes, it fully intends to continue supplying its customers for years to come.

Consumption of mercury has declined significantly in the EU and other OECD countries, but this decline is not evident in the rest of the world, to which mercury trade is increasingly being shifted.

From about 1990 to 2003, mercury prices were at a lower level in real terms than at any time in history, and there are no market reasons for any increase in prices in the foreseeable future. It should be mentioned, however, that 2004 witnessed a significant increase in the mercury price, reportedly due to decreased production from the usual suppliers, as well as rumored efforts by some traders to further restrict supplies.

For at least two decades, the mercury commodity market has been strongly influenced by public policies and the release of government inventories, and has not operated on purely economic grounds. The term "free market" seems no longer applicable. Depending on other developments, by 2020 there may be no more primary mercury mining (there is little or no economically justified mining now), and it is likely that the cost of "producing" mercury will reflect only recycling and recovery costs, which are driven by regulation, and which could therefore send prices lower.

Regulations tend to give mercury a negative value. Whether it is raw mercury or mercury waste, holding mercury is a liability, and it costs money to dispose of it. The only way to add value to mercury is to put it in a product and sell it, or to use it in an industrial process, with the obvious risk of eventual emissions and exposures. Regulation has already pushed many of these production processes out of the OECD.

Despite the very recent surge in the market price of mercury, the various sources of mercury are easily able to meet and outstrip anticipated demand. Under any reasonable assumptions (i.e., declining global demand, decreased mining activity, increased supply from secondary sources, etc.), therefore, and especially in the absence of any coherent international strategy, a net worldwide oversupply of mercury relative to known industrialized demand is expected to remain a hallmark of mercury markets into the foreseeable future.

As mercury markets have demonstrated a particular ability to adapt to changes in supply and demand, continued low prices and oversupply may encourage uses (and eventual emissions) that would not otherwise have occurred, or at the least, they will undoubtedly discourage any mercury reduction efforts.

Once residual mercury from chlor-alkali plants is transferred to another party, the chlor-alkali industry - however responsible – effectively loses control over its ultimate destination and use. The European producers' association, Euro Chlor, has since 2001 had a specific plan in place to address this concern, whereby Euro Chlor mercury is intended to offset mine production, and to adhere to "best environmental practices ... to avoid emissions and spillage into the environment" during handling and marketing of the residual mercury. However, with no independent method of monitoring the ultimate destination of the mercury in the international market, the adequacy of the Euro Chlor plan remains open to criticism.

The availability of mercury from decommissioned chlor-alkali plants will soon become an important contributor to global mercury supplies, and will largely replace the demand for virgin mercury. Chlor-alkali residual mercury has comprised 10-20 percent of global supplies in some years, and will likely contribute 20-30 percent in the near future.

The quantities of residual mercury that will be most easily recovered from Western European chlor-alkali plant closures up to 2020 are roughly equivalent to the expected mercury market oversupply during the same period. As part of a larger strategy, a non-market solution needs to be identified urgently to deal with this oversupply. To its credit, Euro Chlor has stated that it is studying a range of mercury storage options.

Mercury producers make a reasonable case that as long as legitimate demand for mercury persists, they should have a right to supply it.

Therefore, if we wish to restrict the introduction of new mercury into the biosphere, we have to seriously address mercury demand at the same time. In this regard, the widely cited UNEP Global Mercury Assessment produced a general consensus among a large group of stakeholders and experts that concerted international efforts must be made to reduce the circulation of mercury in the economy and the environment - from the supply side as well as the demand side. Mercury "business-as-usual" will only create further opportunity for misuse and abuse, emissions and exposure.

REFERENCES

ACAP Inventory of Mercury Releases from the Russian Federation - Chemical Industry, Dr. Yuri A. Treger, Scientific Research Institute "Syntez", Arctic Council Action Plan to Eliminate Pollution of the Arctic, 2004.

CADTSC "Draft Mercury Report," California Department of Toxic Substances Control, October, 2001.

Chlorine Institute. "Fifth Annual Report to EPA for the Year 2001," Chlorine Institute, USA, 25 April, 2002.

Comtrade international commercial trade statistics. United Nations Department of Economic and Social Affairs – Statistics Division. URL: http://unstats. un.org/unsd/comtrade/.

Euro Chlor . "Western European chlor-alkali industry plant & production data," 1970-2000." Euro Chlor, Brussels, Dec, 2001.

Euro Chlor. Reduction of Mercury Emissions from the West European Chlor Alkali Industry, 3rd Edition, Euro Chlor, Brussels, June, 2001.

Euro Chlor. "Western European chlor-alkali industry plant & production data, 1970-2001." Euro Chlor, Brussels, 2002.

European Commission. COM 489 - "Report from the Commission to the Council concerning mercury from the chlor-alkali industry," Commission of the European Communities, Brussels, 6 Sept, 2002.

European Commission. EU Mercury Strategy stakeholder consultation, European Commission – DG Environment Directorate G - Sustainable development & integration, Brussels, 31 March, 2004.

Feng, X., Mercury Pollution in China – An Overview, State Key Laboratory of Environmental Geochemistry, Institute of Geochemistry, Chinese Academy of Sciences, Guiyang, P.R. China, 2004.

Hylander & Meili. L.D., Hylander, Meili, M., "500 years of mercury production: global annual inventory by region until 2000 and associated emissions," *Sci. Tot. Env.*, 304 13–27, 2003.

MAYASA Email communications regarding the Almadén mine, from Manuel Ramos to Neil Emmott, European Commission, DG Environment, 6 and 9 February, 2004.

Maxson, P., Verberne, F. Mercury concerns in decommissioning chlor-alkali facilities in Western Europe, Concorde East/West Sprl and ERM for the Netherlands' Ministry of Environment VROM, The Hague, September, 2000.

Maxson, P., Mercury flows in Europe and the world: The impact of decommissioned chlor-alkali plants, Concorde East/West Sprl for the European Commission – DG Environment, February 2004, Brussels. Document available at http://europa.eu.int/comm/environment/chemicals/mercury/index.htm. 2004.

Metallgesellschaft Metallstatistik 1929-1938; Metallstatistik 1957-1966; Metallstatistik 1981-1991; Metallstatistik 1985-1995; Metallstatistik 1987-1997. Annual volumes of metal statistics. Frankfurt-am-Main, 1939-1998 (in German).

NEMA Email from Ric Erdheim, National Electrical Manufacturers' Association, to Frank Anscombe, US EPA, dated 26 September, 2003.

Nordic Council Workshop on mercury & international environmental agreements, Nordic Council of Ministers, 29-30 March 2004, Brussels, 2004.

Reese USA Geological Survey Minerals Yearbook: Mercury. All years, 1991-99.

Roskill Roskill's Metals Databook. London: Roskill Information Services, Ltd., 2000.

UNEP Chemicals Global Mercury Assessment. United Nations Environment Program, Chemicals Directorate, Geneva, December, 2002. (Available at: URL: http://www. chem.unep.ch/mercury/), 2002.

USEPA "Potential Revisions to the Land Disposal Restrictions Mercury Treatment Standards," Appendix H: Mercury Waste RCRA Categories. U.S. Environmental Protection Agency, 40 CFR Part 268, 64 FR No. 102, 28 April, 1999.

USGS US Geological Survey Annual Reports – Mercury, Denver. (Available at http://www.usgs.gov) 1991-98.

USGS "Mineral Commodity Summaries/Minerals Yearbook, Mercury." In: Minerals statistics and information 1994-2000 (annual volumes), US Geological Survey, 1995-2001. (Available at http://minerals.usgs.gov/minerals/pubs/commodity/mercury/).

USGS USGS Minerals Yearbook - Mercury. US Geological Survey. (Available at http://minerals.usgs.gov/minerals/pubs/commodity/mercury/), 2003.

Veiga Personal communications with M. Veiga (UNIDO) and L. Hylander (Univ. of Uppsala), 2004.

NOTES

1. Data published periodically by the Algerian Ministry of Energy and Mines.

2. Typical Kyrgyz output has been 500-600 tonnes annually since the mid-1990s, but mine flooding in 2003 was reported to be the major reason for a decline in production.

3. China's recent domestic mercury production has been generally estimated at about 200 tonnes per year (Hylander & Meili, 2003). However, if Xinbin Feng is correct that China's consumption of mercury may now approach 1500 tonnes (see Chapter-27), then its domestic production of mercury is more likely at least 400 tonnes. This is also consistent with USGS (2004) estimates of Chinese production of 435 tonnes in 2002 and 610 tonnes in 2003.

4. This is an approximate doubling of highly conservative estimates for 2000 (Maxson, 2004), based on contributions and discussions at the Nordic Council Workshop in Brussels (Nordic Council, 2004). In an effort to reduce hazardous wastes and mercury emissions, it appears that more mine operators are recovering more secondary mercury than ever before.

5. No new information has come to light to justify a revision to the data gathered by Maxson (2004).

6. Euro Chlor contribution to Nordic Council Workshop in Brussels (Nordic Council, 2004).

7. Euro Chlor member companies have committed themselves to a 2020 phase-out date for Western European mercury cell chlor-alkali plants. The chlor-alkali industry is also covered by the IPPC Directive, which requires installations to have permit conditions based on best available techniques (BAT). The mercury-cell process is not considered to be BAT for the chlor-alkali sector. The Directive states in Article 5 that existing installations, i.e., installations in operation before 30 October 1999, should operate in accordance with the requirements of the Directive by 30 October 2007. Alternatively, a number of EU countries have announced that their mercury cell chlor-alkali plants will be decommissioned and/or converted to mercury-free technology by 2010, in line with a more flexible interpretation of the IPPC Directive. The latter date is also consistent with the PARCOM decision 90/3 of OSPAR, which recommended closing or converting all mercury cell plants in the OSPAR region by 2010. All of the EU-15 countries with mercury cell plants are parties to the OSPAR Convention except Italy and Greece.

8. European Commission (2002).

9. This 12,000 tonnes does not include tens or hundreds of tonnes more that may be recovered from a single site during decommissioning, from internal process sources such as piping, sumps and equipment, or from external sources such as pools of mercury that sometimes accumulate in the soil beneath the cellroom due to mercury leaking through cracks in the floor of the cellroom over many years (Maxson, 2000).

10. 2003 mercury demand by EU-15 chlor-alkali plants is based on data reported by industry to OSPAR for 2002 (104 tonnes), plus a conservative estimate of 16 tonnes for plants in Italy and Greece, for a total of around 120 tonnes. This estimate remains to be confirmed when industry figures become available for 2003.

11. The Czech Ministry of Environment has reported that its two plants alone consumed in 2001 in excess of 100 tonnes. The web link (in Czech) is: http://www.env.cz/AIS/web.pub.nsf/$pid/MZPLSF4H1VU6/$FILE/-oov_65_Rtut_20040414.pdf

12.   Historical mercury demand through the 1990s is based on mercury production data compiled by Hylander and Meili (2003).

13.   COMTRADE international commercial trade statistics, United Nations Department of Economic and Social Affairs – Statistics Division, http://unstats.un.org/unsd/comtrade/.

14.   In response to this statement, a legal representative of the National Electrical Manufacturers' Association (NEMA), which represents major battery manufacturers in the US, confirmed, "Under US law and the laws of about 20 states, it has been illegal since 1996 (and earlier in some states) to sell a mercuric oxide button cell in the US (NEMA, 2004)." On the other hand, he failed to offer any rational explanation of the trade statistics.

15.   The same NEMA official dismissed this significant trade in mercury batteries as "relatively small numbers of specialty batteries sold in non consumer markets" (NEMA, 2004).

Chapter-3

MERCURY EMISSIONS FROM ANTHROPOGENIC SOURCES: ESTIMATES AND MEASUREMENTS FOR EUROPE

Jozef M. Pacyna[1,2], John Munthe[3], Kari Larjava[4] and Elisabeth G. Pacyna[1]

[1] *Norwegian Institute for Air Research (NILU), Kjeller, Norway*
[2] *Gdansk University of Technology, Gdansk, Poland*
[3] *IVL Swedish Environmental Research Institute, Gothenburg, Sweden*
[4] *VTT Technical Research Centre of Finland, Espoo, Finland*

INTRODUCTION

Processing of mineral resources at high temperatures, such as combustion of fossil fuels, roasting and smelting of ores, kilns operations in cement industry, as well as incineration of wastes and production of certain chemicals result in the release of several volatile trace elements into the atmosphere. Mercury is one of the most important trace elements emitted to the atmosphere due to its toxic effects on the environmental and human health, as well as its role in the chemistry of the atmosphere.

Although substantial information has been collected on environmental effects of mercury and its behaviour in the environment much less data is available on atmospheric emissions of the element. Information on emissions is needed for various policy and modelling purposes. This need has been recognized not only locally where mercury may pose direct problems but also on regional scale because the element is a subject of long-range transport while in the atmosphere.

PARAMETERS AFFECTING EMISSIONS OF MERCURY FROM VARIOUS ANTHROPOGENIC SOURCES

There are four major groups of parameters affecting mercury emissions:

- contamination of raw materials by mercury,
- physico-chemical properties of mercury affecting its behaviour during the industrial processes,
- the technology of industrial processes, and
- the type and efficiency of control equipment.

Contamination of raw materials

Concentrations of mercury in coals and fuel oils vary substantially depending on the type of the fuel and its origin, as well as the affinity of the element for pure coal and mineral matter. The sulphide-forming elements, with mercury included, are consistently found in the inorganic fraction of coal. Although it is very difficult to generalize on the impurities in coal, the literature data seem to indicate that the mercury concentrations in coals vary between 0.01 and 1.5 ppm (a review in Pirrone et al., 2001). These concentrations are presented in Table 1. It should be noted that mercury concentrations within the same mining field might vary by one order of magnitude or more.

Table 1. Concentrations of Hg in various fossil fuels.

Fuel	Unit	Concentration
1. Hard coals	g/tonne	
- Europe		0.01-1.5
- USA		0.01-1.5
- Australia		0.03-0.4
- South Africa		0.01-1.0
- Russia		0.02-0.9
- Brown coals	g/tonne	
- Europe		0.02-1.5
USA		0.02-1.0
3. Crude oil		0.001-0.05
4. Natural gas	mg/m^3	0.01-5.0*

(*) A reduction of Hg to concentrations lower than 10 $\mu g/m^3$ must be obtained before the gas can be used.

There is only limited information on the content of mercury in oils. In general, mercury concentrations in crude oils range from 0.01 to as much as 30.0 ppm (Pacyna, 1987). It is expected that mercury concentrations in residual oil are higher than those in distillate oils being produced at an earlier stage in an oil refinery. Heavier refinery fractions, including residual oil, contain higher quantities of ash containing mercury.

Natural gas may contain small amounts of mercury but the element should be removed from the raw gas during the recovery of liquid constituents, as well as during the removal of hydrogen sulphide. Therefore, it is believed that mercury emissions during the natural gas combustion are insignificant.

Mercury appears as an impurity of copper, zinc, lead, and nickel ores. Obviously, there are also mercury minerals, particularly cinnabar. The element is also present in the gold ores. It is very difficult to discuss the average content of mercury in the copper, zinc, lead, nickel and gold ores as very little information is available in the literature on this subject. Average zinc ores contain larger amounts of the element compared to copper and lead ores (Pacyna, 1983).

Chemical composition of input material for incineration is one of the most important factors affecting the quantity of atmospheric emissions of various pollutants from waste incineration. Very limited information exists on mercury concentrations in various types of wastes. Another difficulty is that it is almost impossible to calculate an average value for these concentrations due to the high variabilities in the content and origin of wastes to be incinerated, even in the same incinerator. Therefore, it is rather difficult to extend the information on the mercury content measured in one incinerator for another one.

Physico Chemical Properties of Mercury Affecting its Behaviour during the Industrial Processes

Most of the processes generating atmospheric emissions of mercury employ high temperature. During these processes, including combustion of fossil fuels, incineration of wastes, roasting and smelting operations in non-ferrous and ferrous metallurgy, and cement production, mercury introduced with input material volatilizes and is converted to the elemental form. It has been confirmed in various investigations that almost 100 % of the element is found in exhaust gases in a gaseous form, as discussed later in this paper.

Technology of Industrial Processes

Various technologies within the same industry may generate different amounts of atmospheric emissions of mercury. It can be generalized for conventional thermal power plants that the plant design, particularly the burner configuration has an impact on the emission quantities. Wet bottom boilers produce the highest emissions among the coal-fired utility boilers, as they need to operate at the temperature above the ash -melting temperature. The load of the burner affects also the emissions of trace elements including mercury in such a way that for low load and full load the emissions are the largest. For a 50 % load the emission rates can be lower by a factor of two.

The influence of plant design or its size on atmospheric emissions of mercury from oil-fired boilers is not as clear as for the coal-fired boilers. Under similar conditions the emission rates for the two major types of oil-fired boilers: tangential and horizontal units are comparable.

Non-conventional methods of combustion, such as fluidized bed combustion (FBC) were found to generate comparable or slightly lower emissions of mercury and other trace elements than the conventional power plants (e.g. a review in Pirrone et al., 2001). However, a long residence time of the bed material may result in increased fine particle production and thus more efficient condensation of gaseous mercury.

Among various steel making technologies the electric arc (EA) process produces the largest amounts of trace elements and their emission factors are about one order of magnitude higher than those for other techniques, e.g., basic oxygen (BO) and open hearth (OH) processes. The EA furnaces are used primarily to produce special alloy steels or to melt large amounts of scrap for the reuse. The scrap which often contains trace elements, and on some occasions mercury, is processed in electric furnaces at very high temperatures resulting in volatilization of trace elements. This process is similar from the point of view of emission generation to the combustion of coal in power plants. Much less scrap is used in other furnaces, where mostly pig iron (molten blast-furnace metal) is charged. It should be noted, however, that the major source of atmospheric mercury related to the iron and steel industry is the production of metallurgical coke.

Quantities of atmospheric emissions from waste incineration depend greatly on the type of combustor and its operating characteristics. The mass burn/waterwall (MB/WW) type of combustor is often used. In this design the waste bed is exposed to fairly uniform high combustion temperatures resulting in high emissions of gaseous mercury and its compounds. Other types of combustors seem to emit lesser amounts of mercury as indicated by the comparison of the best typical mercury emission factors for municipal waste combustors (MRI, 1993). It is also suggested that fluidized-bed combustors (FB) emit smaller amounts of mercury to the atmosphere compared to other sewage sludge incineration techniques, and particularly multiple hearth (MH) techniques.

Type and Efficiency of Control Equipment

The type and efficiency of control equipment is the major parameter affecting the amount of trace elements released to the atmosphere. Unlike other trace elements, mercury enters the atmosphere from various industrial processes in a gas form. The application of flue gas desulfurisation (FGD) has a very important impact on removal of mercury. A number studies have been carried out to assess the extend of this removal and parameters having major impact on this removal. These studies were reviewed in connection with the preparation of the EU Position Paper on Ambient Air Pollution by Mercury (Pirrone et al., 2001). It was concluded that the relatively low temperatures found in wet scrubber systems allow many of the more volatile trace elements to condense from the vapour phase and thus to be removed from the flue gases. In general, removal efficiency for mercury ranges from 30 to 50%. It was also concluded that the overall removal of mercury in various spray dry systems varies from about 35 to 85%. The highest removal efficiencies are achieved from spray dry systems fitted with downstream fabric filters.

Table 2. Emission factors for Hg, used to estimate European emissions of the element to the atmosphere in 2000.

Category	Unit	Emission factor
Coal combustion:	g/tonne coal	
- power plants		0.1-0.3
- residential and commercial boilers		0.5
Oil combustion	g/tonne oil	0.006
Non ferrous metal production		
- Cu smelters	g/tonne Cu produced	5.0-6.0
- Pb smelters	g/tonne Pb produced	3.0
- Zu smelters	g/tonne Zu produced	7.5-8.0
Cement production	g/tonne cement	0.1
Pig iron % steel production	g/tonne steel	0.04
Waste incineration	g/tonne wastes	
Municipal wastes		1.0
Sewage sludge wastes		5.0

EMISSION ESTIMATES

Emission estimates are carried out on the basis of information on emission factors and statistics on the production of industrial goods and the consumption of raw materials.

Emission factors are prepared for each source sector and raw material, separately. They can be evaluated on the basis of measurements or mass balance estimates for mercury during certain industrial process or certain application of the element.

The emission factors used for the estimates of the European emissions of Hg in the reference year 2000 are presented in Table 2.

The European emissions from anthropogenic sources in the year 2000 are presented in Table 3. These estimates were made with the use of emission data provided by national experts from a number of European countries. The emission factors from Table 2 were used for estimates for the rest of the European countries.

It can be concluded that a half of the European emissions in the year 2000 is emitted during combustion of fuels.

The second category consists of several industrial processes, including chlor-alkali production, non-ferrous and ferrous metal production and cement production. Other sources include waste incineration and various uses of mercury.

Table 3. Changes in total anthropogenic emissions of mercury in Europe since 1980 (in tonnes/year).

Source category	1980	1985	1990	1995	2000
Combustion of fuels	350	296	195	186	114
Industrial processes	460	388	390	143	99
Other sources	50	42	42	59	26
Total	860	726	627	338	239

Data in Table 3 also indicate steady decrease of Hg emissions in Europe during the last 2-3 decades. Major decline of Hg emissions in Europe occurred at the end of the 1980's and the beginning of the 1990's.

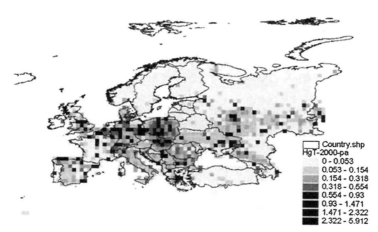

Figure 1. Spatial distribution of Hg emissions from anthropogenic sources in Europe in 2000 within the 50 km x 50 km grid (in t/grid).

This decrease was caused mainly by 1) the implementation of the FGD equipment in large power plants and the other emission controls in other industrial sectors, particularly in Western Europe, and 2) decline of economy in Eastern and Central Europe due to the switch in economies in these countries from centrally planned to market oriented.

Information on emissions from individual point sources and geographical location of these sources was then used to prepare Hg emission maps for Europe. The area source emissions, such as the emissions during combustion of fuels to produce heat in residential boilers were spatially distributed using the population density map as a surrogate parameter. The map for Hg emissions in 2000 from anthropogenic sources in Europe within the grid system of 50 km by 50 km is presented in Figure 1.

The areas where coal combustion is the main source of energy production are the regions with the highest emissions of mercury.

Estimates of emissions of mercury in its major physico-chemical forms were also approached (e.g. Pacyna et al., 2001). A summary of this work is presented in Table 4 in a form of Hg emission profiles for major emission sectors with regard to the three forms: elemental gaseous, bivalent gaseous, and elemental Hg on particles.

Table 4. Emission profiles of Hg from anthropogenic Sources.

Species	Coal combustion		Oil combustion	Non-Ferrous Metals		Pig & Iron	Caustic Soda	Waste Disposal	Other	Average	Source
	Power plants	Res. heat		Pb	Zn						
Hg^0 (gas)	0.5	0.5	0.5	0.8	0.8	0.8	0.7	0.2	0.8	0.64	Modified Pacyna, 1998
Hg^{II}	0.4	0.4	0.4	0.15	0.15	0.15	0.3	0.6	0.15	0.285	
Hg (p)	0.1	0.1	0.1	0.05	0.05	0.05	0	0.2	0.05	0.075	

EMISSION MEASUREMENTS

In order to improve the reliability of current estimates of atmospheric emissions of mercury, source test measurements of individual mercury species are required. With coal-fired power plants and waste incinerators being the dominant stationary anthropogenic sources of atmospheric mercury in Europe including the European Union and Eastern Europe, source test measurements in recent European projects have focused on stationary combustion sources. Concentrations of mercury associated with particulate matter, vapour phase elemental mercury, oxidised mercury and methyl mercury measured in stack gases of 7 fossil fuel-fired utility boilers and 3 thermal waste treatment facilities located in Europe are presented in this paper.

Emissions of mercury species from stationary combustion sources

Plant information

Within the two EU projects: "Optimal Utilisation of coal in modern power plants with respect to control of mass flows and emissions of VOCs/PAHs and mercury " (ECSC Coal Research Agreement No. 7220-ED/089) and "Mercury Species Over Europe (MOE) Mercury Species over Europe. Relative Importance of Depositional Methylmercury Fluxes to Various Ecosystems" (Contract ENV4-CT97-0595) (Munthe et al., 2001), measurements were carried out in ten stationary combustion sources.

Table 5. Main parameters for coal fired power plants reported in this chapter.

Source	Fuel	Output (electricity/total)	Stack gas cleaning technology
Power plant 1	Pulverised hard coal	170 850 MW	ESP
Power plant 2	Pulverised hard coal	150/270 MW	ESP, semi dry de-SOx, fabric filter
Power plant 3	Pulverised hard coal	640/340 MW	ESP, wet de-SOx, SCR
Power plant 4	Lignite	360/- MW	ESP
Power plant 5	Lignite	500/- MW	ESP
Power plant 6	Hard coal	170/ - MW	ESP
Power plant 7	Hard coal	330/ - MW	ESP, de SOx

Table 6. Main parameters for waste incinerators reported in the chapter.

Source	Waste type	Output (electricity/total)	Stack gas cleaning technology
Waste incinerator 1	Municipal waste	24 MW/ 400 000 GJ	ESP
Waste incinerator 2	Municipal waste	20/146 MW	ESP, wash reactor and condensing reactor, SNCR, fabric filter
Waste incinerator 3	Municipal waste	2*10 MW/120 GW	Cyclones, ESP, semi dry lime reactor, activated carbon feeding, fabric filter

The point sources were located in member states of the European Union and also in other European countries. All source test measurements were performed on the premises that the individual stationary combustion source may be kept anonymous. The measurement campaigns were performed in 1998-2000. A description of the point sources is given in Table 5 and 6 for coal fired power plants and waste incinerators, respectively.

Source test measurement methodology

Two different sampling approaches for the measurement of mercury in stack gases were employed. The MESA method and the modified standard test method were used for all stack gas measurements at all ten combustion sources. At two stationary point sources, an on-line speciation method and a denuder sampling method were also employed.

Six tests were usually conducted at each site. Samples were collected for 1-3 hours at gas flow rates of 0.5-2 l/min. During the test periods, operating data of the plant were obtained at each test site. After sampling, the solid and liquid samples were sealed and stored in the dark. The collected mercury was analysed using cold vapour atomic fluorescence spectrometry (CV-AFS) or cold vapour atomic absorption spectrometry (CV-AAS). Based on analytical results and operating data of each plant, mercury mass concentration in the stack was calculated. The concentrations of mercury species in the stack gases of the boilers measured are presented in Table 7 and 8 for coal fired power plants and waste incinerators, respectively.

These results suggest that vapour phase elemental mercury and oxidised mercury (mercury(II) chloride) are the dominant mercury species in stack gases from combustion sources equipped with both conventional and state-of-the-art flue gas cleaning systems. With the MESA method for vapour phase mercury speciation, the relative concentrations levels for vapour phase elemental mercury was found to be 41% for the coal-fired power plant equipped only with electrostatic precipitator and in the range from 0 to 53% for modern hard coal fired utility boilers. With the modified standard method, the relative concentration of elemental mercury vapour was found to be 43% for conventional hard coal fired power plants and from 43 to 75% for modern facilities. Traces of methylmercury were found in a few stack gas samples but the concentration fractions were always below 1% and are considered to be unreliable given the analytical difficulties in these complex samples. The conclusion is that waste incinerators and coal fired power plants are not sources of methylmercury emissions to air. Mercury associated with particulate matter was found to be also below 1% of total mercury in the flue gases. Lignite is an additional domestic fuel used in utility boilers in Europe. Emission estimates for mercury species released from conventional and modern brown coal fired power plants are presented in Table 7. Total mercury emission from a lignite fired power plant equipped only with electrostatic precipitator was found to be 26.9 µg Hg/m³ (13.2 µg Hg/MJ) with the MESA and 10.4 µg Hg/m³ (5.2 µg Hg/MJ) with the modified standard method. With the MESA method and the modified standard method, respectively, the concentration for total mercury in flue gas from a modern lignite fired power plant was found to be 5.1 µg Hg/m³ (1.03 µg Hg/MJ) and 4.9 µg Hg/m³ (0.98 µg Hg/MJ). It is concluded that the elevated concentration of total mercury from the conventional power plant is due to high mercury concentration in the lignite burned.

The calculated relative emissions of vapour phase elemental mercury was found to be from 61 to 99% for the conventional power plants and from 94 to 100% for the modern power plants. The relative concentration levels of mercury associated with particulate matter were found to be well below 1%.

Table 7. Concentrations of Hg in exhaust gas from coal fired power plants (MESA = Mercury Speciation and Adsorption method, MS = Modified standard method).

Source	Hg(II) μgm^{-3}		Hg^0 μgm^{-3}		Hg Tot μgm^{-3}		Emission factor (Tot Hg) $\mu g/mJ$	
	MESA	MS	MESA	MS	MESA	MS	MESA	MS
Power plant 1	2.6	2.7	1.8	2.0	4.4	4.7	1.9	1.2
Power plant 2	0.7	-	0	-	0.7	0.8	0.3	0.3
Power plant 3	2.1	0.8	2.4	2.4	4.5	3.2	1.9	2.0
Power plant 4	10.4	-	16.5	10.4	26.9	10.4	13	5.2
Power plant 5	0.3	0	4.8	4.9	5.1	4.9	1.0	0.98
Power plant 6	2.6	2.7	1.8	2.0	4.4	4.7	1.9	2.0
Power plant 7	0.8	-	0.6	-	1.4	-	-	0.61

Table 8. Concentrations of Hg from waste incinerators.

Source	Hg(II) μgm^{-3}		Hg^0 μgm^{-3}		Hg Tot μgm^{-3}		Emission factor (Tot Hg) mg/tonne waste	
	MESA	MS	MESA	MS	MESA	MS	MESA	MS
Waste incinerator 1	53.1	178	83.3	14	136	194	499	707
Waste incinerator 2	5.1	-	14.7	-	19.8	17.3	81	71
Waste incinerator 3	8	-	15.4	28.4	23.4	28.4	152	183

The concentration levels of total mercury in exhaust gases from waste incinerators were found to be 136 and 194 μg Hg/m^3 (499 and 707 mg Hg/ tonne waste burned) for waste incinerators equipped with only electrostatic precipitator and from 17 to 28 μg Hg/m^3 (71 to 183 mg Hg/tonne waste burned) for thermal waste treatment plant with state-of-the-art air pollution control technology. Measurements performed at thermal waste treatment facilities showed also that both vapour phase elemental mercury and oxidised mercury are the main mercury species released to the atmosphere.

For the vapour phase elemental mercury, the relative concentrations were found to be 61% for conventional waste incinerators and from 66 to 74% for modern facilities. From these results, it is concluded that the speciation pattern of vapour phase mercury deviate considerable from an expected relative proportion of Hg^0 estimated from thermodynamic equilibrium calculations. From the source test measurements, it is also concluded that methylmercury is very unlikely to be present in significant concentration levels in stack gases from waste incinerator equipped with both conventional and state-of-the-art flue gas cleaning systems.

Comparison between measurements and estimates

A comparison of total mercury concentrations in the exhaust gases measured during the EU project on Mercury over Europe (MOE) and estimated during the EU project on the Mediterranean Atmospheric Mercury Cycle System (MAMCS) (http://www.cs.iia.cnr.it/MAMCS/project.htm) is presented in Table 9.

Table 9. Comparison of Hg emission measurements (the MOE project) with estimates (the MAMCS project).

Point source	Country	Emission (Kg/yr)	
		Measured	Emission Factor estimates
Power plant	Poland	2990	2140
Waste incinerator	Sweden	32.4	80
Power plant	Finland	8.2	50
Power plant	Poland	90	145
Waste incinerator	Hungary	191	250
Power plant	Germany	69.8	175

The estimates are generally higher than the measurements, but the difference is within a factor of 2 except for a power plant in Finland. It should be added, however, that many more measurements need to be performed in order to obtain conclusive results with regard to the applicability of currently available emission factors for emission estimates for Hg from anthropogenic sources.

FINAL REMARKS

During the last decade major progress has been made to assess emissions from anthropogenic sources in Europe. This work resulted in the improvement of our knowledge of emission factors and emissions in Europe, as well as in other parts of the world (e.g. Pacyna et al., 2003). It can be concluded that combustion of fossil fuels to produce electricity and heat is the major emission source in Europe and worldwide. Future emissions of Hg from anthropogenic sources can be expected to decrease due to wider application of equipment for desulfurization in major power plants and industrial boilers, as well as wider use of renewable energy sources.

Source test measurement at both coal-fired power plants and waste incinerators in Europe have been performed to determine the concentration levels of mercury associated with particulate matter, and vapour phase elemental mercury, oxidised mercury and methylmercury in stack gases. Vapour phase elemental mercury and oxidised mercury are the main mercury species emitted form conventional and state-of-the-art stationary combustion sources. The relative proportion of mercury associated with particulate matter was found to be low for combustion sources equipped with both conventional and state-of-the-art flue gas cleaning installations. Methylmercury was found generally not occurring in stack gases from combustion sources.

More measurements of Hg emissions from at least major point sources are needed for making emission inventories more accurate. The measurements of Hg species in exhaust gases from these sources are also needed. Results obtained with current measurement methods were found to lie in the same order of magnitude for vapour phase elemental mercury and oxidised mercury. However, it is not conclusive whether the sampling methods determine the mercury species correctly.

REFERENCES

MRI Locating and estimating air emissions from sources of mercury and mercury compounds. Midwest Research Institute Report for the U.S. Environmental Protection Agency, Draft Rep. Research Triangle Park, NC. (EPA-454/R-93-023), 1993.

Munthe, J., Wängberg, I., Iverfeldt, Å. Lindqvist, O., Strömberg, D., Sommar, D., Gårdfeldt, K., Petersen, G., Ebinghaus, R, Prestbo, Larjava, K. and Siemens, V. Distribution of atmospheric mercury species in Northern Europe: Final Results from the MOE Project. *Atmos. Environ.*, 37, S9-S20, 2003.

Pacyna, J.M. Trace element emission from anthropogenic sources in Europe. Lillestrøm, Norwegian Institute for Air Research (NILU Technical Report No. 10/83) 1983.

Pacyna, J.M. Atmospheric emissions of arsenic, cadmium, lead and mercury from high temperature processes in power generation and industry. In: Lead, Mercury, Cadmium and Arsenic in the Environment, T.C. Hutchinson and K.M. Meema (eds.). Chichester, Wiley, pp. 69-87, 1987.

Pacyna E.G., Pacyna J.M. Pirrone N. Atmospheric mercury emissions in Europe from anthropogenic sources. *Atmos. Environ.*, 35, 2987-2996, 2001.

Pacyna, J.M., Pacyna, E.G., Steenhuisen, F., Wilson S. Mapping 1995 global anthropogenic emissions of mercury. *Atmos. Environ.*, 37-S, 109-117, 2003.

Pirrone, N., Munthe, J., Barregard, L., Ehrlich, H.C., Petersen, G., Fernandez, R., Hansen, J.C., Grandjean, P., Horvat, M., Steinnes, E., Ahrens, R., Pacyna, J.M., Borowiak, A., Boffetta, P., Wichman-Fiebig, M. Ambient Air Pollution by Mercury. Position Paper. Office for Official Publications of the European Communities, Brussels, Belgium, 2001.

Chapter-4

LEGISLATION AND POLICY CONCERNING MERCURY IN THE EUROPEAN UNION

Neil Emmott[1] and Martin Slayne[2]

[1] *European Commission, Directorate-General for Environment, Brussels, Belgium*
[2] *European Commission, Directorate-General for Health and Consumer Protection Brussels, Belgium*

INTRODUCTION

This paper presents an introductory overview of legislation and policy concerning mercury in the European Union (EU). It only summarises actions taken at the EU level, and does not attempt to cover the many actions taken individually by Member States or at other levels. Moreover, there is a very broad range of EU-level action, and the paper therefore only covers the main elements. Finally, where the paper attempts to summarise the requirements of EU legislation, some loss of precision is inevitable. Therefore, readers wishing to know the exact legal requirements should refer to the specific legal texts referenced[2].

INTEGRATED POLLUTION PREVENTION AND CONTROL (IPPC)

The purpose of the IPPC Directive[3] is to achieve integrated prevention and control of pollution arising from activities listed in Annex I of the Directive (energy industries, production and processing of metals, mineral industry, chemical industry, waste management and other activities like intense livestock farming, pulp and paper industry and tanneries). These include some of the major sources of mercury emissions.

The Directive lays down the requirement to prevent or, where that is not practicable, to reduce pollution of the air, water and land, including from mercury and its compounds, from the above-mentioned activities, including measures concerning waste, in order to achieve a high level of protection of the environment taken as a whole. Control is to be achieved by way of a permitting regime whereby the operator of an installation applies for a permit and a competent authority determines whether or not a permit is to be issued. Among other requirements, permits are to include emission limit values (or equivalent parameters or technical measures) which are to be based on the "Best Available Techniques" (BAT) for the sector.

The Directive entered into force on 30 October 1999. New installations, and substantial changes to existing installations, require a permit issued in accordance with the Directive before they are brought into operation. Existing installations must be brought into compliance with the requirements of the Directive no later than 30 October 2007.

In order to support the implementation of the Directive the Commission is producing a series of BAT Reference documents (BREFs) for the main industry sectors under the Directive. An important document concerning mercury is the BREF on chlor-alkali manufacturing[4]. This concludes that mercury cells are not BAT.

As the basis for a "European Pollutant Emission Register" (EPER)[5], Member States are also required to submit reports to the Commission on emissions from all individual facilities with one or more activities mentioned in Annex I to the IPPC Directive. The reports must include details of emissions to air and water for all pollutants for which the thresholds specified in an Annex are exceeded. The reporting thresholds for mercury and its compounds are 10 kg/year for emissions to air and 1 kg/year for emissions to water. The data from the first reporting cycle, for 2001, were published in February 2004[6]. The next reporting year is 2004, for publication in 2006.

WATER QUALITY

The water framework Directive[7] establishes a framework for the protection of inland surface waters, transitional waters, coastal waters and groundwater. Article 16 provides for the adoption of EU measures for substances included in a list of priority substances, i.e. those which present a significant risk to or via the aquatic environment. Mercury is identified as one of the "priority hazardous substances"[8] that are subject to cessation or phasing-out of emissions, discharges and losses within 20 years after adoption of measures. The Commission is currently developing proposals for emission controls for point sources and environmental quality standards for mercury and other substances.

The water framework Directive also provides for the review, revision and possible repeal of a number of pre-existing Directives, including Directives dealing with discharges of dangerous substances to water, and with protection of groundwater. The dangerous substances Directive[9] requires Member States to take appropriate steps to eliminate pollution in inland surface, territorial and internal coastal waters by various substances including mercury and its compounds. Articles 5 and 6 lay down the provisions for authorisation of discharges and provide that Member States can choose whether to base their authorisations on emission limit values or quality objectives. Two specific "daughter" Directives deal with mercury in more detail. One[10] provides for specific emission limit values and quality objectives applicable to discharges of mercury from the chlor-alkali electrolysis industry. The second[11] does the same for other industry sectors, and also stipulates the requirement to draw up programmes to avoid or eliminate pollution caused by discharges of mercury from diffuse sources.

The groundwater Directive[12] aims to prevent the pollution of groundwater by substances set out in two lists in an Annex, and as far as possible to check or eliminate the consequences of pollution which has already occurred. Mercury and its compounds are included in the "List I of Families and Groups of Substances", to which the most stringent requirements apply. Direct discharges (introduction without percolation through the ground or subsoil) of substances in List I into groundwater are prohibited. Any disposal or tipping of List I substances which might lead to indirect discharge (introduction after percolation through the ground or subsoil) must be subject to prior investigation. Member States must then prohibit such activity, or authorise it provided that all the technical precautions necessary to prevent such discharge are observed. In addition, all appropriate measures deemed necessary must be taken to prevent any indirect discharge of List I substances due to activities on or in the ground other than disposal or tipping.

Article 11 of the water framework Directive repeats the prohibition of direct discharges of pollutants into groundwater. However, the former is concerned with protecting groundwater not just against pollution from discharges and disposals, but also from other activities. Article 17 calls for the adoption of specific measures to prevent and control groundwater pollution, with the aim of achieving "good groundwater chemical status".

The Commission has therefore proposed a further Directive[13] which includes criteria for the assessment of good groundwater chemical status, and for the identification and reversal of significant and sustained upward trends and the definition of starting points for trend reversals. For mercury, the proposed Directive would not itself set any threshold values, but rather it would require Member States to do so.

AIR QUALITY

The air quality framework Directive[14] defines the basic principles for a common approach for the assessment and management of ambient air quality in the EU. Details of the specific requirements for particular pollutants are set out in daughter directives.

An agreement has been achieved with the European Parliament and Council on a proposal from the Commission[15] for the fourth daughter Directive, relating to arsenic, cadmium, nickel, mercury and polycyclic aromatic hydrocarbons (PAH). Formal adoption is foreseen to take place in the course of autumn 2004. Concentrations of mercury in ambient air in the EU generally are below a level believed to have adverse effects on human health. Therefore, mercury in ambient air is not regulated via a target value in the fourth daughter Directive. However, regardless of the concentration level, all substances covered by the measure, including mercury, are to be measured at background sampling points with a spatial resolution of 100,000 km^2 in order to provide information on geographical variation and long-term trends. The same requirements are laid down for deposition measurements of heavy metals and PAH. Monitoring of particulate and gaseous divalent mercury is also recommended.

USE OF MERCURY

Electrical and Electronic Equipment

The Directive on the restriction of hazardous substances (RoHS) in electrical and electronic equipment[16] requires the substitution of mercury, among other substances, in new electrical and electronic equipment by 1 July 2006. Applications of mercury in fluorescent lamps up to certain levels are exempted. Each exemption must be reviewed at least every four years with the aim of considering deletion.

Closely related to the RoHS Directive is the Directive on waste electrical and electronic equipment (WEEE)[17]. This aims to prevent the generation of WEEE and to support the reuse, recycling and other forms of recovery of such waste. It also seeks to improve the environmental performance of all operators involved in the life cycle of electrical and electronic equipment. In particular, it provides that producers, or third parties acting on their behalf, must set up systems by 13 August 2004 to provide for the treatment of WEEE using best available treatment, recovery and recycling techniques. Member States must achieve a high level of separate collection for WEEE, and any mercury-containing components must be removed from any separately-collected WEEE.

Vehicles

The end-of-life vehicles (ELV) Directive[18] aims, as a first priority, at the prevention of waste from vehicles. It also lays down measures relating to the reuse, recycling and other forms of recovery of ELVs and their components so as to reduce the disposal of waste. According to Article 4 of this Directive mercury, among other substances, is restricted in materials and components of vehicles. In particular, Member States must ensure that materials and components of vehicles put on the market after 1 July 2003 do not contain mercury other than in bulbs and instrument display panels. In addition, under Article 6 Member States must ensure that ELVs are stored and treated in accordance with minimum specified technical requirements, including the removal, as far as possible, of all components identified as containing mercury.

Batteries

The batteries Directive[19] prohibits the marketing of batteries and accumulators containing more than 0.0005% of mercury by weight. Button cells with a mercury content of no more than 2% by weight are exempted. The Directive also requires Member States to take appropriate steps to ensure that spent batteries and accumulators are collected separately with a view to their recovery or disposal, and that batteries and accumulators are marked with information on separate collection, recycling and heavy metal content.

In November 2003 the Commission adopted a proposal for a Directive that would replace and repeal the current batteries Directive. The limit on mercury content by weight of 0.0005%, and the exemption for button cells, would be retained. The explanatory memorandum that accompanies the new proposal notes that mercury consumption in batteries has declined significantly in the EU, but that many mercury batteries produced before the restrictions of the current Directive entered into force are still in use. The new proposal aims to establish a closed loop system for all batteries to avoid their disposal by incineration or landfill. It would also require Member States to set up national collection systems so that consumers can return spent portable batteries free of charge.

Pesticides and Biocides

According to Article 3 of Council Directive 79/117/EEC[21], which took effect in 1981, plant protection products containing one or more of the following active substances may be neither placed on the market nor used: mercury oxide, mercurous chloride (calomel), other inorganic mercury compounds, alkyl mercury compounds, alkoxyalkyl and aryl mercury compounds. An amendment[22] in 1991 deleted some limited exemptions from these restrictions which had previously been allowed.

Biocidal product cannot be placed on the market and used in the territory of the Member States unless authorised in accordance with Directive 98/8/EC[23]. No biocidal products containing mercury have been authorised and accordingly they are banned in the EU.

Cosmetics

Under Directive 76/768/EEC[24], mercury and its compounds may not be present as ingredients in cosmetics, including soaps, lotions, shampoos, skin bleaching products, etc. (except for phenyl mercuric salts for conservation of eye makeup and products for removal of eye make-up in concentrations not exceeding 0.007 percent weight-to-weight) marketed within the EU.

Other Uses

Directive 76/769/EEC[25] creates a framework legislative procedure under which the EU may ban or restrict the use of hazardous chemicals by adding the substances and controls to an Annex. Additions of chemicals have been done in several amendments. The following uses of mercury compounds were prohibited by Directive 89/677/EEC[26]: marine anti-fouling agents, wood preservatives, impregnation of heavy-duty industrial textiles and yarn, and treatment of industrial waters.

More broadly, the Commission has recently proposed a major new EU chemicals regime[27]. This will eventually repeal the framework Directive 76/769, encompass the various controls adopted under it, and provide a more streamlined procedure for the adoption of any further restrictions.

EXPORT AND IMPORT

Regulation 304/2003[28] implements the Rotterdam Convention on the Prior Informed Consent (PIC) Procedure for Certain Hazardous Chemicals and Pesticides in International Trade. The Convention provides for an exchange of information between its parties on restrictions on hazardous chemicals and pesticides and their import and export. The trigger for action is when a party takes regulatory action to ban or severely restrict a hazardous chemical or pesticide in its own territory in order to protect human health and/or the environment. The party must then notify the Secretariat of the Convention of that ban or restriction. It should also make export of the substance subject to a notification procedure, whereby the first export annually to any party would have to be notified in advance to the designated authority in that country of destination. This obligation ends when the substance becomes subject to the PIC procedure and the importing party has given an import decision (see below).

When two notifications of bans or severe restrictions for the same substance are received under the Convention from two geographic regions, a chemical review committee will consider whether these meet the criteria of Annex II to the Convention. The committee may recommend that the substance be added to the PIC procedure and prepare a decision guidance document (DGD), containing relevant information to help parties take informed decisions on whether or not to accept imports. If the Conference of the Parties decides that the chemical should be included in the PIC procedure, the DGD is circulated and all parties should communicate an import decision to the Secretariat on whether and under what circumstances they wish to receive imports of the substance. Exporting parties are then obliged to ensure that their exporters comply with these wishes.

Mercury compounds are listed in Annex I, Part 1 to the Regulation as banned or severely restricted within the EU and are thus subject to the export notification requirements, which are laid down in Article 7 of the Regulation. These requirements apply to exports to all countries. Mercury compounds used as pesticides, including inorganic mercury compounds, alkyl mercury compounds and alkyloxyalkyl and aryl mercury compounds, are also included in Part 3 of Annex I to the Regulation as chemicals subject to the PIC procedure. Thus, in accordance with Article 13 of the Regulation, *inter alia*, EU exporters must comply with the import decisions taken by third countries. The Regulation in fact goes further than the Convention in this respect, in that it requires exports of PIC substances to have the explicit consent of the importing country (whereas under the Convention exports would, after a certain period of time, be permitted to a country that has failed to communicate an import decision). Like export notification, this requirement extends to exports to all countries, irrespective of whether or not they are parties to the Convention.

The Regulation also bans the export from the EU of certain chemicals and articles, listed in Annex V. Cosmetic soaps containing mercury are subject to this ban.

WASTE MANAGEMENT

In addition to the waste management provisions linked to certain product groups described above (e.g. ELVs, batteries, WEEE), there are also various broader requirements in EU legislation.

The Waste Framework Directive and Hazardous Waste

The main basis for waste management in the EU is the waste framework Directive[29]. This requires that Member States take the necessary measures to ensure waste is recovered or disposed of without endangering human health and without using processes or methods that could harm the environment. It includes various provisions – for example relating to the control of waste management facilities – to support this requirement.

Directive 91/689/EEC[30] introduces an additional, more stringent, layer of controls applicable to hazardous waste on top of those that apply under the waste framework Directive. Wastes are identified as hazardous based on properties listed in Annex III of the Directive. By Decision 2000/532/EC[31] a list of waste was adopted, which includes "Sludges containing mercury" (as a waste from natural gas purification), "Waste containing mercury", "Mercury containing batteries", "Amalgam waste from dental care" and "Fluorescent tubes and other mercury containing waste". The determination that waste is hazardous has implications in respect of the application of other

EU measures. For example, Regulation 259/93[32] prevents hazardous waste being exported to non-OECD countries.

Waste Incineration

The waste incineration Directive[33] aims to prevent or to limit as far as practicable negative effects on the environment, in particular pollution by emissions into air, soil, surface water and groundwater, and the resulting risks to human health, from the incineration and co-incineration of waste. Emission limit values for discharges of waste water from exhaust gas cleaning at incineration plants are established in Annex IV of the Directive. The limit value for mercury is 0.03 mg/l. Air emission limit values for incineration plants are set out in Annex V. The limit value for mercury is 0.05 mg/m^3, as an average value over a minimum period of 30 minutes and a maximum of 8 hours (a limit of 0.1 mg/m^3 applies until 1 January 2007 for existing plants for which the permit to operate was granted before 31 December 1996). Mercury in emissions to air has to be measured at least twice per year; mercury in emissions to water at least once per month.

Most waste incineration facilities will also fall under the scope of the IPPC Directive (see above). Where the application of the IPPC Directive would entail stricter requirements than those of the waste incineration Directive, then these stricter requirements take precedence. Work on a BREF document on waste incineration is underway.

Landfills

The landfill Directive[34] aims to prevent or reduce negative effects on the environment and risk to human health from the landfilling of waste. Article 4 requires that Member States classify landfills into those for hazardous waste, those for non-hazardous waste and those for inert waste. Member States must also ensure that certain wastes are not accepted in a landfill. These include liquid waste, and any other waste that does not fulfil the "acceptance criteria" determined in accordance with an Annex. These acceptance criteria were set out in Decision 2003/33/EC[35]. They include specific mercury leaching values for wastes acceptable at the different classes of landfill.

Articles 7 and 8 of the landfill Directive require that operators of landfills apply for permits and that competent authorities ensure that certain conditions will be met in those cases where landfilling is authorised. One such condition is that landfills comply with certain technical standards set in an Annex, for example concerning protection of soil and water. Another is that operators maintain adequate financial security to meet their obligations, including after-care.

Sewage Sludge

Directive 86/278/EEC[36] aims to regulate the use of sewage sludge in agriculture in such a way as to prevent harmful effects on soil, vegetation, animals and humans, while encouraging its correct use. Member States must prohibit the application of sewage sludge to soil where the concentration of one or more metals in the soil exceeds the limit values laid down in a first Annex. For mercury, the soil limit value is 1 to 1.5 mg/kg of dry matter for soils with a pH higher than 6 and lower than 7. Member States must also regulate the use of sludge such that the accumulation of heavy metals in soil does not exceed the limit values. A possible revision of Directive 86/218/EEC is being considered as part of the development of the EU's broader thematic strategy on soil[37].

LIMITING HUMAN EXPOSURE

Drinking Water

Directive 98/83/EEC[38] sets standards for the quality of drinking water. According to Article 5 and Annex I, a maximum level of 1.0 µg/l is specified for mercury in drinking water.

Worker Health and Safety

Directive 98/24/EC[39] lays down minimum requirements for the protection of workers from risks to their safety and health arising, or likely to arise, from the effects of all chemical agents that are present at the workplace or as a result of any work activity involving chemical agents. Consequently this framework Directive regulates all substances including mercury and its compounds.

The EU Scientific Committee on Occupational Exposure Limits (SCOEL) has held extensive discussions on mercury and mercury compounds in order to come up with a Recommendation to the Commission for an occupational exposure limit value. The Committee has proposed levels of 0.02 mg/m^3 as an 8-hour time-weighted average, and 0.01 mg/l in blood and 0.03 mg/g creatinine in urine as biological limit values.

Mercury Contamination in Food

Under Commission Regulation 466/2001[40], a maximum level of 0.5 mg/kg wet weight is set for mercury in fishery products, with the exception of certain fish species for which a separate maximum level of 1 mg/kg wet weight applies.

On 24 February 2004, responding to a request from the Commission, the Scientific Panel on Contaminants in the Food Chain of the European Food Safety Authority (EFSA) adopted an opinion on mercury and methylmercury in food[41]. This took into account the decision in June 2003 of the FAO/WHO Joint Expert Committee on Food Additives to revise its Provisional Tolerable Weekly Intake for methylmercury from 3.3 to 1.6 µg/kg body weight[42]. EFSA also took account of a lower Reference Dose of 0.7 µg/kg body weight per week established by the US National Research Council[43]. It compared these levels against data gathered by the EU Member States and Norway on levels of mercury in foods and estimates of dietary exposure as part of a scientific co-operation (SCOOP) task[44]. The EFSA opinion and the SCOOP report should be referred to directly to see their full analyses and conclusions in context. Some selected findings are given below.

The EFSA opinion concluded that the reduction of the PTWI for methylmercury by JECFA, from 3.3 to 1.6 µg/kg body weight, was justified because rather than focusing on risks to the general population it was based on the most susceptible lifestage, i.e. the developing foetus and intake during pregnancy. Comparison with the lower US NRC recommendation may offer additional guidance.

The estimated intakes of mercury in Europe varied by country, depending on the amount and type of fish consumed. Based on the SCOOP document, national average exposures to methylmercury[45] from fish and seafood products were between 1.3 and 97.3 µg/week, corresponding to <0.1 to 1.6 µg/kg body weight per week (assuming a 60 kg adult body weight). Hence the highest average intake estimates were just at the PTWI, thereby exceeding the US NRC recommendation.

In general, EU consumers who eat average amounts of varied fishery products are not likely to be exposed to unsafe levels of methyl mercury. However, people who eat more than average amounts of fish are more likely to exceed these recommended safety thresholds. In particular, population groups who frequently consume top predatory fish, such as swordfish and tuna, may have a considerably higher intake of methylmercury and exceed the PTWI. The range of high exposure[46] was estimated to be between 0.4 and 2.2 µg/kg body weight per week of methylmercury.

The SCOOP data showed that, although the population in Norway had the highest total consumption of fish and seafood products, the estimated high intake of methylmercury from these foods was lower in Norway than in southern European countries. The reason for this is probably that the type of fish consumed in Norway consists of species, such as cod and saithe, containing relatively low levels of methylmercury. The consumption of top predatory fish, such as swordfish and tuna, which can contain higher levels of methylmercury, may be significantly greater in countries in southern Europe.

A probabilistic analysis carried out by EFSA using the French data from the SCOOP report suggested that, based on the distribution of consumption and fish contamination, in France 11.3% of 293 children aged 3 - 6 years would exceed the JECFA PTWI for methylmercury and 44% would exceed the US NRC recommendation. The figures for 248 adults were 1.2% and 17% respectively. However, the figures for children exceeding the PTWI are likely to represent an overestimate, because young children often tend to eat fish from species that are more likely to contain only low levels of methyl mercury, such as the white fish in fish fingers/ fish sticks. It is also important to note that some of the calculated high intakes may be overestimates in view of limitations on the available data, as indicated in the SCOOP report.

Specific intake data for pregnant women were not available for the EU risk assessment, although the EFSA has highlighted the need to generate reliable intake data from studies focused on women of childbearing age.

In view of the revised safety thresholds and risk assessment advice on dietary intake of methyl mercury, the EU maximum levels for mercury in fishery products are being reviewed and other risk management options are being considered. Initial assessment indicates that it might be difficult to further lower the maximum levels without either making them unachievable for many fish species or overcomplicating the legislation.

As an alternative approach, the European Commission has issued an information note on methyl mercury in fish and fishery products. [47] This has been distributed via consumer and public health networks, to help ensure that the information reaches the targeted vulnerable groups. The note contains advice on fish consumption for women who might become pregnant, who are pregnant or breastfeeding and for young children. It advises that these consumers should not eat more than one small portion (<100 g) per week of large predatory fish, such as swordfish, shark, marlin and pike, and that if they eat this portion, they should not eat any other fish during this period. It also advises that they should not eat tuna more than twice per week. (It is easy to calculate that the PTWI of 1.6 µg/kg body weight equals 96 µg/week for an average 60 kg adult and this amount would be present in 96 g of fish if it contains 1 mg/kg methyl mercury. Swordfish, shark, marlin and pike can often contain such a level.) EU consumers are also advised to pay attention to any more specific advice given by national authorities in light of local or regional consumption characteristics. This advice is roughly in line with advice issued in the USA[48] and Australia and New Zealand. [49] Further discussions on the world-wide approaches to risk management of methyl mercury are planned. The European Community is leading a working group to prepare a discussion paper for the 37th session of the Codex Committee on Food Contaminants in 2005.

DEVELOPMENT OF AN EU MERCURY STRATEGY

The largest present user of mercury in the EU is the chlor-alkali industry. However, the use of mercury in this industry sector is being phased out as "mercury cell" technology is replaced with mercury-free processes. At the EU Environment Council meeting of 7 June 2001, the Council called upon the Commission to clarify the legal situation regarding the conversion of the chlor-alkali industry, identify the possible consequences for the use of mercury and report to the Council on the potential need for co-ordinated action in the EU and the accession countries.

In response to the Council's request, in December 2002 the Commission presented a report to the Council concerning mercury from the chlor-alkali industry[50]. This reviewed mercury production and use generally, use of mercury in the chlor-alkali industry, legal issues concerning the conversion to mercury-free technology and consequences of the mercury-cell phase-out. In relation to the consequences of the mercury cell phase-out, the report analysed different scenarios concerning the fate of the surplus mercury expected to arise in the EU[51].

The Council reacted to the report by inviting the Commission to present "a coherent strategy with measures to protect human health and the environment from the release of mercury based on a life-cycle approach, taking into account production, use, waste treatment and emissions". The EU mercury strategy will therefore look at all aspects of the mercury problem – production and supply, trade, use in products and processes, emissions, recovery and disposal, and exposure – rather than just those aspects relating to the chlor-alkali industry[52]. The strategy is due to be published in 2004.

NOTES

1. The authors are officials in the Commission services who work on aspects of legislation and policy concerning mercury. However, the paper does not necessarily represent the views of the Commission.
2. Legislative texts can be accessed online at: http://europa.eu.int/eur-lex/en/index.html.
3. Council Directive 96/61/EC of 24 September 1996 concerning integrated pollution prevention and control (OJ L 257, 10.10.1996).
4. See http://eippcb.jrc.es/pages/FActivities.htm
5. Commission Decision 2000/479/EC of 17 July 2000 on the implementation of a European pollutant emission register (EPER) according to article 15 of Council Directive 96/61 concerning integrated pollution prevention and control (OJ L192, 28.7.2000).
6. See www.eper.cec.eu.int.
7. Directive 2000/60/EC of the European Parliament and of the Council of 23 October 2000 establishing a framework for Community action in the field of water policy (OJ L 327, 22.12.2000).
8. Decision 2001/2455/EC of the European Parliament and of the Council of 20 November 2001 establishing the list of priority substances in the field of water policy (OJ L 331, 15.12.2001).
9. Council Directive 76/464/EEC of 4 May 1976 on pollution caused by certain dangerous substances discharged into the aquatic environment of the Community (OJ L 129, 18.5.1976).
10. Council Directive 82/176/EEC of 22 March 1982 on limit values and quality objectives for mercury discharges by the chlor-alkali electrolysis industry (OJ L 81, 27.3.1982).
11. Council Directive 84/156/EEC of 8 March 1984 on limit values and quality objectives for mercury discharges by sectors other than the chlor-alkali electrolysis industry (OJ L 74, 17.3.1984).
12. Directive 80/68/EEC on the protection of groundwater against pollution caused by certain dangerous substances (OJ L 20, 26.1.1980).
13. Proposal for a Directive of the European Parliament and of the Council on the protection of groundwater against pollution, COM (2003) 50 final, 19.9.2003.
14. Council Directive 96/62/EC of 27 September 1996 on ambient air quality assessment and management (OJ L 296, 21.11.1996).
15. Proposed Directive of the European Parliament and of the Council relating to arsenic, cadmium, mercury, nickel and polycyclic aromatic hydrocarbons in ambient air, COM (2003) 423 final, 16.7.2003.
16. Directive 2002/95/EC of the European parliament and of the Council of 27 January 2003 on the restriction of the use of certain hazardous substances in electrical and electronic equipment (RoHS) (OJ L 37, 13.2.2003).
17. Directive 2002/96/EC of the European parliament and of the Council of 27 January 2003 on waste electrical and electronic equipment (WEEE) (OJ L 37, 13.2.2003).
18. Directive 2000/53/EC of the European Parliament and of the Council of 18 September 2000 on end-of-life vehicles (OJ L 269, 21.10.2000).
19. Council Directive 91/157/EEC of 18 March 1991 on batteries and accumulators containing certain dangerous substances (OJ L 078, 26.3.1991) adapted to technical progress by Commission Directives 93/86/EEC of 4 October 1993 (OJ L 264, 23.10.1993) and 98/101/EC of 22 December 1998 (OJ L 1, 5.1.1999).

20. Proposal for a Directive of the European Parliament and of the Council on batteries and accumulators and spent batteries and accumulators, COM (2003) 723 final, 21.11.2003.

21. Council Directive 79/117/EEC of 21 December 1978 prohibiting the placing on the market and use of plant protection products containing certain active substances (OJ L 33, 8.2.1979).

22. Commission Directive 91/188/EEC of 19 March 1991 amending for the fifth time the Annex to Council Directive 79/117/EEC prohibiting the placing on the market and use of plant protection products containing certain active substances (OJ L 92, 13.4.1991).

23. Directive 98/8/EC of the European Parliament and of the Council of 16 February 1998 concerning the placing of biocidal products on the market (OJ L 123, 24.4.1998).

24. Council Directive of 27 July 1976 on the approximation of the laws of the Member States relating to cosmetic products (OJ L 262, 27.9.1976).

25. Council Directive 76/769/EEC of 27 July 1976 on the approximation of the laws, regulations and administrative provisions of the Member States relating to restrictions on the marketing and use of certain dangerous substances and preparations (OJ L 262, 27.9.1976).

26. Council Directive 89/677/EEC of 21 December 1989 amending for the 8th time Directive 76/769/EEC on the approximation of the laws, regulations and administrative provisions of the Member States relating to restrictions on the marketing and use of certain dangerous substances and preparations (OJ L 398, 30.12.1989).

27. Proposal for a Regulation of the European Parliament and of the Council concerning the Registration, Evaluation, Authorisation and Restriction of Chemicals (REACH), establishing a European Chemicals Agency and amending Directive 1999/45/EC and Regulation (EC) {on Persistent Organic Pollutants}, COM(2003) 644 final, 29.10.2003.

28. Regulation (EC) No. 304/2003 of the European Parliament and of the Council of 28 January 2003 concerning the export and import of dangerous chemicals (OJ L 63, 6.3.2003).

29. Council Directive 91/156/EEC of 18 March 1991 (OJ L 78, 26.3.1991), amending Directive 75/442/EEC on waste (OJ L 194, 25.7.1975).

30. Council Directive 91/689/EEC on hazardous waste (OJ L 47, 16.2.2001).

31. Commission Decision 2000/532/EC (OJ L 226, 6.9.2000) as amended by Council Decision 2001/532/EC (OJ L 203, 28.7.2001) as regards the list of wastes pursuant to Article 1(4) of Council Directive 91/689/EEC on hazardous waste.

32. Council Regulation (EEC) No 259/93 of 1 February 1993 on the supervision and control of shipments of waste within, into and out of the European Community (OJ L 30, 6.2.1993).

33. Directive 2000/76/EC of the European Parliament and of the Council of 4 December 2000 on the incineration of waste (OJ L 332, 28.12.2000).

34. Council Directive 1999/31/EC of 26 April 1999 on the landfill of waste (OJ L182, 16.7.1999).

35. Council Decision 2003/33/EC of 19 December 2002 establishing criteria and procedures for the acceptance of waste at landfills pursuant to Article 16 of and Annex II to Directive 1999/31/EC (OJ L11, 16.1.2003).

36. Council Directive 86/278/EEC of 12 June 1986 on the protection of the environment, and in particular of the soil, when sewage sludge is used in agriculture (OJ L181, 4.7.1986).

37. See the Commission's Communication towards a Thematic Strategy for Soil Protection, COM (2002) 179 final, 16.4.2002.

38. Council Directive 98/83/EEC of 3 November 1998 on the quality of water intended for human consumption (OJ L 330 of 5.12.1998).

39. Council Directive 98/24/EC of 7 April 1998 on the protection of the health and safety of workers from the risks related to chemical agents at work (OJ L 131, 5.5.1998).

40. Commission Regulation (EC) No 466/2001 of 8 March 2001 setting maximum levels for certain contaminants in foodstuffs (OJ L 77, 16.3.2001) as amended by Commission Regulation (EC) No 221/2002 of 6 February 2002 (OJ L 37, 7.2.2002).

41. See http://www.efsa.eu.int/science/contam_ panel/contam_opinions/259_ en.html.

42. See http://who.int/pcs/jecfa/Summary61.pdf

43. See *Toxicological effects of methylmercury*. Committee on the Toxicological Effects of Methylmercury, National Research Council, National Academy Press, Washington, D.C., 2000.

44. See http://europa.eu.int/comm/food/food/ chemicalsafety/contaminants/scoop _3-2-11_heavy_metals_report_en.pdf.

45. The SCOOP data recorded total mercury rather than methylmercury. Methyl mercury is the chemical form of concern and can make up to more than 90% of the total mercury in fish and seafood. The EFSA opinion based its calculations on the conservative assumption that all the mercury in fish and seafood products is methylmercury.

46. High exposure is measured at the 95th or 97.5^{th} percentile of the distribution for fish- and seafood product consumption depending on the country considered.

47. See http://europa.eu.int/comm/food/food/chemicalsafety/contaminants/information_note_mercury-fish_12-05-04.pdf

48. See http://www.cfsan.fda.gov/~ dms/admehg3.html

49. See http://www.foodstandards.gov.au/_ srcfiles/brochure_mercury _ in_fish_0304v2.pdf

50. COM (2002) 489 final, 6.9.2002.

51. The amount of remaining mercury expected to be decommissioned as a result of the phase-out of mercury cells is estimated by Euro-Chlor at 11,600 tonnes.

52. Details of the development of the strategy can be found on the Commission's website: http://europa.eu.int/comm/environment/chemicals/mercury/index.htm.

Chapter-5

THE DG RESEARCH PERSPECTIVE – RESEARCH ON MERCURY SUPPORTED BY THE EUROPEAN COMMISSION

Hartmut Barth

European Commission, Directorate General Research, Directorate I: Environment Brussels, Belgium

PAST AND CURRENT EU FUNDED RESEARCH

The current development and future implementation of the EU Strategy on mercury has to be based on scientific knowledge in order to provide an informed policy action, but is also indicating gaps of knowledge which will need to be addressed by research, including EU-supported research. The research element of the Mercury Strategy will have to reply to the following questions:

- Which research on mercury has the EU already funded?
- What further research issues are already under consideration that are relevant to mercury?
- What are the key gaps to be identified as research priorities in support to the EU Mercury Strategy?
- How could the EU assist to fill these gaps?

This chapter will be limited to address environment oriented research only, both highlighting key results from past and ongoing EU-funded projects on mercury and will conclude with perspectives regarding future research options.

Over the past 20 years European Commission funded research specifically addressing the origin, transformation, fate, bioaccumulations, and environmental effects of mercury, in particular regarding its most toxic forms mono – and dimethylmercury, was concentrated to a relatively small number of projects which were however well designed and successful in regard to their contributions to European and international mercury assessments. Therefore, I like to present a brief overview of those projects and attempt to highlight a few results and/or conclusions from those "historical" projects which seem to me still valid and important.

During the mid to late 1980s a group of marine chemists, marine biologists and chemical analysts, coordinated by Dr. Michael Bernhard of the marine branch of the Italian research centre ENEA investigated the *"Origin, fate and environmental effects of methylmercury"* through laboratory and field research. This project illustrated that:

" *The actual mean methylmercury (MeHg) concentrations in coastal marine waters of the Mediterranean Sea can not only be explained by microbial methylation processes, because the more than 100 bacterial strains which were known from the scientific literature as strong mercury methylaters did actually methylate inorganic mercury in laboratory experiments only at very low rates or not at all, when they were cultured with realistically low inorganic mercury concentrations as they are typically found in marine coastal waters.* "

The biomagnification process of methylmercury in long-living marine predators, such as tuna, dolphins, Mediterranean harbour seals etc was confirmed as well as the positive correlation between the MeHg concentrations in muscle issue with length/age class.

A screening exercise on MeHg concentrations in the upper 5-10 cm of various soil profiles and in wild terrestrial animals revealed that methylmercury may accumulate to as high values than in the marine foodchain. A striking example was found in red deer muscle tissue reaching equal or higher MeHg concentrations than tuna. However, the small number of species and probes tested did not allow any statistical confirmation of those results and hence they have never been published (except in progress reports to the Commission).

Based on the above results, Dr. Bernhard and his colleagues speculated that an inorganic methylation mechanism might contribute to the relatively high MeHg concentration in coastal marine waters. They proposed an UV energy driven inorganic methylation at the nepheloid layer above the sediments of shallow waters.

During the same period Prof. Nürnberg of the Nuclear and Environment Research Centre in Jülich (KFA-Jülich, Germany) investigated the metal concentrations in the water column and sediments of the North Sea and English Channel, including total mercury concentrations, which was paralleled by investigations by the German Fisheries Research Centre on malformations, skin lesions and other illnesses of bottom-living flatfish (e.g. flounder) in the North Sea. These two independent investigations revealed striking correlations between areas of high metal concentrations in the sediment and the highest number of flatfish with illnesses. Clearly however, those results cannot only be attributed to mercury, but to "cocktails" of high metal concentrations. More importantly, Prof. Nürnberg started to measure the natural background concentrations of total mercury, by analysing deep sea water and sediment probes from the Pacific Ocean with an improved gold-plated stripping voltammetry methodology under ultra-clean conditions in a specific newly constructed laboratory, applying completely automised approaches and respecting extreme precautions against secondary contamination of the samples. Regretfully, he died during one of the project cruises for the EU research project, and I have never seen his potentially highly important results published in the peer-reviewed international scientific literature. From my memory I recall that his "natural background values" were roughly one order of magnitude lower than the published background data until that time. A demonstrated knowledge of such reference values for total mercury concentrations in the marine environment would have had implications for the correct assessment and interpretation of the anthropogenic contribution to the mercury pool.

In the NW-Mediterranean basin the biogeochemical cycle of mercury has attracted much attention, due to the Mediterranean geochemical mercury anomaly, i.e. the leaching from the large cinnabar deposits. In contrast to what might be expected this "anomaly" does not lead to higher ambient mercury levels outside the immediate coastal zone in the vicinity of the ore deposits, and MeHg concentrations in specimen of fish caught off-shore are usually higher than from coastal areas. In addition, mercury concentrations in the striped mullet exhibits a 10-fold increase at an increasing depth of capture (from 60 – 340 m). As part of the "EROS 2000" (European River Ocean System) project in the Northwest Mediterranean Sea, which started in 1988 and ran through various phases until 1995, Dr. Daniel Cossa (IFREMER, Nantes, France) and Prof. Jean-Marie Martin (CNRS-ENS-IBM, Montrouge, France) investigated the biogeochemical cycle and speciation of mercury, i.e. elemental mercury Hg°, reactive mercury Hg(r), dissolved gaseous mercury DGHg, monomethylmercury MMHg, and dimethylmercury DMHg, with particular emphasis on dimethylmercury. From this first speciation study of mercury in the NW-Mediterranean Sea

Cossa and Martin concluded that the low oxygen sub-thermocline water masses constitute an important reservoir of DMHg available for accumulation by deep living pelagic fishes. The direct passive diffusion through the gill of the dissolved gaseous DMHg has been evidenced as a significant pathway for the methylmercury accumulation in fishes, in addition to the metal uptake from their diet. This pathway may explain the very high MeHg levels found in fishes caught in off-shore waters of the Mediterranean compared to those from coastal areas, where – due to its volatility - DMHg is at very low concentrations.

When comparing the riverine, wet and dry deposition of mercury onto the western Mediterranean Sea as well the loads and losses from the exchange of different water masses at the Strait of Gibraltar in order to try to establish a mass balance for mercury, the EROS 2000 project realised that reliable data on Hg concentrations in rain and aerosols was scarce in this area (D. Cossa et al., 1997) and hence the mass balance was mainly based on rough estimates. In addition, the same authors concluded that the Hg concentrations in rain may be underestimated.

The "cold trapping" effect of POPs and mercury in remotely located high mountain lakes, both with increasing height above sea level and with increasing latitudes has been illustrated by the MOLAR project and it evidenced that the mercury concentrations in Arctic charr in lakes of the arctic island Svalbard and in northern Norway regularly exceeded the WHO mercury limit values for consumption.

In addition, during the period 1970-1985 the national Swedish monitoring programme had revealed that the 10 000 lakes under regular control were contaminated by Hg which lead to methylmercury concentrations in fish above the human health guidelines for fish consumption, and for most of those lakes other sources of mercury loads than from atmospheric deposition could be excluded. Furthermore, during the period 1986 – 1997 the mercury emissions in Europe decreased by 40%, but this decrease was only marginally visible in the Swedish lakes where the methymercury levels in fish remained too high.

In order to quantify the sources of atmospheric mercury species and specifically to identify sources of atmospheric methylmercury and their importance for the methylmercury accumulation in aquatic and terrestrial ecosystems the European Commission funded the MOE project "Mercury Species Over Europe", coordinated by Dr. Munthe at the Swedish Environmental Research Institute, whereas in order to assess the spatial and temporal distributions of atmospheric mercury over the Mediterranean Sea, the MAMCS project on the" Mediterranean Atmospheric Mercury Cycle System" (http://www.cs.iia.cnr.it/MAMCS/project.htm), coordinated by Prof. Pirrone at Italian CNR-Institute for Atmospheric Pollution (CNR-IIA) was funded. Both projects were not only complementing each other geographically, covering two different regions of Europe, i.e. central and northern Europe by MOE and southern Europe by MAMCS, but they were also mutually sharing data, measurement and modelling approaches, contributing finally to the establishment of a European mercury emission inventory, a joint atmospheric measurements database and an improved European scale atmospheric model intended for policy support. Major findings of MOE and MAMCS projects have been published in two special issues of Atmospheric Environment (Pirrone et al., 2001; Pirrone et al., 2003).

The most unexpected result of the MOE project was for me the very high gaseous MeHg concentration at the Mace Head station on the west coast of Ireland, where the average contration was around 3.5 pg per qm, more than double as high as the inland stations, including the so-called "black triangle" in central Europe. And, Mace Head had originally been chosen to act as reference station where the lowest mercury concentrations in the air were thought to appear! These results were explained by a strong emission of di-methyl-mercury (DMeHg) from the sea surface of the Atlantic, followed by a chemical degradation to mono-methyl-mercury (MMeHg). In contrast to an earlier hypothesis, evoking that by an inorganic methylation process in cloud droplets inorganic mercury would basically be transformed into MeHg, the MOE project has found this process as insignificant, because too slow to explain the found concentrations in the atmospheric loads.

From the MOE and MAMCS projects it has been concluded that there are no "simple" control options for mercury available in support to European environmental policy., e.g. the implementation of the Council Directive of 1996 regulating the air quality of major persistent pollutants or the Water Framework Directive. Therefore, the European Commission is funding since October 2002 the MERCYMS project "An Integrated Approach to Assess the Mercury Cycle into the Mediterranean Basin", coordinated by Prof. Pirrone at Italian CNR-IIA which aims at:

- improving the predictive capability of the fate of mercury in the marine environment;
- assessing the qualitative and quantitative relationships between atmospheric inputs and direct discharges of mercury to the sea and the cycle of mercury in the marine environment as well as their re-emissions back to the atmosphere;
- developing an integrated modelling system in support of the implementation of EU Directives and international treaties;

- applying this integrated modelling system for environmental and socio-economic scenarios, in order advise on optimised, cost-efficient control strategies and policies for mercury.

The MERCYMS project URL (http://www.cs.iia.cnr.it/MERCYMS/project.htm) provides updated information on the currently achieved results.

Finally, a policy-oriented research project has been launched this year as part of the 6th RTD Framework Programme which will estimate the willingness-to-pay to reduce the risks of exposure to heavy metals and cost-benefit analysis for reducing heavy metals occurrence in Europe, including mercury. This ESPREME project intends to consolidate, improve and provide Europe wide emission data of heavy metals (HM), in particular regarding Hg, Cd, Pb, Ni, As and Cr; to collect systematic data on the possibilities to reduce emissions; to improve process models for HM in the atmosphere, soil and water and to apply them for simulations of the transport of the HM in and though the various media; to collect data on thresholds and information on exposure-response relationships; to estimate the willingness-to-pay to avoid damage from HM exposures by transferring values from available contingent valuation studies; to set up an integrated assessment model (IAM); to carry out runs of the IAM for the identification of cost-effectiveness strategies, i.e. bundles of measures that achieve compliance with thresholds and cost-benefit analyses to identify bundles of measures, where the difference between benefits and costs is maximised; and finally to conduct a feasibility study to identify approaches and further research needs for the macro-economic assessment of HM abatement strategies. The ESPREME project will take into account the whole of Europe, including the EU and Accession Countries. The results are expected to substantially contribute to the policy design in the frame of the CAFÉ strategy and to the ongoing reviews of the EC Air Quality legislation as well as to the UN-ECE Convention on Long-Range Transboundary Air Pollution protocols.

RESEARCH PERSPECTIVES ON MERCURY

Within the "Global Change and Ecosystems" Sub-Priority (1.1.6.3) of the running 6th RTD Framework Programme (2002 – 2006) still two calls for the submission of proposals are foreseen in support of integrating and structuring the European Research Area (ERA). The third call will be published on 16 June this year, with a closure date for proposal submissions on 26 October 2004. The total indicative budget for the 3rd call is 205 million Euros. It is envisaged to fund up to 17 Integrated Projects (IP) or Networks of Excellence (NoE) for a budget of 150 million Euros, to cover 12 sub-topics with Specific Targeted Research Projects (STREP) or C-ordination Actions (CA) for a budget of up to 50 million Euros, and one Area with Specific Support Actions (SSA) for up to 5 million Euros.

None of the areas and topics which are structuring the Work Programme for this third call is specifically addressing research requirements on mercury, but certainly mercury research could be part of proposals addressing several topics open for funding. In the following I will briefly mention those topics where research on mercury might play a role.

In the context of area I. "Impact and mechanisms of greenhouse gas emissions and atmospheric pollutants on climate, ozone depletion and carbon sinks" the sub-topic I.2.1 "Ocean – Atmosphere – Chemistry interactions" calls for up to two Specific Targeted Research Projects (STREP) and/or Co-ordination Actions (CA) which are addressing the feedback mechanisms between climate change and the air-sea exchange of trace gases, involving fluxes (emissions, deposition and ocean uptake); physico-chemical processes occurring within the atmosphere and oceanic boundary layers, including the transfer mechanisms, from the local to the global scale; and production of primary and secondary marine aerosols from natural sources.

In the context of Area VII. "Complementary Research" which is aiming to focus on the development of advanced methodologies for the risk assessment of processes, technologies, measures and policies, the appraisal of environmental quality, including reliable indicators of population health and environmental conditions and risk evaluation in relation to outdoor and indoor exposures, I could imagine that sub-topics VII.1.1.1 "Integrated assessment of environmental stressors, their interactions and development methods and models for the evaluation of combined human exposures", which is open for up to 1 Integrated Project, and VII.1.2.1 "Environmental analysis and monitoring of emerging environmental pollutants", which should be implemented by up to 1 Co-ordination Activity, might incorporate elements on mercury. Sub-topic VII.1.1.1 aims to develop integrated assessment models and methodologies for evaluating the cumulative effects, interactions between various stressors, and their short and long term influence on human health, in support to the EC Environment and Health Initiative. Sub-topic VII.1.2.1 aims to create a network among European reference laboratories and related organisations in order to improve the European capability for large scale monitoring and bio-monitoring of environmental pollutants in the various matrixes (air, water, soil), with emphasis on emerging environmental pollutants for which Europe-wide data are lacking. Relevant pre-normative research on measurement and testing can also be considered.

There are also a number of other topics or sub-topics open in the Areas "Water cycle, including soil-related aspects", "Biodiversity and ecosystems" and "Strategies for sustainable land management, including coastal zones, agricultural land and forests" of interest to mercury research, e.g. regarding "Integrated Coastal Zone Management (ICZM)" or on "Integrated risk-based management of the water-sediment-soil system at river basin scale", but personally I would consider there the involvement of mercury-specific approaches as of lower relevance.

In addition to the Research Priority 6 mentioned above exists also in the Framework Programme a specific "Policy-oriented Research" subject, for which the policy-oriented Directorates General of the European Commission are directly suggesting the priority research themes to be addressed, in order to underpin the implementation of existing policy instruments (directives, regulations, etc) or to prepare new policy initiatives and strategies. The EU Mercury Strategy may formulate relevant policy-oriented research needs for future calls.

The fourth and last major call for proposals within the 6th RTD Framework Programme will possibly be published by the end of 2004, with a deadline in 2005. However, the internal discussions about its priorities are still ongoing and no clear tendency can be deduced from those discussions at this moment.

Also the first thoughts about the orientations of the next, 7th RTD Framework Programme (2006–2010) have been brought forward at political level: more technology research and development in support to the implementation of the objectives formulated at the Lisbon European Council summit, combined with the Göteborg Council conclusions on Sustainable Development; more emphasis on fundamental research, and continuation of the further implementation of the European Research Area, possibly with strengthened approaches of ERA-Nets and other joint national-EU funding mechanisms. In respect to environmental research three thematic areas become already "visible": environmental technologies, natural resources and complex systems, respectively. However, it is premature to speculate at this stage about any more detailed thematic priorities and the implementation mechanisms of the future.

REFERENCES

Cossa, D., Martin, J.M., Takayanagi, K. Sanjuan J. The distribution and cycling of mercury species in the western Mediterranean. In: J.M.Martin (Guest Editor), EROS 2000 (European River Ocean System). The Western Mediterranean. *Deep Sea Research Part II*, 44, 721-740., 1997.

Pacyna, J.M., Barthel, K.G., Barth, H., Pirrone, N. (Editors) Socio-Economic Aspects of Fluxes of Chemicals into the Marine Environment. European Commission-DG Research (Publisher), Brussels, Belgium - EUR 2201, 2000.

Pirrone, N., Pacyna, J.M., Barth, H. Atmospheric Mercury Research in Europe. Special Issue of Atmospheric Environment, volume 35 (17), Elsevier Science, Amsterdam, Netherlands, 2001.

Pirrone, N., Pacyna, J.M., Munthe, J., Barth, H. Dynamic Processes of Mercury and Other Atmospheric Contaminants in the Marine Boundary Layer of European Seas. Special Issue of Atmospheric Environment, volume 37 (S1), Elsevier Science, Amsterdam, Netherlands, 2003.

Chapter-6

PERSPECTIVE ON MERCURY: PROGRESS THROUGH COOPERATION

Marilyn Engle

US-Environmental Protection Agency, Washington D.C., USA

INTRODUCTION

There has been a gradual evolution of how EPA looks at the mercury problem, driven in part by the evolving science linking emissions, transport and fate with wildlife and human exposure and risk. The U.S. has moved from end of pipe/stack controls, to the concepts of pollution prevention and integration of approaches in interdisciplinary fashion, thereby avoiding the transfer the mercury problem from one medium to another, and to the recognition of mercury as a global problem.

Mercury has been of concern to EPA since its inception in December 1970. In fact, concern over mercury was even a factor in the formation of the Agency. At that time, there was much public attention to mercury contamination of the U.S. waters. It was recognized that many industries discharged mercury to surface waters. Likewise, the first worker health standard developed by the new Occupational Health and Safety Administration pertained to inhalation of mercury vapor.

We recognize that many countries, in particular Sweden and other Nordic countries, have had a lengthy acquaintance as well with mercury and are continuing to identify new and important areas in which to focus our attention.

The sad mercury experience in Minimata, Japan in the 1960's contributed to an appreciation of how disastrously mercury entering a water body as methylmercury, or subsequently being methylated by bacteria, can transport up the food chain to man and cause a range of adverse effects linked to dose and time of exposure. The Council on Environmental Quality (CEQ) Second Annual Report of 1971 included mercury in a short list of environmental toxins attributed to being a causal agent in the deaths of the American eagle and stated:

....some problems are emerging with portentous implications. The accumulation of mercury in our waters is a serious problem that may become even worse as organisms continue to concentrate mercury through the food chain. The increases in complex toxic materials-both heavy metals and organic chemicals-may be serious long-term threats to human health and the life cycle upon which all men depend.

While EPA recognized that many industries discharged mercury to surface waters, with some facilities discharging tons each year, and that some sources emitted mercury to air, there was little understanding or appreciation of the dimensions of mercury's behavior in the environment. There was little recognition of its international aspects or of its expected health risks from eating fish, since an adequate health assessment was not available for use in estimating risks of in utero exposures.

Since then, the focus has grown not only to incorporate awareness of the domestic sources and deposition sinks, but also the international scope of this truly globally cycling, persistent bioaccumulative toxic substance, whereby all countries are seen to be both sources and repositories through deposition. Research monitoring and modelling have shown the linkage of coal burning and other emissions sources to atmospheric transport and deposition in water bodies with resultant uptake and bioconcentration in fish and exposure to humans and other living organisms. The U.S. EPA's 1997 *Mercury Study Report to Congress*, wherein this linkage was first reported in the U.S., was a big impetus for the most recent high priority given to mercury at EPA.

It is timely that this workshop will further better information sharing and awareness of developments as well as needs of scientists in the monitoring and modelling sphere and that of the health community. EPA has been committed to enhancing government, industry and public awareness of mercury and its behavior in the environment, and effects to man.

Mercury has been intriguing to man and utilized by him for hundreds, if not thousands, of years: the first Emperor of China died drinking mercury while thinking it was a source of youth and longevity, and of medicinal value; and, beginning over 500 years ago, the Almaden mine in Spain supplied mercury to the new world to help extract the gold and silver contributing to Spain's wealth and stature over a significant span of time. The Almaden mercury was sent as the ballast in ships and the mined metals were the ballast on the return trip. Due to this practice of Almaden mining, it is in the mid 16th century that modern scientists begin to see a widespread distribution of anthropogenic mercury. The U.S.'s California goldrush in the mid nineteenth century also helped distribute mercury all over the globe due to the open burning of mercury used to amalgamate the gold (Schuster et al., 2002). Eighteenth and nineteenth century hat makers in Europe were known as mad hatters due to what later became recognized as the neurological effects of direct inhalation exposure to mercury in the processing.

Fortunately, the new UNEP global mercury program, established in February 2003, provides an international mechanism which can further shape cooperation between counties. It is clear that cooperation between scientific disciplines and between the public and private sectors on a global scale is essential to better understand the linkage of emission sources and receptors, especially most vulnerable populations. This leads to more informed decision-making, buttressed by legislation, which can ultimately lead to reduced uses, emissions and deposition around the world.

All countries contribute to the problem and can aid the solution by reducing use and emissions of mercury within their countries. EPA is committed to a strong mercury program at home and to an ongoing international engagement in concert with the UNEP global program and other bilateral and regional mechanisms.

EPA is increasingly concerned about international as well as domestic sources of mercury. International uses and emissions of mercury account for a significant portion of mercury circulating in the atmosphere. An estimated one-third of all global mercury releases comes from fossil-fuel burning in Asia.

Therefore, in order not only to take actions on mercury emissions at home but also provide leadership abroad, EPA's vision is to reduce uses and releases of mercury as much as possible, and make progress in collaborative monitoring, modelling and health activities, as well as in information sharing across these disciplines. EPA and the U.S. not only have lessons- learned to contribute, but much to benefit from others' experiences as well.

FRAMEWORK: CONCEPTUAL MODEL FOR MERCURY

The following section describes the way in which EPA visualizes the nature of mercury behavior in the environment and its impact upon it.

Mercury is emitted from natural and anthropogenic sources, where it can cycle globally, constantly moving among various environmental compartments through a complex combination of transport and transformation, finally resulting in human and wildlife exposure.

While many mercury sources are well understood, others are poorly characterized. In the United States, the largest sources are coal-fired electric utilities, coal-fired industrial boilers, and gold mining. There are also natural sources of air emissions, e.g., volcanoes. Mercury discharges to surface waters from abandoned gold and mercury mines in the western U.S. have led to many fish advisories, but overall in 2003, there were 47 states, 1 territory, and 3 tribes with such advisories (www.epa.gov/waterscience/fish). Total use of mercury in the U.S. economy has declined by over 80 percent since the late 1980's, but mercury continues to be used in products and industrial processes, including electrical equipment, measurement devices, fluorescent bulbs, certain batteries, dental amalgam, preservatives in pharmaceuticals, as well as in the production of chlorine and caustic soda.

The fate of mercury emissions depends on various factors, including the form (species) of mercury emitted, the location of an emissions source, the topography near the source, and the prevailing circulation patterns. Typical combustion processes, e.g., incinerators, emit three different forms: 1) elemental mercury, believed to have a half life of about one year and to travel globally (see footnote one below); 2) particulate-bound mercury, which, depending upon particle size, can deposit over a variety of distances; and 3) oxidized mercury (sometimes called ionic) mercury, predominantly in water soluble forms, with mercuric chloride being the chief suspect, which can deposit from the atmosphere relatively quickly, even in the absence of precipitation. For these reasons, atmospheric mercury is transported and deposited at the following scales:

1. **Local scale impacts** result from deposition relatively close to an emissions source. For example, a source emitting primarily RGM can be expected to have a relatively high fraction of its mercury emissions deposited within 50 kilometers and have significant local-scale impacts.

2. **Regional scale impacts** result from deposition associated with long-range transport of emissions beyond the local scale, but generally

within a 1000-kilometer range. For example, depending on atmospheric conditions and climate, RGM or elemental mercury emitted from a tall stack might have regional scale impacts.

3. *Continental scale impacts* result from emissions being transported even longer distances and depositing across an entire continental area. For example, mercury that is emitted as elemental and a small amount of RGM lofted high in the atmosphere by convective currents or adsorbing to fine aerosol particulate matter might have continental scale impacts.

4. *Global scale impacts* result from emissions that become part of the global pool, where they can remain for months or years. For example, elemental mercury (or RGM chemically reduced to elemental mercury before deposition) can be transported in air for many months, eventually being oxidized and deposited at any location around the globe.

Once it enters freshwater or marine ecosystems, mercury tends to accumulate in biota, or in sediments where a portion of the mercury can be methylated and subsequently accumulate through food webs into predator fish, mammals, and birds. The degree of methylation varies from one ecosystem to another, depending on biogeochemical factors, and is a controlling factor in the biomagnification of mercury.

Exposure to humans and terrestrial wildlife is primarily through eating fish, fish-eating marine mammals (for various sub-populations, including various tribes in the Arctic), or the eggs of piscivorous birds. In fish, the mercury is almost all methylmercury. (While inorganic mercury may also be present in blood, it is generally below the limit of detection, making total mercury concentration in blood a generally good indicator of methylmercury exposure.) Mercury in urine reflects elemental and inorganic exposure (and not methylmercury exposure).

The oceans and coastal marine ecosystems are recipients of mercury carried by river systems of the world and large scale atmospheric mercury deposition. They are potential sinks for mercury and also sources of mercury to the atmosphere.

The dynamics of the chemistry, fate, and transport of mercury in pelagic (open ocean) and coastal marine ecosystems are different and not well known. The data are not yet clear as to whether consistent reductions have occurred in recent decades, despite a large reduction in mercury use.

While the general outlines of mercury sources, fate and transport, and bioaccumulation are well understood, there remains considerable uncertainty, particularly concerning the detailed processes involved in mercury transport and fate and the rate-controlling factors.

MAJOR MILESTONES IN MONITORING, MODELLING AND HEALTH RESEARCH, AND ASSESSMENT

A number of accomplishments deserve attention, recognizing in each case the importance of EPA's internal coordination and cooperation with others on many levels.

Transport and Fate

Progress to Date in Monitoring

Mercury and Polar Sunrise. One of the key findings presented in the Arctic Monitoring and Assessment Program (AMAP) Phase II Heavy Metals Assessment Report, accepted by the Arctic Council of Ministers in 2003 and expected for publication by December 2004, is the transformation of mercury in the Arctic at polar sunrise. EPA has been instrumental in investigating the nature and geographical extent of the phenomenon termed Arctic Sunrise where atmospheric elemental gaseous mercury levels have been shown to drop drastically during the Arctic Spring when sunlight returns to the region. Indeed, this is seen to be a polar phenomenon, occurring at both poles, based on research findings in the Arctic as well as in the Antarctic, where EPA has collaborated with Italy at their monitoring site there. EPA has a particular focus in the Arctic, due to State of Alaska issues and concerns about mercury generally, which has led EPA to participate in the Arctic Council's Arctic Council Action Plan (ACAP) Mercury Project. The majority of atmospheric mercury is present in elemental form, but transformation to reactive gaseous mercury (RGM), which has much higher wet and dry deposition rates, is now known to occur. Thus, speciation of mercury is of particular interest in the Arctic because of the sunrise phenomenon and the greater local impact of reactive forms.

Since 2000 EPA scientists have designed and implemented a series of atmospheric mercury speciation studies. Successful work first completed in Barrow, Alaska led to implementation of partnership studies during 2002 and 2003 at the Italian South Pole Atmospheric Terra Nova Science Research Base and at the Norwegian Polar Research Base at NyAlesund. EPA scientists provided training to collaborators and helped design and install specialized instrumentation at all three polar monitoring sites. The primary objectives of monitoring studies conducted during polar sunrise were threefold:

1. Measure and speciate the various forms of mercury in air and snow (i.e. elemental mercury (Hg^0), Reactive Gas Phase Mercury (HgX_2- where X is a halide) and fine particle bound mercury ($Hg(p)$)),
2. Obtain snow samples for subsequent chemical analysis, and
3. Obtain air quality data and meteorological measurements.

These measurement campaigns were designed to obtain information on the factors that lead to mercury depletion events (MDEs), to better understand them and model their impacts on the half-life of mercury in the atmosphere.

Progress to Date in Modelling

A special version of the EPA Community Multi-scale Air Quality (CMAQ) model has been developed to include atmospheric mercury reactions, aqueous mercury processes, and wet and dry deposition of mercury. EPA models have been expanded to simulate atmospheric mercury in addition to all other air pollutants. In FY2002 and 2003, the mercury model was applied as part of an Intercomparison Study of Numerical Models for Long-Range Atmospheric Transport of Mercury conducted by the Co-operative Program for Monitoring and Evaluation of the Long-Range Transmission of Air Pollutants in Europe (EMEP) in support of the Convention on Long-Range Transboundary Air Pollution (LRTAP). In FY2003, an initiative was begun to conduct a similar study focused on North America and the likely effectiveness of proposed emission reductions there.

Summary of work: A special version of the mercury model has been developed to include atmospheric reactions, aqueous processes, and wet and dry deposition. These process descriptions continue to evolve as scientific information on mercury behavior is published in the scientific literature. Laboratory investigation of mercury chemistry is currently underway or planned at a variety of scientific organizations around the world, including those sponsored by EPA/ORD's Science To Achieve Results (STAR) program. As these scientific investigations report their results, we will continue to work under this task to further refine the mercury model to ensure that it reflects the best scientific information available about the processes controlling atmospheric mercury transport and fate. This task also focuses on sensitivity testing of the model and assessment of its ability to reproduce observed mercury deposition and air concentration data. It also includes intercomparison of the EPA mercury model results with those obtained from other models using similar initial and boundary conditions, emissions information, and meteorological forcing. The logical structure of this multi-media model linkage is yet to be determined, but preliminary plans have been developed to use individual compartmental model treatments for each of the various surface types underlying the horizontal grid cells of the mercury atmospheric model structure.

EPA plans to collaborate with Italy's Division of Rende of the Institute for Atmospheric Pollution to further shape a modelling framework which incorporates the recent halide chemistry research. This will be a valuable platform for inserting Mauna Loa, Ny Alesund, and other mercury data.

The Everglades Example

It is important to appreciate that in the Florida Everglades, progress in reduction of mercury in fish tissue levels has been measured over a multi-year span through collaborative monitoring activities involving Federal EPA and State of Florida cooperation. This information contributes to our understanding of time scales required for improvement in the environment following management actions. However, ecosystems are unique in their response times, and the Everglades example may not be indicative of other freshwater systems. We know very little at present about marine systems and the time required for improvement in fish tissue levels there.

Health

As reported in the 1997 EPA *Mercury Study Report to Congress* identifying the link between mercury sources emissions and adverse levels of mercury in fish, EPA used a Reference Dose (RfD) of 0.1 micrograms per kilogram of body weight per day, to estimate human health risks in the United States associated with the ingestion of fish. The National Academy of Sciences (NAS) report on the Toxicological Effects of Methylmercury (NRC, 2000) confirmed this reference dose, which is equivalent to eating 3 tuna sandwiches per week, and estimated that more than 60,000 U.S. children are born each year with neurological risk due to exposure to mercury while in the womb and may, as a result, suffer learning disabilities. The NAS Report furthered the drive for interagency cooperation by calling for the need to establish a common scientific basis for exposure guidance among federal agencies in light of the fact that they operate under different legal and regulatory authorities. In 2001, a nation-wide random sample of women of child-bearing age conducted by the U.S. Centers for Disease Control and Prevention (CDC) (NHANES) indicated that over 8 percent have exposures exceeding the RfD.

Joint EPA and FDA Fish Advisory (March 2004)

About 47 states have fish-consumption advisories for mercury. Mercury is the key reason for these advisories in the U.S. EPA is responsible for national freshwater fish advisories, while the Food and Drug Administration (FDA) gives guidance on commercial fish.

EPA and FDA have been consulting on the health issues related to mercury for many years. In March, 2004, EPA and FDA issued AAdvice For: Women Who Might Become Pregnant, Women Who are Pregnant, Nursing Mothers, Young Children. The advisory stated that:

Fish and Shellfish are an important part of a healthy diet. Fish and Shellfish contain high-quality protein and other essential nutrients, are low in saturated fat and contain omega-3 fatty acids. A well balanced diet that includes a variety of fish and shellfish can contribute to heart health and children's proper growth and development. So, women and young children in particular should include fish or shellfish in their diets due to many nutritional benefits..... Yet, some fish and shellfish contain higher levels of mercury that may harm an unborn baby or young child's developing nervous system. The risks from mercury in fish and shellfish depend on the amount of fish and shellfish eaten and the levels of mercury in the fish and shellfish. Therefore, the Food and Drug Administration (FDA) and the Environmental Protection Agency (EPA) are advising women who may become pregnant, pregnant women, nursing mothers, and young children to avoid some types of fish and eat fish and shellfish that are lower in mercury...... "

Information can be obtained at the following web site: http://www.epa.gov/ost/fishadvice/advice.html.

MILESTONE U.S. ACCOMPLISHMENTS IN REDUCING MERCURY USE AND EMISSIONS

Monitoring, modelling and health data factor significantly into EPA's well-integrated multi-media approach involving every program office in the Agency to reduce human and ecological exposure to mercury. EPA regulates many air emissions and direct discharges to water; encourages industries to develop substitutes for mercury in products and industrial processes and collaborates with industry on their voluntary programs; helps states to warn the public about the risks of eating locally caught fish; coordinates with FDA to provide commercial fish advisories; and undertakes an extensive research development and assessment program to better understand how mercury behaves in the environment, its human health effects, and how to reduce releases to the environment.

Some of EPA's earliest actions, concluded in 1971 under the Refuse Act to stop harmful discharges of harmful material, pertained to mercury. At that time, out of 10 facilities, there were nine stipulations for discharge reduction and one plant shutdown, resulting in a total reduction in mercury discharged from 139 to 2 pounds per day, and plans with industry for further reduction were being set into place (CEQ Second Annual Report, 1971).

Reducing Mercury Emissions

The current major U.S. mercury emission sources, according to EPA's latest national emissions inventory (NEI) are indicated in the following Table 1.

The nation has made significant progress towards reducing mercury emissions. As a result of various risk management actions, primarily EPA regulatory controls for incinerators, total US anthropogenic emissions have declined in recent years as follows: The US-EPA estimates that U.S. annual direct emissions from industrial processes totaled 210 short tons (191 metric tons) in 1990, 185 short tons (168 metric tons) in 1996, and, as indicated above, about 118 short tons (107 metric tons) in 1999 (the most recent NEI inventory year).

Table1. U.S. National Emissions Inventory.

Source Category	1999 (tons)
Utility Coal Boilers	47.9
Industrial Boilers	12
Medical Waste Incineration	2.84
Municipal Waste Combustion	5.1
Hazardous Waste Incineration	6.58
Chlorine Production	6.53
Gold Mining	11.5
Other*	25.3
Total	117.76

* Other includes, but is not limited to such items as, cement production, general coal usage, oil, gas and biofuel usage, and miscellaneous industrial processes.

Incinerators

EPA has successfully regulated mercury from municipal waste and medical waste incinerators., which require reductions of emissions by over 90 percent. Prior to implementation of these regulations, these two sources, considered together, were the largest anthropogenic, or man-made, sources in the U.S. after coal-fired electric utility boilers. EPA has also regulated hazardous waste incinerators, with higher emission reductions required.

Chloralkali Mercury Cell Facilities

Regulatory Action for Chloralkali plants: There are water, waste, and air regulations that pertain to the mercury cell sector, dating to the 1970's. EPA has recently promulgated a Maximum Achievable Control Technology (MACT) standard for chloralklai plants that will require actions to further reduce mercury emissions.

The Voluntary Mercury Initiative of U.S. Chloralkali Factories. The Chlorine Institute has developed operational guidance to factories regarding how to curtail mercury air emissions. Information can be found at an EPA web site: http://www.epa.gov/region5/air/mercury/reducing.html#chlor-alkali. In 1996, U.S. mercury-cell chloralkali factories began a voluntary mercury stewardship project. This effort was announced upon proposal, in Spring of 1997, of the Binational Toxics Strategy for collaboration between the U.S. and Canada to address mercury (see Section V).

Through 2001, the 9 U.S. mercury cell factories have reduced their mercury losses by 75 percent (on a capacity-adjusted basis that accounts for factory closures). This represents reduced consumption (loss from the production process) of 120 tons per year, an environmentally significant amount. Some of this mercury likely is trapped within tanks and piping of a factory. Some is likely landfilled, in compliance with environmental regulations. Yet, clearly the U.S. industry's voluntary efforts have paid attention to prevention of fugitive mercury air emissions. Practical steps have included revision to maintenance procedures, to include cool-down of equipment prior to invasive maintenance that could expose mercury to the air; less frequent invasive maintenance; adoption of a UV-light that allows workers to discern mercury vapor leak points at their origin.

The U.S. industry, through The Chlorine Institute, has supported transfer of environmental management lessons to chloralkali factories in developing countries. The Chlorine Institute has provided all of its environmental management materials to counterpart professional associations in South America and India. There are about 30 mercury cell factories in these two regions. The Alkali Manufacturers Association of India reports reduced mercury consumption of about 50 percent since 1999, approaching consumption rates at U.S. factories before onset of their voluntary program (Frank Anscombe, EPA, personal communication).

OTHER SIGNIFICANT USE AND EMISSIONS REDUCTIONS

Mining Voluntary Activities The last mercury ore mine in the U.S. closed in 1990, and gold mining by-products are the only current mining related source of mercury. A voluntary partnership with the gold mines in the State of Nevada has been in place since 2002 to reduce air emissions through process modifications or installing controls. This valuable cooperation may lead to more mining companies to join into the voluntary program, and it is leading to the development of new pollution prevention techniques.

Steel Mills Voluntary Actions Three steel mills in Northwest Indiana – International Steel Group (Burns Harbor), Ispat-Inland (East Chicago), U.S. Steel (Gary) – have produced a progress report on achievements under their voluntary mercury reduction agreement (signed in 1998 with the Lake Michigan Forum, US-EPA, and Indiana Department of Environmental Management (DEM), and coordinated by the Delta Institute). This agreement resulted in a project which called for a 33 percent reduction in usage of mercury-containing equipment within two years, an additional 33 percent in four years, and 90 percent within 10 years. The progress report states that "the project has met with genuine enthusiasm in the plants…and has exceeded expectations." Between 1999 and 2003, the three facilities removed approximately 3700 pounds of mercury for disposal or recycling, roughly 80 percent of the mercury present at the facilities. The progress report also expresses the hope that the Binational Toxics Strategy can be used to "help to recruit other companies and facilities to pursue similar work." The full report is available at: http://www.epa.gov/region5/air/mercury/nwindianareport3-17-04.pdf.

Products There has been a dramatic drop in mercury use in the U.S. over the last two decades due to legislation and a number of regulatory and voluntary programs implemented at the national, state or local level. In the U.S., mercury has been taken out of paints (where it was used as a fungicide), pesticides, certain batteries, and other products as mercury-free alternatives have become available. EPA, states, tribes, local governments and non-governmental organizations are working together to further reduce the use of mercury in products and processes.

NEI EMISSIONS AND TRI RELEASES TO AIR, WATER, AND LAND REPORTING

Overview

Key to understanding progress made over time in mercury emissions reduction has been the U.S. emissions accounting programs, the National Emissions Inventory (NEI) and the Total Release Inventory (TRI).

The NEI is a comprehensive inventory with consistent methodologies covering criteria pollutants and 188 hazardous air pollutants (HAPs), including mercury, whose data can be used for modelling. The current inventory contains speciated emissions for various mercury sources. The latest NEI data are for 1999 and can be found at http://www.epa.gov/ttn/hchief/net/1999 inventory.html#final3haps.

The TRI is a Community Right-to-Know Inventory in which industrial facilities meeting certain criteria are required to report annually on the releases to air, land and water of over 640 chemicals. The latest TRI data are for 2001. TRI data can be found at the following website: http://www.epa.gov/tri/ tridata/tri01/data/index.htm. While TRI provides information on mercury point sources, it is not a modelling inventory and does not contain individual stack emissions for facilities nor stack parameters\, e.g., speciation, necessary for modelling. TRI is not comprehensive for all point and nonpoint sources in the US.

Releases to Land and Water

While deposition of air emissions is the primary route by which mercury is finding its way into fish and ultimately people and wildlife, mercury is also released to water and land. Current data suggest that releases to water are minimal, but that land releases in the United States, primarily from minerals mining in general and that of gold in particular, are large. For example, In 2001, TRI reported that about 1929 metric tons of mercury were released in this manner. While it is believed that most of this mercury is bound up with other substances, the environmental impacts are uncertain.

COOPERATIVE INSTITUTIONAL ARRANGEMENTS: HISTORY AND PROCESS

A number of key examples highlighted in this section illustrate the importance of cooperation and some of the mechanisms under which progress in addressing mercury has been furthered domestically and internationally.

Within EPA

The Persistent Bioaccumulative Toxics (PBT). Initiative EPA has been organized since its inception in 1970 according to the various media, such as air and water, in order to resolve environmental issues in those media. Accordingly, while each individual EPA office and region did a good job in undertaking solutions to reduce pollutants in one place, the holistic picture was sometimes not considered, which created a new issue for another office by moving the pollutant from one compartment to another. Recognition of this problem led to a new approach through creation of the Persistent

Bioaccumulative Toxics (PBT) Initiative in 1998. Mercury was identified as a priority for attention, and the program has aided Agency decision-making, for example in support of trends monitoring.

1997 Mercury Study Report to Congress. In 1997, EPA reported to Congress the magnitude of the mercury emissions and the nature of the problem in the U.S. by describing sources, the health and environmental consequences of those emissions, and opportunities for technologies to address the sources of anthropogenic mercury. All key offices contributed to this *Report to Congress* (EPA, 1997), which serves as a foundation for the EPA risk assessment and risk management vision, contributing to the choices we have made and will make to address mercury. This document is significant in that it is the first time EPA reported the plausible link between human activities that release mercury from industrial and combustion sources across the U.S. and methylmercury concentrations in wildlife and in man. It included the assessment which estimated that between one and three percent of women of childbearing age (between 15 and 44 years of age) in the U.S. eat sufficient amounts of fish for their fetuses to be at risk from methylmercury exposure. Wildlife was also identified to be at risk. It is the most comprehensive human health and environmental assessment of mercury and methylmercury to date.

Mercury Research Strategy. The Mercury Research Strategy, published in 2000, has been the driver for EPA research planning and budgetary decisions to address mercury, and involves cooperation across the agency and with other agencies, the states and the private sector. The Strategy builds on the 1997 *Report to Congress* identified research needs to improve mercury risk assessment and risk management activities, and called for development of a research strategy. In order to guide how EPA addresses remaining uncertainties in risk assessment and risk management of mercury, a team of EPA scientists and engineers from across the Office of Research and Development (ORD) and the agency program offices and regions focused on a range of research topics and posited six core research questions and research needs to be addressed through implementation of the strategy.

1. Transport and fate: How much methylmercury in fish consumed by the U.S. population is contributed by U.S. emissions relative to other sources of mercury (natural sources, those of other countries, and re-emitted mercury from the global pool); and how much and over what time period will levels of methylmercury in fish in the U.S. decrease due to reductions in environmental releases from U.S. sources ?
2. Combustion sources: How much can mercury emissions from coal-

fired utility boilers and other combustion systems be reduced with innovative mercury and multi-pollutant control technologies; 3) Non-combustion sources: What is the magnitude of contribution of mercury releases from non-combustion sources; how can the most significant releases be minimized?

3. Ecological effects and exposure: What are the risks associated with methylmercury exposure to wildlife species and other significant ecological receptors?

4. Human health effects and exposure: What critical changes in human health are associated with exposure to environmental sources of methylmercury in the most susceptible human population; how much methylmercury are humans exposed to, particularly women of child-bearing age and children, what is the magnitude of uncertainty and variability of mercury and methylmercury toxicokinetics in children?

5. Risk communication research: What are the most effective means for informing susceptible populations of the health risks posed by mercury methylmercury contamination of fish and seafood?

ORD Strategic Research Multi-Year Planning To implement the Mercury Research Strategy, EPA's research agenda and multi-year research plan has been prepared, with funding attention in near-term to transport and fate, health and power plant coal combustion sources, supportive of EPA regulatory development for the power sector, and with gradually increasing focus over time to the non-combustion and ecological dimensions. The multi-year plan long-term research goals are: 1) to reduce and prevent releases of mercury into the environment; and 2) to understand the transport and fate of mercury from release to the receptor and its effects on the receptor. The period of time covered is 2002-2010. The Mercury Multi-Year Research Plan Website is: http://intranet.epa.gov/ospintra/Planning/mercury.pdf

Star Grant: To achieve the goals of the research strategy, research is being conducted through in-house resources and through the EPA STAR Grants Program, under the EPA Office of Research and Development. The Science to Achieve Results (STAR). STAR work on mercury began in 1999 when a series of grants were funded to the U.S. and international research community to help the Agency understand the processes that influence human and ecological exposure to mercury. With this information, risk management strategies for mercury can be developed using sound science. Scientists are working with information around the globe to develop a better understanding of the natural and manmade emissions of mercury to the air, and the atmospheric processes that affect the transport, transformation and deposition of those emissions. The following are Mercury STAR Grants websites: http://es.epa.gov/ncer/ publications/ topical/mercury.html; URL: http://cfpub.epa.gov/ncer_abstracts/ index.cfm/fuseaction/recipients.display/rfa_id/109; http://cfpub.epa.gov/ncer_ abstracts/index.cfm/fuseaction/recipients.display/rfa_id/297.

ORD-Sponsored International Workshop In addition to ongoing research, a series of five international technology transfer workshops were convened by EPA's Office of Research and Development (ORD). Recognizing the importance of collaboration with other Federal and State agencies and industry to address research needs and meet goals of the ORD Research Strategy, these meetings involved other agencies, such as U.S. Geological Survey (USGS) and the Department of Energy (DOE), various States, academia and the public. The five workshops and their report references are:

1. Workshop on Source Emission and Ambient Air Monitoring of Mercury, Bloomington, Minnesota, September 13-14, 1999. (EPA/625/R-00/002);
2. Workshop on Mercury in Products, Processes, Waste and the Environment, Baltimore, Maryland, March 22-23, 2000. (EPA/625/R-00/014);
3. Workshop on Mercury in Mining, San Francisco, California, November 28-December1, 2000 (draft proceedings);
4. Workshop on the Fate, Transport, and Transformations of Mercury in Aquatic and Terrestrial Environments, West Palm Beach, Florida, May 8, 2001. (EPA/625/R-02.005);
5. Workshop on Breaking the Mercury Cycle: Long-Term Management of Surplus Mercury and Mercury-Bearing Waste, Boston, Massachusetts, April 2002.

Cooperation with Other Agencies, States and Tribes

U.S. Geological Survey (USGS): In June 2000, EPA and USGS formed the EPA-USGS Mercury Roundtable as a means to communicate on a staff level via teleconference on a regular basis every three months about the most important research and policy developments of interest to both agencies. It is open to all interested government agency representatives, including Federal, state, and local agencies and Tribes, as well as governments of Canada and Mexico. It now draws large numbers of participants from across the country, with approximately 100 persons at any particular meeting.

National Oceanic and Atmospheric Administration (NOAA) and Department of Energy (DOE): EPA has collaborated with NOAA and DOE in conducting Arctic monitoring for mercury at Barrow, Alaska where the polar phenomenon has been investigated. EPA has benefited from NOAA's monitoring infrastructure and at the Mauna Loa, Hawaii high altitude site where EPA has added research and trends mercury measurements. NOAA's long-term carbon dioxide record since the late 1950's there, as well as other measurements, augment our data and help in its interpretation.

Food and Drug Administration (FDA): EPA and the Federal Drug Administration have been coordinating recently in order to shape a joint health advisory for mercury and statement to the public on consumption of fish.

State: Many states have been very active in addressing mercury pollution and exposure and continue to work collaboratively with EPA on a range of mercury issues and programs. For example, EPA and the State of Florida were concerned about high mercury levels in fish and birds in the Florida Everglades and collaborated on monitoring and modelling for mercury. These efforts helped to identify local sources contributing to the mercury deposition in the Everglades and to document observed reductions in fish tissue resulting from emission controls on medical and municipal waste incinerators.

Tribes: Several mechanisms exist for outreach and coordination with tribal governments. The United States government has a trust responsibility to Federally-recognized Indian tribes that arises from Indian treaties, statutes, executive orders and the historical relationships between the United States and Indian tribes, a government-to-government relationship. The Council provides a forum for tribes and EPA to identify priority environmental science issues and collaboratively design effective solutions. The Council seeks to increase tribal involvement in EPA's scientific activities - building bridges between tribal and Agency programs.

Regional North American Cooperation: EPA has been very involved in promoting North American cooperation to address mercury. The Canada-US-Mexico North American Free Trade Agreement (NAFTA) catalyzed the three countries to develop and establish the 1994 North American Agreement on Environmental Cooperation (NAAEC). The NAAEC established the North

American Commission for Environmental Cooperation (NACEC). The NACEC has developed a North American Regional Action Plan (NARAP) for mercury and is also building a monitoring strategy, with assistance being given to Mexico.

Bilateral Collaboration with Canada: Under the Council of Environmental Quality (CEQ) and from the earliest days of EPA, cooperation with Canada and the International Joint Commissions (IJC) in the Great Lakes on mercury issues has been a priority. Since 1972, EPA and Environment Canada have co-led binational cooperation to protect human health and the ecosystem in the Great Lakes Basin. These binational Great Lakes protection and restoration efforts since 1972 have included priority work and goals on heavy metals and mercury contamination in the aquatic food chain. Since 1997, the U.S. and Canada have a Binational Toxics Strategy which seeks a 50 percent reduction nationally in use of mercury and a 50 percent reduction in mercury release to the air and water in the Great Lakes Basin, based on 1990 levels. Monitoring and modelling are tools supporting the goals. A mercury workgroup, consisting of state, provincial, local and tribal governments, industry and non-government organizations, led by EPA and Environment A New England Governors and Eastern Canadian Premiere's Mercury Action Plan, adopted in 1998, focuses on U.S. and Canadian cooperation to address mercury in the Northeast and in addition to setting emissions reduction targets for identified sources and risk communication. The most recent webpage is: www.cap-cpmaca/images/pdf/eng/2003reportmercury.pdf. Additionally, EPA scientists are involved in a Northeastern Ecosystem Research Cooperative (NERC) Mercury Research Group which promotes collaboration among ecosystem research scientists in the NE U.S. and eastern Canada. They monitor and model mercury distribution in this region. Their website is: http://www. ecostudies.org/nerc/

Other Bilateral Cooperation: By 1970, the U.S. and Japan had a long-standing U.S.-Japan Cooperative Program on Natural Resources and joint programs for air and water quality. This has been continued and strengthened through the years. Mercury was on the agenda of discussion at the recent April 2004 visit by the EPA Administrator to Japan. As early as 1971, EPA was shaping liaison on environmental matters, although not necessarily involving mercury, with Mexico, Spain, France and Germany (CEQ Second Annual Report, 1971). More recently, EPA has developed a Letter of Agreement specifically for mercury with the Italy Rende Institute to facilitate cooperation in monitoring, modelling and health issues. This workshop is a result of this collaboration. EPA also gave thought early-on to exploring opportunities for cooperation on environmental matters with developing countries, such as Poland, Yugoslavia and India (CEQ Second Annual Report, 1971). With heightened interest now in EPA regarding the new UNEP Global Mercury Program and the role of international mercury sources to mercury deposition in the U.S., EPA's Office of International Affairs has considered how to build efficiencies into how we coordinate with and assist developing countries to meet their mercury goals for pollution prevention and emissions reductions, as well as improve our understanding, through monitoring and modelling, of apportionment of international and domestic releases of mercury which contribute to U.S. deposition of mercury. See Section VII for more information.

Russia: EPA helped initiate an Arctic Council Action Plan Mercury Project, involving all 8 Arctic countries, whereby special assistance is also given to Russia. This Arctic Council project is described later in this section under Multilateral Global Cooperation.

China: EPA has a new Memorandum of Understanding with China and is in the initial stages of developing joint ambient monitoring at mutually selected coal-fired facilities in China, as well as enhancing awareness of mercury issues. EPA is working with the State Environmental Protection Administration (SEPA), as well as a non-governmental organization (NGO), Global Village Beijing. The ambient mercury speciated measurements activity, which can identify local scale deposition as well as that portion which will travel regionally and globally, can be informative to SEPA's new effort to highlight the importance of mercury domestically and its development of national policy. EPA will seek to coordinate its China measurements with those of the U.S. Department of Energy in China and its liaison with the Chinese Ministry of Science and Technology (MOST) in taking stack measurements. Next year, EPA will coordinate with USGS to provide technical support for mercury for an anticipated, new USGS-led multi-partner initiative in China to bring about more cooperation between the geological sciences and health communities in S.W. China, involving scientist and student exchanges and telemedicine.

India: EPA has a new Memorandum of Understanding with India and this year is identifying a mercury energy sector monitoring activity, to initiate monitoring of mercury at the stack at selected coal-fired facilities.

Also, EPA participated in April 2004 in a chloralkali sound environmental practices workshop in Delhi, sponsored by the private sector, where practical methods to reduce mercury emissions were shared.

Multilateral Global Cooperation

UNEP: In 2000, EPA began to take an interest in a global mechanism to address mercury, as progress was being made in the technical understanding of global behavior and fate of mercury in the environment, and sufficient coordination was taking place on many scales which could facilitate the consultations and activities under a global program for mercury. EPA and Department of State looked to the United Nations Environment Program (UNEP) and its Chemicals Division for a way forward on addressing mercury at this international level, and led the negotiation of a UNEP Governing Council decision in February 2001 that instructed UNEP to conduct a global assessment of mercury and its compounds. The first step was development of a technical assessment document, which was accepted by the February 2003 UNEP Governing Council and formed the basis for its decision to initiate a global program. Countries are asked by UNEP to take immediate actions to assist developing and re-industrializing countries to address mercury, and this is what EPA is working on furthering, in concert with other U.S. agencies and the public and private sector and with other countries.

Other UN Work Co-Led by WHO/UNEP/ILO: Since the 1970s, EPA has contributed to many UN interagency efforts on chemical safety, including scientific and risk assessment work on heavy metals, including mercury. It is worthwhile to note that the first Environmental Health Criteria publication by the joint World Health Organization (WHO) and UNEP, publicly released in 1976, is on mercury. Beginning in the1980s, EPA helped launch and begin implementation of the new United Nations International Programme on Chemical Safety (IPCS) effort involving the WHO, UNEP and International Labor Organization (ILO), in liaison with the Food and Agriculture Organization (FAO). One of the reasons for EPA's priority support and involvement in the IPCS has been to help strengthen regional and global cooperation to increase knowledge about and to better control heavy metals pollution, including mercury. Around 1984, EPA was supportive of the IPCS and the OECD Secretariat establishing a formal, regular, cooperative approach on hazardous and toxic chemicals, for example on mercury.

UNIDO: The UNEP Global Mercury Assessment identified artisanal mining as a key source of mercury emissions to the global pool, and it is recognized that it has been under-represented in global inventories to date. It would be valuable to not only obtain a better perspective on the extent of these practices around the world and the nature of the emissions, but also to help bring about, in concert with UNIDO, application of more efficient technologies which use and emit less mercury, as well as introduction of non-mercury processes. UNIDO's Global Mercury Project last year extended to EPA an invitation, which EPA accepted, to take part on their Advisory Board and provide advisory services and technical field support for their global program assessing environmental and health impacts and promoting reduction of mercury emissions and sustainable development for artisanal miners who use mercury while mining for gold around the world. Six countries are UNIDO's focus: Brazil, Indonesia, Laos PDR, Sudan, Tanzania and Zimbabwe.

Arctic Council Action Plan (ACAP) Mercury Project: In the Arctic, due to meteorological conditions, mercury that travels in the atmosphere from other parts of the globe is deposited and can remain in the Arctic region. The U.S. is participating in an Arctic-wide Arctic Council Action Plan (ACAP) project on mercury, entitled: AReduction of Atmospheric Mercury Emissions from Arctic Countries. A circumpolar perspective on mercury is being obtained by all countries contributing to a regional inventory for mercury, as well as assisting Russia in shaping its own inventory, and then conducting one or more pilot projects in Russia for emissions reduction and possibly workshops for exchange of information about best practices and cleaner technology. During the first phase, all 8 Arctic countries shaped a regional inventory, which has just been finalized. EPA feels that this ACAP project also provides an opportunity to further the objectives of the new UNEP Global Mercury Program, which has identified development of inventories and action plans at the national and regional levels, as well as information sharing and pilot activities to explore ways to control emissions in key source categories in developing or re-industrializing countries.

Arctic Monitoring and Assessment Program (AMAP): The USEPA served on behalf of the United States in its role as Lead Country for Heavy Metals under the Arctic Council, Arctic Monitoring and Assessment Programme. An EPA scientist assumed the role of Lead Country Expert for the U.S., and chaired the Heavy Metals Assessment team from 1999-2004. During that time, EPA sponsored two international science meetings, chaired numerous internationally held team meetings and served as lead author and editor of the assessment. Key findings of the 2003 AMAP assessment for mercury include: 1) In the Arctic, mercury is removed from the atmosphere via unique processes and deposits onto snow in a form that can become bioavailable; 2) Enhanced deposition occurs in the Arctic, which can be an important sink in the global mercury cycle; 3) There is evidence that despite substantial mercury emissions reductions in North America and Western Europe, global mercury emissions may, in fact, be increasing; 4) Mercury levels are increasing in marine birds and mammals in the Canadian Arctic and indication of increases in West Greenland; and 5) Current mercury exposures pose a health risk to some people and animals in the Arctic. Recommendations include: 1) expanding and accelerating research on critical aspects of the mercury cycle and budget in the Arctic; 2) quantifying all sources of mercury and reporting results in a consistent and regular manner to improve emission inventories, with focus on burning of coal at small-scale power plants and in residential heating, as well as waste incineration; 3) continuing temporal trend monitoring and assessment of effects of mercury in key indicator media and biota, to determine effectiveness of measures undertaken through the LRTAP Protocol; 4) ensuring that Arctic concerns are addressed and promoting development of regional and global actions.

TRENDS MONITORING

Strong trends assessment monitoring and modelling programs must address the need for effective coordination and collaborative networking. In order to have an effective and responsive mercury program, an organization must evaluate its mercury efforts and measure progress in meeting its mercury goals. Monitoring of trends in such topics as the amount of mercury used and released, ambient air and water concentrations, deposition rates, and fish/wildlife and human mercury levels is essential to determining the effectiveness of our efforts.

The goals of a trends monitoring strategy should be to provide information to: (1) discern long-terms trends of mercury and other persistent bioaccumulative toxics (PBTs) in the environment and (2) measure the effectiveness of risk management actions. A coordinated multimedia network and assessment program can strengthen and better integrate existing monitoring programs by addressing gaps and promoting better communication and collaboration.

Trends monitoring for mercury

As the conceptual model of the transport of mercury in the environment, discussed earlier in section II, indicates, mercury is released into the air, transported long or short distances, deposited on land or water, transformed in the aquatic environment, accumulated in fish, and ingested by humans or wildlife.

The expected goal of trends monitoring is to find the most efficient points along that flow to determine the trends in environmental levels and whether they are responding to control and reduction measures.

Air Emissions

Atmospheric transport is a primary focus for mercury monitoring and modelling, as it is the dominant means for cycling mercury from anthropogenic sources, such as utility combustion sources, into other media. Ecological and human exposures result primarily from air deposition to water bodies where methylation occurs, followed by uptake and bioaccumulation of the methylated species in fish, and ingestion of fish by humans and piscivorous, i.e., fish eating, marine mammals.

Transboundary Air Transport

Atmospheric mercury concentrations are affected by the emissions of other countries which cycle at various scales. A better understanding of the contribution of global sources to national mercury levels will aid in enacting measures to reduce or prevent such pollution transfer. There currently are some mercury super sites where speciation occurs and sufficient variables (particulates and halides measurements, for example) are measured to interpret sources through back trajectory modelling. For instance, it is possible at these sites, such as in the Ohio River Valley and in Southern Florida, to determine the influence of local sources versus global sources. Currently, speciated mercury measurements are being made and their halide chemistry is being analyzed at a high altitude site at Mauna Loa, Hawaii, in an attempt to better understand the atmospheric mechanisms that are responsible for the transformations of elemental mercury into reactive gaseous mercury (RGM) at high altitudes. An immediate planned enhancement is to establish the Ny Alesund, Norway, site as an international super site with many countries leveraging resources to take samples and share data. An elevated location in

Asia, such as utilization of the Italian K-2 station in the Himalayas between India and China, would be advantageous for mercury measurements. More such strategically-positioned sites are needed. At these sites, it would be valuable to install a sophisticated monitoring capability that will measure, for at least a decade, the 3 forms of mercury EPA is measuring at Mauna Loa, the criteria pollutants, (e.g. sulfur dioxide, ozone, oxides of nitrogen, gas phase halides, fine and coarse particles) and meteorology.

Air Deposition: Air deposition is currently measured through the Mercury Deposition Network (MDN), which monitors wet deposition utilizing a non-probabilistic network design. The MDN provides a good starting point for mercury air deposition monitoring, though some enhancements are needed to fulfill the goals of the Strategy. Analyses have shown that the national coverage of this network could be improved by adding sites in the West. Also, the program could be expanded to include dry deposition.

Sediment and soil: Aquatic sediments play an important role in the mechanism of mercury bioaccumulation. Analysis of sediment core profiles provides the only source of very long-term historical records of air deposition, extending into pre-industrial and pre-European times, although it is slow to reflect the most recent trends. This is a unique form of data that provides an historical perspective on air-deposition rates.

Food: For mercury, unlike some other PBTs, human and wildlife exposure is almost solely through fish consumption. The National Fish Tissue Study, currently being conducted by the EPA Office of Water, provides a one-time probabilistic sample of contaminants in fish tissue in freshwater lakes, including mercury as well as other PBTs. Most human exposure to methylmercury is through the ingestion of fish. Monitoring of commercial fish is within the purview of the Food and Drug Administration (FDA).

Human Tissues: Beginning in 1999, the EPA and other federal agencies funded an add-on to the National Health and Nutrition Examination Study (NHANES) to measure mercury in the blood and hair of a national sample of women of child-bearing age and children. This effort, which is still ongoing, provided the first systematic data on actual mercury exposure levels in the U.S. population. NHANES intends to continue to gather blood mercury data in these populations, making it possible to identify trends over time.

INTERNATIONAL COOPERATION

Vision to Address Mercury Internationally: Mercury is globally cycling, so no one country can achieve its goals unless all participate in solutions. The U.S. has taken a leadership role to provide technical assistance, capacity building and pollution prevention measures to bring along all countries. We want to share our information from lessons learned, and to learn from our international relations.

Identified Research Data Gaps: Key technical data gaps, many identified by EPA for the UNEP Global Mercury Assessment Report, include:

1. atmospheric chemistry related to mercury transport and deposition(source emissions inventory and speciation information; global cycle information and an assessment of air measurement and monitoring techniques);
2. toxicology and human health (to better quantify how large uncertainty factors should be; and better understand effects of mercury in adults);
3. ecological effects (effects of methylmercury on various species; interactions of methylmercury with other chemical and non-chemical stressors on ecological receptors; ecological risk assessment methods;
4. sources and emissions (comprehensive country-specific mercury release inventories from all sources; knowledge of feed stocks, plant designs and operating practices at major sources; new information on measurement methods to aid in inventories);
5. long-term trends in various media (transboundary air transport within and between countries; wet and dry air deposition levels; methylmercury levels in foods, particularly fish tissue for freshwater, estuarine and ocean fish; methylmercury concentrations in human tissues; and methylmercury concentrations in wildlife tissues).

The atmospheric pathway is the most significant for transport and regional and global impacts and is the EPA priority. Based on existing information, the key emissions sources internationally which contribute to global cycling and deposition via air pathways and are of particular interest to the Office of International Affairs (OIA) to address are: *1) coal-fired combustion sources; 2) mining and metals production, including smelting; and 3) chloralkali facilities.* Russia, China and India have coal-fired facilities and also could benefit significantly from attention to mining and chloralkali sectors. In addition to these 3 countries, certain South American (Brazil) and West African countries should receive attention to mining (in concert with EPA's cooperative program with UNIDO).

Coal-fired Combustion: Pacyna (2003) estimated that 75% of current global emissions of mercury come from fuel combustion in China, India and other Asian countries. While many facilities in European Russia are changing to natural gas, the Russian Far East and other areas are still dependent upon coal. The current estimate of coal-fired facility emissions in Russia is 5-7 tons, based on an ongoing ACAP inventory for mercury in Russia, and overall combustion is thought to be 11 tons per year, as presented at the recent EU, Nordic Council of Ministers Meeting in Brussels, Belgium in April 2004.

Mining and Metals Production: Lacerda and Salomons (1997) estimates that mercury emissions from gold mining may be as large as 300 metric tons per year (about 6% of anthropogenic emissions), but others make estimates ranging from 500-2000 tons on a global basis each year (Veiga, 2002; Veiga, 2004).

Chloralkali and the Chemical Industry: There may be 100 chloralkali factories in developing countries. While these factories tend to be smaller in size than U.S. factories, many may never have been given any formal regulations pertaining to their environmental performance. Where this is the case, a factory can potentially release quantities of mercury. EPA feels it is valuable to ensure that both environmental authorities in developing countries and chloralkali factories in these nations are informed, in practical terms, of sound management practices.

CLOSING THOUGHTS

Stemming from EPA's original mandate to have a coordinated approach when dealing with environmental problems, in order to guarantee that as we deal with one difficulty we do not aggravate others (W.D. Ruckelshaus, Environmental News, December 1970), it is through this path of coordination and cooperation, both here and abroad, that we have taken to address the Agency priority toxic pollutant, mercury. We are working closely with all levels of government and with industry and the private sector nationally and internationally, and enhance exchange of information so that further progress can be made in concert with others. The monitoring, modelling and health communities of scientists are mutually dependent upon knowledge each discipline can provide, in order to develop information leading to better exposure and risk assessment, as well as improved risk management. This workshop is one of the first to bring these three communities together internationally to share technical information and collectively plan the next steps for further cooperation. Bringing these technical groups together is something to replicate in the future.

ACKNOWLEDGMENTS

The author gratefully acknowledges assistance by Stan Durkee (EPA, ORD) and Arnold Kuzmack (EPA, OW) for providing information, and for frequent reviews of organization and presentation of materials in this chapter. Other welcome assistance was provided by Ellen Brown (EPA, OAR), Robert Stevens, William Stelz and Douglas Grosse (EPA, ORD), Marianne Bailey and Pete Christich (EPA, OIA), and Frank Anscombe (EPA, Region 5).

DISCLAIMER

This is a staff paper summary of ongoing activities and perspectives about environmental aspects of mercury, and does not itself represent official Agency communication.

REFERENCES

de Lacerda, L., Salomons, W. Mercury from Gold and Silver Mining: a chemical time bomb? Springer Press. 1997.

Environmental Quality, The Second Annual Report of the Council on Environmental Quality, August 1971.

Schuster, P.F., Krabbenhoft, D., Naftz, D., Cecil, L., Olson, M., Dewild, J., Susong, D., Green, J., Abbott M. Atmospheric Mercury Deposition during the Last 270 Years: A Glacial Ice Core Record of Natural and Anthropogenic Sources. *Env. Sci. Technol.* 36, 2303-2310, 2002

The National Academy of Sciences (NAS) report on the Toxicological Effects of Methylmercury (NRC, 2000)

United Nations Environment Programme (UNEP). Chemicals, Global Mercury Assessment, Geneva, 2002.

U.S. Environmental Protection Agency (USEPA). Mercury Study Report to Congress: EPA-452-R-97-003. December 1997. Online at http://www.epa.gov/ttn/oarpg/t3/reports.

U.S. Environmental Protection Agency (USEPA). Mercury Research Strategy: EPA/600/September 2000. Washington, D.C. Online at http://cfpub.epa.gov/ncea/cfm/recordisplay.cfm?deid=2085.

Veiga, M., Hinton J. Abandoned Artisan Gold Mines in the Brazilian Amazon: A legacy of Mercury Pollution. Natural Resources Forum, 2002.

PART-II:
MONITORING AND ANALYTICAL METHODS

Chapter-7

THE MONITORING AND MODELLING OF MERCURY SPECIES IN SUPPORT OF LOCAL, REGIONAL AND GLOBAL MODELLING

Matthew S. Landis[1], Mary M. Lynam[1] and Robert K. Stevens[2]

[1] US Environmental Protection Agency, ORD, RTP, NC 27711, USA
[2] Florida Dept. Environmental Protection at USEPA, RTP, NC 27711, USA

INTRODUCTION

The first major recorded human cases of mercury (Hg) poisoning from eating contaminated fish occurred in Minamata, Japan in 1956 as a result of direct discharges of MeHg into Minamata Bay. "Minamata disease" provided a warning to the rest of the world concerning the human toxicity of Hg. Over the next several decades, legislation was enacted by many countries to protect their citizens against exposure to Hg contaminated fish. Most regulations were focused on the direct discharge of Hg contaminated wastewater. However, fish taken from remote lakes around the world that were known to receive no direct anthropogenic Hg discharges were subsequently found to contain high concentrations of Hg (Helwig et al., 1985; Schroeder et al., 1998). Scientists then discovered that atmospheric deposition was the predominant source of Hg into many aquatic ecosystems (Helwig et al., 1985).

In 1989 the United States Environmental Protection Agency (EPA) classified Hg as a Hazardous Air Pollutant. Congress promulgated the Clean Air Act amendments of 1990 that required EPA to draft a state-of-the-science report for Hg. In 1998 EPA submitted its Mercury Report to Congress (EPA-452/R-97-003, 1998) which identified a plausible link between anthropogenic releases of Hg to the atmosphere and bioaccumulation in fish as MeHg. EPA has recently enacted new regulations on chlor-alkali plants, medical waste incinerators, and municipal waste incinerators in part to cut the emissions of Hg to the atmosphere. Regulations for Hg emissions from coal-fired utility boilers are currently under review. Complicating this evaluation of various Hg emission reduction scenarios are uncertainties in the Hg emission inventories, atmospheric chemistry, and deposition parameterizations in contemporary deterministic Hg models.

The EPA Office of Research and Development (ORD) developed a Hg research strategy that endeavors to answer relevant scientific questions such as:

"How much methyl mercury in fish consumed by the U.S. population is contributed by U.S. emissions relative to other sources of mercury (such as natural sources, emissions from sources in other countries, and re-emissions from the global pool): how much and over what time period will levels of methyl mercury in fish in the U.S. decrease due to reductions in environmental releases from U.S. sources?"

The ability to answer these kinds of policy questions requires innovative research into atmospheric, aquatic, and bioaccumulation processes. In terms of atmospheric processes, our ability to improve the performance of contemporary deterministic models is directly linked to our ability to make reliable automated high-resolution speciated Hg measurements. Over the last decade, improvements in ambient Hg measurement technologies have lead to dramatic new insights into atmospheric Hg dynamics and dry deposition mechanisms. These discoveries are currently reshaping our contemporary atmospheric Hg conceptual model.

The purpose of this chapter is to describe the state-of-the-science with regard to ambient Hg measurement and modelling; their strengths, limitations, and recommendations for further work.

MONITORING

Monitoring of atmospheric Hg has been ongoing for more than three decades. Measurement methodologies have evolved from relatively crude daily manual collection of "total gaseous Hg" to semi-continuous measurement of the three environmentally relevant forms of Hg in the ambient air. These species include: elemental gaseous Hg (Hg^0), divalent reactive gaseous mercury (RGM), and particulate phase Hg (Hg(p)). The behavior of each of the three species in the atmosphere is unique and dependent on their physical and chemical properties. Hg^0 has a high vapor pressure, is relatively insoluble in water (4.9 x 10^{-5} g L^{-1} (Schroeder et al., 1998)), and has a low deposition velocity on the order of 0.05-0.1 cm s^{-1}. As a result, Hg^0 has a long atmospheric half life (weeks to months) and therefore can be transported on a global scale. RGM has a lower vapor pressure, is water soluble (66 g L^{-1} (5)), and is estimated to have a deposition velocity on the order of 1-5 cm s^{-1} (EPA-452/R-97-003, 1998; Lindberg et al., 1998). Brosset, as early as 1983 suggested that RGM was in most cases in the form of mercuric chloride emitted from coal-fired utility boilers and incinerators (Bloom et al., 1996). Hg(p) has a deposition velocity which is particle size dependent and ranges from 0.1-1 cm s^{-1} (Landis et al., 2002). RGM and Hg(p), although usually present in ambient air at a concentration three orders of magnitude less than Hg^0, have higher deposition velocities and undergo both wet and dry deposition on local and regional scales.

Elemental gaseous mercury (Hg^0)

Sampling and analysis of Hg^0 has historically capitalized on the fact that noble metals can be used to amalgamate Hg (Schroeder et al., 1985). Gold (Au) is widely used to preconcentrate Hg^0 during ambient monitoring that is subsequently released for quantification by thermally desorbing the amalgam. When Au traps are employed in monitoring a particulate filter is typically placed upstream to prevent aerosols from collecting in the trap. After a specified sampling interval, the Au trap is then heated to approximately 500°C to release Hg^0 which is detected by either cold vapor atomic fluorescence spectroscopy (CVAFS) (e.g., Tekran model 2537A) or cold vapor atomic absorption spectroscopy (CVAAS) (e.g., Opsis model 200, Nippon model WA-4). These types of instruments typically provide 5 minute integrated samples with detection limits of 0.1 ng m^{-3}.

In monitoring situations with elevated Hg^0 concentrations, fast response (1 Hz) methodologies using no sample preconcentration have been successfully applied. Zeeman background corrected CVAAS (Ohio Lumex model RA-915+) can provide continuous real time Hg^0 results with a detection limit of 2 ng m^{-3}. Differential optical absorption spectroscopy (DOAS, Opsis, Inc.) is an open path method that can provide continuous real time Hg^0 results integrated over a path length ranging from 25 to 1000 meters with a detection limit of 9 ng m^{-3}. Sampling for Hg^0 is now considered routine and data on ambient Hg^0 concentrations are widely available.

Reactive gaseous mercury (RGM)

Source sampling at a medical and a municipal waste incinerator in Florida (Schrock et al., 2000) and numerous coal-fired utility boilers across the U.S. (Prestbo et al., 1995) in the mid 1990's revealed that 95±5%, 78±8%, and 67±27% of the total Hg emitted from these sources was RGM, respectively. Subsequent modelling for the Lake Michigan Mass Balance Study (8) found that RGM dominated Hg dry deposition flux and contributed approximately 40% of the total atmospheric deposition. The overall uncertainty in these RGM dry deposition contributions are unknown because at the time these data were collected there were no reliable ambient RGM measurement methodologies available and estimates of RGM emissions were utilized. The findings from these studies did illustrate the critical need to develop robust methods to measure ambient RGM. Quantifying RGM in ambient air proved to be a challenge because methodologies must be extremely specific since Hg^0 is typically present in the 1.4-3.0 ng m^{-3} range and RGM is typically present in the 1-100 pg m^{-3} range. Nevertheless, during the last five years much effort has been expended in developing suitable methodologies for sampling and analysis of RGM and considerable progress in methods development has been made. Currently used methods include:

Cation-exchange membranes

Cation-exchange membranes have been employed at several locations to sample ambient RGM (Bloom et al., 1996; Ebinghaus et al., 1999). Aerosols are collected by a Teflon pre-filter and RGM is collected onto two cation-exchange membranes in series. After sampling the cation-exchange membranes are extracted with BrCl, reduced to Hg^0 with $SnCl_2$, purged from solution onto a Au trap, and analyzed using CVAFS. While the cation-exchange membranes can be employed in many environmental conditions, their use is currently limited because of (i) concerns over RGM adsorption to aerosols collected on the pre-filter, and (ii) relatively high filter blanks (~200 pg) that necessitate long sampling times (~24 hours) under ambient conditions and can result in high associated uncertainties.

Refluxing mist chambers

Refluxing mist chambers were originally developed and successfully used to measure ambient nitric acid (Cofer et al., 1985).

The mist chamber method was adapted and provided some of the first high resolution (1-2 hour) ambient RGM measurements (Schroeder et al., 1985; Stratton et al., 2001). In this method ambient air is drawn into a glass chamber where it comes in contact with a fine mist of absorbing solution (dilute HCl and NaCl) of aerosol-sized droplets that scrub soluble Hg species. A Teflon filter is employed as a hydrophobic barrier to prevent liquid droplets from escaping the chamber with the exhausted air stream. After sampling, the collected solution is reduced with $SnCl_2$, the collected Hg is purged onto a Au trap, and analyzed using CVAFS. Mist chambers have limited widespread use because they cannot be used in cold environments (<0°C), sampling durations are limited by the evaporation rate of the absorbing solution, and sampling typically must be attended. In at least one study, a comparison between the mist chamber and annular denuder methodologies showed that the former yields significantly higher RGM concentrations than the annular denuder method. The present consensus is that this method may be prone to a positive artifact from the aqueous oxidation of Hg^0 present in the sampled ambient air by the absorbing solution and the entrainment of Hg(p) (Stratton et al., 2001; Landis et al., 2002).

KCl-coated annular denuders

Tubular denuders were first developed and used to remove gaseous ammonia to prevent neutralization of fine (<2.5 μm) acid aerosols (17). Ferm (1979) first used $NaCO_3$-coated tubular denuders for quantification of gaseous nitric acid. A tubular denuder is a cylindrical tube that is chemically coated to remove selected gas-phase species that diffuse to the walls and react or adhere to the active surface during the passage of a laminar air stream. KCl-coated tubular denuders were first used to quantify RGM from incinerator stacks (Stevens et al., 1978) and were then adapted for ambient measurements (Xiao et al., 1997). The tubular denuder method is limited to a low flow rate (~1 L min^{-1}) and requires the RGM to be extracted with an acidic solution, reduced to Hg^0 with $SnCl_2$, purged onto a Au trap, and analyzed. (Possanzini et al., 1983) developed annular denuders to increase collection efficiencies for ambient nitric acid and sulfur dioxide that allowed higher sampling flow rates (10 L min^{-1}). Higher sampling flow rates necessitated the incorporation of a size selective impactor inlet to remove coarse aerosols (<2.5μm) that could deposit to the denuder surface. Fine particles pass through the denuder in the laminar air stream unaffected because they have insufficient diffusion coefficients to deposit to the walls of the denuder (Dzubay et al., 1991). A KCl-coated annular denuder (Figure 1) methodology that used thermal decomposition to quantify the collected RGM was then developed (Landis et al., 2002). Etching both the outer surface of the annulus and the inner surface of the denuder tube dramatically increased its KCl holding capacity and produced a uniform coating. Thermally desorbing the quartz denuder at 500°C in a Hg-free air stream decomposes RGM and allows for quantification as Hg^0 without chemical extraction or sample preparation. Thermal desorption provides for rapid analysis in the field, minimizes the possibility of contamination, and allows the denuder to be immediately reused without further preparation. Currently, the KCl-coated quartz annular denuder method is the most commonly used and is generally accepted as the most appropriate procedure for collection of RGM.

During sampling the annular denuders are heated to 50°C to prevent water vapor from hydrolyzing the KCl coating. Studies carried out to assess denuder performance have shown that the collocated precision of manual denuders is ±15% (Landis et al., 2002). It is currently believed that RGM is physically sorbed to the surface of the KCl coating rather than chemically bound. This is borne out by the fact that long sampling durations will result in diffusion of RGM along the axis of the coating and off the coated surface. Sampling durations up to 12 hours were found to have a collection efficiency of 93% while sampling between 12 and 24 hours resulted in a lowering of the collection efficiency to 83%. Studies in Barrow, Alaska have also revealed that adequate heating of the manual denuders during sampling in arctic environments is imperative to maintain optimum RGM collection efficiency.

Particulate phase mercury (Hg(p))

Quantification of Hg(p) in ambient air is generally accomplished by collection onto quartz filter media followed by analysis by CVAFS or CVAAS.

The quartz filters can either be extracted by microwave digestion in a HNO_3 solution that is subsequently oxidized with BrCl, reduced with $SnCl_2$, and purged onto Au traps (Landis et al., 2002) or thermally decomposed and collected onto Au traps (Landis et al., 2002; Lu et al., 1998).

The collection of Hg(p) samples has been the subject of debate over the last decade because of observed positive artifact formation (RGM sorption onto the particles or the filter) or negative artifacts (volatilization of Hg(p)) during sampling.

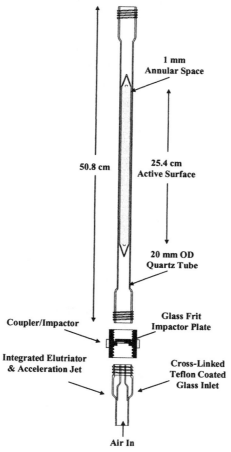

Figure 1. Annular denuder and inlet system for manual collection of RGM.

The use of KCl coated annular denuders to remove positive RGM artifact in the measurement of fine particle (<2.5 µm) Hg(p) has been demonstrated (Landis et al., 2002; Lynam et al., 2002). Both studies reported significantly higher Hg(p) concentrations on undenuded quartz filters when compared to denuded filters. This positive artifact was not quantitative and both studies indicated that the removal of RGM was imperative for the accurate measurement of Hg(p). The occurrence of negative Hg(p) artifacts has been reported (Malcolm et al., 2003) in the marine boundary layer when comparing a series of short term Hg(p) samples (2-3 hour) to longer term integrated samples (6-12 hours). The study recommends shorter sample durations to minimize the negative Hg(p) artifacts.

Semi-Continuous Measurement Methodologies for Mercury

The development of semi-continuous measurement methods was spawned in response to the need for higher time resolved speciation data for understanding atmospheric Hg dynamics, quantifying atmospheric dry deposition, and investigating the impact of Hg source(s).

Cold regions pyrolysis unit

The cold regions pyrolysis unit was designed by Environment Canada scientists and first deployed at Alert to assess if any Hg remained in the ambient air during Hg^0 depletion events observed after polar sunrise, or whether deposition to the snow pack was occurring (Banic et al., 2003). The cold regions pyrolysis unit simultaneously monitors Hg^0 and total atmospheric Hg (TAM; comprised of Hg^0, RGM, and Hg(p)) using two collocated Tekran 2537A mercury analyzers configured with a common inlet. A pyrolysis unit heats the incoming ambient air to 900°C is installed at the front end of one of the Tekran 2537A instruments converting inorganic and organometallic mercury species to Hg^0. The collocated Tekran analyzer samples Hg^0 in the ambient air. One can then compare the concentrations of TAM and Hg^0 from both instruments. During a Hg depletion event Hg^0 in the ambient air decreases and if measurable quantities of TAM are detected with the pyrolysis unit these are likely to be oxidized Hg species which were converted to Hg^0 during pyrolysis. This unit is useful for limited applications mainly because it operates at a flow rate of 3 L min^{-1} which is insufficient to sample larger aerosols thereby leading to an unknown certainty in the Hg(p) portion of the measurement.

Tekran speciation system

The development of the Tekran 1130/1135/2537A system has revolutionized our ability to study Hg speciation in the ambient environment. The system can be operated under a wide temperature range (-40°C to 50°C) and can be tailored to sample ambient air for as little as 1 hour with method detection limits (MDL) that permit quantification of background ambient levels of RGM, Hg(p) and Hg^0. A detailed description of the instrument and how it functions may be found in Landis et al. (2002).

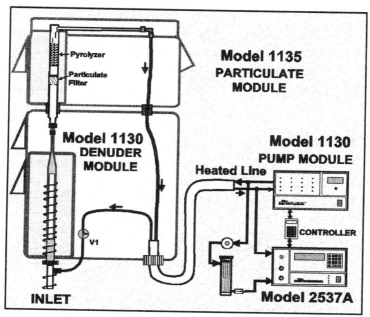

Figure 2. Schematic of Tekran 1130/1135 mercury speciation unit.

Figure 2 illustrates the instrumentation which operates as follows; ambient air is sampled at a flow rate of 10 L min^{-1} through a heated impactor inlet and passes through a heated (50°C) KCl-coated annular denuder for RGM collection, a particulate filter unit for fine (>2.5 μm) Hg(p) collection, and through a heated umbilical line to the model 2537A for Hg^0 collection and analysis. The minimum recommended sampling duration is 1-hour, which permits the collection of an integrated sample for RGM and Hg(p). During sample collection, Hg^0 is sampled and analyzed every 5-min using the Tekran 2537A. Sequentially, after completion of sampling, Hg(p) is desorbed from the filter and RGM is desorbed from the denuder by heating to 800°C and 500°C, respectively. As the Hg(p) and RGM are thermally decomposed they are detected as Hg^0 by the 2537A analyzer. The system is then cooled for 10 minutes and the sampling cycle is resumes. Thus, the Tekran 1130/1135/2537A instrumentation permits semi-continuous monitoring of the three Hg species which are currently believed to be the most important in the atmospheric cycling of Hg. This semi-continuous method is now considered state-of-the-art and is currently employed at numerous monitoring locations world wide in order to glean much needed information on Hg speciation.

What have we learned from Hg speciation measurements?

The ability to measure semi-continuous Hg speciation has resulted in a dramatic increase in our knowledge of atmospheric Hg dynamics and source characterization.

Polar Hg chemistry

The existence of contamination by persistent toxic substances in the Arctic has prompted research on Hg in that ecosystem. In 1995, surface-level measurements at Alert in the Northwest Canadian Territories showed frequent episodic depletions in Hg^0 concentrations during Arctic spring following polar sunrise (Schroeder et al., 1998). At that time it was assumed that Hg^0 was being oxidized by an unknown mechanism to either Hg(p) or RGM, but the technology to continuously monitor these Hg species was not available. In 2000 scientists at the German research station Neumayer also observed Hg^0 depletion events in Antarctica (Ebinghaus et al., 2002).

Semi-continuous Hg speciation monitoring was subsequently carried out in Alaska, Norway and Antarctica. Studies implemented in Barrow, Alaska in 2001 (Lindberg et al., 2001) and 2002 (Lindberg et al., 2002) used the Tekran 1130/1135/2537A speciation system to measure Hg depletion events. These studies confirmed the rapid transformation of Hg^0 to predominantly RGM after polar sunrise. The oxidation mechanism was hypothesized to be mediated by UV light and reactive halogens. RGM then deposited to the snow pack where it accumulated and was released in a bioavailable form during the spring snow melt. Monitoring at Ny-Alesund, Norway (Berg et al., 2003), and Neumayer (Temme et al., 2003) and Terra Nova Bay (Sprovieri et al., 2002) Antarctica stations highlighted the extent of the Hg depletion events and their importance to the global Hg cycle.

Speciation studies in the marine free troposphere

EPA conducted two measurement campaigns off the Atlantic coast of Florida in 2000 using a DeHavilland Twin Otter aircraft to measure Hg^0, RGM, and Hg(p) from 60 to 3,500 meters (Landis et al., 2001).

One sampling campaign was conducted in January when the synoptic flow was from west to east and another campaign was conducted in June when the synoptic flow was from east to west. A total of 21 research flights were undertaken to evaluate the potential for long range transport of Hg to Southern Florida. Figure 3 presents results from the January 2000 campaign that show Hg^0 concentrations decreased with altitude (even when corrected for decreased sensitivity of the Tekran 2537A instruments with altitude) while RGM increased with altitude in the marine free troposphere. Elevated concentrations of RGM in the free troposphere were thought to be a result of the oxidation of Hg^0 to RGM at high altitude. This finding led EPA to initiate automated Hg speciation measurements at the Mauna Loa Observatory (MLO), Hawaii (~3,400 meter elevation) in 2001, in part to investigate the potential for either homo- or heterogeneous oxidation of Hg^0 in the marine free troposphere (Landis et al., 2001). Some initial results from the Tekran 1130/1135/2537A Hg speciation system are depicted in Figure 4 and reveal periods of anticorrelation between Hg^0 concentrations and RGM/Hg(p) during which concentrations of Hg^0 decrease while those of RGM and Hg(p) increase. The MLO observations are consistent with the results from the aircraft measurements indicating a significant Hg^0 oxidation mechanism(s) in the marine free troposphere. Future studies at MLO will incorporate semi-continuous measurements of both gas and particle phase halides in order to gain more insight into potential mechanism(s) for Hg^0 oxidation in this environment.

Mobile sources

In the Hg report to Congress (EPA-452/R-97-003) EPA listed mobile sources as an important category for which Hg emissions estimates were unavailable. Hg is known to be a trace contaminant in crude oil (1505 ± 3278 ng g^{-1}; mean ± standard deviation), light distillates (1.3 ± 2.8 ng g^{-1}), and utility fuel oil (0.7 ± 1.0 ng g^{-1}) (Liang et al., 1996; Wilhelm et al., 2000).

EPA and the Florida Department of Environmental Protection (FLDEP) conducted a study in 1998 at the Ft. McHenry tunnel in Baltimore, Maryland to develop source profiles from gasoline and diesel powered vehicles (Landis et al., 2004). The tunnel measurements indicate that gasoline powered vehicles were a source of Hg^0, RGM, and Hg(p). The average Hg concentrations in the diesel tunnel were similar to ambient background

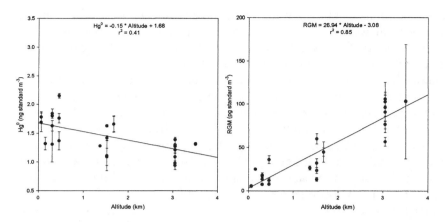

Figure 3. Hg⁰ and RGM aircraft profiles off the Atlantic Coast of South Florida (June, 2000).

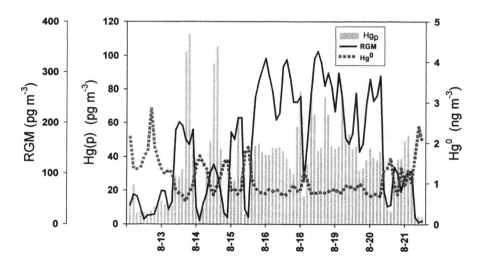

Figure 4. Mercury Speciation data from Mauna Loa Observatory, Hawaii (2001).

concentrations, while the gasoline tunnel concentrations were elevated for Hg^0, RGM, and Hg(p). The tunnel profile observations are consistent with the fuel sample results from this study. Regular (87 octane) gasoline was found to have an average total Hg concentration of 284 ± 109 ng L^{-1} compared to the average total Hg concentration of 62 ± 36 ng L^{-1} in diesel fuel. EPA is currently working to develop quantitative Hg emission factors for gasoline and light duty diesel powered vehicles.

Chlor-alkali emissions

A comprehensive Hg emission study from a Hg cell chlor-alkali plant (MCAP) was conducted in 2001 by EPA and a team of collaborators that determined the emission rate of Hg^0 (518 g day^{-1}) and RGM (10 g day^{-1}) from the facility using a combination of continuous and semi-continuous instrumentation (Kinsey et al., 2004; Landis et al., 2004). The overall Hg emission rate was significantly lower than the contemporary EPA emissions factors and the percentage of Hg emitted as RGM measured during this study was 2%, considerably lower than the estimate of 30% contained in many emission inventories (EPA-452/R-97-003, 1998, Petersen et al., 1995). Nevertheless, the Hg emissions from the MCCAP had a considerable impact on near field Hg atmospheric dry deposition. A Lagrangian transport and deposition model estimated the mean annualized dry deposition within a 10 km radius of the facility due to emissions solely from the MCCAP to be 4.6 μg m^{-3} (Landis et al., 2004). A subsequent study of a European MCCAP found the relative emission fraction of RGM to be 1% (Wängberg et al., 2003).

Medical waste incineration

Stack testing conducted by EPA and FLDEP in 1995 found that Hg emissions from a large medical waste incinerator (MWI) were primarily in the form of RGM ($95\pm5\%$) (Schrock et al., 2000). A comprehensive multimedia analysis subsequently prepared by FLDEP implicated medical and municipal waste incineration as the primary contributors to Hg contamination in the Florida Everglades (Atkeson et al., 2003). New EPA regulations on MWI facilities have drastically cut the number of operating units and mandates installation of "maximum achievable control technology" requiring 80% reduction in Hg emissions on most remaining units. Monitoring by the North Carolina Department of Environmental Protection (NCDEP) observed that a relatively small MWI could dramatically impact near field ambient Hg concentrations (Crawford et al., 2004). NCDEP deployed a Tekran 1130/2537A monitoring system in Matthews, North Carolina approximately 0.5 Km from a MWI for 12 months in 2003. NCDEP reported a 2-hour annual Hg^0 average concentration of 3.9 ± 25.9 ng m^{-3} with a maximum observed value of 623.8 ng m^{-3} and a 2-hour annual RGM average concentration of 122 ± 634 pg m^{-3} with a maximum observed value of 13.5 ng m^{-3}. These are among the highest RGM concentrations ever observed.

Future methods development

In spite of steady advances in ambient Hg monitoring methods accomplished over the last two decades, RGM and Hg(p) remain operationally defined "species" based upon simple fractionation and trapping of groups of related compounds. Identification of individual compounds that constitute RGM and Hg(p) would provide new insights and dramatically improve our understanding of atmospheric Hg chemistry, gas/aerosol interactions, wet and dry deposition phenomenon, and source/receptor relationships.

Particulate phase mercury

As described previously in this chapter, chemical extraction and thermal desorption of quartz filters followed by CVAFS or CVAAS are currently the preferred methods for Hg(p) analysis. The thermal method is attractive since it is faster and far less complicated than chemical extraction. In addition, the thermal method may also permit chemical identification of the component Hg compounds in the sample. Thermal decomposition methodologies have historically been used to quantify Hg content in ores and Hg-contaminated soils (Henry et al., 1972; Biester et al., 1999). In the latter study, temperature controlled continuous heating of the samples in a furnace coupled to a CVAAS allowed the identification of the different redox states of Hg species based on their characteristic release temperatures. However, in contrast to ores and soils where Hg is present in ppm concentrations and can therefore be easily quantified, Hg associated with ambient aerosols is typically present in the ppb range (Landis et al., 2002) and are prone to measurement artifacts. This makes quantification of true Hg(p) challenging and chemical identification impossible with conventional CVAFS or CVAAS analytical techniques.

A new method has been developed for quantification of Hg(p) collected on quartz filters using High Resolution Inductively Coupled Plasma Mass Spectrometry (HRICP-MS) (Klaue et al., 2004). Thermal release of Hg(p) into an argon carrier gas at a heating rate of 1°C s^{-1} from 25-750°C was measured by HRICP-MS. The released Hg(p) was quantified online by isotope dilution using an enriched double-spike of ^{200}Hg and ^{201}Hg generated by a continuous-flow high efficiency cold vapor generator. The accuracy of this method was evaluated using a NIST standard reference material 1633b, a bituminous coal fly ash and found to be in very good agreement with the certified value 0.1389 ± 0.0089 mg kg^{-1} (HRICP-MS) versus 0.141 ± 0.019 mg kg^{-1} (NIST). The detection limit for this method was found to be 20 fg.

The temperature release pattern of Hg from a PM$_{2.5-10}$ sample obtained in Detroit, MI using this technique is shown in Figure 5. This release pattern shows that all of the Hg from this sample was evolved at a temperature below 350°C in a bimodal release pattern. Based on kinetic, thermodynamic considerations and mechanisms of vaporization, as well as previous work on contaminated soils; one can tentatively assign the first release peak as adsorbed Hg vapor and the second peak as an ionic form of mercury likely Hg(II). Thermal release of pure salts (e.g., HgCl$_2$, HgSO$_4$, HgS, and HgO) has been attempted using this method and the temperature of peak release varied from 136°C-600°C (Lauretta et al., 2001). EPA is currently working to (i) refine and optimize the HRICP-MS ambient Hg(p) analysis method and (ii) develop a thermal decomposition library for known Hg compounds.

MODELLING

Deterministic modelling

Deterministic models are routinely used as a tool to investigate the transport, transformation, and deposition of atmospheric Hg (Chapters 13 & 14 this volume). The ability of these models (e.g., CMAQ, GEOS-CHEM, HYSPLIT, REMSAD) to reliably simulate Hg fate are dependent on the (i) quality of emission inventories, (ii) completeness and accuracy of the chemical kinetics module, and (iii) proper parameterization of gas/particle interaction and wet and dry deposition phenomena.

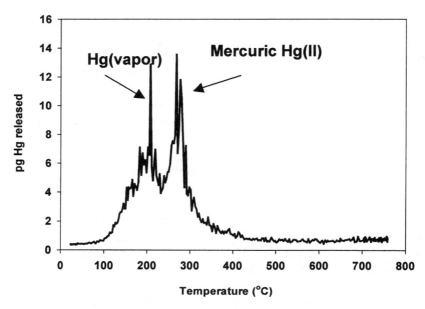

Figure 5. Thermal release profile for a PM$_{2.5-10}$ sample obtained in Detroit, Michigan.

Other chapters in this volume delve into these relevant research areas. It is clear that ambient Hg monitoring has both helped to improve contemporary deterministic models as well as highlight their deficiencies.

Emission inventories

Deterministic models require detailed emissions inventories for all major Hg sources. In the EPA Mercury Report to Congress (EPA-452/R-97-003, 1998) coal-fired utility boilers were identified as the largest contributor to Hg emissions in the US. In 1999, EPA issued an information collection request (ICR) for Hg emissions from these plants. As a result, Hg concentrations in coal used at 450 coal-fired utility boilers in the U.S. were quantified and speciated stack tests were performed at 80 of the plants. While these activities have yielded valuable information on the amount and speciation characteristics of Hg in stacks, recent ambient monitoring data suggest that rapid reduction of RGM to Hg^0 may occur in the exhaust plume (Edgerton et al., 2004) but this is yet to be replicated: if so, this would cast some doubt on the use of in stack Hg speciation data for emissions modelling. Previously discussed monitoring studies focused on mobile sources, MCCAP, and MWI has also highlighted the need for additional work on anthropogenic Hg source characterization in the US. Emissions from Hg-enriched geological formations (Gustin et al., 2003; Nacht et al., 2004), volcanoes (Pyle et al., 2003; Nriagu et al., 2003), and forest fires (Sigler et al., 2003) must also be better quantified to elucidate the relative magnitude and speciation of natural as well as re-emitted anthropogenic Hg sources.

While Hg emissions from most developed countries are decreasing, emissions from developing countries have risen steadily from 1983-1992 at a rate of 4.5-5.5% (Pirrone et al., 1996). In 1995, 75% of total global anthropogenic Hg emissions were attributable to fuel combustion in China, India, and North and South Korea (Pacyna, E.G.; Pacyna, J.M.; 2002). As these developing countries economies continue to grow, Hg emissions are also expected to increase proportionately and will need to be quantified especially since Hg emissions have the potential for long range transport that can contribute to the depositional burden in other countries (Weiss-Penzias 2003).

Chemical kinetics

Atmospheric Hg chemistry is complex. The earliest kinetic studies suggested that Hg^0 was not very reactive in the gas phase ($k \sim 10^{-20}$ cm^3 molecule^{-1} s^{-1}) and therefore likely had a lifetime of upwards of a year in the atmosphere (Hall, B.; 1995). Hg^0 depletion events measured at the Poles and Hg monitoring studies in the marine free troposphere have challenged the contemporary atmospheric Hg conceptual model, and have prompted a resurgence in laboratory kinetics studies (Sumner et al., chapter 9). Rate constants for reaction of Hg^0 with OH radical ($k = 8.7 \pm 2.8$ x 10^{-14} cm^3 molecule^{-1} s^{-1}), (Sommar et al., 2001) with Cl radical ($k = 1.0 \pm 0.2$ x 10^{-11} cm^3 molecule^{-1} s^{-1}) and Br radical ($k = 3.2 \pm 0.3$ x 10^{-12} cm^3 molecule^{-1} s^{-1}) (Ariya et al., 2002) in laboratory settings have been reported. Furthermore, evidence for the existence of HgBr, HgBrO/HgOBr and HgO has been observed upon BrO-initiated oxidation of Hg^0 (Raofie et al., 2004). Smog chamber experiments have also been carried out where Hg^0 has been reacted with O_3 Cl_2, Br_2, BrO, and ClO. The fastest rates were observed with Cl radical, ($k = 6.6$ x 10^{-14} cm^3 molecule^{-1} s^{-1}) and Br radical ($k = 3.0 - 9.7$ x 10^{-13} cm^3 molecule^{-1} s^{-1}) (Spicer et al., 2002). Two rate constants for the reaction of Hg^0 with O_3 were also found, $k \leq 2.5$ x 10^{-18} cm^3 molecule^{-1} s^{-1} and 9 x 10^{-19} cm^3 molecule^{-1} s^{-1} suggesting that further kinetics studies are necessary in order to obtain a better-resolved rate constant for this reaction. These lower reported rate constants suggest an atmospheric lifetime for Hg^0

of days to weeks with respect to reaction with ozone adding credence to the observations of depletion of Hg^0 in anthropogenic plumes during transport in the Pacific Northwest (Weiss-Penzias et al., 2003). Further monitoring and laboratory kinetics work is necessary to identify important oxidation/reduction reactants and to generate relevant rate constants.

Deposition parameterization

Wet and dry deposition mechanisms are the removal pathways for atmospheric Hg. Deterministic models of wet deposition estimates have historically been compared to collected precipitation samples (e.g., Mercury Deposition Network (EPA-452/R-97-003, 1998)) and have performed relatively well. Modeled aerosol and gaseous dry deposition estimates are currently highly uncertain because of a dearth of information on dry deposition modelling parameterizations, little semi-continuous ambient monitoring data, and a lack of validated direct measurement methods to evaluate model performance. Dry deposition of RGM is still a largely unexplored component in Hg monitoring. Much of this stems from the fact that the surrogate surface methodology commonly used for dry deposition measurements are not appropriate for use with RGM. Clearly, the development of new dry deposition monitoring techniques in conjunction with automated Hg species monitoring are required to successfully validate a deterministic dry deposition model.

Source Apportionment Modelling

The application of receptor models provides another approach to understanding the impacts of Hg sources on specific receptor locations. Receptor models quantify the impact of air pollution sources by using multivariate statistical methods on a matrix of elements or compounds in atmospheric samples as tracers for the presence of materials from specific source emissions (Gordon, G.E.; 1985, Hopke, P.K., 1985). Receptor models can be applied to a variety of air pollution studies since the measured properties of aerosols or gases at any monitoring site can be related to the sum of independently contributing causal factors. The goal of receptor modelling is to "apportion" the sources into specific categories (e.g., combustion, incineration, metals smelting, etc.) and quantify their relative importances. Receptor models can be used to constrain the uncertainty in deterministic modelling estimates and help identify sources that may not be accurately represented in emission inventories. The following discussion will describe source apportionment models currently being developed by EPA and its research collaborators that can be applied to Hg fate and transport research studies.

Statistical receptor modelling

The chemical mass balance model version 8.2 (CMB8.2) is now available from EPA. CMB8.2 is an effective variance-weighted least squares model that apportions sources using conserved source profiles. Inputs to the model are source profile abundances and the concentrations of species at a receptor site. It utilizes any number of samples as input. A site specific emissions inventory is necessary in order to determine which source profiles must be selected as well as which chemical parameters must be measured at local source emissions and in the ambient air at receptor locations.

The UNMIX2 model is a multivariate receptor model which seeks to solve a general mixture problem where the data are assumed to be a linear combination of an unknown number of sources of unknown composition which contribute an unknown amount to each sample (Henry, R.C.; 2003). This model usually requires information on 100 samples for input; it does not use data that is below the method detection limit and it generates source profiles and their associated uncertainties.

Positive matrix factorization (PMF) is a multi-linear model that usually requires at least 100 samples and uses data below the method detection limit (Paatero, P., 1997). Using this technique, sources are constrained to have non-negative species concentration and no sample can have a negative source contribution. The technique permits the incorporation of uncertainties for each data point, however, these uncertainties must be chosen judiciously so as to reflect the quality and reliability of each of the data points. This technique also generates source profiles as part of its output.

Hybrid receptor modelling

This type of modelling combines statistical receptor models with meteorological models to resolve types and geographical location of sources. EPA is currently collaborating with the University of Michigan to refine the Quantitative Transport Bias Analysis (QTBA) model (Keeler et al., 1989). QTBA utilizes air mass back trajectories and observed meteorological data along with the measured pollutant concentrations to identify the important source areas contributing to elevated Hg concentrations for a particular location or region. Alternatively, the contribution from various source types identified by traditional receptor models can be combined with the trajectories to estimate the impact of each source type on a location or region. An example of QTBA model results are depicted in Figure 6. Source contribution probability fields for Hg(p) concentrations in the Midwest US were calculated from data collected at 3 monitoring sites -- Eagle Harbor, Michigan; Taquamenon Falls, Michigan; and Brule River, Wisconsin (Keeler et al. 2003).

Ancillary measurements

To successfully apply receptor models to Hg monitoring studies requires a paradigm shift in approach to measurement. It is insufficient to embark on any Hg monitoring activity without also measuring a host of other gas and aerosol species.

Event based sampling of Hg, trace elements, and major ions are recommended for wet deposition source apportionment (Dvonch et al., 1999). SO_2, CO, NO, NO_X, O_3, particulate matter mass, trace element aerosols, elemental carbon (EC), and organic carbon (OC) are recommended for Hg^0, RGM, and Hg(p) source apportionment.

Historically most receptor modelling studies involved daily integrated aerosol samples and criteria gas measurements (Stevens et al., 1980; Dzubay et al., 1982).

However, with the advent of a host of new semi-continuous monitoring equipment it is now possible to conduct ambient source apportionment studies using 1-hour measurements. High resolution particulate mass measurements can be made using a tapered element oscillating microbalance (TEOM) (Rupprecht and Patashnick), a nephelometer (Radiance Research, TSI), or a beta guage (Kimoto).

Semi-continuous OC/EC measurements can be made using nondispersive infrared analyzer (Sunset Laboratories), semi-continuous EC measurements can be made using a spectrum aethelometer (Magee Scientific), and semi-continuous $PM_{2.5}$ sulfate, nitrate, nitrite, phosphate, ammonium, sodium, calcium, potassium and magnesium measurements can be made using a ambient ion monitor sampler (AIMS, URG Corp.) that uses a liquid diffusion denuder to remove interfering gases (e.g., SO_2 and HNO_3).

Perhaps one of the most intriguing new monitoring instruments is the semi-continuous elements in aerosol sampler (SEAS) (Ondov Enterprizes) (Kidwell and Ondov, 2001; Kidwell and Ondov, 2004). The SEAS samples air at 90 L min^{-1}, aerosols are grown by condensation of water vapor thereby effecting their separation from the air stream. Samples are collected as a liquid slurry over 30-min sampling durations into a fraction collector. Unlike filter samples, the resulting slurry samples are suitable for analysis by HRICP-MS with minimal sample preparation.

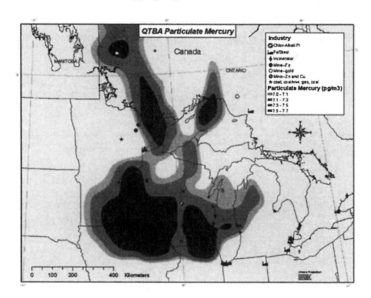

Figure 6. Atmospheric Hg(p) source region probability fields in the Southern Lake Superior region calculated using Quantitative Transport Bias Analysis.

We have been able to quantify up to 40 trace elements in these 30-min integrated samples. Temporally resolved data for some selected trace elements obtained using the SEAS in a study by FLDEP in Tampa, Florida are illustrated in Figure 7. It is clear that multiple sources impacted the sampling site on May 13 that are easily discernable and will not require a statistical receptor model to apportion. This kind of high resolution data when used with meteorological data and limited source emission information (e.g., SO_2) can actually be used to calculate comprehensive source emission profiles (Park, S.S.).

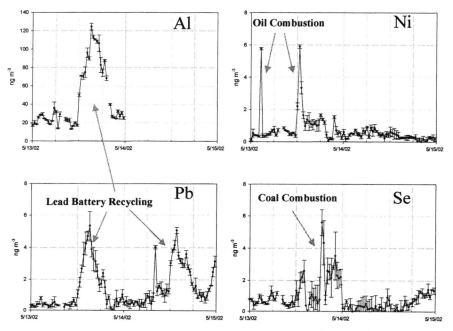

Figure 7. Time resolved data for selected trace elements in Tampa, Fl obtained using the SEAS.

SUMMARY

Ambient Hg measurement studies have played an important role in the development of our current understanding of the atmospheric Hg cycle by (i) characterizing emissions from numerous natural and anthropogenic sources, (ii) identifying important polar and free tropospheric Hg^0 oxidation mechanisms, (iii) correlating observed deposition and ambient concentrations to specific Hg source types, and (iv) providing data for deterministic model evaluation. Recent Hg measurements have highlighted deficiencies in the emission inventories, chemical kinetics modules, and deposition parameterizations used by contemporary deterministic models. Improvements in measurement methods to identify specific compounds contained in RGM and Hg(p), and to directly quantify Hg dry deposition will be required to help guide the development of improved atmospheric Hg chemical kinetics and deposition model parameterizations, respectively.

Statistical and hybrid receptor models currently offer the ability to investigate the impact of specific source types on a receptor locations without regard to current uncertainties is emission inventories, atmospheric Hg chemistry, and deposition phenomena. However, these models do require comprehensive measurement data to develop a data matrix capable of resolving and identifying the contributing causal factors (sources). Receptor modelling results also provide an opportunity to evaluate the accuracy of deterministic model results for specific source categories.

The ability to globally deploy Hg speciation measurement technologies in a thoughtful and coordinated manner is presently lacking. There are no national or international ambient Hg speciation monitoring networks where receptor sites are selected at remote or anthropogenically impacted locations. Every attempt should be made to generate a comprehensive set of measurements through collaborative efforts and leveraging of resources and expertise. These monitoring networks are necessary to assess the success or failure of control strategies, to validate global Hg models, and to observe atmospheric Hg trends.

DISCLAIMER

The United States Environmental Protection Agency through its Office of Research and Development partially funded and collaborated in the research described. It has been subject to Agency Review and approved for publication. Mention of trade names or commercial products does not constitute endorsement or recommendation for use.

REFERENCES

Ariya, P.A., Khalizov, A., Gidas, A. Reactions of Gaseous Mercury with Atomic and Molecular Halogens: Kinetics, Product Studies and Atmospheric Implications. *J. Phys. Chem. A*, 106, 7310-7320, 2002.

Atkeson, T., Axelrad, D., Pollman, C., Keeler, G.J. Integrating atmospheric mercury deposition and aquatic cycling in the Florida Everglades, Florida Department of Environmental Protection, Tallahassee, Florida, 2003.

Banic, C.M., Beauchamp, S.T., Tordon, R.J., Schroeder, W.H., Steffen, A, Anlauf, K.A., Wong, H.K.T. Vertical Distribution of Gaseous Elemental Mercury in Canada. *J. Geophys. Res.,* 108, 4264, 2003.

Berg, T., Sekkesaeter, S., Steinnes, E., Valdal, A.K., Wibetoe, G. Springtime depletion of mercury in the European Arctic as observed at Svalbard. *Sci. Tot. Env.*, 304, 43-51, 2003.

Biester, H., Gosar, M., Muller, G., Mercury Speciation in Tailings of the Idrija Mercury Mine, *J. Geochem. Explor.,* 65, 195-204, 1999.

Bloom, N.S., Prestbo, E.M., Vondergeest, E. Determination of Atmospheric Gaseous Hg(II) at the pg/m3 Level by Collection onto Cation Exchange Membranes, followed by Dual Amalgamation/Cold Vapor Atomic Fluorescence Spectrometry, 4th International Conference on Mercury as a Global Pollutant, Hamburg, August 4-8, 1996.

Brosset, C. Transport of Airborne Mercury Emitted by Coal Burning into Aquatic Systems, *Water Science Technol.*, 5, 59-66, 1983.

Cofer, W.R.III, Collins, V.G., Talbot, R.W. Improved aqueous scrubber for collection of soluble atmospheric trace gases, *Env. Sci. Technol.*, 19, 557-560, 1985.

Crawford, T., Charlotte/Mecklenburg County Air Quality Study, X97426201-0, North Carolina Department of Environment & Natural Resources, Raleigh, North Carolina, 2004.

Dvonch, J.T., Graney, J.R., Keeler, G.J., Stevens, R.K. Uses of Elemental Tracers to Source Apportion Mercury in South Florida Precipitation, *Env. Sci. Technol.*, 33, 4522-4527, 1999.

Dzubay, T.G., Stevens, R.K., Lewis, C.W., Hern, D.H., Courtney, W.J., Tesch, J.W., Mason, M. Visibility and Aerosol Composition in Houston, Texas, *Env. Sci. Technol.*, 16, 514-525, 1982.

Dzubay, T.G., Stevens, R.K. Sampling methods for ambient PM_{10} aerosols. In Receptor Modeling for Air quality Management, Hopke, P. K., Ed., Elsevier Science Publishers, 7, 11-44, 1991.

Ebinghaus, R., Jennings, S.G., Schroeder, W.H., Berg, T., Donaghy, T., Guentzel, J., Kenny, C., Kock, H.H, Kvietkus, K, Landing, W., Muhleck, T., Munthe, J., Prestbo, E.M., Schneeberger, D., Slemr, F., Sommar, J., Urba, A., Wallschlager, D., Xiao, Z. International Field Intercomparison of Atmospheric Mercury Species at Mace Head, Ireland, *Atmos. Environ.*, 33, 3063-3073, 1999.

Ebinghaus, R, Kock, H.H., Temme, C., Einax, J.W., Lowe, A.G., Richter, A., Burrows, J.P., Schroeder, W.H. Antarctic springtime depletion of atmospheric mercury, *Env. Sci. Technol.*, 36, 1238-1244, 2002.

Edgerton, E.S., Jansen, J.J., Hartsell, B.E. Speciated mercury measurements at a rural site near atlanta, GA, USA, *RMZ-Mat. Geoenv.*, 51, 1539, 2004.

Ferm, M. Method for determination of atmospheric ammonia, *Atmos. Environ.,* 13, 1385-1393, 1979.

Fitzgerald, W.F., Engstrom, D.R., Mason, R.P., Nater, E.A. The Case for Atmospheric Mercury Contamination in Remote Areas, *Env. Sci Technol*, 32, 1-7, 1998.

Gordon, G.E. Receptor Models, *Env Sci. Technol.*, 22, 1132-1142, 1985.

Gustin, M.S. Are Mercury Emissions from Geologic Sources Significant? A Status Report, *Sci. Tot.Env.*, 304, 153-167, 2003.

Hall, B. The Gas Phase Oxidation of Elemental Mercury by Ozone, *Water, Air Soil Pollut.* 80, 301-315, 1995.

Helwig, D., Hiskary S. Fish Mercury in Northern Minnesota Lakes, Minnesota Pollution Control Agency, St. Paul, MN., 1985.

Henry H.G., Heady, H.H., Stever K.R., Barry, W.L. Determination of Mercury in Low Grade Ores, *Appl. Spectrosc.*, 26, 2, 288-293, 1972.

Henry, R.C. Multivariate receptor modeling by *N*-dimensional edge detection, *Chemom. Intell. Lab Syst.*, 65, 179-189, 2003.

Hopke, P.K. Receptor modeling in environmental chemistry, 1985, J.W. Wiley & Sons, Hoboken, New Jersey.

Keeler, G.J., Samson, P.J. Spatial Representativeness of Trace Element Ratios, *Env. Sci. Technol.*, 23, 1358-1364, 1989.

Keeler, G.J., Gildemeister, A.E., Rea, A.W., Vette, A.F., Malcolm, E., Marsik, F., Lyon, M., Barres, J, The Lake Superior Basin Trust Final Report, 2003, Submitted to the Michigan DEQ, Lansing, Michigan, 2003.

Kidwell, C.B., and Ondov, J.M. Development and Evaluation of a Prototype System for Collecting Sub-Hourly Ambient Aerosol for Chemical Analysis, *J. Aero. Sci Technol.* 35, 596-601.

Kidwell, C.B., Ondov, J.M. Elemental analysis of sub-hourly ambient aerosol collections, *J. Aero. Sci. Technol.*, 38, 1-14, 2004.

Kinsey, J.S., Swift, J., Bursey, J. Characterization of fugitive mercury emissions from the cell building at a US chlor-alkali plant, *Atmos. Environ.*, 38, 623-631, 2004.

Klaue, B., Lynam, M.M., Keeler, G.J., Blum, J. Thermal Characterization of Mercury Compounds in Atmospheric Particulate Filter Samples by Thermal Analysis Isotope Dilution Cold Vapor Generation Inductively Coupled Plasma Mass Spectrometry (TA-ID-CV-ICP-MS), *RMZ-Mat. Geoenv.*, 51, 1956-1957, 2004.

Klockow, D., Siemens, V., Larjava, K. Application of diffusion separators for measurement of metal emissions, *VDI-Berichte.*, 838, 389-400, 1990.

Landis, M.S., Stevens, R.K. Preliminary results from the USEPA mercury speciation network and aircraft measurements campaigns. Paper presented at the 6th International Conference of Mercury as a Global Pollutant, Minamata, Japan, 2001.

Landis, M.S., Keeler, G.J. Atmospheric mercury deposition to Lake Michigan during the Lake Michigan Mass Balance Study, *Env Sci. Technol.* 36, 4518-4524, 2002.

Landis, M.S., Stevens, R.K., Schaedlich, F., Prestbo, E. Development and characterization of an annular denuder methodology for the measurement of divalent inorganic reactive gaseous mercury, *Env. Sci. Technol.*, 36, 3000-3009, 2002.

Landis, M.S., Vette. A.F., Keeler, G.J. Atmospheric Mercury in the Lake Michigan Basin: Influence of the Chicago/Gary Urban Area, *Env. Sci. Technol*, 36, 4508-4517, 2002.

Landis, M.S., Keeler, G.J., Al-Wali, K.I., Stevens, R.K. Divalent Inorganic reactive Gaseous Mercury Emissions From A Mercury Chlor-Alkali Plant And Its Impact on Near-Field Atmospheric Dry Deposition, *Atmos. Environ.*, 38, 613-622, 2004.

Landis, M.S., Lewis, C.W., Stevens, R.K., Keeler, G.J., Dvonch, J.T., Tremblay, R., Ft. McHenry Tunnel Study: Source Profiles and Mercury Emissions from Diesel and Gasoline Powered Vehicles, submitted to *Atmos. Environ.*, 2004.

Lauretta, D.S., Klaue, B., Blum, J.D., Buseck, P.R. Mercury Abundances and Isotopic Compositions in the Murchison (CM) and Allende (CV) Carbonaceous Chondrites, *Geochim. Cosmochim. Acta*, 65, 2807-2818, 2001.

Liang, L., Horvat, M., Danilchik, P. A novel analytical method for determination of pictogram levels of total mercury in gasoline and other petroleum based products, *Sci. Tot. Env.*, 187, 57-64, 1996.

Lindberg, S. E., Stratton, W.J. Atmospheric Mercury Speciation: Concentrations and Behavior of Reactive Gaseous Mercury in Ambient Air, *Env. Sci. Technol.* 32, 49-57, 1998.

Lindberg, S., Brooks, S., Lin, C-J., Scott, K., Meyers, T., Chambers, L., Landis, M.S., Stevens, R.K. Formation of reactive gaseous mercury in the arctic: evidence of oxidation of Hg^0 to gas-phase Hg-II compounds after arctic sunrise. *Water Air Soil Pollut. Focus*, 1, 295-302, 2001.

Lindberg, S.E., Brooks, S., Lin, C-J., Scott, K., Landis, M.S., Stevens, R.K., Goodsite, M., Richter, A. Dynamic oxidation of gaseous mercury in the arctic troposphere at polar sunrise. *Env. Sci. Technol.* 36, 1245-1256, 2002.

Lu, J., Schroeder, W.H., Berg, T., Munthe, J., Schneeberger, D., Schaedlich, F. A Device for Sampling and Determination of Total Particulate Mercury in Ambient Air, *Anal. Chem.*, 70, 2403-2408, 1998.

Lynam, M.M., Keeler, G.J. Comparison of Methods for Particulate Phase Mercury Analysis: Sampling and Analysis, *Anal. Bioanal. Chem.*, 374, 1009-1014, 2002.

Malcolm, E.G., Keeler, G.J., Landis, M.S. The effects of the coastal environment on the atmospheric mercury cycle, *J. Geophys. Res.*, 108, 4357, 2003.

Mercury Study Report to Congress, EPA-452/R-97-003: U.S. Environmental Protection Agency, Office of Air Quality Planning and Standards, Office of research and Development, U.S. Government printing Office: Washington DC, June 1998.

Nacht, D.M., Gustin, M.S., Engle, M.A., Zehner, R.E., Giglini, A.D. Atmospheric mercury emissions and speciation at the Sulphur Bank mercury mine superfund site, Northern California, *Env. Sci. Technol.*, 38, 1977-1983, 2004.

Nacht, D.M., Gustin, M.S. Mercury Emissions from Background and Altered Geologic Units throughout Nevada, *Water Air Soil Pollut.*, 151, 179-193, 2004.

Nriagu, J, Becker, C., Volcanic Emissions of Mercury to the Atmosphere: Global and Regional Inventories, *Sci. Tot. Env.*, 304, 3-12, 2003.

Ontario Ministry of the Environment, Guide to Eating Ontario Sport Fish, Communications Branch, Toronto, Canada. 1998

Paatero, P. Least squares formulation of robust non-negative factor analysis, *Chemom. Intell. Lab Syst.*, 37, 23-35, 1997.

Pacyna, E.G., Pacyna, J.M. Global Emission of Mercury from Anthropogenic Sources in 1995, *Water Air Soil Pollut.*, 137, 149-165, 2002.

Park, S.S., Pancras P.J., Ondov, J.M., Poor, N. A new pseudo-deterministic multivariate receptor model for accurate individual source apportionment using highly time-resolved ambient concentrations measurements, *J. Geophys. Res.* (In Press).

Petersen, G., Iverfeldt, Å., Munthe, J. Atmospheric mercury species over central and Northern Europe. Model calculations and Nordic air and precipitation network for 1987 and 1988, *Atmos. Environ.*, 29, 47-67, 1995.

Pirrone, N,, Keeler, G.J., Nriagu, J.O. Regional Differences in Worldwide Emissions of Mercury to the Atmosphere, *Atmos. Environ.*, 30, 2981-2987, 1996.

Possanzini, M., Febo, A., Liberti, A. New design of a high-performance denuder for the sampling of atmospheric pollutants, *Atmos. Environ.*, 17, 2605-2610, 1983.

Prestbo, E., Bloom, N. Mercury speciation adsorption (MESA) method for combustion flue gas: methodology, artifacts, intercomparison, and atmospheric implications, *Water Air Soil Pollut.*, 80, 145-158, 1995.

Pyle, D.M., Mather, T.A., The Importance of Volcanic Emissions for the Global Atmospheric Mercury Cycle, *Atmos. Env.*, 37, 5115-5124, 2003.

Raofie, F., Ariya, P.A., Product Study of the Gas-Phase BrO-Initiated Oxidation of Hg^0: Evidence for Stable Hg^{1+} Compounds, *Env. Sci. Technol.* 38, 4319-4326, 2004.

Schrock, J., Bowser, J., Mayhew, W., Stevens, R. South Florida Mercury Monitoring and Modeling Pilot Study Technical Report EPA 600/R-00/102, US EPA National Exposure Research Laboratory, Research Triangle Park, NC, 2000.

Schroeder, W.H., Hamilton, M.C., Stobart, S.R. The Use of Noble Metals as Collection Media for Mercury and its Compounds in the Atmosphere A Review, *Rev. Anal. Chem.*, 179-209, 1985.

Schroeder, W.H., Anlauf, K.G, Barrie, L.A., Lu, J.Y., Steffen, A., Schneeberger, D.R., Berg, T. Arctic Springtime Depletion of Mercury, Nature, 394, 331-332, 1998.

Schroeder, W.H, Munthe, J. Atmospheric Mercury – An Overview, *Atmos. Environ.*, 32, 809-822, 1998.

Sigler, J.M., Lee, X., Munger, W. Emission and Long-range Transport of Gaseous Mercury from a Large-Scale Boreal Forest Fire, *Env. Sci. Technol.* 37, 4343-4347, 2003.

Sommar, J., Gardfeldt, K., Stromberg, D., Feng, X. A Kinetic Study of the Gas-Phase Reaction between the Hydroxyl Radical and Atomic Mercury, *Atmos. Environ.*, 35, 3049-3054, 2001.

Spicer, C.W., Satola, J., Abbgy, A.A., Plastridge, R.A., Cowen, K.A. Kinetics of Gas-Phase Elemental Mercury Reactions with Halogen Species, Ozone, and Nitrate Radical Under Atmopsheric Conditions, 2002, Battelle, Columbus, Final Report to Florida Department of Environmental Protection.

Sprovieri, F., Pirrone, N., Hedgecock, I.M., Landis, M.S., Stevens, R.K. Intensive atmospheric mercury measurements at Terra Nova Bay in Antarctica during November and December 2000. *J. Geophys. Res.,* 107(D23), 4722, 2002.

Stevens, R.K., Dzubay, T.G., Russwurm, G., Rickel, D., Sampling and analysis of atmospheric sulfates and related species, *Atmos. Environ.*, 12, 55-68, 1978.

Stevens, R.K., Dzubay, T.G., Shaw, R.W., McClenny, W.A., Lewis, C.W., Wilson, W.E., Characterization of the Aerosols in the Great Smoky Mountains. *Env. Sci. Technol.* 14, 1491-1498, 1980.

Stratton, W.E., Lindberg, S.E., Perry, C.J. Atmospheric Mercury Speciation: Laboratory and Filed Evaluation of a Mist Chamber Method for Measuring Reactive Gaseous Mercury, *Env. Sci. Technol.*, 35, 170-177, 2001.

Sumner, A.L., Spicer, C.W., Satola, J., Abbgy, A.A.; Plastridge, R.A.; Mangaraj, R.; Cowen, K.A.; Landis, M.S.; Stevens, R.K.; and Atkeson, T.D. Environmental Chamber Studies of Mercury Reactions in the Atmosphere, Chapter 9, this volume.

Temme, C., Einax, J.W., Ebinghaus, R, Schroeder, W.H. Measurements of atmospheric mercury species at a coastal site in the Antarctic and over the South Atlantic during polar summer, *Env. Sci. Technol.*, 37, 22-31, 2003.

U.S. Environmental Protection Agency, Compendium of Methods for the Determination of Inorganic Compounds in Air, Chapter IO-5, EPA-625/R-96/010a: Office of Research and Development, Cincinnati, OH.

Wängberg, I., Edner, H., Ferrara, R., Lanzillotta, E., Munthe, J., Sommar, J., Sjöholm, M., Svanberg, S., Weibring, P. Atmospheric mercury near a chlor-alkali plant in Sweden, *Sci. Tot. Env.*, 304, 29-41, 2003.

Weiss-Penzias, P., Jaffe, D.A., McClintick, A., Prestbo, E., Landis, M.S. Gaseous Elemental Mercury in the Marine Boundary Layer: Evidence for Rapid Removal in Anthropogenic Pollution, *Env. Sci. Technol.* 37, 3755-3763, 2003.

Wilhelm, S.M., Bloom, N, Mercury in petroleum, *Fuel Processing Technology*, 63, 1-27, 2000.

Windmoller, C.C., Wilker R.D., Wilson de Figueiredo J. Mercury Speciation in Contaminated Soils by Thermal Release Analysis, *Water, Air Soil Pollut.* 89, 399-416, 1996.

Xiao, Z.F., Sommar , J., Wei , S., Lindqvist, O., Sampling and determination of gas phase divalent mercury in the air using KCl-coated denuder, Fresenius *J. Anal. Chem.,* 1997. 358, 386-391.

Chapter-8

DETERMINATION OF MERCURY AND ITS COMPOUNDS IN WATER, SEDIMENT, SOIL AND BIOLOGICAL SAMPLES

Milena Horvat

Department of Environmental Sciences, Institute Jožef Stefan, Jamova 39, 1000 Ljubljana, Slovenia

INTRODUCTION

Mercury occurs naturally in the Earth's crust principally as the ore, cinnabar, HgS. Mercury is quite different from other metals in several respects: (i) it is the only metal that is liquid at room temperature; (ii) it is the only metal that boils below 650°C; (iii) it is quite inert chemically, having a higher ionization potential than any other electropositive element with the sole exception of hydrogen; (iv) it exists in oxidation states of zero (Hg^0) and 1 (Hg^{2+}) in addition to the expected state of 2 (Hg^{2+}). Mercury forms alloys ("amalgams") with many metals. Mercury and its chemical derivatives are extremely hazardous. Since the early 1960s, the growing awareness of environmental mercury pollution (e.g. the Minamata tragedy resulting from methyl-mercury poisoning) has stimulated the development of more accurate, precise and efficient methods of determining mercury and its compounds in a wide variety of matrices.

Many of the environmental aspects of mercury and its compounds have been reviewed (UNEP 2002; Drasch et al., 2004; Pirrone et al., 2002). In recent years, new analytical techniques have become available and have been used in environmental studies and consequently the understanding of mercury chemistry in natural systems has improved significantly. Mercury can exist in a large number of different physical and chemical forms with a wide range of properties. Conversion between these different forms provides the basis for mercury's complex distribution pattern in local and global cycles and for its biological enrichment and effects. The most important chemical forms are: elemental mercury (Hg^0), divalent inorganic mercury (Hg^{2+}), methylmercury (CH_3Hg^+), and dimethylmercury ($(CH_3)_2Hg$).

There is a general biogeochemical cycle by which monomethyl and mercury (II) compounds, dimethylmercury and mercury (0) may interchange in the atmospheric, aquatic, and terrestrial environments. Mercury vapour is released into the atmosphere from a number of natural sources and through anthropogenic emissions (mainly from combustion of fossil fuels). A small portion of Hg^0 is converted into water soluble species (probably Hg^{2+}) which can, in part, be re-emitted to the atmosphere as Hg^0 by deposition on land or exchange at the air/water boundary. The atmospheric cycle entails retention of Hg° in the atmosphere for long periods and consequently it is transported over very long distances. The bottom sediment of oceans is thought to be ultimate sink where mercury is deposited in the form of highly insoluble HgS. Changes in speciation from inorganic to methylated forms is the first step in aquatic bioaccumulation processes. These processes are considered to occur in both the water column and sediments. The mechanism of synthesis of methylmercury is not very well understood. Although methylmercury is the dominant form of mercury in higher organisms, it represents only a very small amount of the total mercury in aquatic ecosystems and in the atmosphere. Methylation-demethylation reactions are assumed to be widespread in the environment and each ecosystem attains its own steady state equilibrium with respect to the individual species of mercury. However, owing to the bioaccumulation of methylmercury, methylation is more prevalent in the aquatic environment than demethylation.

Once methylmercury is formed, it enters the food chain by rapid diffusion and tight binding to proteins in aquatic biota and attains its highest concentrations in the tissues of fish at the top of the aquatic food chain due to biomagnification through the trophic levels. The main factors that affect the levels of methylmercury in fish are the diet/trophic level of the species, age of the fish, microbial activity and mercury concentration in the upper layer of the local sediment, dissolved organic carbon content, salinity, pH, and redox potential.

Speciation is therefore a term frequently used and in the case of mercury, (i) it rules the way Hg is transported from its sources to the local environment of man and wildlife, (ii) it rules how mercury is bound in the environment and it therefore more or less available to cause adverse effects, and (iii) it rules one of the most important mercury transformation and build-up of MeHg in fish and other aquatic and terrestrial foods.

The term "speciation" and "fractionation" in analytical chemistry were addressed by the International Union for Pure and Applied Chemistry (IUPAC) which published guidelines (Templeton et al., 2000) or recommendations for the definition of speciation analysis:

- **Speciation analysis** *is the analytical activity of identifying and/or measuring the quantities of one or more individual chemical species in a sample.*
- *The* **chemical species** *are specific forms of an element defined as to isotopic composition, electronic or oxidation state, and/or complex or molecular structure.*
- *The* **speciation** *of an element is the distribution of an element amongst defined chemical species in a system. In case that it is not possible to determine the concentration of the different individual chemical species that sum up the total concentration of an element in a given matrix, that means it is impossible to determine the speciation, it is a useful practice to do* **fractionation** *instead.*
- **Fractionation** *is the process of classification of an analyte or a group of analytes from a certain sample according to physical (e.g. size, solubility) or chemical (e.g. bonding, reactivity) properties.*

During recent years new analytical techniques have become available that have contributed significantly to the understanding of mercury chemistry in natural systems. In particular, these include ultra sensitive and specific analytical equipment and contamination-free methodologies. These improvements eventually allow for the determination of total and major species of mercury to be made in air, water, sediments, and biota.

Analytical methods are selected depending on the nature of the sample and, in particular, the concentration levels of mercury. The present review is an updated and extended version of a review published in 1996 (Horvat, 1996) in order to meet the opbjectives of the present book.

PHYSICAL AND CHEMICAL PROPERTIES OF MERCURY SPECIES

Metallic Mercury

Elemental Mercury (Hg^o) is usually referred to as mercury vapour when present in the atmosphere, or as metallic mercury in liquid form. Hg^o is of considerable toxicological as well as of environmental importance because it has a relatively high vapour pressure (14 mg m^{-1} at 20 °C, 31 mg m^{-3} at 30 °C) and appreciable water solubility ($\sim$ 60 μg.L^{-1} at room temperature). Due to its high lipophilicity, elemental mercury dissolves readily in fatty compartments. Of equal significance is the fact that the vapour exists in a monatomic state.

Inorganic ions of mercury

Many salts of divalent mercury (Hg^{2+}) are readily soluble in water, such as mercury sublimate ($HgCl_2$: 62 g L^{-1} at 20°C), and, thereby, highly toxic. In contrast, the water solubility of HgS (cinnabar) is extremely low ($\sim$ 10 ng L^{-1}), and, correspondingly, HgS is much less toxic than $HgCl_2$ (Simon and Wuhl-Couturier, 2002). The extremely high affinity of Hg^{2+} for sulfhydryl groups of amino acids such as cysteine and methionine in enzymes explains its high toxicity. However, its affinity to SeH-groups is even greater, which may explain the protective role of selenium from mercury intoxication (Yaneda and Suzuki, 1997).

Monovalent mercury is found only in dimeric salts such as Hg_2Cl_2 (calomel), which is sparingly soluble in water and, again correspondingly much less toxic than $HgCl_2$ (sublimate).

Organic mercury compounds

Organic mercury compounds consist of diverse chemical structures in which divalent mercury forms one covalent bond (R-Hg-X) or two (R-Hg-R) with carbon. For all practical purposes, organic mercury compounds are limited to the alkylmercurials monomethyl-Hg, monoethyl-Hg and dimethyl-Hg, to the alkoxymercury compounds, and to the arylmercurials (phenylmercury). Organic mercury cations (R-Hg^{1+}) form salts with inorganic and organic acids (e. g., chlorides and acetates), and react readily with biologically important ligands, notably sulfhydryl groups. Organic mercurials also pass easily across biological membranes, perhaps since the halides (e. g., CH_3HgCl) and dialkylmercury are lipid-soluble. The major difference among these various organomercury compounds is that the stability of carbon-mercury bonds in vivo varies considerably. Thus, alkylmercury compounds are much more resistant to biodegradation than either arylmercury or alkoxymercury compounds.

Monomethylmercury compounds are of greatest concern today as these highly toxic compounds are formed by micro-organisms in sediments and bio-accumulated and bio-magnified in aquatic food chains, thus resulting in exposures of fish eating populations, often at levels exceeding what is regarded as a safe. The term "methylmercury" is used throughout this text to represent monomethylmercury compounds. In many cases the complete identity of these compounds is not known except for the monomethylmercury cation, CH_3Hg^+, which is associated either with a simple anion, like chloride, or a large charged molecule (e.g. a protein).

A specific source of exposure is the use of thiomersal for preservation of vaccines and immunoglobulins (usually 25-50 µg of mercury per injection) (Knežević et al., 2004; Pichichero et al., 2002). Its metabolite, ethylmercury, behaves toxicologically much like methylmercury, but is less stable. Because of the possible health significance to infants, the use of thiomersal is being phased out.

ANALYTICAL METHODS

In general, determination of mercury involves the following steps:

- (a) sample collection;
- (b) sample pretreatment/preservation/storage;
- (c) liberation of mercury from its matrix;
- (d) extraction/clean-up/pre-concentration;
- (e) separation of mercury species of interest;
- (f) quantification.

Over the last twenty years, thousands of papers dealing with determination of mercury and organomercury compounds in environmental samples have been published. This chapter is organized in such a manner that the principal information on mercury distribution in the environment is given with emphasis on mercury speciation, with the identification of major gaps. However, cleaning of laboratory ware and calibration are common to all matrices and are described below.

Cleaning procedures

Rigorous cleaning procedures must be used for all laboratory ware and other equipment which comes into contact with samples. Reagents that are used for the analyses of total and organomercury species must be of suitable quality. The best materials for sample storage and sample processing are Pyrex and silica (quartz) glass, and Teflon (PTFE or FEP). Plastics such as polypropylene are not recommended since these materials can contribute to either contamination or losses of mercury. There are several cleaning procedures which are suitable: (1) aqua regia treatment followed by soaking in dilute (~5-10%) nitric acid for a week; (2) soaking in a hot oxidizing mixture of $KmnO_4$ and $K_2S_2O_8$, followed by NH_4OCl rinsing and soaking for a week in 5M HNO_3; (3) soaking in a 1:1 mixture of concentrated chromic and nitric acids for a few days; (4) soaking in BrCl (mixture of HCl and $KbrO_3$); (5) Teflon is usually cleaned in hot concentrated HNO_3 for 48 hours, followed by soaking in dilute HNO_3 (5%) (which is repeated twice). After such treatments, laboratory ware is usually rinsed with mercury-free deionized water or double distilled water, and stored in a mercury-free place, preferably sealed in mercury-free plastic bags. Some authors recommend storage of laboratory ware in dilute nitric or hydrochloric acids until use. Laboratory ware that is used for methylmercury analyses should be prepared with extreme caution. It has been shown that final soaking of laboratory ware, particularly Teflon, in hot (70°C) 1% HCl removes any traces of oxidizing compounds (e.g. chlorine) which may subsequently destroy methylmercury in solution.

Calibration

Inorganic Mercury

The difficulty of preventing losses of low level mercury from aqueous solutions is well known. Losses during storage are due to adsorption on container walls and volatilization losses due to reduction of inorganic mercury to elemental mercury. Numerous papers have been written describing various treatments of samples to prevent losses of inorganic mercury during storage (Carr and Wilkniss, 1973, Lo and Wai, 1975, Coyn and Callins, 1972) Normally, strong acids and oxidants (such as HNO_3, $H_2SO_4 + KmnO_4$, $K_2Cr_2O_7$, HCl and H_2O_2 and BrCl) are added as preservatives. Losses are very much dependent on the container materials used, the best materials being quartz, Pyrex glass and Teflon. Any working standard solution with a concentration below 10 ng/ml should be carefully prepared freshly on a daily basis.

In a number of studies using the double amalgamation method the calibration of the analytical instrument is performed using the mercury saturated air calibration method (Dumarey et al., 1985). The elemental mercury is placed in a closed glass container, and the temperature is kept at a constant value, below room temperature. The temperature dependence of the saturated mercury vapour concentration is calculated from the ideal gas law. An appropriate volume of Hg vapour is taken by gas tight syringe and injected into the gas train through the chromatographic septum on the sampling gold trap. This approach has the advantage of preventing problems associated with the stability of aqueous mercury solutions and is ideal for calibration of methods for determination of total mercury in air. However, when total mercury is measured by the reduction-aeration method, calibration of the analytical instrument with aqueous standards is preferable because it acts as a control for the reduction amalgamation conditions and represents less danger for possible contamination in the case of damage of the calibration vessel.

Organomercury standards

There have been quite a few studies performed concerning the stability of organomercurials in standard solutions (Lansens et al., 1990; Meulemann et al., 1993). A decrease of methylmercury in aqueous solutions can be caused by adsorption onto the container walls. Losses of methylmercury chloride due to volatilization is unlikely to occur [K_d (gas-liquid distribution coefficient: C_{gas}/C_{H2O}) is 1.07×10^{-5} at 20°C]. The stability is strongly dependent on the concentration, the container materials and the storage temperature. An aqueous methylmercury solution with a concentration of 10 $\mu g.L^{-1}$, stored in Pyrex at low temperature (e.g. in a refrigerator) is stable for approximately one month. If Teflon containers are used, the solution is stable for several months if stored in the dark at room temperature. Some authors recommend storage of methylmercury solutions in 1% HCl.

Methylmercury standard solutions in organic solvents stored in Pyrex glass bottles (for GC analyses) are even more stable than aqueous solutions. Recently, a new CRM for calibration of Me^{202}Hg was produced by the European Institute for Reference Materials and Measurements (Snell et al., 2004) for calibration and quality control using ICP MS detection and isotope dilution analysis.

Volatile organomercury compounds (in particular dimethlymercury) can also be prepared in the gas phase. This is of importance for the optimization of methodologies for organomercury speciation in air and dimethylmercury in water. An aliquot of vapour is removed from a temperature-stabilized vessel using a gas tight syringe. The concentration can be calculated from data on the partial pressure of the individual compound and the gas-law equation.

Water

Concentrations of total mercury in water samples are very low (at the ng L^{-1} level or below), so that accurate analysis is still a major problem. The theoretical approach via stability calculations can be of great help in making rough estimates of the predominant mercury species under various conditions. Mercury compounds occurring in natural waters (Figure 1) are most often defined by their ability to be reduced to elemental mercury. In lake waters methylmercury species account for 1-30 % of total mercury. Most of the methylmercury is probably associated with dissolved organic carbon (DOC). Limited data are available on the formation constants between the methylmercury cation and DOC (Martell et al., 1998). Thiol groups (-RSH) have been shown, however, to have a higher capability to bind methylmercury in comparison with ligands containing oxygen and nitrogen donor atoms and the inorganic ions (CN$^-$, Cl$^-$, OH$^-$). Methyl mercury compounds in surface runoff waters, soil pore waters and ground waters are similar to the species in lake waters and are generally quite strongly associated with DOC. Dimethylmercury has rarely been reported in surface waters except in the deep ocean (Horvat et al., 2003; Cossa et al., 1994; Vandal et al., 1998) and during some seasons in the slurry of salt marshes (Weber et al., 1998). Mercury in sea water exists mainly in the form of Hg^{2+} complexed with Cl$^-$ ions. Methylmercury concentrations in seawater are generally lower than in lake waters. Dissolved gaseous mercury vapour is also present in ocean waters (Horvat et al., 2003). The presence of organomercury species, including dimethylmercury, was also detected in geothermal gases and waters (Hirner et al., 1998).

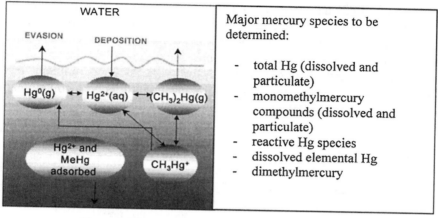

Figure 1. Mercury species and transformation in water.

Collection and handling of aqueous samples for low level determination of mercury must address factors such as whether or not the sample is representative of the sample, possible inter-conversion processes, contamination, as well as preservation and storage of the matrix before analysis. The measurement (sampling and analysis) protocol must be even more carefully designed if speciation of mercury forms in the aqueous samples is intended. There have been remarkable improvements in sampling and analytical techniques which have resulted in a dramatic increase in the reliability of data for mercury levels in water samples over the past 15 years. The stability of mercury in solution is affected by many factors. These include: (a) the concentration of mercury and its compounds, (b) the type of water sample, (c) the type of containers used, (d) the cleaning and pre-treatment of the containers, and (e) the preservative added.

Sampling and storage

The main steps for water sample pretreatment, preservation, and storage for determination of various mercury compounds are shown in Figure 2. Contamination-free sampling devices (e.g. Teflon Go-Flo samplers) are commonly used. Alternatively, the water can be pumped through Teflon tubing using a peristaltic pump. Collection of surface waters is usually performed by hand, using arm-lenght plastic gloves. Samples are taken upwind of a rubber raft or a fiberglass boat. Precipitation samples can be collected by automatic samplers, with in-line filtration if desired. Wide-mouth Teflon jars have been favoured for sampling waters with low mercury concentrations. Containers and other sampling equipment which come into contact with water samples should be made of borosilicate glass, Teflon or silica glass. These materials have been found to be free from mercury contamination and therefore suitable for work at low, ambient levels. However, Teflon showed the best performance regarding both contamination and loss-free storage of aqueous samples.

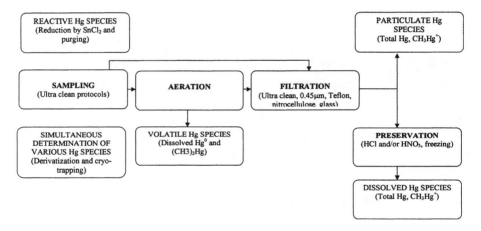

Figure 2. Most common steps for determination of total mercury and its compounds in natural water samples.

The most volatile mercury forms possibly present in the water are Hg° and dimethylmercury. They should be removed from the samples immediately after the sampling step by aeration with collection on Carbotrap or other suitable adsorption media for subsequent analyses in the laboratory (Mason, Fitzgerald 1993; Horvat et al., 2003).

Measurement of "dissolved" mercury compounds requires removal of "particulate" materials from the sample by either filtration or centrifugation. Various types of filters can be employed: 0.45 μm membrane filters (precipitation, sea water), disposable polystyrene units (Nalgen) with nitrocellulose membrane filters, and pretreated glass fibre filters.

In order to store samples prior to analyses, samples should be preserved, generally by acidification, even though it has been shown that the Hg concentration in unpreserved precipitation samples is more stable than previously assumed (Bloom 1989). However, acidification is subject to two limitations: (a) suspended matter has to be removed prior to acidification and (b) dimethylmercury and Hg° have to be removed, otherwise conversion of these species into methylmercury and mercury(II) may occur. In general, samples are processed immediately after removal of the particulate matter. However, if water samples do have to be stored for longer periods the mercury present may be stabilized by adding acid. For the analysis of organomercurials, preservation with oxidative reagents (as advised for total mercury analysis) should be avoided, since organomercurials are converted into inorganic mercury. Stabilization by HNO_3 results in a decrease in methylmercury, while Hg(II) remains stable in the presence of this acid.

HCl was found the most appropriate acid for storing aqueous methylmercury solutions (Ahmed et al., 1987). Sulphuric acid can also be used for preservation of methylmercury solutions in distilled water although it is not suitable for natural water samples. Some authors claim that for methylmercury determinations, storage of unpreserved samples at low temperatures (or even deep-frozen) is better than adding acid (Bloom 1989; Horvat et al., 2003).

Determination of total mercury

As generally agreed, problems arising in the analysis of total mercury in natural water samples are not connected with the final measurement, but rather with difficulties associated with contamination-free sampling and potential losses due to volatilization and adsorption during storage. The best hope to overcome these problems would seem to be in immediate measurement at the sampling site.

Analytical techniques suitable for total Hg determination in natural waters at the picogram level are either based on cold vapour atomic absorption (CV AAS), ICP-MS, plasma atomic emission (plasma-AES) or atomic fluorescence spectrometry (AFS). Of these, CV AAS is the most widespread method. In recent years, CV AFS and ICP-MS/ICP-AES techniques have become increasingly important, since the detection limits of the instruments is about 1 picogram, which is an order of magnitude better than CV AAS. Since the development of a simple, very sensitive (0.3 pg) and inexpensive CV AFS detector (Kvietkus et al., 1983; Bloom and Fitzgerald, 1988) many research groups are using it for mercury measurement of low level air and natural water samples.

In nearly all analytical procedures used for measurement of Hg in natural waters the method of preconcentration on gold trap/amalgamatio is applied.

A schematic flow chart for determination of total mercury in natural water samples is shown in Figure 3.

The sampling, filtration and preservation steps have been discussed in previous Chapter. Decomposition of all mercury species into Hg^{2+} is necessary if total mercury is to be measured. In a large number of studies this was achieved by the oxidizing agent BrCl in HCl (Bloom 1989; Bloom and Crecelius 1983). Also, techniques based on UV oxidation in HCl media have successfully been applied (Ahmed et al., 1987; May et al., 1987). The use of other oxidative mixtures are limited due to relatively high reagent blanks. In the case of humic rich water samples, a combination of BrCl and UV oxidation is very effective and results in complete recovery.

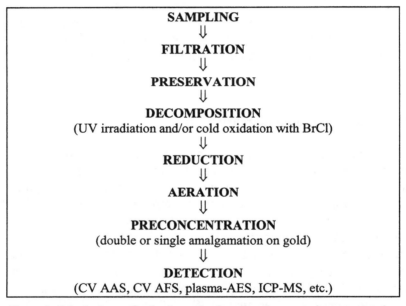

Figure 3. Principal steps for the determination of total mercury in water samples.

After the decomposition step, an aliquot of the sample is transferred into a reduction cell and treated with a reducing agent and purged with a Hg-free gas flow. As reductant, $SnCl_2$ solution is used, but an aqueous solution of $NaBH_4$ may also be employed (Heraldsson et al., 1989, Iverfeld et al., 1988).

Speciation of mercury in water

A large number of articles dealing with the determination of methylmercury compounds in biological and sediment samples have been published. Only a few analytical techniques, however, have been developed for the reliable determination of organomercury species in water samples. In many studies, mercury compounds in aquatic environments are speciated according to their ability to be reduced to the elemental state. Dimethylmercury and Hg° (the most volatile mercury compounds) can be isolated by aeration and adsorption on a suitable adsorbent (Carbotrap or Tenax), coupled with noble metal amalgamation (for Hg°), immediately after sampling (Horvat et al., 2003). Alternatively, they can be directly analyzed by cryogenic trapping, separation on GC columns and detected by one or more suitable mercury detectors. ICP-MS is frequently used in recent years (Hintelmann and Ogrinc 2003; Stoichev et al., 2004; Monperrus et al., 2004). Samples should not be acidified prior to such separations since dimethylmercury and Hg° are transformed into methylmercury and Hg(II), respectively.

For specific organomercury compound determinations, a pre-separation and pre-concentration method, followed by a very sensitive detection system is necessary. In a typical extraction method, methylmercury halide (Br⁻, Cl⁻ or I⁻) is extracted into an organic solvent (benzene or toluene) after acidification. This is followed by derivatization to a water-soluble adduct of methylmercury-cysteine, which is extracted into the aqueous phase. After acidification, CH_3HgX (X is a halide ion) is back-extracted into a small amount of organic solvent. An aliquot is then injected onto a GLC column and detected by ECD or any other suitably sensitive detector (such as a plasma emission detector). Packed or capillary columns can be used. More precise descriptions of chromatographic conditions are given in the section 3.3 of this manuscript describing methylmercury determination in other environmental samples. There are quite a few modifications to this extraction procedure and these have been reviewed by Craig (Craig 1986). For example, the methylmercury compound may be transferred into organic solvent as dithizonates followed by clean-up steps and detection by GC-ECD (Akagi et al., 1991). Inorganic and organic mercury species can be pre-concentrated on dithiocarbamate or sulphydril cotton-fibre adsorbent which is then extracted as described above (Lee and Mowrer 1989; Jones et al., 1995). However, in some water samples artifact formation of methylmercury was observed during solid-phase extraction of water samples (Celo et al., 2004). The common drawbacks of most of these extraction procedures are the large sample requirements, low extraction yields, and non-specific separation of dimethylmercury, if present.

There are also methods for determination of "total" organomercury compounds. Inorganic and organic mercury are pre-concentrated on a dithiocarbamate resin and are subsequently eluted with thiourea. Separation of organic and inorganic mercury is achieved by differential reduction and detection by CV AAS (Minagawa et al., 1979). Inorganic and organic mercury can also be separated using anion exchange resins]. Organic mercury is then decomposed (by UV irradiation) and measured by CV AAS. However, it has been shown that the levels obtained by this method do not necessarily correspond to methylmercury (owing to the lack of specificity of the protocol). The method has recently been improved by the introduction of more specific separations of organic and inorganic mercury species by water vapour distillation (Padberg and Stoeppler 1991).

Another method based on aqueous-phase ethylation, room temperature pre-collection, separation by GC with CV AFS detection has been described (Bloom 1989). The method has been frequently adopted in laboratories involved in studies of the biogeochemical cycle of mercury. Ionic mercury species in the sample are ethylated according to the following reactions:

$$[CH_3Hg^+] + [Hg^{2+}] + 3NaB(C_2H_5)_4 \rightarrow [CH_3HgC_2H_5] + [Hg(C_2H_5)_2] + 3Na^+ + 3B(C_2H_5)_3$$

Ethylated mercury species are volatile and therefore can be purged from solution at room-temperature and then collected on adsorbent materials such as Carbotrap or Tenax. After thermal release, individual mercury compounds are separated by cryogenic or isothermal GC. As the species are eluted they are thermally decomposed (pyrolized) at 900°C and are measured as Hg° using a CV AFS detector, which achieves very low detection limits (< 10 pg). A CV AAS detector can also be used, but its detection limit is much higher, 167 pg (Rapsomanikis and Craig, 1991).

In recent years ICP-MS is being more and more frequently used as it offers numerous advantages over AAS and AFS detectors. Instead of ethylation, propylation was recently shown to be an even more suitable derivatization procedure, being free from interferences caused by halide ions (Demuth and Heumann, 2001). Hydration was also proven to be a useful derivatization method, in particular when coupled with preconcentration by cryotrapping (Tseng et al., 1998; Stoichev et al., 2004, Monperrus et al., 2004).

In any case, the critical part of this procedure is the preparation of samples prior to derivatization. Methylmercury compounds must be removed from bound sites to facilitate the ethylation reaction. Interfering compounds (such as sulphides) must also be removed. Two approaches have been used so far. The first is based on extraction of methylmercury compounds into methylene chloride and then back-extraction into water by solvent evaporation (Bloom 1989). The second is based on water vapour distillation. Distillation has advantages since it quantitatively releases methylmercury from sulphur and organic-rich water samples (Horvat et al., 1993).

The analysis of different mercury compounds at environmental concentrations should be developed further. Reliable data can be obtained for monomethyl mercury compounds and dimethylmercury. There is, however, a need to identify biogeochemically important mercury fractions which are currently measured by operationally-defined rather than direct ("analytically rigorous") protocols.

Other environmental matrices

Sampling and storage

Sediments and soils: Mercury in these two compartments of the environment is associated with humic matter. In contaminated sites (particularly mining areas) it is mostly present as HgS. Following methylation, methylmercury does not usually build up in sediments to more than about 1.5% of the total mercury present. This appears to be an approximate equilibrium level between formation and removal. Methylation-demethylation reactions are assumed to be widespread in the environment and each ecosystem attains its own steady state with respect to the individual species of mercury. Dimethylmercury is considered to be unstable in sediments, but is assumed to be stabilized by a conjunction of factors, such as high sulphide levels, salinity, anoxic conditions and constant inputs of methane into the media (Weber et al., 1998).

Sediment and soil samples should also be prepared with caution since the percentage of methylmercury in these samples is very low (e.g. <2% of total mercury) and improper handling and storage may lead to inaccurate results. Moreover, changes of pH, redox potential, moisture, etc. may significantly influence the stability of methylmercury in sediments. Due to changes during sampling, conversion of mercury species may occur (methylation, demethylation, reduction), particularly in the case of sediments taken from the oxygen-depleted bottom of water bodies (Horvat et al., 2004). These samples are better analyzed fresh or, if long-term storage is unavoidable, samples should be kept in the dark at low temperature, in an inert atmosphere and deep frozen.

Biological samples: The highest levels found in fresh and marine organisms are found at the highest trophic levels, where mercury levels can exceed the "black list limit" of 0.5 mg kg-1 (IPCS/WHO, 1990). The percentage of methylmercury to total mercury in fish muscle varies from 80 to 100 %, but in other organs its concentration is smaller (in liver and kidney up to 20 %). In other aquatic organisms, the percentage of methylmercury is more variable, depending on water depth, location, and the type of organism. Relatively high mercury and methylmercury concentrations have been reported for fish-eating marine birds. Birds feeding on wild vegetation generally have much lower mercury in their bodies. There have recently been many studies performed on terrestrial ecosystems (particularly in Canada, Sweden and the USA). Mercury also passes from vegetation into the food chains of fauna (Gnamuš et al., 2000).

Relatively little is known about the effects of storage on the stability of methylmercury in biological samples. Significant external contamination of samples with methylmercury is unlikely to occur; however extreme precautions are necessary to avoid contamination by inorganic mercury.

Blood and hair samples are often analyzed in order to estimate exposure of humans to mercury and its compounds. Blood is a suspension of cellular components in a protein-containing medium in a ratio of approximately 43:57. It should be taken by venipuncture. It is mandatory to collect blood without addition of any preservative, or of any coagulant, and this for two reasons. First of all because the preservative may contain a mercury compound or be contaminated with mercury impurities. Secondly because any preservative and/or anticoagulant is liable to break up the original mercury species. Most anticoagulants are either polyanions (e.g. heparin) or metal chelators (e.g. EDTA, citrate) and therefore have a high affinity for metal species. As a rule neither heparinised samples, nor EDTA , citrate or any other anticoagulant doped samples should be used. Add to this that when blood is collected with an anticoagulant it separates into plasma and red blood cells, which is different from serum and packed cells. Serum contains no fibrinogen. Packed cells consist not only of red blood cells but include all other cellular material as well.

As speciation of the mercury species is the aim of the analysis, it is of little mean to analyze total blood, because of the very different nature of its constituents (serum and packed cells). The same applies to many real samples, which are complex mixtures.

After blood has been sampled without an anti-coagulant, it will clot spontaneously and separate into serum and packed cells. The process lasts 15 to 30 minutes at room temperature. It is then submitted to centrifugation, what should be completed within 1 hour. This separation should be done as soon as possible to avoid haemolysis of blood. Haemolysed samples cannot be considered for analysis, the more so not for speciation analysis. The distribution of the different trace element species between serum and cells may vary by several order of magnitudes. Concentrations in the two phases are controlled by different mechanisms. The determination of trace element species in whole blood is, therefore, far less informative than their analysis in serum and packed cells separately.

Serum and packed cells may be deep-frozen only once, as repeatedly frozen and thawed samples showed a remarkable decrease in methylmercury concentrations. There is also some evidence that methylmercury may be destroyed during lyophilization of blood samples (Horvat and Byrne 1992, la Fleur 1973).

Analysis of human hair offers several advantages over analysis of blood samples: e.g. ease of sampling and sample storage, the concentration of methylmercury is approximately 250 times higher than in blood, and analysis of different longitudinal sections of hair can give information on the history of the exposure to methylmercury ingested through food. Adhering dust and grease should be removed by one of the following solvents: hexane, alcohol, acetone, water, diethylether, or detergents. IAEA and WHO (UNEP/WHO/IAEA 1987) recommend the use of only water and acetone. Long-term storage of human hair samples has shown that methylmercury is stable for a period of a few years if stored dry and in darkness at room temperature.

Biological samples are preferably analyzed fresh or after lyophilization. Deep-freezing of fresh samples, especially with long storage, should be avoided, since it has been noticed that in some organisms methylmercury may decompose with repeated freezing and unfreezing (particularly in bivalves) (Horvat and Byrne 1992). Methylmercury and total mercury in lyophilized biological samples, such as biological certified reference materials (CRMs), are stable for years. CRMs are, however, sterilized either by autoclaving or by γ-irradiation. This important step prevents bacteriological activity, which may otherwise lead to methylation and demethylation processes. In general, very little is known of the effects of sterilization on the stability of methylmercury compounds. More studies are needed to investigate the stability of organomercury compounds in biological samples, particularly under various sample preparation and long-term storage conditions.

Determination of total mercury

Sample pretreatment

Most of the methods for the determination of total mercury in solid samples require preliminary digestion of the sample. They are classified as wet (oxidizing digestion) and dry (combustion/pyrolysis) decomposition methods.

In order to quantitatively release mercury from the sample, wet oxidation procedures require one or more oxidizing agents in an acidic medium. Most common reagents that have been used include HNO_3, HCl, BrCl, H_2SO_4, $HClO_4$, H_2O_2, V_2O_5, $KmnO_4$, $K_2Cr_2O_7$. Mixtures of various other reagents have also been used (Horvat 1989). It is important to note that these reagents should be of proper quality (low in mercury). Samples are normally digested in closed, semi-closed, and/or sealed containers at elevated temperatures (max. 90-100°C). Particular attention should be paid to preventing losses of mercury at elevated digestion temperatures. Therefore, closed or sealed digestion containers should be employed (Horvat et al., 1991). Nowadays, several laboratories follow procedures based on oxidative digestion (HNO_3, H_2SO_4 and BrCl) in semi-closed Teflon containers at 90°C for 12 hours. Frequently, incomplete digestion of the matrix has led to erroneous results. In the case of solid samples such as soli and sediments, it is strongly recommended to use acid digestion including HF in order to completely remove Hg from the inorganic matrix (Kocman et al., 2004).

Combustion or pyrolysis procedures (dry ashing) are often advised in the literature as an alternative decomposition method. They can be performed under reductive (Aston and Riley 1972) or oxidative conditions (Nicholson 1977; Dumarey and Dams 1984; Kosta and Byrne 1969; Byrne and Kosta 1974). Normally, they are combined with a noble metal amalgamation or pre-concentration in oxidizing solution and/or other absorbent materials.

The general advantage of these approaches is a rapid quantitative separation by a physicochemical process not requiring any chemical manipulation and the avoidance of chemical interferences that are connected with the reduction/aeration step (Horvat et al., 1991; Bartha and Krenyi, 1982).

Instrumental methods

For the determination of low level mercury concentrations a number of instrumental analytical methods can be used (Schroeder 1995, Drasch et al., 2004). Among the most frequently employed are the following: CV AAS, CV AFS, ICP-MS, electrochemical methods, and neutron activation analyses (NAA). During the last two decades CV AAS and CV AFS have replaced most of other techniques. The relative detection limits of some techniques are presented in Table 1. It should be noted, however, that the detection limits reported are dependent on the overall analytical procedure including sample preparation prior to the final quantification step.

Cold Vapour Atomic Absorption Spectrometry. Most of the procedures are based on the principle developed by Poluektov et al. (1964) (see also Hatch and Ott 1968). A reducing agent ($SnCl_2$ or $NaBH_4$) is added to a reaction vessel containing the prepared sample. The $Hg°$ vapour is then liberated from the sample solution and swept either directly into the cell of the spectrometer, or is pre-concentrated on a gold surface before being thermally desorbed prior to analysis. Such reduction-aeration procedures are easy to perform, rapid, selective, and accurate by comparison with many other techniques. A typical detection limit of a CV AAS system, with pre-concentration on gold, is 0.05 ng/g. The method is fairly specific. However, a number of spectral interferences can occur due the presence of water vapour, NO_2, SO_2, ozone and other organic and inorganic compounds (Schroeder 1982). Some of these interferences can be successfully removed by the Zeeman background correction. All mercury forms should be converted into $Hg°$, as the AAS can only detects mercury in its elemental form.

Numerous analytical protocols have been developed and optimized by instrument producers world-wide. For example, solid sampling CV AAS has been increasingly used for the determination of total Hg in solid and liquid samples (Roos-Barraclough et al., 2002). CV AAS procedures are also frequently based on flow injection analysis (Murphy et al., 1996).

Cold Vapour Atomic Fluorescence Spectrometry. With improvements in light sources in recent years, AFS has become increasingly popular. The basis of AFS determination of mercury is detection of the radiant energy emitted, perpendicular to the incident light beam. In this way the signal is measured relative to "zero" rather than as a small difference in a large signal (as in the case of AAS). Therefore, AFS achieves much better sensitivities (less than 1 picogram) and linearity over a wider concentration range (Kvietkus et al., 1983). Care is needed to avoid quenching of the fluorescence signal by gaseous substances such as CO/CO_2, O_2, or N_2. As in the case of AAS, it detects mercury in its elemental form only.

In both methods, CV AAS and CV AFS, a reduction/aeration step is used. Apart from spectral interferences, there are a number of interferences dependent on the composition of the sample. They may cause non-specific absorption (volatile organic compounds), interfere with the reduction (bind ionic mercury in complexes or amalgamate $Hg°$), and interfere with the preconcentration of Hg on the adsorption trap (volatile halides and hydrides). A number of these interferences can be avoided by careful optimization of the analytical procedure. The most serious interferences may occur with the determination of mercury in geological samples (Horvat 1989, Horvat et al., 1991, Bartha and Ikrenyi 1982), due to high concentrations of palladium, platinum, gold, silver, antimony, copper, zinc, or lead. By proper selection of the pH and the reducing agent ($SnCl_2$ or $NaBH_4$) these interferences can be minimized or completely removed.

Inductively-coupled plasma - mass spectrometry. ICP-MS has become increasingly used in mercury research studies and has been demonstrated to be a very powerful tool (Hintelmann and Ogrinc 2003). Introducing mercury in the form of gaseous species into a dry plasma greatly reduced occurrence of memory effects, which was one of the major problems for the effective use of ICP-MS initially. ICP-MS can achieve absolute detection limits of less than 100 pg of Hg. Moreover, the capability of ICP-MS to take advantage of special isotope dilution methods makes this technique suitable for very precise and accurate measurements.

In addition, multiple stable tracer experiments to study the fate of Hg species in the environment and biological systems are available for investigation of multiple transformation processes simultaneously (Domuth and Heumann 2001; Stoichev et al., 2004; Tseng et al., 1998).

Neutron Activation Analysis. This can be performed as non-destructive instrumental NAA (Das and van der Sloot 1976; Dams et al., 1970) or radiochemical NAA (Kosta and Byrne 1969; Byrne and Kosta 1974).). k_0 standardization instrumental NAA is now available and can be used on a routine basis (Jacimovic and Horvat 2003).

Table 1. Most frequently used methods for quantification of mercury and their relative detection limits, adopted from Horvat, 1996.

Method		Detection limits
Colorimetric methods		$0.01 - 0.1$ µg g^{-1}
AAS	graphite furnace (GF AAS)	1 ng g^{-1}
	cold vapour (CV AAS)	$0.01 - 1$ ng g^{-1}
AFS	cold vapour (CV AFS)	$0.001 - 0.01$ ng g^{-1}
NAA	instrumental (INAA)	$1 - 10$ ng g^{-1}
	radiochemical (RNAA)	$0.01 - 1$ ng g^{-1}
GC	Electron capture detector	$0.01 - 0.05$ ng g^{-1}
	Atomic emission detector	0.05 ng g^{-1}
	Mass spectrometer	0.01 ng g^{-1}
	CV AAS/AFS	$0.01 - 0.05$ ng g^{-1}
HPLC	UV	0.1 ng mL^{-1}
	CV AAS	0.5 ng mL^{-1}
	CV AFS	0.08 ng mL^{-1}
	Electrochemical detectors	$0.1 - 1$ ng mL^{-1}
ICP MS		0.01 ng mL^{-1}
ICP AES		2 ng mL^{-1}
Photo-acoustic spectroscopy		0.05 ng
X ray fluorescence		5 ng g^{-1} – 1 ng g^{-1}
Gold-film analyzer		0.05 µg g^{-1}

Good agreement of the results obtained by k0-INAA with other methods was observed in environmental samples such as soil, sediments and sewage sludge with elevated mercury values (> 1 mg/kg), while at lower concentration agreement is good in the absence of major interferences in k$_0$-INAA. In biological samples (plants, algae and tissues) the agreement is satisfactory at concentrations higher that 0.05 mg/kg. The sensitivity of k$_0$-INAA largely depends on the presence of other elements which interfere with the gamma line of ^{203}Hg. Instrumental k$_0$-NAA may suffer from spectral interferences and, when plastic irradiation vials are used, from volatilization losses, therefore the use of RMs with known values and chemical composition close to those of the samples analysed should be used for validation purposes. Because sample preparation and handling steps are minimal before the irradiation of the sample (almost no contamination problems), NAA has often been used as a reference method against which other methods were checked and compared. (Jacimovic and Horvat 2003). However, it requires very expensive facilities, well-trained personnel and lengthy procedures, and it is not suitable for use in the field.

Atomic Emission Spectrometry. In recent years several types of plasma sources, including direct current, inductively-coupled, and microwave - induced gas (helium and argon) plasmas have been used for the determination of mercury (Fukushi et al., 1993). These methods are very sensitive, but compared to AAS and AFS they are too complex and expensive for routine work.

Photo-acoustic spectroscopy. Mercury is first preconcentrated on a gold trap and after thermal release it is quantified by measuring the sound produced from fluorescent quenching when the sample vapour is irradiated with a modulated mercury vapour lamp. The detection limit is 0.05 ng. The method has been successfully used for detection of ultratrace levels of Hg in air and snow(de Mora et al., 1993; Patterson 1984).

X-*Ray Spectoscopy.* X-ray fluorescence is convenient as the sample preparation is minimal, analysis is quick and non-destructive and it is indifferent to the chemical or physical state of the analyte. However, it is less sensitive than AAS and NAA and only detection limits in the µg range can be achieved if the sample is directly measured. The sensitivity can be improved by pre-separation and pre-concentration of Hg (D'Silva and Fassel 1972, Bennnun and Gomez, 1998). In vivo determination of mercury was investigated and applied (O'Meara et al., 2000). Synchrotron radiation XRF has successfully been applied to biological monitoring using hair. Its advantage is in studying mercury dynamics in a small sample (Shimijo et al., 1997)

Recently, X-ray absorption spectroscopy (XAS), in particular extended X-ray absorption fine structure (EXAFS) spectroscopy has been applied for mercury speciation in mercury–bearing mine wastes (Kim et al., 2000).

Electrochemical Methods. The use of these methods is less popular and has been replaced by the other measurement techniques mentioned above. One of the important advantages is that, for example, using anodic stripping voltammetry (ASV) it is possible to separate Hg(I) and Hg(II) in aqueous solutions. However, the sensitivity is poor compared to other techniques for determination of total mercury (Sipos et al., 1980).

Mercury speciation

In general, methods are classified according to the isolation technique and the detection system (Drasch et al., 2004, Sanchez Uria and Sanz Medel 1998, Horvat and Schroeder 1995). Most methods for the isolation/separation of organomercury compounds have been based on solvent extraction, differential reduction, difference calculations between "total" and "ionic" mercury, derivatization, or on paper- and thin layer chromatography. The most common approaches to organomercury separation and detection are schematically presented in Figure 4.

Separation and detection systems

During the last twenty years hundreds of papers dealing with determination of organomercury compounds in environmental samples have been published. Most of them are based on the method originally developed by Westöö (Westoo 1966). In recent years, however, significant improvements of analytical methods in terms of specificity and sensitivity have been achieved. This has allowed the determination of mercury speciation in all environmental compartments. Only a brief overview is given here of methods which have been reported in review articles (Drach 2004, Horvat, 1996). Instead, particular emphasis hasis placed on more recent analytical developments and future needs.

The basis of most present methods was introduced by Japanese and Scandinavian workers (Westoo 1966, Sumino 1968). It involves the extraction of organomercury chloride from acidified homogeneous samples into benzene (however, the use of toluene is strongly recommended, for health and safety reasons). Organomercury compounds are then back-extracted into an aqueous cysteine solution. The aqueous solution is then acidified and organomercury compounds are re-extracted with benzene or toluene. This double partitioning enables removal of many interferences (e.g. benzene-soluble thiols). Finally methylmercury is analyzed by gas chromatography with electron capture detection. Several modifications have been made to this protocol for the separation and identification of organic mercury in biological and other samples. For example, in the initial step the addition of copper (II) ions (or mercury(II)) enhances the removal of mercury bound to sulphur. Copper (II) was found superior to mercury(II) since it avoided the problem of decomposition of dimethylmercury, if present. The method has also been modified in terms of the quantity of chemicals used. A semi-micro scale method developed by Uthe and co-workers (Uthe et al., 1972) has been widely applied. However, inorganic mercury cannot be determined by this procedure, unless a reagent is added to form, for example, alkyl- and aryl- derivatives, which can then be extracted and determined by GLC (Zarnegar and Muschak 1974). In general, solvent extraction procedures are time consuming, corrections for the recovery of the procedure vary from sample to sample, and with some sample types (e.g. those rich in lipids) phases are difficult to separate due to the presence of persistent emulsions, particularly during the separation of the aqueous cysteine phase. To overcome these problems methylmercury can be adsorbed on cysteine paper (instead of into cysteine solution) during the clean-up stage 8 Horvat et al. (1988). Using additional pre-separations prior to extraction such as volatilization of methylmercury in a microdiffusion cell (Zelenko and Kosta 1973) and distillation (Horvat et al., 1988; 1994) may also facilitate separation of phases during extraction.

When speciation is required with insoluble samples (such as sediments and soils), it is difficult to estimate recovery. In such samples, recovery of spiked methylmercury is not equivalent to the methylmercury originally present. By comparing various isolation techniques for methylmercury compounds in sediment samples and soils it has been shown that conventional methods based on acid leaching of organomercury compounds prior to their extraction into an organic solvent are inadequate in most cases for releasing methylmercury from sediment samples. Improved recoveries have been achieved by extraction of methylmercury with nitric acid at elevated temperature or assisted by microwave energy (Lang et al., 2004). It is important to mention that some protocols may lead to artifact methylmercury production, especially in procedures where methylmercury is isolated at higher temperatures (Falter 1999). The quality of the results should therefore be regularly checked by the use of appropriate reference materials, if available, or by comparison of the results from different laboratories and/or the use of different analytical approaches.

It is important to mention that the use of ICP-MS and isotope dilution analysis (IDA) overcomes problems associated with incomplete recoveries of organomercury species, particularly in biological samples. The key stage in the IDA procedure is the equilibration of the isotopically modified spike and the sample MeHg; if this is achiveved the spike material acts as an ideal internal standard. So far, such a protocol has been successfully applied to numerous environmental and biological samples (Clough et al., 2003; Hintelmann 1999; Falter 1999; Snell et al., 2000).

Sample collection
↓
Sample pretreatment
↓
Liberation of MeHg from its matrix
(acid leaching, alkaline dissolution, volatilization, distillation, super fluid extraction, microwave assistance)
↓
Extraction/clean-up/preconcentration
(Solvent extraction, derivatization such as ethylation, butylation, hydration and iodination; cryogenic trapping; preconcentration on solid phases)
↓
Separation of mercury species of interest
(gas chromatography; HPLC; ion-exchange)
↓
Quantification
(CV AAS, CVAFS, GC-ECD, AED, MS, ICP-MS)

CV AAS – cold vapour atomic absorption spectrometry
CV AFS – cold vapour atomic fluorescence spectrometry
GC-ECD – gas chromatography – electron capture detector
AED – atomic emission detector
ICP-MS – inductively coupled mass spectrometry
HPLC – high performance liquid chromatography

Figure 4. Steps for determination of organomercury compounds.

Chromatographic conditions: Apart from the above mentioned problems associated with the extraction of organomercurials, problems also exist in the chromatography of organomercurial halides. Many investigators have recommended that columns packed with 5 % DEGS-PS on 100-120 mesh Supelcoport be used. Some other polar stationary phases have also been employed, e.g. PEGS, Carbowax 20M, Durapak, Carbowax 400, PDEAS, HIEFF-2AP, etc.. In order to prevent ion-exchange and adsorption processes on the column (which cause undesirable effects such as tailing, changing of the retention time and decrease of peak areas/heights) passivation of the packing material is needed with Hg (II) chloride in benzene (O'Reilly 1982). Although the more inert nature of capillary columns would be expected to minimize such effects, improved chromatographic performance over packed columns cannot be readily achieved. Some workers still prefer to use packed columns since the analytical protocols using capillary columns require additional research to optimize performance. The following capillary columns have so far been reported to give good results: OV-17 WCOT, Beijing Chemical Industry Works; Superox 20M FSOT, and OV 275. Several workers have chosen to derivatize mercury species to their corresponding non-polar, alkylated analogues such as butyl derivatives, which can then be separated on non-polar packed or capillary columns (Bulska et al., 1992).

Detectors: Various detectors can be used in combination with GLC for the determination of mercury species. The electron capture detector (ECD) is a very sensitive detector with an absolute detection limit of approximately a few picograms. It does not, however, measure mercury directly, but responds to the halide ion attached to the CH_3Hg^+ ion. The identification of small methylmercury peaks can sometimes be subject to positive systematic error owing to co-eluting contaminants. The use of a plasma atomic emission detector, a mass spectrometric detector, CV AAS, CV AFS, or ICP - MS can avoid such problems, since mercury is measured directly. Miniaturised automated speciation analyzers have recently been developed for the determination of organomercury compounds, based on microwave induced plasma emission detector (Slaets and Adams 2000).

Derivatization methods

Many methods use the formation of volatile organomercury derivatives (through ethylation, propylation, butylation, hydration and iodination) in order to separate them from the bulk of the sample by simple room-temperature aeration. The same ethylation method as described for water samples has also been applied to biological and sediment samples (Bloom 1989). An aliquot of sample is subjected to ethylation by sodium tetraethylborate. Methylmercury is transformed into methylethylmercury and mercury (II) is transformed into diethylmercury. The two species can be determined simultaneously (Liang et al., 1994). Volatile ethylated mercury compounds, as well as elemental mercury and dimethylmercury, are removed from solution by aeration and are then trapped on an adsorbent (Carbotrap or Tenax). Mercury compounds are separated on a GC column, and pyrolized to elemental $Hg°$ at 900°C for subsequent mercury determination by CV AFS, CV AAS, or ICP MS. As mentioned previously, very low detection limits may be achieved by CV AFS and ICP MS (6 pg/L for water and 1pg/g for biota and sediment samples). Instead of sodium tetraethyl borate, sodium borohydride may also be used to form volatile methylmercury hydride, which is then quantified by gas chromatography in line with a Fourier transform infrared spectrophotometer (Fillipelli et al., 1992). CH_3I formed in a headspace vial may also be introduced onto a GC column and detected by microwave-induced plasma atomic emission spectrometry (MIP-AES) or AFS detectors. Propylation and hydration have also been applied with great success as described above (Domuth and Heumann 2001; Logar et al., 2004).

Differential reduction

There are also a few methods that are based on differential reduction. In the method developed by Magos (Magos 1971) the inorganic mercury in an alkaline digested sample is selectively reduced by stannous chloride, while organomercury compounds are reduced to elemental mercury by a stannous chloride-cadmium chloride combination. Elemental mercury released can be measured by CV AAS. The method has been successfully applied to biological samples in toxicological, epidemiological and clinical studies. CV AAS has also been used for detection of organomercury compounds after pre-separation of organomercury by (a) anion exchange (May et al., 1987), (b) volatilization and trapping on cysteine paper (Zelenko and Kosta 1973) and (c) water vapour distillation (Horvat et al., 1986; 1993). Organomercury compounds must be destroyed by either UV-irradiation or acid digestion prior to detection by CV AAS. In most biological samples, the organomercury concentrations usually correspond to methylmercury. In some environmental samples such as sediment, soil, and water samples, the concentrations of organic mercury (particularly if separated by anion-exchange) have been found to be much higher than those of methylmercury compounds. This is probably due to presence of some other organic mercury compounds which have not, as yet, been identified.

Miscellaneous Methods

The first practical method for differentiating between organic and inorganic mercury was a colorimetric method developed by Gage (1961). Organomercury compounds were extracted into an organic solvent and determined spectrophotometrically as dithizone complexes. The method basically suffers from low sensitivity. Simple extraction procedures have also been successfully used followed by AAS. High performance liquid chromatography (HPLC) has proven of use with reductive amperometric electrochemical detection, ultraviolet detection, inductively coupled plasma emission spectrometric detection, or AAS detection. NAA has been used for methylmercury determinations in fish, blood and hair samples after suitable preseparation procedures. Graphite furnace AAS has also been used for the final determination of methylmercury in toluene extracts to which dithizone was added. An anodic stripping voltammetry technique has been developed for determination of methylmercury. However, the method has never been used for environmental samples. Methylmercury has also been extracted into dichloromethane (CH^2Cl^2). This was then evaporated down to 0.1 ml and subjected to GC with an atmospheric pressure active nitrogen detector (Horvat 1996).

An enzymatic method for specific detection of organomercurials in bacterial cultures has been developed. It is based on the specific conversion of methylmercury (no other methyl-metallo groups are enzymatically converted) to methane by organomercurial lyase. Ethyl and phenylmercury can also be detected (Baldi and Fillipelli 1991).

Determination of other organomercurials

Among organomercury species currently of interest, ethylmercury (EtHg) is a compound that requires further attention as it is still used in Thiomersal for preservations of vaccines. It is important to analyze ethylmercury in vaccines, in wastewater from waste treatment plants in industries using ethylmercury, as well biological samples in order to understand ethylmercury uptake, distribution, excretion, and effects. In principle, methods developed for methylmercury can also be used for ethylmercury, except in protocols using derivatization by ethylation. In such cases propylation is recommended (Logar et al., 2004).

Only a few investigations concerning the determination of other organomercurials used in agriculture and for other purposes (Horvat and Schroeder 1995) have been reported. Methoxyethyl- and ethoxyethylmercury have been examined by thin layer chromatography (TLC) and gas-liquid chromatography (GLC). It would appear that the only method that can separate and measure many of the compounds simultaneously is high performance liquid chromatography (HPLC) with UV detection (Hintelmann and Wilken 1993, Hempel et al., 1992). It offers several advantages. The separation of the compounds is performed at ambient temperatures, hence thermal decomposition does not occur. It offers the possibility to separate less volatile or non-volatile species such as mersalylic acid or the aromatic organomercurials, which usually present a problem for GLC. It is, however, very important to isolate these compounds from environmental samples quantitatively. Methyl- and ethylmercury can easily be isolated from soils by extraction from acidified samples. Several extraction agents have been tested in order to release organomercurials from soils. Methyl- and phenylmercury can be extracted by potassium iodide-ascorbic acid and oxalic acid with satisfactory yields, whereas ethylmercury is only partly extracted. No suitable extraction techniques have been found for methoxyethyl- and ethoxyethylmercury in soils, (due to decomposition of these compounds under acidic conditions).

Fractionation of mercury in soils and sediments

The biogeochemical and especially the ecotoxicological significance of Hg input is determined by its specific binding form and coupled reactivity rather than by its accumulation rate in the solid material. Consequently, these are the parameters that have to be determined in order to assess the potential for Hg transformation processes (such as methylation, reduction, demethylation), and to improve data for environmental risk assessment. Hg pyrolysis followed by AAS detection was developed to distinguish among cinnabar bound Hg, metallic Hg and matrix bound Hg (Biester et al., 2000, Bloom et al., 2003). Alternative approaches used for mercury fractionation are based on sequential extractions and leaching to provide information on the solubility and reactivity of Hg. Sequential extraction schemes developed by Bloom (Bloom et al., 2003) consists of six steps, including water soluble, 'human stomach acid' soluble, organo-chelated, elemental Hg, mercuric sulfide and residual fraction. An additional step was incorporate into this scheme in order to provide information on the volatilization potential of mercury present in soil (Kocman et al., 2004). It is important to note that these schemes are based on the analytical protocols used and slight changes in may result in a different relative distribution of mercury fractions.

Methylation/demethylation/reduction potential using tracers. In order to assess the potential for mercury transformation rates under various environmental conditions and matrices, analytical protocols using stable and radioactive isotopes of mercury were applied by different groups. As mentioned above enriched stable isotopes in combination with the ICP MS are increasingly used (Monperrus et al., 2004) . Multiple stable tracer experiments allow studies of the fate of Hg species in the environment and in biological systems. This concept allows the investigation of multiple transformation processes simultaneously (Monperrus et al., 2004; Hintelmann and Ogrinc 2003; Domuth and Heumann 2001). The use of radioisotopes to trace different transport and transformation processes is also widespread; in the case of mercury the most frequently used radiotracer is ^{203}Hg ($t_{1/2}$ = 46 d) (Guimares et al., 1995; Mauro et al., 2002). However, when adequate facilities are available ^{197}Hg ($t_{1/2}$ = 64.14 h) can be also employed successfully, as it was demonstrated in mercury methylation/de-methylation studies in soils and sediments (Guevara et al., 2004). Demethylation was studied by the use of ^{14}CH$_3$Hg$^+$, where the ^{14}CH$_4$ produced indicates reductive demethylation, and ^{14}CO$_2$ oxidative pathways of detoxification mechanisms (Hines et al., 2000; Oremland et al., 1991).

QUALITY ASSURANCE AND QUALITY CONTROL

Quality assurance refers to those procedures that ensure that analytical results are valid, traceable, reproducible, representative, complete and accurate, i.e. close to the "true value". It also includes measures developed to assess performance. It is generally accepted that mercury analylsis and speciatiom must be done by well trained staff who, in principle, should be involved in the measurement process from sampling to the production of final results, particularly if speciation of mercury is intended. The use of reference materials certified for mercury and its compounds plays an important role in method validation and demonstration of traceability. At present there are many reference materials certified for total mercury concentrations in various matrices (sediment, soil, ash, water, plants, and tissues) of different origin. Unfortunately, only a few reference materials are certified for methylmercury compounds (Horvat, 1999).

It is understood that these materials are not sufficient to satisfy the quality assurance requirements in many laboratories performing methylmercury compounds analyses. Therefore, apart from the analysis of CRMs, the accuracy of analytical procedures for determination of methylmercury was tested by several intercomparison exercises on biological, soil, and sediment and water samples.. A review of these exercises has shown that the determination of total methylmercury compounds in samples such as soil, sediment, and water is difficult and is also method dependent.

It is generally accepted that the use of CRMs represents only one aspect of the QA/QC programme and can only cover a limited number of environmental samples. For example, concentration levels of mercury in air and water are extremely low and even highly sophisticated equipment cannot guarantee accurate measurements. The reliability of the results depends on the overall procedure including sampling, storage, and laboratory handling. One way to check the accuracy of the results is to participate in field intercomparison exercises or by comparison of the results obtained by various methods. Such exercises are now regularly organized by different international agencies, RM producers, and programmes. The results obtained are encouraging, demonstrating the comparability of the data sets being generated by diverse groups around the world.

Further development and optimisation is needed for mercury analyses and speciation/fractionation in soils and sediments and "dynamic" measurements (transformation and transport measurements). It is suggested that "method specific" techniques should be avoided unless they provide information which is biogeochemically important.

In conclusions, chemical metrology in mercury analysis and speciation needs to develop further in order to achieve comparability of results.

Currently available matrix CRMs are not sufficient to establish comparability of chemical measurements due to poor coverage of concentrations and matrix matching. In order to demonstrate traceability to international standards calibration standards for Hg speciation with small uncertainties are urgently needed. Questions related to operationally defined parameters (RGM, reactive Hg in water etc.) need to be addressed from the metrological point of view in order to demonstrate comparability of results.

REFERENCES

Ahmed, R., May, K., Stoppler, M. Ultratrace analysis of mercury and methylmercury in rainwater using cold vapour absorption spectrometry, *Fres. J. Anal. Chem.*, 326, 510-516, 1987.

Akagi, H., Nishimura, H. Speciation of mercury in the environment. In Advances in mercury toxicology (Eds. Suzuki T, Nobumassa I, Clarkson TW.) Plenum, new York, 1991.

Baldi, F., Fillipelli, M., New method for detecting methylmercury by its enzymatic conversion to methane. *Env. Sci. Technol.*, 25, 302-305, 1991.

Bartha, A., Ikrenyi, K. Interfering effects on the determination of low concentration of mercury in geological materials by cold vapor atomic absorption spectrometry, *Anal. Chim. Acta*, 139, 329-332, 1982.

Bennun, L., Gomez, J. Determination of mercury by total reflection X-ray fluorescence using amlgamation with gold. *Spectochim. Acta B.*, 52, 1195-1200, 1997.

Biester, H., Gosar, M., Müller, G. Mercury speciation in tailings of the Idrija mercury mine *J. Geochem. Explor.,* 65, 195-204, 1999.

Bloom, N.S., Crecelius, E.A. Determination of mercury in sea water at sub-nanogram per liter levels, *Mar. Chem.* , 14: 49-59, 1983.

Bloom, N.S., Fitzgerald, W,F. Determination of volatile mercury species at the picogram level by low-temperature gas chromatography with cold vapour atomic fluorescence detection, *Anal. Chim. Acta*, 208,151-161, 1988.

Bloom, N.S. Determination of picogram levels of methylmercury by aqueous phase ethylation, followed by cryogenic gas chromatography with cold vapour atomic fluorescence detection, *Can. J. Fish. Aquat. Sci.*, 46, 1131-1140, 1989.

Bloom, N.S., Preus, E., Katon, J., Hiltner, M. Selective extractions to assess the biogeochemically relevant fractionation of inorganic mercury in sediments and soils. *Anal. Chim Acta.*, 79, 33-248, 2003.

Bulska, E., Emteborg, H., Baxter, D.C., Frech, W. Speciation of mercury in human whole blood by capillary gas chromatography with a microwave-induced plasma emission detector system following complexometric extraction and butylation, *Analyst*, 117, 657-665, 1992.

Byrne, A.R., Kosta, L. Simultaneous neutron activation determination of selenium and mercury in biological samples by volatilization, *Talanta*, 211, 1083, 1974.

Carr, R.A., Wilkniss, P,E. Mercury: Short-term storage of natural waters, *Env. Sci. Technol.*, 7,1, 62-63, 1973.

Celo, V., Ananth, R.V., Scott, S.L., Lean, D.R.S. Methylmercury artifact fiormation during solid-phase extraction of water samples using sulfhydryl cotton fiber adsorbent. *Anal. Chim Acta*, 516, 171-177, 2004.

Clought, R., Belt, S., Evans, E.H., Fairman, B., Catterick, T. Investigation of equilibration and uncertainty contributions for the determination of inorganic mercury and methylmercury by isotope dilution inductively coupled plasma mass spectrometry *Anal. Chim. Acta*, 500, 155-170, 2003.

Cossa, D., Martin, J.M., Sanjuan, J. Dimethylmercury formation in the Alboran Sea *Mar. Poll. Bull.*, 28: 381-384, 1994.

Coyn, R.V., Collins, J. Loss of mercury from water during storage, *Anal. Chem.*, 44, 1093-1096, 1972.

Craig, P.J., 'Organometallic Compounds in the Environment - Principles and Reactions', Longman Group Limited, England, 1986.

Dams, R., Robbins, J.A., Rahn, K.A., Winchester, J.W. Nondestructive neutron activation analysis of air pollution particulates *Anal. Chem.*, 42, 861-867, 1970.

Das, H.A., van der Sloot, H.A. Sampling problems and the determination of mercury in surface water, seawater, and air. NBS Special Publication No.422, Washington, 1976.

de Mora, S.J., Patterson, J.E., Bibby, D.M. (3) Baseline atmospheric mercury studies at Ross Island, Antartctica, *Antar. Sci.*, 7, 323-326, 1995.

Demuth, N., Heuman, K.G. Validation of Methylmercury Determinations in Aquatic Systems by Alkyl Derivatization Methods for GC Analysis Using ICP-IDMS *Anal.Chem.*, 73, 4020-4027, 2001.

Drasch, G., Horvat, M., Stoeppler, M. 'Mercury' In "Elements and their compounds in Environment", E. Merian (Ed.), M. Anke (Ed.), M. Ihnat (Ed.), M. Stoeppler (Ed.), Wiley-VCH Verlag GmBH&Co. KgaA, Weinheim, 2nd edition, 2004.

D'Silva, A.P., Fassel, V.A. Ultratrace level detection of mercury by an x-ray excited optical fluorescence technique, *Anal. Chem.*, 44, 2115-2116, 1972.

Dumarey, R., Dams, R. Pyrolysis/CV AAS for determination of mercury in solid environmental samples, *Microchim. Acta*, 111, 191-198, 1984.

Dumarey, R., Temmerman, E., Dams, R., Hoste, J. The accuracy of the vapour-injection calibration method for the determination of mercury by amalgamation/cold vapour atomic absorption spectrometry, *Anal. Chim. Acta*, 170: 337-340, 1985.

Falter, R. Experimental study on the unintentional abiotic methylation of inorganic mercury during analysis: Part 1: Localisation of the compounds effecting the abiotic mercury methylation, *Chemosphere*, 39, 1051-1073, 1999.

Fillipelli, M., Baldi, F., Brinckman, F.E., Olson, G.J. Methylmercury determination as volatile methylmercury hydride by purge and trap gas chromatography in line with Fourier Transform Infrared Spectroscopy, *Env. Sci. Technol.*, 25, 1457-1462, 1992.

Fukushi, K., Willie, N.S., Sturgeon, R.E. Subnanograme determination of inorganic and organic mercury by helium-microwave induced plasma-atomic emission spectrometry, *Anal. Lett.*, 26, 325-340, 1993.

Gage, J.C. The trace determination of phenyl- and methylmercury salts in biological material, *Analyst*, 86: 457-459,1961.

Gnamuš, A., Byrne, A.R., Horvat, M. mercury in the soil-plant-deer-predator food chain of a temeprate forest in Slovenia. *Env. Sci. Technol.*, 34, 3337, 2000.

Guevara, S., Jereb, V., Arribere, M., Perez Catan, S., Horvat, M. The production and use of 197Hg radiotracer to study mercury transformation processes in environmental matrices. *Mat. Geoenviron.*, 51, 1928-1931, 2004.

Guimaraes, J.R.D., MalmO, Pfeiffer WC. A simplified radiochemical technique for measuremnt of mercury methylation rates in aquatic systems near goldmining areas, Amazon, Brazil. *Sci. Tot. Env.*, 175: 151-162, 1995.

Hatch, W.R., Ott, W.L. Determination of submicrogram quantities of mercury by atomic absorption spectrophotometry *Anal. Chem.*, 40, 2085-2087, 1968.

Hempel, M., Hintelman, H., Wilken R.D. Determination of organic mercury species in soils by high-performance liquid chromatography with ultraviolet detection, *Analyst*, 117, 669-674, 1992.

Heraldsson, C., Westerlund, S., Öhman, P. Determination of mercury in natural samples in the sub-nanogram level using inductively coupled plasma/mass spectrometry after reduction to elemental mercury, *Anal. Chim. Acta*, 221, 77-84, 1989.

Hines, M.E., Horvat, M., Faganeli, J., Bonzongo, J.C.J., Barkay, T., Major, E.B., Scott, K.J., Bailey, E.A., Warwick, J.J., Lyons, W.B. Mercury biogeochemistry in the

Idrija River, Slovenia, from above the mine into the Gulf of Trieste. *Environ. Res.,* 83, 129-139, 2000.

Hintelmann, H., Wilken, R.D. The analysis of organic mercury compounds using liquid chromatography with on-line atomic fluorescence spectrometric detection, *Appl. Organomet. Chem.* 7, 173-180. 1993.

Hintelmann, H. Comparison of different extraction techniques used for methylmercury analysis with respect to accidental formation of methylmercury during sample preparation *Chemosphere*, 39, 1093-1105, 1999.

Hintelmann, H. Ogrinc N. Determination of stable mercury isotopes by ICP-MS and their applicatuion in environmental studies' in "Biogeochemistry of Environmetally Important Trace Elements", Yong Cai (Ed.), Olin C. Braids (Ed.), ACS Symposium Series, 835, 321-338, 2003.

Hirner, A.V., Feldmann, J., Krupp, E., Grumping, R., Guguel, R., Cullen, W.R., metal(loid)organic compounds and geothermal gases and waters. *Org. Geochem.,* 29, 1765-1778, 1998.

Horvat, M., May, K., Stoeppler, M., Byrne, A.R. Comparative studies of methylmercury determination in biological and environmental samples *Appl. Organometal. Chem.,* 2, 515-524, 1988.

Horvat, M. Development and study of analytical methods for determination of low level mercury concentration and its application in analysis of biological and other environmental samples, Ph.D. Thesis, University of Ljubljana, Slovenia, 1989.

Horvat, M., Lupsina, V., Pihlar, B. Determination of total mercury in coal fly ash by atomic absorption spectrometry, *Anal. Chim. Acta*, 243, 71-79, 1991.

Horvat, M. Byrne, A.R. Preliminary study of the effects of some physical parameters on the stability of methylmercury in biological samples, *Analyst*, 117, 665-668, 1992.

Horvat, M., Liang, L., Bloom, N.S. Comparison of distillation with other current isolation methods for the determination of methylmercury compounds in low level environmental samples; Part I. Sediments, *Anal. Chim. Acta*, 281, 135-152, 1993.

Horvat, M., Liang, L., Bloom, N.S. Comparison of distillation with other current isolation methods for the determination of methylmercury compounds in low level environmental samples Part II. Water, *Anal. Chim. Acta*, 282: 153-168, 1993.

Horvat, M., Mandic, V., Liang, L., Bloom, N.S., Padberg, S., Lee, Y.H., Hintelmann, H., Benoit, J. Certification of methylmercury compounds concentration in marine sediment reference material, IAEA-356. *Appl. Organomet. Chem.*, 8, 533-540, 1994.

Horvat, M., Schroeder, W.H. Mercury (a) Determination of organomercurials. Encyclopedia of Analytical Science, Harcourt Brace & Company Limited, London 1995.

Horvat, M. 'Global and Regional Mercury Cycles: Sources, Fluxes and Mass Balances', W. Baeyens (Ed.), R. Ebinghaus (Ed.), O. Vasiliev (Ed.), Kluwer Academic Publishers, Netherlands, 1996.

Horvat, M. Current status and future needs for biological and environmental reference materials certified for methylmercury compounds. *Chemosphere*, 39, 1167-1179, 1999.

Horvat, M., Kotnik, J., Logar, M., Fajon, V., Zvonaric, T. Pirrone, N., Speciation of mercury in surface and deep sea waters in the mediterranean sea. *Atmos. Environ.,* 37, 93-108, 2003.

Horvat, M., Logar, M., Ogrinc, N., Fajon, V., Lojen, S., Akagi, H., Ando, T., Tomiyasu, T., Matsuyama, A. The effect of sampling and sample pretreatment on MeHg concentration in coastal marine sediments. *Mat. Geoenviron.*, 51, 1939-1943, 2004.

IPCS/WHO, 'Environmental Health Criteria 101', "Methylmercury", WHO, Geneva 1990.

IPCS/WHO, 'Environmental health criteria 118', "Inorganic Mercury", WHO, Geneva, Switzerland, 1991.

Iverfeldt, Å. Mercury in the Norwegian fjord Framvaren, Mar. Chem. 1988; 23: 441-445.

Jaćimović, R., Horvat, M. Determination of total mercury in environemtal and biological samples using k0-INAA, RNAA and CVAAS/AFS techniques: advantages and disadvantages. *J. Rad. Nucl. Chem.*, 259, 385-390, 2004.

Jones, R., Jocobson, M.F., Jaffe, R., West-Thomas, A.C, Alli, A. Method development and sample processing of water, soil, and tissue for the analyses of total and organic mercury by CV AFS, *Water Air Soil Pollut.*, 80(1-4) 1285-1294, 1995.

Kim, C.S., Rytuba, J.J., Brown, G.E., Jr. EXAFS study of mercury(II) sorption to Fe- and Al-(hydr)oxides: I. Effects of pH. *J. Colloid Interf. Sci.*, 27, 1-15, 2004.

Knezevic, I., Griffiths, E., Reigel, F., Dobbleaer, R. Vaccine, (in press), 2004.

Kocman, D., Horvat, M., Kotnik, J., Mercury fractionation in contaminated soils of the Idrija mercury mine region, *J. Environ. Monitor.*, (in press), 2004.

Kosta, L., Byrne, A.R. Activation analyses for mercury in biological samples at nanogram levels *Talanta*, 16, 1297, 1969.

Kvietkus, K., Sakalys, J., Sopauskas, K. The application of the atomic fluorescence method for determining mercury concentrations by a photon counter, *Atmos. Phys.*, 8: 127-135, 1983.

LaFleur, P.D. Retention of mercury when freeze-drying biological materials *Anal. Chem.*, 45, 1534-1536, 1973.

Lansens, P., Meuleman, C., Baeyens, W. Long-term stability of methylmercury standard solutions in distilled, deionized water, *Anal. Chim. Acta*, 229, 281-285, 1990.

Lee, Y.H., Mowrer, J. Determination of methylmercury in natural waters at sub-nanogram per liter level by capillary gas chromatography after adsorbent preconcentration, *Anal. Chim. Acta*, 221, 259-264, 1989.

Liang, L., Bloom, N., Horvat, M. Simultaneous determination of mercury speciation in biological materials by GC/CVAFS after ethylation and room temperature precollection. *Clin Chem.*, 40, 602-607, 1994.

Liang, L., Horvat, M., Feng, X., Shang, L., Li, H., Pang, P. Re-evaluation of distillation and comparison with HNO$_3$ leaching/solvent extraction for isolation of methylmercury compounds from sediment/soil samples, *Appl. Organomet. Chem.*, 18, 264-270, 2004.

Lo, J. Wai, C. Mercury loss from water during storage: Mechanisms and preservation *Anal. Chem.* 47, 1869-1871, 1975.

Logar, M., Horvat, M., Horvat, N., Benedik, M., Marn-Pirnat, A., Ponikvar, R., Osredkar, J. Determination of ethyl mercury and methylmercury in blood samples. *Mat. Geoenviron.* 51, 1976-1978, 2004.

Magos, L., Selective Atomic-absorption determination of inorganic mercury and methylmercury in undigested biological samples, *Analyst*, 96, 847-852, 1971.

Martell, A.E., Smith, R.M., Motekaitis, R.J. 'NIST critically selected stability constants of metal complexes data base', NIST Std. Ref. Database, No. 46, Department of Commerce, Gaithersburg, MD, 1998.

Mason, R.P., Fitzgerald, W.F. The distribution and biogeochemical cycling of mercury in the equatorial Pacific ocean, *Deep-Sea Res.*, 40, 1897-1924, 1993.

Mauro, J.B.N., Guimaraes, J.R.D., Hintelman, H., Watras, C.J., Haack, E.A., Coelho, S.A. -Souza. Mercury methylation in mycrophytes, periphyton, and water – comparative studies with stable and radio-mercury additions, *Anal. Bioanal. Chem.*, 374, 983-989, 2002.

May, K., Stoeppler, M., Reisinger, K. Studies in the Ratio Total Mercury/Methylmercury in the Aquatic Food Chain, *Toxicol. Environ. Chem.*, 13, 153-159, 1987.

Meuleman, C., Laino, C.C., Lansens, P., Baeyens, W. A study of the behaviour of methylmercury compounds in aqueous solutions, and of gas/liquid distribution coefficients, using had space analysis, *Wat. Res.*, 27, 1431-1446, 1993.

Minagawa, K., Takizawa, Y., Kifune, I. Determination of very low levels of inorganic and organic mercury in natural waters by CV AAS after preconcentration on a chelating resin, *Anal. Chim. Acta*, 115, 103-110,1979.

Monperrus, M., Krupp, E., Amouroux, D., Donard, O.F.X., Rodríguez Martín-Doimeadios, R.C. Potential and limits of speciated isotope-dilution analysis for metrology and assessing environmental reactivity *TrAC*, 23, 261-272, 2004.

Murphy, J., Jones, P., Hill, S.J. Determination of total mercury in environmental and biological samples by flow injection cold vapour atomic absorption spectrometry *Spectrochim. Acta B*, 51, 1867-1873, 1996.

Nicholson, R.A. Rapid thermal decomposition for the atomic absorption determination of mercury in rocks, soils, and sediments, *Analyst*, 102, 399-403, 1977.

O'Meara, J.M., Borjesson, J., Chettle, R. Improving the in vivo X-ray fluorescence (XRF) measuremnt of renal mercury. *Appl- Rad. Isotop.*, 53, 639-646, 2000.

O'Reilly, J.E. Gas chromatographic determination of methyl and ethyl mercury: "passivation" of the chromatographic column, *J. Chromatogr.*, 238, 433, 1982.

Oremland, R.S., Culbertons, C.W., Winfrey, M.R. Methylamercury decomposition ins ediments and bacterial cultures. Involvement of methanogens and sulfate reducers in oxidative demethylation. *Appl. Environ. Microbiol.*, 57, 130-137, 1991.

Padberg, S. and Stoeppler, M. Studies of transport and turnover of mercury and methylmercury. Metal Compounds in Environment and Life: (Interrelation between Chemistry and Biology), 4, 329-340, 1991.

Patterson, J.E. A differential photoaccoustic mercury detector, *Anal. Chim. Acta*, 164, 119-126, 1984.

Pichichero, M.E., Cernichiari, E., Lopreiato, J., Treanor, J. *Lancet*, 360, 1737, 2002.

Pirrone, N., Wichmann-Fiebig, M., Pacyna, J.M., Boffetta, P., Hansen, J.C., Grandjean, P., Horvat, M., Barregard, L., Munthe, J. Ambient Air Pollution by mercury (Hg) Position Paper, EC, 2002.

Poluektov, N.S., Vitkun, Y.V. Zelyukova, Y. Zh. *Anal. Khim.*, 18, 937-948, 1964.

Rapsomanikis, S. Craig, P.J. Speciation of mercury and methylmercury compounds in aqueous samples bycChromatography - AAS after ethylation with sodium tetraethylborate, *Anal. Chim. Acta*, 248, 563-567, 1991.

Rodriquez-Vazquez, J.A. Gas-Chromatographic determination of organomercury(II) Compounds, *Talanta*, 25, 299-310, 1978.

Roos-Barraclough, F., Givelet, N., Martinez-Cortizas, A., Goodsite ME, Biester H and Shotyk W, An analytical protocol for the determination of total mercury concentrations in solid peat samples. *Sci. Tot. Env.*, 292, 129-139, 2002.

Sanchez Uria, J.E., Sanz-Medel, A. Inorganic and methylmercury speciation in environmental samples *Talanta*, 47, 509-524, 1998.

Schroeder, W.H. Sampling and Analysis of Mercury and its Compounds in the Atmosphere, *Env. Sci. Technol.*, 16, 394, 1982.

Schroeder, W.H. 'Mercury: Inorganic (and total) determination' in "Encyclopedia of Analytical Science", Academic Press Limited, 3050-3059, 1995.

Schroeder, W.H. Sampling and analysis of mercury and its compounds in the atmosphere, *Env. Sci. Technol.*, 1982; 16, 362A-400A.

Schroeder, W.H. Developments in the speciation of mercury in natural waters, Trends in *Anal. Chem.*, 8, 339-347, 1989.

Shimojo, N. Homma-Takeda S, Ohuchi K, Shinyashiki M, Sun GF, Kumagai Y. Mercury dynamics in hair of rats exposed to methylmercury by synchrotron radiation X-ray fluorescence imaging. *Life Sci.*, 60, 2129-2137, 1997.

Simon, M. Wuhl-Couturier, G. 'Mercury' in "Ullmann's Encyclopedia of Industrial Chemistry, 6th Completely Revised Edition", F. Bohnet (Ed.), Wiley-VCH, Weinheim, Germany, 2002.

Sipos, L., Nurnberg, H.W., Valenta, P. Branica. M. The reliable determination of mercury tracers in sea water by subtractive differential pulse voltammetry at the twin gold electrodes, *Anal. Chim. Acta*, 115, 25-42, 1980.

Slaets, S., Adams, F.C. Determination of organomercury compounds with a miniaturised automated speciation analyser *Anal. Chim. Acta.*, 414; 141-149, 2000.

Snell, J., Bjorn, E., Frech, W., Investigation of errors introduced by the species distribution of mercury in organic solutions on total mercury detrmination by electro vaporization-inductively coupled plas ma mass spectrometry *J. Anal. At. Spectrom.*, 15, 397-402, 2000.

Snell, J.P., Quetel, C.R., Lambertsson, L., Qvarnstrom, J. A new 202 Hg istopically enriched methylmercury spike material with SI-traceable reference values for isotope dilution measurements in biological and environmental samples. *Mat. Geoenviron.*, 51, 2026-2029, 2004.

Stoichev, T., Rodriguez, Martin-Doimeadios, R.C., Tessier, E., Amouroux, D., Donard, O.F.X. Imrouvement of analytical performance of mercury speciation by on-line derivatization, cryofocussing and atomic fluorescence. *Talanta*,. 62, 433-438, 2004.

Sumino, K. Analysis of organic mercury compounds by gas chromatography - Part II. Determination of organic mercury compounds in various samples *Kobe J. Med. Sci.*, 14, 131, 1968.

Templeton, D.M., Ariese, F., Cornelis, R., Danielsson, L.G., Muntau, H., van Leeuwen, H.P., Lobinski, R., IUPAC Guidelines for Terms Related to Speciation of Trace Elements, *Pure Appl. Chem.*, 72, 1453-1470, 2000.

Tseng, C.M., de Diego, A., Pinaly, H., Amoroux, D., Donard, O.F.X. Field cryofocussing hydride generation applied to the simultaneous multi-elemental determination of alkyl-metal(loid) species in natural waters using ICP-MS detection *J. Environ. Monitor.*, 2: 603-612, 2000.

UNEP/WHO/IAEA, 'The Determination of Methylmercury, Total Mercury and Total Selenium in Human Hair' in "Reference Methods for Marine Pollution Studies No. 46" (Draft), UNEP 1987.

UNEP/WHO/IAEA, The Determination of Methylmercury, Total Mercury and Total Selenium in Human Hair, Reference Methods for Marine Pollution Studies No. 46 (Draft), UNEP 1987.

UNEP/IOMC, 'Global Mercury Assessment', UNEP Chemicals, Geneva, 2002.

Uthe, J.F., Solomon, J., Grift, B. A rapid semi-micro method for the determination of methylmercury in fish tissue, *J. Assoc. Offic. Anal. Chem.*, 55, 583-594, 1972.

Vandal, G.M., Mason, R.P., McKnight, D., Fitzgerald, W. Mercury speciation and distribution in a polar desert lake (Lake Hoare, Antarctica) and two glacial meltwater streams *Sci. Tot. Environ.*, 213, 229-237, 1998.

Weber, J., Evans, R., Jones, S.H., Hines, M.E., Conversion of mercury(II) into mercury(0), monomethylmercury cation, and dimethylmercury in saltmarsh sediment slurries *Chemosphere*, 6, 1669-1687, 1998.

Westöö, G. Determination of Methylmercury Compounds in Foodstuffs I. Methylmercury Compounds in Fish, Identification and Determination", *Acta Chem. Scand.*, 20, 2131-2137, 1966.

Yaneda, S., Suzuki, K.T. Equimolar Hg-Se complex binds to Selenoprotein P. Biochem. *Biophys. Res. Comm.*, 231, 7-11, 1997.

Zarnegar, P., Mushak, P. Quantitative measurements of inorganic mercury and organomercurials in water and biological media by gas liquid chromatography, *Anal. Chim. Acta*, 69, 389-407, 1974.

Zelenko, V., Kosta, L. A new method for the isolation of methylmercury from biological tissues and its determination at the parts-per-milliard level by gas chromatography", *Talanta*, 20, 115-123, 1973.

PART-III:
CHEMICAL AND PHYSICAL PROCESSES

Chapter-9

ENVIRONMENTAL CHAMBER STUDIES OF MERCURY REACTIONS IN THE ATMOSPHERE

Ann L. Sumner[1], Chester W. Spicer[1], Jan Satola[1], Raj Mangaraj[1], Kenneth A. Cowen[1], Matthew S. Landis[2], Robert K. Stevens[3] and Thomas D. Atkeson[4]

[1] *Battelle Memorial Institute, 505 King Ave., Columbus, Oh 43201, USA;*
[2] *N.S. Environmental Protection Agency, 109 T.W. Alexander Dr., Research Triangle Park, Nc 27709, USA;*
[3] *Florida Department Of Environmental Protection At U.S. Environmental Protection Agency 109 T.W. Alexander Dr., Research Triangle Park, Nc 27709, USA;*
[4] *Florida Department Of Environmental Protection, 2600 Blair Stone Rd., Tallahassee, Fl 32399-2400, USA*

INTRODUCTION

Mercury (Hg) is released to the atmosphere by both natural and anthropogenic processes. Natural sources include evasion from enriched geologic materials and volcanoes. Combustion processes, including burning of fossil fuels, non-ferrous metal production, and waste incineration account for most of the anthropogenic input. The concern over Hg in the atmosphere stems from its eventual deposition into aquatic ecosystems and subsequent conversion to methylated Hg. Methyl Hg can bioaccumulate, leading to high Hg levels in fish that are consumed by humans. This is a major concern because Hg is a human neurotoxin that has been linked to poisonings through contaminated food at many locations throughout the world. EPA has recently issued a report, "The Mercury Research Strategy", EPA/600/R-00/73 (2000) that sets an aggressive agenda to improve knowledge about Hg's sources and behavior in the environment, the risks it poses to humans and ecosystems, and mitigation of those risks. This research strategy builds on the findings of an earlier report to Congress EPA-452R-97-003 (1997) that summarized knowledge and uncertainties regarding Hg in the environment.

These and other reports suggest the need for improved understanding of the chemical processes affecting Hg in the atmosphere. Lin and Pehkonen (Lin, C. and Pehkonen, 1999) reviewed atmospheric Hg chemistry in 1999 and reported that "... many questions still remain unanswered". They point specifically to unidentified gas-phase transformation pathways as an important area for future research.

Hg is generally thought to exist in three states in the atmosphere: elemental gaseous Hg (Hg^0), reactive gaseous Hg (Hg(II)) and particle-phase Hg (Hg(p); in either oxidation state). Hg^0 is thought to dominate the atmospheric mercury burden.[4,5] Hg(p) is usually a small fraction of the total atmospheric Hg load, with the possible exception of the vicinity of some emission sources. Because the deposition of Hg is dependent on its atmospheric form, knowledge of the reactions which interconvert the various Hg forms is critical to understand and model Hg deposition. Knowledge of the reactions which transform Hg, and their rates, is also important in assessing the sources of Hg that affect deposition to sensitive areas. For example, Hg has been found at high levels in fish in the Everglades National Park in South Florida, the Great Lakes, and the Gulf of Mexico. Policy makers need to know if Hg deposition to such sensitive regions is due to reactive forms of Hg emitted locally, or to transformation of long-lived Hg^0 and deposition of the products accelerated by some specific local conditions. To address such questions, improved information on Hg reaction kinetics is needed.

Here we report on a laboratory study that examined the reactions of Hg^0 with several potential atmospheric reactants, including halogen species, ozone (O_3) and nitrate radical (NO_3). The reactions with halogen species may be relevant in polar regions, mid-latitude coastal areas, and in the upper troposphere. Additional studies of Hg reaction with bromine and chlorine atoms (Spicer et al., 2004) and the radicals BrO and ClO (Spicer, et al., 2002) are reported elsewhere and underway in this laboratory.

EXPERIMENTAL METHODS

Reactions were studied in a 17.3 m^3 environmental chamber that is equipped with fluorescent black lamps and sun lamps. The chamber has a surface to volume ratio of 2.5 m^{-1} and the internal surfaces are aluminum and Teflon. For some experiments, the chamber was fully lined with Teflon film to reduce interactions with the walls. One wall of the chamber is made of Teflon film, to allow the light from the radiation source (115 black lamps and sun lamps) to fully irradiate the chamber volume. Because of the Teflon film wall, the chamber pressure must be maintained close to atmospheric pressure. As air is withdrawn from the chamber by the measurement systems, it is replaced by ultra high purity (UHP) air from two high capacity clean air generators (Aadco, Inc., model 737). The rate of dilution of the chamber contents is measured using the inert tracer sulfur hexafluoride (SF_6), which is determined by gas chromatography (Shimadzu Mini II GC with electron capture detector). The chamber contains a mixing fan which was operated at low speed to assure uniform mixing of the gases in the chamber.

Hg^0 was measured by two different approaches. For some experiments, it was observed that one Hg^0 measurement method produced more reliable results then another, as described in more detail elsewhere. (Spicer, C. et. al., 2004) In these cases, the kinetic data were analyzed using the more reliable Hg^0 measurement method. Hg^0 was measured every five minutes using a Tekran 2537A vapor phase mercury analyzer. This instrument preconcentrates Hg from the sample air stream on one of two parallel gold traps and thermally desorbs the trap to a cold vapor atomic fluorescence spectrometer (CVAFS). The measurement approach has been described by Lindberg et al. (2000). We incorporated a Teflon filter (Whatman 7592-104) and a soda lime trap in the inlet line to ensure the instrument measured only Hg^0 during these experiments.

Hg^0 was also monitored continuously using a Lumex Model RA-915$^+$. This instrument employs differential Zeeman atomic absorption spectroscopy with high frequency modulation of the polarized light source. It provides rapid real-time response, with a range to 50,000 ng m^{-3} and a detection limit of 2 ng m^{-3} (ng m^{-3} = 0.1 part per trillion (ppt)). The sampling rate of the Lumex instrument was reduced to 3-5 L min^{-1} to lower the rate at which the chamber was diluted with clean air. The calibrations of the Tekran 2537A and the Lumex RA-915$^+$ were checked before most experiments by a Tekran Model 2505 primary calibration unit and an air stream containing a known amount of Hg from a low rate permeation device (VICI Metronics), respectively. The permeation rate of the device was confirmed by sampling emissions from the tube for one hour into dilute $KMnO_4$ solution and analyzing the solution for Hg by ICP-MS.

In addition to measurements of Hg^0, RGM was collected at either 5 or 10 L min^{-1} using a KCl-coated annular denuder with an impactor inlet and thermally desorbed into the Tekran 2537A at 500°C as described by Landis (2002). For some experiments, Hg(p) was collected onto pre-fired quartz filters positioned downstream of the denuder and similarly thermally desorbed at 800°C into the Tekran 2537A. Particles larger than 2.5 micron in diameter were removed by the impactor inlet and not included in the Hg(p) measurements. RGM and Hg^0 deposition to the chamber surfaces were estimated by exposing 47 mm diameter quartz filters to the reaction mixtures in the environmental chamber during each experiment. The filters were thermally desorbed into the Tekran 2537A at 800°C and the mercury content extrapolated to the surface area of the environmental chamber. It should be noted that only the geometric surface area of the filter was considered. This treatment excludes any additional surface area due to the fibrous surface and pores of the filter. It is also expected that the Hg deposition rate to the Teflon wall material would differ from that to the quartz filter; as such, the measure of deposited Hg to the quartz filters is a first order approximation of the Hg deposited to the chamber walls.

Br₂ and Cl₂ were monitored using atmospheric pressure chemical ionization tandem mass spectrometry (Perkin Elmer-Sciex API 365) in the negative ionization mode. The measurement approach for the halogens has been described elsewhere. (Spicer et al., 1998; Spicer et al., 2002a) The instrument was calibrated using emissions from certified permeation devices for Br_2 and Cl_2 (VICI Metronics). Dimethyl sulfide was also measured by APCI-MS/MS for some experiments, but using benzene as a charge transfer agent in the positive ionization mode (Kelly 1991). A certified permeation tube (VICI Metronics) was used for calibration. For most experiments in which dimethyl sulfide was monitored, a Nafion drier was used at the inlet to the APCI-MS/MS instrument to dry the sample to minimize the effect of water vapor on the sensitivity for dimethyl sulfide. For experiments conducted in 2004, dimethyl sulfide was measured in canister samples by GC-MSD.

Ozone was monitored by the chemiluminescence from the reaction of O_3 with ethene (Bendix Model 8002) or by UV photometry (Thermo Electron Model 49). Although the UV photometry analyzer, which measures O_3 absorbance at a mercury emission line (254 nm), can suffer from Hg interference, the interference effect was negligible under the conditions of these experiments. Both O_3 monitors were calibrated using an O_3 generator and independent UV photometer (Dasibi Model 1008).

Dimethyl sulfide and n-butane were measured by collection of air samples from the chamber in evacuated 1- or 6-L canisters followed by GC-MSD analysis per Method TO−15 EPA/625/R-96/010b, (1999). Nitric oxide was measured by chemiluminescence using a Monitor Labs Model 8840 or Thermo Electron Model 42-S. Temperature and relative humidity (RH) were monitored by a thermocouple and condensation on a chilled mirror respectively (EG&G Model 911). Size-segregated particles were counted by optical light scattering (Climet Innovation CI-500).

Test atmospheres were prepared in the 17.3 m³ environmental chamber after purging overnight with UHP air. Hg concentrations between approximately 50 and 180 ng m^{-3} were achieved prior to each experimental run. For some experiments, water vapor was added to the chamber to achieve the desired RH by passing UHP air through a baffled water bath containing heated, distilled, deionized water, before entering the chamber. Hg^0 was introduced into the chamber via a timed injection from a 5 cm high emission rate permeation tube (VICI Metronics) at 100°C. Br_2, was injected into the chamber in an air stream that passed over a 10 cm high emission rate permeation tube (VICI Metronics) at 30°C. Cl_2 (5830 parts per million (ppm), Scott Specialty Gas) and F_2 (0.1% F_2 in argon, Nova Gas Technologies) were injected using gas-tight syringes or known-volume transfer vessels. Sulfur hexafluoride (99.8%, Scott Specialty Gas) and n-butane (99.8%, Matheson) were injected using a gas-tight syringe. A gas tight syringe was also used to transfer a known volume of dilute dimethyl sulfide to the chamber from a compressed gas cylinder (50 ppm, Scott Specialty Gas). Nitric oxide was injected from a 1000 ppm NO in N_2 cylinder (Matheson). The dilute NO was passed through a rotameter and into a stream of N_2 to provide further dilution before mixing with the air in the chamber. Ozone was produced by an electrostatic generator (PSI Corp.) and introduced into the chamber in a flowing air stream. All chemicals were used as received.

RESULTS AND DISCUSSION

Experiments were carried out to estimate rate constants for the reactions of Hg^0 with the molecules O_3, Cl_2, Br_2, F_2, and with NO_3. Experiments with the free radical NO_3 employed a relative rate approach to determine the rate constant. We used a reference compound whose rate constant for reaction with NO_3 is known. The removal rate of Hg^0 relative to the reference compound is used to estimate the rate constant of Hg^0 with the atom or free radical. The removal rate of Hg^0 and the reference compound may include loss due to the chemical reaction of interest, as well as loss due to dilution and deposition on the chamber surfaces. Dilution affects all species equally, but wall loss rates can vary from chemical to chemical and must be evaluated for both the target reactant and the reference compound. Dimethyl sulfide $((CH_3)_2S)$ was used as the reference compound in these experiments. Under the conditions of these experiments loss rates for Hg^0 and $(CH_3)_2S$ measured in clean air in the dark chamber were indistinguishable from the dilution rate.

The relative rate approach requires knowledge of the rate constant of the reference chemical with the reactants of interest. The rate constant used for this purpose for the reaction of NO_3 with $(CH_3)_2S$ was 1.0×10^{-12} $cm^3molecule^{-1}s^{-1}$.(DeMore et al., 1997).

For the other reactants, the absolute reaction rate was determined using either measured (O_3 and Cl_2) or calculated (Br_2 and F_2) initial reactant concentrations. An example of data obtained during this study is shown in Figure 1. The figure shows the removal rate of the dilution tracer SF_6 and Hg^0 as monitored by the Lumex instrument. For each species, the natural logarithm of the instrument signal is plotted as a function of time. In this example, the initial slope of the Hg^0 line after fluorine addition is used to define the overall loss rate of Hg^0, and the slope of SF_6 determines the dilution rate. Uncertainty in the measured rate constants were calculated as the 95% confidence interval for a single experiment or as the standard deviation of the mean rate constant.

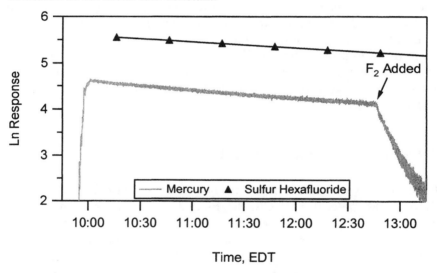

Figure 1. Example plot of Ln SF_6 and Ln Hg^0 for reaction of Hg^0 and F_2.

Reaction of Hg^0 with O_3

The reaction of O_3 with Hg^0 (Reaction 1) was examined in four experiments, two of which were conducted when the chamber surfaces were fully lined with Teflon. An electrostatic O_3 generator was used to inject O_3 into the chamber. Because these generators are known to produce trace concentrations of nitrate radical (NO_3) which can react with Hg^0, (Spicer, C. W., 2002) 1 ppm of n-butane was added to the chamber before adding O_3, to scavenge any NO_3 radicals. In the first experiment, 48 ng/m^3 of Hg^0 reacted with 75 parts per billion (ppb) O_3 in clean dry air in the dark. No change in Hg^0 removal rate was detected after O_3 was mixed into the chamber, even after four hours of reaction time. Based on our ability to detect small changes in the rate of Hg^0 removal above the dilution rate, we can estimate an upper limit for Reaction 1 of $k_1 \leq 2.5 \times 10^{-18}$ cm^3-molecule^{-1}s^{-1}. Using a slightly higher O_3 concentration (1 ppm) and comparable Hg^0 (45 ng/m^3), a k_1 of 9×10^{-19} cm^3-molecule^{-1} s^{-1} was measured.

$$Hg^0 + O_3 \rightarrow \text{Products} \qquad\qquad 1$$

Two additional investigations of Reaction 1 were carried out with higher O_3 mixing ratios (9.5 and 14 ppm) in order to drive Reaction 1 at a rate that could be clearly distinguished above the dilution rate. The conditions and results of these experiments are summarized in Table 1. For the first experiment, O_3 was added to the chamber followed by mercury vapor to start the reaction. An initial denuder sample was taken and the quartz filters (surrogate surfaces) were installed in the chamber before the addition of Hg^0. Upon completion of the experiment, a final denuder sample was collected and the quartz filters removed from the chamber.

Table 1. Experimental conditions and measured rate constants for the reaction of Hg^0 with O_3.

Initial Hg^0 (ng/m^3)	Initial O_3 (ppmv)	k_1 (cm^{3}mol^{-1}s^{-1})	Reacted Hg(ng)	RGM (ng)	Deposited Hg (ng)	Hg(p) (ng/m^3)	Difference (ng)[a]
48	0.075	$\leq 2.5 \times 10^{-18}$	NM[b]	NM	NM	NM	NM
45	1	9×10^{-19}	NM	NM	NM	NM	NM
122[c]	14	4.7×10^{-19}	583	90	152	NM	341
130[c]	9.5	5.6×10^{-19}	1401	24	405	1	971

[a] Difference = Reacted Hg - [RGM + Deposited Hg + Hg(p)]
[b] NM = not measured.
[c] Chamber walls fully lined with Teflon during this experiment.

The final experiment was designed to minimize the Hg analyzer's exposure to the high O_3 concentrations employed in these experiments since the high O_3 concentrations in the chamber appeared to negatively impact the collection efficiency of the gold traps in one of the Tekran analyzers (passivation), as confirmed by the addition of Hg vapor to the injection port of the analyzer. For this experiment, Hg^0 and n-butane (NO_3 sink) were added to the chamber and the initial denuder sample collected. Tekran data were obtained both with and without soda lime traps. The mercury analyzers were disconnected from the chamber, the dilution air flow reduced, and an

O_3 monitor connected to the chamber. Ozone was added to the chamber to a concentration of 9.5 ppm; once the concentration had stabilized, the O_3 monitor was disconnected from the chamber and all dilution air stopped. After a reaction time of approximately three hours, the Hg^0, O_3, and SF_6 analyzers were connected to the chamber. Tekran data were again obtained both with and without the soda lime traps.

The data in Table 1 also show that, for the final two experiments, a significant amount of Hg^0 was lost during the experiment that cannot be accounted for by dilution. Of that "reacted Hg," a fraction (30-40%) could be accounted for as RGM, deposited RGM, or Hg(p); 60-70% of the reacted mercury could not be accounted for using the available measurement data. It should be noted that the value for deposited RGM was extrapolated from the small surface area of the filters to the large surface area of the chamber walls and that differences in the sticking coefficient of Hg to quartz and Teflon could account for this discrepancy. In any case, the data do indicate that deposition of the reaction products (RGM) is significant during these experiments.

The recommended value for k_1 measured during these experiments is $6.4 \pm 2.3 \times 10^{-19}$ cm^{3}molecule^{-1}s^{-1}, where the estimated uncertainty is the standard deviation of the mean rate constant. Our findings are compared to previously reported values for k_1 in Table 2. Within the broad range of reported values, our results are consistent with previous studies, particularly with the values reported by Slemr et al. (1985) and Pal and Ariya (2004).

Table 2. Reported values of $Hg^0 + O_3$ rate constant.

k_1 Estimate (cm^3 molecule^{-1}s^{-1})	Reference
4.2×10^{-19}	Slemr et al., 1985
1.7×10^{-18}	Iverfeldt and Lindqvist 1986
4.9×10^{-18}	Schroeder et al., 1991
3×10^{-20}	Hall 1995
$7.5 \pm 0.9 \times 10^{-19}$	Pal and Ariya 2004
$6.4 \pm 2.3 \times 10^{-19}$[a]	This work

[a] Estimated uncertainty reported as the standard deviation of the mean rate constant.

Reaction of Hg^0 with Br_2

Reaction 2 was investigated through two chamber experiments. The first experiment employed initial conditions of 93 ng m^{-3} Hg^0 and a nominal Br_2 concentration of 2 ppb at 35% RH. No loss of Hg^0 was observed above the dilution rate. After two hours of reaction, the concentration of RGM in the chamber was the same as background levels within experimental uncertainty.

$$Hg^0 + Br_2 \rightarrow Products \qquad\qquad 2$$

The second experiment was carried out with initial conditions of 55 ng m^{-3} Hg^0 and approximately 1 ppb Br$_2$ at 30% RH. One hour of reaction showed no evidence of Hg^0 loss beyond that due to dilution. Measurements of RGM, Hg(p), and deposited Hg were not conducted for the reaction of Br$_2$ with Hg^0.

The results of these experiments are consistent with the findings of Ariya et al. (2002) who report an upper limit for the room temperature rate constant, k_2, for the reaction of Br$_2$ with Hg^0 ($<0.9 \pm 0.2) \times 10^{-16}$ cm^{3}molecule^{-1}s^{-1}; these experiments were conducted in small (0.1 to 5 L) cells for which surface effects are expected to be important. The conclusion that the gas phase reaction of Br$_2$ with Hg^0 is atmospherically unimportant is consistent with the results of this study and those reported by Ariya et al. (2002).

Reaction of Hg^0 with Cl$_2$

During a preliminary scoping study, (Spicer et al., 2002b) one environmental chamber experiment was conducted to explore Reaction 3:

$$Hg^0 + Cl_2 \rightarrow Products \qquad\qquad 3$$

A rate constant of 5×10^{-17} cm^{3}molecule^{-1}s^{-1} was estimated for this reaction, whereas values for this rate constant from unpublished work by Calhoun and Presbo cited in Seigneur et al. 2001 and Hall et al., 1991 as interpreted by Sliger et al., 2000 ranged from 4.0 10^{-18} cm^{3}molecule^{-1}s^{-1} to 5.6 $\times$ 10^{-15} cm^{3}molecule^{-1}s^{-1}. Recently, (Ariya et al., 2001) reported a value for this rate constant of 2.6 $\pm$ 0.2 $\times$ 10^{-18} cm^{3}molecule^{-1}s^{-1}. The present study has attempted to reduce the uncertainty in this range of rate constants by conducting experiments in a much larger reaction vessel at much lower concentrations. For the present study, the chamber was lined with FEP Teflon film to minimize removal of halogens on the walls, and a surrogate surface was employed to estimate mercury deposition to the chamber surface.

Three experiments were carried out to study the Reaction 3. In addition, three other experiments that employed Cl$_2$ as a precursor of chlorine atoms can provide some information on the Cl_2 + Hg^0 reaction. Results from the first three experiments are given in Table 3. Initial Hg^0 concentrations ranged from 75 to 180 ng/m^3 (9 to 22 ppt) and initial Cl_2 from 10 to 20 ppb. In separate experiments the Cl_2 loss rate in the Teflon-lined chamber was shown to be 0.198 h^{-1}, including a dilution rate of ~0.13 h^{-1}, so the stability of Cl_2 in the dark chamber was much improved over the scoping study. Reaction times for the three experiments listed in Table 3 were between 4 and 5 hours.

Table 3. Experimental conditions and measured rate constants for the reaction of Hg^0 with Cl_2.

Initial Hg^0 (ng/m^3)	Initial Cl_2 (ppbv)	k_1 (cm^{3}mol^{-1}s^{-1})	Reacted Hg (ng)	RGM (ng)	Deposited Hg (ng)	Hg(p) (ng/m^3)	Difference (ng)[a]
50	3	5×10^{-17}	NA	NA	NA	NA	NA
180[b]	10	9.8×10^{-18}	16	0.3	BDL	NA	15.3
80[b]	10	$<1.6 \times 10^{-17}$	0	—	BDL	NA	—
75[b]	20	[c]	0	0.2	16	NA	—

[a] Difference = Reacted Hg - [RGM + Deposited Hg + Hg(p)]
[b] Chamber walls fully lined with Teflon during this experiment.
[c] Mercury monitors yielded inconsistent results.

Two features of the data in Table 3 suggest that the reaction of Hg^0 with Cl_2 is very slow. First, the rate constants that could be estimated were 9.8×10^{-18} cm^{3}molecule^{-1}s^{-1} and an upper limit estimate of $<1.6 \times 10^{-17}$ cm^{3}molecule^{-1}s^{-1}. A rate constant could not be determined for the third experiment because of inconsistencies with the mercury monitors.

The second factor that indicates a very slow reaction is the mass of reacted Hg shown in the table. This is calculated using the initial Hg^0 concentration and the measured dilution rate to estimate the mass of Hg (in any form) that should be present at the end of the 4-5 hour reaction period, and subtracting the residual Hg measured at the end of the reaction period. In two of the three experiments all of the Hg was accounted for as residual gaseous Hg^0, and in the other experiment only a trace amount of Hg (16 ng or 0.5% of the starting mass) may have reacted.

One of the significant and previously unrecognized difficulties in conducting these experiments with Cl_2 is a problem with the Tekran analyzers in the configuration we utilized. Figure 2 shows the Hg concentrations for one of the experiments, reported by the Tekran and the Lumex instruments. Cl_2 was mixed into the chamber in the dark at 13:12 EST.

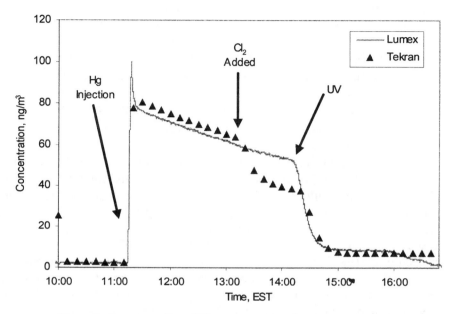

Figure 2. Response of two different mercury analyzers to mercury
in the presence of ppb Cl_2 levels in air.

The mercury concentration reported by a Lumex instrument showed no dramatic change when Cl_2 was injected into the chamber, but the Hg^0 level measured by the Tekran dropped from 63 ng m^{-3} to below 50 ng m^{-3} in just a few minutes. Additional studies of the effect of Cl_2 on the mercury response of the Tekran analyzer are reported elsewhere. (Spicer et al., 2004). Because of this phenomenon, the results from the Lumex instrument were given greater credence than the Tekran analyzer observations for experiments with Cl_2.

In addition to the three experiments dedicated to the Cl_2 reaction with Hg^0, two other experiments provided data that can help elucidate this reaction. These experiments were carried out to study the reaction of atomic chlorine with Hg^0 reported elsewhere. (Spicer et al., 2004) The photolysis of Cl_2 was employed to generate Cl atoms, and we were able to measure the rate of Hg^0 loss in the dark prior to the initiation of photolysis. For this set of experiments, higher Cl_2 concentrations were employed, and the reaction times were shorter, approximately 1 hour. Also, because the focus of these experiments was on the Cl reaction, there was not time to collect the denuder samples and surrogate surface samples necessary to characterize RGM and deposited Hg for the Cl_2 reaction. The results from the two experiments of opportunity for $Hg^0 + Cl_2$ are shown in Table 4. The two estimated rate constants are 2.5×10^{-18} and $<5 \times 10^{-18}$ cm^{3}molecule^{-1}s^{-1}.

Table 4. Results for $Hg^0 + Cl_2$ reaction from the Two Experiments of Opportunity.

Initial Hg^0 (ng/m^3)	Initial Cl_2 (ppbv)	k_3 (cm^{3}mol^{-1}s^{-1})
65	101	$<5 \times 10^{-18}$
105	67	$2.5 \pm 0.9 \times 10^{-18(a)}$

[a] Estimated uncertainty reported as the 95% confidence interval.

In summary, all of the rate constants obtained in these recent experiments are lower than the value estimated during the preliminary experiment (5×10^{-17} cm^{3}molecule^{-1}s^{-1}). The new rate constants were obtained under experimental conditions that were better controlled than the earlier test, but there is still considerable scatter in the rate constant values reported in Tables 3 and 4. The estimates range from 2.5×10^{-18} to $<1.6 \times 10^{-17}$ cm^3 molecule^{-1}s^{-1} (more than a factor of 6). In the current experiments, the combination of low reactant concentrations and the slow rate constant resulted in a very slow reaction rate that could not be measured with great precision under these conditions. There is considerable scatter in our estimate of k_3, but the result from our most reliable experiment (Table 4) with a value of $2.5 \pm 0.9 \times 10^{-18}$ cm^{3}molecule^{-1}s^{-1} (uncertainty reported as the 95% confidence interval) is reasonably consistent with the Ariya et al. (2002) estimate of $2.6 \pm 0.2 \times 10^{-18}$ cm^3 molecule^{-1} s^{-1}.

Reaction of Hg^0 with F_2

The reaction of Hg^0 with F_2, Reaction 4, was studied in four experiments in the fully Teflon-lined chamber under dry conditions. Once the desired quantities of Hg and SF_6 were added to the chamber, a 0.1% F_2 (in argon) mixture was added to the chamber using a known-volume transfer vessel. The Lumex Hg^0 measurement data for the first experiment are shown in Figure 3. The Hg^0 decay resulting from the addition of F_2 is clearly visible in the figure. We were not able to monitor the F_2 concentration using APCI-MS/MS, so the initial F_2 concentrations were calculated based on the volume of the F_2 mixture added to the chamber. The experimental conditions and measured rate constants for studies of Reaction 4 are summarized in Table 5.

$Hg^0 + F_2 \rightarrow$ Products   4

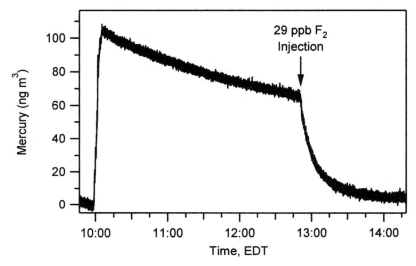

Figure 3. Lumex Hg^0 measurement showing a 29 ppb F_2 injection.

Table 5. Results from Environmental Chamber Study of $Hg^0 + F_2$ Reaction.

Initial Hg^0 (ng/m^3)	Initial F_2 (ppbv)	k_1 (cm^{3}mol^{-1}s^{-1})	Reacted Hg (ng)	RGM (ng)	Deposited Hg (ng)	Hg(p) (ng/m^3)	Difference (ng)[a]
78	29	1.9×10^{-15}	721	12	146	10	553
66	4	1.3×10^{-15}	910	2	177	BDL	731
44	4	1.6×10^{-15}	[b, c]				
140	$(3 \times) \sim 4$ [d]	[d]	$716^{(e)}$	6	1078	2	-370
45	29	2.2×10^{-15}					
122	4	2.1×10^{-15}	688 $^{(c,e,f)}$	3	1034	28	-377
90.55	$(2 \times) 29$	[e]					

[a] Difference = Reacted Hg - [RGM + Deposited Hg + Hg(p)]
[b] The denuder sample and quartz filters were collected before the second F_2 injection.
[c] UV irradiation occurred during this experiment.
[d] Injections were only partly successful. No rate constant could be measured.
[e] Multiple F_2 injections were made during this experiment.
[f] Too few data points available for reliable measurement

During several of the experiments, F_2 was injected more than once. For each injection, it was assumed that all F_2 from the previous injection was no longer present in the chamber (either due to chemical transformation or deposition) since the Hg decay quickly returned to approximately the dilution rate after each injection. The measured rate constants from each experiment are quite consistent, with an average k_4 of 1.8×10^{-15}

$cm^3 molecule^{-1} s^{-1}$. There are no known measurements of this rate constant in the literature to which this value may be compared. For each experiment, a substantial quantity of Hg was removed due to reaction, as shown in Table 5. Most of the recovered Hg was in the deposited form, but for the first two experiments, only 15-20% of the reacted Hg could be accounted for.

At the end of the last two experiments, the reaction of fluorine (F) atoms with Hg was probed by irradiating the chamber with Hg and F_2 present. One of these reactions is shown in Figure 4, which shows the Lumex trace during an injection of 29 ppb F_2 followed quickly by UV irradiation. As soon as the UV lamps were turned on, the Hg decay stopped. We believe that this is due to the very rapid reaction of F atoms with water vapor. The experiment shown in Figure 4 was conducted under very dry conditions (RH < 10%), but even under these conditions the lifetime of F atoms with respect to reaction with water vapor is less than 10^{-6} seconds. Consequently, even at very low ambient humidities, and F atoms that may be formed in the atmosphere will react with waver vapor and the reaction of F atoms with Hg^0 in the atmosphere cannot be important.

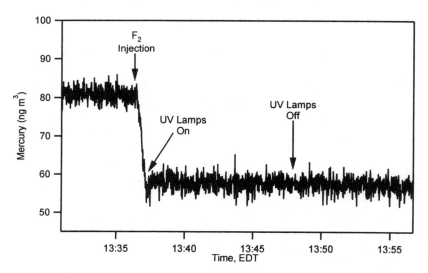

Figure 4. Lumex data during a F_2 injection followed by UV irradiation.

Reaction of Hg^0 with NO_3

During some initial tests in which O_3 was added to the chamber, the dimethyl sulfide concentration decreased more rapidly than by dilution in the dark without halogen species present. The cause was the presence of trace levels of NO_x in the chamber, which react with O_3 to produce NO_3. The NO_3 in turn reacts rapidly with dimethyl sulfide via reaction 5:

$$(CH_3)_2S + NO_3 \rightarrow C_2H_5S + HNO_3 \qquad 5$$

This presented the opportunity to estimate the rate constant for the reaction of Hg^0 with NO_3 (Reaction 6) utilizing dimethyl sulfide as the reference compound and the recommended rate constant[12] for reaction 5 of $k_5 = 1.0 \times 10^{-12}$ cm^{3}molecule^{-1}s^{-1}.

$$Hg^0 + NO_3 \rightarrow Products \qquad 6$$

We have used the relative rate approach, with dimethyl sulfide as the reference compound, to assess the reaction of NO_3 with Hg^0 for portions of experiments in which NO_3 was present. The three determinations of k_6 (all upper limits) are $< 7 \times 10^{-15}$, $< 1.3 \times 10^{-14}$, and $< 3 \times 10^{-14}$ cm^{3}molecule^{-1}s^{-1}. The lowest upper limit for k_6 of $< 7 \times 10^{-15}$ is within a factor of two of the upper limit reported by Sommar et al. (1997) for k_6 of $< 4.0 \times 10^{-15}$ cm^{3}molecule^{-1}s^{-1}. Additional experiments are recommended to decrease the uncertainty in k_6.

SUMMARY

This project has added to the current knowledge of atmospheric reactions of mercury through studies of selected reactions in a large environmental chamber using low concentrations of Hg^0 and other reactants. The reactants targeted for study include O_3, the molecular halogens Br_2, Cl_2, and F_2, and NO_3. The experiments with NO_3 made use of the relative rate method for estimating the rate constant for reaction with mercury. Table 6 provides the rate constants estimated in this study using our best judgment of the most reliable experiments to include in the estimate.

Increases in RGM were observed in several of our experiments. This provides clear evidence that RGM can be produced by atmospheric transformations under ambient conditions.

Table 6. Rate constants estimated during this study.

Reaction	Best Value for k $(cm^3 molecule^{-1} s^{-1})^{(a)}$
$Hg^0 + O_3$	$6.4 \pm 2.3 \times 10^{-19}$
$Hg^0 + Br_2$	No reaction detected
$Hg^0 + Cl_2$	$2.5 \pm 0.9 \times 10^{-18(b)}$
$Hg^0 + F_2$	$1.8 \pm 0.4 \times 10^{-15}$
$Hg^0 + NO_3$	$< 7 \times 10^{-15}$

(a) Estimated uncertainty reported as the standard deviation of the mean rate constant.
(b) Estimated uncertainty reported as the 95% confidence interval from the most reliable experiment.

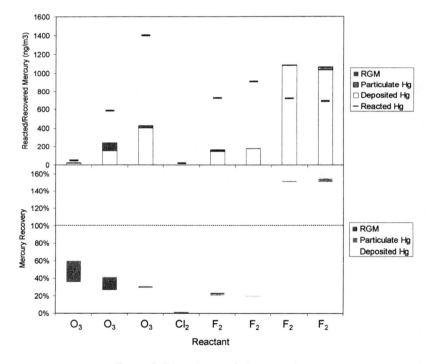

Figure 5. Mercury mass balance results.

RGM is known to be adsorptive and the losses due to deposition to the walls of our chamber were estimated in several of the experiments.

In addition, Hg(p) was measured for some experiments.

Because the amount of RGM that is lost during an experiment and during the subsequent RGM sampling period is unknown, our measurements should be considered as somewhat qualitative reports of the fraction of reacted Hg^0 that was converted to RGM. The question of RGM yield and the identification of RGM constituents from the different reactions is an important area for future research. Figure 5 shows results of Hg mass balance calculations based on the experimental data presented in Section 3 for experiments in which RGM, deposited, and in some cases Hg(p) were measured. The top panel shows the absolute reacted Hg (black line) and the recovered Hg (RGM, Hg(p), and deposited) are stacked bars to indicate the fraction of the reacted Hg that was recovered in each experiment. The bottom panel shows the same data, but as the percent Hg recovered relative to the reacted Hg. For some of these experiments, Hg(p) was also measured. Except for two of the F_2 experiments, not all of the initial Hg mass could be identified at completion of the experiment.

The importance of the gas phase reactions studied here to the oxidation of Hg^0 can be addressed by estimating the Hg lifetime with respect to each reaction. Table 7 shows estimated Hg^0 lifetime for each reaction along with the assumed oxidant concentration used for the estimate. Based on these results, of the reactants studied here, ozone and the nitrate radical are the most likely to influence the Hg^0 lifetime in the atmosphere and play a role in the transformation of Hg^0 to RGM, or other forms (i.e., deposited or Hg(p)). The relative importance of each reactant may change significantly in remote locations, such as polar regions, and other reactants (e.g. Br, BrO) may become important. We have reported preliminary results (Spicer et al., 2002b; Spicer et al., 2004, for the reaction of Hg^0 with Br, Cl, BrO, and ClO and additional studies are underway.

Table 7. Estimated Hg0 Lifetime due to Reaction.

Reactant	Atmospheric Mixing Ratio	k $(cm^3 molecule^{-1}s^{-1})$	Lifetime
O_3	35 ppb	6.4×10^{-19}	23 days
Cl_2	10 ppt	2.5×10^{-18}	56 years
F_2[a]	1 ppt	1.8×10^{-15}	0.8 years
NO_3[b]	20 ppt	$<7 \times 10^{-15}$	>4 days

[a] No measurements of tropospheric F_2 are known.
[b] Only important in the absence of sunlight due to fast NO_3 photolysis.

ACKNOWLEDGEMENTS

The United States Environmental Protection Agency through its Office of Research and Development partially funded and collaborated in the research described here under contract 68-D-99-011, task order #25 to Battelle Memorial Institute. It has been subject to Agency Review and approved for publication. We gratefully acknowledge the Florida Department of Environmental Protection for partial funding of this work.

REFERENCES

Ariya, P.A., Khalizov, A., Gidas, A., Reactions of Gaseous Mercury with Atomic and Moleculer Halogens: Kinetics, Product Studies, and Atmospheric Implications, *J. Phys. Chem.*, 106, 7310-7320, 2002.

DeMore, W.B., Sander, S.P., Golden, D.M., Hampson, R.F., Kurylo, M.J., Howard, C.J., Ravishankara, A.R., Kolb, C.E., Molina, M.J. Chemical Kinetics and Photochemical Data for Use in Stratospheric Modeling. Evaluation number 12, JPL Publication 97-4, 1-266, 1997.

Hall, B., Schager, P., Lindqvist, O. Chemical Reactions of Mercury in Combustion Flue Gases, *Water, Air Soil Pollut.*, 56, 3-14, 1991.

Hall, B. The Gas Phase Oxidation of Elemental Mercury by Ozone, *Water, Air Soil Pollut.*, 80, 301-315, 1995.

Iverfeldt, A. Lindqvist, O., Atmospheric Oxidation of Elemental Mercury in the Aqueous Phase, *Atmos. Environ.*, 20, 1567-1573, 1986.

Kelly, T.J., Kenny, D.V. Continuous Determination of Dimethyl Sulfide at Part-Per-Trillion Concentrations in Air by Atmospheric Pressure Chemical Ionization Mass Spectrometry, *Atmos. Environ.*, 25A, 2155, 1991.

Landis, M.S., Stevens, R.K., Schaedlich, F., Prestbo, E. Development and Characterization of an Annular Denuder Methodology for the Determination of Divalent Inorganic Reactive Gaseous Mercury in Ambient Air, *Env. Sci. Technol.*, 36, 3000-3009, 2002.

Lin, C., Pehkonen, S.O. The Chemistry of Atmospheric Mercury: A Review, *Atmos. Environ.*, 33, 2067, 1999.

Lindberg, S.E., Vette, A., Miles, C., Schaedlich, F. Application of an automated mercury analizer to field speciation measurements: Results for dissolved gaseous mercury in natural waters. *Biogeochemistry*, 48, 237, 2000.

Pal, B., Ariya, P.A. Studies of ozone initiated reactions of gaseous mercury: Kinetics, product studies, and atmospheric implications. *Phys. Chem. Chem. Phys.*, 6, 572-579 2004.

Schroeder, W.H., Yarwood, G., Niki, H. Transformation Processes Involving Hg Species in the Atmosphere—Results from a Literature Survey, *Water, Air Soil Pollut.*, 56, 653, 1991.

Seigneur, C., Karamchandani, P., Lohman, K., Vijayaraghavan, K. Multiscale Modeling of the Atmospheric Fate and Transport of Mercury, *J. Geophys. Res.*, 106, 27797, 2001.

Slemr, F., Schuster, G., Seiler, W. Distribution, Speciation and Budget of Atmospheric Mercury, *J. Atmos. Chem.*, 3, 407, 1985.

Sliger, R.N., Kramlich, J.C., Marinov, N.M. Development of an Elementary Homogeneous Mercury Oxidation Mechanism, Towards the development of a chemical kinetic model for the homogeneous oxidation of mercury by chlorine species. *Fuel Process. Technol.*, 65, 423-438, 2000.

Sommar, J., Hallquist, M., Ljungstrom, E., Lindqvist, O. On the Gas Phase Reactions Between Volatile Biogenic Mercury Species and the Nitrate Radical, *J. Atmos. Chem.*, 27, 233-247, 1997.

Spicer, C.W., Chapman, E.G., Finlayson-Pitts, B.J., Plastridge, R.A., Hubbe, J.M., Fast, J.D., Berkowitz, C.M. Unexpectedly High Concentrations of Molecular Chlorine in Coastal Air, *Nature*, 394, 353, 1998.

Spicer, C.W., Plastridge, R.A., Foster, K.L., Finlayson-Pitts, B.J. Bottenheim, J.W., Grannas, A.M., Shepson, P.B. Molecular Halogens Before and During Ozone Depletion Events in the Arctic at Polar Sunrise: Concentrations and Sources, *Atmos. Environ.*, 36, 2721-2731, 2002a.

Spicer, C.W., Satola, J., Abbgy, A.A., Plastridge, R.A., Cowen, K.A. Kinetics of Gas-Phase Elemental Mercury Reactions with Halogen Species, Ozone, and Nitrate Radical Under Atmospheric Conditions, Battelle Columbus. Final Report to Florida Department of Environmental Protection under Contract AQ174, 2002b.

Spicer, C.W., Sumner, A.L., Satola, J., Mangaraj, R. Kinetics of Mercury Reactions in the Atmosphere. Battelle Columbus. Final Report to the U. S. Environmental Protection Agency under Contract 68-D-99-016, Task 0025, 2004.

U.S. Environmental Protection Agency, Mercury Study Report to Congress, EPA-452R-97-003, 1997.

U.S. Environmental Protection Agency Method TO-15, in Compendium of Methods for the Determination of Toxic Organic Compounds in Ambient Air, Second Edition. EPA/625/R-96/010b, 1999.

U.S. Environmental Protection Agency, The Mercury Research Strategy, EPA/600/R-00/73 2000.

Chapter-10

AIR-SEA EXCHANGE AND MARINE BOUNDARY LAYER ATMOSPHERIC TRANSFORMATIONS OF MERCURY AND THEIR IMPORTANCE IN THE GLOBAL MERCURY CYCLE

Robert P. Mason

Chesapeake Biological Laboratory, University of Maryland Center for Environmental Science, Solomons, MD 20688, USA

INTRODUCTION

The atmosphere is the most important pathway for the worldwide dispersion and transport of Hg (Fitzgerald et al., 1998; Mason et al., 1994). Understanding the global transport and atmospheric transformations of Hg is important because of the ability of Hg deposited to aquatic systems, to be converted to methylmercury (MeHg) and to bioaccumulate through all levels of the food chain. Most of the Hg in the atmosphere is elemental Hg (Hg^0), which is relatively unreactive with the net average atmospheric residence time of around one year. In addition to Hg^0, two other atmospheric Hg fractions have been operationally defined based on physicochemical properties - the gaseous ionic Hg^{II} fraction, which has been termed reactive gaseous Hg (RGHg), and particulate-bound Hg (Hg-P). The speciation of RGHg is not known in detail but based on laboratory studies and the methods of its collection (Landis et al., 2002; Sheu and Mason, 2001; Lindberg and Stratton, 1998; Ariya et al., 2002; Sheu and Mason, 2004), it is assumed to consist of gaseous neutral Hg^{II} complexes such as $HgCl_2$, $HgBr_2$, and HgOBr (Balabanov and Peterson, 2003; Kalizov et al., 2003). Such compounds are highly surface-reactive and substantially more water-soluble than Hg^0. Estimates of dry deposition velocities for RGHg for the open ocean range from 0.5-4 cm s^{-1} (Laurier et al., 2003), estimated using the formulation of Shahin et al. (2002) and are much higher than those for Hg-P (0.1-0.5 cm s^{-1}), which being mostly derived from high combustion sources, is associated with the fine particulate fraction (Bullock et al., 1997). In many locations, dry deposition could be as important as wet deposition in terms of being a Hg source to the Earth's surface. Global Hg models have identified wet and dry particle deposition and evasion of dissolved gaseous Hg from the ocean as critical pathways for global Hg cycling (Mason et al., 1994; Hudson et al., 1995; Lamborg et al., 1999; Shia et al., 1999).

Natural sources of Hg to the atmosphere are mainly in the form of Hg^0 although emissions of Hg-P also occur (e.g., volcanoes, dust) while anthropogenic sources contribute all forms of Hg to the atmosphere (Ebinghaus et al., 1999). In addition, *in situ* oxidation of atmospheric Hg^0 in the gas phase could be a source of RGHg as Hg^0 can be oxidized by hydrogen peroxide (H_2O_2), albeit slowly (Tokos et al., 1998), the nitrate (NO_3) radical (Sommar et al., 1997) and other reactive nitrogen intermediates, ozone (O_3) (Hall, 1995), and the hydroxyl (OH) radical (Sommar et al., 2001). Given the typical atmospheric Hg concentrations of these oxidants, it is unlikely that these homogeneous reactions dominate the Hg^0 oxidation in general, based on estimates of the rate that is required for its removal via wet and dry deposition. However, the gas phase oxidation of Hg^0 by halogen atoms and molecules (Cl, Br, Br_2, $Cl_{2;}$ Lin and Pehkonen, 1999; Sliger et al., 2000; Ariya et al., 2002), and potentially by other halogen compounds (e.g., BrCl, BrO), has been recently demonstrated. The reactions with atomic Cl and Br have the larger rate constants (Ariya et al., 2002) and therefore oxidation by these mechanisms may proceed much faster than oxidation by O_3 and OH, in certain locations such as the polar region, the marine boundary layer, and at high altitudes (Laurier et al., 2003; Lindberg et al., 2002; Landis and Stevens, see Chapter-7 in this book).

Measurements of Hg^0 depletion events in surface-level Arctic air during the three month period following polar sunrise (Schroeder et al., 1998) provided the first indication of the importance of halogen-mediated reactions in Hg^0 oxidation. The fluctuation of total atmospheric Hg (primarily Hg^0) strongly resembled the fluctuation of ambient O_3 concentrations during the same period (Schroeder et al., 1998), suggesting that both species were removed by similar mechanisms. Many more recent studies (Ebinghaus et al., 2002; Temme et al., 2003; Lindberg et al., 2002) have confirmed that such rapid Hg depletion events commonly occur during polar sunrise, being driven by photochemical reactions, in both the Arctic and Antarctic and that the depletion of Hg^0 coincides with the increase in RGHg concentrations (Lindberg et al., 2002). The destruction of O_3 during polar sunrise correlates with the production of reactive halogen species (RXS). It is currently thought that O_3 destruction is initiated by Br atoms and BrO, and to a lesser extent, by Cl atoms (McConnell et al., 1992; Barrie et al., 1988; Foster et al., 2001; Figure 1). with the sea-salt particles that deposit to and accumulate in the snow pack during winter being the major source of the precursors of Br and Cl atoms in polar regions, e.g., Br_2 and BrCl (Finlayson-Pitts et al., 1990; Foster et al., 2001). Similar reactions could oxidize Hg^0 to RGHg.

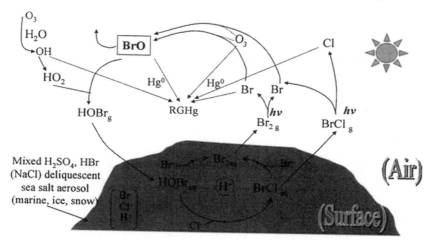

Figure 1. Proposed mechanisms for elemental oxidation in the marine boundary.

Sources of reaction Initiation include: Photolysis of CH_3br; dark reaction of O_3 with Br; and photoconversion of Br to gas phase Br_2 (Caro's acid oxidation).
Adapted from Vogt et al. (1996) and Sheu and Mason (2004).

The X_2 precursors of the RXS (Br, Cl, BrO, ClO), such as Br_2, BrCl, and Cl_2, are also liberated from sea-salt particles in the marine boundary layer (MBL), and the subsequent O_3 destruction, has also been measured and described (Mozurkewich, 1995; Vogt et al., 1996; Knipping et al., 2000; Galbally et al., 2000; Hirokawa et al., 1998) and observed in laboratory studies (Oum et al., 1998; Gabriel et al., 2002; Fickert et al., 1999). Additionally, high concentrations of Cl_2 have also been detected in coastal air (Spicer et al., 1998). An important intermediate product of the O_3 destruction chain reaction, BrO, was detected in the ambient air over the Dead Sea and showed an inverse correlation with O_3 (Hebestreit et al., 1999). BrO is the precursor to HOBr, which directly interacts with salt surfaces to regenerate RXS (von Glasow et al., 2002; Vogt et al., 1996; Figure 1). Dickerson et al. (1999) reported large diurnal variations in O_3

concentration in the marine boundary layer over the tropical Indian Ocean and found, in modelling their results, that the temporal trend of, and the magnitude of variation in, O_3 concentrations were more compatible with the field data when halogen chemistry and RXS formation was included. The presence of RGHg in the MBL in remote locations, and evidence of a diurnal variation in concentration (e.g., Laurier et al., 2003; Hedgecock et al., 2001; 2003) suggests that RGHg is being produced photochemically by reaction with RXS in conjunction with ozone destruction. Laboratory studies also confirm the potential importance of such reactions (Sheu and Mason, 2004; Ariya et al., layer, or in other regimes where there is the presence of halogen-containing aerosol. Principal oxidation reactions involve reactive halogen species such as Br and Cl. The hydoxyl radical is also directly or indirectly involved in mercury oxidation.

In addition to halogen-mediated Hg^0 oxidation at the Earth's surface, there is also accumulating evidence for Hg^0 depletion events in the upper atmosphere. Landis and Steven (see Chapter-7 in this book) discuss results from studies at Mauna Loa (3500m) and from aircraft studies off the coast of Florida (60-3500m). In earlier mass balance models it was assumed that dry deposition was not an important component. This was because, as discussed above, dry particulate Hg deposition is much less than wet deposition for most remote environments. However, scavenging of particulate Hg is likely to be an important contributor to wet deposition. For dry deposition of gaseous Hg, it is known that Hg^0 deposition does not normally occur. The more recent studies that have measured RGHg using accepted protocols have shown that, because of its high deposition velocity, RGHg deposition can rival that of wet deposition in locations where RGHg concentrations are in excess of about 20 pg m^{-3}. The relative importance of wet versus dry deposition will be discussed further in this chapter.

At the air/water interface, gas exchange of dissolved gaseous mercury (DGHg) is the main "sink", *via* evasion, for surface ocean Hg. DGHg consists of elemental Hg (Hg^0) and dimethylmercury (DMHg) (Kim and Fitzgerald, 1988; Mason et al., 1995; Cossa et al., 1997; Lamborg et al., 1999) with Hg^0 being the dominant form of DGHg in the surface ocean. Earlier studies focused on air-sea exchange for the open ocean, both the Atlantic (Cossa et al., 1997; Mason and Sullivan, 1999; Lamborg et al., 1999) and Pacific Oceans (Kim and Fitzgerald, 1988; Mason and Fitzgerald 1993). More recent studies have focused on coastal regions and the Mediterranean (Gardfeldt et al., 2003; Ferrara et al., 2003; Baeyens and Leermakers, 1998; Mason et al., 1999). Many studies have suggested that the estimated evasion rates for Hg from the ocean substantially exceed the current wet plus particulate dry deposition estimates and riverine inputs, suggesting another potential source for upper ocean Hg. One potential reason for this lack of balance is the relative paucity (both temporally and spatially) of DGHg data for the ocean and thus its potentially unrepresentative nature. Accordingly, it has recently been hypothesized that there is a substantial input of Hg to the ocean via dry deposition of RGHg (Mason and Sheu, 2002).

Based on the available evidence, it is proposed that salt particle-mediated halogen chemistry plays an important role in Hg chemistry in marine air, and in the deposition of Hg to open ocean and coastal waters. This chapter will review the details of our recent studies of Hg^0 oxidation in the marine boundary layer and compare and contrast these to data obtained by others in similar environments. The implications of such processes on the input of Hg to the ocean's surface waters is also discussed in the context of the overall air-sea exchange of Hg and the global Hg cycle.

MERCURY SPECIATION IN THE MARINE ATMOSPHERE AND OCEAN SURFACE WATERS

Mercury in the atmosphere is present in two oxidation states, as elemental Hg (Hg^0) and as ionic Hg^{II} species. The dominant form is Hg^0 which is typically >95% of the total (Schroeder and Munthe, 1998). Concentrations range from around 1-2.5 ng m^{-3}, typically, with higher concentrations present in impacted regions, and historically in some locations. Hg^0 is relatively insoluble (49.4 x 10^{-6} g/L at 20°C) and its major loss pathway from the atmosphere is through oxidation and subsequent removal by wet and dry deposition. As most water bodies are typically saturated with Hg^0 relative to the atmosphere, and the Henry's Law coefficient (H) for Hg^0 (729 Pa m^3 mol-1 at 20°C) is relatively high, dry deposition of Hg^0 to water is not an important removal mechanism. However, there is evidence for the uptake of Hg^0 by vegetation (St. Louis et al., 2001). Overall, based on global inter-hemispheric Hg concentration differences, the residence time of Hg^0 in the atmosphere is about a year (Fitzgerald and Mason, 1997). However, given the recent identification of the rapid transformation of Hg^0 in some locations, as discussed above, this estimated residence time likely reflects the net lifetime of Hg^0 which could be oxidized and deposited and then re-emitted to the atmosphere numerous times prior to final removal to the other reservoirs.

Particulate mercury (Hg-P) consists mainly of Hg(II) species adsorbed to surfaces or incorporated into particles during high level combustion and other processes. As a result, Hg-P resides predominantly in the fine particulate fraction (< 2.5 μm) fraction. The deposition velocity has been estimated to be <0.5 cm s^{-1} typically and thus given the relatively low Hg-P concentrations (<30 pg m^{-3} for locations away from point source inputs) (Fitzgerald and Mason, 1997; Bullock, 2000), the input of Hg via dry deposition of Hg-P is not a major component of the overall flux. However, Hg-P is also scavenged by wet deposition. The residence time of Hg-P has been estimated to be from days to weeks (Bullock, 2000). It should be noted that earlier measurements of Hg-P using filters could have potentially over-estimated the Hg-P concentration if substantial amounts of RGHg were trapped by the Hg-P filter. However, since measured Hg-P concentrations over the ocean (< 5 pg m^{-3} typically) are low and less than those of RGHg, such an artifact does not appear to occur to any significant degree (Mason et al., 1992; Lamborg et al., 1999).

Gaseous ionic mercury (Hg^{II}) or reactive gaseous mercury (RGHg) refers to a group of compounds, such as HgCl$_2$ and HgBr$_2$, that can exist in the atmosphere in relatively low concentrations. Such compounds are highly soluble (e.g., HgCl$_2$, 66 g/L at 20°C) and have a high deposition velocity – typically >1 cm s^{-1}. Given the solubility and the low H (3.7 x 10^{-5} Pa m^3 mol^{-1} at 20°C) for HgCl$_2$, it is possible to estimate the dry deposition velocity using the modelling approach of Shahin et al. (2002), which relates the deposition velocity (k, cm s^{-1}) to the diffusion coefficient of the species of interest (D$_A$; cm^2 s^{-1}) and the wind speed at 10 m (u$_{10}$, m s^{-1}), with k = D$_A^{0.5}$(0.98u$_{10}$ + 1.26). Values estimated for the Pacific Ocean during the 2002 cruise using this approach ranged from <1 to 4 cm s^{-1} (Laurier et al., 2003).

While RGHg has been demonstrated to be emitted from point sources (Porcella et al., 1997) it can also be produced in situ via oxidation reactions (Sheu and Mason, 2004; Ariya et al., 2002). Concentrations are typically <20 pg m^{-3} except in the region of point sources (Sheu et al., 2002), and in locations where active oxidation is occurring (polar sunrise, open ocean boundary layer, upper atmosphere) where concentrations >100 pg m^{-3} have been recorded (Lindberg et al., 2002; Landis and Steven in Chapter-7 of this book). Finally, MeHg exists in the atmosphere although its concentration in the air is low, and below the detection limit of most analytical methods (Mason and Benoit, 2003). It has been measured in wet deposition at low concentrations (<1% of total Hg, typically) and likely exists in the atmosphere at sub-pg m^{-3} levels, based on its concentration in rain.

Mercury in surface waters also exists in two oxidation states, Hg^0, as a dissolved gas, and Hg^{II}. The ionic Hg exists both in the dissolved phase and attached to particulate matter, both living (phytoplankton, zooplankton) and detritus. In addition, in most surface waters, a small fraction of the total Hg is MeHg^{II}, which also can be dissolved or particulate-associated. In the dissolved phase, both Hg^{II} and MeHg^{II} are complexed with inorganic and organic ligands with only a small fraction of each existing as the free metal ion (Mason and Benoit, 2003). The reactions leading to the transformation of Hg between the two oxidation states in surface waters can be both abiotically and biotically mediated, with the abiotic processes being mostly photochemically driven. Details of the reactions and the processes mediating them will be discussed in more detail below.

MEASUREMENT METHODS

Sampling Locations

The open ocean atmospheric data discussed in this chapter was primarily collected on two ocean cruises. The first cruise on the RV *Melville* was in the North and equatorial Pacific Ocean in May/June 2002 (Laurier et al., 2003). It started from Osaka, Japan ($34^0 65$'N, $135^0 42$'E) on May 1st 2002 and ended in Honolulu, Hawaii ($24^0 15$'N, $153^0 84$'W) on June 4th 2002. All the data discussed here were collected aboard from $47^0 83$'N, $162^0 50$'E to $22^0 75$'N, $158^0 00$'W during a two week sampling period (May 14th, May 30th). The cruise track during this time consisted of a north-south transect around 170^0E and then a west-east transect along 22-23^{0}N (Laurier et al., 2003). The second cruise took place in August 2003 aboard the *RV Cape Hatteras* from Norfolk, Virginia to Bermuda and then from Bermuda to Barbados. The cruise track was modified on route due to hurricane activity. Data from one period during the second leg of this cruise are presented.

In addition to the ocean atmospheric sampling, atmospheric sampling has been done at the Chesapeake Biological Laboratory (CBL), which is situated on the mouth of the Patuxent River, a tributary to the Chesapeake Bay, in Maryland, USA. This site has been used for prior atmospheric measurements and for wet deposition collection (Mason et al., 2000; Mason et al., 1997a; Sheu et al., 2002; Sheu and Mason, 2001). Additional databases that are used to help constrain concentrations for box modelling purposes are from previous studies off Bermuda (Mason and Sheu, 2002) and in other locations on the shores of the Chesapeake Bay (Sheu et al., 2002; Mason et al., 1997b).

Water samples were collected on the open ocean cruises discussed above as well as on previous ocean cruises, as summarized by Mason et al. (2001). For the current paper, the focus will be on surface water dissolved gaseous mercury (DGHg) measurements, which for surface waters comprise mostly

Hg^0. These data are used to constrain estimates of Hg gas exchange. Total Hg and methylated Hg measurements were also made during these cruises.

Sampling and Analytical Techniques

Ancillary data on the cruises were measured by others, as detailed below, and by the ship's on-board equipment. For example, meteorological parameters, air temperature, relative humidity, barometric pressure, wind speed and wind direction were measured by the RV *Melville*'s continuous underway data monitoring system with integration steps every 30 seconds. Ozone was measured using a Thermo Environmental Instruments, Model 49C analyzer which was calibrated by an ozone standard generator (Model 49CPS). Atmospheric Hg^0 and RGHg measurements were performed using the Tekran mercury speciation unit (Tekran 1130) coupled to a Tekran 2537A analyzer (Tekran Inc., Toronto, Canada), as described by Landis et al. (2002). The precision of the instrument based on field comparisons of paired 1130 speciation units is 15% (Landis et al., 2002). On the *Melville*, the speciation unit was set up on the front of the ship at ~15m above sea level. On the *Cape Hatteras*, the speciation denuder unit was set up on the top of the bridge also at ~15m, and the analytical equipment housed inside the bridge. At CBL, the Tekran is mounted on a pier over the water, 5-10 m above water level.

With the Tekran instrument, Hg^0 is determined using a vapor-phase mercury analyzer (Model 2537A) which consists of a gold amalgamation system coupled to a cold-vapor atomic fluorescence spectrometer (Bloom and Fitzgerald, 1988). The Tekran model 1130 denuder speciation unit was used for the RGHg measurements. The sampling time resolution was 5-min for Hg^0 and 2-h for RGHg. By pumping ambient air through the denuder, RGHg is adsorbed onto the KCl coated denuder while Hg^0 is quantified in the analyzer. After a 2-h sampling period, the denuder is heated to 500^0C and RGHg is thermally decomposed into mercury-free air and analyzed as Hg^0. The particulate mercury was collected on a downstream quartz fiber filter and was not analyzed. In the earlier collections at CBL and on Bermuda, a cation exchange filter was used to collect RGHg (Sheu and Mason, 2001). The cation exchange sampling approach and the KCl denuder approach were carefully tested at CBL, and on Bermuda, and shown to give similar results. Overall, results were within a factor of two which is acceptable given the low concentrations found during the comparisons (<50 pg m^{-3}) and the differences in collection period (2 hours for dender and 12 hours for the filter packs) (Sheu and Mason, 2001). The acid-cleaned 5-stage Teflon filter holders used for collection contained 2 Teflon filters, mounted in front to trap Hg-P, and 3 cation exchange membranes, used to collect the RGHg. All filters were analyzed for total Hg using conventional techniques i.e., BrCl oxidation, followed by $SnCl_2$ reduction, purge and trap collection and cold-vapor atomic fluorescence spectrometry (CVAFS) detection (Sheu and Mason, 2001).

The underway surface-water samples in the Pacific were collected using a "fish" sampler, which is a device used to sample in the surface water beside the ship while it is moving forward with the inlet away from the ship's contamination. Water was pumped on board the ship into a laminar flow hood using a peristaltic pump through acid-cleaned tubing. Such a system was also used on the previous cruises in the Atlantic (Mason et al., 2001) but not on the 2003 Atlantic cruise. For the 2003 cruise, the on-board seawater sampling system of the ship was used as it had been previously demonstrated to be suitable for contaminant-free Hg sampling (Mason et al., 2001). Samples were collected into 2 L Teflon bottles and were processed on board for DGHg immediately after collection, or at least within 3 hours after collection, in an onboard clean room. Sample collection and treatment were performed using ultra-clean techniques. All Teflon and plastic-ware had been acid washed prior to use. Cleaned Teflon bottles were stored filled with HCl (1% v/v) and double-bagged until use. DGHg concentrations were measured using a Teflon bubbler head attachment that fitted directly onto the 2 L Teflon bottles. Samples were bubbled for 40-min at 500 ml min^{-1} with Hg-free argon and the released DGHg was trapped on gold columns. Quantification, by CVAFS, was achieved by heating the gold trap in a stream of argon.

Laboratory Experiments

Photochemical oxidation experiments, both indoors and outdoors, were carried out, under static conditions, at CBL (Sheu and Mason, 2004) using high quality quartz cells as reactors. A 75W O_3 free Xenon lamp (ORIEL 6263) was used as the light source for the indoor experiments. For the outdoor experiments, the ambient sunlight was the light source. All the outdoor experiments were conducted on cloudless sunny days with moderate to high UV indices in June, 2001. Chemicals used in the experiments were prepared as discussed in Sheu and Mason 2004. A volume of 100 μL of gaseous Hg^0 was injected into the prepared cell at the start of the experiment. The cell was then irradiated. Each experiment therefore generated one data point. Dark controls involved wrapping the cells with aluminum foil.

Once the irradiation was stopped, the cell was flushed with high purity Argon (Ar) gas to collect the remaining gaseous Hg^0 in the reactor on a gold trap. Initial tests showed that none of the RGHg formed by oxidation was in the gas phase at the end of the experiment and so the cell was rinsed twice with 1M HCl solution to remove any Hg adsorbed to the glass (Sheu and Mason, 2004). Mercury was quantified by CVAFS using dual gold trap amalgamation (Bloom and Fitzgerald, 1988). The gold traps were analyzed directly. The operational detection limit (DL) for the gas phase was 0.05 pmole or <1% of the added Hg^0. The Hg content in the HCl rinses were quantified by $SnCl_2$-reduction CVAFS (Mason et al., 2001). Given typical DLs for the measurements of dissolved Hg of 0.1 pmole, this corresponds to a DL for oxidation of 1.5% conversion of the added Hg^0.

MERCURY OXIDATION IN THE MARINE BOUNDARY LAYER

Open Ocean Studies

The results from the cruise in the North Pacific in May-June 2002 demonstrated that elevated RGHg concentrations (up to 100 pg m^{-3}) exist in

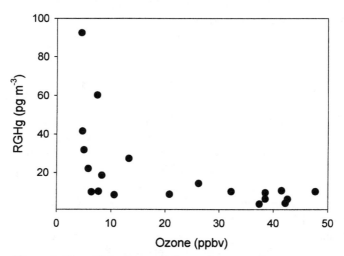

Figure 2. Plot of 2 hr average daily maximum reactive gaseous mercury concentration.

the marine boundary layer of the subtropical North Pacific and that the diurnal cycle in RGHg concentration coincides with that of UV radiation (Laurier et al., 2003).

The maxima in RGHg concentration also occurred under conditions where O_3 concentrations were low (<10 ppb), and coincided with the maximum of photochemical processes and maximum UV.

The relationship between the maximum daily (sunrise to sunset) two hour average RGHg concentration and the corresponding ozone concentration is shown in Figure 2.

The non-linear negative correlation suggests strongly that the factors leading to ozone depletion are those that result in significant RGHg formation. Furthermore, it appears that significant RGHg formation occurred only when ozone concentration was less than 10 ppbv (Laurier et al., 2003).

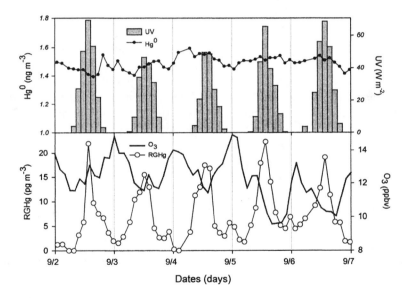

Figure 3. Concentrations of reactive gaseous mercury, elemental mercury and ozone, and UV radiation for 5 days (September 2-7, 2003) during the cruise on the RV Cape Hatteras in the North Atlantic.

Estimated rates of RGHg formation from the model and estimated dry deposition velocities (Laurier et al., 2003) are of the same order. The model results showed that formation rates of RGHg during the day were up to 1.2 x 10^{-2} ng m^{-3} hr^{-1}. For a 1000 m boundary layer, this corresponds to a production of 12 ng m^{-2} hr^{-1}. Given dry deposition velocities of 1-4 cm s^{-1} during the cruise (Laurier et al., 2003), and the observed concentrations of RGHg, up to 90 pg m^{-3}, the corresponding deposition flux is <1-10 ng m^{-2} hr^{-1}. Thus, dry deposition of RGHg would rapidly deplete the MBL if no formation was occurring. Clearly, deposition alone can account for the nightly decrease in observed concentration when formation is not occurring to any significant extent and reduction of RGHg species does not need to be invoked to explain the field results, nor has it been demonstrated in the lab or field. Furthermore, since deposition rates are related to both production rates and wind speed, then low concentrations in the atmosphere and/or a small magnitude change in RGHg concentrations over a diurnal cycle does not necessarily mean that there is not substantial formation of RGHg occurring, but that the rate of formation is similar to the rate of deposition.

In August 2003, samples were collected in a further field campaign on the *RV Cape Hatteras* from Norfolk, Virginia to Barbados, covering the temperate to subtropical North Atlantic. During the cruise there was evidence of a similar diurnal RGHg formation from with higher concentrations during the day. However, ozone concentrations were not as low as those encountered in the North Pacific 2002 cruise. Data for five days at the end of the cruise when the lowest ozone concentrations were encountered are shown in Figure 3. The UV radiation levels were similar to those found in the subtropical North Atlantic, but as comparison of Figures 2 and 3 demonstrates, ozone levels were not as low and never fell below 10 ppbv. There was, however, a daily minimum in ozone that corresponded to the daily maximum in RGHg. Overall, the highest RGHg concentration was about 4 times less than the maximum found in the Pacific. As mentioned above, the highest RGHg concentrations found in the North Pacific were during time of very low ozone (< 10 ppb) and indeed the magnitude of the RGHg values for the North Atlantic are comparable, for comparable ozone, with the North Pacific data (Figures 2 and 3). The average value was, however, about half that found during the period of maximum formation in the North Pacific.

In contrast, Mason et al. (2001) and Sheu et al. (2002) found higher levels of RGHg during cruises off Bermuda and from weekly integrated measurements on Bermuda using the filter pack collection techniques. While these values were somewhat higher than those measured more recently using the Tekran (Fig. 3), these results do provide additional data to support the presence of RGHg in the MBL. In another study, Ebinghaus et al. (1999) found an average value for RGHg of 18 ± 4 pg m^{-3} sampling air of oceanic origin at Mace Head, Ireland. Clearly, more studies are needed to examine in more detail the factors controlling the formation of RGHg in marine air. However, in summary, while the open ocean measurements are limited, they do provide distinct evidence for photochemical oxidation of Hg^0 in the MBL and, in conjunction with the results for the Mediterranean and those in coastal regions, discussed below, suggest that RGHg formation in the MBL is enhanced under conditions of extremely low ozone, which is similar to the results found in the polar regions (Lindberg et al., 2002).

Coastal Ocean and Enclosed Seas

Similar diurnal changes in RGHg in marine boundary layer air have been noted over the Mediterranean Sea during a cruise from 14 July to 5 August, 2000 in both the Eastern and Western Basins on the *R.V. Urania* (Hedgecock et al., 2003; Sprovieri et al., 2003). As discussed in detail elsewhere in this book, the first leg of the Mediterranean study was stormy with rain and high seas, and RGHg concentrations were less than 15 pg m^{-3}. In contrast, the second leg was much more typical of summertime Mediterranean weather. For this section of the cruise, higher RGHg concentrations were found, above 30 pg m^{-3}, and there was an obvious diurnal cycle in concentration with a maximum during the period of highest photochemical production (Sprovieri et al., 2003).

The southeast coast of the USA, as represented by the studies at CBL, on the shores of the Chesapeake Bay in Maryland, may represent another location where the importance of anthropogenic impacts cannot be ignored. Data now exists for a number of campaigns at CBL, starting with studies in 1997 where RGHg concentrations were measured using ion exchange membranes (Mason et al., 1997; Table 1). The filter pack method required typically 12 hours of deployment per sample so that the resolution possible with the automated Tekran instrument was not obtained but it is clear that the average values and ranges found for the two techniques are similar. The large standard deviation of the average found for the more recent measurements using the Tekran reflect the fact that a diurnal cycle in RGHg concentration is often found, and that the variability is over a longer time frame than the measurement length (2 hours). Overall, there appears to be a decrease in both RGHg and Hg-P concentrations since measurements began in 1995. Clear diurnal trends were found during all studies – in February/March 2002 (maximum 2 hr average around 80 pg m^{-3}; Mason and Sheu, 2002); during an intensive study in October 2002 (maximum 2 hr average around 90 pg m^{-3}; Laurier, pers. comm.) and in continual and on-going collections since November 2003 (maximum 2 hr average on occasion between 100-120 pg m^{-3}).

In October 2002, rain samples were collected during the period of RGHg measurement and the results confirm the notion that RGHg is rapidly removed from the atmosphere by precipitation (Laurier et al., in prep.). Similar results have been found by others (Malcolm and Keeler, 2002). Four periods of rain occurred and it was found in all cases that the concentration of total Hg in the first rain sample collected during the event was about twice that of subsequent rain samples. Also, after this initial decrease, the concentration in the rain appeared to be relatively constant. Rain concentrations for the initial samples ranged from 20 to 35 pM, with the latter concentrations being less than 20 pM. During the periods prior to the rain event, RGHg concentrations varied but were generally above 20 pg m^{-3}.

Table 1. Concentrations of reactive gaseous mercury and particulate mercury measured at the Chesapeake Biological Laboratory between 1995 and 2004.

Date	Hg-P (pg m^{-3})	RGHg (pg m^{-3})	Method	Reference
Summer 1995	16 ± 15	-	Filters	Mason et al., 1997
Winter 1995	21 ± 20	-	Filters	Mason et al., 1997
1997*	9 (2.3)	10 (2.3)	Filters	Sheu and Mason, 2001
1998*	21 (1.0)	24 (1.0)	Filters	Sheu and Mason, 2001
1999*	9.0 (0.7)	29 (0.6)	Filters	Sheu, 2001
2000*	9.0 (1.3)	12 (0.7)	Filters	Sheu, 2001
2-3/02	-	10.6 ± 11.2	Tekran	Mason and Sheu, 2002
10/02	-	7.9 ± 12	Tekran	Unpublished data
11/03-5/04	-	9.5 ± 11.4	Tekran	Unpublished data

* These data are for the median and the relative error (in brackets).

The potential importance of scavenging in contributing to Hg in precipitation can be estimated, as done previously by Mason et al. (1997), although in that analysis they did not include scavenging of RGHg from the atmosphere by precipitation. The previous evaluation concluded that the measured amounts of Hg-P in air did not appear to be sufficient to account for the concentrations of Hg found in precipitation, indicating other sources and/or in-cloud processes as being important. That analysis is updated here, considering the total amount of Hg in the air that can be scavenged to be the sum of the RGHg and Hg-P concentrations. As seen in Table 1, RGHg and Hg-P concentrations at CBL are comparable, on average, and the combined total ranges from about 20-40 pg m^{-3}. Based on this, the data (lines on the graph) in Figure 4 was generated for values of scavenged Hg of 15, 30 and 40 pg m^{-3}, and a cloud height of 10 km. In addition, the measured concentrations of total Hg in rain for CBL are also plotted (Mason et al., 2000).

Overall, it appears reasonable to conclude that scavenging of RGHg and Hg-P from the atmosphere is sufficient to account for much of the Hg in rain, with the remaining Hg likely supplied by oxidation processes during cloud formation. Thus, RGHg is contributing Hg through both wet and dry deposition to surface waters. The relative importance of RGHg in contributing to Hg deposition is discussed more below. Suffice to note that for all the open ocean and coastal regions that we have studied, it is an important part of the Hg speciation in the atmosphere and in deposition and cannot be ignored.

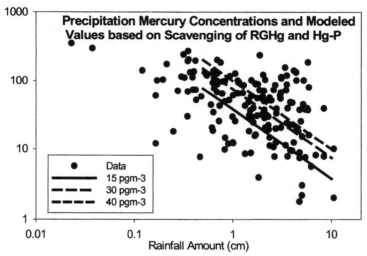

Figure 4. Concentrations of mercury in wet deposition at the Chesapeake Biological Laboratory and modeled scenarios of the contribution of mercury scavenging from the atmosphere to mercury in wet deposition.

Laboratory Studies

While there have been a number of studies that have examined the oxidation of Hg^0 through homogeneous gas phase reactions (e.g., Lin and Pehkonen, 1999; Sliger et al., 2000; Ariya et al., 2002, Sommar et al., 1997; 2001) there have been few studies of Hg^0 oxidation in the presence of halide surfaces. Sheu and Mason 2004 performed such experiments, as well as examining the oxidation of Hg^0 in the presence of various potential oxidants. There was essentially no reaction in the dark but reaction occurred in the light in all treatments. As shown in Table 2, the rates of oxidation were relatively low in the presence of light, with or without the addition of water and nitrite solution (NO_2^-). Nitrite decomposes in the presence of light to form the hydroxyl radical (OH), which is known to oxidize Hg^0, and was added to the chamber for this purpose. However, NO_2^- can also be a sink for OH and the complex interaction between N-containing and O-containing species can result, depending on the conditions, in the net formation or net loss of ozone, and the OH radical (Seinfeld and Pandis, 1998). Given the results in Table 2, it is possible to conclude that the addition of NO_2^- did not enhance Hg^0 oxidation, but hindered it. More work should be done to examine these interactions, especially in the presence of salt surfaces. As seen in Table 2, in the presence of either a NaCl or NaBr salt surface coating in the reactor, the rate of oxidation was enhanced by one to two orders of magnitude, both in indoor experiments with artificial light and outdoors (Table 2). The rate of oxidation was slower in the presence of outdoor light and this likely reflects the lower light intensity during the experiments. Finally, Sheu and Mason 2004 showed that the removal of light of <324 nm completely stopped the reactions, even in the presence of the NaCl salt surface. The most important reactions (shown in Figure 1) that rely on the presence of short wavelength light are those involved in the formation of the OH radical, and in ozone destruction, and this likely accounts for the effect of wavelength on Hg^0 oxidation.

Table 2. Summary of the pseudo-first order reaction rates for the indoor Xenon lamp-irradiated and outside (natural sunlight) gaseous Hg^0 oxidation experiments. Taken from Sheu and Mason 2004.

Treatment	Indoor Experiments Reaction Rate	Outdoor Experiments Reaction Rate (min^{-1})
Light alone	7×10^{-4}	$<10^{-4}$
Light+H_2O	2.4×10^{-3}	8×10^{-4}
Light+NO_2^-	7×10^{-4}	3×10^{-4}
Light+NaCl	9.4×10^{-2}	-
Light+NaCl+H_2O	1.0×10^{-1}	$3 \times 10^{-4} - 5 \times 10^{-3}$
Light+NaCl+NO_2^-	2.2×10^{-2}	$1.1 - 1.5 \times 10^{-2}$
Light+NaBr+NO_2^-	5.5×10^{-1}	-

While the experiments performed and discussed in Sheu and Mason (2004) were insufficient to examine in detail all the potential interactions that are known to exist with the photochemical reactions of nitrogen, oxygen and halide reactive species and intermediates, it did appear from the experiments that the role of nitrogen species was complex in that it could either enhance or hinder oxidation. It is known that NO_x can react with ozone, OH and with halogens, and thus given the complex reaction cycle proposed in Figure 1, it is not unreasonable to expect the impact of NO_x to be variable, as likely is the presence and concentration of other oxidants (Seinfeld and Pandis, 1998). Clearly, more studies are required to accurately assess the role of various processes and oxidants - in the marine boundary layer, - in the coastal atmosphere, and - in other locations in the oxidation of Hg^0 as this process could be considered as a "short-cycling" process that alters the rate of the global distribution of Hg^0.

MERCURY IN SURFACE OCEAN WATERS

Mercury exists in surface waters as both dissolved Hg^0 and as Hg^{II} complexes. It has been shown that reduction of Hg^{II} to Hg^0 occurs in surface waters under a variety of conditions and that both biotic and abiotic processes are important. Recent studies, in both freshwater and saline waters, have focused on understanding the photochemical processes influencing DGHg concentrations and while progress has been made, the understanding is still incomplete (Amyot et al., 1997; Rolfhus and Fitzgerald, 2001). However, there is now sufficient evidence to suggest that in most situations while there is net Hg^0 formation, there is also Hg^0 oxidation occurring (Amyot et al., 1997; Mason et al., 2001; Lalonde et al., 2001). Much of the lack of understanding of the oxidation processes is driven by the difficulty in performing the experiments with a relatively insoluble trace gas which has substantial gas exchange losses under most experimental conditions.

A number of studies have shown that Hg^{II} reduction occurs in the presence of algae and bacteria (e.g., Mason et al., 1995; Lanzillotta et al., 2004). It has been demonstrated that the extent of reduction can be affected by the addition of chemicals that block photosynthetic organisms and biochemical processes and Mason et al. (1995) concluded that bacteria and picoplankton (cyanobacteria) were the more important organisms involved in biotic reduction in environmental waters. Lanzillotta et al. (2004) showed that Hg^{II} reduction occurred in the presence of a marine diatom and that reduction also occurred in the filtered medium suggesting that these organisms release organic matter into the water that can play a role in the reduction processes. Thus, the role of organisms is to either directly reduce the Hg^{II} or to release organic matter into solution that leads to enhanced reduction. The reduction of Hg^{II} in the presence of organic matter has been shown by others. Costa and Liss 1999 found reduction enhanced in the light in the presence of humic acid although the increase in Hg^0 formation with increasing humic acid concentration was non-linear. This is likely a result of the adsorption of light by the humic acid present at high concentrations, and by the impact of complexation. It has been suggested that the reduction of Hg^{II} may occur through charge transfer reactions involving Hg-DOC complexes but while this may be a mechanism it is not the only, or perhaps most important, mechanism as organic matter complexation can also lead to a decrease in Hg^{II} reduction (Rolfhus and Fitzgerald, 2001). Overall, higher DOC concentrations remains one factor that likely explains the lower concentrations and rates of Hg^0 evasion from coastal and estuarine environments compared to the open ocean (Table 3). The potential for photochemical processes to dominate Hg^{II} reduction is demonstrated by the work of Lanzillotta and Ferrara 2001 which measured short-term changes in Hg^0 concentration at a coastal site and showed a strong diurnal cycle linked to photosynthetically active radiation with a linear relationship between the two variables both in the field and in laboratory studies. Similar results have been found in lakes (e.g., Amyot et al., 1997).

Two primary recent studies of Hg^0 oxidation in seawater have documented this process and demonstrated the same overall result. Lalonde et al. (2001) showed that Hg^0 oxidation did not occur in the dark or in the light in distilled water, in the presence of Cl^- ions only, or in the presence of semiquinones alone. However, in the presence of both Cl^- and semiquinones, Hg^0 oxidation was relatively rapid. Furthermore, in the presence of Cl^- and fulvic acid oxidation also occurred, but the rate of reaction was slower. Mason et al. (2001) showed that in the presence of Cl^- and NO_2^-, Hg^0 oxidation was rapid in the light and that the rate of reaction was enhanced in the presence of Br^- compared to Cl^-. All these experimental results are consistent with the reaction mechanisms described above in detail for atmospheric oxidation of Hg^0. For example, NO_2^- is photolysed to form OH (Seinfeld and Pandis, 1998) which can then initiate the reaction sequence with halogen ions to form RXS in solution, which then oxidizes Hg^0. Furthermore, it is likely that the halide ions in solution stabilize the reaction products and intermediates, as the overall oxidation is a two electron transfer process and thus must involve the formation of a transient Hg^I intermediate. Lin and Pehkonen (1999) showed that Hg^0 is also oxidized by $HOCl/OCl^-$ suggesting that a variety of dissolved RXS species could be responsible for oxidation in natural waters. In the studies of Lalonde et al. (2001), it is likely that the photochemical degradation of the seimquinone, or the fulvic acid, released reactive oxygen intermediates which through a series of reactions likely led to the formation of RXS species. Thus, in both instances, an initiation reaction was required to provide the RXS species that oxidize Hg^0 and the halide present likely stabilized the reaction products through complexation. However, the HgX_2 species formed as products are also likely to be readily reduced unless further complexed by DOC, and thus the net rate of formation would depend on this. The presence of DOC will lead to Hg^{II} complexation but this has been shown to enhance Hg^{II} reduction on occasion, rather than diminish it. Overall, in the high Cl^-, low DOC environments of the open ocean it is likely that both oxidation and reduction occur rapidly, and as stated above, based on data gathered to date, it appears that overall there is net reduction.

If both reactions occur, and can be considered pseudo first order because of the general excess of oxidant, then the ratio of the rate of reduction to that of oxidation at steady state is equal to the ratio of the reactive amount of Hg^0 to that of Hg^{II}. It is reasonable to assume that all the Hg^0 is available for oxidation, but it is likely that not all the Hg^{II} is easily reduced. In fact, the method of measurement of "reactive Hg", based on $SnCl_2$ reduction of unamended seawater samples, have shown that only a small fraction of the total Hg in surface ocean and estuarine waters is easily reduced (<20%; Mason et al., 1998; 1999; 2001). This is similar to the concentration of Hg^0. During the recent North Pacific Ocean cruise, the reactive Hg concentration was similar to that of Hg^0 (Laurier et al., 2003) suggesting that the rates of oxidation and reduction were similar. In contrast, in the North Atlantic during the August 1993 cruise, the fraction of the total dissolved Hg as Hg^0 was 20-100% suggesting that under these conditions, net reduction was occurring and that oxidation must have been minimal during the sampling period.

Estimations of fluxes of Hg^0 from the ocean to the atmosphere have generally been made based on available data, which is still limited. These data, summarized for some coastal and ocean studies in Table 3, have been used to extrapolate fluxes to the global scale. Exchange rates estimated from the Hg^0 concentration measured during cruises in the Atlantic Ocean and elsewhere suggest that fluxes can be much higher than estimated atmospheric wet deposition rates (Mason et al., 1995; 2001; Mason and Sullivan, 1999; Table 3). For example, for the South and equatorial Atlantic, estimated evasion rates during the cruise in May/June 1996 were substantially greater that estimates of wet deposition for this region (Lamborg et al., 1999; Mason et al., 1994; 2001). In the North Atlantic near Bermuda, estimated evasional fluxes were lower and showed no strong seasonal trend but were still greater than wet deposition (Mason et al., 2001). Clearly, these rates must represent a transient situation.

While Hg^0 oxidation is possible in surface waters (Amyot et al., 1997; Lalonde et al., 2001; Mason et al., 2001), the flux estimates are based on actual measurements of Hg^0, and as recent sample collections have been in the top meters of the ocean, these represent the steady state concentration between oxidation and reduction processes. It is likely that evasion rates are lower in winter, as suggested by the model developed for the far North Atlantic (Mason et al., 1998), although there is little open ocean data to support this notion. For coastal environments, such as the Scheldt estuary and Long Island Sound, there is evidence for a lower flux in winter compared to summer (Baeyens and Leermakers, 1998; Rolfhus and Fitzgerald, 2001). In their recent overall mass balance, Mason and Sheu (2002) increased the magnitude of oceanic Hg evasion compared to that proposed earlier (Mason et al., 1994) but the magnitude of the increase was constrained by that of the emission fluxes for the terrestrial environment, and by the magnitude of the dry deposition flux to the ocean (primarily that of RGHg). Mason and Sheu (2002) estimated an average flux of 13 Mmol yr^{-1} (7.2 µg m^{-2} yr^{-1}) for evasion from the ocean, and this value was constrained for seasonal and latitudinal differences. Recent data (e.g., Laurier et al., 2003) does not suggest that this estimate needs to be substantially changed. While Mason and Sheu (2002) proposed a 30% increase over the estimate of Mason et al. (1994), it still may be an underestimate as the average (0.6 µg m^{-2} mnth^{-1}) is at the low end of the flux estimates in Table 3.

Table 3. The concentrations and estimated fluxes of elemental mercury for various water bodies. As fluxes are mostly from short-term measurements, they are scaled to a monthly rather than a yearly basis.

Location	DGHg (pM)	Flux (µg m^{-2} mth^{-1})	References
Equatorial Pacific	0.05-0.36	0.7-7	1
North Pacific	0.13 ± 0.07	0.7-1.8	9
Equatorial Pacific 2002	0.06 ± 0.03		
N. Atlantic – summer	0.25-1.25	12	2
S. Atlantic – summer	1.24 ± 0.81	36	3
Bermuda	0.08-0.2	2.7	4
Long Island Sound, USA	0.04-0.55	2.1	5
Scheldt Estuary, Belgium	0.1-0.65	1.2-2.4	6
North Sea	0.06-0.8	0.35-6.7	
Chesapeake Bay, USA	0.02-0.2	0.8	7
Lakes/Wetlands	-	0.2-2	8

References: #1= Mason and Fitzgerald, 1993; #2= Mason et al., 1998; #3= Lamborg et al., 1999; 4 = Mason et al., 2001; #5= Rolfhus and Fitzgerald, 2001; #6= Baeyens and Leermakers, 1998; #7= Mason et al., 1999; #8= Zhang and Lindberg, 1999 and reference therein; #9 = Laurier et al., 2003.

As discussed below, this estimate reflects evasion from the surface ocean but because of Hg^0 oxidation and RGHg formation in the MBL, the next evasion to the global atmosphere is less than this.

AIR-WATER EXCHANGE AND THE GLOBAL MERCURY CYCLE

It can be concluded based on the studies discussed above that Hg^0 is relatively easily oxidized in the atmosphere under the right conditions, and that these conditions often exist in the marine boundary layer. In general, it appears that conditions conducive to ozone destruction, specifically those that involve the formation of RXS, can also lead to Hg^0 oxidation, although the data available to date suggest that other secondary reactions, such as those with N-containing radicals, can also be important in modulating Hg^0 oxidation. Overall, current modelling efforts which include the primary reactions for which rate constants have been measured, can reproduce field data with relative accuracy suggesting that the major aspects of the atmospheric chemistry is now understood.

Furthermore, it appears that the formation of RGHg that occurs in the boundary layer over the ocean, and in coastal areas, can be an important flux of mercury to these systems, rivaling that of wet deposition. The data from the studies in Maryland suggest that for coastal sites such as CBL, the RGHg dry deposition flux is comparable to that of wet deposition, and that the particulate flux is about 10% of the total input. For example, Sheu et al. (2002) estimated that the dry RGHg flux was 11.5 µg m^{-2} yr^{-1} for 1998 and wet deposition was 12.5 µg m^{-2} yr^{-1}. The more recent measurements (2003/04) suggest that the dry deposition flux is currently lower, around 6 µg m^{-2} yr^{-1}. It is not known if wet deposition fluxes have also decreased over the same period but given the importance of RGHg scavenging to Hg in rain, this is to be expected. Based on 5 years of data between 1995 and 2000, however, Mason et al. (2000) concluded that there was not a statistically significant trend in the yearly wet deposition flux. However, longer term studies at other sites in the USA have noted a decrease in Hg in wet deposition in the last decade.

Another coastal site which has data on atmospheric Hg speciation and Hg in wet deposition is Mace Head in Ireland. Pirrone et al. (2001) report median concentrations of RGHg and Hg-P of 19.1 and 2.9 pg m^{-3} and a median concentration in wet deposition of 3.8 ng/L. Given these data, it can be concluded that dry deposition of RGHg is higher than that of wet deposition at this location and that dry particulate deposition is a small fraction of the total flux. In contrast to the data from CBL, where the concentrations of RGHg and Hg-P are similar, the RGHg concentration is about six times that of Hg-P at Mace Head. This contrasts the other background European stations where concentrations of RGHg and Hg-P are comparable, or where Hg-P is higher than RGHg. Thus, given our understanding of the fate and transport of RGHg and Hg-P, there must be some non-terrestrial source of RGHg at Mace Head. It is reasonable to conclude that the ocean boundary layer is the source given that the wind direction at Mace Head is predominantly oceanic.

The results from the cruise in the North Pacific can be combined with those for the North Atlantic to estimate the flux of RGHg to the ocean surface. In the temperate North Pacific RGHg concentrations were lower but because of higher wind speeds, deposition velocities were higher. Overall, however, fluxes were highest during the day in the equatorial region. The average RGHg flux for the equatorial region was around 16 ng m^{-2} d^{-1} while further north the flux was around 6 ng m^{-2} d^{-1}. Based on these data, and the data from the North Atlantic 2003 cruise, and the estimated coastal fluxes, and taking into account seasonal variability in flux, an overall global estimate of 4.8 µg m^{-2} yr^{-1} is derived, or 8.6 Mmol yr^{-1}. A more simplistic estimation can be based on the average open ocean RGHg concentration from the studies to date (10 pg m^{-3}) and an average dry deposition velocity of 1.5 cm s^{-1}. This yields a global flux of 8.5 Mmol yr^{-1}, which is comparable to the above estimate. Mason and Sheu 2002 concluded on the basis of more limited data that the flux was 5.7 Mmol yr^{-1} and the current analysis suggests that this earlier value is an underestimate. If the formation and deposition of RGHg is higher than previously estimated, then, as a result of mass balance estimations, the net flux of Hg^0 to the global atmosphere via gas evasion from the ocean surface may be less than previously estimated (7.3 Mmol yr^{-1}; Mason and Sheu, 2002). The overall outcome of such a scenario is that the overall residence time of Hg in the upper ocean is actually longer than previously estimated. If so, then the response time of the upper ocean to changes in atmospheric Hg concentrations will be slower, and it may not be reasonable to expect the surface ocean concentration to respond rapidly to changes in atmospheric Hg concentration.

Given a wet deposition flux of about 10 Mmol yr^{-1} (Mason and Sheu, 2002), it can be concluded that the dry deposition of RGHg to the ocean is an important flux, being almost as large as the wet deposition flux, as found for coastal sites (Sheu et al., 2002). Finally, it is clear that formation of reactive gaseous mercury via atmospheric reactions over the ocean rivals the input of reactive gaseous mercury from anthropogenic activities (~4.4 Mmol/yr; Mason and Sheu, 2002). Substantial conversion of elemental mercury in remote regions of the world, and subsequent deposition to sensitive ecosystems has important managerial implications for the regulation of Hg emissions from anthropogenic sources.

REFERENCES

Amyot, M., Gill, G.A., Morel, F.M.M. Production and loss of dissolved gaseous mercury in coastal seawater. *Env. Sci. Technol.*, 31, 3606-3611, 1997.

Ariya, P.A., Khalizov, A., Gidas, A. Reactions of gaseous mercury with atomic and molecular halogens: Kinetics, product studies, and atmospheric implications *J. Phys. Chem. A.*, 106, 7310-7320, 2002.

Baeyens, W., Leermakers, M. Elemental mercury concentrations and formation rates in the Scheldt estuary and the North Sea. *Mar. Chem.*, 60(3-4) 257-266, 1998.

Balabanov, N.B., Peterson, K.A. Mercury and Reactive Halogens: The Thermochemistry of Hg + {Cl2, Br2, BrCl, ClO, and BrO}. *J. Phys. Chem. A*, 107, 7465, 2003.

Barrie, L.A., Bottenheim, J.W., Schnell, R.C., Crutzen, P. J., Rasmussen, R.A. Ozone destruction and photochemical reactions at polar sunrise in the lower Arctic atmosphere, *Nature*, 334, 8-141, 1988.

Bloom, N.S., Fitzgerald, W.F. Determination of volatile species at the pictogram level by low temperature gas chromatography with cold-vapor atomic fluorescence detection, *Anal. Chim. Acta*, 208, 151-161, 1988.

Bullock, O.R. Modeling assessment of transport and deposition patterns of anthropogenic mercury air emissions in the United States and Canada. *Sci. Total Environ.*, 259(1-3) 45-157, 2000.

Cossa, D., Martin, J.M., Takayanagi, K. , SanJuan, J. The distribution and cycling of mercury species in the western Mediterranean, *Deep-Sea Res. II*, 44, 721-740, 1997.

Dickerson, R.R., Rhoads, K.P., Carsey, T.P., Oltmans, S.J., Burrows, J.P., Cruzten, P.J., Ozone in the remote marine boundary layer: a possible role for halogens, *J. Geophys. Res.*, 104, 21,385-21,395, 1999.

Ebinghaus, R., Tripathi, R.M., Wallschläger, D. Lindberg, S.E. Natural and Anthropogenic Mercury Sources and their Impact on the air-surface exchange of mercury on regional and global scales, In: Ebinghaus, R., Turner, R.R., Lacerda, D., Vasiliev, O., Salomons, W. (eds.): Mercury Contaminated Sites—Characterization, Risk Assessment and Remediation, Springer Verlag Berlin Heidelberg New York, ISBN 3-540-63731-1, 3-50, 1999.

Ebinghaus, R., Knock, H.H., Temme, C., Einax, J.W., Lowe, A.G., Richter, A., Burrows, J.P., Schroeder, W.H. Antarctic springtime depletion of atmospheric mercury, *Environ. Sci. Technol.*, 36, 1238-1244, 2002.

Ferrara, R., Ceccarini, C., Lanzillotta, E., Gardfeldt, K., Sommar, J., Horvat, M., Logar, M., Fajon, V., Kotnik, J. Profiles of dissolved gaseous mercury in Mediterranean seawater. *Atmos Environ.*, 37, 85 – 92, 2002.

Fickert, S., Adams, J.W., Crowley, J.M. Activation of Br2 and BrCl via uptake of HOBr onto aqueous salt solutions. *J. Geophys. Res.*, 104, 23719-23727, 1999.

Finlayson-Pitts, B.J., Livingston, F.E., Berko, H.N. Ozone destruction and bromine photochemistry at ground level in the Arctic spring. *Nature*, 343, 622-625, 1990.

Fitzgerald, W.F., Mason, R.P. Biogeochemical cycling of mercury in the marine environment. In: Sigel, A., Sigel, H. (Eds.), Mercury and Its Effects on Environment Biology. Marcel Dekker, New York, 53-111, 1997.

Fitzgerald, W.F., Engstrom, D.R., Mason, R.P., Nater, E.A. The case for atmospheric mercury contamination in remote areas. *Env. Sci. Technol.*, 32, 1-7, 1998.

Foster, K.L., Plastridge, R.A., Bottenheim, J.W., Shepson, P.B., Finlayson-Pitts, B.J., Spicer, C.W. The role of Br2 and BrCl in surface ozone destruction at polar sunrise. *Science*, 291, 471-474, 2001.

Gabriel, R., von Glasow, R., Sander, R., Andrea, M.O., Crutzen, P.J. Bromide content of sea-salt aerosol particles collected over the Indian Ocean during INDOEX 1999. *J. Geophys. Res.*, 107(D19), 2002.

Galbally, I.E., Bentley, S.T., Meyer, C.P. Mid-latitude marine boundary-layer ozone destruction at visible sunrise observed at Cape Grim, Tasmania, 41°S. *Geophys. Res. Lett.*, 27, 3841-3844, 2000.

Gardfeldt, K., Sommar, R., Ferrara, R., Ceccarini, C., Lanzillotta, E., Munthe, J., Wangberg, I., Lindqvist, O., Pirrone, N., Sprovieri, F., Pesenti, E. Evasion of mercury from Atlantic coastal water and the Mediterranean Sea, coastal and open water, *Atmos. Environ.*, 37, S73-S84.

Hall, B. The phase oxidation of elemental mercury by ozone. *Water, Air Soil Pollut.*, 80, 301-315, 1995,

Hebestreit, K., Stutz, J., Rosen, D., Matveiv, V., Pelg, M., Luria, M., Platt, U., DOAS measurements of tropospheric bromine oxide in mid-latitudes, *Science*, 283, 55-57, 1999.

Hedgecock, I.M. Pirrone, N. Mercury photochemistry in the marine boundary layer-modeling studies suggest the in situ production of reactive gas phase mercury. *Atmos. Environ.*, 35, 3055-3062, 2001.

Hedgecock, I.M., Pirrone, N., Sprovieri, F., Pesenti, E. Reactive gaseous mercury in the marine boundary layer, Modeling and experimental evidence of its formation in the Mediterranean region. *Atmos. Environ.*, 37, S41-S49, 2003.

Hirokawa, J., Onaka, K., Kajii, Y. Akimoto, H. Heterogeneous processes involving sodium halide particles and ozone: Molecular bromine release in the marine boundary layer in the absence of nitrogen oxides. *Geophys. Res. Lett.*, 25, 2449-2452, 1998.

Hudson, R.J.M., Gherini, S.A., Fitzgerald W.F., Porcella D.B., Anthropogenic influences on the global mercury cycle: a model-based analysis. *Water Air Soil Pollut.*, 80, 265-272, 1995.

Khalizov, A.F., Viswanathan, B., Larregaray, P., Ariya, P.A. A theoretical study on the reactions of Hg with halogens: atmospheric implications. *J. Phys. Chem. A*, 107, 6360-6365, 2003.

Kim, J.P., Fitzgerald, W.F. Gaseous mercury profile in the tropical Pacific Ocean. *Geophys. Res. Lett.*, 15, 40-43, 1988.

Knipping, E.M., Lakin, M.J., Foster, K.L., Jungwrith, P., Tobias, D.J., Gerber, R.B., Dabdub, D., Finlayson-Pitts, B.J. Experiments and simulations of ion-enhanced interfacial chemistry on aqueous NaCl aerosols. *Science*, 288, 301-306, 2000.

Lalonde, J.D., Amyot, M., Kraepiel, A.M.L. Morel, F.M.M. Photooxidation of Hg(0) in artificial and natural waters. *Env. Sci. Technol.*, 35,1367-1372, 2001.

Lamborg, C.H., Rolfhus, K.R., Fitzgerald, W.F., Kim, G. The atmospheric cycling and air-sea exchange of mercury species in the South and Equatorial Atlantic. *Deep-Sea Res. II*, 46, 957-977, 1999.

Landis, M.S., Stevens, R.K., Shaedlich, F., Prestbo, D.E.M. Development and characterization of an annular denuder methodology for the measurement of divalent inorganic reactive gaseous mercury in ambient air. *Env. Sci. Technol.*, 36, 3000-3009, 2002.

Lanzillotta, E. Ceccarini, C., Ferrara, R. Photo-induced formation of dissolved gaseous mercury in coastal and offshore seawater of the Mediterranean basin. *Sci Tot. Env.*, 300, 179-187, 2002.

Laurier, F.J.G., Mason, R.P., Whalin, L., Kato, S. Reactive gaseous mercury formation in the North Pacific Ocean's marine boundary layer: A potential role of halogen chemistry. *J. Geophys. Res.*, 108, 4529, doi:10.1029/2003JD003625, 2003.

Lin, C., Pehkonen, S.O. The chemistry of atmospheric mercury, a review. *Atmos. Environ.*, 33, 2067-2079, 1999.

Lindberg, S.E. Stratton, W.J. Atmospheric mercury speciation, concentration and behavior of reactive gaseous mercury in ambient air. *Env. Sci. Technol.*, 32, 49-57, 1998.

Lindberg, S.E., Brooks, S., Lin, C.J., Scott, K.J., Landis, M.S., Stevens, R.K., Goodsite,M., Richter, A. Dynamic Oxidation of Gaseous Mercury in the Arctic Troposphere at Polar Sunrise. *Env. Sci. Technol.*, 36, 1245-1256, 2002.

Malcolm, E.G., Keeler, G.J. Measurements of mercury in dew: atmospheric removal of mercury species to a water surface. *Env. Sci. Technol.*, 36, 2815-2821, 2002.

Mason, R.P., Fitzgerald, W.F., Vandal, G.M. The sources and composition of mercury in Pacific Ocean rain. *J. Atm. Chem.*, 14, 489-500, 1992.

Mason, R.P., Fitzgerald, W.F. The distribution and biogeochemical cycling of mercury in the euatorial Pacific Ocean. *Deep-Sea Res.*, 40, 1897-1924, 1993.

Mason, R.P., Fitzgerald, W.F., Morel, F.M.M. The biogeochemical cycling of elemental mercury: Anthropogenic influences. *Geochim. Cosmochim. Acta*, 58, 3191-3198, 1994.

Mason, R.P., Rolfhus, K.R., Fitzgerald, W.F. Methylated and elemental mercury in the surface and deep ocean waters of the North Atlantic, *Water Air Soil Pollut.*, 80, 775-787, 1995.

Mason, R.P., Lawson, N.M., Sullivan, K.A. Atmospheric deposition to the Chesapeake Bay watershed–regional and local sources. *Atmos. Environ.*, 31, 3531-3540, 1997.

Mason, R.P., Rolfhus, K.R., Fitzgerald, W.F., Mercury in the North Atlantic. *Mar. Chem.*, 61, 37-53, 1998.

Mason, R.P., Lawson, N.M., Lawrence, A.L., Leaner, J.J., Lee, J.G., Sheu, G.R. Mercury in the Chesapeake Bay. *Mar. Chem.*, 65, 77-96, 1999.

Mason, R.P., Sullivan, K.A. The distribution and speciation of mercury in the South and equatorial Atlantic. *Deep-Sea Res. II*, 46, 937-956, 1999.

Mason, R.P., Lawson, N.M., Sheu, G-R. Annual and seasonal trends in mercury deposition in Maryland. *Atmos. Environ.*, 34, 1691-1701, 2000.

Mason, R.P., Lawson, N.M., Sheu, G-R. Mercury in the Atlantic Ocean: factors controlling air–sea exchange of mercury and its distribution in the upper waters. *Deep-Sea Res. II*, 48, 2829-2853, 2001.

Mason, R.P., Sheu, G-R. The role of the ocean in the global mercury cycle. *Global Biogeochem. Cyc.*, 16, 1093, 2002.

Mason, R.P., Benoit, J.M. Organomercury compounds in the environment. In: P.J. Craig [ed.], Organometallics in the Environment, 2nd Edition, John Wiley & Sons, New York,57-99, 2003.

McConnell, J.C., Henderson, G.S., Barrie, L., Bottenheim, J., Niki, H., Langford, C.H., Templeton, E.M.J. Photochemical bromine production implicated in Arctic boundary-layer ozone depletion. *Nature*, 355, 150-152, 1992.

Mozurkewich, M. Mechanism for the release of halogens from sea-salt particles by free-radical reactions. *J. Geophys. Res.*, 100, 14199-14207, 1995.

Oum, K.W., Lakin, M.J., DeHann, D.O., Brauers, T., Finlayson-Pitts, B.J. Formation of molecular chlorine from the photolysis of ozone and aqueous sea-salt particles. *Science*, 279, 74-77, 1998.

Pirrone, N., Costa, P., Pacyna, J.M., Ferrara, R. Atmospheric mercury emissions from anthropogenic and natural sources in the Mediterranean region. *Atmos. Environ.*, 35, 2999-3006, 2001.

Porcella, D. B., Ramel, C., Jernelov, A. Global mercury pollution and the role of gold mining: An Overview. Water, Air and Soil Pollution 1997; 97, 205-207.

Rolfhus, K.R., Fitzgerald, W.F. The evasion and spatial/temporal distribution of mercury species in Long Island Sound, CT-NY. *Geochim. Cosmochim. Acta*, 65, 407-418, 2001.

Schroeder, W.H., Anlauf, K.G., Barrie, L.A., Lu, J.Y., Steffen, A., Schneeberger, D.R., Berg, T. Arctic springtime depletion of mercury. *Nature*, 394, 331-332, 1998.

Schroeder, W. H., Munthe, J. Atmospheric Mercury – an overview. *Atmos. Environ.*, 32, 809-822, 1998.

Seinfeld, J.H. Pandis, S.N. Atmospheric Chemistry and Physics from Air Pollution to Climate Change. John Wiley and Sons, New York, 1998.

Shahin, U.M., Holsen, T.M. Odabasi, M. Dry deposition with a surface water surface sampler a comparison to modeled results, *Atmos. Environ.*, 36, 3267-3276, 2002.

Sheu, G.-R Mason, R.P. An examination of methods for the measurements of reactive gaseous mercury in the atmosphere *Env. Sci. Technol.*, 35, 1209-1216, 2001.

Sheu, G.-R, Mason, R.P., Lawson, N.M. In Fate, Impacts and Remediation R.L. Lipnick R.P. Mason, M.L. Phillips, and C.U. Pittman, Jr. eds. Chemicals in the Environment. ACS Symposium Series 806, ACS, Washington, DC, 223-242, 2002.

Sheu, G-R. Mason, R.P. An Examination of the Oxidation of Elemental Mercury in the Presence of Halide Surfaces. *J. Atmos. Chem.*, 48, 107-130. 2004.

Shia, R-L., Seigneur, C., Pai, P., Ko, M., Sze, N.-D. Global simulation of atmospheric mercury concentrations and deposition fluxes. *J. Geophys. Res.*, 104, 747-760, 1999.

Sliger, R.N., Kramlich, J.C., Marinov, N.M. Towards the development of a chemical kinetic model for the homogeneous oxidation of mercury by chlorine species. *Fuel Process. Technol.*, 65-66, 423-438, 2000.

Sommar, J., Hallquist, M., Ljungström, E., Lindqvist, O. On the gas phase reactions between volatile biogenic mercury species and the nitrate radical. *J. Atmos. Chem.* 27, 233-247, 1997.

Sommar, J., Gårdfeldt, K., Strömberg, D., Feng, X. A kinetic study of the gas-phase reaction between the hydroxyl radical and atomic mercury. *Atmos. Environ.*, 35, 3049-3054, 2001.

Spicer, C.W., Chapman, E.G., Finlayson-Pitts, B.J., Plastridge, R.A., Hubbe, J.M., Fast, J.D., Berkowitz, C.M. Unexpectedly high concentrations of molecular chlorine in coastal air. *Nature*, 394, 353-356, 1998.

Sprovieri, F., Pirrone, N., Gärrdfeldt, K., Sommar, J. Mercury speciation in the marine boundary layer along a 6000 km cruise path around the Mediterranean Sea. *Atmos. Environ.*, 37, S63-S71, 2003.

Temme, Ch., Einax, J.W., Ebinghaus, R., Schroeder, W.H. Measurements of Atmospheric Mercury Species at a Coastal Site in the Antarctic and over the South Atlantic Ocean during Polar Summer. *Env. Sci. Technol.*, 37, 22-31, 2003.

Tokos, J.J.S., Hall, B., Calhoun, J.A., Prestbo, E. M. Homogeneous gas phase reaction of Hg^0 with H_2O_2, O_3, CH_3I, and $(CH_3)_2S$: implications for atmospheric Hg cycling, *Atmos. Environ.*, 32, 823-827, 1998.

Vogt, R., Crutzen, P.J. Sander, R. A mechanism for halogen release from sea salt aerosol in the remote marine boundary layer. *Nature*, 383, 327-330, 1996.

von Glasow, R., Sander, R. Bott, A. Crutzen, P.J. Modeling halogen chemistry in the marine boundary layer 1. Cloud free MBL, *J. Geophys. Res.*, 107, 4341, 2002.

Chapter-11

TERRESTRIAL MERCURY FLUXES: IS THE NET EXCHANGE UP, DOWN, OR NEITHER?

Mae S. Gustin[1] and Steven E. Lindberg[2]

[1] *Department of Natural Resources and Environmental Science University of Nevada-Reno, NV*
[2] *Environmental Science Division, Oak Ridge National Laboratory, Oak Ridge, TN, USA*

INTRODUCTION

Natural terrestrial sources of atmospheric mercury (Hg) include geologically enriched substrate, volcanoes, geothermal areas, forest fires, vegetation, and "background soils" (soils that have low concentrations of Hg ($<0.1 \mu g$ Hg/g) and have not been enriched by geologic processes). Emissions from the latter three are probably dominated by re-emission of previously deposited atmospheric Hg derived from anthropogenic and natural sources. Natural terrestrial sinks include soils, plant foliage, and regions where the atmospheric chemistry facilitates formation of reactive gaseous Hg (RGM) (i.e. Polar Regions).

Modeled estimates of natural source emissions in the literature range from 800 to 3000 Mg/y (Nriagu, 1989; Lindqvist et al., 1991; Mason et al., 1994; Mason and Sheu, 2002; Lamborg et al., 2002; Seigneur et al., 2001; Bergan et al., 1999) (Table 1). Most of these were derived by difference, using measured air Hg concentrations, wet deposition rates, and anthropogenic emissions estimates in mass balance models, and are based on precious few (if any) actual flux measurements from substrate. Mercury is also known to be dry deposited although this is often not considered. Lindberg et al. (1998) used measured terrestrial flux data to develop an estimate of emissions from forests and background soils ranging from 1400 to 3400 Mg/y. However, their estimate did not include Hg fluxes from naturally geologically enriched substrates, volcanoes, geothermal systems and fires; which, as new data described below demonstrates, will add significantly to the emissions.

Point source anthropogenic emission estimates remain highly uncertain (up to 50%) (Pai et al., 1998; Pacyna et al., 2001); yet, the range applied in global mass balance models is more limited than that for natural sources (~2000 to 2400 Mg/y) (Bergan et al., 1999; Mason and Sheu, 2002; Lamborg et al., 2002; Seigneur et al., 2004) (Table 1). Dastoor and Larocque (2004) concluded that known anthropogenic emissions can account for only ~1/3 of the measured concentration of gaseous elemental Hg (Hg^0) at ground level, and suggested that natural sources and re-emission account for 2/3 of the total estimated 6400 Mg emitted each year. Pirrone et al. (2001) suggested that natural and anthropogenic emissions in Europe were approximately equal.

Table 1. Summary of recent Global flux estimates for the sources and sinks of atmospheric mercury.

Source	Range in emissions Mg/y	References
Anthropogenic	2000 to 2400	Bergan et al., 1999; Mason and Sheu, 2002; Lamborg et al., 2002; Seigneur et al., 2004
Natural terrestrial	800 to 3000	Nriagu, 1989; Lindqvist et al., 1991; Masonet al., 1994; Mason and Sheu, 2002; Lamborg et al., 2002; Seigneur et al., 2001; Bergan et al., 1999
Re-emission	?	
Total emissions	6000 to 6600	Bergan et al., 1999; Mason and Sheu, 2002; Lamborg et al., 2002; Seigneur et al., 2004
Sink	Deposition, Mg/y	
Oceans	500, 1230	Lamborg et al., 2002; Mason and Sheu; 2002
Wet and dry deposition	3000	Mason and Sheu, 2002

Global estimates for wet deposition to the continents are 2000 Mg/yr with dry deposition estimates at 1000 Mg/yr (Mason and Sheu, 2002) (Table 1). This is not enough to balance the estimated 6000 to 6600 Mg/y (Bergan et al., 1999; Mason and Sheu, 2002; Lamborg et al., 2002; Seigneur et al., 2004) emitted.

The arctic sink (see below) appears to account for only a few hundred Mg/y (Schroeder et al., 2003). Recent models indicate that the ocean could be a sink; however, as Mason and Sheu (2002) point out, there is limited and variable data for assessing the role of the ocean in the Hg biogeochemical cycle. Mason et al. (1994) first suggested that the net ocean flux was zero; however, Mason and Sheu (2002) revised this estimate, based on new data on concentrations of RGM in the marine boundary layer, and concluded that the ocean is a net sink (~500 Mg/y). Lamborg et al. (2002) assumed significantly lower ocean evasion rates than Mason and Sheu (2002) (820 Mg/y versus 2665 Mg/y, respectively), and suggested the ocean is a net sink (1230 Mg/y). These differences in numbers illustrate that the quantitative role of the ocean remains uncertain.

Based on current data and the uncertainty associated with flux estimates, we appear to have too many sources and not enough sinks for atmospheric Hg. If air concentrations are not increasing, then we must be missing or underestimating some important sinks, or perhaps Hg is constantly recycled between terrestrial ecosystems and the atmosphere. The latter process would give Hg the overall appearance of exhibiting a long residence time when it actually is rather short for any particular atom. In this paper we will summarize our current understanding of natural terrestrial sources and sinks for atmospheric Hg. Advances in our ability to measure atmospheric Hg concentrations and speciation have allowed us to improve the estimates of emissions from geologically enriched substrates, assess the potential significance of atmospheric Hg exchange with background soils, and allowed us to generate data that will help us understand the significance of vegetation in Hg cycling.

ASSESSMENT OF THE SIGNIFICANCE OF GEOLOGIC SOURCES

Natural geologic sources of Hg include volcanic emissions, enhanced Hg flux from zones of high crustal heat flow, areas of fossil and current geothermal activity, and volatilization from substrate naturally enriched in Hg (which may include areas of precious and base metal mineralization) (Table 2). These natural sources tend to be concentrated within, but are not limited to, three broad global belts that include plate tectonic boundaries (cf. Pennington, 1959; Jonhanssan and Boyle, 1972).

Volcanic emissions. Volcanic emission estimates are typically based on a few measured Hg/SO_2 mass ratios, which are then scaled up using SO_2

concentrations derived for volcanic systems. Hg/SO2 ratios span four orders of magnitude (Pyle and Mather, 2003).

The wide range in the data is due to the fact that Hg content of volcanic gases can vary as a function of the eruptive phase of the system and geologic setting (Pyle and Mather, 2003; Varenkamp and Buseck, 1981; 1984; 1986).

Table 2. Summary of estimated fluxes of Hg for terrestrial sources and sinks.

Source	Range in emissions, Mg/y	References
Volcanoes	94-700	Nriagu and Becker, 2003; Mather and Pyle, 2003
Geothermal	60	Varenkamp and Busek, 1984
Naturally enriched substrate	>1500	This paper
Background soils (re-emission and natural)	?	
Vegetated ecosystems	Forests-850 to 2000 Grasslands- 800 to 2300	Lindberg et al., 1998 Obrist et al., 2004
Forest fires	200 to 1000	Brunke et al., 2001; Friedli et al., 2001
Sink		
Litterfall sink	2400 to 6000	Lindberg et al., 2004
Arctic sink	100-300	Lindberg et al., 2002b; Schroeder et al., 2003; Skov et al., 2004
Background soils	?	

Estimates for volcanic emissions range from 1 to ~700 Mg/yr (Nriagu, 1989; Fitzgerald and Lamborg, 2003; Varenkamp and Buseck, 1986; Ferrara et al., 2000; Nriagu and Becker, 2003; Pyle and Mather, 2003). Pyle and Mather (2003) suggested that lower estimates were not accurate due to inappropriate extrapolation of data collected from low temperature degassing systems to active volcanoes. They suggested that 700 Mg/y was a more reasonable estimate of emissions with continuous degassing accounting for 10% of the flux, 75% being accounted for by small sporadic eruptions, and rare large events accounting for 15%. In contrast, Nriagu and Becker (2003) suggested that 94 Mg/y were emitted. As demonstrated by recent discussion (cf. Nriagu and Becker, 2004; Mather and Pyle, 2004) these estimates are highly uncertain.

Emissions from geothermal areas

In geothermal areas, Hg releases to the atmosphere will occur via three processes: evaporation from substrate enriched in Hg, volatile loss driven by high heat flow, and emission as a gaseous phase from fumaroles and hot springs. Emissions from geothermal systems are highly variable because factors such as the age and type of geothermal system as well as the host rock type will influence the amount of Hg released. For example, at the Steamboat Springs geothermal area, Nevada, USA, average Hg emissions from substrate (over the 8 km^2) were quite high (180 ng/m^{2}h) although soil Hg concentrations for the dominant substrate type were fairly low (1.9 $\pm$ 2.0 μg Hg/g) (Coolbaugh et al., 2002). The elevated soil flux was hypothesized to be due to the high heat flow in the area. Limited data for fumarole gases in this area indicated that concentrations ranged from 3040 to 100,000 ng/m^3 and were responsible for a small component of the total emissions (0.3 kg/y out of ~12 kg/y). Engle et al. (2004) found substrate Hg fluxes at Lassen National Park, California, USA, to be elevated (mean area flux from background soils was ~10 ng/m^{2}h) and suggested that subsurface heat flux was driving Hg through the soils. Based on measurement of fumarole gas concentrations, they estimated geothermal gas emissions of 2 to 9 kg Hg/yr, which are slightly less than estimated substrate emissions of 10 to 20 kg/y from 637 km^2. In contrast, recent work at Yellowstone National Park, WY, USA, indicated that in the caldera (2284 km^2), soil fluxes were only high in association with areas altered by acidic hydrothermal fluids (Engle et al., 2004). Fluxes measured from substrate with background Hg concentrations were extremely low, with no evidence of heat driving additional Hg from the substrate. The Yellowstone hydrothermal system is associated with a mantle hot spot and heat has been moving through substrate in the area for hundreds of thousands of years. It is possible here that much of the Hg may have been released in previous eruptions and during earlier heating events, or that there is not much Hg associated with the host rocks (Engle et al., 2004).

Varenkamp and Busek (1984) estimated that geothermal sources contributed roughly 60 Mg/y to the atmosphere based on the average Hg content in hot springs and global convective heat transport. Their estimate did not include volatilization from naturally enriched soils or fumaroles, which could be a more significant source than hot springs. Because of the variability from system to system, their estimate is highly uncertain. Scaling up geothermal emissions to obtain a global estimate is difficult without additional flux data from a variety of representative systems.

Emissions from naturally Hg-enriched soils

In order to scale up emissions from substrate geologically enriched in Hg, the factors controlling emissions must be understood. Process level information is now being developed with this objective. Gustin (2003) summarized the current understanding of the controlling mechanisms, and separated them into two groups: those that control the magnitude of the flux, and those that control seasonal and diel variations in flux. Substrate Hg concentration, rock type, the presence and type of hydrothermal alteration, and the presence of heat sources and geologic structures will influence the overall magnitude of emissions. Meteorological parameters, especially increases in light (Gustin et al., 2002), temperature (Lindberg et al., 1979; Gustin et al., 1997), and precipitation (Lindberg et al., 1999), as well as soil moisture (Gustin et al., 2004), will enhance Hg emissions and dictate diel trends. The potential for constituents of ambient air, to enhance Hg release from soils by a simple exchange process was suggested by Zhang and Lindberg (1999). Recent research by Engle et al. (2004) has demonstrated that atmospheric ozone will enhance Hg release from enriched soils.

Scaling up emissions for areas naturally enriched in Hg is usually done by making in situ flux measurements from representative substrate in an area and then developing algorithms between Hg flux and environmental parameters that allow for extrapolation of Hg fluxes to larger areas. Reported area average Hg fluxes, determined for naturally Hg-enriched terrains ranging in size from ~1 to 900 km^2, are 2 to 440 ng/m^2 hr (2 to 110 kg/yr) (Gustin, 2003). The magnitude of these flux estimates depends on the proportion of the surface area with high Hg-enrichment relative to that of the entire area studied. Mercury flux from substrate has been found to the strongly correlated with substrate Hg concentrations (Rasmussen et al., 1998; Coolbaugh et al., 2002; Zehner and Gustin, 2002). Typically, areas with high Hg concentrations in substrate are surrounded by areas with lower levels of enrichment that grade into regions with substrate of background concentrations. When emissions are scaled up for areas including predominantly the mineralized zone with high Hg concentrations, area average fluxes are quite high (i.e. Knoxville District, CA, USA, 114 ng/m^2 h, 37 km^2 (Gustin et al., 2003); Almaden, Spain, ~200 to 1500 ng/m^{2}h, ~1 km^2 (Ferrara et al., 1998)). If the area of scaling includes zones of concentrated Hg mineralization, the surrounding area with low levels of Hg enrichment and some background soils, the area average flux becomes less due to the lower fluxes from the background terrain (i.e. Medicine Lake, CA, USA, 2 ng/m^2 h, 242 m^2 (Coolbaugh et al., 2002); Flowery Peak, NV, USA, 18.5 ng/m^2 h, 251 m^2 (Engle and Gustin, 2002)). In areas typical of the Hg global belts there are zones of concentrated Hg enrichment surrounded by vast areas of substrate with low and background Hg concentrations. When scaling most of the emissions occurs from the terrain covered by the latter two substrates because they cover a larger surface area (Gustin, 2003).

In order to understand the significance of natural source emissions we need to develop defensible approaches to scale the spatially limited measured fluxes to larger areas. This is difficult, and only one rigorous attempt has been reported in the literature. Zehner and Gustin (2002) estimated area average Hg emissions for the state of Nevada, USA, which is located within one of the global mercuriferous belts. Their GIS approach entailed the use of a detailed geologic map of the state, the average Hg concentration for specific rock types published in the literature, a database of ~71,000 Hg concentrations in substrate, an algorithm developed from field data (n=303) relating Hg flux to soil Hg concentration (r^2=0.71 for a log-log plot, p<0.005), and delineation of areas of hydrothermal alteration using LANDSAT imagery. Based on their compiled data, the mean area average flux for the state ranged from ~3 to 4 ng/m^{2}h depending on assumptions made regarding the proportion of wet and dry deposition, and re-emission amounts (30-50%). They estimated an overall net emission of 10 Mg/y which is ~1/4 of that thought to be contributed to the atmosphere by coal-fired utility plants in the United States (EPA, 1997). If one scales up their minimum estimate for Nevada to the land area of the global mecuriferous belts given in Lindqvist et al. (1991), a global geologic substrate emission estimate of ~1500 Mg/y is obtained. This should be considered a minimum estimate, however, because not all Hg enriched areas are located within these global belts. Early global models applied a value of ~1 ng/m^{2}h to these broad belts which resulted in a global estimate of 500 Mg Hg/y (cf. Lindqvist et al., 1991). Based on measured fluxes reported by Zhang et al. (2001) and Nacht and Gustin (2004) for background substrates, the flux of 1 ng/m^{2}h is more representative of emissions from terrestrial landscapes with background Hg concentrations.

THE ROLE OF BACKGROUND SOILS

In areas without natural Hg-enrichment by geologic processes after rock formation, Hg in the surface soils consists of a fraction that exists in the parent rock material (USGS, 1970), plus that deposited from the atmosphere over time. For example, Obrist et al. (2004) reported that in tall grass prairie soils the top 1 cm had Hg concentrations of 20 $\pm$ 1.4 ng Hg/g while concentrations at 170 cm were 3.7 ± 0.2 ng Hg/g. The concentration at depth represents the natural geologic signature of the substrate, and the surface concentration, the natural signature plus enrichment from atmospheric deposition directly to the soil and to the overlying vegetation. Atmospheric deposition can include dry deposition of Hg^0 and RGM and wet deposition of what is thought to be predominantly RGM (Schroeder and Munthe, 1998). In the past the contribution from wet deposition was considered the major input. It has been suggested that dry deposition of gaseous elemental Hg^0 to soils could be important (cf. Engle et al., 2001; Kim et al., 1995) and with the developed capability to measure RGM (cf. Landis et al., 2002), which has a high deposition velocity (Lindberg and Stratton, 1998), we are beginning to realize that the oxidation of Hg^0 to RGM, and its subsequent dry deposition could be an important process.

Only recently has work been systematically done to investigate the potential for background soils to be a source or sink of Hg^0. Laboratory studies, using cleaned air (stripped constituents other than nitrogen and oxygen), have revealed a very low range in Hg^0 compensation point air concentrations for soil types with different characteristics (0.1 to 2.4 ng/m^3) (Xin et al., in prep). At air concentrations above the compensation point Hg^0 is deposited to substrate, while below that concentration Hg^0 is emitted. However, when similar experiments are done using ambient air, the data are noisy, resulting in poor regression coefficients for air Hg concentration versus flux, and a higher range for the compensation point (3 to 9 ng/m^3). These observations suggest that in cleaned air, soil physicochemical properties are influencing the compensation point, while in ambient air the compensation point is also influenced by atmospheric chemistry. Engle et al. (2004) have demonstrated in the laboratory that Hg^0 is liberated from soils when they interact with air containing ozone. In addition to heterogeneous reactions between soils and the atmosphere, precipitation, light and increased temperature promote the release of Hg from background soils, as they do for Hg enriched substrates (Gustin et al., 2002; Zhang et al., 2001; Gustin et al., 2004). Once deposited, atmospheric Hg can also become sequestered in surface soils, and its residence time will vary and cannot be generalized. It can be re-emitted by processes described above, removed from the soil surface by processes such as transport via erosion or burial, or it can become strongly adsorbed to soil mineral or organic constituents. The potential for sequestration in the soil and re-emission is being investigated as part of the METALLICUS (Mercury Experiment to Assess Atmospheric Loading in Canada and the United States) project (cf. Hintleman et al. 2002). This project, now in its 5th year, uses stable Hg isotope spikes to different ecosystem compartments to see how Hg moves through the system. Lindberg et al. (2003) estimated that short-term re-emission, immediately following deposition, removed ~10 to 40% of the Hg added as $HgCl_2$, with the percentage dependant upon the surface (the highest re-emission rates occurred for deposition to water surfaces).

Thus, these new data indicate that background soils can accumulate Hg^0 by direct dry deposition, and a variety of physicochemical parameters, including light, moisture, and atmospheric chemistry can facilitate re-emission. Data from the METALLICUS project suggests that RGM deposited in precipitation, as Hg (II), may be rapidly re-emitted once deposited. This process of cycling of Hg between background substrate and the atmosphere would significantly affect the ability of simple mass balance models to successfully model the global Hg cycle.

THE ROLE OF VEGETATION

Vegetation has been demonstrated to function as a sink and source for atmospheric Hg. Foliar uptake of Hg^0 has been suggested to be an important pathway for atmospheric Hg to enter terrestrial ecosystems, and may represent a significant unrecognized sink within the biogeochemical cycle (Lindberg et al., 1992; Ericksen et al., 2003; Lindberg, 1996). Frescholtz et al. (2003), using multiple plant growth chambers and different air and soil Hg exposure concentrations, demonstrated that Hg^0 was accumulated by aspen foliage as a function of air concentration and time. They showed that air concentrations were the dominant factor controlling foliage concentrations and that soil Hg concentration exerted a minor influence. Mosbaek et al. (1988) also demonstrated that the atmosphere was the primary source of Hg in three crop species.

Atmospheric Hg accumulated in foliage could then be transferred to terrestrial and aquatic ecosystems by way of litterfall. Mass balance data collected in forested ecosystems suggests that litterfall is the largest single flux of Hg and methyl Hg to forested systems (Iverfeldt, 1991; Johnson and Lindberg, 1995; Munthe et al., 1995; St Louis et al., 2001). Some authors (e.g. Driscoll et al., 1994; Johnson and Lindberg, 1995; Rea et al., 2002; Ericksen et al., 2003) have suggested that dry deposition to the forested landscape can be estimated from simple measures of throughfall (includes dry deposited RGM+Hg_p) plus litterfall. A very crude estimate of the global flux of Hg in litterfall was made by Lindberg et al. (2004) assuming a range of Hg concentrations in background foliage of ~20-50 ng/g (e.g. Friedli et al., 2001; Lindberg, 1996; Rea et al., 2002; Frescholtz et al., 2003) and global litterfall rates (e.g. Matthews, 1997). This Hg dry deposition flux, if real, could represent an additional Hg sink on the order of ~2400 to 6000 T/y. Some of this additional sink is temporary, of course, with the ultimate net sink depending on the rate of Hg sequestration in soils. However, since Hg sequestered in soils resides initially in the organic horizon, it too is subject to some degree of re-emission, either slowly by evasion, or quickly by fire or other types of disturbance.

Laboratory studies have demonstrated, depending on air and soil Hg concentrations, plants can emit Hg to the atmosphere. To predict whether Hg will be deposited or emitted from foliar surfaces, compensation point concentrations need to be identified for various plant species under different conditions. At concentrations above the compensation point atmospheric Hg^0 will be deposited to the leaf surface and below this concentration Hg^0 will be emitted. Ericksen and Gustin (2003) demonstrated a compensation point of 3 to 4 ng/m^3 in the light and 2 to 3 ng/m^3 in the dark for aspen saplings using a single plant gas exchange chamber. A higher range was determined for red maple, white oak and Norway spruce (10 to 25 ng/m^3) by Hanson et al. (1995). Work is ongoing to determine the compensation point for a variety of other plant species.

Deposition and emission fluxes of Hg to terrestrial and aquatic vegetated systems have been measured in field studies (Lindberg et al., 1998; 2002a; Lee et al., 2002). Lindberg et al. (1998) determined that the net flux was emission for a mature deciduous forest and a pine plantation (mean 20-30 ng/m^2 hr) growing in soils with low Hg concentrations (~50 – 500 ng/g). Lindberg et al. (1998), 2002a suggested that emissions from forests alone could range from 850 to 2000 Mg/y (Table 2). Obrist et al. (2004) measured Hg fluxes over a year in large mesocosms containing intact monoliths of tall grass prairie (soil Hg = 4 to 20 ng/g), and determined that the annual net exchange was 61 ± 25 ug/m^2 with some deposition occurring only in the winter months. They scaled up measured Hg^0 fluxes to the area of the world's grasslands (~3.75 x 10^7 km^2 (DeFries and Townshend, 1994)), and found that Hg emissions to the atmosphere could be ~ 800 to 2300 tons year[1]. It is unclear however, for these two estimates how much Hg was moved from the soil through the plant to the atmosphere for fluxes associated with foliar surfaces were not directly measured. Mercury being moved from the soil to the atmosphere by plants can be new Hg derived from the soil pool and re-emission of previously deposited atmospheric Hg.

The presence of a foliar canopy can reduce substrate emissions. In controlled mesocosm studies, using Hg contaminated soils (12.3 µg Hg/g), during leaf-out, it was clearly demonstrated that Hg flux declined as the soil was shaded by the developing leaf canopy (Gustin et al., in review). Under the full canopy, Hg flux was reduced 1.2 to 1.5 times relative to that occurring from bare soil. Zhang et al. (2001) compared Hg emissions occurring from soils in shaded versus open field sites and found that fluxes were reduced in the shaded site. Gustin et al. (in review) determined that reduction of light on the soil due to the plant canopy was the primary factor causing the reduced flux.

Thus, plants can assimilate atmospheric Hg^0 in their tissue, they can move gaseous Hg^0 from soil pools to the atmosphere (thereby enhancing the net emission over that of bare soils), and they can reduce Hg^0 emissions from soils by way of shading the soil surface. The relative significance of these processes is highly uncertain, and cannot readily be considered in global models.

Because the volatility of Hg is well known and its accumulation in foliage has been demonstrated, it should come as no surprise that biomass burning has recently been suggested to be a significant source of atmospheric Hg. New aircraft measurements of Hg, CO and CO_2 in plumes have been applied for making estimates of Hg released during biomass burning. Mercury emissions of 200 to 1000 Mg/y, were estimated by Brunke et al. (2001) and Friedli et al (2001). Biwas et al. (2003) suggested that soil burning released a more significant amount than biomass, and estimated that 100 Mg/y was removed by this process.

THE POLES AND OTHER NEW SINKS

New laboratory and modelling studies of the role of reactive halogens, especially bromine (e.g. Ariya et al., 2002; Calvert and Lindberg, 2003) clearly indicate their ability to oxidize Hg^0 from the global pool to more reactive and shorter-lived airborne Hg (II) species. This process has been observed during the polar winter and spring after polar sunrise where Hg^0 becomes depleted in the atmosphere, and simultaneously, RGM and particulate Hg concentrations increase with subsequent increases in Hg concentrations in snow (Lindberg et al., 2002b). Initially the atmospheric Hg depletion events measured in the Arctic were postulated to be an important Hg sink (e.g. Schroeder et al., 1998; Lindberg et al., 2002b; Ebinghaus et al., 2002). However, estimates of the size of this net sink (perhaps a few hundred T/y after accounting for re-emissions (e.g. Lindberg et al., 2002b; Schroeder et. al., 2003; Skov et al., 2004) suggest that this may not be a large enough sink to offset the various new sources recently quantified. New data suggests that Polar warming could alter this cycle dramatically (e.g. Schroeder et. al.,

2003), increasing the area over which depletion event occur (Lindberg et al., 2002b). Other evidence suggests that similar halogen reactions, mediated by oxidants such as hydroxyl radical (e.g. Weiss-Penzias et. al., 2003), and oxidation and dry deposition of gaseous Hg (II) over the oceans in more temperate regions of the globe (e.g. Mason and Sheu, 2002) may contribute to a shorter residence time of Hg than predicted. The magnitude of the latter potentially important sink throughout the marine boundary layer as suggested by Mason and Sheu (2002) remains to be quantified.

Finally, the recent reporting of a highly enriched layer of aerosol Hg at the tropopause (associated with elevated halogens, e.g. Murphy and Thompson 2000), makes one question whether stratosphere/troposphere cycling should be included in the global cycle.

CONCLUSIONS

Current emissions estimates for Hg inputs to the atmosphere are on the order of 6000 to 6600 Mg/y, with natural source estimates ranging from 800 to 3400 Mg/y, and anthropogenic sources estimates ranging from 2000 to 2400 Mg/y (Table 1). Re-emission of previously deposited Hg derived from natural and anthropogenic sources is an unquantified source that would add to the atmospheric pool. As discussed above, there is significant uncertainty associated with these estimates. If air concentrations are not increasing, dry and wet Hg deposition estimates (~3000 Mg/y) do not balance the estimated inputs. The role of the ocean as a source or sink is uncertain and based on limited and variable data. Litterfall could be an important unaccounted for sink (though temporary); however, plants may also move Hg from the soil to the atmosphere, and the relative significance of these two processes is not known. In addition, the significance of the role of background soils as a sink, source or temporary resting place for Hg is not known. Based on recent work it appears that Hg can be fairly rapidly cycled between terrestrial surfaces and the atmosphere. This would cause us to have to rethink the year long residence time that is usually attributed to Hg and revolutionize our thinking with the respect to the whole Hg biogeochemical cycle. Investigations of soil compensation points suggest that the balance between deposition and emission can occur at low air concentrations. This work along with the ongoing work to characterize the significance of re-emission is critical for understanding whether simple mass balance models are adequate for characterizing global Hg fluxes.

Because of the large range of uncertainties in estimates of the significance of natural sources or sinks (Table 2), and our incomplete understanding of processes controlling Hg cycling, the question of whether the true net flux of Hg is directed towards or away from the Earth's surface may not yet be answered. Recent research is pointing towards a more rapid recycling between terrestrial surfaces and the atmosphere than has previously been realized, leading us to speculate that the atmospheric residence time is much shorter than models predict. If this is true, global reductions of Hg emissions to the atmosphere should have significant impacts on Hg cycling. However, reductions of industrial emissions will be offset by the degree to which previous deposition is being re-emitted. New isotope manipulation data suggest that deposited Hg is rapidly re-emitted at rates, which decrease over time (Lindberg et al. 2003). New point source controls will initially be most effective in reducing the deposition of the most reactive forms of Hg emitted, and it may require years after major controls to see significant reductions of the overall global pool of Hg^0 because of recycling.

ACKNOWLEDGEMENTS

The authors thank the following for support for research which has lead to a better understanding of natural terrestrial sources and sinks of Hg: U. S. EPA STAR and EPSCoR programs, EPRI, National Science Foundation, U. S. Geological Survey and U.S. Department of Energy. We also thank our many collaborators and graduate and undergraduate students whose diligent work has helped move the research forward. Oak Ridge National Laboratory is operated by UT-Battelle for the US Department of Energy.

REFERENCES

Ariya, P.A., Khalizov, A., Gidas, A. Reactions of gaseous mercury with atomic and molecular halogens: Kinetics, product studies, and atmospheric implications. *J. Phys. Chem. A*, 106, 7310-7320, 2002.

Bergan, T., Gallardo, L., Rohde, H. Mercury in the global trophosphere: A three-dimensional model study. *Atmos. Environ.*, 33, 1575-1585, 1999.

Biswas, A., Blum, J.D., Keeler, G.J. Forest fire effects on mercury and other trace metal concentrations in a Rocky Mountain forest ecosystem. EOS Transaction AGU 2003; 84(46), Fall Meeting Supplement, Abstract B32C-05.

Brunke E.G., Labuschagne D., Slemr F. Gaseous mercury emissions from a fire in the Cape Peninsula, South Africa. *Geophys. Res. Lett.*, 28, 1483-1486, 2001.

Calvert, J.G. Lindberg S.E. A modelling study of the mechanism of the halogen-ozone-mercury homogeneous reactions in the troposphere during the polar spring. *Atmos. Environ.*, 37, 4467-4481, 2003.

Coolbaugh, M.F., Gustin, M.S., Rytuba, J.J. Annual emissions of mercury to the atmosphere from three natural source areas in Nevada and California. *Env. Geol.*, 42, 338-349, 2002.

Dastoor, A.P. Larocque Y. Global circulation of atmospheric mercury: A modelling study. *Atmos. Environ.*, 38, 147-161, 2004.

DeFries, R.S., Townshend, J.R.G. NDVI-Derived land-cover classifications at a global-scale. *Int. Remote Sens.*, 15, 3567-3586, 1994.

Driscoll, C.T., Otton, J.K., Iverfeldt, A. "Trace metals speciation and cycling." In Biogeochemistry of Small Catchments: A Tool for Environmental Research, B. Moldan J. Cerný, eds. Chichester, England: John Wiley & Sons Ltd., 299-322, 1994.

Ebinghaus, R., Kock, H.H., Temme, C., Einax, J.W., Löwe A.G., Richter A., Burrows J.P., Schroeder W.H. Antarctic springtime depletion of atmospheric mercury. *Env.Sci. Technol.*, 36, 1238-1244, 2002.

Engle, M.A., Gustin, M. S., Zhang, H. Quantifying natural source mercury emissions from the Ivanhoe Mining District, north-central Nevada, USA, *Atmos. Environ.*, 35, 3987-3997, 2001.

Engle, M.A., Gustin, M.S., Lindberg, S.E., Gertler, A.W. Investigation of the effect of tropospheric oxidants on Hg emissions from substrates, Symposium paper, 7th International Conference on Mercury as a Global Pollutant; Ljubliana, Slovenia, June 2004.

Engle, M.A. Gustin, M.S. Scaling up atmospheric mercury emissions from three naturally enriched areas: Flowery Peak, Nevada, Peavine Peak, Nevada and Long Valley Caldera, California. *Sci. Tot. Env.*, 290, 91-104, 2002.

EPA. Mercury Study Report to Congress. U. S. Environmental Protection Agency, Office of Air Quality Planning Standards and Office of Research and Development 1997: EPA-452/R-97 004.

Ericksen, J.A. Gustin, M.S. Foliar exchange of mercury as function of soil and air concentration. *Sci. Tot. Env.*, 324, 271-279, 2003.

Ericksen, J.A., Gustin, M.S., Schorran, D.E., Johnson, D.W., Lindberg, S.E. Coleman, J.S. Accumulation of atmospheric mercury in forest foliage. *Atmos. Environ.*, 37, 1612-1622, 2003.

Ferrara, R., Maserti, B.E., Anderson, M., Edner, H., Ragnarson, P., Svanberg, S., Hernandez, A. Atmospheric mercury concentrations and fluxes in the Almaden district (Spain). *Atmos. Environ.*, 32, 3897-3917, 1998.

Ferrara, R., Mazzolai, B., Lanzillotta, E., Nucaro, E., Pirrone, N. Volcanoes as emission sources of atmospheric mercury in the Mediterranean Basin. *Sci. Tot. Env.*, 259, 115-121, 2000.

Fitzgerald, W.F Lamborg, C.H. "Geochemistry of mercury in the environment." In Treatise on Geochemistry, H.D., Holland, K.K., Turekian, eds. Amsterdam: Elsevier, 107-148, 2003.

Frescholtz, T.F., Gustin, M.S., Schorran, D.E., Fernandez, G.C. Assessing the source of mercury in foliar tissue of quaking aspen. *Env. Toxicol. Chem.*, 22, 2114-2119, 2003.

Friedli, H.R., Radke, L.R., Lu, J.Y. Mercury in smoke from biomass fires. *Geophys. Res. Lett.*, 28, 3223-3226, 2001.

Gustin, M.S., Taylor, G.E., Jr., Maxey, R.A. Effect of temperature, wind velocity and concentration on the flux of elemental mercury from mill tailings to the atmosphere. *J. Geophys. Res.*, 102, 3891-3898, 1997.

Gustin, M.S., Biester, H., Kim, C. Investigation of light enhanced emission of mercury from naturally enriched substrate. *Atmos. Environ.*, 36, 3241-3254, 2002.

Gustin, M.S. Are mercury emissions from geologic sources significant? A status report. *Sci. Tot. Env.*, 304, 153-167, 2003.

Gustin, M.S., Coolbaugh, M., Engle, M., Fitzgerald, B., Keislar, R., Lindberg, S., Nacht, D., Quashnick, J., Rytuba, J., Sladek, C., Zhang, H., Zehner, R. Atmospheric mercury emissions from mine wastes and surrounding geologically enriched terranes. *Env. Geol.*, 43, 339-351, 2003.

Gustin, M.S., Ericksen, J.A., Schorran, D.E., Johnson, D.W., Lindberg, S.E., Coleman, J.S. Application of controlled mesocosms for understanding mercury air-soil-plant exchange, *Env. Sci. Technol.*, 38, 6044-6050, 2004.

Gustin, M.S., Zehner, R., Stamenkovic, J. Experimental examination of the influence of precipitation and moisture content on mercury emissions from soils. Symposium paper, 7th International Conference on Mercury as a Global Pollutant; Ljubliana, Slovenia, June 2004.

Hanson, P.J., Lindberg, S.E., Tabberer, T.A., Owens, J.G., Kim, K.-H. Foliar exchange of mercury vapor: Evidence for a compensation point. *Water, Air Soil Pollut.*, 56, 553-564, 1995.

Hintelmann, H., St.Louis, V., Scott, K., Rudd, J., Lindberg, S.E., Krabbenhoft, D., Kelly, C., Heyes, A., Harris, R., Hurley, J. Reactivity and mobility of newly deposited mercury in a Boreal catchment. *Env Sci Technol.*, 36, 5034-5040, 2002.

Iverfeldt, A. Mercury in forest canopy throughfall water and its relation to atmospheric deposition. *Water, Air Soil Pollut.*, 56, 553-564, 1991.

Johnson, D.W. Lindberg S.E. Sources, sinks, and cycling of mercury in forested ecosystems. *Water, Air Soil Pollut.*, 80, 1069-1077, 1995.

Jonhanson, I.R. Boyle, R.W. Geochemistry of mercury and origins of natural contamination in the environment. CIM Transactions 1972; v. LXXV, 8-15.

Kim, K.-H., Lindberg, S.E., Meyers T.P. Micrometeorological measurements of mercury vapor fluxes over background forest soils in eastern Tennessee. *Atmos. Environ.*, 29, 267-282, 1995.

Lamborg, C.H., Fitzgerald, W.F., O'Donnell, J., Torgerson, T. A nonsteady state compartmental model of global-scale mercury biogeochemistry with interhemispheric atmospheric gradients. *Geochim. Cosmochi. Acta*, 66, 1105-1118, 2002.

Landis, M.A., Stevens, R.K., Schaedlich, F., Prestbo, E. Development and characterization of an annular denuder methodology for the measurement of divalent inorganic reactive gaseous mercury in ambient air. *Env Sci Technol.*, 36, 3000-3009, 2002.

Lee, X., Benoit, G., Hu, X.Z. Total gaseous mercury concentration and flux over saltmarsh vegetation in Connecticut, USA. *Atmos. Environ.*, 34, 4205-4213, 2002.

Lindberg, S.E., Jackson, D.R., Huckabee, J.W., Janzen, S.A., Levin, M.J., Lund, J.R. Atmospheric emission and plant uptake of mercury from agricultural soils near the Almaden mercury mine. *J. Envi. Quality*, 8, 572-578, 1979.

Lindberg, S.E., Meyers, T.P., Taylor, G.E., Turner, R.R., Schroeder, W.H. Atmosphere / surface exchange of mercury in a forest: Results of modeling and gradient approaches. *J. Geophys. Res.*, 97, 2519-2528, 1992.

Lindberg, S.E. "Forests and the global biogeochemical cycle of mercury: the importance of understanding air/vegetation exchange processes." In Regional and Global Mercury Cycles, W. Baeyens, ed. Dordrecht, Holland: Kluwer Academic Publisher, 359-380, 1996.

Lindberg, S.E., Hanson, P.J., Meyers, T.P., Kim, K.-H. Air/surface exchange of mercury vapor over forests: The need for a reassessment of continental biogenic emissions. *Atmos. Environ.*, 32, 895-908, 1998.

Lindberg, S.E. Stratton, W.J. Atmospheric mercury speciation: Concentrations and behavior of reactive gaseous mercury in ambient air. *Env Sci Technol.*, 32, 49-57, 1998.

Lindberg, S.E., Zhang, H., Gustin, M.S., Casimir, A., Ebinghaus, R., Edwards, G., Fitzgerald, C., Kemp, J., Kock, H., Leonard, T., Majewski, M., Marsik, F., Owens, J., Poissant, L. Rasmussen, P., Schaedlich, F., Schneeberger, D., Sommar, J., Turner, R., Vette, A., Wallschlaeger, D., Xiao, Z. The role of rainfall and soil moisture on mercury emissions from mercuriferous desert soils. *J. Geophys. Res.*, 104, 21,879-21,888, 1999.

Lindberg, S.E., Dong, W., Meyers, T. Transpiration of gaseous mercury through vegetation in a subtropical wetland in Florida. *Atmos. Environ.*, 36, 5200-5219, 2002a.

Lindberg, S.E., Brooks, S.B., Lin, C-J., Scott, K.J., Landis, M.S., Stevens, R.K., Goodsite, M., Richter, A. The Dynamic Oxidation of Gaseous Mercury in the Arctic Atmosphere at Polar Sunrise. *Env Sci Technol.*, 36, 1245-1256, 2002b.

Lindberg, S.E., Southworth, G., Peterson, M., Hintelmann, H., Graydon, J., St Louis, V., Amyot, M., Krabbenhoft, D. Quantifying reemission of mercury from terrestrial and aquatic systems using stable isotopes: Results from the Experimental Lakes Area METAALICUS study. *EOS Trans.* 84(46) Fall Meeting Supplement, Abstract B31E-0364, 2003.

Lindberg, S.E., Porcella, D., Prestbo, E., Friedli, H., Radke, L. The problem with mercury: too many sources, not enough sinks. Symposium paper, 7th International Conference on Mercury as a Global Pollutant; Ljubliana, Slovenia, June 2004.

Lindqvist, O., Johanssson, K., Aastrup, M., Andersson, A., Bringmark, L., Hovsenisus, G., Hakanson, L., Iverfelt, A., Meili, M., Timm, B. Mercury in the Swedish environment: Recent research on causes, consequences and corrective methods. *Water, Air Soil Pollut.*, 55, 1-261, 1991.

Mason, R.P., Fitzgerald, W.F., Morel, F.M.M. The biogeochemical cycling of elemental mercury: anthropogenic influences. *Geochim. Cosmochim. Acta*, 58, 3193-3198, 1994.

Mason, R.P. Sheu, G.-R. Role of the ocean in the global mercury cycle. *Global Biogeochem. Cyc.*, 16, 401-414, 2002.

Mather, T.A. Pyle, D. M. Comment on Volcanic emissions of mercury to the atmosphere: global and regional inventories *Sci. Tot. Env.*, 327, 323-329, 2004.

Matthews, E. Global litter production, pools, and turnover times: Estimates from measurement data and models. *J. Geophys. Res.*, 102, 18771-18800, 1997.

Mosbaek, H., Tjel,l J.C., Sevel, T. Plant uptake of airborne mercury in background areas. *Chemosphere*, 17, 1277-1236, 1988.

Munthe, J., Hultberg, H., Iverfeldt, A. Mechanisms of deposition of methyl mercury and mercury to coniferous forests. *Water, Air Soil Pollut.*, 80, 363-371, 1995.

Murphy, D.M. Thomson, D.S. Halogen ions and NO^+ in the mass spectra of aerosols in the upper troposphere and lower stratosphere. *Geophy. Res. Lett.*, 27, 3217, 2000.

Nacht, D.M. Gustin, M.S. Mercury emissions from background and altered geologic units throughout Nevada. *Water, Air Soil Pollut.*, 151, 179-193, 2004.

Nriagu, J.O. A global assessment of natural sources of atmospheric trace metals. *Nature*, 338, 47-49, 1989.

Nriagu, J.O. Becker, C. Volcanic emissions of mercury to the atmosphere: Global and regional inventories. *Sci. Tot. Env.*, 304, 3-12, 2003.

Nriagu, J.O. Becker, C. Reply to Comment on Volcanic emissions of mercury to the atmosphere: global and regional inventories. *Sci. Tot. Env.*, 327, 331-333, 2004.

Obrist, D., Gustin, M.S., Arnone, J.A., Johnson, D.W., Schorran, D.E., Verburg, P.J. Large annual mercury emissions to the atmosphere measured over Tallgrass Prairie ecosystems. *Atmos. Environ.*, (Submitted) 2004.

Pacyna, E.G., Pacyna, J.M., Pirrone, N. European emissions of atmospheric mercury from anthropogenic sources in 1995. *Atmos. Environ.*, 35, 2987-2996, 2001.

Pai, P., Heisler, S., Joshi, A. An emission inventory for regional atmospheric modeling of mercury. *Water, Air Soil Pollut.*, 101, 289-308, 1998.

Pennington, J.W. Mercury: A materials survey. Bureau of Mines Information Circular 7941, United States Government Printing Office, Washington D.C., 28, 1959.

Pirrone, N., Costa, P., Pacyna, J.M., Ferrara, R. Mercury emissions to the atmosphere from natural and anthropogenic sources in the Mediterranean region. *Atmos. Environ.*, 35, 2997-3006, 2001.

Pyle, D.M. Mather, T.A. The importance of volcanic emissions for the global atmospheric mercury cycle. *Atmos. Environ.*, 37, 5115-5124, 2003.

Rasmussen, P.E., Edwards, G.C., Kemp, J.R., Fitzgerald-Hubble C.R., Schroeder, W.H. Towards an improved natural sources inventory for mercury. In: Skeaff J, editor. Proceedings on the Metals in the Environment: Candian Institute of Mining, International Symposium 74-82, 1998.

Rea, A.W., Lindberg, S.E., Scherbatskoy, T. Mercury accumulation in foliage over time in two northern mixed-hardwood forests. *Water, Air Soil Pollut.*, 133, 49-67, 2002.

Schroeder, W.H. Munthe, J. Atmospheric Mercury-An Overview. *Atmos. Environ.*, 32, 809-822, 1998.

Schroeder, W.H., Anlauf, K.G., Barrie, L.A., Lu, J.Y., Steffen, A., Schneeberger, D.R., Berg, T. Artic springtime depletion of mercury. *Nature*, 394, 331-332, 1998.

Schroeder, W.H., Steffen, A., Scott, K., Bender, T., Prestbo, E., Ebinghaus, R., Lu, J.Y., Lindberg, S.E. Summary report: First international Arctic atmospheric mercury research workshop. *Atmos. Environ.*, 37, 2551-2555, 2003.

Seigneur, C., Karamchandandi, P., Lohman, K., Vijayarahavan, K., Shia, R.L. Multiscale modeling of atmospheric fate and transport of mercury. *J. Geophys. Res.*, 106, 27795-27809, 2001.

Seigneur, C., Vijayaraghavan, K., Lohman, K., Karamchandani, P., Scott, C. Global source attribution for mercury deposition in the United States. *Env Sci Technol.*, 38, 555-569, 2004.

Skov H., Christensen J.H., Goodsite M.E., Heidam N.Z., Jensen B., Wahlin P., Geernaert G. Fate of elemental mercury in the artic during atmospheric mercury depletion episodes and the load of atmospheric Hg to the Artic. *Env Sci Technol.*, 38, 2372-2382, 2004.

St. Louis, V.L., Rudd, J.W.M., Kelly, C.A., Hall, B.D., Rolfus, K.R., Scott, K.J., Lindberg, S.E., Dong, W. Importance of forest canopy to fluxes of methyl mercury and total mercury to boreal ecosystems. *Env. Sci. Toxicol.*, 35, 3089-3098, 2001.

U.S. Geological Survey. Mercury in the Environment. U.S.G.S. Professional Paper 713, 67, 1970.

Varekamp, J.C., Buseck, P.R. Mercury emissions from Mount St. Helens during September 1980. *Nature*, 293, 555-556, 1981.

Varekamp, J.C. Buseck, P.R. The speciation of mercury in hydrothermal systems, with applications to ore deposition. *Geochim. Cosmochim. Acta*, 48, 177-185, 1984.

Varenkamp J. C., Buseck P.R. Global mercury flux from volcanic and geothermal sources. *Appl. Geochem.*, 1, 65-73, 1986.

Weiss-Penzias, P., Jaffe, D.A., McClintick, A., Prestbo, E.M., Landis, M.S. Gaseous elemental mercury in the marine boundary layer: Evidence for rapid removal in anthropogenic pollution. *Env. Sci. Technol.*, 37, 3755-3763, 2003.

Xin, M., Gustin, M. S., Johnson, D.W. Elemental Hg deposition to soil: Development of compensation points and an understanding of controlling factors. (In preparation).

Zehner, R., Gustin M.S. Estimation of mercury vapor flux emissions from natural geologic sources in Nevada. *Env. Sci. Technol.*, 36, 4039-4045, 2002.

Zhang H. Lindberg S.E. Processes influencing the emission of mercury from soils: A conceptual model. *J. Geophys. Res.*, 104, 21,899-21,896, 1999.

Zhang H., Lindberg S.E., Marsik F.J., Keeler G.J. Mercury air/surface exchange kinetics of background soils of the Tahquamenon River watershed in the Michigan upper peninsula. *Water, Air Soil Pollut.*, 126, 151-169, 2001.

Chapter-12

CHEMICAL TRANSFORMATION OF GASEOUS ELEMENTAL MERCURY IN THE ATMOSPHERE

Parisa A. Ariya[1] and Kirk A. Peterson[2]

[1] *Departments of Chemistry and Atmospheric and Oceanic Sciences McGill University, 801 Sherbrooke St. W.,Montreal, PQ, Canada, H3A 2K6*
[2] *Department of Chemistry, Washington State University Pullman, WA, USA, 99164-4630*

INTRODUCTION

Mercury is a fascinating fluid metal that exists as liquid at 25° C and has the lowest known critical temperature (1478 ° C) (Hensel and Warren, 1999). The dominant form of mercury in atmosphere is the gaseous elemental mercury ($Hg^0_{(g)}$) and it is assumed to have a relatively long lifetime (Seiler et al., 1980; Slemr et al., 1985; Lindqvist and Rodhe, 1985). It is speculated that there is about 6,000 tonnes of mercury in the atmosphere, predominantly from anthropogenic origin, such as fuel and coal combustion and waste incineration (UNEP, 2002). Natural emissions, including those from volcanic eruptions, soils, lakes, open water and forest fires, contribute less significantly than anthropogenic sources (*ca.*, 2000 tonnes/yr (Mason et al., 1994) in contrast to 4000 tonnes/yr (Porcella et al., 1997). However, there are significant uncertainties in natural emission inventories (Mason et al., 1994; Gardfeldt, 2003). Atmospheric transformation of mercury can indeed play an important role in the global cycling of this toxic element, as the atmosphere is the fastest moving fluid in the Earth's ecosystem. The atmosphere provides an efficient platform for chemical and physical transformation, as well as short and long-term transport of this pollutant around the globe. The background concentration of atmospheric Hg, which is mainly elemental mercury, in the lower troposphere of the Northern Hemisphere and Southern Hemisphere is around 1.7 and 1.3 ng m^{-3}, respectively (Slemr, 1996; Slemr et al., 2002; Ebinghaus et al., 2002a). Recently it has been reported (Temme et al., 2003) that the fast oxidation of gaseous elemental mercury leads to variable Hg^0 concentrations during the Antarctic summer, accompanied by elevated concentrations, up to more than 300 pg m^{-3}, of reactive gaseous mercury.

A major interest in the understanding of atmospheric transformation stems from its potential impact on mercury bioaccumulation. Mercury speciation in the atmosphere has a significant influence on its deposition on environmental surfaces. Solubility and deposition of elemental mercury is quite distinct from Hg^{+2} (see Table 1), and thus deposition rates on the Earth's surface vary substantially.

Table 1. Selected physical and chemical properties of mercury compounds.

	Hg^0*	HgCl$_2$	Hg(CH$_3$)$_2$	HgBr$_2$	HgBr	HgO
Physical state (colour, phase, etc)	Silver coloured liquid	White crystalline solid	Colourless liquid	White crystalline solid	White crystalline solid	Yellow or red crystalline solid
ΔH (enthalpy of formation)	0.0	-224.3 kJ/mol	162.2 kJ/mol	-169.45 kJ/mol	-206.90 kJ/mol	-90.79 kJ/mol
ΔG	0.0	-178.6 kJ/mol		-153.10 kJ/mol	-181.075 kJ/mol	
Melting point (°C)	-38.72°C	278°C	-43°C	238°C	345°C (sublimation point)	500°C
Water solubility	Insoluble	Soluble	Insoluble	Slightly soluble		Insoluble
Density	13579.04 kg/m^3	5600 kg/m^3	2961 kg/ m^3	6050 kg/m^3	7307 kg/m^3	11140 kg/m^3
Ionization energy	1007.1 kJ/mol	11.38 eV (1 098.0 kJ/mol)	9.10 eV (878.0 kJ/mol)	10.560 eV (1 018.9 kJ/mol)		
Oxidation States	2,1	1		1	0	0
Atomic mass	200.59	271.495	230.66	360.398	280.49	216.589
Vapour pressure	0.0002Pa @ -38.72°C	1.3mmHg (173.32 Pa) @236°C	62.3 mmHg (8 305.98 Pa) @ 25 °C	120.3 mm Hg (16040 Pa) @ 238°C		

* Electronic configuration 5s^{2}p^{6}d^{10} 6s^2

Methyl mercury has been shown to be biomagnified along food chains. The extent of methylation depends on a constant supply of inorganic mercury from the atmosphere (Mason and Sheu, 2002). Indeed, atmospheric deposition is considered to be a major source of mercury in most remote aquatic systems (Mason et al., 1994; Nriagu and Pacyna, 1988). It is noteworthy that the physicochemical-biological processes that dictate the bioaccumulation of mercury in the food chain have yet to be fully characterized (Morel et al., 1998). Consequently, the extent of incorporation of oxidized mercury produced via atmospheric chemical reactions into the food chain has yet to be evaluated. Figure 1 depicts a simplified schematic of mercury cycling in the Earth's ecosystem and Table 1 illustrates some general physical and chemical properties of mercury compounds of atmospheric interest.

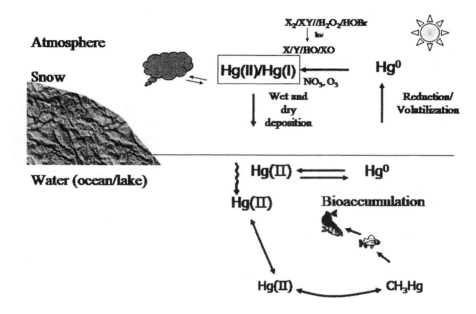

Figure 1. Simplified schematic of Hg cycling in the Earth's ecosystem.

Hg^0 is emitted to the atmosphere from a variety of natural and anthropogenic sources and is removed through dry and wet deposition processes (Lindqvist and Rodhe, 1985). Hg^0 exists in ambient air both in the vapour and particle phase associated with aerosols. Interestingly, in the high-Arctic region (Schroeder et al., 1998), Arctic (Lindberg et al., 2002), and sub-Arctic (Poissant, 2001), the rapid depletion of mercury has been observed. Nearly complete depletions of ozone in the boundary layer occurred over large areas, and evidence of reactive halogens have been observed during most mercury depletion events (MDEs) (Barrie et al., 1988; Jobson et al., 1994; Ariya et al., 1998; 1999). Upon reaction with atmospheric oxidants, elemental mercury can be transformed to its oxidized forms, which are also more bio-accumulative than elemental mercury (Gardfeldt et al., 2001). Observed ozone depletion events at the ground are suggested to be driven by sunlight and bromine atoms derived from reactions of atmospheric reactive halogens with marine sea salt in surface snow and ice (Barrie et al., 1988; Jobson et al., 1994; Ariya et al., 1998; 1999 and 2004). Soon after, mercury depletion was found to be wide spread. Such depletion events have also been observed in the Antarctic (Ebinghaus et al., 2002b), where they are influenced by the photochemical oxidation of elemental mercury in the troposphere involving sea salt on snow/icepack or aerosols (Schroeder and Barrie, 1998; Schroeder et al., 1999; Lu et al., 2001).

Unlike the reactions of Hg^0 in solution, experimental data of the gas-phase reactions of elemental mercury with some atmospheric oxidants are limited due to the low concentrations of species at atmospheric conditions, the low volatility of products, and the strong effects of water vapour and surface on kinetics. Existence of a dark heterogeneous reaction leading to the formation of Br_2 via the oxidation of Br^- by O_3 has been investigated (Oum et al., 1998). Molecular halogens are photochemically dissociated during polar sunrise, resulting in the production of corresponding atomic species. The halogen atoms further react with ozone, forming halogen oxide radicals (BrO^-/ClO^-), which undergo reactions with elemental mercury and transfer them to the oxidized Hg^{+2} species (Raofie and Ariya, 2004). The primary halogen atoms (Br/Cl) and molecular halogens (Br_2/Cl_2) may also oxidize elemental mercury to $HgBr_2$, and $HgCl_2$, respectively (Ariya et al., 2002a). The possible reactions of elemental Hg^0 with a variety of atmospheric oxidants and reductants were evaluated using their thermodynamic data by researchers (Schroeder et al., 1991), who then suggested that O_3 and Cl_2 may be important oxidants of Hg^0, while SO_2 and CO may be important reductants of Hg^{+2}. Both oxidation of Hg^0 and reduction of Hg^{+2} by H_2O_2 are thermodynamically favourable, although literature (Wigfield and Perkins, 1985) indicates the incapability of the oxidation of Hg^0 by H_2O_2.

There have been several excellent review articles on mercury transformation in atmosphere (Linqivist et al., 1999; Schroeder et al., 1991; Lin and Pehkonen, 1999), particularly on its properties, sources, sinks, and fluxes of mercury. As such, in the light of recent laboratory and computational studies, we will attempt to focus instead on a comprehensive review of the kinetic, product studies and thermochemical calculations of gaseous elemental mercury reactions with atmospheric oxidants. We will outline major gaps and some future research directions.

Kinetic and product analysis

The rate of the atmospheric chemical transformation of elemental mercury towards a given oxidant is dependent on two factors. The first factor is the reactivity of mercury toward a given oxidant at environmentally relevant conditions, such as temperature, pressure, oxygen concentration, and relative humidity. The second factor is the concentration (or mixing ratio) of the given oxidant in atmosphere. The existing mercury kinetic reactions have been obtained using steady state reaction chamber or fast flow tubes. Both relative and absolute techniques were used in laboratory studies. The latter refers to an experimental setup in which the reactant destruction or product formation is monitored as a function of time. There are several absolute studies of gas-phase reactions of mercury with various atmospheric oxidants namely for $Cl + Hg$ (Horne et al., 1968) for Cl_2 and $Br_2 + Hg$ (Ariya et al., 2002a), for $O_3 + Hg$ (Pal and Ariya, 2004a) and for $HO + Hg$ (Bauer et al., 2003). An alternative to absolute measurement is the relative technique. It refers to two competing reactions, one of which is known and used as reference. If there are no secondary reactions leading to further decay reactants or reformation of them, one then can express the kinetics of the reaction of interest using a very simple kinetic expression:

$$X + Hg^0 \rightarrow \text{products} \qquad \qquad \text{(R1)}$$
$$X + Ref \rightarrow \text{products} \qquad \qquad \text{(R2)}$$

If both mercury and reference molecules are removed solely by the reaction with oxidant "X", the integrated relative rate equation is expressed as:

$$\ln\{[Hg^0]_0/[Hg^0]_t\} = k_{Hg}/ k_{Ref} \ln\{[Ref]_0/[Ref]_t\} \qquad \qquad \text{(I)}$$

where $[\ ]_0$ and $[\ ]_t$ are the corresponding initial concentrations and the concentrations after time t of Hg^0 and the reference compounds, k_{Hg} and k_{Ref} are the rate constants for reactions (R1) and (R2), respectively. A plot of $\ln\{[Hg^0]_0/[Hg^0]_t\}$ vs. $\ln\{[Ref]_0/[Ref]_t\}$ should give a straight line of slope k_{Hg} / k_{Ref} and a zero intercept. Both absolute and relative techniques have advantages and disadvantages. The disadvantage of the relative rate is that the calculated absolute rate is only as good as the original value of the reaction rate constant for the reference molecule used, and this why most detailed relative rate studies include several reference molecules to overcome this challenge. In addition, one should perform experiments to assure that the extent of side or secondary reactions are negligible and indeed the calculated rates are accurate. To do so, several experiments should be performed under various experimental conditions (e.g., various concentration/pressure of reactants, reactions in different bath gases and at different pressure, different temperature, using scavengers, etc). An advantage of a detailed relative study is the fact that one can readily perform the experiments under tropospheric conditions, and also the reaction chambers can be coupled for simultaneous analysis to several state-of-the-art instruments that allows detailed product analysis as well as kinetics. The advantage of the absolute method is clearly the fact that there is no need for incorporation of errors due to the reference molecule. However, in many absolute studies, one can follow merely one or two reactants, and considering the complexity of mercury reactions, and the extent of secondary reactions, the calculated values may be biased. Another challenge is that some absolute studies are performed at lower pressure than tropospheric boundary layer pressure ($\sim$ 740 Torr), and hence the data obtained under such conditions should be properly corrected for the ambient tropospheric situation. Particularly in the case of complex mercury adduct reactions, and given the lack of detailed product analysis, and different carrier gases, the absolute kinetic studies following one reactant or product, and not having a detailed chemical behaviour of the reaction mixture, may not be the best way to perform experiments. However, in the ideal scenario, both relative and absolute techniques should yield identical values within experimental uncertainties (Pal and Ariya, 2004a).

REACTIONS WITH ATMOSPHERIC OXIDANTS

Most atmospheric oxidation reactions are likely to react via multi-step reaction mechanisms involving the formation of various intermediates that may be thermodynamically favourable even if formation of products is unfavourable, leading to reasonably fast reaction rates. Figure 2 represents the calculated lifetime of the major atmospheric oxidants with elemental mercury, based on the estimated oxidant concentrations. It is of note that significant concentrations of halogens are observed predominantly over the marine boundary layer and not generally over the continents, with the exception of high halide source regions such as salt lakes, and some industrial regions where halogens are widely used.

Hg⁰ + HO

Hydroxyl radical (HO) is considered to be the dominant daytime cleanser of the atmosphere.

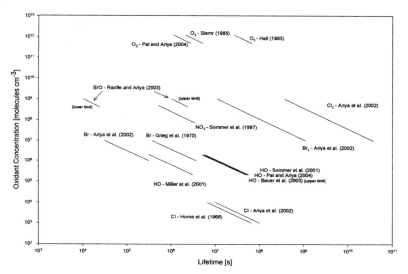

Figure 2. Approximate lifetime of mercury at different oxidant concentrations.*

* *Note:* That to date halogens and halogen oxides have been mainly observed in the marine boundary layer or coastal regions, nitrate radicals are observed only at nighttime.

The major formation pathway in the troposphere is considered to involve photolysis of ozone followed by the reaction with water vapour:

$$O_3 + h\upsilon \text{ (predominantly } \lambda \leq 320 \text{ nm)} \rightarrow O(^1D) + O_2 \tag{1}$$
$$O(^1D) + H_2O \rightarrow 2 \text{ HO} \tag{2}$$

In addition to the photolysis pathway, there are some dark reactions (ozonolysis) that have been proposed to be of significance at night or during the winter (Ariya, et al., 2000). Typical background concentrations of ozone range from 20-30 ppbv (1 ppbv = 2.45×10^{10} cm^{-3}) and can peak to a few hundred ppb during smog situations (Finalyson-Pitts and Pitts, 1997).

To date, there is very limited kinetic data on HO + Hg⁰$_{(g)}$ as shown in Table 2. The results of Sommar et al. (2001) are in excellent agreement with the recent results of Pal and Ariya (2004), and in contrast with Miller et al.

(2001). The values of pre-exponential factors were found to be 3.55×10^{-14} and 2.83×10^{-14} using ethane and cyclopropane, respectively. The (E/R) values were also calculated to be $-(294 \pm 16)$ and $-(322 \pm 24)$ for this reaction, using ethane and cyclopropane, correspondingly. Hence, hydroxyl initiated reactions of elemental mercury seem to possess a slight negative temperature dependence. Hence, assuming a typical tropospheric HO concentration of 1×10^6 radicals cm^{-3}, the tropospheric lifetime of mercury against reaction with HO radical was calculated to be over 3 months using the estimated rate constant (9.3×10^{-14} cm^3 molecule^{-1} s^{-1}) at 298 K.

Table 2. Available literature room temperature (296 ± 2 K) rate constant for selected atmospheric reactions.

Reaction	Rate constants (molecule cm^{-3} s^{-1})	Reference	Comments
$Hg^0 + O_3 \rightarrow$ products	$(3 \pm 2) \times 10^{-20}$ $(7.5 \pm 0.9) \times 10^{-19}$	Hall (1995) Pal and Ariya, (2004)	Temperature dependence is evaluated
$Hg^0 + HO \rightarrow$ products	$(8.7 \pm 2.8) \times 10^{-14}$ $(1.6 \pm 0.2) \times 10^{-12}$ $(9.3 \pm 1.3) \times 10^{-14}$ $<10^{-13}$	Sommar et al. (2001) Miller et al. (2001) Pal and Ariya, (2004) Bauer et al. (2003)	70 °C Temperature dependence is evaluated at 100 and 400 Torr He and air
$Hg^0 + Cl \rightarrow$ products	$(1.0 \pm 0.2) \times 10^{-11}$ $(1.5) \times 10^{-11}$	Ariya et al. (2002) Horne et al. (1968)	120 – 170 °C
$Hg^0 + Cl_2 \rightarrow$ products	$(2.7 \pm 0.2) \times 10^{-18}$	Ariya et al. (2002)	
$Hg^0 + Br \rightarrow$ products	$(3.2 \pm 0.3) \times 10^{-12}$ $(2.7) \times 10^{-13}$	Ariya et al. (2002) Grieg et al. (1970)	120 – 170 °C
$Hg^0 + BrO \rightarrow$ products	$1 \times 10^{-15} < k$ $< 1 \times 10^{-13}$	Raofie and Ariya (2003)	
$Hg^0 + Br_2 \rightarrow$ products	$\leq (9 \pm 2) \times 10^{-17}$	Ariya et al. (2002)	
$Hg^0 + NO_3 \rightarrow$ products	$\leq 4 \times 10^{-15}$	Sommar et al. (1997)	
$Hg^0 + H_2O_2 \rightarrow$ products	$\leq 8 \times 10^{-19}$	Tokos et al. (1998)	
$Hg(CH_3)_2 + HO \rightarrow$ products	$(1.97 \pm 0.23) \times 10^{-11}$	Niki et al. (1983a)	
$Hg(CH_3)_2 + Cl \rightarrow$ products	$(2.75 \pm 0.3) \times 10^{-10}$	Niki et al. (1983b)	
$Hg(CH_3)_2 + NO_3 \rightarrow$ products	$(7.4 \pm 2.6) \times 10^{-14}$	Sommar et al. (1996)	

This calculated lifetime hints at the importance of HO radicals as an effective oxidant of tropospheric mercury, however it suggests that the observed rapid Arctic mercury depletion cannot be explained by HO chemistry.

$O_3 + Hg^0_{(g)}$

Ozone is an important atmospheric constituent and due to its atmospheric abundance, $O_3 + Hg^0$ has been a target of several laboratory studies. In all existing studies relative rate techniques have been used. Therefore, it is important to evaluate the direct impact of O_3-initiated oxidation reactions of mercury on human health and the chemistry of global atmosphere. The kinetics of several aqueous phase reactions of elemental mercury with O_3 were identified (Iverfeldt and Lindqvist, 1986; Munthe, 1992). The corresponding gas phase reaction has also been reported (Hall, 1995; P'yankov, 1949). The kinetic rate constant of 4.9 x 10^{-18} to 4.2 x 10^{-19} $cm^3 molec^{-1}s^{-1}$ (at 293 K) for the gas phase reaction of Hg^0 and O_3 has been estimated (Slemr et al., 1985; Schroeder et al., 1991; Iverfeldt and Lindqvist, 1986) using the experimental data of P'yankov (1949). In 1995, Hall re-investigated the kinetics of gas-phase oxidation of Hg^0 by O_3 and the rate constant found was 3.0 x 10^{-20} $cm^3 molec^{-1}s^{-1}$. Recent studies of Pal and Ariya (2004) provided temperature dependent kinetics of this reaction, with a room temperature rate constant of 7.5 × 10^{-19} cm^3 $moelcule^{-1}$ s^{-1}. Although the existing range of the rate value is indeed large, they all point to the fact that the lifetime of mercury upon reactions with ozone is indeed large (> one month) and is sufficient for long-range transport. However, one should not overrule the possibility of a shorter lifetime when ozone concentration is substantially higher (for instance close to urban sites).

$NO_3 + Hg^0_{(g)}$

The nitrate radical, NO_3, is an important intermediate in the night time chemistry of the atmosphere. Upon sunrise, nitrate ions undergo photolysis to NO_2 or NO (Finlayson-Pitts and Pitts, 1997). Temperature dependent kinetics of elemental mercury, as well as dimethyl mercury, with NO_3 have been studied (Table 2 and Figure 2). Sommar et al. (1997) employed a fast flow-discharge technique to study these reactions and obtained a second order rate constant value of 4 × 10^{-15} cm^3 $molec^{-1}$ s^{-1}. It is of note that since nitrate reactions are only valid at night, this estimated lifetime should be considered as the upper limit. The observed nitrate concentrations are much more erratic and variable than HO and ozone and hence more detailed research in measurement, laboratory and modelling are required prior to proper evaluation of the importance of NO_3 in the oxidation of elemental mercury.

$X_2/X/XO$ (X = Cl, Br, and I) + $Hg^0_{(g)}$

There is very limited data on reactions of halogens and halogen oxides with mercury, due to the complexity of these reactions in terms of side and heterogeneous reactions. In this section, we mainly deal with the experimental studies. In the following section ab-initio studies of halogen and halogen oxides with mercury will be discussed. Very few experimental studies dealing with the gas phase reactions between elemental mercury and halogen-containing molecules have been reported (Table 2). For instance, methyl iodide has been shown to be non-reactive toward Hg^0 under atmospheric conditions ($k < 1 \times 10^{-21}$ cm^3 molecules^{-1} s^{-1}) (Tokos et al., 1998). In others, molecular chlorine was suggested to have a relatively modest reaction rate, 4×10^{-16} cm^3 molecules^{-1} s^{-1} (Menke and Wallis, 1980; Medhekar et al., 1979; Skare and Johansson, 1992; Schroeder et al., 1991; Seigneur et al., 1994), though the reaction was found to be strongly surface catalysed, (Medhekar et al., 1979; Skare and Johansson, 1992) and this value should be considered as an upper limit. Recently, lower values ($(2.7 \pm 0.2) \times 10^{-18}$) for reactions of molecular chlorine (Ariya et al., 2002a) have been reported. The early studies of atomic halogens are limited to one, where the kinetics of the reaction between gaseous mercury and chlorine atoms was followed by monitoring HgCl using time resolved absorption spectroscopy in the temperature range 383 – 443 K (Horne et al., 1968):

$$Hg^* + CFCl_3 \rightarrow Hg + CF_3 + Cl \qquad (3)$$
$$Hg + Cl \rightarrow HgCl \qquad (4)$$
$$HgCl + HgCl \rightarrow Hg_2Cl_2 \qquad (5)$$

The extinction coefficient of HgCl was evaluated first and the recombination rate constant, k_3 was determined to be 5.0×10^{-10} and 3.2×10^{-10} cm^3 molecules^{-1} s^{-1} in 720 Torr CF$_3$Cl and in a mixture of 10 Torr CF$_3$Cl+710 Torr Ar as a bath gas, respectively. The rate constant, k_2 for the reaction of mercury with chlorine atoms was then derived to be 5.0×10^{-11}

cm^3 molecules^{-1} s^{-1} (720 Torr CF$_3$Cl) and 1.5×10^{-11} cm^3 molecules^{-1} s^{-1} (10 Torr CF$_3$Cl+710 Torr Ar) using corresponding values of k_3. The authors (Horne et al., 1968) have mentioned that k_2 has an uncertainty of a factor of three because of the accumulation of experimental errors in evaluating the separate terms. Recently extensive kinetic and product studies on the reactions of gaseous Hg^0 with molecular and atomic halogens (X/X$_2$ where X = Cl, Br) have been performed at atmospheric pressure (750±1 Torr) and room temperature (298±1 K) in air and N$_2$ (Ariya et al., 2002a). Kinetics of the reactions with X/X$_2$ were studied using relative and absolute techniques by cold vapour atomic absorption spectroscopy (CVAAS) and gas chromatography with mass spectroscopic detection (GC-MS). The measured rate constants for the reactions of Hg^0 with Cl$_2$, Cl, Br$_2$, and Br were $(2.6\pm0.2)\times10^{-18}$, $(1.0\pm0.2)\times10^{-11}$, $<(0.9\pm0.2)\times10^{-16}$ and $(3.2\pm0.3)\times10^{-12}$ cm^3 molecule^{-1} s^{-1}, respectively. Chlorine and bromine radicals were generated using UV and visible photolysis of molecular chlorine and bromine, respectively, in addition to UV (300 ≤ λ ≤ 400 nm) photolyis of chloroacetyl chloride and dibromomethane, for generation of Cl and Br. The reaction products were analyzed in the gas-phase from the suspended aerosols and from the wall of the reactor using MS, GC-MS and inductively coupled plasma mass spectrometry (ICP-MS). The major products identified were HgCl$_2$ and HgBr$_2$ adsorbed on the wall. Suspended aerosols, collected on the micron filters, contributed less than 0.5% of the reaction products under our experimental conditions.

Reactions of mercury with halogen oxide radicals drew major attention in the light of satellite "BrO" column measurements as well as simultaneous mercury and ozone depletion in the planetary boundary layer (Richter et al., 1998; 2002; Müller et al; 2002; van Roozendael et al., 2002). Experimental studies of XO reactions are very scarce. To our knowledge there is only one published laboratory kinetic study on the reaction of BrO with elemental mercury (Raofie and Ariya, 2003) during which, using the relative rate methods, the room temperature bimolecular rate constant for BrO + Hg$^0_{(g)}$ was estimated to lie within the range $10^{-15} < k < 10^{-13}$ cm^3 molecule^{-1} s^{-1}. This estimated value renders BrO a significant potential contributor to mercury depletion events in the Arctic. Recently, a report was published on the first experimental product study of BrO-initiated oxidation of elemental mercury at atmospheric pressure of ~ 740 Torr and T = 296±2 K (Raofie and Ariya, 2004). The authors used chemical ionization and electron impact mass spectrometry, gas chromatography coupled to a mass spectrometer, a MALDI-TOF mass spectrometer, a cold vapour atomic fluorescence spectrometer, and high-resolution transmission electron microscopy coupled to energy dispersive spectrometry. BrO radicals were formed using visible and UV photolysis of Br_2 and CH_2Br_2 in the presence of ozone. They analyzed the products in the gas phase, on suspended aerosols and on wall deposits, and identified HgBr, HgOBr or HgBrO, and HgO as reaction products. Experimentally, they were unable to distinguish between HgBrO and HgOBr. Mercury aerosols with a characteristic width of approximately 0.2 micrometers were observed as products (Figure 3). The existence of stable Hg^{1+} in form of HgBr, along with Hg^{2+} upon BrO-initiated oxidation of Hg^0, suggests that in field studies it is fundamental to selectively quantify various mercury species in mercury aerosols and deposits.

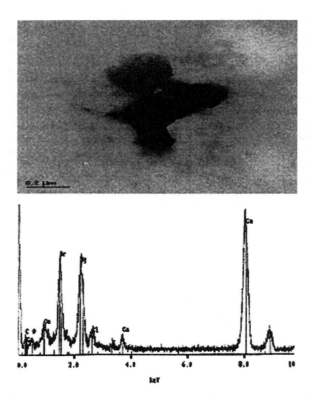

Figure 3. Mercury aerosols: (top) HRTEM (high resolution transmission electron microscopy) image of air-dried product collected from the gas-aerosol, dispersed onto carbon supported Cu grid; (bottom) EDS spectrum showing the chemical composition of the product collected from the gas-aerosol (adapted from Raofie and Ariya, EST, 2004).

The majority of mercury containing products were identified as deposits, however, aerosols accounted for a substantial portion of products. It is noteworthy that we anticipate the possibility of transformation of Hg^{1+} to Hg^{2+} at high humidity levels. Since the hygroscopic nature of Hg^{1+} is different from Hg^{2+}, one suspects a difference in methylation rates for Hg^{1+} in comparison to Hg^{2+} in aquatic systems that may potentially affect mercury bioaccumulation.

There is new modelling work that has incorporated the recent kinetic data on halogens to evaluate the observations (Ariya and Ryhzkhov, 2003; Ariya et al., 2004). These studies, which include the novel kinetic data and multi-scale modelling, show that Br atoms and BrO radicals are the most effective halogens driving mercury oxidation, even considering the reduction back to Hg^0 of oxidized mercury deposited in snow pack and subsequent diffusion to the atmosphere. They found that the reduction cannot compensate for the total deposition and thus a net accumulation occurs. Using a unique global atmospheric mercury model to estimate halogen driven mercury depletion events, they estimate a 44% increase in the net deposition of mercury in the Arctic. Over a one-year cycle, we estimate an accumulation of 325 Tons of mercury in the Arctic (Ariya et al., 2004). Interesting, some modelling studies have been developed on the importance of iodine chemistry and its potential impact on mercury depletion events (Calvert and Lindberg, 2004a,b). Since there is no existing laboratory study on kinetics and products of I_2, I and IO with elemental mercury, we encourage additional studies in this domain to evaluate further the implications of iodine chemistry in the troposphere.

REACTIONS OF ELEMENTAL MERCURY WITH OTHER ATMOSPHERIC OXIDANTS

The oxidation of Hg^0 by H_2O_2 has been performed by Tokos et al. (1998), who found the reaction to be thermodynamically favourable. They obtained a value of 8.5×10^{-19} cm^3 molecule^{-1} s^{-1} for such bimolecular reactions. The reaction of Hg^0 with O_2 is unlikely to proceed at a significant rate ($k \leq 1 \times 10^{-23}$ cm^3 molecule^{-1} s^{-1}), as Hg^0 was found to be the dominant form of mercury in the ambient atmosphere. Hall and Bloom (1993) have investigated the reaction of Hg^0 with HCl by performing experiments in darkness and sunlight at different temperatures. The above-mentioned values render these reactions unlikely drivers of the rapid atmospheric chemistry, however, these reactions and their products can be of significance in condensed phase atmospheric chemistry, and hence we recommend further detailed studies.

METHODOLOGY USED IN MERCURY KINETIC STUDIES

Atomic absorption and fluorescence spectroscopy

Cold vapour atomic absorption spectroscopy (CVAAS) (e.g. Goulden and Anthony, 1980) and cold vapour atomic fluorescence spectroscopy (CVAFS) are the most common instruments employed in both field and laboratory mercury studies. Several types of coupling including Zeeman-modulated CVAAS have been used in mercury kinetic studies (Sommar et al., 1996). CVAFS, in contrast to CVAAS, has not gained widespread general use for routine elemental analysis, but it has been a great tool for elemental mercury due its high sensitivity for this element.

Mass spectroscopy techniques

Gas chromatograph with mass selective detection (GC-MSD) studies

Concentrations of mercury can be monitored using quadrupole mass spectrometer separation on a gas chromatograph using a variety of gas chromatography columns (e.g, cross-linked phenyl-methyl-siloxane column). Analysis can be performed using the scanning or single ion monitoring (SIM) modes (Ariya et al, 2001; Pal and Ariya, 2004; Raofie and Ariya 2004).

Direct mass spectrometry

Reaction products can be collected in a tube, which will then be evaporated to the chemical ionization (CI), electron spray (ES), or electron impact (EI) ion source of mass spectrometry, depending on the targeted experiment (Ariya et al., 2002a; Pal and Ariya, 2004 a, b; Raofie and Ariya, 2004). Reaction products are generally pre-concentrated in a trap cooled by liquid nitrogen or dissolved in small volume of solvent before analysis.

Inductively coupled plasma mass spectrometer (ICP-MS)

ICP-MS is a common technique for mercury analysis, including mercury isotopic studies (Hintelmann et al., 1997). For kinetic studies, reaction products of oxidant initiated oxidation of elemental mercury can be collected using mixtures of HNO_3 or H_2O_2. The samples can be diluted and heated to high temperature (ca. 350-400 K) for a period of time to decompose the washing agents. Mercury containing compounds can then be analyzed by ICP-MS. Coupling of CVAFS with high resolution ICP-MS can provide a powerful technique for in-depth analysis of reaction products.

Derivatization following identification

This method is targeted specifically to separate divalent mercury species. It is based on the qualitative conversion of Hg^{+2} to $HgCl_2$, and then to the more volatile organomercury compound, n-Bu_2Hg that can be analyzed by most mass spectroscopy techniques or even with gas chromatography coupled to mass spectrometry or even flame ionization detection on selective columns:

$$Hg^{2+} + HCl \rightarrow HgCl_2 + 2 H^+ \tag{6}$$

$$HgCl_2 + 2 \text{ n-}Bu_2MgCl \rightarrow \text{n-}Bu_2Hg + 2MgCl_2 \tag{7}$$

It should be noted that Hg_2X_2 is not derivatized under these conditions, and hence it is a good way to distinguish between Hg^{+1} and Hg^{+2}.

Transmission electron microscopy coupled to energy dispersive spectroscopy

This method is targeted to study the size distribution and morphology of mercury condensed products, including aerosols, in addition to some elemental analysis characterization. Reaction products can be collected on microscope grids and products can be analyzed using high-resolution transmission electron microscope. The chemical composition can be qualitatively determined by energy dispersive spectroscopy (Raofie and Ariya, 2004)

Matrix-assisted laser desorption ionization time of flight mass spectrometry (MALDI-TOF-MS)

The reaction products can be collected using different trapping systems (e.g., Raofie and Ariya, 2004). The collected products can be analyzed using MALDI-TOF-MS. Several matrices and ionizations can be employed (Karas and Kruger, 2003). The matrix generally has resonance absorption at laser energy and thus absorbs the energy, causing rapid heating of the matrix. This rapid heating results in expulsion and soft ionization of the sample molecules without fragmentation.

Laser induced fluorescence (LIF) spectroscopy

LIF technique was used to perform kinetic studies of HO + $Hg^0_{(g)}$ (Bauer et al., 2003). They used single and sequential two-photon laser induced fluorescence (LIF) techniques for the detection of elemental mercury. Single photon LIF involves excitation of the 6^3P_1-6^1S_0 transition at 253.7 nm, followed by observation of resonance fluorescence. The second technique includes sequential two-photons, following the initial 6^3P_1-6^1S_0 excitation with a second excitation to the 7^1S_0 or 7^3S_1 levels. To do so, they used two independently tuneable diode laser systems and simultaneous detection of two fluorescence wavelengths.

Other techniques

UV absorption spectrophotometery (e.g., Keeler et al. 1995), atomic emission spectrometry (Pirrone et al., 2001; Sommar et al., 1998; 1999) coupled to CVAFS, as well as Gas chromatography techniques are among techniques used in the field measurement of mercury species or aqueous phase kinetic reactions.

EXPERIMENTAL CHALLENGES OF MERCURY REACTIONS IN THE GAS PHASE

Indeed, many gaseous reactions of mercury with atmospherically important oxidants are difficult to investigate experimentally due to the low concentrations of species at atmospheric conditions, the low volatility of products, side reactions and the strong effects of water vapour and surfaces on kinetics. In the following sections, we separated surface catalyzed reactions from other secondary reactions, due to their particular importance in providing biased kinetic data.

Surface catalyzed reactions

The term heterogeneous reaction refers to a reaction in which reactants react in two or more phases. Thus, in the case of gaseous reactions, if there is a reaction with the wall the reaction product deposits on walls and on aerosols (condensed matter suspended in air or diluent gas). One of the major challenges with a mercury system is to ensure that one studies entirely the unique gas phase reaction of interest, as opposed to a mixture of several reactions. Reactions on surfaces, or catalyzed by surfaces, have been a long-time challenge of gas phase experimental physical chemists. There are several means to overcome this challenge, including using less reactive surfaces and changing the surface-to-volume ratio of reaction chambers, generally using large reaction vessels. Inactive surfaces, such as Teflon or Teflon coated chambers, preferably with large volumes, are among the most common setups. It is of note that unfortunately, Teflon surfaces are porous and in the case of toxic molecules, require careful handling. One should hence pay particular attention to assure homogeneity of Teflon coated surfaces that undergo substantial damage and non-homogeneity after continuous usage of the reaction chamber. An alternative pathway to minimize undesired wall reactions corresponds to the deactivation of walls using various types of coatings (Coquet and Ariya, 2000; Ariya et al., 2002a). The loss of reactants in the absence of initiators is more pronounced in non-treated Pyrex flasks than treated walls. For instance, applying the DMDCS coating led to more than a two-fold decrease of the adsorption rate of mercury (Ariya et al., 2002a). In addition, another potential problem is desorption of reactants that are adsorbed on the surfaces back to the gas phase, biasing kinetic analysis. Hence prior to, and during the kinetic analysis, one needs to examine the setup very carefully to assure that adsorption/desorption processes are significantly slower than the reaction of interest. Another challenge with mercury and oxidant systems lies with the fact that some of the reaction products can produce reactive surfaces for further catalysis leading to gas phase, and gas-phase + heterogeneous reaction rates. Careful analysis of experimental data, and variation of experimental conditions including concentration, surface to volume ratio, photolysis time, and aging of samples are required to properly obtain kinetic values.

Secondary reactions

A challenge in generation of any radical is to avoid production, or at least minimize the scope, of other side reactions. For instance, in generation of HO using CH_3ONO as the HO initiator, one will also generate HO_2, and depending on the NO condition, one may also generate O_3, which in turn reacts further with elemental mercury (Pal and Ariya, 2004a). There are several interesting experimental procedures that have been suggested to assure that one investigates only the reaction of interest. It is of note, however, that it is unlikely to disregard all the side reactions, but we argue that it may be sufficient, if one performs the reactions under conditions where side reactions are significantly less important than the reaction of interest (<< experimental uncertainties).

Impact of water vapour

Hg^0 oxidation processes are observed to be accelerated in the presence of water, hinting at some catalysis reactions, although no specific pathways were suggested (Lindqvist and Rodhe, 1985). We encourage further detailed experimental studies on the impact of water on the kinetics as well as the product distribution of oxidant initiated reactions of mercury.

AB INITIO THERMOCHEMISTRY

The possibility of theoretically predicting the thermochemistry of mercury-containing species of atmospheric interest is of strong importance due to the paucity of accurate experimental information. Obtaining accurate ab initio results for quantities like heats of formation, reaction enthalpies, and activation energies are particularly challenging in these cases, however, due to the large nuclear charge and number of electrons intrinsic to mercury. In particular, the treatment of the extensive relativistic effects due to the former can present major additional sources of error not generally present in calculations involving only elements of the first few rows of the periodic table. Scalar relativistic effects (e.g., the mass velocity and Darwin terms in the Breit-Pauli Hamiltonian) can be accurately recovered using all-electron methods, such as implementations of the Douglas-Kroll-Hess (DK) Hamiltonian (Jansen and Hess, 1989), which is available in several commonly used electronic structure programs. For heavy elements such as mercury, however, these methods can lead to the use of very large 1-particle basis sets, which result in relatively high computational costs for accurate results. The accurate treatment of spin-orbit relativistic effects in all-electron work is especially difficult for heavy elements. On the other hand, a nearly effortless way to recover relativistic effects involves the use of relativistic effective core potentials (RECPs) or pseudopotentials (PPs). For this reason nearly all of the previous ab initio work on atmospheric mercury compounds have utilized relativistic PPs. The accuracy of this approximation does depend, however, on the method of PP adjustment, and there are several choices available for Hg. These include the older Los Alamos ECP implemented in the small LANL2DZ basis set (Hay and Wadt, 1985), the so-called SKBJ ECPs (Stevens et al., 1992), and the energy consistent pseudopotentials developed by the Stuttgart/Köln groups (Haussermann et al., 1993). The latter have been shown to lead to very accurate results for various mercury-containing species. In addition, the accuracy of any PP can also be critically dependent on the number of electrons replaced in the core (Dolg, 1996). With large-core PP's only a minimal number of valence electrons are retained, while small-core PP's retain both the valence and semi-core electrons. Fortunately, for mercury nearly all of the commonly used PPs are of the more accurate small-core variety, i.e., 60 electrons are replaced in the core leaving 20 electrons (5s, 5p, 5d, and 6s) explicitly treated in the valence space.

Unlike all-electron work, the choice of a particular relativistic PP often dictates the basis set that will be used in the calculations, i.e., the one that accompanies the PP. Unfortunately these are often only of double-ζ quality and thus not appropriate for accurate thermochemical studies. Some larger basis sets have been reported for Hg in conjunction with Stuttgart/Köln PPs (Schwerdtfeger et al., 2001), but these lack the important property of being systematically extendable towards the complete basis set (CBS) limit. Recently a new series of basis sets have been developed for Hg (as well as the main group elements) in conjunction with Stuttgart/Köln relativistic PPs (Peterson et al., 2003; Peterson, 2004). These are analogous to the all-electron correlation consistent family of Gaussian basis sets developed by Dunning and co-workers (Dunning, 1989; Wilson et al., 1999; Dunning et al., 2003) and are designed to systematically approach the CBS limit. This property provides the possibility to decouple the errors due to using incomplete basis sets from the errors intrinsic to the chosen electron correlation method. As shown below, the systematic convergence of these sets is very important for accurate thermochemistry studies involving Hg.

In the case of compounds containing only elements in the first few periods of the periodic table, ab initio calculations of thermochemical quantities to within chemical accuracy (1 kcal/mol) have been reported in numerous studies (c.f., Feller et al., 2003 and references therein) Some of these methods, e.g., the Gaussian-X model chemistry approaches (Curtiss et al., 1998), rely on a large training set of molecules with experimentally determined thermochemical properties in order to apply empirical corrections. Unfortunately in the case of mercury chemistry, this type of parameterization is not currently possible due mainly to the lack of sufficient experimental data. This also has negative implications for the effective use of isodesmic-type approaches. Another method, however, that has been used extensively in one of our laboratories (c.f., Feller at al., 2003; Balabanov and Peterson, 2003), as well as the closely related work by Martin and co-workers (Boese et al., 2004) involves a composite theoretical approach without recourse to empirical parameters, and as such while computationally more expensive, is completely amenable to mercury-containing species. This approach utilizes high-level ab initio electronic structure methods, typically singles and doubles coupled cluster theory with a perturbative treatment of connected triple excitations, CCSD(T) (Raghavachari et al.,1998), with explicit extrapolations of the 1-particle basis set to the CBS limit. All other major sources of error in the calculations, e.g., core-valence electron correlation, relativistic effects, etc., are then systematically and accurately accounted for.

The results are thermochemical properties with a very high inherent accuracy only limited essentially by the intrinsic errors in the CCSD(T) method, which for systems dominated by a single electronic configuration (which include all of the molecules noted below) closely mimics that of full configuration interaction (FCI).

Recently this composite approach has been used to accurately predict the thermochemical properties of a variety of mercury oxides and halides, including HgO, $HgCl_2$, $HgBr_2$, and BrHgO (Balabanov and Peterson, 2003; Shepler and Peterson, 2003). In each case the equilibrium geometries and total energies were computed with the CCSD(T) method with a series of correlation consistent-type basis sets ranging from double- to quintuple-ζ quality. Small-core relativistic PPs of the Stuttgart/Köln variety were used on both Br and Hg (Haussermann et al., 1993; Peterson et al., 2003), which were matched to newly developed cc-pVnZ-PP basis sets (n=D – 5) (Peterson et al., 2003; Peterson, 2004). Since the species studied were fairly ionic in nature, extra diffuse-functions were also included that were taken from the aug-cc-pVnZ-PP basis sets. Standard all-electron aug-cc-pVnZ basis sets for oxygen (Dunning, 1989; Kendall et al., 1992) and aug-cc-pV(n+d)Z sets for Cl were used (Dunning et al., 2001). The CBS limits of the total energies were then obtained using two different basis set extrapolation formulas (Peterson et al., 1999):

$$E(n) = E_{CBS} + Ae^{-(n-1)} + Be^{-(n-1)^2} \qquad \text{(Eq. 1)}$$

$$E(n) = E_{CBS} + \frac{A}{n^3} \qquad \text{(Eq. 2)}$$

The final CBS limits [using n=3-5 for Eq.(1) and n=4-5 for Eq.(2)] were obtained by averaging the results of Eqs.(1) and (2) and the spread in these two results provided an estimate of the uncertainty in the basis set extrapolation. Table 3 shows some typical results from Balabanov and Peterson (2003) and Shepler and Peterson (2003) for HgO, HgCl$_2$, HgBr$_2$, and HgBrO. It is readily apparent from these results that basis set incompleteness can represent a significant source of error. In particular, the large basis set superposition error (BSSE) present in the double-ζ (DZ) basis sets can lead to anomolously good agreement with the CBS limit in some cases and very poor agreement in others.

The results shown in Table 3 also demonstrate that the choice of Eq. (1) or (2), i.e., CBS1 or CBS2, respectively, do not differ by more than $0.1 - 0.3$ kcal/mol for these species when large basis sets are used.

Table 3. Dependence on basis set of CCSD(T) electronic energy differences for selected mercury reactions.[a]

	ΔE_e(DZ)	ΔE_e(TZ)	ΔE_e (QZ)	ΔE_e (5Z)	ΔE_e (CBS1)	ΔE_e (CBS2)
HgO ($^1\Sigma^+$) → Hg + O(^{3}P)	−5.75	0.04	1.59	2.17	2.5	2.8
Hg+Br$_2$ → HgBr+Br	26.51	31.64	32.68	33.19	33.5	33.7
→ HgBr$_2$	−42.30	−40.11	−41.35.	−41.64	−41.8	−42.0
Hg +Cl$_2$ → HgCl +Cl	27.71	33.97	35.08	35.73	36.1	36.4
→ HgCl$_2$	−49.84	−46.84	−47.96	−48.25	−48.4	−48.6
Hg+BrO → HgBr+O	33.15	38.56	39.11	39.40	39.6	39.7
→ HgO+Br	52.79	54.20	54.84	55.04	55.2	55.3
→ BrHgO	−19.34	−17.97	−18.97	−19.25	−19.4	−19.6

[a] Relativistic pseudopotentials for Hg and Br were taken from Häussermann et al. (1993) and Peterson et al. (2003), respectively. See Shepler and Peterson (2003) and Balabanov and Peterson (2003) for details.

The resulting CCSD(T)/CBS limit electronic energy differences were then combined with contributions from:

(a) ΔE_{ZPE} : zero-point vibrational energy,

(b) ΔE_{CV} : core-valence correlation,

(c) ΔE_{SR} : residual scalar relativistic effects from O and Cl,

(d) ΔE_{PP} : corrections for the pseudopotential approximation, and

(e) ΔE_{SO} : molecular and atomic spin-orbit coupling.

Representative results are shown in Table 4. The zero-point vibrational corrections were obtained from anharmonic (diatomic) and harmonic (triatomic) large basis set CCSD(T) calculations. For the majority of mercury compounds, however, calculating this correction at lower levels of theory, e.g., density functional theory (DFT) or even Hartree-Fock (HF), would not be expected to result in appreciable errors in the final thermochemical properties. From Table 4 it can be observed that accounting for core-valence electron correlation can impact the final calculated enthalpies by nearly 1 kcal/mol.

It should be noted that these calculations require that the basis set be augmented to provide additional basis functions appropriate for core or semi-

Table 4. Calculated energetic contributions[a] and resulting (0 K) enthalpies of reaction, ΔH_r (kcal/mol), for selected mercury reactions.

	ΔE_e (CBS)	ΔE_{ZPE}	ΔE_{CV}	ΔE_{SR}	ΔE_{PP}	ΔE_{SO}	ΔH_r	Expt.[b]
HgO ($^1\Sigma^+$) → Hg + O(^{3}P)	2.66	−0.64	0.02	---	−0.52	2.4	3.9	64±15
Hg+Br$_2$ → HgBr+Br	33.62	−0.20	0.68	---	−0.53	−3.01	30.6	29.44±9.13
→ HgBr$_2$	−41.89	0.47	0.00	---	−1.31	0.31	−42.4	−42.42±2.05
Hg +Cl$_2$ → HgCl +Cl	36.27	−0.39	0.73	0.05	−0.70	−0.84	35.1	32.89±2.29
→ HgCl$_2$	−48.49	0.59	0.31	0.30	−1.38	0.02	−48.7	−49.44±1.52
Hg+BrO → HgBr+O	39.64	−0.78	0.26	0.09	−0.76	1.00	39.5	39.21±9.52
→ HgO+Br	55.22	−0.40	0.04	0.12	−0.56	−4.27	50.2	−8.1±15.5
→ BrHgO	−19.49	0.52	−0.40	0.22	−1.55	0.55	−20.2	---

[a]ΔE_e = CCSD(T) electronic energy difference, ΔE_{ZPE} = zero-point energy contribution,

ΔE_{CV} = core-valence correlation contribution, ΔE_{SR} = scalar relativity correction (MVD),

ΔE_{PP} = correction for pseudopotential approximation, ΔE_{SO} = molecular and atomic spin-orbit coupling contributions. See the text and Shepler and Peterson (2003) and Balabanov and Peterson (2003).

[b] Chase (1998)

core correlation since most basis sets are only designed for valence electron correlation only.

The present work utilized core-valence basis sets analogous to the cc-pwCVnZ sets reported recently (Peterson and Dunning, 2002) for the second row main group elements.

Of the remaining corrections, the largest in magnitude are ΔE_{PP} and ΔE_{SO}. The former is obtained by comparing CCSD(T) calculations on each species with both the PP approach and all-electron DK calculations with analogous triple-ζ quality basis sets developed in our laboratory. As seen in Table 4, the older Häussermann et al. (1993) PP for Hg that was used in this work results in errors ranging from 0.5 – 1.6 kcal/mol. Obviously it is important to take this into account for accurate studies. Last, the effect of spin-orbit coupling is observed to also be relatively large in magnitude. Most of ΔE_{SO} can be attributed to the zero-field splittings in the atomic asymptotes, which can be obtained from available experimental data (Moore, 1971), but large molecular SO effects were also calculated, e.g., over 2 kcal/mol for HgO and over 1 kcal/mol for BrO. In the present work the effects of SO coupling were obtained using the SO parameters from the PPs and diagonalizing a small SO Hamiltonian in a basis of several pure spin electronic eigenstates of differing multiplicities as implemented in the MOLPRO program suite (Berning et al., 2000). In addition, larger spin-orbit CI calculations have been carried out with the COLUMBUS program (Yabushita et al., 1999), which is also based on one-component molecular orbitals.

The final calculated reaction enthalpies displayed in Table 4 are generally well within the experimental uncertainties, except for the cases involving HgO, whose experimental value has previously been called into question (Shepler and Peterson, 2003). The enthalpies for the reactions leading to HgBr have the largest experimental uncertainties due primarily to the large uncertainty in the heat of formation of HgBr (Chase, 1998). The calculated values in Table 4 for these reactions are in excellent agreement with experiment and have expected uncertainties that are smaller by nearly an order of magnitude.

Of the other recent studies involving the ab initio thermochemistry of atmospheric mercury species, the works of Tossell (2003), Wilcox et al. (2003), and Khalizov et al. (2003) are fairly representative. References to earlier work can be found in these papers, as well as in Balabanov and Peterson (2003) and Shepler and Peterson (2003). Tossell has reported reaction enthalpies for several reactions involving Hg using the double-ζ quality basis sets accompanying the SKBJ RECPs (Stevens et al., 1992) and the MP2, QCISD, and CCSD(T) electron correlation methods. Compared to the accurate results shown in Table 4, this choice of basis set/ECP results in an underestimation of the reaction enthalpies by about 10 kcal/mol at the CCSD(T) level of theory. The differences between the relatively inexpensive MP2 method and CCSD(T) range from 2 – 11 kcal/mol, while the effect of the triple excitations, i.e., the difference between CCSD and CCSD(T), also ranges from 2 – 11 kcal/mol. The basis set and correlation errors are often in different directions, which can make an accuracy assessment very difficult when using small basis sets and low levels of theory. The work of Khalizov et al (2003) demonstrates the inadequacy of the B3LYP method for the reliable prediction of the thermochemistry of reactions involving mercury; the B3LYP reaction enthalpies calculated with triple- and quadruple-ζ quality basis sets were shown to differ from the analogous CCSD(T) results by nearly a factor of two for reactions such as $Hg + Br_2$ and $Hg + Br$. The B3LYP/LANL2DZ level of theory exhibited very large deviations from the large basis set CCSD(T) results. Finally, the study of Wilcox et al. (2003) on HgCl and HgCl + HCl also clearly demonstrates how the choice of PP/basis set can affect the accuracy of the results. They reported calculated enthalpies using both the SKBJ and Häussermann et al. PPs with their accompanying basis sets. For the dissociation of HgCl, the QCISD values differed by nearly 8 kcal/mol. On the other hand, the reaction enthalpies for the HgCl+HCl reaction differed by just 4 kcal/mol. Of course, some of these effects are due to the different basis sets accompanying each PP as well as the different methods of adjusting the relativistic, spin-averaged PP parameters themselves. In particular, for HgCl the QCISD/SKBJ results compared better to experiment while in the case of the HgCl+HCl reaction, the Häussermann PP results appeared to be more accurate. Obviously from these examples, without a careful study of the major contributions to an ab initio enthalpy, e.g., correlation method, basis set, relativistic effects, etc., it is very difficult to assess the accuracy of a given result. In particular, a fortuitous cancellation of errors that is present for one reaction system should not be relied upon to occur for another reaction without a careful analysis.

Ab initio kinetics

From an ab initio point of view, most of the factors important for accurate thermochemistry, e.g., choice of electron correlation method, treatment of relativistic effects, basis set truncation errors, etc., are also responsible for obtaining accurate kinetic data since the latter depends intimately on the underlying potential energy surface. A rigorous calculation of the rate coefficient for a given reaction generally involves either quantum scattering or classical trajectory calculations, which in turn require a global or semiglobal potential energy surface (PES) calculated by ab initio methods. While these treatments are feasible for relatively small systems (depending on the required accuracy of the underlying PES), most studies employ more approximate treatments of the reaction dynamics, e.g., transition state theory (TST) or RRKM theory (Rice-Ramsberger-Kassel-Marcus theory).

For bimolecular reactions involving a barrier, transition state theory is often used. The basic tenet is that there exists a critical configuration lying between reactants and products where at that configuration all trajectories arising from reactants are assumed to irreversibly lead to products (Steinfeld et al., 1989; Truhlar et al., 1996). In conventional transition state theory (TST), this critical configuration is assigned to the top of the reaction barrier, i.e., the transition state. In addition to the "no recrossing" assumption, the reaction coordinate is generally assumed to be separable from the other degrees of freedom and is treated classically as a translation. In TST calculations the only ab initio data required are the equilibrium structures and harmonic vibrational frequencies of the reactants and the transition state, as well as the barrier height. Simple corrections for tunneling can also be applied that do not require any additional information. The critical quantity of interest, the barrier height (ΔE_b), is as much a challenge to ab initio theory (if not more) as calculating accurate reaction or bond dissociation enthalpies discussed above. Often the sensitivity to the choice of electron correlation method is even more critical since transition states can involve highly stretched bonds where the electronic wave function is no longer strongly dominated by a single electronic configuration. Obtaining an accurate ΔE_b is essential for calculating reliable rate constants since even a 1 kcal/mol error in the barrier height can lead to nearly an order of magnitude error in the resulting rate constant. The composite approach described in detail above for ab initio thermochemistry can be applied straightforwardly in the calculation of reliable reaction barrier heights. Even with an accurate calculation of the barrier height, however, conventional TST can still strongly overestimate the rate constant since the occurence of trajectories that originate with reactants, pass through the transition state, but return back to reactants, i.e., recrossing trajectories, can be common. Thus the TST rate constant is an upper bound to the exact classical rate constant. In variational transition state theory (VTST), the position of the dividing surface between reactants and products is varied to yield the smallest rate constant, thus minimizing the number of recrossing trajectories. In canonical VTST, this procedure corresponds to placing the transition state where the free energy is a maximum, while in microcanonical VTST the location of the dividing surface (for a particular energy of the reactants) corresponds to a minimum in the transition state's sum of states. The additional computational burden of carrying out VTST calculations, however, is the calculation of the reaction path (structures) between the TS and reactants, with vibrational frequencies at each or selected points along this path. This can be a significant additional computational expense, but often the vibrational frequencies along the reaction path can be calculated at a lower level of theory than what was used for the reactants and transition state, which can still yield much improved rate constants compared to just the conventional TST results. The easily accessible POLYRATE program (Corchado et al., POLYRATE 8.5.1) can be used for a large variety of VTST calculations.

For reactions that proceed without a barrier, e.g., unimolecular dissociation or recombination reactions, RRKM theory is often employed. The use of RRKM involves two central approximations (c.f., Steinfeld et al. 1989 and Gilbert and Smith 1990 and references therein):

(i) as with transition state theory, RRKM assumes the existence of a critical configuration between reactants and products which is not recrossed and

(ii) the energy of the excited reactant is distributed randomly throughout all the available molecular states.

To satisfy the first approximation, it is generally very important to employ the variational version, which is equivalent to a microcanonical VTST calculation. So as above, one needs to calculate structures and vibrational frequencies along the reaction path. In order to satisfy the second criterion, the reactant must be a molecule large enough to provide efficient intramolecular vibrational energy redistribution. Hence, the use of RRKM for atom-atom recombination reactions is probably quite suspect. In these cases, quasiclassical trajectory calculations would seem to be the most reliable method. For reaction rate constant calculations for barrierless reactions using RRKM and VTST-like methods, the Variflex program (Klippenstien et al., 1999) is a convenient choice for polyatomic systems, since it also allows several options for the calculation of pressure effects on the the rate constant (standard VTST yields only a high pressure limit rate constant).

FUTURE DIRECTIONS AND CONCLUDING REMARKS

Recent kinetic data indicate that the atmospheric lifetime of elemental mercury over marine boundary layer (such as the Arctic) can be shorter than previously believed. However, over continental regions, in absence of rapid halogen reactions, the regional and global transport of mercury seem to be feasible even in the light of recent kinetic studies, which confirm mercury as a global pollutant. Despite the novel positive trend in laboratory and theoretical studies of gas-phase elemental mercury, chemical reaction studies of elemental mercury are still relatively scarce. It is essential to provide kinetic, product, and thermochemical studies on complex reactions, particularly in the presence of important atmospheric partners including organic radicals. There are some limited studies on the kinetics of gas-phase elemental mercury oxidation on surfaces, e.g., on titania particles (Lee et al., 2004), and on powdered activated carbon (Flora et al., 1998). There are also limited available literatures on uptake kinetics of elemental mercury by surfaces such as virgin and sulfur-impregnated activated carbons (Vidic et al., 1998). However, experimental studies on uptake or kinetics of heterogeneous reactions of mercury on various environmentally relevant surfaces such as ice, snow, and aerosols, are desired. Microbiological transformations of mercury in aquatic systems have been suggested. In the light of new evidence on the role of biological aerosols (Ariya et al., 2002b), the redox reaction of mercury on and in bioaerosols can be an area of interest for future research.

ACKNOWLEDGMENTS

PAA is grateful to NSERC and McGill University for financial support. KAP acknowledges the support of the U.S. National Science Foundation. Patrick Costelo and Jackie Johnstone are acknowledged for their help in the preparation of this manuscript.

REFERENCES

Ariya, P.A., Jobson, B.T., Sander, R., Niki, H., Harris, G.W., Hopper, J. F., Anlauf, K.G. J. Measurements of C2-C8 hydrocarbons during the Polar Sunrise Experiment 1994: Further evidence for halogen chemistry in the Arctic troposphere. *Geophys. Res.*, 103, 169-13180, 1998.

Ariya, P.A., Niki, H., Harris, G. W., Hopper, F., Worthy, D. Polar Sunrise Experiment 1995: Hydrocarbon measurements and halogen chemistry in the Arctic troposphere. *Atmos. Environ.*, 33, 931-938, 1999.

Ariya, P.A., Khalizov, A., Gidas, A. Reaction of Gaseous Mercury with Atomic and Molecular Halogens: Kinetics, Product Studies, and Atmospheric Implications. *J. Phys. Chem. A*, 106, 7310-7320, 2002a.

Ariya, P.A., Nepotchatykh, O., Ignatova, O., Amyot, M. Microbiological degradation of organic compounds in the atmosphere. *Geophy. Res. Lett.*, 29, 341-4, 2002b.

Ariya, P. A., Ryzhkov, A., Modelling of mercury in the Arctic. *J. Phys. IV*, 107, 57-60, 2003.

Ariya, P.A., Dastoor, A.P., Amyot, M., Schroeder, W.H., Barrie, L., Anlauf, K., Raofie, F., Ryzhkov, A., Davignon, D., Lalonde, J., Steffen, A. Arctic a sink for mercury, *Tellus B*, 56, 397-403, 2004.

Balabanov, N.B., Peterson, K.A. Mercury and Reactive Halogens: The Thermochemistry of Hg + {Cl_2, Br_2, BrCl, ClO, and BrO} *J. Phys. Chem. A*, 107, 7465, 2003.

Barrie, L.A., Bottenheim, J.W., Schnell, R.C., Crutzen, P.J., Rasmussen, R.A. Ozone destruction and photochemical reactions at polar sunrise in the lower Arctic atmosphere. *Nature*, 334,138-141, 1988.

Bauer, D., D'Ottone, L., Campuzaon-Jos, P., Hynes, A.J. Gas phase elemental mercury: a comparison of LIF detection techniques and study of the kinetics of reaction with the hydroxyl radical. *J. Photochem. Photobiol.*, 157, 247-256, 2003.

Berning, A., Schweizer, M., Werner, H.-J., Knowles, P.J., Palmieri, P. Spin-orbit matrix elements for internally contracted multireference configuration interaction wavefunctions. *Mol. Phys.*, 98, 1823, 2000.

Calvert, J.G., Lindberg, S.E. A study of the potential influence of iodine containing compounds in the extent of ozone depletion in the troposphere in the polar spring I. *Atmos. Environ.*, 38, 5087 –5104, 2004a.

Calvert, J.G., Lindberg, S.E. The potential influences of iodine containing compounds on the chemistry of the troposphere in the polar spring II. *Atmos. Environ.*, 38, 5105-5116, 2004b.

Chase, M.W., Jr., ed. NIST-JANAF thermochemical tables, 4th edition, 1998.

Coquet, S., Ariya, P. A. Kinetics of the gas-phase reactions of Cl atom with selected C_2-C_5 unsaturated hydrocarbons at $283 < T < 323$ K . *Int. J. Chem. Kinet.*, 32, 478-484, 2000.

Corchado, J.C., Chuang, Y.-Y., Fast, P.L., Villà, J., Hu, W.-P., Liu, Y.-P., Lynch, G.C., Nguyen, K.A., Jackels, C.F., Melissas, V.B., Lynch, B.J., Rossi, I., Coitiño, E.L., Fernandez-Ramos, A., Steckler, R., Garrett , B.C., Isaacson, A.D., Truhlar, D.G., POLYRATE 8.5.1: a new version of a computer program for the calculation of chemical reaction rates for polyatomics.

Curtiss, L.A., Raghavachari, K., Redfern, P.C., Pople, J.A. Gaussian-3 (G3) theory for molecules containing first and second-row atoms. *J. Chem. Phys.*, 109, 7764, 1998.

Dolg, M. On the accuracy of valence correlation energies in pseudopotential calculations. *J. Chem. Phys.*, 104, 4061, 1996.

Dunning, T.H., Jr. Gaussian basis sets for use in correlated molecular calculations. I. The atoms boron through neon and hydrogen. *J. Chem. Phys.*, 90, 1007, 1989.

Dunning, T.H., Jr., Wilson, A.K., Peterson, K.A. Gaussian basis sets for use in correlated molecular calculations. X. The atoms aluminum through argon revisited. *J. Chem. Phys.*, 114, 9244, 2001.

Ebinghaus, R., Kock, H.H., Coggins, A.M., Spain, T.G., Jennings, S.G., Temme, Ch. Long-term measurements of atmospheric mercury at Mace Head, Irish west coast, between 1995 and 2001. *Atmos. Environ.*, 36, 5267-5276, 2002a.

Ebinghaus, R., Kock, H.H., Temme, C., Einax, J. W., Löwe, A.G., Richter, A., Burrows, J.P., Schroeder, W.H. Antarctic Springtime Depletion of Atmospheric Mercury. *Env. Sci. Technol.*, 36, 1238-1244, 2002b.

Feller, D., Peterson, K.A. Re-examination of atomization energies for the Gaussian-2 set of molecules. *J. Chem. Phys.*, 110, 8384, 1999.

Feller, D., Peterson, K.A., de Jong, W.A., Dixon, D.A. Performance of coupled cluster theory in thermochemical calculations of small halogenated compounds. *J. Chem. Phys.*, 118, 3510, 2003.

Finlayson-Pitts, B. J., Pitts, J., Jr. Atmospheric Chemistry: Fundamentals and Experimental Techniques. New York: John Wiley & Sons, 1986.

Flora, J.R.V., Vidic, R.D., Liu, W., Thurnau, R.C. Modelling powdered activated carbon injection for the uptake of elemental mercury vapours. *J. A.&W.M.A.*, 8, 1051-1059, 1998.

Gårdfeldt, K., Sommar, J., Strömberg, D., Feng X. Oxidation of atomic mercury by hydroxyl radicals and photoinduced decomposition of methylmercury in the aqueous phase. *Atmos. Environ.*, 35, 3039, 2001.

Gardfeldt, K. The transformation of mercury species in the aquatic phase. Ph.D. thesis 2003.

Gilbert, R.G., Smith, S.C., Theory of Unimolecular and Recombination Reactions. Oxford: Blackwell Scientific, 1990.

Goulden, P.D., Anthony, D.H.J. Chemical speciation of mercury in natural waters. *Anal. Chem. Acta*, 120,129-139, 1980.

Hall, B., Bloom, N. S. Annual Report to the Electric Power Research Institute, Palo Alto, CA 1993.

Hall, B. The gas phase oxidation of elemental mercury with ozone, *Water, Air, Soil Pollut.*, 80,301-315, 1995.

Häussermann, U., Dolg, M., Stoll, H., Preuss, H., Schwerdtfeger, P., Pitzer, R.M., Accuracy of energy-adjusted quasi-relativistic ab-initio pseudo-potentials all electron and pseudo-potential benchmark calculations for Hg, HgH and their cations, *Mol. Phys.*, 78, 211, 1993.

Hay, P.J., Wadt, W.R. Ab initio effective core potentials for molecular calculations. Potentials for the transition metal atoms Sc to Hg. *J. Chem. Phys.*, 82:270, 1985.

Hensel, Fredrich Warren, William W., Jr., Fluid Metals. Princeton: Princeton University in press, 1999.

Hintelmann H., Falter R., Ilgen G., Evans, R. D. Determination of artifactual methylmercury (CH_3Hg^+) formation in environmental samples using stable Hg^{2+} isotopes with ICP-MS detection: Calculation of contents applying species specific isotope addition. *Fres. J. Anal. Chem.*, 358, 363-370, 1997.

Horne, D.G., Gosavi, R., Strausz, O.P. Reactions of Metal Atoms. I. The Combination of Mercury and Chlorine Atoms and the Dimerization of HgCl. *J. Chem. Phys.*, 48, 4758-4764, 1968.

Iverfeldt, Å., Lindqvist, O. Atmospheric oxidation of elemental mercury by ozone in the aqueous phase. *Atmos. Environ.*, 20, 1567-1573, 1986.

Jacob, Daniel, Introduction to Atmospheric Chemistry. Princeton: Princeton Press, 1999.

Jansen, G., Hess, B.A. Revision of the Douglas-Kroll transformation. *Phys. Rev.*, 39, 6016, 1989.

Jobson, B.T., Niki, H., Yokouchi, Y., Bottenheim, J., Hopper, F., Leaitch, R. Measurements of C2-C6 hydrocarbons during the Polar Sunrise 1992 Experiment: Evidence for Cl atom and Br atom chemistry [C_2-C_6]. *J. Geophys. Res.*, 99, 25355-25368, 1994.

Keeler, G., Glinsorn, G., Pirrone, N. Particulate mercury in the atmosphere: its significance, transport, transformation and sources. *Water, Air, Soil Pollut.*, 80, 159-168, 1995.

Kendall, R.A., Dunning, T.H., Jr., Harrison, R.J. Electron affinities of the first-row atoms revisited. Systematic basis sets and wave functions. *J. Chem. Phys.*, 96, 6796, 1992.

Khalizov, A.F., Viswanathan, B., Larregaray, P., Ariya, P.A. Theoretical Study on the Reactions of Hg with Halogens: Atmospheric Implications. *J. Phys. Chem. A*, 107, 6360, 2003;

Klippenstein, S.J., Wagner, A.F., Robertson, S.H., Dunbar, R.C., Wardlaw, D.M.VARIFLEX-version 1.0 (VARIFLEX is a freeware program package for calculating gas phase reaction rates) http://chemistry.anl.gov/variflex , 1999.

Klopper, W., Bak, K.L., Jørgensen, P., Olsen, J., Helgaker, T. Highly accurate calculations of molecular electronic structure. *J. Phys. B.*, 32, R103, 1999.

Karas, M., Krüger, R. Ion Formation in MALDI: The Cluster Ionization Mechanism. *Chem. Rev.*, 103, 427, 2003.

Lee, T.G., Biswas, P., Hedrick, E. Overall Kinetics of Heterogeneous Elemental Mercury Reactions on TiO2 Sorbent Particles with UV Irradiation. *Ind. & Eng. Chem. Res.*, 43,1411-1417, 2004.

Lin, C.-J., Pehkonen, S.O. The chemistry of atmospheric mercury: a review. *Atmos. Environ.*, 33, 2067-2079, 1999.

Lindberg, S.E., Brooks, S., Lin, C.J., Scott, K.J., Landis, M.S., Stevens, R.K., Goodsite,M., Richter, A. Dynamic Oxidation of Gaseous Mercury in the Arctic Troposphere at Polar Sunrise. *Environ. Sci. Technol.*, 36, 1245-1256, 2002.

Lindqvist, O., Rodhe, H. Atmospheric mercury - a review. *Tellus*, 37B, 136-159, 1985.

Lindqivist, O. Sommar, J., Feng, X., Gardfeldt K. Measurements of fractionated gaseous mercury concentrations over northwestern and central Europe, 1995-99. *J. Env. Monit.*, 1, 435-9, 1999.

Lu, J.Y., Schroeder, W.H., Barrie, L.A., Steffen, A., Welch, H.E., Martin, K., Lockhart, L., Hunt, R.V., Biola, G., Richter, A. Magnification of atmospheric mercury deposition to polar regions in springtime: the link to tropospheric ozone depletion chemistry. *Geophys. Res. Lett.*, 28, 3219-3222, 2001.

Mason, R.P., Fitzgerald, W.F., Morel, F.M.M. The biogeochemical cycling of elemental mercury: Anthroposenic influences. *Goechim. Cosmochim. Acta*, 58, 3191-3198, 1994.

Mason, R.P., Sheu, G.-R. Role of the ocean in the global mercury cycle. *Glob. Biogeoch. Cyc.*, 16, 1093, 2002.

Medhekar, A.K., Rokni, M., Trainor, D.W., Jacob, J.H. Surface catalyzed reaction of Hg + Cl_2. *Chem. Phys. Lett.*, 65, 600-604, 1979.

Menke, R., Wallis, G. Detection of Mercury in Air in the Presence of Chlorine and Water Vapor. *Am. Ind. Hyg. Assoc. J.*, 41, 120-124, 1980.

Miller, G.C., Quashnick, J., Hebert, V. Abstr. Pap. - Am. Chem. Soc. 221st:AGRO-016, 2001.

Moore, C.E., Atomic Energy Levels. NSRDS-NBS 35, Office of Standard Reference Data, National Bureau of Standards, Washington, D.C., 1971.

Morel, F.M.M., Kraepiel, A.M.L., Amyot, M. The Chemical Cycle and Bioaccumulation of Mercury.Annual. *Rev. Ecol. System*, 29, 543-566, 1998.

Müller, R.W., Bovensmann, H., Kaiser, J.W., Richter, A., Rozanov, A., Wittrock, F., Burrows, J. P. Consistent Interpretation of Ground based and GOME BrO Slant Column Data. *Adv. Space Res.*, 29, 1655-1660, 2002.

Munthe, J. The aqueous oxidation of elemental mercury by ozone. *Atmos. Environ.*, 26A, 1461-1468, 1992.

Nriagu, J.G., Pacyna, J.M. Quantitative assessment of worldwide contamination of air, water, and soils with trace metals. *Nature*, 333, 134-139, 1988.

Oum, K.W., Lakin, M.J., Finlayson-Pitts, B.J. *Geophys. Res. Lett.*, 25, 3923-3926, 1998.

P'yankov, V. A. Zhurmal Obscej Chemii Akatemijaneuu SSSR, 19, 224-229, 1949.

Pal, B., Ariya, P.A., Kinetics and mechanism of O_3-initiated reaction of Hg^0: atmospheric implication, *Phys. Chem. Chem. Phys.*, 6, 752, 2004.

Pal, B., Ariya, P.A. Gas-phase Reaction of Hydroxyl Initiated Reaction of Elemental Mercury: Kinetics and Product Studies, *Env. Sci. Technol.*, 38, 5555, 2004.

Pehkonen, S.O., Lin, C.-J. Aqueous Photochemistry of Divalent Mercury with Organic Acids. *Journal of AWMA*, 48, 144-150, 1998.

Peterson, K.A., Dunning, T.H., Jr. Accurate correlation consistent basis sets for molecular core–valence correlation effects: The second row atoms Al–Ar, and the first row atoms B–Ne revisited. *J. Chem. Phys.*, 117, 10548, 2002.

Peterson, K.A. In Recent Advances in Electron Correlation Methodology. Wilson, A.K., Peterson, K.A. eds. *A.C.S.*, (in press) 2004.

Peterson, K.A., Figgen, D., Goll, E., Stoll, H., Dolg, M. Systematically convergent basis sets with relativistic pseudopotentials. II. Small-core pseudopotentials and correlation consistent basis sets for the post-d group 16–18 elements. *J. Chem. Phys.*, 119, 11113, 2003.

Pirrone, N., Sprovieri, F., Pesenti, E. MED-OCEANOR Project-technical report. Workshop on data analysis of the URANIA oceanographic Mediterranean campaign-summer 2000; 2001 February 26-27; Pisa, Italy, 2001.

Poissant, L. 2001, personal communication.

Porcella, D.B., Ramel, C., Jernelov, A., Global mercury pollution and the role of gold mining: An Overview. *Water, Air Soil Pollut.*, 97, 205-207, 1997.

Raghavachari, K., Trucks, G.W., Pople, J.A., Head-Gordon, M. A fifth-order perturbation comparison of electron correlation theories. *Chem. Phys. Lett.*, 157, 479, 1989.

Raofie, F., Ariya, P. A. Reactions of BrO with mercury: kinetic studies, *J. Phys. IV*, 107, 1119-1121, 2003.

Raofie, F., Ariya, P.A. First evidence for stable Hg +1 in the aerosols. *Env. Sci. Technol.*, 38, 4319, 2004.

Richter, A., Wittrock, F., Eisinger, M., Burrows, J. P. GOME observations of tropospheric BrO in Northern Hemispheric spring and summer 1997. *Geophys. Res. Lett.*, 25, 2683-2686, 1998.

Richter, A., Wittrock, F., Ladstätter-Weißenmayer, A., Burrows, J. P. GOME measurements of stratospheric and tropospheric BrO. *Adv. Space Res.*, 29, 1667-1672, 2002.

Schroeder, W.H., Yarwood, G., Niki, H. Involving Mercury Species in the Atmosphere - Results from a Literature Survey. *Water, Air, Soil Pollut.*, 56, 653-666, 1991.

Schroeder, W.H., Anlauf, K.G., Barrie, L.A., Lu, J.Y., Steffen, A., Schneeberger, D.R., Berg, T. Arctic springtime depletion of mercury. *Nature*, 394, 331-332, 1998.

Schroeder, W.H., Barrie L.A. Is mercury input to polar ecosystems enhanced by springtime ozone depletion chemistry? IGAC Newsletter. 14, 7-8, 1998.

Schroeder, W.H.; Anlauf, K.G.; Barrie, L.A.; Steffen, A.; Lu, J.Y. In Proceedings of EUROTRAC-2 Symposium '98, Vol. 2, pp 358-368. Borrell, P.M., Borrell, P., eds. Southampton, U. K.: WIT Press, 1999.

Schwerdtfeger, P., Wesendrup, R., Moyano, G.E., Sadlej, A.J., Greif, J., Hensel, F. The potential energy curve and dipole polarizability tensor of mercury dimer. *J. Chem. Phys.*, 115, 7401, 2001.

Seigneur, C., Wrobel, J., Constantinou, E. A chemical kinetic mechanism for atmospheric inorganic mercury. *Env. Sci. Technol.*, 28, 1589-1597, 1994.

Seiler, W., Eberling, C., Slemr, F. Global distribution of gaseous mercury in troposphere, *Pageoph*, 118, 963-973, 1980.

Shepler, B.C., Peterson, K.A. Mercury monoxide: A systematic investigation of its ground electronic state, *J. Phys. Chem. A*, 107, 1783-1787, 2003.

Skare, I., Johansson, R. Reactions between mercury vapor and chlorine gas at occupational exposure levels. *Chemosphere*, 24, 1633-1644, 1992.

Slemr, F., Schuster, G., Seiler, W. Distribution, speciation, and budget of atmospheric mercury *J. Atmos. Chem.*, 3, 407-434, 1985.

Slemr, F. In NATO-ASI-Series, 2. Environment, Vol. 21, Baeyens, W., Ebinghaus, R., Vasiliev, O., eds. Dordrecht, The Netherlands: Kluwer Academic Publishers, 1996.

Slemr, F., Brunke, E., Labuschagne, C., Ebinghaus, R., Munthe, J. Worldwide Trends in Atmospheric Mercury Concentrations. Poster at EUROTRAC-2 Symposium; March 11-15; Garmisch-Partenkirchen, Germany, 2002.

Sommar, J., Hallquist, M. Ljungstrom, E. Rate of reaction between the nitrate radical and dimethyl mercury in the gas phase. *Chem. Phys. Lett.*, 257, 434-438, 1996.

Sommar, J., Hallquist, M., Ljungström, E., Lindqvist, O. On the Gas Phase Reactions Between Volatile Biogenic Mercury Species and the Nitrate Radical. *J. Atmos. Chem.*, 27, 233-247, 1997.

Sommar, J., Feng, X., Garfeldt, K., Lindqvist, O. Measurements of fractionated gaserous mercury concentrations in Northwestern and central Europe, 1995-1999. *J. Environ. Monitor.*, 1, 435-439, 1999.

Sommar, J., Gardfeldt, K., Stromberg, D., Feng, X. A kinetic study of the gas-phase reaction between the hydroxyl radical and atomic mercury. *Atmos. Environ.*, 35, 3049-3054, 2001.

Steinfeld, J.I., Francisco, J.S., Hase, W.L., Chemical Kinetics and Dynamics. Englewood Cliffs, N.J.: Prentice-Hall, 1989.

Stevens, W.J., Krauss, M., Basch, H., Jasien, P.G. Relativistic compact effective potentials and efficient, shared exponent basis sets for 3rd row and 4th row, and 5th row atoms, *Can. J. Chem.*, 70, 612, 1992.

Temme, Ch., Einax, J.W., Ebinghaus, R., Schroeder, W.H. Measurements of Atmospheric Mercury Species at a Coastal Site in the Antarctic and over the South Atlantic Ocean during Polar Summer. *Environ. Sci. Technol.*, 37, 22-31, 2003.

Tokos, J.J.S., Hall, B., Calhoun, J.A., Prestbo, E.M. Homogeneous gas-phase reaction of Hg° with H_2O_2, O_3, CH_3I, AND $(CH_3)_2S$: Implications for atmospheric Hg cycling. *Atmos. Environ.*, 32, 823-827, 1998.

Tossell, J.A. Calculation of the Energetics for Oxidation of Gas-Phase Elemental Hg by Br and BrO. *J. Phys. Chem. A*, 107, 7804, 2003.

Truhlar, D.G., Garrett, B.C., Klippenstein, S.J. Current Status of Transition-State Theory. *J. Phys. Chem.*, 100, 12771, 1996.

UNEP Global assessment report for mercury. http://www.chem.unep.ch/mercury/report/final%20assessment%20report.htm, 2002.

van Roozendael, M., Wagner, T., Richter, A., Pundt, I., Arlander, D. W., Burrows, J. P. Chipperfield, M., Fayt, C., Johnston, P. V., Lambert, J.-C., Kreher, K., Pfeilsticker, K. Platt, U., Pommereau, J.-P., Sinnhuber, B.-M., Toernkvist, K. K., Wittrock, F Intercomparison of BrO Measurements from ERS-2 GOME, ground-based and Balloor Platforms. *Adv. Space Res.*, 29, 1661-1666, 2002.

Vidic, R.D., Chang, M.-T., Thurnau, R.C. Kinetics of vapor-phase mercury uptake by virgir and sulfur-impregnated activated carbons, *J.A.&W.M.*, 48, 247-255, 1998.

Wigfield, D.C., Perkins, S.L., Oxidation of elemental mercury by hyudroperoxides in aqueous solution, *Canadian J. Chem.*, 63, 275-278, 1985.

Wilcox, J., Robles, J., Marsden, D.C.J., Blowers, P. Theoretically Predicted Rate Constants for Mercury Oxidation by Hydrogen Chloride in Coal Combustion Flue Gases. *Environ. Sci. Technol.*, 37, 4199, 2003.

Wilson, A.K., Peterson, K.A., Woon, D.E., Dunning, T.H., Jr. Gaussian basis sets for use in correlated molecular calculations. IX. The atoms gallium through krypton. *J. Chem. Phys.*, 110, 7667, 1999.

Yabushita, S., Zhang, Z., Pitzer, R.M. Spin-Orbit Configuration Interaction Using the Graphical Unitary Group Approach and Relativistic Core Potential and Spin-Orbit Operators. *J. Phys. Chem.*, 103, 5791, 1999.

Chapter-13

MODELLING CHEMICAL AND PHYSICAL PROCESSES OF HG COMPOUNDS IN THE MARINE BOUNDARY LAYER

Ian M. Hedgecock and Nicola Pirrone

CNR-Institute for Atmospheric Pollution, Division of Rende, 87036 Rende, Italy

INTRODUCTION

Only five years ago a chapter in a book such as this with the above title would either have been extremely short, or simply not included. The inclusion of this chapter is evidence of the progress made in fields as diverse as analytical methods and instrumentation, chemical kinetics and atmospheric chemistry modelling, as well as of the ever increasing interest in mercury (Hg) within both the scientific and environmental policy communities. The reason for the current interest in Marine Boundary Layer (MBL) processes and their influence on Hg was the discovery of higher than expected concentrations of Reactive Gaseous Mercury (RGM or RGHg or $Hg^{II}_{(g)}$) at coastal sites (Mason et al., 2001, Wangberg et al., 2001) and in the MBL of both seas (Sprovieri et al., 2003) and oceans (Mason et al., 2001, Laurier et al., 2003). The $Hg^{II}_{(g)}$ concentration was also found to vary diurnally with a maximum occurring with maximum solar radiation intensity and a minimum at night. $Hg^{II}_{(g)}$ is not a single compound, it is an operationally defined quantity which refers to the oxidised inorganic mercury compounds present in the gas phase which are collected on KCl denuders, see chapter 7. The most probable components of $Hg^{II}_{(g)}$ are $HgCl_2$ and $HgBr_2$, possibly with HgO and $Hg(OH)_2$. $Hg^{II}_{(g)}$ is fundamental to Hg cycling because its chemical and physical characteristics differ so greatly from those of $Hg^0_{(g)}$. Where $Hg^0_{(g)}$ is volatile and sparingly soluble, the compounds which make up $Hg^{II}_{(g)}$ are far less volatile and far more soluble, thus Hg deposition is almost totally dominated by $Hg^{II}_{(g)}$ whilst Hg emission, even industrial emission, is predominantly $Hg^0_{(g)}$. Modelling studies had suggested that the sea salt aerosol could be important in Hg cycling in the MBL because of its ability to form a range of Hg^{II} complexes due to its high chloride ion content (Pirrone et al., 2000). Further modelling studies suggested that Hg^{II} produced in the gas phase (Hg^0 from the reaction of Hg^0 + O_3) could be scavenged and then cycled via the sea salt aerosol to $HgCl_2$, which is more volatile than HgO, thus providing an MBL source of $Hg^{II}_{(g)}$, (Hedgecock and Pirrone, 2001).

The development of our understanding of Hg chemical and physical processes in the MBL, is intertwined with the study of Hg in the Arctic, and has gained very much from the study, both in the field and modelling, of tropospheric O_3 in remote areas, specifically the Arctic and the remote MBL.

Taking things in chronological order, it was in the late 80s that ozone depletion events were first reported during polar spring, (Bottenheim et al, 1986), these events were linked to photochemical processes, (Barrie et al., 1988). In the years that followed there were a number of field and theoretical investigations into the mechanism behind the rapid O_3 destruction, which pointed to the involvement of bromine containing radicals, (Fan and Jacob, 1992, Barrie and Platt, 1997, Foster et al, 2001). Measurements of $Hg^0_{(g)}$ over a number of years at Alert in Canada showed that when O_3 depletion events occurred, the concentration of $Hg^0_{(g)}$ also decreased rapidly, often to below the detection limit of the measurement techniques employed (Schroeder et al, 1998). This observation led to the suggestion that it was likely to be the Br radical compounds which were responsible for the Hg depletion events. At around the same time model studies were performed to see what effect, if any, the halogen radical chemistry seen in the Arctic may have in the MBL, (Vogt et al, 1996, Sander and Crutzen, 1996). Model predictions suggested that halogen compounds (Br_2 and BrCl) would be released to the atmosphere as a result of the acidification of the sea salt aerosol, and that the radicals produced by their photolysis would destroy O_3.

Since then the phenomenon known as 'Sunrise Ozone Destruction' has been observed in the sub-tropical Pacific (Nagao et al., 1999), and a number of studies have observed depletion of Br containing species in supermicrometer sea-salt aerosol, and enrichment of the same species in submicron marine aerosol, (Sander et al., 2003). The link between O_3 destruction at polar dawn with $Hg^0_{(g)}$ depletion and the measurement of elevated RGM concentrations in marine air masses hinted that Hg^0 oxidation processes were also at work in the MBL. The recent measurement of the reaction rates of $Hg^0_{(g)}$ + $OH_{(g)}$ (Sommar et al., 2001) and subsequently of $Hg^0_{(g)}$ + $Br_{(g)}$ (Ariya et al., 2002), have been included in a box model, AMCOTS (Atmospheric Mercury Chemistry Over The Sea, Hedgecock and Pirrone, 2004)) of multiphase MBL photochemistry along with previously known Hg gas and aqueous phase chemistry (Pleijel and Munthe, 1995, Lin and Pekhonen, 1999). The AMCOTS results can be compared directly to results from measurement campaigns, (Hedgecock et al., 2003, Hedgecock and Pirrone, 2004). The high measured concentrations of RGM coupled with the modelling support have meant that the cycle of Hg in the MBL, its emission, transport, chemistry and deposition have had to be reconsidered, and the role of the ocean-atmosphere in the global Hg cycle and the global Hg budget reassessed (Mason and Sheu, 2002).

The MBL chemistry of Hg therefore cannot be removed from the transport and physical processes which ultimately have an influence on the concentrations of Hg species in the MBL. Where Hg in the MBL comes from, and which chemical species are emitted or transported with it, can directly or indirectly affect Hg chemistry. The scavenging of Hg compounds by deliquesced aerosol particles or rain droplets, and the dry deposition of oxidised Hg compounds not only exert a major influence over the concentrations of Hg species in the MBL but also on the marine emission/deposition budget. The chemistry of Hg is however not only the central theme of this chapter, but also the central process which links emissions to deposition, governs the lifetime of $Hg^0_{(g)}$ in the MBL, and therefore the influence that marine emissions may have on the global Hg budget.

THE CHEMISTRY OF Hg IN THE MBL

The Gas Phase

One of gas phase elemental mercury's defining characteristics under typical atmospheric conditions of temperature and pressure, and in the presence of typical concentrations of the common tropospheric oxidants, is its general lack of reactivity. $Hg^0_{(g)}$ reacts extremely slowly with O_3, H_2O_2, HCl, and Cl_2, the rates range from 3×10^{-20} to 2.6×10^{-18} $cm^3 molecules^{-1} s^{-1}$ (see table 1). These gases are present at trace concentrations in the parts per billion (ppb = nmol mol^{-1}) to parts per trillion (ppt = pmol mol^{-1}) range. Recently however it has been found that $Hg^0_{(g)}$ reacts much faster with certain radicals present in the troposphere, notably OH, Br and Cl, the concentrations of which are generally very low indeed, so low in fact that direct measurement is very difficult in the field. The concentrations of these radicals are inferred to range from effectively zero to $10^6 - 10^7$ molecules cm^{-3}; at 25°C and 1 atmosphere there are in total roughly 2.5×10^{19} molecules cm^{-3}. These maximum radical concentrations, because of their photolytic production mechanisms, occur with maximum incident solar radiation, and their average concentrations are therefore far lower. The reaction between $Hg^0_{(g)}$ and NO_3 has also been measured (Sommar et al., 1997) but only an upper limit (k = 4×10^{-15} $cm^3 molecule^{-1}s^{-1}$) was reported. NO_3 is a night time oxidant and given the experimental evidence which shows a daytime $Hg^{II}_{(g)}$ maximum it seems unlikely that his reaction plays an important role in the MBL. The reactions with OH, Br and Cl radicals are nonetheless important in Hg cycling because even though their concentrations are low, the relative rapidity with which they oxidise $Hg^0_{(g)}$ compared to other more abundant atmospheric oxidants, mean that these reactions account for the major part of the tropospheric oxidation of $Hg^0_{(g)}$ where they are present. The halogen radicals are found in greater than average abundance during the polar springtime 'Bromine explosion', and also, but to a lesser extent in the MBL and above salt lakes. The OH radical on the other hand is generally more abundant where O_3 concentrations are high, as it is produced mainly by the reaction between water vapour and the O (^{1}D) atom produced by one of the photolysis pathways of O_3. The measurement of the rates of reaction between $Hg^0_{(g)}$ and radical species has only begun recently, and it is only right to point out that not all the measured reaction rates have been independently confirmed, and that there is not unanimous agreement over the results obtained. Comparison between modelling studies and field campaign measurements have in some cases provided some corroboration using the measured reaction rates, but in others have called into question the accuracy of some of the rate constants.

The Aqueous Phase

The atmospheric aqueous phase, cloud and fog droplets and deliquescent aerosol particles play an extremely important role in Hg chemistry and not just in the MBL. As mentioned in the introduction $Hg^{II}_{(g)}$ is much more soluble and less volatile than $Hg^0_{(g)}$, therefore most deposited Hg is Hg^{II}, and wet deposition is often the major Hg deposition process. $Hg^{II}_{(g)}$ is rapidly scavenged in the presence of an atmospheric aqueous phase, $Hg^0_{(g)}$ is also scavenged to a certain extent and both oxidation and reduction processes occur in the aqueous phase. In the absence of the radicals which oxidise Hg in the gas phase, Hg oxidation proceeds much more rapidly in the aqueous phase than the gas phase, predominantly via the reaction with $HOCl_{(aq)}/OCl^-_{(aq)}$ and $O_{3(aq)}$, see Table 1. There is some debate whether oxidised Hg compounds can be reduced in environmental aqueous systems by reaction with $HO_{2(aq)}/O_2^-_{(aq)}$; Pehkonen and Lin (1997) sustained that reduction occurred with a rate constant of 1.1×10^4 M^{-1} s^{-1}, while Gårdfeldt, and Jonsson (2003) suggest that the reaction does not proceed under environmental conditions. The influence of the inclusion or not of this reduction pathway on modelled Hg concentrations in rain has been studied by Bullock (2005) (see Chapter-14). The thermal decomposition of $HgSO_3$ also yields elemental Hg (Table 1). Apart from the redox chemistry which occurs in the aqueous phase, the ionic composition of the droplets or aerosols make a difference to the quantity of $Hg^{II}_{(aq)}$ which can feasibly be associated with the particles. Modelling studies showed that because of the greater number of potential complexes between $Hg^{++}_{(aq)}$ and $Cl^-_{(aq)}$ (and other halides) compared to $OH^-_{(aq)}$ or $SO_3^{2-}_{(aq)}$, that is: $HgCl^+_{(aq)}$, $HgCl_{2(aq)}$, $HgCl_3^-_{(aq)}$ and $HgCl_4^{2-}_{(aq)}$, that higher $Hg^{II}_{(aq)}$ concentrations, by a factor of 100, could be found in droplets with high halide ion concentrations (Forlano et al., 2000). Although this is not generally important in fog and cloud droplets it is potentially very important for sea salt aerosol particles.

The high halide ion concentration found in sea salt particles potentially also has an effect on the speciation of $Hg^{II}_{(g)}$ found in the MBL. HgO has a very high Henry's Law constant and HgO produced by the gas phase oxidation of $Hg^0_{(g)}$ by $OH_{(g)}$ or $O_{3(g)}$, (if it is HgO produced in the reaction, product studies are difficult with $Hg^{II}_{(g)}$ species because they tend to condense on to any available surface), would be rapidly scavenged by the sea salt particles. It would then react with $H^+_{(aq)}$ to give $Hg^{++}_{(aq)}$ which would be complexed as discussed above. Modelling studies show that as the Hg chloride complex concentrations increase the imbalance between the aqueous and gas phase concentrations of $HgCl_2$ results in net outgassing of $HgCl_2$ from the particles.

Mass Transfer

Mass transfer of any species between droplets or deliquesced aerosol particles depends on the Henry's Law constant, the temperature, the droplet radius, the accommodation coefficient of the species involved and the liquid water content (LWC) of the atmosphere. For $Hg^{II}_{(g)}$ species the general direction of the transfer, because of high their Henry's Law constants, is from the gas phase to the aqueous phase. This is certainly the case when the LWC is high, such as in fogs or clouds. However, as described above when the only liquid water is deliquesced aerosol particles, and they high concentrations of halide ions (sea salt aerosol can have $Cl^-_{(aq)}$ concentrations over 5M) it is possible that Hg^{II} species such as $HgCl_2$ pass from the aqueous to the gas phase. $Hg^0_{(g)}$ however is relatively insoluble and has a low Henry's Law constant, see Table 1, and its aqueous phase concentration is low. The equilibrium concentration in the aqueous phase is determined by the Henry's Law constant, the rate at which equilibrium is reached, however, depends on

Table 1. The Hg chemistry included in AMCOTS.

Mercury Reactions and Equilibria	k or K (298K)	Reference
$Hg^0_{(g)} + O_{3(g)} \rightarrow HgO_{(g)}$	3.0×10^{-20} cm^3 molec^{-1} s^{-1}	(a)
$Hg^0_{(g)} + O_{3(g)} \rightarrow HgO_{(g)}$	7.5×10^{-19} cm^3 molec^{-1} s^{-1}	(b)
$Hg^0_{(g)} + H_2O_{2(g)} \rightarrow HgO_{(g)}$	8.5×10^{-19} cm^3 molec^{-1} s^{-1}	(c)
$Hg^0_{(g)} + OH_{(g)} \rightarrow HgO_{(g)}$	8.7×10^{-14} cm^3 molec^{-1} s^{-1}	(d)
$Hg^0_{(g)} + HCl_{(g)} \rightarrow \rightarrow HgCl_{2(g)}$	1×10^{-19} cm^3 molec^{-1} s^{-1}	(e)
$Hg^0_{(g)} + Cl_{2(g)} \rightarrow HgCl_{2(g)}$	2.6×10^{-18} cm^3 molec^{-1} s^{-1}	(f)
$Hg^0_{(g)} + Cl_{(g)} \rightarrow \rightarrow HgCl_{2(g)}$	1×10^{-11} cm^3 molec^{-1} s^{-1}	(f)
$Hg^0_{(g)} + Br_{2(g)} \rightarrow HgBr_{2(g)}$	9×10^{-17} cm^3 molec^{-1} s^{-1}	(f)
$Hg^0_{(g)} + Br_{(g)} \rightarrow \rightarrow HgBr_{2(g)}$	3.2×10^{-12} cm^3 molec^{-1} s^{-1}	(f)
$Hg^0_{(aq)} + O_{3(aq)} \rightarrow HgO_{(aq)}$	4.7×10^7 M^{-1} s^{-1}	(g)
$HgO_{(aq)} + H^+_{(aq)} \rightarrow Hg^{++}_{(aq)} + OH^-_{(aq)}$	1×10^{10} M^{-1} s^{-1}	(h)
$Hg^{++}_{(aq)} + OH^-_{(aq)} \rightarrow HgOH^+_{(aq)}$	3.9×10^{10} M^{-1}	(h)
$HgOH^+_{(aq)} + OH^-_{(aq)} \rightarrow Hg(OH)_{2(aq)}$	1.6×10^{11} M^{-1}	(h)
$HgOH^+_{(aq)} + Cl^-_{(aq)} \rightarrow HgOHCl_{(aq)}$	2.7×10^7 M^{-1}	(h)
$Hg^{++}_{(aq)} + Cl^-_{(aq)} \rightarrow HgCl^+_{(aq)}$	5.8×10^6 M^{-1}	(h)
$HgCl^+_{(aq)} + Cl^-_{(aq)} \rightarrow HgCl_{2(aq)}$	2.5×10^6 M^{-1}	(h)
$HgCl_{2(aq)} + Cl^-_{(aq)} \rightarrow HgCl_3^-{}_{(aq)}$	6.7 M^{-1}	(i)
$HgCl_3^-{}_{(aq)} + Cl^-_{(aq)} \rightarrow HgCl_4^{--}{}_{(aq)}$	13 M^{-1}	(i)
$Hg^{++}_{(aq)} + Br^-_{(aq)} \rightarrow HgBr^+_{(aq)}$	1.1×10^9 M^{-1}	(i)
$HgBr^+_{(aq)} + Br^-_{(aq)} \rightarrow HgBr_{2(aq)}$	2.5×10^8 M^{-1}	(i)
$HgBr_{2(aq)} + Br^-_{(aq)} \rightarrow HgBr_3^-{}_{(aq)}$	1.5×10^2 M^{-1}	(i)
$HgBr_3^-{}_{(aq)} + Br^-_{(aq)} \rightarrow HgBr_4^-{}_{(aq)}$	23 M^{-1}	(i)
$Hg^{++}_{(aq)} + SO_3^{--}{}_{(aq)} \rightarrow HgSO_{3(aq)}$	2.1×10^{13} M^{-1}	(j)
$HgSO_{3(aq)} + SO_3^{--}{}_{(aq)} \rightarrow Hg(SO_3)_2^{--}{}_{(aq)}$	1.0×10^{10} M^{-1}	(j)
$HgSO_{3(aq)} \rightarrow Hg^0_{(aq)}$ + products	0.0106 s^{-1}	(k)
$Hg^0_{(aq)} + OH_{(aq)} \rightarrow Hg^+_{(aq)} + OH^-_{(aq)}$	2.0×10^9 M^{-1} s^{-1}	(l)
$Hg^+_{(aq)} + OH_{(aq)} \rightarrow Hg^{++}_{(aq)} + OH^-_{(aq)}$	1.0×10^{10} M^{-1} s^{-1}	(l)
$Hg^{II}_{(aq)} + O_2^-{}_{(aq)} \rightarrow Hg^+_{(aq)} + O_{2(aq)}$	1.1×10^4 M^{-1} s^{-1}	(m)
$Hg^{II}_{(aq)} + HO_{2(aq)} \rightarrow Hg^+_{(aq)} + O_{2(aq)} + H^+_{(aq)}$	1.1×10^4 M^{-1} s^{-1}	(m)
$Hg^+_{(aq)} + O_2^-{}_{(aq)} \rightarrow Hg_{(aq)} + O_{2(aq)}$	fast	(m)
$Hg^+_{(aq)} + HO_{2(aq)} \rightarrow Hg^0_{(aq)} + O_{2(aq)} + H^+_{(aq)}$	fast	(m)
$Hg^{II}_{(aq)} + O_2^-{}_{(aq)} \rightarrow Hg^+_{(aq)} + O_{2(aq)}$	0	(n)
$Hg^{II}_{(aq)} + HO_{2(aq)} \rightarrow Hg^+_{(aq)} + O_{2(aq)} + H^+_{(aq)}$	0	(n)
$Hg_{(aq)} + HOCl_{(aq)} \rightarrow Hg^{++}_{(aq)} + Cl^-_{(aq)} + OH^-_{(aq)}$	2.09×10^6 M^{-1} s^{-1}	(o)
$Hg_{(aq)} + ClO^-_{(aq)} \rightarrow Hg^{++}_{(aq)} + Cl^-_{(aq)} + OH^-_{(aq)}$	1.99×10^6 M^{-1} s^{-1}	(o)
$Hg^0_{(g)} \leftrightarrow Hg^0_{(aq)}$	0.13 M atm^{-1}	(p)
$HgO_{(g)} \leftrightarrow HgO_{(aq)}$	2.69×10^{12} M atm^{-1}	(p)
$HgCl_{2(g)} \leftrightarrow HgCl_{2(aq)}$	2.75×10^6 M atm^{-1}	(p)
$HgBr_{2(g)} \leftrightarrow HgBr_{2(aq)}$	2.75×10^6 M atm^{-1}	(q)
$HgO_{(g)} \rightarrow$ deposition	2.0 cm s^{-1}	(r)
$HgCl_{2(g)} \rightarrow$ deposition	2.0 cm s^{-1}	(r)
$HgBr_{2(g)} \rightarrow$ deposition	2.0 cm s^{-1}	(q)

(a) Hall (1995), (b) Pal and Ariya (2004), (c) Tokos et al. (1998), (d) Sommar et al., (2001), (e) Hall and Bloom (1993), (f) Ariya et al. (2002), (g) Munthe (1992), (h) Pleijel and Munthe (1995), (i) Clever et al. (1985), Van Loon et al. (2001), (k) Van Loon et al. (2000), (l) Lin and Pehkonen (1997), (m) Pehkonen and Lin (1997), (n) Gårdfeldt, and Jonsson (2003), (o) Lin and Pehkonen (1999), (p) Schroeder and Munthe (1998), (q) Hedgecock and Pirrone, 2004, (r) Hedgecock and Pirrone, 2001.

a number of factors as shown below. The rate of change of the aqueous phase concentration is given in equation 1, following the notation of Sander (1999),

$$dc_a/dt = k_{mt} (c_{g,\infty} - c_{a,surf} / k_H RT) \tag{1}$$

where c_a is the aqueous phase concentration of the species (mol m$^{-3}_{aq}$),
$c_{a,surf}$ the aqueous phase concentration at the surface (mol m$^{-3}_{aq}$), $c_{g,\infty}$ the background gas phase concentration (mol m$^{-3}_{air}$), k_H the Henry's Law constant (mol m$^{-3}_{aq}$ Pa^{-1}), R the gas constant (J mol^{-1} K^{-1}) and T the temperature (K). k_{mt} is the mass transfer coefficient, defined as:

$$k_{mt} = (r^2/3D_g + 4r/3\alpha v)^{-1} \tag{2}$$

where r is the droplet radius (m), D_g the gas phase diffusivity (m^{2}s^{-1}), v the mean molecular velocity (ms^{-1}) and α the accommodation coefficient (dimensionless). k_{mt} is expressed in s^{-1}, but contains the conversion m$^3_{air}$ to m$^3_{aq}$, Sander (1999).

As can be seen from equation 2, the mass transfer coefficient is inversely proportional to the square of the droplet radius, hence the very small aerosol droplets reach aqueous phase equilibrium rapidly. Larger droplets such as cloud or rain droplets take significantly longer. Some droplets may not reach equilibrium during the lifetime of a cloud or if they are precipitating, thus chemical models tend to model mass transfer as a kinetic rather than equilibrium process.

General MBL Photochemistry

Mass transfer between phases turns out to be very important in MBL photochemistry in general. The chemistry of the MBL is distinguished from other parts of the atmosphere by its constantly high relatively humidity, usually above the deliquescence point of sea salt and non-sea-salt (nss) sulphate aerosols, and almost always higher than the crystallization point, this means that aerosol emitted from the sea surface as liquid droplets remain liquid and do not crystallize. The sea-salt aerosol droplets are characterised by high concentrations of Cl$^-$ and Br$^-$ ions; the acidification of freshly produced sea salt aerosol, which is slightly alkaline, triggers a series of chemical responses in the aerosol, such as the acid displacement of HCl and the production of neutral, relatively insoluble Br containing species in the aqueous phase which then pass from the aqueous to the gas phase according to their Henry's Law constant.

The release of halogen containing species from aerosol particles is the key to the difference between Hg chemistry in the MBL and elsewhere in the troposphere. The conversion of bromide ions present in sea salt particles to less soluble and volatile Br containing species, particularly Br_2 and BrCl, which degas from the aerosol, is initiated by its reaction with $O_{3(aq)}$, $OH_{(aq)}$, $NO_{3(aq)}$, $N_2O_{5(g)}$ or $HSO_5^-{}_{(aq)}$, (Sander et al., 2003; von Glasow et al., 2002). Vogt et al., 1996 proposed an autocatalytic cycle starting from the reaction of HOBr with Cl⁻ in the presence of acid as shown below (from Sander et al., 2003);

$$HOBr_{(aq)} + Cl^-{}_{(aq)} + H^+{}_{(aq)} \rightarrow BrCl_{(aq)} + H_2O_{(aq)} \quad (3)$$
$$BrCl_{(aq)} + Br{-}_{(aq)} \rightarrow Br_{2(aq)} + Cl{-}_{(aq)} \quad (4)$$
$$Br_{2(aq)} \rightarrow Br_{2(g)} \quad (5)$$
$$Br_{2(g)} + h\nu \rightarrow Br_{(g)} + Br_{(g)} \quad (6)$$
$$Br_{(g)} + O_{3(g)} \rightarrow BrO_{(g)} + O_{2(g)} \quad (7)$$
$$BrO_{(g)} + HO_{2(g)} \rightarrow HOBr_{(g)} + O_{2(g)} \quad (8)$$
$$HOBr_{(g)} \rightarrow HOBr_{(aq)} \quad (9)$$

which leads to the net reaction;

$$Br^-{}_{(aq)} + H^+{}_{(aq)} + O_{3(g)} + HO_{2(g)} + h\nu \rightarrow Br_{(g)} + 2O_{2(g)} + H_2O_{(aq)} \quad (10)$$

or alternatively expressed, see von Glasow et al. (2002);

$$2HO_{2(g)} + H^+{}_{(aq)} + 2O_{3(g)} + Br^-{}_{(aq)} + h\nu \rightarrow HOBr_{(g)} + 4O_{2(g)} + H_2O_{(aq)} \quad (11)$$

It is the release of Br containing compounds which is of most interest for Hg chemistry. The difference in the release mechanisms of Br and Cl from the sea salt aerosol results in Cl being released to a large extent as HCl, whilst Br is emitted mostly as Br_2 or BrCl. HCl is stable chemically and photolytically, whereas Br_2 and BrCl are rapidly photolysed to their atomic constituents. The gas phase concentration of reactive Br compounds is therefore higher than the concentration of reactive Cl compounds, with the result that the Br compounds have more influence on Hg chemistry than Cl compounds, even thought the reaction with the Cl atom is the fastest of the known reactions between $Hg^0{}_{(g)}$ and the halogen atoms and compounds (Arya et al., 2002).

Modelling MBL Chemistry

A number of box and 1-dimensional modelling studies of MBL chemistry have been published in recent years, most of which concentrate on the role of the marine aerosol in the production of reactive halogen compounds (Sander and Crutzen,1996, Vogt et al., 1996; Toyota et al., 2001; von Glasow et al., 2002; Toyota et al., 2003). These models have increased in complexity and now take into account both sea salt and non-sea-salt sulphate particles, some including particle size distributions, and sea salt aerosol production. The chemistry in these models focuses on halogen gas and aqueous phase chemistry but also include NO_x, HO_x and SO_x chemistry. Non-methane hydrocarbons are included in all the models with varying levels of detail. The most complex treatment is given in Toyota et al. (2003). The AMCOTS model uses the reaction database from von Glasow et al. (2002) which includes CH_4, C_2H_6, C_2H_4, and their reaction products in the gas phase and where the products are soluble, the aqueous phase.

The rate and extent of halogen release from sea salt aerosols under varying atmospheric conditions is discussed in more detail below, but in terms of Hg chemistry; for more general information the reader is referred to von Glasow et al. (2002), Sander et al. (2003), Toyota et al. (2003), and references therein.

FACTORS INFLUENCING THE RATE OF $Hg^0_{(g)}$ OXIDATION IN THE MBL

Br production

In clean MBL air, with low levels of NO_x and O_3, the concentration of OH is low and the major $Hg^0_{(g)}$ oxidant species will be $Br_{(g)}$. Therefore the factors which influence $Br_{(g)}$ concentrations will have a direct effect on the rate at which $Hg^0_{(g)}$ is oxidised and thus the concentration of $Hg^{II}_{(g)}$. Among the more important factors influencing the rate and extent of halogen activation in the MBL are the wind speed, the temperature, the intensity of solar radiation and latitude.

Wind speed, which determines the production of sea salt aerosol particles, has a direct effect on atmospheric sea salt LWC. Increasing sea salt LWC increases the scavenging rate of $Hg^{II}_{(g)}$ and can therefore lower the gas phase concentration. However, increased sea salt LWC also potentially supplies more $Br_{(g)}$ to the MBL increasing its capacity to oxidise $Hg^0_{(g)}$, but increases in the aerosol numbers, and/or size, increase the buffering capacity of the MBL because freshly produced sea salt aerosol is alkaline, and therefore the acid dependent halogen activation mechanism takes longer to start. A theoretical box model study of the lifetime of $Hg^0_{(g)}$ in the MBL showed that up to a threefold increase in sea salt LWC (from 3×10^{-11} to 9×10^{-11} m^3_{aq}/m^3_{air}) increased the rate of $Hg^0_{(g)}$ oxidation, whereas decreasing the sea salt LWC decreased the rate (Hedgecock and Pirrone, 2004). One of the problems with a box model study is that a constant aerosol loading is assumed to the top of the model volume, which is not the case given that production occurs at the surface. Both model and experimental studies need to confront this aspect of MBL chemistry.

From the net equations describing Br release from the sea salt aerosol above, it can be seen that acid is consumed, and therefore the originally alkaline sea salt particles need to take up acidic gases from the atmosphere to initiate the process. The air temperature influences the solubility of gases, thus lower temperatures favour the more rapid uptake of acidic gases, accelerating the rate of acidification of the sea salt particles and halogen activation (von Glasow et al., 2002).

The intensity of solar irradiation is important both for the photolysis of Br containing compounds, but also for the production of $HO_{2(g)}$ which is the major sink for $BrO_{(g)}$ in the MBL. Because the oxidation of $Hg^0_{(g)}$ depends on photolytically produced oxidants the rate naturally depends on both latitude and time of year, Figures 1a and b, (Hedgecock and Pirrone, 2004).

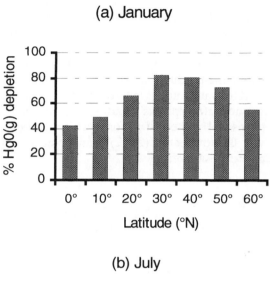

(a) January

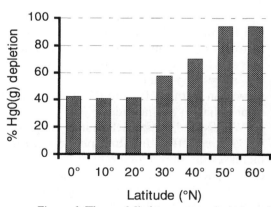

(b) July

Figure 1. The modelled percentage depletion of $Hg^0_{(g)}$ after 1 week as a function of latitude for the months of January and July.

Hedgecock and Pirrone (2004) also showed that some cloud cover could actually enhance the rate at which $Hg^0_{(g)}$ is oxidised in the MBL. Figure 2 shows an example of this phenomenon, the $Hg^0_{(g)}$ depletion was modelled for January at a latitude of $10°$ N, and it can be clearly seen that increasing the cloud cover up to an optical depth of 20 increases $Hg^0_{(g)}$ depletion with respect to the clear sky simulations.

The reason for this is that the attenuation of the incoming solar radiation produced by clouds is not equal over the UV-visible spectrum, it is more marked at shorter than at longer wavelengths. Br_2 and BrCl photolysis begins earlier in the morning and ceases later in the evening than O_3 photolysis because the solar spectrum in the lower troposphere is shifted to the red at high solar zenith angles.

This accounts for the earlier build-up of $Br_{(g)}$ in the MBL, because in the early morning the longer path which sunlight travels through the atmosphere means that Br_2 photolysis commences earlier. In fact the Br produced reacts with O_3 to produce BrO, a phenomenon called 'sunrise ozone destruction', (Nagao et al., 1999). The concentration of BrO in fact peaks at dawn

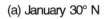

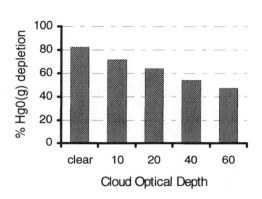

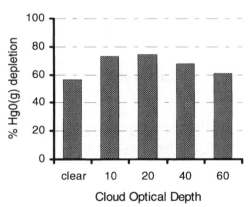

Figure 2. The influence of cloud optical depth on $Hg^0_{(g)}$ depletion at 10 ° N during January, after one week of model simulation.

and dusk, it has a minimum around midday when $HO_{2(g)}$ concentrations are at their maximum (von Glasow et al., 2002).

This increase in $Hg^0_{(g)}$ oxidation rate occurs only at certain latitudes and at certain times of the year, and only up to a certain cloud optical depth, after which the rate of oxidation decreases.

OH production

As well as $Br_{(g)}$, $OH_{(g)}$ is potentially a major contributor to the production of $Hg^{II}_{(g)}$ in the MBL. In the clean MBL it is likely to be significantly less important than $Br_{(g)}$ due to the low O_3 concentrations found and the low concentrations of possible O_3 precursors such as NO_x. However, in regions where the MBL is influenced by continental air masses which contain urban or industrial NO_x emissions, it is possible that conditions are amenable to $OH_{(g)}$ production. Such conditions would also result in higher $HO_{2(g)}$ concentrations than in the clean MBL, thus suppressing $BrO_{(g)}$, which, because BrO photolysis is the main source of $Br_{(g)}$, reduces the role of $Br_{(g)}$ at the same time.

A prime example of a region where the MBL is directly affected by continental emissions is the Mediterranean Sea. A combination of meteorological factors, boundary layer air flow is slow but generally from north to south, high levels of solar irradiation, stable anticyclonic conditions and little rain, provide all the ingredients in the Mediterranean summer for O_3 production. Concentrations of O_3 at ground level across the whole Mediterranean regularly exceed the European 8 hour exposure limit (Nolle et al., 2002; Lelieveld et al., 2002). It is probable that the contribution of $OH_{(g)}$ to the MBL oxidation of $Hg^0_{(g)}$ under such conditions is greater than that of $Br_{(g)}$, partly because of the higher $OH_{(g)}$ concentration and partly because of the effect that higher $HO_{2(g)}$ concentrations would have on the production of $Br_{(g)}$ as described above. This hypothesis is supported by recent modelling studies as described below.

$Hg^{II}_{(g)}$ IN THE MBL: MODELLING vs. MEASUREMENTS

The first measurements of RGM in the MBL (rather than at coastal sites) were made during three cruises between 1999 and 2000 near to the Bermuda Time Series Station (BATS; 31.678N, 64.178W), in the North Atlantic, (Mason et al, 2001), using the filter pack technique (Sheu and Mason, 2001) and in the Mediterranean in 2000 (Sprovieri et al., 2003), using the annular denuder technique (Landis et al., 2002). Since then there have been further measurement campaigns in the North Pacific in 2002 (Laurier et al., 2003), and in the Mediterranean in 2003 and 2004 (F. Sprovieri, private communication), during these cruises $Hg^0_{(g)}$ and $Hg^{II}_{(g)}$ were measured using the same techniques and the same type of instrument: the Tekran 1130 speciation unit coupled with to the Tekran 2537A analyser (Tekran Inc.,

Toronto, Canada). The time resolution of the measurements was two or four hours.

The first attempt to model MBL $Hg^{II}_{(g)}$ concentrations used measured $Hg^0_{(g)}$ concentrations and ancillary data obtained during from the first Mediterranean cruise campaign in 2000 as input and using a box model of MBL photochemistry with Hg chemistry included, sought to reproduce the measured $Hg^{II}_{(g)}$ concentrations. The box model itself is described in Hedgecock and Pirrone (2001), and the results of the comparison between modelled and measured $Hg^{II}_{(g)}$ concentrations in Hedgecock et al. (2003).

That original model did not include the reaction of $Hg^0_{(g)}$ with $Br_{(g)}$, as the rate constant was determined after the modelling studies were performed and the model unsurprisingly underestimated somewhat the $Hg^{II}_{(g)}$ concentrations, although the minima and the diurnal variation in $Hg^{II}_{(g)}$ was reproduced reasonably well.

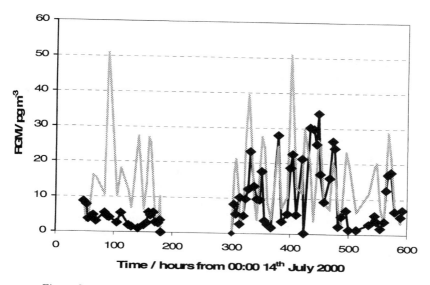

Figure 3. Measured (dark line with diamonds) and modelled with $z_{(surface\ mixed\ layer)}$ set at 300m (light line) RGM values (pg m-3) from the Mediterranean campaign.

Later studies included the $Hg^0_{(g)} + Br_{(g)}$ reaction and, in order to simulate halogen activation more accurately, both the sea salt and non-sea-salt sulphate aerosols were included (Hedgecock and Pirrone, 2004); the previous model included only the sea salt aerosol.

This latest version of the box model, AMCOTS, (Atmospheric Mercury Chemistry Over The Sea) is being used (work in progress) to re-analyse the data from the Mediterranean campaign aboard the R.V. Urania in 2000, and the north Pacific campaign aboard the R.V. Melville in 2002. This new version of the model uses the measured $Hg^0_{(g)}$ concentrations and ancillary chemical and meteorological data, to model the $Hg^{II}_{(g)}$ concentration, similarly to Hedgecock et al. (2003). Some preliminary results for the Mediterranean are shown in Figure 3.

Two things need to be mentioned regarding the application of a box model to measured data, the first is that precipitation and sea surface conditions, are not taken into account. Both increased sea salt aerosol and precipitation increase the rate at which $Hg^{II}_{(g)}$ is scavenged, due to the increased atmospheric liquid water content. The first period of the oceanographic campaign was characterised by rough seas and rain, which accounts for the model's overestimation of the $Hg^{II}_{(g)}$ concentration (Figure 3). The second, is that a box model requires that the box height is included as input in order to calculate the rate of dry deposition and emission to and from the sea surface. Typical values of the MBL height (z_{MBL}) are 1000m for the remote MBL (such as the Pacific, Sander et al., 1996) and 400m for the summertime Mediterranean (Kallos et al., 1998) respectively. However the height of the capping inversion over the MBL does not always correspond to the height of the surface mixed layer, (Garratt, 1992), and in a model of MBL aerosol formation from biogenic iodine emissions, O'Dowd et al. (2002) used a globally averaged value of 300m for the height of the surface mixed layer. The mixed layer height also varies near the coast when advection of air across the coastline results in a thermal internal boundary layer (Garratt, 1992). The height of the boundary layer over the Mediterranean Sea is relatively stable but variations between day and night do occur. The actual height of the surface mixed layer is not simple to determine, (Seibert et al., 2000), and in the absence of appropriate measurements meteorological models which calculate advection and transport could be one means of providing an estimate from the surface mixed layer height. Unfortunately complex chemical models such as AMCOTS with over 200 chemical species and 900 reactions are too computationally expensive to be linked directly to meteorological / dispersion models to perform regional simulations over the periods of measurement campaigns. However, models such as the Regional Atmospheric Modelling System (RAMS) (ATMET, 2004) can be used to estimate the height of the mixing layer, using variation in Turbulent Kinetic Energy (TKE), the virtual potential temperature (θ_v) or the relative humidity. Simulations using nested grids for the period and area of the last half of the Urania cruise suggest that the mixing layer height varied between less than 150m to around 350m.

Figure 4 shows the results assuming that the mixing layer height ($z_{surface\ mixed\ layer}$) varies as indicated by the RAMS TKE values for the second period of the Mediterranean campaign in 2000. The model output is averaged over four hours to correspond to the sampling time used in the measurements.

One of the main reasons for attempting to model measured $Hg^{II}_{(g)}$ is partly to see if known Hg chemistry is comprehensive enough to make sense of the observations. In fact the first modelling studies (Hedgecock and Pirrone, 2001; Hedgecock et al., 2003) $Hg^{II}_{(g)}$ concentrations were underestimated.

Modelling studies also help to identify the predominant oxidation pathways, because the model uses measured data to recreate the chemistry, allowing reaction rates to be followed individually during the simulation. This last approach has recently identified what appears to be a difference between open ocean and closed sea oxidation of $Hg^0_{(g)}$. Preliminary simulations for the Pacific data (Laurier et al, 2003) and Mediterranean data (Sprovieri et al, 2003) indicate that although Br was the most important oxidant in the Pacific whilst in the Mediterranean most oxidation was the result of the reaction with $OH_{(g)}$. This is most likely to be due to the higher $O_{3(g)}$ concentrations and temperatures in the Mediterranean, however it is too soon to draw hard and fast conclusions.

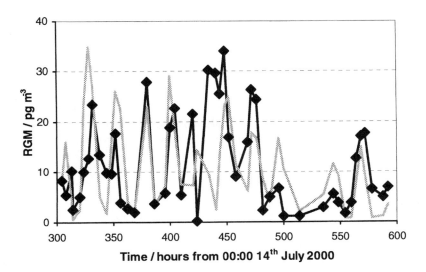

Figure 4. Measured (dark line with diamonds) and modelled (light line) RGM values (pg m^{-3}) from the Mediterranean campaign, with z$_{(surface\ mixed\ layer)}$ temporal variation calculated using RAMS.

IMPLICATIONS OF Hg CHEMISTRY IN THE MBL

The relatively new discovery of $Hg^{II}_{(g)}$ formation in the MBL is changing some of the assumptions previously made about atmospheric Hg, its cycling between air and sea, its emission from the world's oceans and the global Hg budget. A few of the more important of these are briefly described here, a fuller discussion may be found in the relevant chapters of this book and the literature cited below.

The lifetime of $Hg^0_{(g)}$ in the MBL

The AMCOTS model has been used to estimate the atmospheric lifetime of $Hg^0_{(g)}$ in the MBL, (Hedgecock and Pirrone, 2004). Using the approach of Seinfield and Pandis, (1998) the atmospheric lifetime (τ) of $Hg^0_{(g)}$ is the sum of the inverse of the oxidation reaction rates (rate constant multiplied by concentration :

$$\tau = \{k(Hg^0+O_3).[O_3] + k(Hg^0+OH).[OH] + k(Hg^0+Br).[Br]\}^{-1} \text{ s} \qquad (12)$$

thus

$$\tau^{-1} = 3\times10^{-20}\,[O_3] + 8.7\times10^{-14}\,[OH] + 3.2\times10^{-12}\,[Br]\,\text{s}^{-1} \qquad (13)$$

from AMCOTS, the modelled one week average concentrations of the oxidants found using average cloud optical depths for summer, at 10, 40 and 60° N were 21 ppb O_3, OH in the range $5-20\times10^5$ molecules cm^{-3}, and Br around 3×10^5 molecules cm^{-3},

$$\tau^{-1} = 2\times10^{-8} + 8.7\times10^{-8} + 1\times10^{-6}$$
$$\text{thus, } \tau = 10.5 \text{ days,} \qquad (14)$$

alternatively the lifetime with respect to the individual reactions under the model conditions are: (Hg^0+O_3) = 578 days, (Hg^0+OH) = 133 days and (Hg^0+Br) = 11.5 days days. Using the newly measured value for Hg^0+O_3 (Pal and Ariya, 2004) the calculation gives:

$$\tau^{-1} = 7.5\times10^{-19}\,[O_3] + 8.7\times10^{-14}\,[OH] + 3.2\times10^{-12}\,[Br]\,\text{s}^{-1} \qquad (15)$$
$$\tau^{-1} = 3.9\times10^{-7} + 8.7\times10^{-8} + 1\times10^{-6} \qquad (16)$$

thus, τ = 7.8 days, (Hg^0+O_3) = 29 days, (Hg^0+OH) = 133 days and (Hg^0+Br) = 11.5 days.

Such a short lifetime for $Hg^0_{(g)}$ is not found in other parts of the atmosphere and has implications for the cycle of Hg in the marine troposphere as discussed below.

Marine emissions of Hg^0 and the global Hg budget

The world's oceans are a source of Hg^0 as well as a receptor for Hg^{II}. The potentially rapid cycling of Hg in the MBL seems to conflict with the well known fact that the hemispherical background concentration of $Hg^0_{(g)}$ is both uniform and constant. The depletion in $Hg^0_{(g)}$ concentrations simulated by Hedgecock and Pirrone (2004) are not seen in the field and imply that the Hg being lost via oxidation and deposition is constantly being replaced. Therefore Hedgecock and Pirrone (2004) included emissions in their model to maintain a stable $Hg^0_{(g)}$ concentration and calculated the emission rates under typical conditions of cloud cover, and temperature for different seasons at different latitudes. The emission rates obtained were of the same order of magnitude as the measured data available in the literature, illustrating that oxidation in the MBL is not incompatible with stable background concentrations. It should also be borne in mind that replenishment of $Hg^0_{(g)}$ in the MBL may be partially due to exchange with the free troposphere as is the case for O_3.

Model improvements required

The next steps in improving MBL Hg chemistry models will include moving to 1 or more dimensional models which allow the variation with height of atmospheric LWC, aerosol pH, and particularly halogen species concentrations. This should serve to avoid the problems of varying surface mixed layer height, and ideally if combined with height resolved measurements of $Hg^0_{(g)}$ and $Hg^{II}_{(g)}$, allow the determination of the relative influences of Hg replenishment from ocean surface emissions and entrainment of Hg from the free troposphere. It will also require that there be some representation of sea salt aerosol production and its advection in the MBL.

Currently AMCOTS uses two mono-disperse distributions to describe the sea salt and sulphate aerosol particles. This is another point where MBL photochemical models have progressed by beginning to use size segregated bins to describe the aerosol population (Toyota et al., 2001, von Glasow et al., 2002), although even this method averages aqueous phase concentrations over size bins, it is nonetheless better than averaging concentrations over the whole aerosol population. Inclusion of a full description of MBL and Hg chemistry in 3-d meteorological / dispersion models is hindered by the time that the calculations would require. They are also a less appropriate tool for sensitivity studies because the meteorological model input data cannot be changed simply parameter by parameter in the same way that it can with box and 1-d models.

Useful data

There are two specific sets of data which would improve the understanding of MBL Hg chemistry in particular but also Hg chemistry in general. The first is reaction rate data. There is still no real consensus on the rates, or mechanisms, or products of the potentially most important $Hg^0_{(g)}$ oxidation reactions. This is partly because interest in determining these rates is recent, as the number of experimental and theoretical studies published in the last two to three years shows, and partly due to practical difficulties in performing the experiments. It should be mentioned also that $Hg^{II}_{(g)}$ and $Hg^{II}_{(aq)}$ reduction reactions, chemical and photolytic could play an important role in atmospheric Hg chemistry, but few data are available. The aqueous phase photo-reduction of Hg^{II} compounds in particular is important for modelling emissions and to establish the magnitude of sea-air exchange resulting from Hg^{II} deposition and Hg^0 emission. The other type of data which would prove extremely useful is speciated $Hg^{II}_{(g)}$ measurements. At present it is not possible to distinguish HgO, from $HgCl_2$ and $HgBr_2$ using highly time resolved techniques in the field. Should this become practically feasible one of the major obstacles to being certain that atmospheric Hg chemistry is understood would be overcome.

Planning measurement campaigns

The ancillary data obtained during measurement campaigns is of fundamental importance to the interpretation and modelling of Hg compound concentration measurements. The current knowledge of atmospheric Hg chemistry means that without O_3 concentration data, UV-A and UV-B intensities, relative humidity, and temperature, the data pertaining to Hg is almost very difficult to interpret and therefore effectively useless. Measurements of SO_2 and CO as indicators of anthropogenic influence on air masses is also useful. Specific data that would be of particular use in interpreting data from the MBL are the atmospheric LWC and size-resolved aerosol Hg concentrations. If these measurements could be performed at different altitudes the results would almost certainly be fascinating. The practical complexities involved however are numerous.

CONCLUSIONS

Hg chemistry in the MBL is now rightly a part of the wider research effort to understand the atmospheric cycling of Hg, as it is via the atmosphere that Hg is transported from emission source to receptor site. The deposition and subsequent uptake of Hg by plants and animals can represent a major problem to their well-being, particularly to higher animals, as discussed elsewhere in this book. The transport of $Hg^0_{(g)}$, the overall impact of marine Hg^0 emissions, the deposition of Hg^{II} to open oceans, and perhaps more importantly to coastal areas, are all directly influenced by Hg chemistry in the MBL.

The lifetime of $Hg^0_{(g)}$ in the MBL depends on sea salt aerosol production and loading, and therefore wind speed, on incoming solar radiation thus on latitude, time of year and cloud cover, air temperature and the chemical composition of the MBL. There may therefore be great variability in the lifetime of $Hg^0_{(g)}$, but it can be short which would preclude long range transport, and suggest that a multi-hop mechanism is more likely.

Emissions (or re-emissions) of Hg^0 from the sea contribute to the atmospheric burden of Hg, however oxidation in the MBL and re-deposition could mitigate the effect of these emissions. It should not be forgotten however that the conditions which favour atmospheric oxidation may also favour photo-reduction in marine waters.

The potential impact on coastal environments that the deposition of Hg^{II} resulting from the confluence of more polluted continental air and cleaner but possibly more highly oxidising marine air is an aspect of MBL Hg chemistry which merits further investigation.

The increased understanding of Hg chemistry in the MBL, just as in so many fields of study, increases the number of questions for which an answer may soon be found. In spite of rapid progress many of the conclusions reached with the help of a better understanding of MBL Hg chemistry are still qualitative; clearly the next steps will be towards quantitative answers, particularly regarding deposition / emission in the open ocean, and deposition in coastal areas.

ACKNOWLEDGEMENTS

The authors wish to acknowledge the financial contributions received from the European Commission in the framework of several European projects including MAMCS (ENV4-CT97-0593) and MERCYMS (EVK3-2002-00070), the Italian Ministry of Research and the Italian National Research Council in the framework of MED-OCEANOR 2000, 2003 and 2004 projects. Special thanks to all students and technicians for their valuable contribution provided during the field campaigns.

REFERENCES

Ariya, P.A., Khalizov, A., Gidas, A. Reaction of gaseous mercury with atomic and molecular halogens: kinetics, products studies, and atmospheric implications, *J. Phys. Chem. A*, 106, 7310-7320, 2002.

ATMET, Atmospheric, Meteorological and Environmental Technologies, 2004 http://atmet.com.

Barrie, L.A., Bottenheim, J.W., Schnell, R.C., Crutzen, P.J., Rasmussen, R.A. Ozone destruction and photochemical reactions at polar sunrise in the lower Arctic atmosphere, *Nature*, 334, 138-141, 1988.

Barrie, L., Platt, U. Arctic tropospheric chemistry: an overview, *Tellus*, 49B, 450-454, 1997.

Bauer, D., D'Ottone, L., Campuzano-Jost, P., Hynes, A.J. Gas phase elemental mercury: a comparison of LIF techniques and study of the kinetics of reaction with the hydroxyl radical, *J. Photochem. Photobiol. A*, 157, 247-256, 2003.

Bottenheim J.W., Gallant A.C., Brice K.A. Measurements of NO_y species and O_3 at 82° N latitude, *Geophys. Res. Lett.*, 13, 113-116, 1986.

Clever, H.L., Johnson, S.A., Derrick, M.E. The solubility of Mercury and Some Sparingly Soluble Mercury Salts in Water and Aqueous Electrolyte Solutions. *J. Phys. Chem. Refer. Data*, 14, 631-680, 1985.

Fan, S-M. Jacob, D.J. Surface ozone depletion in Arctic spring sustained by bromine reactions on aerosols, *Nature*, 359, 522-524, 1992.

Foster, K.L., Plastridge, R.A., Bottenheim, J.A., Shepson, P.B., Finlayson-Pitts, B.J., Spicer, C.W. The role of Br_2 and BrCl in surface ozone destruction at polar sunrise, *Science*, 291, 471-474, 2001.

Gårdfeldt, K., Jonsson, M. Is bimolecular reduction of divalent mercury complexes possible in aqueous systems of environmental importance? *J. Phys. Chem. A*, 107, 4478-4482, 2003.

Hall, B., Bloom N., Report to EPRI, Palo Alto, CA, USA, 1993.

Hall, B. The phase oxidation of elemental mercury by ozone. *Water, Air Soil Pollut.*, 80, 301-315, 1995.

Hedgecock, I.M., Pirrone, N. Mercury photochemistry in the marine boundary layer modeling studies suggest the in situ production of reactive gas phase mercury, *Atmos. Environ.*, 35, 3055-3062, 2001.

Hedgecock, I.M., Pirrone, N., Sprovieri, F., Pesenti, E. Reactive gaseous mercury in the marine boundary layer: Modeling and experimental evidence of its formation in the Mediterranean region, *Atmos. Environ.*, 37, S41-49, 2003.

Hedgecock, I.M., Pirrone, N. Chasing Quicksilver: Modeling the Atmospheric Lifetime of $Hg^0_{(g)}$ in the Marine Boundary Layer at Various Latitudes. *Environ. Sci. Technol.*, 38, 69-76, DOI: 10.1021/es034623z, 2004.

Kallos, G., Kotroni, V., Lagouvardos K., Papadopoulos, A. On the long-range transport of air pollutants from Europe to Africa, *Geophys. Res. Lett.*, 25, 619-622, 1998.

Laurier, F.J.G., Mason, R.P., Whalin, L., Kato, S. Reactive gaseous mercury formation in the North Pacific Ocean's marine boundary layer: A potential role of halogen chemistry, *J. Geophys. Res.*, 108, 4529. doi:10. 1029/ 2003JD003625, 2003.

Lelieveld, J., Berresheim, H., Borrmann, S., Crutzen, P.J., Dentener, F.J., Fischer, H., Feichter, J., Flatau, P.J., Heland, J., Holzinger, R., Korrmann, R., Lawrence, M.G., Levin, Z., Markowicz, K.M., Mihalopoulos, N., Minikin, A., Ramanathan, V., De Reus, M., Roelofs, G.J., Scheeren, H.A., Sciare, J., Schlager, H., Schultz, M., Siegmund, P., Steil, B., Stephanou, E.G., Stier, P., Traub, M., Warneke, C., Williams, J., Ziereis, H. Global air pollution crossroads over the Mediterranean,

Science, 298, 794-799, 2002.

Lin, C-J., Pehkonen, S.O. Aqueous Free Radical Chemistry of Mercury in the Presence of Iron Oxides and Ambient Aerosol. *Atmos. Environ.*, 31, 4125-4137, 1997.

Lin, C-J., Pehkonen, S.O. Two-Phase Model of Mercury Chemistry in the Atmosphere. *Atmos. Environ.*, 32, 2543-2558, 1998.

Lin, C-J., Pehkonen, S.O. Aqueous Phase Reactions of Mercury with Free Radicals and Chlorine: Implications for Atmospheric Mercury Chemistry. *Chemosphere*, 38, 1253-1263, 1999.

Mason, R.P., Lawson, N.M., Sheu, G.-R. Mercury in the Atlantic Ocean: factors controlling air-sea exchange of mercury and its distribution in the upper waters. *Deep-Sea Res. II*, 48, 2829-2853, 2001.

Mason, R.P., G.-R. Sheu, The role of the ocean in the global mercury cycle, *Global Biogeochem. Cyc.*, 16, 1093. doi:10.1029/2001GB001440, 2002.

Munthe J. The aqueous oxidation of elemental mercury by ozone. *Atmos. Environ.*, 26A, 1461-1468, 1992.

Nagao, I., Matsumoto, K., Tanaka, H. Sunrise ozone destruction found in the sub-tropical marine boundary layer, *Geophys. Res. Lett.*, 26, 3377- 3380, 1999.

Nolle, M., Ellul, R., Heinrich, G., Gusten, H., A Long term of Background Ozone Concentrations in the Central Mediterranean - Diurnal and Seasonal Variations on the Island of Gozo, *Atmos. Environ.*, 36, 1391-1402, 2002.

O'Dowd, C., Jimenez, J.L., Bahreini, R., Flagan, R.C., Seinfeld, J.H., Hameri, K., Pirjola, L, Kulmala, M., Jennings, S.G., Hoffmann, T. Marine aerosol formation from biogenic iodine emissions, *Nature*, 417, 632-636, 2002.

Pal, B., Ariya, P.A. Studies of ozone initiated reactions of gaseous mercury: kinetics, product studies, and atmospheric implications, *Phys. Chem. Chem. Phys.*, 6, 572-579, 2004.

Pehkonen, S.O., Lin, C-J. Aqueous Photochemistry of Mercury with Organic Acids. *J. A&WM Assoc.*, 48, 144-150, 1997.

Pirrone, N., Hedgecock, I.M., Forlano, L., The role of the ambient aerosol in the atmospheric processing of semi-volatile contaminants: A parametrised numerical model (GASPAR), *J. Geophys. Res.*, 105, 9773-9790, 2000.

Pleijel, K., Munthe, J. Modelling the Atmospheric Mercury Cycle - Chemistry in Fog Droplets. *Atmos. Environ.*, 29, 1441-1457, 1995.

Sander, R., Keene, W.C., Pszenny, A.A.P., Arimoto, R., Ayers, G.P., Baboukas, E., Cainey, J.M., Crutzen, P.J., Duce, R.A., Hönninger, G., Huebert, B.J., Maenhaut, W., Mihalopoulos, N., Turekian, V.C., Van Dingenen, R. Inorganic bromine in the marine boundary layer: a critical review. *Atmos. Chem. Phys. Discuss.*, 3, 3050, 2003.

Sander, R. Modeling atmospheric chemistry: Interactions between gas-phase species and liquid cloud/aerosol particles. *Surv. Geophys.*, 20, 1-31, 1999.

Sander, R., Cruzten, P.J. Model study indicating halogen activation and ozone destruction in polluted air masses transported to the sea, *J. Geophys. Res.*, 101, 9121- 9138, 1996.

Schroeder, W.H., Anlauf, K.G., Barrie, L.A., Lu, J.Y., Steffen, A., Schneeberger, D.R., Berg, TArtic springtime depletion of mercury, *Nature*, 394, 331-333, 1998.

Schroeder, W. H., Munthe, J., Atmospheric Mercury – an overview, *Atmos. Environ.*, 32, 809-822, 1998.

Seibert, P., Beyrich, F., Gryning, S-E., Joffre, S., Rasmussen, A., Tercier, P. Review of operational methods for the determination of the mixing height, *Atmos. Environ.*, 34, 1001-1027, 2000.

Seinfield, J. H., Pandis, S. N. Atmospheric Chemistry and Physics - From Air Pollution to Climate Change, John Wiley and Sons, Inc., New York., 1998.

Sommar, J., Hallquist, M., Ljungstrom, E., Lindqvist, O. On the gas phase reactions between volatile biogenic mercury species and the nitrate radical, *J. Atmos. Chem.*, 27, 233-247, 1997.

Sommar, J., Gårdfeldt, K., Strömberg, D., Feng, X. A kinetic study of the gas-phase reaction between the hydroxyl radical and atomic mercury, *Atmos. Environ.*, 37, 3049-3054, 2001.

Sprovieri, F., Pirrone, N., Gårdfeldt, K., Sommar, J. Mercury speciation in the marine boundary layer along a 6000 km cruise path around the Mediterranean Sea, *Atmos. Environ.*, 37, S63-71, 2003.

Tokos, J. J. S., Hall, B., Calhoun, J. A., Prestbo, E. M. Homogeneous gas phase reaction of Hg^0 with H_2O_2, O_3, CH_3I, and $(CH_3)_2S$: implications for atmospheric Hg cycling, *Atmos. Environ.*, 32, 823-827, 1998.

Toyota K., Takahashi, M., Akimoto, H., Modeling multiphase halogen chemistry in the marine boundary layer with size segregated aerosol module: Implication for quasi-size dependent approach, *Geophys. Res. Lett.*, 28, 2899-2902, 2001.

Toyota K., Kanaya, Y., Takahashi, M., Akimoto, H., A box model study on photochemical interactions between VOCs and reactive halogen species in the marine boundary layer, *Atmos. Chem. Phys. Discuss.*, 3, 4549-4632, 2003.

Van Loon, L. L., Mader, E. A., Scott, S. L. Sulfite Stabilization and Reduction of the Aqueous Mercuric Ion: Kinetic Determination of Sequential Formation Constants, *J. Phys. Chem. A.*, 105, 3190-3195, 2001.

Van Loon, L. L., Mader, E. A., Scott, S. L. Reduction of the Aqueous Mercuric Ion by Sulfite: UV Spectrum of $HgSO_3$ and Its Intramolecular Redox Reaction, *J. Phys. Chem. A.*, 104, 1621-1626, 2000.

Vogt, R., Crutzen, P. J., Sander, R. A mechanism for halogen release from sea-salt aerosol in the remote marine boundary layer, *Nature*, 383, 327-330, 1996.

Wängberg, I., Munthe, J., Pirrone, N., Iverfeldt, Å., Bahlman, E., Costa, P., Ebinghaus, R., Feng, X., Ferrara, R., Gårdfeldt, K., Kock, H., Lanzillotta, E., Mamane, Y., Mas, F., Melamed, E., Osnat, Y., Prestbo, E., Sommar, J., Schmolke, S., Spain, G., Sprovieri, F., Tuncel, G. Atmospheric mercury distribution in Northern Europe and in the Mediterranean region, *Atmos. Environ.*, 35, 3019-3025, 2001.

Chapter-14

MODELLING TRANSPORT AND TRANSFORMATION OF HG AND ITS COMPOUNDS IN CONTINENTAL AIR MASSES

Russell Bullock

NOAA Air Resources Laboratory (on assignment to the U.S. EPA) U.S. EPA Mail Drop E243-03, Research Triangle Park, NC 27711, USA

INTRODUCTION

It is known that atmospheric deposition contributes the majority of the Hg found in nearly all contaminated ecosystems, and in some cases it is the only source. In recent years, a number of numerical simulation models of atmospheric mercury have been developed to help understand the atmospheric pathway of mercury and to help formulate effective emission reduction strategies to reduce atmospheric deposition to specific nations, regions and watersheds (Cohen et al., 2004; Dastoor and Larocque, 2004; Christensen et al., 2003; Bullock and Brehme, 2002; Travnikov and Ryaboshapko, 2002; Berg et al., 2001; Lee et al., 2001; Petersen et al., 2001; Seigneur et al., 2001; Xu et al., 2000). Mercury deposition to specific fresh-water systems has been the subject of most interest as compared to deposition to oceanic areas, and the models used thus far for emission reduction strategy development have tended to be continental-scale models with oceanic areas at the model boundary.

Elemental Hg (Hg^0) emitted to air is believed to remain in the atmosphere for a period of time on the order of one year (Schroeder and Munthe, 1998). There are also a number of oxidized compounds of mercury known to exist in air. Most atmospheric mercury models simulate two additional species:

reactive gaseous mercury (RGM) and particulate Hg. These oxidized Hg species are more subject to atmospheric deposition than Hg^0 and they have significantly shorter atmospheric lifetimes (Schroeder and Munthe, 1998). Recent research has discovered certain phenomena in the Arctic (Schroeder et al., 1998) and at high altitude (Landis, 2004) that appear to involve atmospheric oxidation of Hg^0 on time scales of days or even hours, which would suggest the average atmospheric lifetime of Hg^0 emitted to air might be somewhat shorter than one year if that Hg^0 is transported to certain locations. Nonetheless, a lifetime of only 5 to 7 days would be required for significant intercontinental transport of Hg^0 to occur.

Thus, any model of atmospheric mercury in continental air masses must also consider the various forms of mercury being transported in from the marine environments at its lateral boundaries.

There are a few special considerations for modelling the behavior of Hg in continental air masses. First, the continents are where most humans live. Nearly all direct anthropogenic emissions of mercury occur in continental areas. Oxidized forms of mercury are more rapidly deposited from the atmosphere and a significant fraction of anthropogenic Hg emissions are in the oxidized form, especially those from combustion sources (Pacyna and Pacyna, 2002; Pacyna et al., 2003). Thus, the continents are where our industrial emissions have the most concentrated and significant impacts. Also, the surface underlying a continental air mass can be land or water, and land surfaces usually include vegetation which can have widely varying effects on atmospheric deposition processes depending on the season of the year. The effect of vegetation on atmospheric processes can even vary depending on current meteorological conditions or those in the recent past (e.g., leaf area index and evapotranspiration). Continental water bodies are not as deep as most oceans and their water temperatures can vary more widely from season to season. They also freeze and thaw more rapidly. Finally, unlike oceans, continents have topographic relief. Mountains and valleys can create complex atmospheric flow patterns that require fine-scale (spatial and temporal) modelling to resolve.

Since oceans cover two-thirds of the earth's surface, the behavior of mercury in marine environments is an important factor in its atmospheric lifetime and the concentration of the various forms of mercury in air and cloud water that might be expected at the horizontal boundaries of the continental-scale models. This chapter will focus on modelling mercury in continental air masses, but the importance of the marine environment to the global distribution of mercury and its eventual deposition to fresh-water systems should not be discounted.

GENERAL MODELLING CONCEPTS

Numerical simulation models for air pollutants are developed with the intent that they address all of the current knowledge regarding emission sources, transport and diffusion, chemical and physical transformations, and removal processes. However, many of the meteorological processes known to affect mercury occur on rather small spatial and temporal scales (e.g., convective clouds) and it is currently impractical to model them in full detail over continental model domains, much less the entire global mercury cycle. A compromise between model domain size and temporal/spatial resolution is required. As the calculating power of computing equipment has increased, this compromise has become less difficult. However, it is still necessary to use limited-area models in order to simulate atmospheric mercury at the level of spatial detail required to resolve the atmospheric flow patterns (e.g., land/sea-breeze circulations) and cloud elements in which important mercury transport and transformations are known to occur. Albert Einstein is quoted as having said "Make everything as simple as possible, but not simpler."

This advice certainly applies to modelling atmospheric mercury. But, with our scientific understanding of atmospheric mercury continuing to evolve, the level of spatial/temporal detail necessary to resolve all important processes may change with each new discovery.

The individual atmospheric mercury models that have been developed thus far fall across a wide range of the simple-complex scale. Models that encompass the entire global domain of the mercury cycle must simulate several years in order to accumulate a global background air concentration of Hg^0 and they tend to neglect small-scale air circulations, planetary boundary layer (PBL) structure, and cloud-water physicochemistry in their simulations. Those models that manage to resolve these complex atmospheric processes in high detail are only applicable up to the regional spatial scale, at best, and can only be used to simulate a few weeks or months of time on the most powerful computing systems. This compromise between model complexity and domain size is made based on a desire to minimize "modelling uncertainty", which occurs when the model is not accurately treating accepted scientific fact in its calculation of a numerical simulation.

A different type of uncertainty that must always be considered when atmospheric mercury modelling is applied, especially for assessment purposes, is "scientific uncertainty". This is true regardless of the complexity of the models being used or the quality of their input data. Some future scientific discovery regarding sources of atmospheric mercury, its atmospheric composition or behavior, or its modes of deposition could make the results of any modelling assessment performed today scientifically indefensible and obsolete. Modelling for the purposes of scientific research is performed in anticipation of such discoveries, and it plays an important role in the support and direction of the field research through which these discoveries may be made. Model simulations are also used to develop public policy, and this implies a certain level of confidence in the current science that should not be misunderstood. Policy decisions are made with the best information available at the time, and mercury policy decisions are no exception. Our current scientific understanding of atmospheric mercury is incomplete and improved knowledge will lead to more complete scientific certainty. As the noted historian Daniel Boorstin once said, "The greatest obstacle to discovery is not ignorance - it is the illusion of knowledge."

BASIC SIMULATION MODEL STRUCTURES

There are two basic types of numerical model structures currently being used for the simulation of long-range atmospheric mercury transport and deposition, Lagrangian and Eulerian.

Lagrangian modelling

In Lagrangian modelling, the model's reference frame moves with the entity or substance being modeled. Individual parcels of air and the mercury contained in them are resolved and their motion and physicochemical processes are simulated based on pre-defined meteorological information. Usually, these parcels are defined as originating at the location of each air emission source in the inventory being used. They are "released" from each source at specific time intervals with the mercury mass loading in each parcel upon release determined by multiplying the time interval by the mercury emission flux rate for the source in question. Modelers commonly refer to these as "puff" models. Rather than the continuous pollution elements that are simulated by Gaussian-plume models for short-range single-source applications, Lagrangian models simulate a train of discrete emission parcels that, when totaled together, resemble a continuous plume.

Chemical and physical reactions of mercury are simulated in most Lagrangian models one parcel at a time with no interaction with the other pollutant parcels present in the simulation. If one parcel contains twice the amount of mercury as a second one under the same meteorological conditions and with the same air concentrations for all other reactants, the simulated rate of all reactions and depositions for the first parcel will be exactly twice those for the second one. This "linear" chemistry assumption is currently acceptable for atmospheric mercury simulation because; 1) the chemical and physical reactions of mercury that have been identified thus far have all been found to be either first-order or zero-order reactions, and 2) the concentrations of mercury are normally so much smaller than the concentrations of its reactants that significant depletion of reactants does not occur. However, many of the chemical species known to react with mercury are the products of photochemical reactions that are known to be non-linear. Therefore, Lagrangian modelling is dependent on some sort of previous determination of the air concentrations of these reactants. Also, limited-area modelling with a Lagrangian modelling framework requires the simulation of synthetic sources to account for mercury transporting or diffusing into the model domain across the lateral and top boundaries.

Eulerian modelling

In Eulerian modelling, the model's reference frame is fixed and the entity or substance being modeled moves through this reference frame. Eulerian models are often referred to as "fixed grid" models. The atmosphere is defined as a three-dimensional stationary array of finite volume elements. Physical and chemical variables in the model are simulated as if they each have only one value all throughout a particular volume element at any particular point in time.

Transport and diffusion of atmospheric constituents are simulated by transfers from one volume element to another, normally to an adjacent element except when special techniques are employed to negate artificial numerical diffusion. Eulerian modelling has been used since the mid 20th century to simulate physical meteorology. Nearly all current weather forecast models use this type of modelling framework. Over the years, meteorological modelers have striven to reduce the size of the volume elements to better resolve pressure, wind, temperature, humidity and clouds as they exist in the real atmosphere, knowing that volume averaging of these physical parameters was a primary source of modelling uncertainty in their simulations. The same is true for atmospheric chemistry modelling. Simulating a single air concentration for each chemical constituent throughout the entire expanse of a large volume element obviously deviates from reality.

The numerical methods used to simulate the motion of a substance from one volume element to another require that this motion or diffusion be calculated in small increments of time. These "time steps" must be kept below a certain length to avoid instabilities in the mathematical calculations that can cause artificial numerical errors to amplify. Reducing all three dimensions of the volume elements by one-half creates eight times as many volume elements to simulate. It also requires that the time step for pollutant transport be reduced by one-half, resulting in another doubling of the calculations required to simulate a particular period of time. In essence, doubling the fineness of the model's spatial resolution produces a 16-fold increase in the number of calculations that have to be made.

Regardless of these difficulties, an Eulerian modelling framework is required for the realistic simulation of complex, non-linear chemistry. Even where simple chemistry is involved, if a pollutant is released to air from a large number of sources, or is released from a diffuse source over a large area (e.g., oceanic emissions), an Eulerian modelling framework may be the most efficient option due to the large number of discrete pollutant parcels that a Lagrangian model would have to track and simulate.

MERCURY EMISSIONS INTO CONTINENTAL AIR MASSES

As mentioned before, most direct anthropogenic emissions of mercury occur in continental areas, and a significant fraction of anthropogenic emissions are in the oxidized forms that deposit from the atmosphere most rapidly. Model simulations of the atmospheric fate and transport of mercury suggest that the speciation of the mercury emissions governs mercury deposition patterns in industrial source areas, while in remote areas the atmospheric chemistry of mercury determines those deposition patterns (Pai et al., 1999). Clearly, it is important to ensure that numerical models of atmospheric mercury include a proper treatment of mercury transformations, but in continental air masses the speciation of emissions is more important than anywhere else. When modelling mercury deposition from continental air masses, an accurate inventory of industrial emissions of oxidized mercury is most critical to achieving accuracy in the simulated wet and dry deposition of total mercury to nearby sensitive watersheds.

Because of the strong diel (day-night cycle) fluctuations in the PBL structure of continental air masses, the transport and diffusion characteristics of any air pollutant emitted into them are highly dependent on the time of day and height of the emission. When daytime surface heating is at a maximum, strong convective vertical mixing occurs between the surface and the PBL top which may be 2 km or more above the surface. Under these conditions, even pollutant emissions from tall stacks can rapidly mix down to the surface and lead to localized increases in air contaminant concentrations. However, vertical homogenization of the air within this deep PBL tends to rapidly dilute pollutant air concentrations. Therefore, deep and rapid mixing with the PBL formed under conditions of strong solar heating does not necessarily lead to the highest surface-level air concentrations near emission sources.

Under the opposite conditions of nighttime surface cooling, a vertically stratified nocturnal boundary layer develops over the surface. Emissions near the surface within this shallow layer of temperature inversion (increasing temperature with height) will not be mixed vertically and will remain near the surface. Winds also tend to be rather light within this nocturnal boundary layer, leading to high pollutant concentrations in the locality of an emission source. Elevated and buoyant emissions which are released or rise above the top of this nocturnal boundary layer will exhibit vertical mixing to some degree dependent on the PBL structure left behind from the previous day, but downward mixing will be limited by the top of the nocturnal inversion layer. Under this condition, pollutant emissions can travel very long distances aloft during the night with little or no surface-level air concentration from the source in question.

Similar vertical thermal structures can exist in marine air masses, but they are rarely as intense and do not have such strong diel cycles as those found in continental air masses. For this reason, simulations of pollutant transport and diffusion in continental air masses depend more strongly on detailed information regarding the time of day and effective stack height, which is the sum of physical stack height and plume rise, of pollutant emissions.

A variety of mercury emission inventories have been developed, but only a portion of them are useful for numerical simulation modelling applications. Inventories where the emission flux is defined as national totals, or as totals for large political regions like state or provinces, are rarely useful for atmospheric simulation modelling. The spatial resolution of emissions information provided in this form is simply insufficient. In a technical sense, these emission inventories are simply data sets which are used as inputs to numerical simulation models and are not really a part of the models themselves. Nonetheless, the various assumptions and simplifications that are made during the course of compiling these emission inventories can have profound effects on the results of model simulations. Emission sources are nearly always grouped or categorized based on industrial activity or physical characteristics of the emission source. But the definitions for these groups and categories of sources can vary quite significantly between individual inventories. Instead of describing the emissions of Hg^0, RGM and particulate Hg for each source, most current inventories provide only total mercury emission rates. The critical estimation of the speciation of those mercury emissions are usually left for the modeler to perform based on the type or category of the source in question. Beginning with their 1999 base year, mercury emission inventory developers at the U.S. EPA began to prescribe speciation profiles for each source type and category in their inventory.

To be used for simulation purposes, mercury emission inventories must contain descriptions of each major source of mercury providing its geographic coordinates (e.g., latitude and longitude) for locating the source within the modelling domain and exhaust characteristics (e.g., stack height, exhaust temperature) for estimating the effective height of emission in addition to basic mercury emission flux rates. Most emission inventories provide no temporal resolution at all for the emission flux rates and simply provide a single mass per time quantity for each source element in the inventory. Rarely do these inventories provide an indication of the fraction of the total emissions applicable to each hour of the day, day of the week, or month/season of the year. Smaller anthropogenic sources of mercury that are not accounted for individually during inventory development are typically described as "area sources" and are accounted for as a group for each state, province, or other political division. Since the geographic size and shape of these political divisions may vary widely, the modeler is faced with a spatially allocation problem that has no standard solution. Sometimes the geometric centroid of the political division is used as the simulated emission point for a single integrated area source. Sometimes the flux rate indicated for the area source is spatially distributed among the model grid points or grid cells falling within the boundaries of the political division. Whatever the case, the spatial resolution of the emission inventory is usually inferior to that of the model and additional modelling uncertainty results from the spatial allocation.

There is one strictly terrestrial mercury air emission source type that can be significant in comparison to industrial sources and is sometimes considered natural, but is in fact largely anthropogenic; that being forest fires (Friedli et al., 2003; Friedli et al., 2001). Not only are many forest fires intentionally set, but much of the mercury emitted from them is actually anthropogenic mercury accumulated in forest soils from atmospheric deposition over the decades or centuries since the last forest fire at that location. Intentional agricultural fires (prescribed burning) also fall into this category, but in this case there is good reason to expect the amount of mercury accumulated in the underlying soils to be less important than in forest fires due to the shorter accumulation period since the last burning. Being a combustion source, one might expect a significant fraction of the mercury emitted from these fires to be in oxidized form. However, recent measurements of Hg emitted from forest fires in the Cascade Mountains in Washington State showed 95% or more to be in the form of Hg^0 gas with the remainder found in the particulate fraction of the smoke (Friedli et al., 2003).

There is currently no comprehensive mercury emission inventory for modelers to account for individual forest and agricultural fires in their simulations, and no generally-accepted method to estimate long-term average emission rates from such sources. This is certainly a subject in need of additional research.

Emissions of mercury to continental air from water bodies, soils, and vegetation are also anthropogenic to a large degree since much of the mercury in those media was deposited over time from anthropogenic sources. However, nearly all of these emissions are believed to be in the form of Hg^0 (Vette et al., 2002; Grigal, 2002). Truly natural emissions of mercury from geologic formations are also believed to be mostly Hg^0 (Gustin et al., 2002). Continental-scale modelling has shown that industrial Hg^0 emissions do not add significantly to the Hg^0 air concentration except in intense source regions and the effect of Hg^0 air concentrations on the simulated deposition of total mercury is mostly influenced by the air concentration at the lateral model boundaries (U.S. EPA, 1997). Except where natural substrates are exceptionally rich in mercury, it follows that natural emissions of Hg^0 may also be factored into continental-scale modelling by the background Hg^0 air concentration. Nonetheless, the processes that govern the exchange of mercury between the atmosphere and the various natural surfaces below it need to be studied and understood so that, eventually, these processes and sources of natural and recycled mercury can be included more explicitly in atmospheric modelling.

ATMOSPHERIC MERCURY CHEMISTRY

Nearly all contemporary models of atmospheric mercury attempt to simulate generally the same gas-phase reactions in air and aqueous-phase reactions in cloud water that have been identified and studied through laboratory investigation. Some of these chemical reactions were first identified and studied many years ago, but modern laboratory techniques developed only in past few years have allowed much improved determination of the rate of these reactions under various conditions. Some models ignore the slower, less significant reactions of mercury in order to concentrate efforts on simulating the reactions that are believed to be most important (Ryaboshapko et al., 2002). Laboratory investigation of mercury chemistry is continuing and it sometimes results in significantly revised estimates of reaction rate for particular reactions and considerable debate among the modelling community as to which rate determination to accept as most accurate.

In just the past few years, a number of scientific papers have been published describing chemical reactions of mercury in air and in cloud water. Some provide information about chemical reactions that were suspected for some time to occur in the atmosphere, but for which no reaction rate determinations had been published before (Ariya et al., 2002). Some suggest significantly different values for reactions rates that had been published previously (Van Loon et al., 2000). One even suggests the existence of an unusual form of water-soluble mercury that had never been considered before, a Hg^0-SO_2 complex (Van Loon et al., 2001), while another provides evidence that a chemical reaction previously included in many atmospheric mercury models does not actually occur in atmospheric water (Gårdfeldt and Jonsson, 2003).

Atmospheric measurements taken in recent years have identified previously unknown chemical phenomena regarding mercury. It is only since the late 1990's that gaseous elemental mercury depletion events (GEMDEs) have been recognized to occur in northern Canada (Schroeder et al., 1998). Subsequent research has improved our understanding of GEMDEs and has found them to occur at various circumpolar locations (Lindberg et al., 2002; Ebinghaus et al., 2002; Sommar et al., 2004). We are only just now beginning to understand how GEMDEs occur and how they may influence the global cycling of mercury (Bottenheim et al., 2002). Ongoing observations of RGM and particulate Hg at high altitude, both aloft and on mountain tops, suggest that chemical oxidation of mercury is occurring on in the middle and/or upper atmosphere for reasons that are not yet understood (Landis, 2004).

These new and revised chemical rate determinations and observations of unexplained atmospheric phenomena are certainly keeping the chemistry formulations of atmospheric mercury simulation models in a state of uncertainty and change. This should be considered a good thing, as it indicates progress is being made in developing our understanding of the behavior of atmospheric mercury and the sources most likely responsible for observed atmospheric mercury deposition and ecosystem contamination in continental areas.

MODELLING MERCURY WITH THE COMMUNITY MULTI-SCALE AIR QUALITY MODEL

The Community Multiscale Air Quality model (CMAQ) is an Eulerian type model used to simulate the transport, transformation and deposition of air pollutants and their precursors (Byun and Ching, 1999). The pollutants simulated by the standard version of CMAQ include tropospheric ozone, acidic and nutrient substances, and aerosol matter of various composition and particle size. The standard version was used as the basis for development of the mercury version of CMAQ (CMAQ-Hg) which has been applied to simulate continental domains in North America and Europe. A complete description of all special CMAQ model formulations for the simulation of atmospheric mercury is given in Bullock and Brehme (2002). Basic transport and diffusion simulation in the CMAQ-Hg was kept the same as in the standard CMAQ. All pollutant species normally simulated by the standard CMAQ are also simulated in CMAQ-Hg along with mercury and molecular chlorine gas. Descriptions of the standard CMAQ are found in Byun and Ching (1999).

The CMAQ-Hg modelling domain is variable in size and spatial resolution. Horizontal spacing of grid points in mercury simulations performed thus far has been 36 km in both dimensions. Two vertical layer structures have been employed using 14 and 21 layers of varying thickness from the surface to the 10 kPa pressure level, with the finest resolution near the surface. All CMAQ-Hg simulations performed thus far have used the Carbon Bond-IV (CB-IV) gaseous chemistry mechanism from the standard model plus special chemistry treatments for mercury and molecular chlorine gas which is a significant oxidizing agent for mercury in air and cloud water. Meteorology has been defined using the Fifth-Generation Pennsylvania State University / National Center for Atmospheric Research (NCAR) Mesoscale Model (Grell et al., 1994), commonly referred to as "MM5", employing the surface energy flux and the planetary boundary layer model of Pleim and Xiu (1995).

The CMAQ-Hg simulates mercury in three separate forms; (1) gaseous Hg^0, (2) RGM, and (3) particulate Hg. Particulate Hg is resolved in two particle size modes; the super-fine Aitken mode and the fine-particle accumulation mode. A treatment for coarse-mode particulate Hg is currently under development. Four gas-phase oxidation reactions for mercury are simulated in the CMAQ-Hg as shown in Table 1. The Hg^{2+} products of these reactions are modeled in CMAQ-Hg as either RGM or particulate Hg based on the vapor pressure of the compounds that are assumed to be produced.

Reduction of Hg^{2+} species to Hg^0 is simulated only in the cloud chemistry mechanism. While some chemical reduction of Hg^{2+} to Hg^0 may occur in the gas phase, the mechanism and rate for this type of reaction remains poorly understood. In general, gas-phase reactions of mercury appear to be of minor importance to its oxidation state as compared to its aqueous-phase reactions.

Chlorine photolysis is known to occur at a rather rapid rate, similar to nitrogen dioxide, a gas-phase species simulated in the standard CMAQ and in CMAQ-Hg. The Cl_2 photolysis rate in CMAQ-Hg is referenced to the CMAQ photolysis rate for NO_2 with a proportionality factor of 0.295. This referencing is based on efforts at the University of Texas to expand the CB-IV mechanism to include Cl_2 and is based on their analysis of updated actinic flux data for Cl_2 and NO_2 (Finlayson-Pitts and Pitts, 2000).

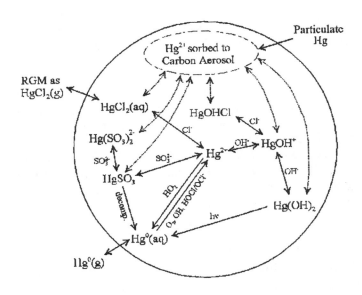

Figure 1. Schematic representation of the aqueous mercury chemical mechanisms.

The cloud chemistry mechanism for mercury in CMAQ-Hg is illustrated in Figure 1. It simulates gas/liquid partitioning, aqueous chemistry and reversible sorption of Hg^{2+} complexes to elemental carbon aerosol (ECA) suspended in cloud water, and was developed based on the approach of Pleijel and Munthe (1995). This aqueous chemistry mechanism involving Hg^0, the Hg^{2+} ion, and six Hg^{2+} compounds is simulated simultaneously with the pre-existing aqueous chemistry from the original CMAQ. Gas/liquid partitioning of Hg^0 and RGM is simulated using Henry's law equilibrium assumptions. RGM is assumed to partition based on the Henry's constant for mercuric chloride ($HgCl_2$).

Particulate mercury is assumed to be completely incorporated into cloud water and composed of Hg^{2+} sorbed to ECA at the start of the CMAQ operator splitting technique that is employed to simulate cloud chemistry and wet deposition (Byun and Ching, 1999). All aqueous chemical reactions for mercury and their rate constants are shown in Table 1.

Table 1. Chemical reactions for mercury and their rate constants used in CMAQ model.

No.	Reaction	k or K	Ref.
Gaseous-phase reaction of Hg			
RG1	$Hg^0_{(g)} + O_{3(g)}$ TPM	3.0×10^{-20} cm^3 molecules^{-1} s^{-1}	1
RG2	$Hg^0_{(g)} + Cl_{2(g)}$ RGM	4.8×10^{-18} cm^3 molecules^{-1} s^{-1}	2
RG3	$Hg^0_{(g)} + H_2O_{2(g)}$ TPM	8.5×10^{-19} cm^3 molecules^{-1} s^{-1}	3
RG4	$Hg^0_{(g)} + OH_{(g)} \rightarrow$ TPM	8.7×10^{-14} cm^3 molecules^{-1} s^{-1}	4
Aqueous-phase reactions of Hg			
RA1	$Hg^0_{(aq)} + O_{3(aq)} \rightarrow Hg^{2+}_{(aq)}$ + products	4.7×10^7 M^{-1} s^{-1}	5
RA2	$HgSO_{3(aq)} \rightarrow Hg^0_{(aq)}$ + products	$T \times e^{((31.971 \times T)-12595)/T}$ s^{-1}	6
RA3	$Hg(OH)_{2(aq)} + h\nu \rightarrow Hg^0_{(aq)}$ + products	6.0×10^{-7} s^{-1} (maximum) †	7
RA4	$Hg^0_{(aq)} + OH_{(aq)} \rightarrow Hg^{2+}_{(aq)}$ + products	2.0×10^9 M^{-1} s^{-1}	8
RA5	$Hg^{2+}_{(aq)} + HO_{2(aq)} \rightarrow Hg^0_{(aq)}$ + products	1.1×10^4 M^{-1} s^{-1}	9
RA6	$Hg^0_{(aq)} + HOCl_{(aq)} \rightarrow Hg^{2+}_{(aq)}$ + products	2.09×10^6 M^{-1} s^{-1}	10
RA7	$Hg^0_{(aq)} + OCl^-_{(aq)} \rightarrow Hg^{2+}_{(aq)}$ + products	1.99×10^6 M^{-1} s^{-1}	10
Aqueous-phase chemical equilibria for Hg			
E1	$Hg^{2+} + SO_3^{2-} \leftrightarrow HgSO_3$	2.0×10^{-13} M	11
E2	$HgSO_3 + SO_3^{2-} \leftrightarrow Hg(SO_3)_2^{2-}$	4.0×10^{-12} M	11
E3	$Hg^{2+} + 2Cl^- \leftrightarrow HgCl_2$	1.0×10^{-14} M^2	12
E4	$Hg^{2+} + OH^- \leftrightarrow HgOH^+$	2.51×10^{-11} M	11
E5	$HgOH^+ + OH^- \leftrightarrow Hg(OH)_2$	6.31×10^{-12} M	11
E6	$HgOH^+ + Cl^- \leftrightarrow HgOHCl$	3.72×10^{-8} M	11
Henry's equilibria for Hg			
H1	$Hg^0_{(g)} \Leftrightarrow Hg^0_{(aq)}$	1.1×10^{-1} M atm^{-1}	13
H2	$HgCl_{2(g)} \Leftrightarrow HgCl_{2(aq)}$	1.4×10^6 M atm^{-1}	14

† Rate constant for RA3 is scaled to the cosine of solar zenith angle

References: (1) Hall (1995); (2) Calhoun and Prestbo (2001); (3) Tokos et al. (1998); (4) Sommar et al. (2001); (5) Munthe (1992); (6) Van Loon et al. (2000); (7) adapted from Xiao et al. (1994); (8) Lin and Pehkonen (1997); (9) Pehkonen and Lin (1998); (10) Lin and Pehkonen (1998); (11) Smith and Martell (1976); (12) Lin and Pehkonen (1999); (13) Sanemasa (1975); (14) Lindqvist and Rodhe (1985)

Chemical equilibria used in the aqueous mechanism are also in Table 1, as are the Henry's constants used for gas/liquid partitioning. Sorbed aqueous Hg^{2+} complexes are not subject to aqueous chemical reduction to the elemental form and subsequent out gassing from cloud droplets (Seigneur et al., 1998). Thus, sorption to ECA can affect the amount of mercury in cloud water subject to removal by precipitation.

Upon completion of the CMAQ operator splitting function for cloud chemistry and wet deposition, all aqueous chemical species are transferred back to the gas phase for the simulation of transport and dry deposition. This transfer is necessary because the current versions of the CMAQ and CMAQ-Hg have no explicit simulation of cloud water transport. Cloud water concentration is estimated at the beginning of each aqueous chemistry time loop based on the previous MM5 meteorological model simulation. Given the assumptions of Henry's equilibrium for all gaseous species and complete incorporation of particulate Hg into cloud water at the beginning of the cloud chemistry time loop, the effect on the aqueous chemistry simulation from this transfer to the gas phase is minimal. Methods for explicit simulation of the transport of cloud water and its chemical constituents by the CMAQ modelling system are currently under development. In the mean time, dissolved Hg^0 is transferred as gaseous Hg^0, all dissolved Hg^{2+} species are transferred collectively as RGM, and all Hg^{2+} species sorbed to ECA are transferred collectively as particulate Hg.

It should be noted that emissions of sea salt aerosol and other sources of chloride ion (Cl⁻) in cloud water are not yet defined for the standard version of CMAQ. A minimum aqueous Cl⁻ concentration of 1.0×10^{-3} g l⁻¹ is assumed in the current form of CMAQ-Hg based on the value of 2.5×10^{-3} g l⁻¹ previously adopted for long-range mercury transport model intercomparison (Ryaboshapko et al., 2002). The speciation of dissolved Hg^{2+} compounds in cloud water is dependent on Cl⁻ concentration. A top priority for future development of CMAQ is the addition of sea salt and crustal aerosol emissions.

The CMAQ-Hg model parameterization for the sorption and desorption of aqueous Hg^{2+} species to ECA suspended in cloud water is adapted from previous work by Seigneur et al. (1998). A complete description of the CMAQ-Hg treatment of Hg^{2+} sorption in the aqueous media is available in Bullock and Brehme 2002.

The CMAQ-Hg model simulates the wet deposition of Hg^0, RGM, and particulate Hg in the same manner as for all other aqueous pollutant species previously resolved in the standard CMAQ model. The cloud-water concentration of each pollutant is deposited to the surface based on the simulated rate of precipitation falling from each clouded grid volume during the cloud chemistry time splitting operation. The cloud water concentration of Hg^0 is relatively low compared to the total dissolved and sorbed Hg^{2+}, and the simulated wet deposition of Hg^0 is minor compared to that of RGM and particulate Hg.

The cloud-water concentration of RGM is calculated as the sum of the dissolved-phase concentrations of all seven Hg^{2+} species in the aqueous chemistry mechanism. The cloud-water concentration of particulate Hg is calculated as the sum of the concentrations of all sorbed Hg^{2+} species.

Dry deposition of Hg^0 is assumed to be negligible in comparison to that of RGM and particulate Hg. It may actually be a minor compensation to emission fluxes of Hg^0 from geology, vegetation and water bodies such that dry evasion of Hg^0 might well be occurring instead. The dry deposition velocity for Hg^0 is currently set to zero the CMAQ-Hg model. RGM dry deposition is known to occur very readily, especially to lush vegetation and to water surfaces. The standard CMAQ dry deposition parameterization for gaseous nitric acid is also used for RGM. Dry deposition of particulate Hg is simulated based on pre-existing deposition velocity formulations in the standard CMAQ for the two modeled particle size modes.

MODELLING ACCURACY

An inter-comparison of mercury cloud-chemistry models was organized in 2000 at the Meteorological Synthesizing Center – East (MSC-East) in Moscow, Russia, with the participation of atmospheric mercury model developers from Germany, Russia, Sweden, and the U.S. (Ryaboshapko et al., 2001). Five mercury chemistry models, including CMAQ-Hg, were compared using the same hypothetical conditions for a cloud/fog volume. The same cloud microphysical parameters (i.e., liquid water content, droplet size, temperature and pressure) and initial conditions for chemistry in air and cloud water were used by all models, including identical initial concentrations in air and cloud water for three forms of mercury, Hg^0, RGM (assumed to $HgCl_2$), and particulate Hg. The results showed considerable variation among the models in their simulated accumulation of these three forms of mercury in cloud water during the 48-hour simulation period. With the exception of one model which used a rather different overall chemical system for mercury, it was not expected at the outset of the study that there would be a large amount of variation between the simulations. However, further investigation found a number of differences between the models in the exact set of reactions they simulated and the rate constants they employed for each reaction. There was also considerable difference in the way each model treated the association of aqueous Hg^{2+} with suspended particulate matter in cloud water. Overall, the results suggested that further basic research is critically needed to reduce uncertainties in the formulation of Hg chemistry models.

This international atmospheric mercury model inter-comparison activity is continuing, and a second phase of study involving full-scale model simulations for Europe has been conducted by participants from Bulgaria, Canada, Denmark, Germany, Russia, and the U.S. (Ryaboshapko et al., 2003). In addition to comparing results between the various models, these full-scale model simulations have been compared to observations of surface-level air concentration of total gaseous mercury (TPM), RGM and particulate Hg taken at five locations during 11 days in June/July of 1995 and 14 days in November 1999 (Ryaboshapko et al., 2003). Six of the models applied were Eulerian in structure and one was Lagrangian. All models used the same mercury emission inventories for input.

The air concentrations of Hg^0 assumed to be present at the lateral boundaries of the Eulerian models and super-imposed across the model domain of the Lagrangian model were a strong factor in each model's simulated total gaseous mercury (TGM) air concentration pattern over Europe, and each simulation of this species agreed quite well with observations in a statistical sense. However, for the RGM fraction of TGM, it was found that the models were only able to match observations of air concentration within a factor of 4 on average, and the difference was found to exceed an order of magnitude for some individual samples. The agreement between models and measurements for TPM was generally within a factor of 2, and most of the models were able to reproduce observed peaks in TPM air concentration at the proper time, if not at the proper amplitude. Ryaboshapko et al. (2003) concluded that the models had shown some skill in reproducing general TGM conditions, but our knowledge of the physico-chemical processes of RGM is still incomplete and sufficiently accurate parameterizations for this species have not yet been incorporated into the models.

A third phase of the MSC-East model inter-comparison study is now in the final stages of completion where simulations of monthly wet deposition flux are being compared to observations at nine locations in Europe. These results are not yet available for publication, but a comparison of simulated weekly wet deposition flux to observations at eleven sites in the U.S. has been conducted by the U.S. EPA using the same full-scale model it later applied for the European study (Bullock and Brehme, 2002). In the earlier study, the Community Multi-scale Air Quality (CMAQ) model was used to

Figures 2 and 3 show comparisons of simulated weekly Hg wet deposition versus observations from the Mercury Deposition Network (MDN) taken during the Spring and Summer test periods, respectively.

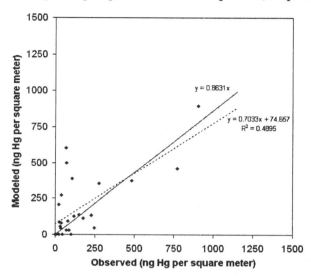

Figure 2. Scatter of the CMAQ – simulated versus MDN-observed weekly wet deposition flux of total mercury for the period 4 April to 2 May 1995.

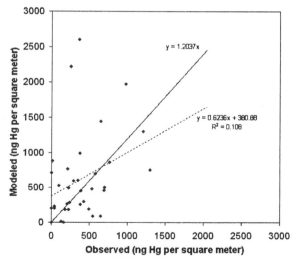

Figure 3. Scatter plot of the CMAQ-simulated versus MDN-observed weekly wet deposition flux of total mercury for the period 20 June to 18 July 1995.

simulate mercury for two 4-week periods in 1995 over the central and eastern U.S. and adjacent parts of southern Canada.

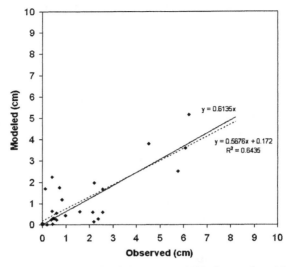

Figure 4. Scatter plot of the MM5-simulated versus MDN-observed weekly precipitation amount the period 4 April to 2 May 1995.

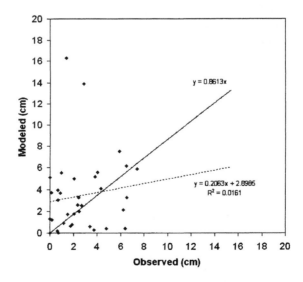

Figure 5. Scatter plot of the MM5-simulated versus MDN-observed weekly precipitation amount for the period 20 June to 18 July 1995.

The results in Figure 2 differ slightly from the Spring-period results published in Bullock and Brehme (2002) and show a slight increase in model accuracy due to correction of a slight error in the results analysis. Figure 2 indicates moderate model skill, but Figure 3 shows rather poor model performance.

Figures 4 and 5 show comparisons of simulated versus observed weekly precipitation amount during the Spring and Summer periods, respectively.

Precipitation rate and all other meteorological variables for CMAQ simulations are derived from previous applications of the MM5 meteorological model.

Figure 5 shows that the precipitation inputs for the CMAQ atmospheric mercury simulation of the Summer period were highly inaccurate. Obviously, the accuracy of CMAQ simulations of the wet deposition of Hg is dependent on an accurate definition of precipitation.

Comprehensive evaluation of this or any other atmospheric mercury model is currently not feasible due to a lack of measurements of the dry deposition flux of mercury. Without a confident definition of the true atmospheric deposition flux of Hg by both wet and dry processes, we cannot determine with confidence that any model is an accurate reflection of real atmospheric processes. A model could be simulating the wet deposition flux in the correct amount only because it has grossly over-represented the dry flux and has therefore left too little pollutant in air to be wet deposited. Likewise, it could be simulating too much wet deposition because the dry deposition flux was under-represented and the simulated air concentrations are too high. Air concentration measurements for each of the pertinent mercury species could certainly help to constrain this problem. However, we have no reason to believe that mercury contamination of aquatic ecosystems is due only to the wet deposition of atmospheric mercury.

Until we can assess model performance regarding the *total* atmospheric deposition of mercury, we gain insight into only a fraction of the problem, and we cannot know the size of that fraction.

SUMMARY

Atmospheric simulation models are being used widely to inform governmental policy makers about current scientific indications regarding the most effective mercury emission control options to reduce atmospheric deposition. Unfortunately, these models do not always agree as to the sources responsible for the deposition observed in particular areas. This disagreement between models does not appear to come only from differences in the sets of chemical and physical reactions of mercury they simulate. Uncertainty regarding actual air concentrations of Hg^0, RGM and particulate Hg at high altitudes and over remote locations requires continental-scale modelers to adopt boundary air concentration values that are based either on lower-resolution global-scale models or on very sparse observational data. Uncertainty about exchanges of mercury between the atmosphere and the various surfaces beneath it requires still more assumptions to be made. Obviously, these types of assumptions provide ample opportunity for model simulations to differ. The current lack of measurement data to support comprehensive model evaluation will certainly need to be addressed if we are to know which models, with all of their attendant assumptions, are the most accurate reflections of reality.

DISCLAIMER

The research presented here was performed under the Memorandum of Understanding between the U.S. Environmental Protection Agency (EPA) and the U.S. Department of Commerce's National Oceanic and Atmospheric Administration (NOAA) and under agreement number DW13921548. Although it has been reviewed by EPA and NOAA and approved for publication, it does not necessarily reflect their policies or views.

REFERENCES

Ariya, P.A., Khalizov, A., Gidas, A. . Reactions of gaseous mercury with atomic and molecular halogens: kinetics, product studies, and atmospheric implications. *J. Phys. Chem. A*, 106, 7310-7320, 2002.

Berg, T., Bartnicki, J., Munthe, J., Lattila, H., Hrehoruk, J., Mazur, A. Atmospheric mercury species in the European Arctic: measurements and modelling. *Atmos. Environ.*, 35, 2569-2582, 2001.

Bottenheim, J.W., Shepson, P.B., Sturges, B. (Eds.). Special Issue: Air/Snow/Ice Interactions in the Arctic: Results from ALERT 2000 and SUMMIT 2000. *Atmos. Environ.*, 36, 2467–2798, 2002.

Bullock, Jr. O.R. Brehme, K.A. Atmospheric mercury simulation using the CMAQ model: formulation description and analysis of wet deposition results. *Atmos. Environ.*, 36, 2135-2146, 2002.

Byun, D.W. Ching, J.K.S. Science Algorithms of the EPA Models-3 Community Multiscale Air Quality (CMAQ) Modeling System. EPA-600/R-99/030, US Environmental Protection Agency, US Government Printing Office, Washington, DC. 1999.

Calhoun, J.A. Prestbo, E. Kinetic study of the gas phase oxidation of elemental mercury by molecular chlorine. Report available from Frontier Geosciences, Inc., 414 Pontius Avenue N., Seattle, WA 98109, 2001.

Christensen, J.H., Brandt J., Frohn, L.M., Skov, H. Modelling of mercury with the Danish Eulerian Hemispheric Model. *Atmos. Chem. Phys. Disc.*, 3, 3525-3541, 2003.

Cohen, M., Artz, R., Draxler, R., Miller, P., Poissant, L., Niemi, D., Ratte, D., Deslauriers, M., Duval, R., Laurin, R., Slotnick, J., Nettesheim, T., McDonald, J. Modeling the atmospheric transport and deposition of mercury to the Great Lakes. *Env. Research*, 95, 247-265, 2004.

Dastoor, A.P., Larocque, Y. Global circulation of atmospheric mercury: a modelling study. *Atmos. Environ.*, 38, 147-161, 2004.

Ebinghaus, R., Kock, H.H., Temme, C., Einax, J.W., Löwe, A.G., Richter, A., Burrows, J.P., Schroeder, W.H. Antarctic springtime depletion of atmospheric mercury. *Env. Sci. Technol.*, 36, 1238-1244, 2002.

Finlayson-Pitts, B.J. Pitts, Jr. J.N. Chemistry of the Upper and Lower Atmosphere. Academic Press, New York., 2000.

Friedli, H.R., Radke, L.F., Lu, J.Y. Mercury in smoke from biomass fires. *J. Geophys. Res. Lett.*, 28. 3223-3226, 2001.

Friedli, H. R., Radke, L. F., Prescott, R., Hobbs, P. V., Sinha, P. Mercury emissions from the August 2001 wildfires in Washington State and an agricultural waste fire in Oregon and atmospheric mercury budget estimates, *Global Biogeochem. Cyc.*, 17, 1039, 2003.

Gårdfeldt, K. Jonsson, M. Is bimolecular reduction of Hg(II) complexes possible in aqueous systems of environmental importance. *J. Phys. Chem. A*, 107, 4478-4482, 2003.

Grell, G.A., Dudhia, J., Stauffer, D.R. A description of the fifth-generation Penn State/NCAR mesoscale model (MM5). NCAR Technical Note, NCAR/TN-398+STR.

Grigal, D.F. Inputs and outputs of mercury from terrestrial watersheds: a review. *Environ. Rev.*, 10, 1-39, 2002.

Gustin, M.S., Biester, H., Kim, C.S. Investigation of the light-enhanced emission of mercury from naturally enriched substrates. *Atmos. Environ.*, 36, 3241-3254, 2002.

Hall, B. The gas phase oxidation of elemental mercury by ozone. *Water, Air Soil Pollut.*, 80, 301–315, 1995.

Landis, M.S. Personal Communication in May. U.S. EPA, National Exposure Research Laboratory, Mail Drop E205-03, Research Triangle Park, NC 27711, 2004.

Lee, D.S., Nemitz, E., Fowler, D., Kingdon, R.D. Modelling atmospheric mercury transport and deposition across Europe and the UK. *Atmos. Environ.*, 35, 5455-5466, 2001.

Lin, C.-J., Pehkonen, S.O. Aqueous free radical chemistry of mercury in the presence of iron oxides and ambient aerosol. *Atmos. Environ.*, 31, 4125–4137, 1997.

Lin, C.-J., Pehkonen, S.O. Oxidation of elemental mercury by aqueous chlorine (HOCl/OCl⁻): implication for tropospheric mercury chemistry. *J. Geophys. Res.*, 103, 28093–28201, 1998.

Lin, C.-J., Pehkonen, S.O. The chemistry of atmospheric mercury: a review. *Atmos. Environ.*, 33, 2067–2079, 1999.

Lindberg, S.E., Brooks, S., Lin, C.-J., Scott, K.J., Landis, M.S., Stevens, R.K., Goodsite, M., Richter, A. Dynamic oxidation of gaseous mercury in the Arctic troposphere at polar sunrise. *Env. Sci. Technol.*, 36, 1245-1256, 2002.

Lindqvist, O., Rodhe, H. Atmospheric mercury—a review. *Tellus*, 37B, 136–159, 1985.

Munthe, J. Aqueous oxidation of elemental Hg by O3. *Atmos. Environ.*, 26A, 1461–1468, 1992.

Pacyna, E.G. Pacyna, J.M. Global emission of mercury from anthropogenic sources in 1995. *Water, Air Soil Pollut.*, 137, 149–165, 2002.

Pacyna, J.M., Pacyna, E.G., Steenhuisen F., Wilson S. Mapping 1995 global anthropogenic emissions of mercury. *Atmos. Environ.*, 37 (S1), S109-S117, 2003.

Pai, P., Karamchandani P., Seigneur C., Allan M. Sensitivity of simulated atmospheric mercury concentrations and deposition to model input parameters. *J. Geophys. Res.*, 104, 13855-13868, 1999.

Pehkonen, S.O., Lin, C.-J. Aqueous photochemistry of mercury with organic acids. *J. A. W. M. A.*, 48, 144–150, 1998.

Petersen, G., Bloxam, R., Wong, S., Munthe, J., Krüger, O., Schmolke, S.R., Kumar, A.V. A comprehensive Eulerian modelling framework for airborne mercury species: Model development and applications in Europe. *Atmos. Environ.*, 35, 3063-3074, 2001.

Pleijel, K., Munthe, J. Modelling the atmospheric mercury cycle — chemistry in fog droplets. *Atmos. Environ.*, 29, 1441–1457, 1995.

Pleim, J.E., Xiu, A. Development and testing of a surface flux and planetary boundary layer model for application in mesoscale models. *J. Appl. Met.*, 34, 16–32, 1995.

Ryaboshapko, A., Ilyin, I., Bullock, R., Ebinghaus, R., Lohman, K., Munthe, J., Petersen, G., Seigneur, C., Wängberg, I. Intercomparison Study of Numerical Models for Long-Range Atmospheric Transport of Mercury.- Stage I. Comparison of chemical modules for mercury transformations in a fog/cloud environment. MSC-E Technical Report 2/2001. (on-line at http:/www.msceast.org/publications.html) 2001.

Ryaboshapko, A., Bullock, R., Ebinghaus, R., Ilyin, I., Lohman, K., Munthe, J., Petersen, G., Seigneur, C., Wängberg, I. Comparison of mercury chemistry models. *Atmos. Environ.*, 36, 3881-3898, 2002.

Ryaboshapko, A., Artz, R., Bullock, R., Christensen, J., Cohen, M., Dastoor, A., Davignon, D., Draxler, R., Ebinghaus, R., Ilyin, I., Munthe, J., Petersen, G.,

Syrakov, D. Intercomparison Study of Numerical Models for Long-Range Atmospheric Transport of Mercury.- Stage II. Comparison of modeling results with observations obtained during short-term measuring campaigns. MSC-E Technical Report 1/2003. (on-line at http:/www.msceast.org/publications.html) 2003

Sanemasa, I. The solubility of elemental mercury vapor in water. *Bull. Chem. Soc. Japan*, 48, 1795–1798, 1975.

Schroeder, W.H., Anlauf K.G., Barrie L.A., Lu J.Y., Steffon A., Schneeberger D.R., Berg T. Arctic springtime depletion of mercury. *Nature*, 394, 331–332, 1998.

Schroeder, W.H., Munthe, J., Atmospheric mercury - an overview. *Atmos. Environ.*, 32, 809-822, 1998.

Seigneur, C., Abeck, H., Chia, G., Reinhard, M., Bloom, N.S., Prestbo, E., Saxena, P. Mercury adsorption to elemental carbon (soot) particles and atmospheric particulate matter. *Atmos. Environ.*, 32, 2649–2657, 1998.

Seigneur, C., Karamchandani, P., Lohman, K., Vijayaraghavan, K., Shia, R.-L. Multiscale modeling of the atmospheric fate and transport of mercury. *J. Geophys. Res.*, 106, 27795-27809, 2001.

Smith, R.M., Martell, A.E. Critical Stability Constants Inorganic Complexes, Vol. 4. Plenum Press, New York., 1976.

Sommar, J., Gärdfeldt, K., Strömberg, D., Feng, X. A kinetic study of the gas-phase reaction between hydroxyl radical and atomic mercury. *Atmos. Environ.*, 35, 3049–3054, 2001.

Sommar, J., Wängberg, I., Berg, T., Gärdfeldt, K., Munthe, J., Richter, A., Urba, A., Wittrock F., Schroeder W.H. Circumpolar transport and air-surface exchange of atmospheric mercury at Ny-Ålesund (79°N), Svalbard, spring 2002. *Atmos. Chem. Phys. Disc.*, 4, 1727-1771, 2004.

Tokos, J.J.S., Hall, B., Calhoun, J.A., Prestbo, E.M. Homogeneous gas-phase reaction of Hg0 with H2O2, O3, CH3I, and (CH3)2S: implications for atmospheric Hg cycling. *Atmos. Environ.*, 32, 823–827, 1998.

Travnikov, O. Ryaboshapko, A. Modelling of Mercury Hemispheric Transport and Depositions. MSC-E Technical Report 6/2002. 67 pages. (http://www.msceast.org/publications.html) 2002.

U.S. EPA. Mercury Study Report to Congress. Volume III: Fate and Transport of Mercury in the Environment. Report number EPA-452/R-97-005, 1997.

Van Loon, L., Mader, E., Scott, S.L. (2000). Reduction of the aqueous mercuric ion by sulfite: UV spectrum of $HgSO_3$ and its intramolecular redox reaction. *J. Phys. Chem. A*, 104, 1621-1626, 2000.

Van Loon, L., Mader, E., Scott, S.L. Sulfite stabilization and reduction of the aqueous mercuric ion: kinetic determination of sequential formation constants. *J. Phys. Chem. A*, 105, 3190-3195, 2001.

Vette, A.F., Landis, M.S., Keeler, G.J. Deposition and emission of gaseous mercury to and from Lake Michigan during the Lake Michigan Mass Balance Study (July, 1994-October, 1995). *Env. Sci. Technol.*, 36, 4525-4532, 2002.

Xiao, Z.F., Munthe, J., Strömberg D., Lindqvist O. Photochemical behavior of inorganic mercury compounds in aqueous solution. In: Watras, C.J., Huckabee, J.W. (Eds.), Mercury as a Global Pollutant—Integration and Synthesis. Lewis Publishers, New York, 581–592, 1994.

Xu, X., Yang, X., Miller, D.R., Helble, J.J., Carley, R.J. A regional scale modeling study of atmospheric transport and transformation of mercury. I. Model development and evaluation. *Atmos. Environ.*, 34, 4933-4944, 2000.

PART-IV:
HUMAN EXPOSURE

Chapter-15

EXPOSURE TO MERCURY IN THE AMERICAS

Kathryn R. Mahaffey

Office of Science Coordination and Policy, United States Environmental Protection Agency Washington, D.C. 20640, USA

INTRODUCTION

Non-occupational exposures of the human population to mercury occur predominantly as methylmercury from fish and shellfish consumption and as inorganic mercury among the segment of the population for whom silver-mercury amalgams have been used in dental restorations. These sources of exposure are sufficiently common as to constitute the background onto which additional sporadic exposures are added. Examples of the latter include mercury in cosmetics, folk remedies, cultural practices, accidents, and forensic or use of mercury as a poison.

Occupational exposures which cover a wide range of types of work (e.g., chloralkali factory workers, dentists, dental technicians, gold miners) are predominantly to inorganic mercury, mercury vapors, and various mercury salts. The rare occupational exposure to organo-mercurials has proven extremely dangerous. However, organo-mercurial residues have been identified in environmental samples at former mercury plants where organo-mercurials for pesticide use were produced (Hempel et al., 1995).

This chapter provides an overview of reports on sources of mercury in the Americas during the past 30 years. The descriptions are for specified subgroups and usually provide no indication of the prevalence of the exposures in the general population. The usual situation relies on reported mercury exposures in a geographic location, a cohort, an occupation, a poisoning episode, or in some other subgroup.

Based on this collage of data describing the magnitude of mercury exposures in the Americas, an estimate of the sources and pervasiveness of the mercury exposures emerges. This chapter does not compare exposure estimates with risk values for methylmercury (e.g., NAS/NRC, 2000; Rice et al., 2003; JECFA, 2003).

BIOLOGICAL INDICATORS OF MERCURY EXPOSURE

Biomarkers of Mercury Exposure

Environmental data describing mercury concentrations in media including air, water, sediments, biota, etc. indicate the magnitude of contamination of various routes of mercury exposure. In risk assessments relating mercury exposure to health effects, dose-response for a group of individuals may include consideration of biomarkers of exposure to the prevalence of adverse health effects. Chemical analyses of body fluids and tissues provide an indication of the chemical species of mercury to which the person was exposed, the magnitude, the time and duration of the exposure. Whole blood, urine, and hair are the body fluids and products typically analyzed to determine their mercury concentration to provide an indication of the person's mercury exposures. These three tissues provide diverse information. For example, blood mercury typically is used to indicate exposure to organic mercury, almost always dominated by methylmercury. Urinary mercury is an indicator of exposure to inorganic mercury. Hair mercury is dominated by exposure to methylmercury. Because mercury can deposit on hair from surface contamination, total mercury content of hair may not be a reflection of mercury excreted from the body, but may be dominated by mercury from surface contamination. In depth review articles on this topic include: Risher et al. (2002), Mason et al. (2001). Veiga (2004) provides guidance on practical aspects of collection of environmental and human biological samples for evaluation of exposure to mercury vapor, inorganic mercury and methylmercury. Veiga (2004) also reviews in-depth methodologies for analyses that provide both total and chemically speciated mercury concentrations. A brief overview of the relationship of chemical species of mercury to mercury concentrations in biological materials used as biomarkers of exposure follows.

Chemical Forms of Mercury to Which People Are Commonly Exposed

Methylmercury. More than 95% of methylmercury present in diet is absorbed from the gastrointestinal tract. The half-life of total blood mercury in human adults is generally given as approximately 70 days, however, individual values as long as 250 days have been reported by Birke et al. (1972). Based on data from 48 patients al-Shahristani and Shihab (1974) reported that biological half-times of blood mercury varied between 37 and 189 days, with an average of 72 days. The half-life of methylmercury in blood has been estimated to be 44 days based on single exposure intravenous injection of radio-labeled methylmercury (Smith et al., 1994). Based on results from Kershaw et al. (1980) elimination of methylmercury from blood had a biphasic half-time of 7.6 hours (range 4.9 hours to 9.5 hours) and 52 days (range 39 days to 66 days) following consumption of single serving of high-mercury fish (Kershaw et al., 1980). Sherlock et al. 1984 reported a half-time of 50 days (range 42 to 70 days) for methylmercury from consumption of fish over a period of several weeks in a group of adult volunteers.

Among the general population in the United States when total blood mercury exceeded ~ 4 µg/L more than 90% of the mercury present was organic mercury (Mahaffey et al., 2004). The source of organic mercury for the general population is methylmercury from the consumption of fish and shellfish.

Inorganic mercury. Blood inorganic mercury concentrations have been used to detect acute, high dose exposures. At high exposures the cell-to-plasma ratio ranges from a high of two to less than one (Goyer and Clarkson, 2001; Mason et al., 2001). Alternately monitoring of urine is the indicator for exposure to chronic, low-to-moderate inorganic mercury exposure (Mason et al., 2001). Health risks associated with various values are discussed elsewhere. The distribution of inorganic mercury concentrations is discussed below.

The half-life of inorganic mercury in blood is about three days based on experimental studies with radio-labeled mercury (Cherian et al., 1978). A second slower half-life of about two to three weeks has been indicated based on studies in chloralkali workers whose exposures were terminated (Barregard et al., 1992). Urinary levels are used to monitor sustained exposure to inorganic mercury.

Mercury vapor. An excellent recent review of the health effects of mercury vapor is provided in the UNIDO Protocols for Environmental and Health Assessment of Mercury Released by Artisanal and Small-Scale Gold Miners (Veiga, 2004). Inhalation of mercury vapor is the most significant form of mercury exposure for mining and gold shop workers engaged in these types of mining operations. Once inside the lungs mercury is oxidized forming Hg (II) complexes which are soluble in many body fluids.

The half-time of Hg in blood absorbed as a vapor is 2 to 4 days after which 90% is excreted through urine and feces followed by a second phase with a half-time of 15-30 days (WHO, 1991). Between passage of elemental mercury through the alveolar membrane and complete oxidation, mercury accumulates in the central nervous system (Mitra, 1986). During this process mercury can irreversibly damage the central nervous system. At exposures of moderate duration, the kidneys are also affected. Short-term exposure to high levels of mercury vapor produces chest pain, dyspnea, cough impaired pulmonary function, interstitial pneumonitis (Veiga, 2004; WHO, 1991). Occupational exposures to mercury vapor have caused psychiatric symptom, hallucinations, erethism (exaggerated emotional response), insomnia, and muscular tremors (WHO, 1991).

Ethylmercury, Phenylmercury and Other Mercury Compounds Ethylmercury has been added as a preservative to vaccines and biologicals since the 1930s (Midthun, 2004). The metabolism of ethylmercury administered by intramuscular injection (as it would be used in vaccines) differs from methylmercury based on studies with primates (Burbacher et al, 2004). Compared with methlymercury, ethylmercury derived from thimerosal has a shorter terminal half-life in both blood and brain. In young monkeys, Burbacher et al. (2004) found minimal accumulation of total blood mercury from ethylmercury derived from thimerosal, but found that the brain-to-blood partition ratio for ethyl mercury was higher than that for methylmercury. Earlier work by Pichichero et al. (2002) had reported that the blood half-life for ethylmercury in human infants was 7 days, substantially shorter than the half-life of methylmercury. Measurement of blood mercury levels in infants 48 to 72 hours after hepatitis B vaccinations at birth showed a substantial elevation in their blood mercury concentration (Stajich et al., 2000).

Phenylmercury is absorbed from the gastrointestinal tract and excreted in the urine based on studies with adult volunteers. This form of mercury has been used as a preservative in cosmetics and has been used as an anti-infection agent in products such as mouth washes (Lauwerys et al., 1977). A study with adult human volunteers (Lauwerys et al., 1977) showed use of these products resulted in a marked increased urinary mercury excretion and an increase in blood mercury concentration with even very short-term use. Use of hand soap containing phenylmercuric borate was associated with increased urinary mercury among adult male and female health professionals using the soap as a disinfectant in hand cleaning (Peters-Haefeli et al., 1976). It was estimated that use of hand soap containing 0.04% phenylmercuric borate increased daily mercury absorption between 30 to 100 µg Hg/24 hours, in part due to hand-to-mouth transfer of mercury.

Biological Media Used to Indicate Mercury Exposure

Blood Mercury. Blood mercury concentration is the preferred biomarker for exposure to methylmercury. When blood mercury exceeded ~ 4 µg/L, more than 90% of mercury in blood was present as organic or methylmercury in adult women in a sample representative of the United States population (Mahaffey et al., 2004). At lower blood mercury concentrations in the range of 1 µg/L to 2 µg/L blood mercury indicates exposures to both inorganic mercury and methylmercury (Mahaffey et al., 2004; Morrissette et al., 2004). Whether the inorganic form or methyl form predominates, depends on the relative contribution to the total mercury burden of fish consumption vs. dental amalgams.

In occupational settings where there is high exposure to mercury vapor, blood inorganic mercury has been found to be elevated and closely correlated with urinary mercury (e.g., data on chlor-alkali workers - see Smith et al., 1970). These findings were reported from subjects whose exposures ranged to levels far higher than observed in the general population in the Americas; i.e., > 20-30 µg/L whole blood. Exposure to mercury vapor may occur non-occupationally in mercury spills or in ritualistic use of mercury (see 4.7 below).

Urinary mercury. Urine mercury concentration is usually expressed in µg/L. Frequently mercury concentrations are adjusted for creatinine excretion. Creatinine is excreted by the kidneys and is a breakdown produce of creatine which is a constituent of muscle. Creatinine excretion depends on muscle mass so values for a person are relatively constant depending on age and lean body mass. Creatinine concentrations are usually expressed in mg/dL. Urinary values expressed simply in volume (per L) are subject to dilution and consequently more variable with the quantity of fluids consumed.

The half-life of urinary excretion of mercury has been reported to range from as short as 20 days to as long as 90 days (Mason et al., 2001). Based on likely half-lives of 40 to 90 days, urinary mercury is an integrated marker of exposure over previous months (Mason et al., 2001).

Reference published for urinary mercury, e.g., Iyenger and Woittiez (1988) indicated the 95% upper confidence limit was < 20 µg/L. Among the 1748 women aged 16-49 years who participated in the 1999-2000 NHANES, the 95th percentile value was 5 µg/L or 3.27 µg Hg/g creatinine (CDC, 2003).

Urinary mercury is the preferred biomarker for exposure to inorganic mercury and over time increases in response to exposure to low levels of mercury vapor. Mercury concentrations in the urine will also increase with exposure to other mercurials including phenylmercury. Inorganic mercury can also arise from demethylation of methylmercury. An increase in urinary inorganic mercury can be shown in subjects with a high methylmercury intake (Carta et al., 2003; Johnsson et al., 2004).

Hair mercury. Growth, morphology and histochemistry of human hair have been reviewed in detail (Swift, 1997). Hair is approximately 95% proteinaceous and 5% a mixture of lipids, glycoproteins, remnants of nucleic acids, and in the case of pigmented hairs of melanin and phaeomelanin. Hair contains a central core of closely packed spindle-shaped cortical cells, each filled with macrofibrils which, in turn, consists of microfibril/matrix composit. The amino acid composition of hair is high in those amino acids with side-chains (particularly, those containing cystine, cysteine, tyrosine, and tryptophan). The cortical core is covered by some sheet-like cells of the cuticle. The surfaces of all of the hair shaft have a thin layer of lipids which is covalently attached to the underlying protein.

As hair grows methylmercury is incorporated into hair. There is a general view that hair grows about 1 cm per month, although there is evidently substantial variability around this value. Hair mercury analyses are complicated by the problem of deposition of mercury onto the hair after it has been formed from external sources including cosmetics (e.g., dyes and Hg^0 vapors).

Hair mercury is predominantly methylmercury, but the percent of total hair mercury that is methylmercury has been reported to be 80% (WHO, 1990) to values ranging from 90% to 98% (Dolbec et al., 2000). Generally hair is thought to be 250 to 300 times more concentrated in mercury than is blood (WHO, 1990; Veiga, 2004), although lower values have also been given including a range of 150 to 200 (Gill et al., 2002). A far wider range of individual values exists. For example, Dolbec et al. (2000) reported hair to blood ratios ranging between 81 and 624. The extent to which this ratio is applicable across all age and ethnic/racial groups remain to be confirmed. Seidel et al. (2001) have noted many problems with commercial laboratories performing hair analyses for trace elements.

Inorganic mercury is not considered to be excreted in hair, although inorganic mercury can be a surface contaminant on hair. Hair is not considered as good an indicator of exposure to mercury vapor (Veiga, 2004) or to inorganic mercury (Veiga, 2004) as is urine.

Biological Variation by Age, Gender and Physiological Status

Because concentrations of mercury in human tissues are determined by environmental exposures to mercury, it is difficult to describe biological variation by age, gender, and physiological status, with a few exceptions. The generalities are that exposures depend more on the work habits and food consumption habits of the population being described. If mercury exposures are unusually high in a particular occupation, for example dentistry, the sex distribution of professionals in that occupation will, of course, greatly influence sex-related differences in that particular occupational exposure. With regard to methylmercury, if methylmercury were closely tied to caloric intakes per kilogram of body weight, men would consistently have higher exposures to methylmercury than women. However, methylmercury exposure is closely linked to consumption of fish and shellfish, in general, and to consumption of high-trophic level piscivorous fish, in particular. Consequently eating patterns of particular species of fish and shellfish predict dietary intakes of methylmercury.

Generally men have higher intakes of methylmercury on a per kilogram body weight basis than do women, as well as having higher hair and blood mercury concentrations. Higher mercury exposures among men have been identified in some data sets (e.g., Dolbec, et al., 2000). However, the gender-associated difference is not consistent. For example, Kosatsky et al. (2000) found no association between gender and mercury exposure among sports fish consumers.

Some major data sets report an increase in blood mercury concentrations with age among women of reproductive age. Examples of data sets in which blood mercury levels increased with age are results from the 1709 women ages 16 through 49 who were examinees in the 1999-2000 National Health and Nutrition Examination Survey in the United States (Mahaffey et al., 2004) and the cord blood mercury data for a group of ~1100 southern Quebec, Canadian women reported by Rhiands et al. (1999). Other data sets indicate no increase in concentration of mercury with age (among others see Kosatsky et al., 2000).

One pattern that has consistently been identified in data from multiple sources is the ratio of cord blood mercury concentration to maternal blood mercury concentration. Cord blood averages 70% to 80% higher than maternal blood based on analyses of matched cord maternal pairs (Sterns and Smith, 2003). In addition to the data sets included in the analyses by Sterns and Smith (2003), several published subsequently from diverse parts of the world have provided subsequent confirmation of this ratio. Among others see: Muckle et al. (2001) among the northern Quebec Inuits; Morrisette et al. (2004), among women living along the St. Lawrence River; Sakamoto et al. (2004) among

Japanese women.

Blood mercury levels among children have to be interpreted with knowledge of the child's age and degree of mercury exposure. At birth newborns have higher blood mercury concentrations than their mothers. These blood mercury concentrations appear to decline rapidly in populations with low mercury exposures.

For example, in the NHANES data (Schober et al., 2003), children's blood concentrations were approximately one-fourth as high as those of adult female examinees. Hair mercury concentrations among 1-through-5 year-old children were less than one-half those of the women ages 16-through-49 years (McDowell et al., 2004). By contrast, under conditions when methylmercury exposures are substantially higher, children's mercury concentrations do not differ very much from those of adults. For example, Santos et al. (2002) found mean hair mercury concentrations of children aged 7 to 12 years were 14.4 µg/g, compared with 15.7 µg/g for adult women aged 14 to 44 years, and 14.1 µg/g for all other people in their sample.

HUMAN DATA INDICATING MERCURY EXPOSURE IN THE AMERICAS

Most reports on mercury exposure have aimed to evaluate populations thought to be at risk of exposure to amounts of mercury thought to be problematic. Efforts to identify data that provide reference ranges for the population not considered at unusual risk of exposure and located in the Americas are few. General population data for the entire age and gender range appear not to be available for any country in the Americas.

United States of America

The recent United States data on blood mercury and hair mercury concentrations are limited to women of childbearing age and children. Data on adult males will be reported only beginning with examinees who participated in the National Health and Examination Surveys (NHANES) in 2003 and later. These data are unlikely to be published until late 2005 or 2006. Estimates of exposures to people considered to have no unusual risk of mercury exposure can be inferred by looking at data from groups selected as "controls" when evaluating "exposed" groups.

Surveys among the General Population

The largest study of men from the general population is that of Kingman et al. (1998) who analyzed urine and blood mercury concentration among 1127 Vietnam-era United states Air Force pilots (all men, average age 53 years at the time of blood collection) for whom extensive dental records were available. Blood values were determined for total mercury, inorganic mercury and organic/methylmercury. The mean total blood mercury concentration was 3.1 µg/L with a range of "zero" (i.e., detection limit of 0.2 µg/L) to 44 µg/L. Overall, 75% of total blood mercury was present as organic or methylmercury. Less than 1% of the variability in total blood mercury was attributable to variation in the number and size of silver-mercury amalgam dental restorations. Dietary data on the former pilots were very limited, so typical patterns of fish consumption were not reported.

Table 1. Total Blood Mercury (µg/L) for Women Ages 16-49 Years among the 1999-2000 NHANES Examinees (Mahaffey et al., 2004).

	Sample Persons	Geometric Mean	Percentiles Selected			
			95% CI	50th	90th	95th
Total	1709	1.02	0.85-1.20	0.94	4.84	7.13
Age (years)						
16-19	513	0.63	0.49-0.76	0.58	2.47	3.28
20-29	437	0.87	0.68-1.06	0.79	4.79	6.43
30-39	405	1.09	0.84-1.35	0.97	5.07	8.95
40-49	354	1.32	1.04-1.60	1.24	5.31	7.13

Table 2. Blood Organic Mercury Concentrations (µg/L) for Women Ages 16-49 Years among the 1999-2000 NHANES Examinees (Mahaffey et al., 2004).

	Sample Persons	Geometric Mean	Selected Percentiles			
			95% CI	50th	90th	95th
Total	1707	0.80	0.34-0.66	0.60	4.44	6.73
Age (Years)						
16-19	513	0.49	0.40-0.58	ND	2.03	2.88
20-29	436	0.70	0.54-0.87	0.44	4.50	6.10
30-39	405	0.83	0.61-1.04	0.61	4.65	8.62
40-49	363	1.02	0.79-1.24	0.90	4.93	6.73

In the United States the Centers for Disease Control conducts the NHANES which provides medical examinations, biochemistry assessments for exposure to environmental contaminants, dietary and medical histories, clinical chemistry profiles, and a large number of other specialized tests to approximately 8,000 persons per year. This survey is conducted in approximately 25 to 30 communities per year and these communities are selected so that if treated by appropriate population statistics, the data can provide a profile that is representative of the United States as a whole.

Among adult women of childbearing age (considered being 16 years through 49 years) and young children ages one through six years (beginning with the years 1999 and 2000), biomarkers of mercury exposure were measured. These included hair mercury (total and inorganic), blood mercury (total and inorganic), and urinary mercury (total only and among adult women only). Organic mercury was calculated by differences and chemical speciation of samples indicated that the predominant chemical species was methylmercury (Schober et al., 2003; Mahaffey et al., 2004). Although adult men, children older than six years, and women older than 49 years were included among the NHANES examinees, mercury measurements were not included for the age and gender groups during the 1999 and 2000 years of the survey. These groups have been added beginning in 2003. Blood mercury concentration data are shown in Table 1 which indicates the geometric mean, 95% CI, 50th, 90th and 95th percentiles for blood total mercury by age group. Organic blood mercury concentrations, almost entirely methylmercury, are shown in Table 2. Children in the age 1-to-6 year age range have much lower blood mercury concentrations than did the adult women. Table 3 shows hair mercury concentrations for adult women and children who were examinees in the 1999 and 2000 NHANES.

Table 3. Mean and Selected Percentiles of Hair Mercury (Hg) Concentrations for Children Aged 1-6 Years and Women Aged 16-49 years - National Health and Nutrition Examination Survey, United States, 1999 (McDowell et al., 2004).

			Selected Percentiles					
Hair Hg	No.	Mean (95% CI)	10th (95% CI)	25th (95% CI)	50th (95% CI)	75th (95% CI)	90th (95% CI)	95th (95% CI)
Children	838	0.22 (0.18,0.25)	0.03 (0.01,0.05	0.06 (0.05,0.07)	0.11 (0.10,0.13)	0.21 (0.15,0.27)	0.41 (0.33,0.49)	0.65 (0.52,0.76)
Women	1726	0.47 (0.35,0.58)	0.04 (0.02,0.05)	0.09 (0.07,0.11)	0.19 (0.14,0.24)	0.42 (0.29,0.55)	1.11 (0.54,1.68)	1.73 (1.44,2.02)

Surveys from Individual States

Analyses of blood and hair samples for mercury indicate a range of exposures present under localized conditions. The upper end of this exposure distribution extends into concentrations reported to be 50 µg/L to 70 µg/L for blood; hair values occasionally > 10 ppm with average values typically in the 1 ppm to 3 ppm range. This information has been provided in detail in US EPA's Mercury Study Report to Congress (US EPA, 1997), Volume 4, Section 6.4.4 (blood mercury) and Section 6.5.2 (hair mercury). Blood mercury concentration was associated with increased consumption of fish, particularly species known to be high in methylmercury or species containing < 1 ppm mercury but consumed routinely.

More recent data are not dissimilar, although from geographically diverse areas. For example, hair mercury concentration of a group of 16 rural Alaskan subsistence food users from Napakiak, a small Yup'ik Eskimo in southwest Alaska was compared with a group of non-subsistence non-Yup'ik adults from the urban interior city of Fairbanks, Alaska (Rothschild and Duffy, 2002). The mean methylmercury concentration in hair for the Yup'iks was 1.45 ppm (range 0.32-4.00 ppm), which was higher than that of the Fairbank dwellers whose mean was 0.19 ppm (range 0.03 to 0.43 ppm). Pregnant women (mainly during their first trimester) from the eastern United States' State of New Jersey had hair and blood samples analyzed for total mercury (Stern et al., 2001). Between 85% and 90% of women had hair mercury levels < 1.0 µg/g, and 1% to 2% had values > 4.0 µg/g.

Exposures at the Higher End of the Continuum of the General Population in the United States and Territories

There is a continuum of exposure to mercury with higher exposures found among people with particular characteristics including geographic locations (e.g., island populations), ethnic groups with food habits which prefer fish (e.g., Asian populations), occupations (e.g., dentists, gold miners), and life styles (e.g., affluent patients who have been consuming a high-fish diet for promotion of cardiovascular health). These groups' mercury exposures will be quite different from those identified by either the NHANES data or general data reported in state surveys.

Data from the NHANES provide an estimate for the United States population. It may be important, however, that frequency of fish consumption - the primary determinant of methylmercury exposure - was not a selection factor for inclusion in the survey. This may or may not affect the validity of extrapolation of the NHANES data in making population estimates. The initial reports from 1999-2000 NHANES were based on 26 communities. Additional two-year cycles of the NHANES will add between 25 and 30 communities with each cycle. Currently the 2001/2002 data are being statistically analyzed and the

2003 and early 2004 samples have been completed with the 2004 survey underway.

These additional years of data from NHANES will strengthen our data base and provide greater ability to better describe subpopulations. Nonetheless, based on reports from case series, communities and medical practices, it is already known that much higher exposures to methylmercury in the United States have been documented than were identified among the 1999-2000 NHANES examinees. As physicians have become more aware of mercury as a health concern for the general population, obtaining mercury measurements on patients' blood samples have increased resulting in expanded awareness of the magnitude of mercury exposure in groups not typically considered to be at elevated risk of exposure to methylmercury.

Some health departments in the United States (e.g., the State of North Carolina) have begun to offer screening for blood mercury to people who may be concerned that their mercury exposures are elevated. For example, in Louisiana the State Office of Public Health began offering to measure blood mercury levels and reported screening values for 313 participants (Bellanger et al., 2000) whose blood values ranged from < 0.3 μg/l to 35 μg/L with 1.9% > 20 μg/L. Higher values were reported among commercial fishermen and their families. Blood mercury values were twice as high among people who ate fish at least once a week compared with people who ate fish twice a month or less. Elevated blood mercury concentrations have been found among gamefish consumers in Arkansas (Burge and Evans, 1994).

High blood mercury concentrations (e.g., > 80 μg/L) have been reported among high-income individuals who choose to consume diets that contain large amounts of fish in the view that such diets will be advantageous to their health. Others pursue these diets for culinary preference. For example, Hightower and Moore (2003) reported blood mercury concentrations from a private practice patient population in San Francisco. From a total of 116 patients evaluated, 89% had blood mercury concentration ≥5 μg/L, and 16% had concentrations ≥20 μg/L. Four individuals in this group had blood mercury levels more than 50 μg/L. Following the 2003 publication Hightower tested an additional 107 patients whose average blood mercury concentration was 21 μg/L (Hightower, 2004). Saint-Phard et al. (2004) reported a case series from University of Colorado in which patients with elevated blood mercury (range 27 μg/L to 96 μg/L) associated with fish consumption had neurological symptoms (i.e., including paresthesias of the extremities and or electrodiagnostic evidence of sensorimotor peripheral neuropathy).

In addition to these cases, other clinics and individual physicians have identified individuals with blood, hair, and urinary mercury values considerably higher than those reported among NHANES examinees. Among others see Kales and Goldman (2002). In addition to methylmercury ingested from frequent consumption of fish containing high concentrations of methylmercury, other medical reports indicate exposures to inorganic mercury from folk remedies, cosmetics, gold smelting. It is important to know that the highest blood value reported from 1999-2000 NHANES was 35 µg/L of which only about 50% of the total mercury present was organic mercury (Mahaffey et al., 2004).

Residents of those states and territories of the United States (e.g., Hawaii, Puerto Rico) who have geographic proximity to a steady supply of fresh fish have greater intake of methylmercury than the more inland populations. For example, Ortiz-Roque and Lopez-Rivera (2004) investigated blood mercury concentrations and seafood consumption frequency among reproductive-age women in two areas of Puerto Rico.

Overall-United-States data from NHANES indicated 30-day consumption of fish and shellfish at the 50th percentile was 1.54 meals (95% CI: 1.25-1.82) and at the 90th percentile 10.81 meals (95% CI: 7.15-14.47) (Schoeber et al., 2003). Average Puerto Rican total seafood consumption was 12.1 meals/30 days in NE Puerto Rico and 21.2 meals/30 days in Vierques, an island municipality. That is to say that average Puerto Rican fish and shellfish consumption was higher than 90th percentile intake for the overall United States. The percent of women whose mercury intake exceeded the RfD was 6.6 in NE Puerto Rico and 26.8 in Vieques. Among Viequenses three of the 41 women had hair mercury concentrations higher than 12 ppm which is the lower bound of the benchmark dose for mercury, an effect level (Ortiz-Roque and Lopez-Rivera, 2004).

Canada

Northern Canada

Canadian data on mercury exposure can generally be separated into studies of northern Canada and studies from more southern Canadian regions. A comprehensive review of the Arctic studies by Van Oostdam et al. (1999) provided an in-depth analysis of the cultural and nutritional significance of wildlife as food for indigenous people, as well as a consideration of a range of contaminants including organochlorines, heavy metals, and radionucleides. A monitoring program providing data on mercury levels among the Cree population of James Bay, Quebec in 1988 and in 1993/1994 indicated that the proportion of the Cree population with total hair mercury concentrations > 15.0 µg/g decreased from 14.2% in 1988 to 2.7% in 1993/1994 (Dumont et al., 1998).

Inuit inhabitants of northern Canada in a region called Nanavik consume in their traditional diet large amounts of marine foods. Muckle et al. (2001) reported cord blood, maternal blood, and maternal hair mercury concentrations averaged 18.5 µg/L, 10.4 µg/L, and 3.7 µg/g. Time trends in cord blood mercury reported by Dallaire et al. (2003) describe concentrations of Inuit infants born between 1994 and 2001, indicate mean cord blood concentrations declined from approximately 20 µg/L in 1993 to approximately 8 µgl/L in 1999. This decrease does not appear to depend on a decline in fish consumption based on ω-3 fatty acid level, an indicator of fish intake.

Southern Canada

Reports on blood and hair mercury concentrations in southern Canada indicate much lower mercury exposures in general than observed in northern Canada (see Table 4).

Rhainds et al. (1999) data on newborns' after adjustment of mercury concentration for differences in the cord blood to maternal blood ratio using 1.7 to 1.0 (Sterns and Smith, 2003) indicated maternal blood mercury concentrations averaged < 1 µg/L. Comparing regions of Quebec, blood mercury concentrations were associated with residency area with coastal > urban > suburban > rural. Morrissette et al. (2004) identified a similar exposure range for women early in pregnancy. Speciation of blood samples indicated about half of the mercury in blood at this exposure range was inorganic mercury.

Table 4. Mercury Concentrations (µg/g wet weight) in Traditional Foods Consumed by Canadian Aboriginal Peoples (modified from Chan, 1998).

Food Group	Number of Sites	Number of Samples	Arithmetic Mean	Standard Deviation	Maximum
Marine Mammal Meat	32	764	0.85	1.05	33.4
Marine Mammal Blubber	6	71	0.08	0.05	0.13
Terrestrial Mammal Meat	6	19	0.03	0.02	0.17
Terrestrial Mammal Organs	14	254	0.86	0.90	3.06
Fish	799	31,441	0.46	0.52	12.3
Birds	24	216	0.38	0.59	4.4
Plants	8	14	0.02	0.02	0.05

Mahaffey and Mergler (1998) found that increased consumption of lake fish was associated with higher blood mercury concentrations among 289 residents of Southwest Quebec, but with one exception total blood mercury concentrations were < 5 µg/L. One individual, however, had a blood mercury concentration of ~70 µg/L greatly exceeding all of the other subjects.

Among groups who consume sports fish, methylmercury exposures are somewhat higher. Kosatsky et al. (2000) compared blood mercury concentrations among those who ate sports fish at least once a week and reported a geometric mean blood mercury concentration of 3.03 ±2.43 µg/L compared with blood mercury of 1.44±2.23 µg/L for those eating sports fish less than once a week. The highest blood mercury concentrations were associated with consumption of high-food-chain piscivorous fish.

The wide variation in blood mercury concentrations, however, indicates some consumers of fish with higher mercury levels. Hair mercury concentrations also reflected these differences. Evaluation of the patterns of fish consumption among the sports fishers (Kosatsky et al., 2000) showed a strong correlation between blood mercury and hair mercury. A small group of sports fishers of recent Asian immigration (from Vietnam and Bangladesh) had higher hair mercury concentrations than the high-level, native-born Canadians (Kosatsky et al., 1999b).

Brazil

Amazon River Basin and Tributaries

Widespread use of mercury to amalgam gold by miners in the Amazon River Basin and its tributaries over past decades has resulted in severe contamination with inorganic mercury. The process of producing a mercury-gold amalgam and using torches to drive off the mercury in open air with no worker protection has resulted in severe mercury poisoning among miners. The condensed mercury vapor and inorganic residue are then dumped into the Amazon River Basin and its tributaries. This topic is covered in detailed reviews by many others including Eisler (2004).

To summarize briefly twenty-five years of significant data on the impact of use of mercury in gold mining and its impact on the people living in the region, hair mercury concentrations among riparian river people are highly elevated, but variable depending on the amount and species of fish consumed. Seasonal variation is also found because the species of fish that are available changes with the season (Dolbec et al., 2001).

Mean or median hair mercury concentrations between 2 µg/g and 20 µg/g have been reported (Akagi et al., 1995; Barbosa et al., 1998; Lebel et al., 1998; Dolbec et al., 2000). Harada et al. (2001) evaluated three fishing villages located several hundred kilometers downstream from the gold mining areas. Eighty percent of subjects had hair mercury concentrations > 10 ppm. Similar exposures are reported by Dorea et al. (2003) who evaluated adult women of childbearing age living in locations of the Rio Negro basin that were not impacted by gold mining. These women consumed an estimated fish intake of 170 grams per day.

The mean mercury concentration in the individual fish species varied between 0.04 and 0.59 µg/g, but 27% of samples had concentrations > 0.50 µg/g and 7% were > 1.00 µg/g. The women's hair mercury concentrations ranged between 6.5 and 32.6 µg/g, with 82% of women's values > 10 µg/g.

Sing et al. (2003) have studied Yanomama Indian villagers living near mined and unmined rivers beginning in 1994. Re-evaluating these villagers indicated average blood mercury concentration between 21.2 µg/L and 43.1 µg/L. Mercury concentrations in piscivorous fish (piranha) from the mined Catrimani River ranged from 0.23 to 1.08 ppm. High mercury concentrations were also observed in fish and villagers along the unmined Ajarani and Pacu Rivers.

Santos et al. (2002) conducted a cross-sectional study among the Munduruku Indians living in the State of Para, Brazil. Mean hair mercury concentrations were 14.4 µg/g for children aged 7 to 12 years, 15.7 µg/g for women aged 14 to 44 years, and 14.1 µg/g for all other people. Although the mercury concentrations of the fish were < 0.5 µg/g, the amount of fish consumed resulted in hair mercury concentrations for the women of childbearing age being higher than considered without risk by the World Health Organization 1990.

Hair mercury in these Amazonian fish-eating populations has been chemically speciated between inorganic mercury and organic mercury. In a Negro River fish-eating population who consumes fish at least twice a day, total hair mercury ranged from 1.5 µg/g to 59.0 µg/g (Barbosa et al., 2001). The mean percentage of methylmercury was 71.3% (range 34% to 100%) of total mercury in hair. The percent methylmercury was comparable across age and sex groups. Among women of reproductive age, 65% had hair mercury concentrations > 10 µg/g.

Other Regions of Brazil

Regions of Brazil away from the Amazon have reported far lower mercury exposures. For example, a coastal region of Brazil was evaluated because of possible mercury contamination due to a chlor-alkali plant installed along one of the tributaries of the channel (Nilson et al., 2001). Hair mercury concentrations in this area were 1.9 µg/g total mercury and 1.2 µg/g methylmercury. Local fish contained 0.026 total mercury and 0.019 methylmercury.

Argentina

Study of mercury in marine mammals in coastal waters indicates mercury contamination (Marcovecchio et al., 1994). Importance of local sources of contamination illustrated by the distribution of mercury from a mercury cell chlor-alkali factory that had operated for more than 40 years and dischaged plant effluents near the main irrigation channel for agricultural land showed accumulation of mercury in river bed sediments (Arribere et al., 2003).

Bolivia

Higher hair mercury concentrations among examinees in Japanese immigrant settlements in Bolivia compared with other groups in Brazil and Paraguay related to higher fish consumption among the Japanese immigrants (Tsugane and Knodo, 1987).

Chile

An overview of mercury contamination in Chile is recently available (Barrios-Guerra, 2004) that includes a description of industrial sources of mercury in Chile. Hair mercury concentrations among 59 pregnant and nursing women with normal to high fish and seafood consumption living in fishing villages distributed throughout the coastal zone of Chile were compared with an inland Chilean town (Bruhn et al., 1994). Total hair mercury content of the women in the coastal villages was 2.0± 1.4 µg/g compared with the control group's mean value of 0.4 ± 0.2 µg/g.

French Guiana

An investigation of 500 individuals from 13 health centers indicated that consumption of methylmercury from fish consumption (especially freshwater fish) in diet played a predominant role in the total mercury burden and that in some communities hair mercury concentrations exceeded 10 µg/g (Cordier et al., 1998). In their overall sample 12% exceeded 10 µg/g, but in some Amerindian communities up to 79% of the children had hair mercury concentrations > 10 µg/g. These investigators also identified some high exposures to inorganic mercury associated with use of mercury for religious rituals.

SOURCES OF MERCURY EXPOSURE

Mercury in Foods and Products

Chemical species of Mercury

Methylmercury is bound to the amino acids in fish muscle and cannot be removed by food preparation or cooking (Morgan et al., 1997). Estimates of exposure to mercury through diet can be determined by chemical analyses of food (Gunderson, 1995; Larsen et al., 2002; Nakagawa et al.; 1997; Sanzo et al., 2001; Urieta et al., 1996; Ysart et al., 2000). Within the diet consumed by populations who do not include marine mammals in their diets (marine mammals are discussed separately below), fish and shellfish contain the highest mercury concentrations (Larsen et al. 2002; Urieta et al., 1996; Ysert et al., 2000), although trace amounts of total mercury may be detected in other dietary components (e.g., eggs, organ meats such as kidney (Larsen et al., 2002), or offal (Ysart et al., 2000). Diets are usually analyzed for total mercury without chemical speciation to differentiate between organic mercury, specifically methylmercury, and inorganic mercury. Based on total diet data for the United States, in which only total mercury is measured routinely, mercury is identified consistently only in the fish and shellfish components of diet. MacIntosh et al. (1996) indicated that dietary mercury exposures at the upper end of the established distribution for approximately 120,000 adults were dominated by the consumption of fish products (87% of total mercury intake), principally canned tuna (65%), in the Nurses' Health Study and Health Professional Follow-Up Study. The organic mercury in fish and shellfish has been repeatedly speciated and is methylmercury (Bloom, 1992; Falter and Scholer, 1994; Haxton et al., 1979; Kannan et al., 1998).

Mercury present in fish is approximately 85% methylmercury and higher for fish muscle (e.g., Storelli et al.; 2000, 2002). In nonmuscle tissues (e.g., organs such as kidney or liver), the fraction of total mercury that is methylmercury is substantially less than 80% to 90%. The fraction of mercury in shellfish that is methylmercury varies even within a species. For example, Ipolyi et al. (2004) found that the percent methylmercury in mussels varied between 33% and ~ 90%.

Mercury Concentrations in Fish and Shellfish. Mercury concentration in fish is determined by the feeding habits of the fish, the mercury concentration in tissues of its prey, the fish's age, and place in the food chain. Some highly migratory species such as tuna appear to be more consistent in their mercury concentration at a particular age and size than do species that remain in a much more restricted geographic range that can be more readily influenced by local conditions. Reference to some of the larger or more detailed data sets on mercury concentrations in fish and shellfish includes the following for particular countries in the Americas:

USA	Bahnick et al., 1994
	Hall et al., 1978
	Adams and McMichael, 2001
Canada	Jensen et al., 1997 Canadian Arctic Contaminants Assessment Report
	Chan, 1998 Data Base for Environmental Contaminants in Traditional Foods
Brazil	Bidone et al., 1997 Tapajós River Basin, Pará
	Boischio and Henshel, 2000 Madeira River
	Dolbec et al., 2001 Tapajós River Basin, Pará
	Dos Santos et al., 2000 Tapajós River Basin, Pará
	Niercheski et al., 1997 Patos and Mirim Lagoons
	Passos et al., 2003 Tapajós River Basin, Pará

The concentration of methylmercury in fish and shellfish species ranges from < 0.1 ppm for shellfish species to > 1 ppm for high-end predatory fish including ocean fish [such as tuna (Storelli and Marcotrigiano, 2000), marlin (Schultz et al., 1976), and sharks (Penedo de Pinho et ala., 2002)] and freshwater fish [e.g., walleye and northern pikes (Gilmour and Riedel, 2000; Jewett et al., 2003)]. Consequently mercury intake depends on the species of fish consumed, as well as the quantity of fish eaten.

Mercury in Marine Mammals. Most of the traditionally harvested fish and land/marine mammals consumed by the northern Canadian indigenous peoples are long-lived and are from the higher trophic levels of the food chain which contain greater concentrations of methylmercury than are found in nonpredatory fish (Jensen et al., 1997). Several extensive data sets on mercury concentrations in marine mammals consumed by indigenous populations living in the circumpolar regions have been published (Wagemann et al., 1996 and 1998; Caurant et al., 1996; Dietz et al., 1996). Analyses that determined chemically speciated mercury have shown that mercury present in muscle tissue is largely (>75%) organic mercury (i.e., methylmercury) (Caurant et al., 1996). By contrast, mercury present in organs such as liver and kidney is predominantly in an inorganic form (Caurant et al.,1996). Traditional diets of indigenous peoples living in northern communities differ from one another (e.g., Dene Communities in contrast with Inuit Communities) with substantial differences in their methylmercury content (Table 5). Jensen et al. (1997) in the *Canadian Arctic Contaminants Assessment Report* identified wide variability in the consumption of fish and marine mammals by various aboriginal groups.

Table 5. Estimated Daily Intake of Mercury Using Contaminant Data Base and Dietary Information from Dene and Inuit Communities in Canada (adapted from Chan, 1998).

Food Group	Dene Community		Inuit Community	
	Food (g/day)	Mercury (µg/day)	Food (g/day)	Mercury (µg/day)
Marine Mammal Meat	0	0	199	170
Marine Mammal Blubber	0	0	30	2
Terrestrial Mammal Meat	205	6	147	4
Terrestrial Mammal Organs	23	20	1	1
Fish	80	13	1	1
Birds	8	1	2	1
Plants	2	0	2	0
Total	318	40	423	185

Quantity of Fish, Shellfish and Marine Mammals Consumed Dietary intakes of fish and shellfish are enormously variable across countries in the Americas ranging from no intake to several hundred grams per day.

Consumption of marine mammals appears limited to indigenous peoples living in circumpolar regions. Even within the population of one country, fish and shellfish consumption vary substantially. In the United States less than 15% of the population consumes fish or shellfish on a weekly basis, yet a few percent of people consume fish several times a week. In the Amazon region whose riverine population's mercury exposure has been documented, the fish consumption of the urban areas has also been found to be substantial. Giugliano et al. (1978 - *cited in* Boischio and Henshel, 2000) estimated daily fish consumption per capita to be approximately 100 to 150 grams of fish among an urban population based on 1200 households in Manaus, Amazonas. However, the species of fish consumed differed from those of the riverine peoples.

An analysis of the patterns of fish species consumed (herbivores, omnivores, phanktophagus, piscivore) and the mercury concentration of these species has been used to predict mercury exposure for the Madeira River people by Boischio and Henshel (2000).

Table 6. Hair and Blood Mercury Concentrations in Southern Canadians.

Group	Reference	Findings
During 1993 to 1995 10 hospitals in southern Quebec monitored umbilical cord blood samples from 1109 newborns.	Rhainds et al., 1999	Mean mercury in cord blood 4.82 nmol/L (CI 4.56-5.08) or 0.96 ug/L. Using a cord:maternal ratio of 1.7:1.0 and converting to µg/L this corresponds to a maternal mercury concentration of 0.56.
Montreal area sports fishers with fish consumption of 18.3 kg/yr of sports fish vs. reference group consuming 3.3 kg/yr of sports fish.	Kosatsky et al., 1999a	Hair mercury median concentration of 0.73 µg/g for sports fishers (n=25). Highest value, 4.4 ppm. Vs. 0.23 µg/g (n=15). Highest value 0.82 µg/g.
St. Lawrence River fish consumers of Asian origin (Bangladeshi and Vietnamese) and a reference group of high-level sports fish consumers.	Kosatsky et al., 1999b	Hair mercury concentrations for Vietnamese males (n=9). Median 1.23 µg/g, 90th percentile 4.62 µg/g, and Bangladeshi males, median 1.07 µg/g, 90th percentile 2.29 µg/g, and high-end sports fish consumers, median 0.73 µg/g, 90th percentile 1.88 µg/g.
St. Lawrence River sportsfish consumers comparison between those who ate sports fish at least once a week and those eating sports fish less than once weekly.	Kosatsky et al., 2000	Persons consuming sports fish at least once a week (n=60). Hair mercury geometric mean, 0.82±2.54 µg/g and blood mercury geometric mean of 3.03 ±2.43 µg/L. Consuming sportsfish < once/week (n=72) hair mercury geometric mean 0.38±2.38 µg/g and blood Hg of 1.44±2.23 µg/L. Persons consuming sports fish 2 or more times/week
Population in Southwest Quebec living near the Upper St. Lawrence River System..	Mahaffey and Mergler, 1998	Increased consumption of lake fish associated with higher blood mercury among 289 residents. Total blood Hg < 5 µg/L except for one individual with a blood Hg of 70 µg/L.
Pregnant women living along the St. Lawrence River. 1st trimester of pregnancy. Market fish were the source of mercury exposure.	Morrissette et al., 2004.	First trimester of pregnancy (n=39). Total blood Hg. Geometric mean (GM): 0.85 µg/L. Arithmetic mean (AM)= 0.99 µg/L. 5th - 95th, 0.40-2.20 µg/L. Organic blood Hg. GM: 0.36 µg/L. AM: 0.48 µg/L 5th - 95th: n.d. - 1.20 µg/L Inorganic blood Hg GM: 0.45 µg/L AM: 0.51 µg/L 5th - 95th: n.d. - 1.80

Among the fish species described mean mercury concentrations ranged from 0.06 ppm to 1.44 ppm. In most of the Amazonian studies, piscivorous fish species have mercury concentrations > 0.5 ppm (Lodenius and Malm, 1998).

Other estimates of fish intake in Amazonia provided by Boischio and Hensel (2000) ranged between 90 grams/day and 370 grams/day. In view of the wide range in quantities of fish consumed and differences in the mercury concentration in the fish, it is more informative to consider the data on hair and blood mercury concentrations reported from Amazonian communities.

Cosmetics

Use of "beauty" and skin-lightening creams or lotions have been reported worldwide [e.g., Sin and Tsang (2003) from Hong Kong; Al-Saleh et al. (2003) from Saudi Arabia; Adebajo 2002 from Nigeria]. These include various "beauty" lotions containing inorganic mercury which are also found in the Americas. One product has been well described and found to contain between 6% and 10% mercury by weight (Balluz et al., 1997), was distributed across the Mexican-US border, and was associated with an increase in urinary excretion (Weldon et al., 2000) of inorganic mercury to levels > 100 µg/L (reference range: 0 to 20 µg/L) among users of this product which contained "calomel" or mercurous chloride [Hg_2Cl_2] (CDC, 1996). These products are commonly available. For example the Mexican Secretary of Health seized 35,000 containers of this product in one state in Mexico (CDC, 1996).

In the United States mercury compounds can only be used as preservatives in eye-area cosmetics at concentrations not exceeding 65 ppm mercury (CDC 1996). Balluz et al. (1997) point out that standards for production and regulation of cosmetic products vary worldwide. Ingredients that are restricted in one country may be entirely legal in another.

Dental Amalgams

An alloy of silver, copper, tin and 50% inorganic mercury has been used in dental practices as a restorative material to fill teeth. Mercury released from these amalgam fillings occurs in multiple forms: elemental mercury vapor, metallic ions, and/or fine particles. Results indicate that placement of mercury-containing amalgams in teeth result in an increased body burden of mercury in body tissues (US DHHA, 1993; Henderson, 1995; Weiner and Nylander, 1993). Barregard et al. (1995) and Francis et al., (1982) indicate that individual variation in habits can influence the amount of mercury released from mercury amalgams including bruxism and gum chewing.

The controversy regarding whether or not mercury from silver-mercury amalgams produces adverse health effects has been a topic for more than 150 years and has been the subject of a substantial literature including more than 110 review articles based on a search of Pub Med. Among many others see Ekstrand et al. (1998), Dunne et al. (1997), Eley (1997a and b), and Jones et al. (1999). Currently a clinical trial is underway (Children's Amalgam Trial Study Group, 2003) to investigate the effect of amalgam restorations on change in IQ scores and other neuropsychological assessments and renal functioning.

There is a consensus, however, that dental workers need to be protected from exposure to mercury vapors in dental clinics (Chang et al., 1992). Dental associations have provided practice information to provide guidance to dentists and their staff members for safe handling of mercury and dental amalgam (ADA Council on Scientific Affairs, 2003). Guidance is also provided to reduce release of mercury from dental offices into the environment.

Drugs, Biologicals and Folk Remedies

Drugs and Biologicals

Ethylmercury under the trade name thimerosal has been used as a preservative in vaccines since the 1930s (Midthun, 2004) and biological products (ophthalmic solutions, optic suspension, creams) for at least a century. Many vaccines contained a preservative because they were marketed for use in multi-dose vials. Mercury exposures via this route of administration differ because these are administered by injection resulting in a high, bolus dose producing a high blood mercury concentration. Beginning in the mid-1990s in the United States there has been an intense controversy regarding the neurodevelopmental effects of greater exposure to ethylmercury in vaccines as part of an accelerated immunization schedule for infants. This controversy continues. The most recent health assessment was reported from the United States National Academy of Sciences' Institute of Medicine (NAS, 2004). A neurodevelopmental follow-up study is underway in Italy with resulted expected in 2006 (Wharton, 2004).

Thimerosal has been removed from a large number of vaccines (CDC, 1999). It should be noted, however, that mercury continues to be used in biologicals sold in the United States and quite likely also in other countries in the Americas. Mercury containing ingredients currently used in biologics in the United States include thimerosal, phenylmercuric acetate, phenylmercuric nitrate, mercuric acetate, mercuric nitrate, merbromin, and mercuric oxide (http://www.fda.gov/

cder/ fdama/mercury300.htm). This information was derived from submissions made in response to the Food and Drug Administration Modernization Act of 1997 which required US FDA to review the risks of all mercury-containing food and drugs. These products include vaccines, ophthalmic solutions, nasal sprays, and immune globulins.

Sanitizers and Antifungals

Phenylmercury compounds are part of a broad class of aryl and alkoxyl aryl mercurials that have been used world wide as disinfectants and fungicides (Gotelli et al., 1985). A massive exposure to infants was discovered in Buenos Aires in 1980 following use of phenylmercury by a commercial diaper service. Urinary mercury concentrations for the exposed infants were 20-times higher than the control infants and inorganic mercury accounted for more than 90 % of the total mercury (Gotelli et al., 1985). Zaidi et al, (1995) reported mercury compounds included as disinfectants in hospitals in the state of Yucatan , Mexico. Use of phenylmercuric borate (0.04%) along with hexachlorophene as a hand disinfectant in soap resulted in increased urinary mercury compared with control subjects. It was concluded that the absorption of mercury from the phenylmercuric borate-containing soap occurred partly by transfer from the hands to the oral cavity (Peters-Haefeli et al., 1976).

Organo-mercurial Pesticides

Although these are banned from commerce, use of mercury compounds in farming may continue (Camara and Corey, 1994). For example, the Ministry of Agriculture/Plant Health Protection banned the use of all mercury-containing compounds in Brazil in 1980, however, the importance has been emphasized for continuance of programs for surveillance, particularly for problems arising as a result of accidents or from the importation, production, sale, and uses of officially banned products (Camara and Corey, 1994).

Miscellaneous sources of inorganic mercury

Ingestion of small quantities of elemental mercury is commonly reported to poison control centers. For example, nearly 3,000 cases of mercury exposure were reported to the American Association of Poison Control Centers Toxic Exposure Surveillance System in 1996 (Litovitz et al., 1996). However, elemental mercury may be deliberately ingested. Elemental mercury is part of various folk remedies, particularly for gastroenteritis (Geffner and Sandler, 1980), and may produce medial complications depending on the quantity ingested (McKinney, 1999).

In the Americas ethnic and folk uses of mercury are associated with cultural practices known as "Santeria," "Espiritismo" and "voodoo." Santeria, for example, is a religion that developed in Cuba from the sixteenth to the nineteenth century as a syncretism of African religions, Roman Catholicism, and French spiritism (Lefever, 1996). It's practice and beliefs have been transferred to many communities in the Americas including the United States. Espisitismo is a spiritual belief system indigenous to Puerto Rico and other Caribbean islands (Bird and Canino, 1981). Mercury in the form of metallic mercury (Forman et al., 2000) is sold in botanicas, stores that specialize in selling religious items used in Espiritismo, voodoo, and Santeria (Zagas and Ozuah, 1996). Mercury is used in several ways: carried in sealed pouches, sprinkled in the home, carried in pockets, sprinkled in the car, or consumed in small quantities (Zayas and Ozuah, 1996). A cluster of cases in New York has been described in which vapors of metallic mercury from mercury to be used in mercury-filled ampulets prepared for practioners of Santeria were the source of elevated urinary mercury levels (US EPA, 2002; Forman et al., 2000). The extent of this practice is not known. In the United States analyses for inorganic mercury is now being included in housing surveys in addition to inorganic lead used to identify lead-based paints.

Spills of inorganic mercury occur in homes from accidents, measuring devices in homes (e.g., gas meters), and deliberate contamination. The actual number is unknown. An analysis of the number that were reported to hazardous substance authorities was provided by the Agency for Hazardous Substances and Disease Registry (ATSDR) in 2002 (Zeitz et al., 2002). Reporting on 406 events that occurred between 1993 and 1998 in which mercury was the only substance released, there were reports of adverse health effects and elevated blood mercury levels in some individuals. Evacuations were ordered in 22% of the spills causing an economic and public health impact.

Occupational

Mining

In the Americas Artisanal gold mining typically encompasses small, medium, informal, legal, and illegal miners who use rudimentary processes to extract gold ore (Veiga, 2004). The number of Artisanal gold miners in Latin America is estimated to be between ~ 500,000 and just over 1 million who mine between 115 and 188 tonnes of gold annually Table 7 (Veiga, 1997).

Although use of mercury in mineral processing is illegal in most countries, mercury amalgamation is the preferred method employed by small Artisanal miners (Veiga, 2004). Mining operations expose miners directly to mercury vapor from burning of mercury amalgam or gold burning to separate gold from mercury. Secondary exposures to methylmercury from consumption of locally obtained fish contaminated with methylmercury incorporated from mercury discharged into the environment effects miners and the entire community. Consequently the number of people impacted by the mining operations exceeds the number directly involved in mining.

Table 7. Artisanal Gold Miners in Latin America from Veiga (1997).

Country	Gold (tonnes)	Number of Miners
Brazil	30 - 50	200,000 - 400,000
Colombia	20 - 30	100,000 - 200,000
Peru	20 - 30	100,000 - 200,000
Ecuador	10 - 20	50,000 - 80,000
Venezuela	10 - 15	30,000 - 40,000
Suriname	5 - 10	15,000 - 30,000
Bolivia	5 - 7	10,000 - 20,000
Mexico	4 - 5	10,000 - 15,000
Chile	3 - 5	6,000 - 10,000
French Guyana	2 - 4	5,000 - 10,000
Guyana	3 - 4	6,000 - 10,000
Nicaragua	1 - 2	3,000 - 6,000
Dominican Republic	0.5 - 1	2,000 - 3,000
Others	2 - 5	6,000 - 15,000
Total	115.5 - 188	543,000 - 1,039,000

Inorganic mercury is not strongly absorbed through the skin and is minimally absorbed from the gastrointestinal tract. The exposure danger is through inhalation of mercury vapor as the gold-mercury amalgam is retorted or simply burnt openly in pans. Specifics on the methods of retorting and reducing exposures are covered by Veiga (2004).

The gold doré that results when the amalgam is retorted rather than burnt is sold in villages to gold shops that melt the gold doré to rid it of its impurities. This doré contains mercury impurities that are then released into gold shops and distribute mercury vapors into the urban atmosphere (Veiga, 2004).

Chloralkali Factories

The chlor-alkali electrolysis process produces chlorine, hydrogen and sodium hydroxide (caustic) solution. In some countries chlorine is the product of commercial importance and is used as a bleaching agent for the textile and paper industries and for general cleaning and disinfection. Since the 1950s, chlorine has become an increasingly important raw material for synthetic organic chemistry and is an essential component of a multitude of end products. Plants utilize either the diaphragm cell process or mercury cell process. The environmental concern with these plants is release of mercury emissions from the plants, so called "fugitive sources" (Lecloux, 2003; Wangberg et al., 2003; http://www.epa.gov/ttn/oarpg/) and in particular identification of the atmospheric mercury species/fractions measured in the plume from the chlor-alkali plant emissions.

Smith et al. (1970) conducted an extensive assessment of 642 workers in 21 chloralkali plants in the United States and Canada. Time weighted average exposures to mercury ranged from near zero to 270 $\mu g/m^3$ with 85% of the group ≤ 100 $\mu g/m^3$. When exposures exceeded 100 $\mu g/m^3$ symptoms included loss of appetite and weight, tremors, insomnia, and other indicators of early effects on the nervous system. Even among controls urinary mercury concentrations were much higher than general population values reported in current values (i.e.,United States, non-occupational, post 2000). Approximately 65% of the control group had urinary mercury concentrations > 10 $\mu g/L$. Worker values exceeded 1,000 $\mu g/L$ for 7% to 8% of subjects.

Reports from other countries indicate neurological and renal effects of exposure to inorganic mercury and mercury vapor. Specifically cerebral effects of long-term exposures to mercury vapor (Pikivi and Tolonen, 1989) are shown by abnormalities in electroencephalography. Increased urinary excretion of renal enzymes has been found in chlor-alkali workers exposed to inorganic mercury compared with age-matched controls when urinary mercury values averaged 35 $\mu g/g$ creatinine (Barregard et al., 1988). Albers et al. (1982) have described plyneuropathy among chlor-alkali plant workers chronically exposed to inorganic mercury vapor. They concluded that elemental mercury exposure was associated with a sensorimotor polyneuropathy of the axonal type and that the degree of neurologic impairment appeared related to the magnitude of mercury exposure.

A number of chlor-alkali factories have been closed. For example, in the United States a 1992 report indicated 52 chlor-alkali plants in operation in 23 states (US EPA, 1992).

In the announcement of the final rule to reduce toxic air pollutants from mercury cell chlor-alkali plants, there were nine facilities located in eight states (

http://www.epa.gov/ttn/oarpg/t13/fact_sheets/mccap_fs.pdf). However, closed chlor-alkali plants can continue to release inorganic mercury into the environment over time. An environmental assessment in the Upper Negro River valley area in Argentina in the vicinity of a closed chlor-alkali plant (Arribére et al., 2003) indicated, however, that because the drainage areas near the plant had not become permanently flooded biomagnitication of mercury in the food chain was not occurring.

Other Industries

Industries in which occupational exposure to mercury may occur include chemical and drug synthesis, hospitals, laboratories, instrument manufacture, and battery manufacture (USA, National Institute for Occupational Safety and Health, 1977). Jobs and processes involving mercury exposure include manufacture of measuring instruments (barometers, thermometers, etc.), mercury arc lamps, mercury switches, fluorescent lamps, mercury boilers, mirrors, electric rectifiers, paints, explosives, photographs, disinfectants, and fur processing. Occupational mercury exposure can also result from the synthesis and use of metallic mercury, mercury salts, mercury catalysts (in making urethane and epoxy resins) mercury fulminate, Millon's reagent, pharmaceuticals, and antimicrobial agents [USA (Occupational Safety and Health Administration (OSHA, 1989)]. Within the USA Campbell et al. 1992 reported that about 70,000 workers are annually exposed to mercury. Inorganic mercury accounts for nearly all occupational exposures with airborne elemental mercury vapor the main pathway of concern in most industries, in particular those with the greatest number of mercury exposures.

SUMMARY

Mercury exposures in the Americas are highly varied ranging from analytically non-detectable to very high levels producing clinical poisoning. Methylmercury is the dominant chemical form of mercury for people who consume fish and shellfish. Who within the population consumes a high percentage of their dietary protein from fish/shellfish differs with social, economic, cultural and geographic conditions. The highest exposures to methylmercury in the Americas have been found among affluent, urban dwellers who consumed fish in pursuit of health (Hightower and Moore, 2003), among remote villagers in the Amazon who have an environment in which the fish are seriously contaminated with mercury (Dolbec et al., 2001), and among Inuit villagers who consume marine mammals (Chan, 1998; van Oostdam et al., 1999) as part of their cultural traditions.

Among the general population of countries ranging from Canada to the USA to Chile exposure to methylmercury depends on both the quantity and species of fish consumed. General population survey data are available only for the United States and these data indicate that for the overall general population, 9% of women of childbearing age consume fish weekly (Mahaffey et al., 2004). Within the general population, approximately 3% of women consume fish daily. Groups at greater than average risk of methylmercury exposure from routine consumption of fish are persons of Asian or island ethnicity, coastal and/or island populations, individuals following "health" promoting diets, and some indigenous tribal groups. Their higher methylmercury exposure results from the greater overall quantities of fish consumed and/or the consumption of fish with relatively high mercury concentrations.

Within regions such as the Amazon having wide-spread mercury contamination secondary to release of inorganic mercury from mining operations, elevated mercury exposures with hair mercury concentrations frequently exceed 10 ppm hair mercury. In these regions choice of fish species to minimize mercury intake has been shown (Dorea et al., 2003; Passos et al., 2003) to be a possible means of reducing methylmercury exposures in a region where fish is the main routinely available source of protein.

Non-occupational inorganic mercury exposures, for the portion of the population who experience these, occur from mercury released from dental amalgams. Occupational exposures to inorganic mercury include persons working in dental professions, the chlor-alkali industry, and fluorescent bulb production and recycling. Statistics on the number of workers employed in the Americas in occupations that utilize mercury need to be summarized.

Far higher exposures to mercury vapor occur among artisanal miners, their families, and in communities associated with gold mining and gold ore processing. These exposures have been reported to be of a magnitude that produces overt clinical toxicity within a few months to a few years. Because of contamination of the surrounding ecosystems with mercury-laden waste from these mining operations, vast geographic areas surrounding mining activities have been grossly contaminated with mercury. Elevated mercury levels in the food chain result. Because people engaged in mining also may consume fish in their diets, a combined methylmercury and inorganic mercury pattern of exposure has been documented in Latin American countries.

Against this substantial background of methylmercury and inorganic/mercury vapor exposures, there are far less common exposures to organo-mercurials including phenylmercury and ethylmercury. Such compounds are found in "beauty" preparations, creams, topical preparations to treat skin conditions.

These compounds have produced poisoning among the unwary.

Biomonitoring of mercury concentrations in blood, hair, and urine are used to identify the magnitude and patterns of mercury exposure. Combined with dietary, occupational, cultural, and socio-demographic data such information can be used to identify sources of mercury exposure for individuals and populations. Such information can provide governments, public health authorities, medical organizations, physicians, and individuals with the links to stop or reduce exposures to mercury and mercurial compounds.

DISCLAIMER

The statements in this publication are the professional views and opinions of the author and should not be interpreted to be the policies of the United States Environmental Protection Agency.

REFERENCES

ADA Council on Scientific Affairs. Dental mercury hygiene recommendations. *J. Am. Dent. Assoc.*, 134, 1498-9, 2003.

Adams, D.H., McMichael, R.H. Mercury levels in marine and estaurine fishes of Florida. Florida Marine Research Institute Technical Report. TR-6. 35 pp. 2001.

Adebajo, S.B. An epidemiological survey of the use of cosmetic skin lightening cosmetics among traders in Lagos, Nigeria. *West Afr. J. Med.*, 21, 51-5, 2002.

Akagi, H., Malm, O., Kinjo, Y., Harada, M., Branches, F.J.P., Pfeiffer W.C., Kata, H. Methylmercury pollution in the Amazon, Brazil. *Sci. Tot. Env.*, 175, 85-96, 1995.

Albers, J.W., Cavender, G.D., Levine, S.P., Langolf, G.D. Asymptomatic sensorimotor polyneuropathy in workers exposed to elemental mercury. *Neurology*, 32, 1168-74, 1982.

Al-Saleh, I., Khogali, F., Al-Amodi, M., El-Doush, I., Shinwari, N., Al-Baradei, R., Histopathological effects of mercury in skin-lightening cream. *J. Environ. Pathol. Toxicol. Oncol.*, 22, 287-99, 2003.

Al-Shahristani, H., Shihab, K.M. Variation of biological half-life of methylmercury in man. *Arch. Environ. Health*, 28, 342-344, 1974.

Arribere, M.A., Ribeiro Guevara, S., Sanchez , R.S., Gil, M.I., Roman Ross, G., Daurade, L.E., Fajon, V., Horvat, M., Al Calde, R., Kestelman, A.J. Heavy metals in the vicinity of a chlor-alkali factory in the Upper Negro River ecosystem, Northern Patagonia, Argentina. *Sci. Tot. Env.*, 301, 187-203, 2003.

Bahnick, D., Sauer, C., Butterworth, B., Kuehl, D. A national study of mercury contamination of fish. *Chemosphere*, 29, 537-46, 1994.

Balluz, L.S., Philen, R.M., Sewell, C.M., Voorhees, R.E., Falter, K.H., Paschal, D. Mercury toxicity associated with a beauty lotion, New Mexico. *Intern. J. Epidemiol.*, 26, 1131-1132, 1997.

Barbosa, A.C., Jardim, W., Dorea, J.G., Fosberg, B., Souza, J. Hair mercury speciation as a function of gender, age, and body mass index in inhabitants of the Negro River basin, Amazon, Brazil. *Arch. Environ. Contam. Toxicol.*, 40, 439-44, 2001.

Barbosa, A.C., Silva, S.R., Dorea, J.G. Concentration of mercury in hair of indigenous mothers and infants from the Amazon basin. *Arch. Environ. Contam. Toxicol.*, 34, 100-105, 1998.

Barregård, L., Hultberg, B., Schutz, A., Salsten, G. Enzymuria in workers exposed to inorganic mercury. *Int. Arch. Occup. Environ. Health*, 61, 65-9, 1988.

Barregård, L., Sallsten, G., Jarholm, B. People with high mercury uptake from their own dental amalgam fillings. *Occup. Environ. Med.*, 52, 124-28, 1995.

Barregård, L., Sallsten, G., Schutz, A., Attewell, R., Skerfving, S., Jarvholm, B. Kinetics of mercury in blood and urine after brief occupational exposure. Arch. Environ. Health, 47, 176-84, 1992.

Barrios-Guerra, C.A. Mercury contamination in Chile: a chronicle of a problem foretold. *Rev. Environ. Contam. Toxicol.*, 183, 1-19, 2004.

Bellanger, T.M., Caesar, E.M., Trachtman, L. Blood mercury levels and fish consumption in Louisiana. *J. La State Med. Soc.*, 152, 64-73, 2000.

Bidone, E.D., Castilhos, Z.C., Cid de Souza, T.M., Lacerda, L.D. Fish contamination and human exposure to mercury in the Tapajos River Basin, Para State, Amazon, Brazil: a screening approach. *Bull. Environ. Contam. Toxicol.*, 59, 194-201, 1997.

Bird, H.R., Canino, I. The sociopsychiatry of espiritismo: findings of a study in psychiatric populations of Puerto Rican and other Hispanic children. *J. Am. Acad. Child Psychiatry*, 20, 725-40, 1981.

Birke, G., Johnels, A.G., Plantin, L.O., Sjostrand, B., Skerfving, S., Westermark, T. Studies on humans exposed to methyl mercury through fish consumption. *Arch. Environ. Health*, 25, 77-91, 1972.

Bloom, N.S. On the chemical form of mercury in edible fish and marine invertebrate tissue. *Can. J. Fish. Aquat. Sci.*, 49, 1010-1017, 1992.

Boischio, A.A., Cernichiari, E. Longitudinal hair mercury concentration in riverside mothers along the upper Madeira river (Brazil). *Environ. Res.*, 77, 79-83, 1998.

Boischio, A.A., Henshel, D.S. Linear regression models of methyl mercury exposure during prenatal and early postnatal life among riverside people along the upper Madeira river, Amazon. *Environ. Res.*, 83, 150-61, 2000.

Bruhn, C.G., Rodriguez, A.A., Barrios, C., Jaramillo, V.H., Becera, J., Gonzales, U., Gras, N.T., Reyes, O., Seremi-Salud. Determination of total mercury in scalp hair of pregnant and nursing women resident in fishing villages in the Eighth Region of Chile. *J. Trace Elem. Electrolytes Health Diss.*, 8, 79-86, 1994.

Burbacher, T., Shen, D., Clarkson, T. Mercury in Macaque infants following oral ingestion of methylmercury or IM injection of vaccines containing thimerosal. Presented in: "Mercury: Medical and Public Health Issues" Tampa, Florida, USA, 2004.

Burge, P., Evans, S. Mercury contamination in Arkansas game fish. A public health perspective. *J. Ark. Med. Soc.*, 90, 542-4, 1994.

Camara, V.M., Corey, G. Epidemiologic surveillance for substances banned from use in agriculture. *Bull. Pan. Am. Health. Organ.*, 28, 355-9, 1994.

Campbell, D., Gonzales, M., Sullivan, J.B. "Mercury" In: *Hazardous Materials Toxicology. Clinical Principles of Environmental Health*. JB Sullivan and G Rigger, eds. Baltimore, MD, Williams and Wilkins, 1992.

Carta, P., Flore, C., Alinovi, R., Ibba, A., Tocco, M.G., Aru, G., Carta, R., Girei, I., Mutti, A., Lucchini, R., Randaccio, F.S. Sub-clinical neurobehavioral abnormalities associated with low level of mercury exposure through fish consumption. *Neurotoxicology*, 24, 617-623, 2003.

Caurant, F., Navarro, M., Amiard, J-C. Mercury in pilot whales: possible limits to the detoxicifaction process. *Sci Tot. Env.*, 186, 95-104, 1996.

CDC. Centers for Disease Control. Thimerosal in vaccines: a joint statement of the American Academy of Pediatrics and the Public Health Service. *Morb. Mortal Wkly. Rep.*, 48, 563-5, 1999.

CDC. Centers for Disease Control. Summary of the joint statement on thimerosal in vaccines. American Academy of Family Physicians, American Academy of Pediatrics, Advisory Committee on Immunization Practices, Public Health Service. *Morb. Mortal Wkly. Rep.*, 49, 622-631, 2000.

CDC. Centers for Disease Control. Mercury poisoning associated with beauty cream – Texas, New Mexico, and California, 1995-1996. *Morb. Mortal Wkly. Rep.*, 45, 400, 1996.

Centers for Disease Control and Prevention. Second National Report on Human Exposure to Environmental Chemicals. NCEH Pub. No. 03-0022. January 2003. <222.cdc.gov/exposurereport/2nd/pdf/secondner.pdf>

Chan, H.M., A database for environmental contaminants in traditional foods in northern and Arctic Canada: development and applications. *Food Addit. Contam.*, 15, 127-34, 1998.

Chang, S.R., Siew, C., Gruninger, S.E. Factors affecting blood mercury concentrations in practicing dentists. *J. Dent. Res.*, 71, 66-74, 1992.

Cherian, M.G., Hursh, J.R., Clarkson, T.W., Allen, J. Radioactive mercury distribution in biological fluids and excretion in human subjects after inhalation of mercury vapor. *Arch. Environ. Health*, 33, 109-114, 1978.

Children's Amalgam Trial Study Group. The Children's Amalgam Trial: design and methods. *Control Clin. Trials*, 24, 795-814, 2003.

Cordier S., Grasmick C., Paquier-Passelaigue M., Mandereau L., Weber J., Jouan M. Mercury exposure in French Guiana: levels and determinants. *Arch. Environ. Health*, 53, 299-303, 1998.

Dallaire, F., Dewailly, E., Muckle, G., Ayotte, P. Time trends of persistent organic pollutants and heavy metals in umbilical cord blood of Inuit infants born in Nunavik (Quebec, Canada) between 1994 and 2001. Environ. Health Perspect., 111, 1660-4, 2003.

de Kom, J.F., van der Voet, G.B., de Wolff, F.A. Mercury exposure of Maroon workers in the small scale gold mining in Suriname. *Environ. Res.*, 77, 91-97, 1998.

Dietz, R., Riget, F., Johansen, P. Lead, cadmium, mercury, and selenium in Greenland marine animals. *Sci Tot. Env.*, 186, 67-93, 1996.

Dolbec, J., Mergler, D., Larribe, F., Roulet, M., Lebel, J., Lucotte, M. Sequential analysis of hair mercury levels in relation to fish diet of an Amazonian population, Brazil. *Sci. Tot. Env.*, 23, 87-97, 2001.

Dolbec, J., Mergler, D., Sousa Passos, C-J., Sousa de Morais, S., Lebel, J. Methylmercury exposure affects motor performance of a riverine population of the Tapajós river, Brazilian Amazon. *Int. Arch. Occup. Environ. Health*, 73, 195-203, 2000.

Dorea, J., Barbosa, A.C., Ferrari, I., de Souza, J.R. Mercury in hair and in fish consumed by Riparian women of the Rio Negro, Amazon, Brazil. *Int. J. Environ. Health Res.*, 13, 239-48, 2003.

Dos Santos, L.S., Muller, R.C., de Sarkis, J.E., Alves, C.N., Brabo, E.S., Santos, E.C., Bentes, M.H. Evaluation of total mercury concentrations in fish consumed in the municipality of Itaituba, Tapajos River Basin, Para, Brazil. *Sci Tot. Env.*, 261, 1-8, 2000.

Dumont, C., Girard, M., Bellavance, F., Noel, F. Mercury levels in the Cree population of James Bay, Quebec From 1988 to 1993/94. *CMAJ*, 158, 439-45, 1998.

Dunne, S.M., Gainsford, I.D., Wilson, N.H. Current materials and techniques for direct restorations in posterior teeth. Part 1. Silver amalgam. *Int. Dent. J.*, 47, 123-36, 1997.

Eisler, R. Mercury hazards from gold mining to humans, plants, and animals. *Rev. Environ. Contam. Toxicol.*, 181, 139-98, 2004.

Ekstrand, J., Bjorkman, L., Edlund, C., Sandborgh-Englund, G. Toxicological aspects on the release and systemic uptake of mercury from dental amalgam. *Eur. J. Oral. Sci.*, 106, 678-86, 1998.

Eley, B.M. The future of dental amalgam: a review of the literature. Part 4: Mercury exposure hazards and risk assessment. *Br. Dent. J.*, 182, 373-81, 1997a.

Eley B.M. The future of dental amalgam: a review of the literature. Part 6: Possible harmful effects of mercury from dental amalgam. *Br. Dent. J.*, 182, 455-9, 1997b.

Falter, R., Scholer, H.F. Determination of methyl-, ethyl-, phenyl and total mercury in Neckar River fish. *Chemosphere*, 29, 1333-38, 1994.

Forman, J., Moline, J., Cernichiari, E., Sayegh, S., Torres, J.C., Landrigan, M.M., Hudson, J., Adel, H.N., Landrigan, P.J. A cluster of pediatric metallic mercury exposure cases treated with meso-2,3-dimercaptosuccinic acid (DMSA). *Environ. Health Perspect.*, 108, 575-7, 1994.

Francis, P.C., Birge, W.J., Roberts, B.L., Black, J.A. Mercury content of human hair: a survey of dental personnel. *J. Toxicol. Environ. Health*, 10, 667-72, 1982.

Geffner, M.E., Sandler, A. Oral metallic mercury. A folk medicine remedy for gastroenteritis. *Clin. Pediatr. (Phila).*, 19, 435-37, 1980.

Gill, U.S., Schwartz, H.M., Bigras, L. Results of multiyear international interlaboratory comparison program for mercury in human hair. *Arch. Environ. Contam. Toxicol.*, 43, 466-72, 2002.

Gilmour, C.C., Riedel, G.S. A survey of size-specific mercury concentrations in game fish from Maryland fresh and estaurine waters. *Arch. Environ. Contam. Toxicol.*, 39, 53-59,

2000.

Gotelli, C.A., Astolfi, E., Cox, C., Cernichiari, E., Clarkson, T.W. Early biochemical effects of an organic mercury fungicide of infants: "dose makes the poison". *Science*, 227, 638-40, 1985.

Goyer, R.A., Clarkson, T.A. Toxic Effects of Metals pgs. 811 - 867. In: Casarett & Doull's Toxicology The Basic Science of Poisons 6th Edition. Klaassen C.D. (Ed). McGraw-Hill Pubs. New York, 2001.

Gunderson, E.L. FDA Total Diet Study. July 1986-April 1991. Dietary intakes of pesticides, selected elements, and other chemicals. *JAOAC Int.*, 78, 1353-1363, 1995.

Hacon, S., Yokoo, E., Valenta, J., Campos, R.C., da Silva, V.A., de Menezes, A.C., de Moraes, L.O., Ignotti, E. Exposure to mercury in pregnant women from Alta Floresta-Amazon basin, Brazil. *Environ. Res.*, 84, 204-10, 2000.

Hall, R.A., Zook, E.G., Meaburn, G.M. National Marine Fisheries Survey of Trace Elements in the Fishery Resource. NOAA Technical Report NMFS SSRF-721. Washington, D.C:US Department of Commerce. 1998.

Harada, M., Nakanishi, J., Yasoda, E., Pinheiro, M.C., Oikawa, T., De Assis Guimaraes, G., da Silva Cardoso, B., Kizaki, T., Ohno, H. Mercury pollution in the Tapajos River basin, Amazon: mercury level of head hair and health effects. *Environ. Int.*, 27, 285-90, 2001.

Haxton, J., Lindsay, D.G., Hislop, J.S., Salmon, L., Dixon, E.J., Evans, W.H., Reid, J.R., Hewitt, C.J., Jeffries, D.F. Duplicate diet study on fishing communities in the United Kingdom: mercury exposure in a "critical group". *Environ. Res.*, 18, 351-368, 1979.

Hempel, M., Chau, Y.K., Dutka, B.J., McInnis, R., Kwan, K.K., Liu, D. Toxicity of organomercury compounds: bioassay results as a basis for risk assessment. *Analyst*, 120, 721-4, 1995.

Henderson B. Dental amalgam: scientific consensus and CDA policy. *J. Can. Dent. Assoc.*, 61, 429-31, 1995.

Hightower, J.M., Moore, D. Mercury levels in high-end consumers of fish. *Environ. Health Perspect.*, 111, 604-608, 2003.

Hightower, J.M. " Mercury and clinical practice". Presented in: Mercury: Medical and Public Health Issues. April 28-30, 2004. Tampa, Florida, USA.

Ipolyi, I., Massanisso, P., Sposato, S., Fodor, P., Morabito, R. Concentration of levels of total and methylmercury in muscle samples collected along the coasts of Sardinia Island (Italy). *Analytica Chim. Acta*,505, 145-51, 2004.

Iyengar, V., Woittiez, J. Trace elements in human clinical specimens: evaluation of literature data to identify reference values. Clin Chem 1988; 34:474-81.

JECFA (Joint FAO/WHO Expert Committee on Food Additives). Sixty-first meeting. Rome. 10-19. June 2003. Summary and conclusions. Available from: URL:ftp://ftp/fao. org/es/esn/jecfa/jecfa61sc.pdf. Accessed June 8, 2004.

Jensen, J., Adare, K., Shearer, R. Canadian Arctic Contaminants Assessment Report. Department of Indian Affairs and Northern Development. 10 Wellington. Ottawa. Ontario K1A OH4. Canada 1997.

Jewett, S.C., Zhang, X., Naidu, A.S., Kelley, J.J., Dasher, D., Duffy, L.K. Comparison of mercury and methylmercury in northern pike and arctic grayling from western Alaska rivers. *Chemosphere*, 50, 383-92, 2003.

Johnsson, C., Sällsten, G., Schütz, A., Sjörs, A., Barregård, L. Hair mercury levels versus freshwater fish consumption in household members of Swedish angling societies. *Environ. Res.*, 96, 257-63, 2004.

Jones, D.W. Exposure or absorption and the crucial question of limits for mercury. *J. Can. Dent. Assoc.*, 65, 42-46, 1999.

Kales, S.N., Goldman, R.H. Mercury exposure: current concepts, controversies, and a clinic's experience. *J. Occup. Environ. Med.*, 44, 143-54, 1999.

Kannan, K., Smith, R.G.Jr, Lee, R.F., Windon, H.L., Heitmuller, P.T., Macauley, J.M. Distribution of total mercury and methyl mercury in water, sediment, and fish from south Florida estuaries. *Arch. Environ. Contam. Toxicol.*, 34, 109-18, 1998.

Kershaw, T.G., Dhahir, P.H., Clarkson, T.W. The relationship between blood levels and dose of methylmercury in man. Arch. Environ. Health., 35, 28-36, 1980.

Kingman, A., Albertini, T., Brown, L.J. Mercury concentrations in urine and whole blood associated with amalgam exposure in a U S military population. *J. Dent. Res.*, 77, 461-71, 1980.

Kosatsky, T., Przybysz, R., Armstrong, B. Mercury exposure in Montrealers who eat St. Lawrence River sportfish. *Environ. Res.*, 84, 36-43, 2000.

Kosatsky, T., Przybysz, R., Shatenstein, B., Weber ,J-P., Armstrong, B. Fish consumption and contaminant exposure among Montreal-area sport fishers: pilot study. *Environ. Res.*, 80, S150-S158, 1999a.

Kosatsky, T., Przybysz, R., Shatenstein, B., Weber ,J-P., Armstrong, B. Contaminant exposure in Montrealers of Asian origin fishing the St. Lawrence River: Exploratory assessment. *Environ. Res.*, 80, S159-S165, 1999b.

Larsen, E.H., Andersen, N.L., Moller, A., Petersen, A., Mortensen, G.K., Petersen J. Monitoring the content and intake of trace elements from food in Denmark. *Food. Addit. Contam.*, 19, 33-46, 2002.

Lauwerys, R., Roels, H., Buchet, J.P., Bernard, A. Non-job related increased urinary excretion of mercury. *Int. Arch. Occup. Environ. Health*, 39, 33-6, 1977.

Lebel, J., Mergler, D., Branches, F., Lucotte, M., Amorim, M., Larribe, F., Dolbec, J. Neurotoxic effects of low-level methylmercury contamination in the Amazon basin. *Environ. Res.*, 79, 20-32, 1998.

Lecloux, A.J. Scientific activities of Euro Chlor in monitoring and assessing naturally and man-made organohalogens. *Chemosphere*, 52, 521-9, 2003.

Lefever, H.G. When the saints go riding in: Santeria in Cuba and the United States. *J. Sci. Study Religion*, 35, 318-30, 1996.

Litovitz, T.L., Smilkstein, M., Felberg, L., Klein-Felberg, L., Klein-Schwartz, W., Berlin, R., Morgan, J.L. 1996 annual report of the American Association of Poison Control Centers Toxic Exposure Surveillance System. *Am. J. Emerge. Med.*, 15, 447-500, 1997.

Lodenius, M., Malm, O. Mercury in the Amazon. Rev Environ Contam Toxicol 1998; 157:25-52.

MacIntosh D.L., Spengler J.D., Ozkaynak H., Tsai L., Ryan P.B. Dietary exposures to selected metals and pesticides. *Environ. Health. Perspect.*, 104, 202-09, 1996.

Mahaffey, K.R., Clickner, R.P., Bodurow, C.C. Blood organic mercury and dietary mercury intake: National Health and Nutrition Examination Survey, 1999 and 2000. *Environ. Health. Perspect.*, 112, 562-570, 2004.

Mahaffey, K.R., Mergler D. Blood levels of total and organic mercury in residents of the upper St. Lawrence River basin, Quebec: association with age, gender, and fish consumption. *Environ. Res.*, 77, 104-114, 1998.

Marcovecchio, J.E., Gerpe, M.S., Bastida, R.O., Rodriguez, D.H., Moron, S.G. Environmental contamination and marine mammals in coastal waters from Argentina: an overview. *Sci. Tot. Env.*, 154, 141-51, 1994.

Mason, H.J., Hindell, P., Williams, N.R. Biological monitoring and exposure to mercury. *Occup. Med. (Lond.)*, 51, 2-11, 2001.

McDowell, M.A., Dillon, C.F., Osterloh, J., Bolger, P.M., Pellizzari, E., Fernando, R., Montes de Oca, R., Schoeber, S. E., Sinks, T., Jones, R. L., Mahaffey, K. R. Hair mercury levels in US children and women of childbearing age. Reference range data from NHANES 1999-2000. *Environ. Health Perspect.*, 112, 1165-71, 2004.

McKinney, P.E. Elemental mercury in the appendix: an unusual complication of Mexican-American folk remedy. *J. Toxicol. Clin. Toxicol.*, 37, 103-7, 1999.

Midthun, K. Thimerosal as a preservative in vaccines: An FDA perspective. Presented 2004 in "Mercury: Medical and Public Health Issues" Tampa, Florida. April 25-28, 2004.

Mol, J.H., Ramlal, J.S., Lietar, C., Verloo, M. Mercury contamination in freshwater, estuarine, and marine fishes in relation to small-scale gold mining in Suriname, South America. *Environ. Res.*, 82, 183-97, 2001.

Morgan, J.N., Berry, M.R.Jr., Graves, R.L. Effects of commonly used cooking practices on total mercury concentrations in fish and their impact on exposure assessments. *J. Expo. Anal. Environ. Epidemiol.*, 7, 119-133, 1997.

Morrissette, J., Takser, L., St.Amour, G., Smargiassi, A., Lafond, J., Mergler, D. Temporal variation of blood and hair mercury levels in pregnancy in relation to fish consumption history in a population living along the St. Lawrence River. *Environ. Res.*, 95, 363-74, 2004.

Muckle, G., Ayotte, P., Dewailly, E.E., Jacobson, S.W., Jacobson, J.L. Prenatal exposure of the northern Quebec Inuit infants to environmental contaminants. *Environ. Health Perspect.*, 109, 1291-9, 2001.

Nakagawa, R., Yunita, Y., Hiromoto, M. Total mercury intake from fish and shellfish by Japanese people. *Chemosphere*, 35, 2909-2913, 1997.

National Academy of Sciences. National Research Council. Committee on Toxicology of Methylmercury. *Toxicology of Methylmercury*. Washington DC: National Academy Press. 2000.

National Academy of Sciences. Institute of Medicine. Immunization Safety Review Committee. *Immunization Safety Review. Thimerosal-Containing Vaccines and Neurodevelopmental Disorders*. Stratton, K., Gable, A., and McCormick, M.C. Eds. Washington DC: National Academy Press. 2004.

National Institute for Occupational Safety and Health and Health (NIOSH). A Recommended Standard for Occupational Exposure to Inorganic Mercury. Washington, D.C. 1977

Niencheski, L.F., Windom, H.L., Baraj, B., Wells, D., Smith, R. Mercury in fish from Patos and Mirim Lagoons, Southern Brazil. *Mar. Pollut. Bull.*, 42, 1403-6, 2001,

Nilson, S.A. Jr., Costa, M., Akagi, H., Total and methylmercury levels of a coastal human population and of fish from the Brazilian northeast. *Environ. Sci. Pollut. Res. Int.*, 8, 280-4, 2001.

Ortiz-Roque, C., Lopez-Rivera, Y. Mercury contamination in reproductive age women in a Caribbean island: Vieques. *J. Epidemiol. Community Health*, 58, 756-7, 2004.

Palheta, D., Taylor, A. Mercury in environmental and biological samples from a gold mining area in the Amazon region of Brazil. *Sci Tot. Env.*, 19, 63-9, 1995.

Passos, C.J., Mergler, D., Gaspar, E., Morais, S., Lucotte, M., Larribe, F., Davidson, R., de Grosbois, S. Eating tropical fruit reduces mercury exposure from fish consumption in the Brazilian Amazon. *Environ. Res.*, 93, 123-130, 2003.

Penedo de Pinho, A., Daves Guinaraes, J.R., Martins, A.S., Costa, P.A., Olavo, G., Valentin, J. Total mercury in muscle tissue of five shark species from Brazilian offshore waters: effects of feeding habit, sex, and length. *Environ. Res.*, 89, 250-58, 2002.

Peters-Haefeli, L., Michod, J., Aellig, A., Varone, J.J., Schelling, J.L., Peters, G. Urinary mercury excretion in professional users of an antiseptic soap containing 0,04% phenylmercury borate. *Schweiz. Med. Wochenschr.*, 106, 171-8, 1976.

Pichichero, M.E., Cernichiari, E., Lopreiato, J., Treanor, J. Mercury concentrations and metabolism in infants receiving vaccines containing thiomersal: a descriptive study. *Lancet*, 360, 1737-41, 2002.

Piikivi, L., Tolonen, U. EEG findings in chlor-alkali workers subjected to low long term exposure to mercury vapour. *Br. J. Ind. Med.*, 46, 370-5, 1989.

Rhainds, M., Levallois, P., Dewailly, E., Ayotte, P. Lead, mercury, and organchlorine compound levels in cord blood in Quebec, Canada. *Arch. Environ. Health*, 54, 40-47, 1999.

Rice, D.C., Schoeny, R., Mahaffey, K.R. Methods and rationale for derivation of a reference dose for methylmercury by U.S. EPA. *Risk Anal.*, 23, 107-15, 2003.

Risher, J.F., Murray, H.E., Prince, G.R. Organic mercury compounds: human exposure and its relevance to public health. *Toxicol. Ind. Health*, 18, 109-60, 2002.

Rothschild, R.F., Duffy, L.K. Methylmercury in the hair of subsistence food users in a rural Alaskan village. *Alaska Med.*, 44, 2-7, 2002.

Saint-Phard, D., Gonzalez, P.G., Sherman, P. Poster 88. Unsuspected mercury toxicity linked to neurologic symptoms: A case Series. *Arch. Phys. Med. Rehabil.*, 85, E25, 2004.

Sakamoto, M., Kubota, M., Liu, X.J., Murata, K., Nakai, K., Satoh, H. Maternal and mercury and −3 polyunsaturated fatty acids as a risk and benefit of fish consumption to fetus. *Env. Sci Technol.*, 38, 3860-63, 2004.

Santos, E.C., de Jesus, I.M., Camara Vde, M., Brabo, E., Loureiro, E.C., Mascarenhas, A., Weirich, J., Luiz, R.R., Cleary, D. Mercury exposure in Munduruku Indians from the community of Sai Cinza, State of para, Brazil. *Environ. Res.*, 90, 98-103, 2002.

Sanzo, J.M., Dorrosoro, M., Amiano, P., Amurrio, A., Aguinagalde, F.X., Azpiri, M.A. Estimation and validation of mercury intake associated with fish consumption in an EPIC cohort of Spain. *Public Health Nutr.*, 4, 981-988, 2001.

Schober, S.E., Sinks, T.H., Jones, R.L., Bolger, P.M., McDowell, M., Osterloh, J., Garrett, E.S., Canady, R.A., Dillon, C.F., Sun, Y., Joseph, C.B., Mahaffey, K.R. Blood mercury levels in US children and women of childbearing age, 1999-2000. *J. Am. Med. Assoc.*, 289, 1667-74, 2003.

Schultz, C.D., Crer, D., Pearson, J.E., Rivera, J.E., Hylin, J.W. Total and organic mercury in the Pacific blue marlin. *Bull. Environ. Contam. Toxicol.*, 15, 230-234, 1976.

Seidel, S., Kreutzer, R., Smith, D., McNeel, S., Gilliss, D. Assessment of commercial laboratories performing hair mineral analysis. *JAMA*, 285, 67-72, 2001.

Sherlock, J.C., Hislop, J.E., Newton, D., Topping, G., Whittle, K. Elevation of mercury in human blood from controlled chronic ingestion of methylmercury in fish. *Hum. Toxicol.*, 3, 117-31, 1984.

Sherlock, J.C., Lindsay, D.R., Hislop, J.E., Evans, W.H., Collier, T.R. Duplication diet study on mercury by fish consumers in the United Kingdom. *Arch. Environ. Health*, 37, 271-278, 1982.

Sin, K.W., Tsang, H.F. Large-scale mercury exposure due to a cream cosmetic: community-wide case series. *Hong Kong Med. J.*, 95, 329-34, 2003.

Sing, K.A., Hryhorczuk, D., Saffrio, G., Sinks, T., Paschal, D.C., Sorensen, J., Chen, E.H. Organic mercury levels among the Yanomama of the Brazilian Amazon Basin. *Ambio*, 32, 434-9, 2003.

Skerfving, S. Methylmercury exposure, mercury levels in blood and hair, and health status in Swedes consuming contaminated fish. *Toxicology*, 2, 3-23, 1974.

Smith, J.C., Allen, P.V., Turner, M.D., Most, B., Fisher, H.L., Hall, L.L. The kinetics of intravenously administered methyl mercury in man. *Toxicol. Appl. Pharmacol.*, 128, 251-56, 1994.

Smith, R.G., Vorwald, A.J., Patil, L.S., Mooney, T.F.Jr. Effects of exposure to mercury in the manufacture of chlorine. *Am. Ind. Hyg. Assoc. J.*, 31, 687-700, 1970.

Stajich, G.V., Lopez, G.P., Harry, S.W., Sexson, W.R. Iatrogenic exposure to mercury after hepatitis B vaccination in preterm infants. *J. Pediatr.*, 136, 678-81, 2000.

Stern, A.H., Gochfeld, M., Weisel, C., Burger, J. Mercury and methylmercury exposure in the New Jersey pregnant population. *Arch. Environ. Health*, 56, 4-10, 2001.

Stern, A.H., Smith, A.E. An assessment of the cord blood-maternal blood methylmercury ratio: implications for risk assessment. *Environ. Health. Perspect.*, 111, 1465-70, 1970.

Storelli, M.M., Marcotrigiano, G.O. Fish for human consumption: risk of contamination by mercury. *Food Addit. Contam.*, 17, 1007-11, 2000.

Storelli, M.M., Stuffler, R.G., Marcotrigiano, G.O. Total and methylmercury residues in tuna-fish from the Mediterranean sea. *Food. Addit. Contam.*, 19, 715-20, 2002.

Suzuki, T., Miyama, T., Toyama, C. The chemical form and bodily distribution of mercury in marine fish. *Bull. Environ. Contam. Toxicol.*, 10, 347-55, 1973.

Swift, J.A. Morphology and histochemistry of human hair. *EXS*, 78, 149-75, 1997.

Tsugane, S., Kondo, H. The mercury content of hair of Japanese immigrants in various locations in South America. *Sci Tot. Env.*, 63, 69-76, 1987.

Urieta, I., Jalon, M., Eguilero, I. Food surveillance in the Basque Country (Spain). II. Estimation of the dietary intake of organochlorine pesticides, heavy metals, arsenic, aflatoxin M1, Iron, and zinc through the Total Diet Study 1990/91. *Food Addit. Contam.*, 13, 29-52, 1996.

US EPA. Task Force on Ritualistic Use of Mercury Report Mercury is used to attract luck, love, or money. It is also used to protect against evil. http://www.epa.gov/superfund/action/ community/mercury.pdf 2002.

US EPA. Fact Sheet:Proposed Rule to Reduce Toxic Air Pollutants Mercury Cell Chlor-Alkali Plants. http://www.epa.gov/ttn/oarpg/t3/fact_sheets/mccap_fs.pdf

US EPA. Background Report: AP-42 Section 5.5: Chlor-Alkali Industry. An emission factor relates the quantity of pollutants emitted to a unit of activity. URL:http://www.epa. gov/ttn/chief/ap42/ch08/bgdocs/b08s11.pdf 1992.

US EPA. Mercury Study Report to Congress. Volume IV. An Assessment of Exposure to Mercury in the United States. URL: http://www.epa.gov/oar/mercury.html 1997.

US FDA. Center for Drug Evaluation and Research. Mercury in Drug and Biologic Products Retrieved May 2004 from: http://www.fda.gov/cder/fdama/mercury300.htm

US Occupational Safety and Health Administration. Job Health Hazards Series. Mercury. OSHA Report 2234. 1975.

Van Oostdam, J., Gilman, A., Dewailly, E., Usher, P., Wheatley, B., Kuhnlein, H., Neve, S., Walker, J., Trace, B., Feeley, M., Jerome, V., Kwavnick, B. Human health implications of environmental contaminants in Arctic Canada: a review. *Sci. Tot. Env.*, 230, 1-82, 1999.

Veiga, M.M. Protocols for environmental and health assessment of mercury released by Artisanal and small -scale gold miners (ASM). Global Mercury Project. UNIDO Project EG/GLO/01/G#$: Removal of Barriers to Introduction of Cleaner Artisanal Gold Mining and Extraction Technologies. Vienna: United Nations Industrial Development Organization. 2004.

Veiga, M.M. UNIDO/UBC/CETEM. Introducing new technologies for abatement of global mercury pollution in Latin America. Ed. UNIDO/UBC/CETEM. Rio de Janeiro. 94 p. ISBN: 85-7227-100-7, 1997.

Wagemann, R., Innes, S., Richard, P.R. Overview and regional and temporal differences of heavy metals in Arctic whales and ringed seals in the Canadian Arctic. *Sci. Tot. Env.*, 186, 41-66, 1996.

Wagemann, R., Trebacz, E., Boila, G., Lockhart, W.L. Methylmercury and total mercury in tissues of arctic marine mammals. *Sci. Tot. Env.*, 218, 19-31, 1998.

Wangberg, I., Edner, H., Ferrara, R., Lanzillotta, E., Munthe, J., Sommar, J., Sjoholm, M., Svanberg, S., Weibring, P. Atmospheric mercury near a chlor-alkali plant in Sweden. *Sci. Tot. Env.*, 304, 29-41, 2003.

Weiner, J.A., Nylander, M. The relationship between mercury concentration in human organs and different predictor variables. *Sci. Tot. Env.*, 138, 101-15, 1993.

Weldon, M.M., Smolinski, M.S., Maroufi, A., Hasty, B.W., Gilliss, D.L., Boulanger, L.L., Ballus, L.S., Dutton, R.J. Mercury poisoning associated with a Mexican beauty cream. *West J. Med.*, 173, 15-18, 2000.

Wharton, M., "Thimerosal and Vaccines: CDC Perspectives". Mercury: Medical and Public Health Issues. Tampa, Florida. April 25-28, 2004.

WHO - World Health Organization. *Environmental Health Criteria 101. Methylmercury.* Geneva. World Health Organization. 1990.

WHO - World Health Organization. *Environmental Health Criteria 118. Inorganic Mercury.* Geneva: World Health Organization. 1991.

Ysart, G., Miller, P., Crosdale, M., Crews, H., Robb, P., Baxter, M., de L'Argy, C., Harrison, N. 1997 UK Total Diet Study - Dietary exposures to aluminum, arsenic, cadmium, chromium, copper, lead, mercury, nickel selenium, tin, and zinc. *Food Addit. Contam.*, 17, 775-786, 2000.

Zaidi, M., Angulo, M., Sifuentes-Osornio, J. Disinfection and sterilization practices in Mexico. *J Hosp. Infect.*, 31, 25-32, 1995.

Zayas, L.H., Ozuah, O. Mercury use in Espiritismo: a survey of botanicas. *Am. J. Public Health*, 86, 111-112, 1996.

Zeitz, P., Orr, M.F., Kaye, W.E. Public health consequences of mercury spills: Hazardous Substances Emergency Events Surveillance System, 1993-1998. *Environ. Health Perspect.* 110, 129-32, 2002.

Chapter-16

EXPOSURE TO MERCURY IN THE GENERAL POPULATION OF EUROPE AND THE ARCTIC

Lars Barregård

Sahlgrenska University Hospital and Academy, P.B. 414, 405 30 Göteborg, Sweden

INTRODUCTION

Mercury (Hg) is a toxic heavy metal. It occurs in several physical and chemical forms. The most important from a toxicological point of view are the metallic (elemental) form (Hg^0), and methylmercury (MeHg). Mercury is emitted to the atmosphere, mainly as Hg^0 by natural evaporation and from man-made sources. It is converted to soluble forms (e.g. Hg^{2+}) and deposited by rain into soil and water. Mercury is methylated non-enzymatically or by microbes. The methylmercury compounds bioaccumulate up the aquatic food chain and reach the human sea-food diet (WHO 1990; WHO 1991).

Inhalation of Hg^0 vapour, and ingestion of MeHg via fish are the most important routes of human exposure to mercury. In the general population, the major source of inhaled Hg^0 is Hg released from dental amalgam fillings. In European occupational settings, like chloralkali plants, mercury mines, the exposure levels are much higher (Sallsten et al., 1990; Boffetta et al., 1998; Pranjic et al., 2003; Jarosinska et al., 2004).

In addition to Hg^0, low amounts of inorganic divalent Hg may be present in diet. Exposure to ethyl mercury (EtHg) may occur from certain medical products (e.g. vaccines and immunoglobulins containing tiomersal), since EtHg is used as a preservative (Pichichero et al., 2002; Magos et al., 2003).

INORGANIC MERCURY

A major source of exposure to inorganic mercury in the general population is Hg^0 from dental amalgam fillings. Inhaled Hg^0 is rapidly absorbed in blood, and about 80 % is retained. Elimination occurs via urinary and faecal excretion. Urinary mercury (U-Hg) reflects kidney Hg and at steady state also the body burden, and it is widely used for assessment of exposure to inorganic Hg in humans (WHO, 1991). The half time for U-Hg is about two months, thereby reflecting the exposure in the past year. In order to decrease the impact of urinary flow rate, U-Hg is generally expressed in µg/g creatinine. In spite of this there is a normal day-to-day variability. The fact that women have a lower mean creatinine excretion rate results in lower average U-Hg in µg/g than in men, even when men and women have the same Hg excretion rate.

Another source of inorganic Hg in humans is demethylation of MeHg. Therefore, an increase of U-Hg is often shown in subjects with a high MeHg intake (Carta et al., 2003; Johnsson et al., 2004).

URINARY MERCURY CONCENTRATIONS

Some recent studies, stratified by country have been summarized in Table 1. In subjects who have no dental amalgam filling, and eat fish less than once a week, typical levels are only about 0.2 µg/gC. In subjects with a moderate number of amalgam fillings, or a high fish consumption, median U-Hg is about 1 µg/gC. Women tend to have slightly higher concentrations than men. Levels in adults and children are similar,

The increase of U-Hg with the number of amalgam fillings or the number of amalgam surfaces is well known and was shown in a large number of studies in the 1990s (Akesson, 1991). Similar findings are shown in those recent studies in Europe, where amalgam status was examined, Table 1. Thus, there is a tendency towards higher exposure to Hg^0 and higher U-Hg in countries where the average number of amalgam fillings in the population is high. This is the case for example in the Scandinavian countries, where public dental care was free and well organized at schools, and amalgam was the standard treatment for caries.

A typical increase of urinary Hg by amount of dental amalgam is shown in Figure 1.

Table 1. Urinary mercury concentrations in the general population, adults or children (Ch) in various European countries, and the Arctic. In the selection of studies, the aim was to report at least one study from each country in the past 5 years. If no recent study was available, the time period was extended somewhat backwards. When many studies were available from one country, priority was given to large ones. Results from a recent non-published EU study were also included. Controls (C) refers to control subject in occupational studies or similar.

Country	Year	Subjects R, C, Ch, P	N	Sex F	Age mean Or median (range)	U-Hg Mdn/GM µg/gC	U-Hg mean µg/gC	Dental amalgam %	Dental amalgam amount	Reference
Czech Republic		C	46	50% F	36	0.8/24h				Urban et al., 1999
Czech Republic		Ch		M						Benes et al., 2002
France			174	F	(9-12)	0.96				Burbure et al., 2003
France		Ch	165	M	(9-12)	1.0				Burbure et al., 2003
France			91	F	35	0.55				Burbure et al., 2003
France			247	50%F	35	0.64				Burbure et al., 2003
Germany	1991		4000	50%F	25-69	0.4	0.71			Seifert et al., 200ab
Germany	1998	R	1560	50%F	(18-69)	0.2	0.28	0		Becker et al., 2003
Germany	1998	R	1353	50%F	(18-69)	0.3	0.50	100	1-4 teech	Becker et al., 2003
Germany	1998	R	1092	50%F	(18-69)	0.6	0.84	100	5-8 teech	Becker et al., 2003
Germany	1998	R	467	50%F	(18-69)	1.0	1.33	100	>8 teech	Becker et al., 2003
Germany	1996	R Ch	245	50%F	9	0.22		30		Pesch et al., 2002
Germany	2001	R Ch	1800	50%F	10	0.2 µg/L	1.0	80		Gabrio et al., 2003
Germany	2001	R Ch	1500	50%F	10	<0.2µg/L	0.9	0		Gabrio et al., 2003
Italy [1]	2002	R	48	M	37	0.93	1.3	0		Barregard et al., 2003
Italy [1]	2002	R	34	F	38	0.92	1.6	0		Barregard et al., 2003
Italy [1]	2002	R	48	F	40	1.3	2.0	100	1-4 teech	Barregard et al., 2003
Italy [1]	2002	R	62	M	38	1.1	1.4	100	1-4 teech	Barregard et al., 2003

Table 1. cont'd.

Country	Year	Subjects R, C, Ch, P	N	Sex F	Age mean or median (range)	U-Hg Mdn/GM µg/gC	U-Hg mean µg/gC	Dental amalgam %	Dental amalgam amount	Reference
Italy		Tuna eaters	22	M	52	6.5			15mm2	Carta et al., 2003
Italy		C	22	M	51	1.5			15mm2	Carta et al., 2003
Italy	1999		383	49% F	41	0.79	1.4		3.5 fillings	Apostoli et al., 2002
										Soleo et al., 2003
Netherlands – (23 surf.)		C	52	60% F	35	1.5		100	11. fillings	Schuurs et al., 2001
Norway	1997	C	47	M	42		2.3			Ellingsen et al., 2000
Poland	2003	R	49		38	0.18	0.65	0		Jarosinska et al., 2004
Poland	2003	R	60		39	0.22	1.18	0		Jarosinska et al., 2004
Poland	2003	R	19	M	32	0.50	0.65	100	1-4 teech	Jarosinska et al., 2004
Poland	2003	R	43	F	36	0.42	1.01	100	1-4 teech	Jarosinska et al., 2004
Portugal	1997	R Ch	150	50% F	10	1.1µg/L		0		Evans et al., 2001
Portugal	1998	R Ch	150	50% F	11	2.1µg/L		50		Evans et al., 2001
Scotland		C univ. staff	163	47% F	32	0.88	1.18	>0		Ritchie et al., 2002
Spain	1997	R	180	50% F	(18-69)	1.7µg/L				Gonzales et al., 2000
Sweden [1]	2002	R	90	F	21	0.20	0.24	0		Barregard et al., 2003
Sweden [1]	2002	R	70	M	21	0.10	0.13	0		Barregard et al., 2003
Sweden [1]	2002	R	65	M	44	0.61	1.0	100	8 teech	Barregard et al., 2003
Sweden	1995	P	226	F	31	1.6µg/L		90%	5-10 fillings	Vahter et al., 2000

Note: [1] indicates that data from the Italian and Swedish populations were collected in the same study; R = random, C = controls, Ch = children, P = pregnant, F = female

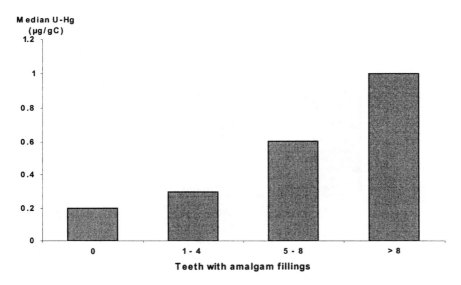

Figure 1. Urinary Hg excretion in µg/g creatinine versus dental amalgam in the GerESIII study. The study included a random sample (proportionately stratified for age, gender and community size) of 4800 Germans, 18-69 years of age.
The data are taken from (Becker, 2003).

What has been less often described is the effect of fish consumption on U-Hg excretion. This is not caused by the small amount of inorganic Hg present in fish (generally about 5 %), but it is rather the result of demethylation of MeHg (Apostoli et al., 2002; Carta et al., 2003; Barregard et al., 2003; Johnsson et al., 2004). The effect of fish intake is shown by the positive associations between reported number of fish meals and U-Hg concentrations in an Italian population study (Figure 2). It also explains differences between countries, as shown in a recent EU-funded study, where sampling was identical and samples from Sweden and Italy were analysed in the same laboratory (Barregard et al., 2003). In that study the higher average MeHg intake from fish in Italy resulted in U-Hg levels of the general population without amalgam fillings that were about 5 times higher compared to Sweden and Poland (Barregard et al., 2003; Jarosinska et al., 2004; see Table 1). This effect is also shown by the clear association between hair-Hg, an indicator of MeHg exposure, and U-Hg in subjects without amalgam fillings (Johnsson et al., 2004).

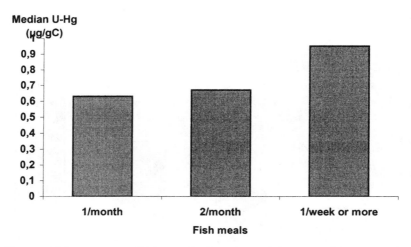

Figure 2. Urinary mercury (Hg) excretion in µg/g creatinine versus fish consumption
in an Italian population study of 374 subjects from from four cities.
Data are taken from (Soleo, 2003).

The contribution from ambient air is generally negligible. Typical outdoor air-Hg concentrations in Europe are 2-5 ng/m³. Assuming no additional indoor sources, inhalation of 15-20 m³ per day, and 80 % absorption, will result in an uptake of less than 0.1 µg/day (Barregard et al., 2003). This is much less than the uptake from amalgam fillings, which is typically at least 10-100 times higher (WHO, 1991).

The time trends for U-Hg in Europe depend on time trends in occurrence of amalgam fillings, time trends in fish consumption, and time trends in MeHg concentrations in fish. Clear declining trends have been shown in Germany, and Sweden (Seifert et al., 2000; Becker et al., 2003; Akesson et al., 1990; Barregard et al., 2003).

OCCURRENCE OF HIGH LEVELS

Most big population studies have shown that high U-Hg levels are found in a small fraction of the general population, as shown by the maximum concentrations indicated in Table 1. Although, in most reports the reasons for extreme levels have not been clarified, some information is available. In a study of German children, very high concentrations were found in some children who used ointments containing mercury (Gabrio et al., 2003). A closer study revealed that U-Hg was high in most of the members of the family, immigrated from Kosovo, where there is less control of such products than in Germany. Some studies showed that U-Hg levels were statistically associated with use of chewing gum in subjects with amalgam filling. A previous reports showed that long term intense chewing on amalgams may result in U-Hg concentrations of 25 µg/gC (Sällsten, 1996). The impact of high intake of MeHg from seafood was clearly shown in a study from Sardinia, where the median U-Hg was 6.5 µg/gC (maximum 21.5 µg/gC) in 22 men with a high consumption of tuna with high Hg content, around 1.5 µg/g (Carta et al., 2002; 2003), see Figure 3.

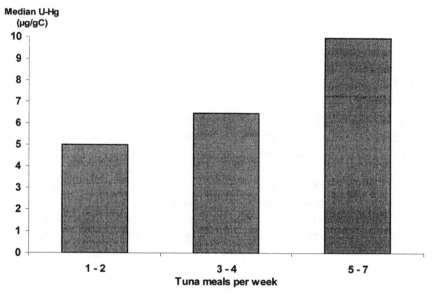

Figure 3. Urinary mercury (Hg) excretion in µg/g creatinine versus fish consumption in 22 consumers of tuna fish with high (about 1.5 µg/g) mercury concentrations. The data are taken from (Carta, 2002; 2003).

The effect of methyl mercury exposure on urinary mercury excretion is also indirectly shown by a close intra-individual association between U-Hg and hair-Hg in Swedish anglers consuming freshwater fish (Johnsson et al., 2004b).

Population studies generally show a lognormal distribution of U-Hg. The large studies give a reasonable estimate of the right tail of the distribution, e.g. the GerESIII Study from Germany including nearly 5000 adults showing a 98-percentile of 2.8 µg/gC in 1998. In a Swedish study, data from about 2000 subjects in 1990, all with amalgam fillings, showed a 99-percentile of 10 µg/gC (Barregard et al., 1995). In Italian subjects the 95-pecentiles were about 5 g/gC, and the 99-percentiles 10-15 µg/gC (Apostoli et al., 2002; Soleo et al., 2003; Barregard et al., 2003). In the large population studies, or in case reports, maximum U-Hg concentrations of 25µg/gC or higher have been reported, i.e levels that have been shown to have subtle effects on humans, e.g. on kidney function. Such cases seem to be rare, however.

In summary, median U-Hg levels in Europe are relatively low, and there is a declining trend. Dental amalgam and demethylation of MeHg from fish are the dominating sources. High levels occur in people chewing intensely on amalgam fillings, having a high MeHg intake, or using skin creams containing mercury.

METHYL MERCURY

For most people the source of exposure to methyl mercury is fish, although for some populations marine mammals are also important. Most of the mercury content in fish, about 90-95 % (Kehrig et al., 1998; Evens et al., 2001) is in the form of MeHg, the rest being inorganic Hg. The Hg content of fish is highly variable, the highest concentrations in fish eaten in Europe being found in predatory species like swordfish, shark, some big tuna species, and in piscivorous freshwater fish like pike and perch. In Europe there is a limit for mercury in fish of 0.5 µg/g, except for certain species like pike and eel, where the limit is 1 µg/g (European Commission, 2001).

MeHg intake is a function of the amount of fish consumed and the Hg content of the fish species eaten. Populations living in coastal areas, rivers or lakes where fishing is common on average have higher fish consumption than inland populations. The same is true for fishermen and anglers. Some intake estimates for mercury from fish have been made for average populations indicating, higher intake in coastal Spain and Croatia than in Germany and Scandinavian countries (European Commission, 2002).

MeHg in food is almost completely (> 90 %) absorbed in the gut. MeHg in blood peaks rapidly, and distributes relatively uniformly to most tissue. It readily passes the blood-brain barrier. In blood, most MeHg is found in red blood cells, with concentrations 10-20 times higher than in plasma (WHO, 1990; Clarkson, 2002). MeHg is secreted into bile and partly reabsorbed into the portal circulation and thereby returned to the liver. Most of the MeHg is eliminated by demethylation and excretion of the inorganic form (mainly Hg^{2+}) in faeces. The demethylation of MeHg takes place in several organs, e.g. liver, spleen, lymph nodes, and in the intestines (by the microflora). The fraction of total mercury present as in tissue as inorganic Hg, after exposure to MeHg depends on the duration of exposure and the time after cessation of exposure to MeHg (Carrier et al., 2001). Excretion of inorganic Hg in urine is small in the early phase after exposure but increase with time. The half time of MeHg in blood and in most organs is about 2 months. Demethylation also takes place in the brain, where the inorganic fraction may have a long half time, binding to SH-groups and selenium (WHO, 1990; Clarkson, 2002).

Hair-Hg is used for biological monitoring of exposure to MeHg, and reflects past exposures, up to a year or more, depending on the length of the hair. Hair incorporates Hg from blood when formed, and grows with about 1 cm per month. Thus, the mercury content in the part of a hair strand closest to the scalp reflects the recent blood Hg levels (WHO, 1990). The possibility of external contamination of hair should be taken into account, as well as the potential leaching of mercury from the hair at permanent treatment.

HAIR MERCURY CONCENTRATIONS

As shown in Table 2, typical hair-Hg concentrations in central and northern Europe are 0.2-0.4 µg/g (Rosborg et al., 2003; Bjornberg et al., 2003; Benes et al., 2003; Lindow et al., 2003; Gebel et al., 1998; Pesch et al., 2002; Hac et al., 2000) in subjects not regularly consuming freshwater fish. Studies from the UK show concentrations around 0.5 µg/g (Ritchie et al., 2002; Lindow et al., 2003). In groups who consume freshwater fish regularly, typical levels are 1-2 µg/g (Salonen et al., 2000; Johnsson et al., 2004a).

In the Mediterranian, hair-Hg levels are typically higher than in central and northern Europe, with a median of 1-2 µg/g in the general population on the Italian coast (n=140), and 4 µg/g (n=19) in fishermen (Horvat et al., 2004; personal communication). In regular tuna eaters from Sardinia the median hair-Hg was 9.6 (n=8, Carta et al., 2003). Older studies on Thyrrenian coast in Italy and on the island of Madeira also showed high hair mercury concentrations in fishermen and their families (Renzoni et al., 1998; Murata et al., 1999). In addition to the data given in Table 2, recent results from pregnant women in the Friuli-Venezia region of Italy show mean hair-Hg concentrations in pregnant women of about 1 µg/g, and population studies on Greek islands in the Eastern Aegean show median hair-Hg concentrations of 1-1.5 µg/g (M. Horvat, personal communication).

Many studies have shown a relation between hair mercury and fish intake (Figure 4).

In the Arctic, the levels are highly variable depending on communities, but in general higher than in Europe.

A study of Inuits in Canada showed a median concentrations of 4 µg/g (Muckle et al., 2001), while Inuits in Greenland had median concentrations of 5-15 µg/g (Weihe et al., 2002).

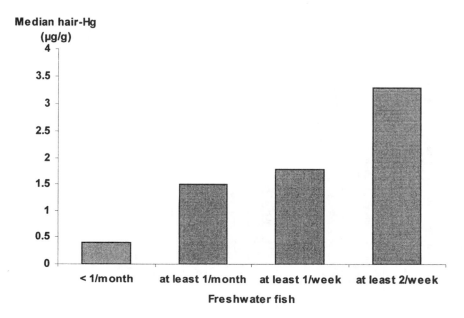

Figure 4. Association between self-reported freshwater fish consumption and hair mercury concentrations in Swedish anglers. Data taken from (Johnsson, 2004).

In the Faroe Islands, the levels in children and adults have decreased to 1-2 µg/g around the year 2000 (Dewailly et al., 2003; Murata et al., 2004).

There is a clear downward trend in hair Hg in the Faroe Islands (Dewailly et al., 2003). For the rest of Europe and the Arctic, the time trend is unclear.

OCCURRENCE OF HIGH LEVELS

Obviously high hair-Hg levels are found in people consuming either large amounts of fish with moderate Hg levels, or at least a weekly fish meal with high Hg content. The groups with the highest exposure seem to be the inutis in the Arctic, the Sardinians regularly consuming large tuna, and the Mediterranian fishermen. The median concentrations in these groups are shown in Table 2.

Table 2. Hair mercury concentrations in the general population, adults or children in various European countries, and the Arctic. In the selection of studies, the aim was to report at least one study from each country in the past 5 years. If no recent study was available, the time period was extended somewhat backwards. When many studies were available from one country, priority was given to large ones. Controls (C) refers to control subject in occupational studies or similar. P=pregnant.

Country	Year	Subjects R, C, Ch	Site	N	Sex	Age mean or median (range)	Hair-Hg median ($\mu g\ g^{-1}$)	Hair-Hg mean ($\mu g\ g^{-1}$)	Fish meals mean	Reference
Albania		C		25			0.40	0.41		Babi et al. 2000
Czech Republic	94-01	RCh	Inland	3500	50%	10	0.19	0.27		Benes et al. 2003
England	1997	Mothers	Mixed	52	F	26		0.44		Lindow et al. 2003
Finland	1986	R	Inland	1000	M	52		1.8		Salonen et al. 2000
Germany	1994		Inland	213	60% F		0.21	0.25		Gebel et al. 1998
Germany	1996	RCh	Inland	245	50% F	9	0.18		< 1	Pesch et al. 2002
Italy	2003	R	Coast	140			1.7			Horvat 2004
Italy	1999	Tuna eaters	Coast	8	M	52	9.6		4/w	Carta et al. 2002; 2003
Poland	1998	Sudn deaths	Inland	46	17%F	50	0.29	0.38		Hac et al. 2000
Portugal [(1)]	1994	mothers	Island	146	F		9.6		3/w	Murata et al. 1999
Portugal	1994	Ch	Island	149	F	7	3.8		3/w	Murata et al. 1999
Scotland		Univ. Staff		163	47% F	32	0.47	0.57		Ritchie et al. 2002
Sweden	1996	Anglers	Inland	143	63% F	61	0.9	1.7	1/w	Ritchie et al. 2002
Sweden [(2)]	1997	Random	Inland	90	F		0.35			Johonsson et al. 2004
Sweden	1997	P	mixed	127	F	(27)	0.35		1.5/w	Bjornberg et al. 2003

Table 2. cont'd.

Country	Year	Subjects R, C, Ch	Site	N	Sex	Age mean Or median (range)	Hair-Hg median (µg g⁻¹)	Hair-Hg mean (µg g⁻¹)	Fish meals mean	Reference
Alaska	2000	Inuits		16	69% F	49	1.1	1.5		Rotschild et al. 2002
Canada	1997	Pregn. Inuits		108	F	24	3.7	4.3	3.3+	Muckle et al. 2001
Canada	70-95			39000				1.2		Van Oostdam et al. 1999
Faroe Islands	1998	P All			F			2.1		Dewailly et al. 2003
Faroe Islands	2001	Ch All		839		14	0.96			Murata et al. 2004
Greenland		R Ch		43			5.5		1+	Wehile et al. 2002
Greenland		R								

Note: R = random, C = controls, Ch = children, F = female, ⁽¹⁾ fishing village on Madeira; ⁽²⁾ freshwater fish

The maximum concentrations obviously depend on the sizes of the groups sampled. The maximum concentration in 108 pregnant iniuts from Canada was 18.5 μg/g (Muckle et al., 2001). In 43 Inuit children from Greenland it was 18.4 μg/g, and in 31 mothers from the same area the maximum hair mercury concentration was 32.9 μg/g (Weihe et al., 2002). Similar maximum hair Hg concentrations were also found in regular consumers of Swedish freswater fish (maximum 18.5 μg/g; Johnsson et al., 2004) or tuna caught outside Sardinia (34.5 μg/g; Carta et al., 2003).

The maximum concentrations found in random samples in Germany (N=500; Gebel et al., 1998; Pesch et al., 2002) and the Czech Republic (N=3500; Benes et al., 2002) populations, respectively were below 2 μg/g.

In adults, the proposed reference dose of 0.1 μg/kg body weight is equivalent to a hair mercury concentrations of about 1.1 μg/g at long term stable exposure. This level is exceeded in most people in coastal areas of the Mediterranian countries. In people who consume large amounts of seafood with high mercury content, hair Hg concentrations may also exceed 10 μg/g, the lower limit of the benchmark dose (BMDL) for toxic effects on the developing brain. This is especially the case for some Inuit populations in the Arctic.

BLOOD MERCURY CONCENTRATIONS

Blood mercury reflects exposure to methyl mercury (most of it in the red blood cells) as well as inorganic mercury (divided about equally between plasma and red blood cells). Many studies have been performed on blood mercury in Europe, but no population studies as large as those on U-Hg or hair-Hg, and few studies on children. One reason is the invasiveness of blood sampling, and the other one is the fact that the analysis of blood Hg is somewhat more difficult than analysis of U-Hg. This is in particular the case if speciation of inorganic/organic Hg or inorganic/MeHg is to be performed. In case of exposure to MeHg, the blood mercury concentrations is typically 1/250 of the hair mercury concentration.

The blood Hg level is an excellent indicator of recent exposure to inorganic Hg or MeHg, if speciation is performed. The same is true for total B-Hg if exposure to inorganic Hg or MeHg is high. In general population samples where there is a low exposure to inorganic Hg as well as MeHg, U-Hg as indicator of inorganic Hg exposure, and hair-Hg as indicator of MeHg exposure have the advantage of better discrimination of the two major types of Hg exposure. For newborns the situation is different. The determination of

Hg in cord blood gives better information about the exposure levels for the foetus during the last period of the pregnancy than does the mother's hair-Hg.

In the present summary, focussed on the general population of Europe, having a mixed exposure to inorganic Hg and MeHg, the main indicators chosen were U-Hg and hair-Hg.

CONCLUSIONS

The major sources of mercury exposure in the European population are elemental mercury from dental amalgam and methyl mercury from seafood. The variability in these exposure sources is well shown in the ranges of urinary and hair mercury levels. In rare cases the exposure to inorganic mercury may affect human health. For methyl mercury, the intake is higher than the recommended reference doses in large parts of the populations in Europe and the Arctic. In some populations it is clearly at a level with exposures that has been shown to affect human health. Measure should be taken to reduce exposure, especially in pregnant women.

REFERENCES

Akesson, I., Schutz, A., Attewell, R., Skerfving, S., Glantz, P.O. Status of mercury and selenium in dental personnel: Impact of amalgam work and fillings. *Arch. Environ. Health*, 46, 102-109, 1991.

Apostoli, P., Cortesi, I., Mangili, A., Elia, G., Drago, I., Gagliardi, T., Soleo, L., Valente, T., Sciarra, G.F., Aprea, C., Ronchi, A., Minoia, C. Assessment of reference values for mercury in urine: the results of an Italian polycentric study. *Sci. Tot. Env.*, 289, 13-24, 2002.

Ask, Bjornberg, K., Vather, M., Petersson-Grawé, K., Glynn, A., Cnattingius, S., Darnerud, P.O., Atuma, S., Aune, M., Becker, W., Berglund, M. Methyl mercury and inorganic mercury in Swedish pregnant women and in cord blood: Influence of fish consumption. *Env. Health Perspect.*, 111, 637-641, 2003.

Babi, D., Vasjari, M., Celo, V., Koroveshi, M. Some results on Hg content in hair in different populations in Albania. *Sci. Tot. Environ.*, 259, 55-60, 2000.

Barregard, L., Sallsten, G., Jarvholm, B. People with high mercury uptake from their own dental amalgam fillings. *Occup. Env. Med.*, 52, 124-128, 1995.

Barregard, L., Sallsten, G., Mazzolai, B., Tripoli, G., Raffa, S., Mattoli, V., Horvat, M., Gibicar, D., Fajon, V., Fontanelli, R., LaManna, S., Di Bona, S. EMECAP – Deliverable D3.1. Work package no. 3. Epidemiological research. Report on urinary Hg in three groups (MCCA workers, potentially exposed citizens and control groups). Report to the EU commission. Department of Occupational and Environmental Medicine, Sahlgrenska University Hospital and Academy, Göteborg, Sweden, 2003.

Becker, K., Schulz, C., Kaus, S., Seiwert, M., Seifert, B. German Environmental Survey 1998 (GerES III): environmental pollutant in the urine of the German population. *Int. J. Hyg. Env. Health*, 206, 15-24, 2003.

Beneš, B., Sladká, J., Spěváčková, V., Šmíd, J. Determination of normal concentration levels of Cd, Cr, Cu, Hg, Pb, Se and Zn in hair of the child population in the Czech Republic. *Cent. Eur. J. Publ. Health*, 4, 184-186, 2003.

Burbure, C., Buchet, J.P., Bernard, A., Leroyer, A., Nisse, C., Haguenoer, J.M., Bergamaschi, E., Mutti, A. Biomarkers of renal effects in children and adults with low environmental exposure to heavy metals. *J. Toxicol. Environ. Health A*, 66, 783-798, 2003.

Carrier, G., Brunet, R.C., Caza, M., Bouchard, M. A toxicokinetic model for predicting tissue distribution and elimination of organic and inorganic mercury following exposure to methyl mercury in animals and humans. I. Development and validation of the model using experimental data in rats. *Toxicol. Appl. Pharmacol.*, 171, 38-49, 2001.

Carrier, G., Brunet, R.C., Caza, M. A toxicokinetic model for predicting tissue distribution and elimination of organic and inorganic mercury following exposure to methyl mercury in animals and humans. II. Application and validation of the model in humans. *Toxico.l Appl. Pharmacol.*, 171, 50-60, 2001.

Carta, P., Flore, C., Alinovi, R., Ibba, A., Tocco, M.G., Aru, G., Carta, R., Girei, E., Mutti, A., Lucchini, R., Randaccio, F.S. Sub-clinical neurobehavioral abnormalities associated with low level of mercury exposure through fish consumption. *Neurotoxicology*, 24, 617-623, 2003.

Clarkson, T.W. The three modern faces of mercury. *Env. Health Perspect.*, 110, 11-23, 2002.

Ellingsen, D., Efskind, J., Berg, K.J., Gaarder, P.I., Thomassen, Y. Renal and immunologic markers for chloralkali workers with low exposure to mercury vapor. *Scand. J. Work Environ. Health*, 26 (5) 427-435, 2000.

European Union Commission Regulation (EC). Maximum levels for certain contaminants in food stuff. No. 466/2001 of March 2001.

European Comission Chapter 5. Exposure to mercury in Europe. In: Position paper on mercury. European Comission, Air Quality, Daughter Directives. Available at www.europa.eu.int/comm/environment/air/background.htm#mercury

Evens, C.C., Martin, M.D., Woods, J.S., Soares, H.L., Bernardo, M., Leitão, J., Simmonds, P.L., Liang, L., DeRouen, T. Examination of dietary methylmercury exposure in the Casa Pia study of the health effects of dental amalgams in children. *J. Toxicol. Environ. Health A*, 64, 521-530, 2001.

Gabrio, T., Benedikt, G., Broser, S., Felder-Kennel, A., Fichtner, G., Horras-Hun, G., Jovanovic, S., Kirsch, H., Kouros, B., Link, B., Maisner, V., Piechotowski, I., Rzonca, E., Schick, K.H., Schrimpf, M., Schröder, S., Schwenk, M., Spöker-Maas, K., Weidner, U., Wuthe, J., Zöllner, I. 10 Jahre Beobachtungsgesundheitsämter in Baden-Württemberg – Beurteilung der Human-Biomonitoring-Untersuchungen bezüglich der Quecksilber-Belastung durch Amalgamfüllungen und andere Quellen. (10 years of observation by Public Health Offices in Baden-Württemberg – Assessment of human biomonitoring för mercury due to dental amalgam fillings and other sources.) *Gesundheitswesen*, 65, 327-335, 2003.

Gebel, T., Suchenwirth, Behmke, C., Plessow, Claussen, K., Schulze, E., Dunkelberg, H. Biomonitoring-Untersuchung bei Personen in Wohngebieten mit erhöhen Bodenwerten an Quecksilber, Arsen und Antimon. *Gesundheitswesen*, 60, 580-585, 1998.

Gonzalez, C.A., Kogevinas, M., Gadea, E., Huici, A., Bosch, A., Bleda, M.J., Päpke, O. Biomonitoring study of people living near or working at a municipal solid-waste incinerator before and after two years of operation. *Arch. Environ. Health*, 55, 259-267, 2000.

Hać, E., Krzyżanowski, M., Krechniak, J. Total mercury in human renal cortex, liver, cerebellum and hair. *Sci, Tot. Env.*, 248, 37-43, 2000.

Jarosińska, D., et al., EMECAP – Deliverable D3.1. Work package 3. Epidemiological research. Report on urinary mercury levels in three study groups: MCCA workers, potentially exposed individuals and population in the reference area. Institute of Occupational Medicine and Environmental Health, Sosnowiec, Poland; 2004.

Johnsson, C., Sallsten, G., Schütz, A., Sjors, A., Barregard, L. Hair mercury levels versus freshwater fish consumption in household members of Swedish angling societies. *Env. Res.*, 96, 257-263, 2004a.

Johnsson, C., Schütz, A., Sallsten, G. Impact of Consumption of Freshwater Fish on Mercury Levels in Hair, Blood, Urine, and Alveolar Air. (Submitted) 2004b.

Lindow, S.W., Knight, R., Batty, J., Haswell, S.J. Maternal and neonatal hair mercury concentrations: the effect of dental amalgam. *Br. J. Obstet. Gynaecol.*, 110, 287-291, 2003.

Magos, L. Neurotoxic character of thiomerosal and the allometric extrapolation of adult clearance half-time to infants. *J. Appl. Toxicol.*, 23, 263-269, 2003.

Muckle, G., Ayotte, P., Dewailly, E., Jacobson, S.W., Jacobson, J.L. Determinant of polychlorinated biphenyls and methylmercury exposure in Inuit women of childbearing age. *Env. Health Perspect.*, 109, 957-963, 2001.

Murata, K., Weihe, P., Renzoni, A., Debes, F., Vasconcelos, R., Zino, F., Araki, S., Jorgensen, P.J. White, R.F., Grandjean, P. Delayed evoked potentials in children exposed to methylmercury from seafood. *Neurotoxicol. Teratol.*, 21, 343-8, 1999.

Murata, K., Weihe, P., Budtz-Jørgensen, E., Jørgensen, P., Grandjean, P. Delayed brainstem auditory evoked potential latencies in 14-year-old children exposed to methylmercury. *J. Pediatr.*, 144, 177-183, 2004.

Pesch, A., Wilhelm, M., Rostek, U., Schmitz, N., Weishoff-Houben, M., Ranft, U., Idel, H. Mercury concentrations in urine, scalp hair, and saliva in children from Germany. *J. Expos. Anal. Environ. Epidemiol.*, 12, 252-258, 2002.

Pichichero, M.E., Cernichiari, E., Lopreiato, J., Treanor, J. Mercury concentrations and metabolism in infants receiving vaccines containing thiomersal: a descriptive study. *Lancet*, 360, 1737-1741, 2002.

Pranjic, N., Sinanovic, O., Jakubovic, R. Chronic psychological effects of exposure to mercury vapour among chlorine-alkali plant workers. *Med. Lav.*, 94, 531-41,2003.

Renzoni, A., Zino, F., Franchi, E. Mercury levels along the food chain and risk for exposed populations. *Env. Res.*, 77, 68-72, 1998.

Ritchie, K.A., Gilmour, W.H., Macdonald, E.B., Burke, F.J.T., McGowan, D.A., Dale, I.M., Hammersley, R., Hamilton, R.M., Binnie, V., Collington, D. Health and neuropsychological functioning of dentists exposed to mercury. *Occup. Env. Med.*, 59, 287-293, 2002.

Rosborg, I., Nihlgård, B., Gerhardsson, L. Hair element concentrations in females in one acid and one alkaline area in southern Sweden. *Ambio*, 32, 440-446, 2003.

Rothschild, R.F.N., Duffy, L.K. Methylmercury in the hair of subsistence food users in a rural Alaskan village. *Alaska Medicine*, 44, 2-7, 2002.

Sallsten, G., Thorén, J., Barregard, L., Schütz, A., Skarping, G. Long-term use of nicotine chewing gum and mercury exposure from dental amalgam fillings. *J. Dent. Res.*, 75, 594-598, 1996.

Salonen, J.T., Seppänen, K., Lakka, T.A., Salonen, R., Kaplan, G.A. Mercury accumulation and accelerated progression of carotid atherosclerosis: a population-based prospective 4-year follow-up study in men in eastern Finland. *Atherosclerosis*, 148, 265-273, 2000.

Schuurs, A., Exterkate, R., ten Cate, B. Biological mercury measurements before and after administration of a chelator (DMPS) and subjective symptoms allegedly due to amalgam. *Eur. J. Oral Sci.*, 108, 511-522, 2000.

Seifert, B., Becker, K., Helm, D., Krause, C., Schulz, C., Seiwert, M. The German Environmental Survey 1990/1992 (GerES II): reference concentrations of selected environmental pollutants in blood, urine, hair, house dust, drinking water and indoor air. *J. Expos. Anal. Environ. Epidemiol.*, 10, 552-565, 2000.

Seifert, B., Becker, K., Hoffman, K., Krause, C., Schulz, C. The German Environmental Survey 1990/1992 (GerES II): a representative population study. *J. Expos. Anal. Env. Epidemiol.*, 10, 103-114, 2000.

Soleo, L., Elia, G., Russo, A., Schiavulli, N., Lasorsa, G., Mangili, A., Gilberti, E., Ronchi, A., Balducci, C., Minoia, C., Aprea, C., Sciarra, G.F., Valente, T., Fenga, C. Reference values of urinary mercury in the Italian population (In Italian). *G. Ital. Med. Lav. Ergon.*, 25, 107-13, 2003.

Urban, P., Lukáš, E., Nerudová, J., Čábelková, Z., Cikrt, M. Neurological and electrophysiological examinations in three groups of workers with different levels of exposure to mercury vapors. *Euro. Neurol.*, 6, 571-577, 1999.

Vahter, M., Akesson, A., Lind, B., Bjors, U., Schütz, A., Berglund, M. Longitudinal study of methylmercury and inorganic mercury in blood and urine of pregnant and lactating women, as well as in umbilical cord blood. *Env. Res., A*, 84, 186-194, 2000.

Van Oostdam, J., Gilman, A., Dewailly, E., Usher, P., Wheatley, B., Kuhnlein, H., Neve S, Walker J, Tracy B, Feeley M, Jerome V, Kwavnick B. Human health implications of environmental contaminants in arctic Canada: a review. *Sci. Tot. Env.*, 230 1-82, 1999.

Weihe, P., Hansen, J.C., Murata, K., Debes, F., Jørgensen, P.J., Steuerwald, U. White RF, Grandjean P. Neurobehavioral performance of Inuit children with increased prenatal exposure to methylmercury. *Int. J. Circumpolar Health*, 61, 41-49, 2002.

WHO IPCS Environmental Health Criteria, Vol. 101, Methylmercury. Geneva: World Health Organization, 1990.

WHO IPCS Environmental Health Criteria, Vol. 118, Inorganic Mercury. Geneva: World Health Organization, 1991.

Chapter-17

METHYLMERCURY EXPOSURE IN GENERAL POPULATIONS OF JAPAN, ASIA AND OCEANIA

Mineshi Sakamoto[1], Akira Yasutake[2] and Hiroshi Satoh[3]

[1] *Department of Epidemiology, National Institute for Minamata Disease, 4058-18 Hama, Minamata, Kumamoto 867-0008, Japan*
[2] *Department of Basic Medical Science, National Institute for Minamata Disease, 4058-18 Hama, Minamata, Kumamoto 867-0008, Japan*
[3] *Department of Environmental Health Sciences, Tohoku University Graduate School of Medicine, 2-1 Seiryomachi, Aoba-ku, Sendai 980-8575, Japan*

INTRODUCTION

Fetuses and neonates are known to be high-risk groups for methylmercury (MeHg) exposure (Amin-Zaki et al., 1981). In the epidemics in Minamata and Niigata, Japan and in Iraq, many infants were congenitally affected by MeHg (Amin-Zaki et al., 1981; National Research Council, 2000). The clinical reports of patients in the Minamata district indicated that the infant victims showed symptoms such as severe cerebral palsy, while their mothers had only mild manifestations of the poisoning (WHO, 1990). The cause was fish contaminated by man-made environmental MeHg pollution. However, the main problem today is MeHg exposure through fish consumption as a widespread environmental neurotoxicant. In the natural course of events, most human exposure to MeHg is through fish/shellfish consumption. Generally, the larger fish and sea mammals at the top of the food chain, such as shark, tuna and whale, contain higher levels of MeHg than the smaller ones. The higher mercury accumulation and higher susceptibility to toxicity in the fetus than in the mothers during gestation is well established (WHO, 1990; National Research Council, 2000). Therefore, the effect of MeHg exposure on pregnant women remains an important issue for elucidation, especially for the Japanese and some Asian populations which consume much fish/shellfish. In this paper we first report on fetal MeHg exposure through the placenta in pregnant Japanese women (Sakamoto et al., 2004) and current methylmercury exposure levels in the population of Japan (Yasutake et al., 2004). The Japanese are known around the world as a population which consumes a large amount of fish/shellfish. Not only Japanese but some other Asian people also depend on marine products for their protein and other nutrients. The MeHg exposure status in Japan will provide important information about MeHg exposure in other Asian populations. Further, we summarized the data on hair mercury concentrations in some populations without any particular exposure to mercury in Asia and Oceania.

MATERIALS AND METHODS

MeHg exposure to fetus through placenta

The mercury concentration in red blood cells (RBC-Hg) is the best biomarker of MeHg exposure (Swedish Expert Group, 1971; WHO, 1990; Svensson et al., 1992). Additionally, more than 90% of Hg in RBC is known to be in the methyl form in high fish-consuming populations (Kreshaw et al., 1980). Further, hematocrit (Htc) values are quite different between mother and fetus at parturition (Table 1). Therefore, we used total Hg concentrations not in whole blood but in RBC to reveal the MeHg levels in mothers and fetuses in the present study. Sixty-three healthy pregnant Japanese women, ranging in age from 21 to 41 yr (average 29.6 ± 4.4), planning to deliver in Munakata Suikokai General Hospital, Munakata City, Fukuoka, Japan, gave informed consent to take part in the present trial. Blood samples were collected from the mothers and the umbilical cord. The samples included 13 ml of venous umbilical cord blood at birth and 10 ml of venous maternal blood 1 day after parturition before breakfast. Both blood samples were obtained by venipuncture with a small amount of heparin-Na and centrifuged at 3000 rpm for 10 min to separate red blood cells (RBC) and plasma. Samples were stored at -80°C until analysis. This study was approved by the Ethics Committee of the National Institute for Minamata Disease (NIMD). Total Hg in 0.5 g of RBC was determined by cold vapor atomic absorption spectrophotometry (CVAAS) according to the method of Akagi and Nishimura (1991). The method involves sample digestion with HNO_3, $HClO_4$ and H_2SO_4 followed by reduction to Hg^0 by $SnCl_2$. The detection limit was 0.1 ng/g. Accuracy was ensured using certified reference material (DORM-2; dogfish muscle prepared by the National Research Council,

Canada) as the quality control material; the Hg concentration obtained averaged 4.53 µg/g, as compared to the assigned value of 4.64 ± 0.26 µg/g. The total analytical precision of this analysis was estimated to be 3.9%.

CURRENT HAIR MERCURY LEVEL IN GENERAL POPULATIONS IN JAPAN

Hair samples were collected in 2000-2002 at beauty parlors and barber shops across Japan: in Miyagi (M=561, F=624), Chiba (M=253, F=232), Nagano (M=342, F=311), Wakayama (M=413, F=299), Tottori (M=611, F=207), Hiroshima (M=440, F=561), Fukuoka (M=474, F=570), Kumamoto (M=385, F=326), Minamata (M=389, F=648), Okinawa (M=406, F=613). Survey questionnaires were distributed at the time on fish consumption (amount and species), age, sex, hair-dye and artificial hair permanent wave treatment. About 60 to 70% of the visitor of a beauty parlors and barbershop approved to take part in the present trial. The hair samples were also collected at a primary school in each district to obtain samples from children. The age of the overall ranged from 0 to 95, with 92.4% ranged between the ages of 5 to 74. This study had been agreed by NIMD Board of Ethics. For mercury analysis, hair samples weighing from 0.1 to 1g each person, were washed well with detergent, and rinsed twice with acetone to dry. The dried hair was then cut into small pieces (<2 mm) with scissors. Aliquots of samples (15 to 20 mg) were dissolved in 0.5 ml of 2N NaOH with heating at 60°C for 1 hour. Ten or twenty µl of the solution was used to analyze the total mercury levels by oxygen combustion-gold amalgamation methods using an atomic absorption detector MD-1 (Nippon Instrument Co., Ltd., Osaka). 2.5 µM mercuric chloride (0.5 µg/ml) in 0.5 M L-cysteine/0.2% bovine serum albumin solution was used as an external standard.

HAIR MERCURY LEVELS IN SOME GENERAL POPULATIONS OF ASIA AND OCEANIA

NIMD has conducted many surveys in mercury contaminated sites. In the surveys, we collected the data in control populations without any particular contamination. We summarized the data by the survey of NIMD and other scientists as general hair mercury levels in some general populations in Asia and Oceania.

Various hair sampling areas. Shen Yang, China; A large and old industrial city, about 200 km from the Yellow Sea. Samples were collected from 55 mothers in a nursery school in 1999 (Liu et al., not published).

Zhoushan islands, China; Islands near Shanghai surrounded by sea with fishing as the main industry. Samples were collected from 28 males and 19 females in 2003. (Liu et al., not published).

Gui Yang, China; An inland medium-sized city located more than 500 km from the sea. Samples were collected from 28 inhabitants as a control at the mercury-contaminated site of Gui Zou Province in 1998 (Yasuda et al., 2002).

Sihanoukville, Cambodia; A small resort town facing the sea. Samples were collected from 18 males in 1999 (NIMD Mission Report, 1999).

Hoa Binh, Vietnam; A small town about 200 km from the Gulf of Tonkin. The samples were collected from 384 males and 329 females in 2000-2003 (Loi, 2003).

Deder Bubu, Kyrgyzstan; A small inland village in an area completely isolated from the sea. Samples were collected from 51 males and 40 females in 1997 (NIMD Mission Report, 1997).

Kojar, Kyrgyzstan; A medium-sized town near Deder Bubu in an area completely cut off from the sea. Samples were collected from 57 males and 99 females in 1997 (NIMD Mission Report, 1997).

Harbin, China and Medan, Indonesia; Harbin is a large northern city approximately 600 km from the coast. Medan is also a big city about 30 km from the coast. Samples were collected from 64 males in Harbin and 55 males in Medan. (Feng et al., 1998).

Hong Kong; Samples were collected from 166 husbands and 35 wives in an IVF(in vitro fertilization) Center (H.K.) from 1994 to 1996. (Dickman and Leung., 1998).

Davao, Philippines; Davao is a central city in a major island of Mindanao, Philippines. Forty two samples were collected in 1999 as a control of the gold mining area of Diwalwal. (Drasch et al., 2001)

Bangladesh; Samples were collected from 219 males from 6 different regions in Bangladesh in 1991. (Holsbeek et al., 1996)

Papua New Guinea; Samples were collected from 210 inhabitants from Lake Murray (n=114), Suki (n=51) and Rumginae (n=45) in Papua New Guinea (Kyle and Ghani, 1982).

RESULTS AND DISCUSSION

MeHg exposure to fetus through placenta

Table 1 presents the Japanese subject characteristics and values, including maternal and fetal mercury concentrations in red blood cells and hematocrit values. In all 63 cases fetal RBC-Hg were higher than maternal RBC-Hg.

Table 1. Japanese subject characteristics (n=63) and values, including maternal and fetal mercury concentrations in red blood cells and hematocrit values.

Category	Mean (Geomean)	SD	Min	Max
Maternal age (years)	29.6	4.37	21	41
Maternal RBCs Hg (ng/g)	9.12 (8.41)	3.63	3.76	19.1
Maternal Htc value	31.5	3.18	23.9	38.4
Fetal RBCs Hg (ng/g)	14.7 (13.4)	6.37	4.92	35.4
Fetal Htc value	45.2	3.64	38.1	53.7
Fetus/Mother RBCs Hg ratio	1.6	0.27	1.08	2.19

The geometric mean of fetal RBC-Hg was 14.7 ng/g for fetuses which was significantly ($p<0.01$) higher than that for mothers (8.41 ng/g). A strong correlation was observed in RBC-Hg between mothers and fetuses ($r = 0.92$, $p<0.001$; Figure 1), the average fetal/maternal RBC ratio was 1.6 (Figure 2). This suggests that MeHg is actively transferred to the fetus across the placenta via a neutral amino acid carrier, as demonstrated by previous studies (Kajiwara et al. 1996; Ashner and Clarkson, 1988).

This higher Hg accumulation in the fetuses than in mothers is widely acknowledged from human and animal studies (Amin-Zaki et al. 1981; WHO, 1990; National Research Council, 2000; Sakamoto et al. 2004; Sakamoto et al. 2002a; Sakamoto et al., 2002b).

In addition, the susceptibility of the developing brain itself is known to be high (WHO, 1990; Yasutake et al., 2004). Thus, intensive studies should be conducted on pregnant women and women of child-bearing age.

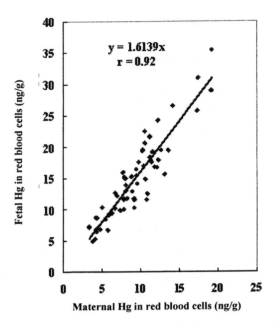

Figure 1. Correlation between maternal and fetal mercury concentrations in red blood cells in 63 maternal-fetal pairs.

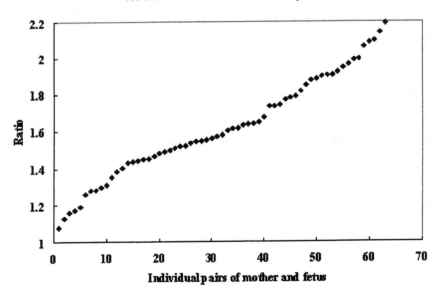

Figure 2. Individual fetus/mother ratio of Hg concentration in red blood cells in 63 maternal-fetal pairs.

However, the individual fetal/maternal RBC-Hg ratio varied from 1.08 to 2.19, indicating individual differences in MeHg concentrations between maternal and fetal circulations at late gestation. This can be partly explained by the individual differences in MeHg transfer from mother to fetus through the placenta. The maternal MeHg level tends to be influenced by the latest meal. On the other hand, the blood-organ ratios of MeHg concentration will be constant at parturition in fetal circulation. Therefore, not the maternal side biomarker, but the fetal side biomarker is essential to evaluate the MeHg exposure of the fetus during gestation.

Current hair mercury level in the general populations of Japan

Regional differences

Sex and age group-dependent hair mercury levels among the sampling districts are shown in Figure 3 (males=4274 and females=4391 in total 8665). The geometric mean mercury levels among the districts varied from 1.67 to 4.75 μg/g for males and from 1.07 to 2.29 μg/g for females.

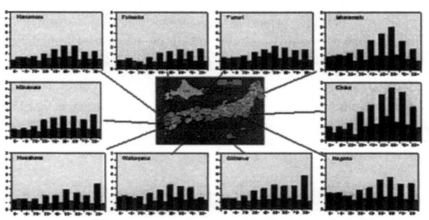

Figure 3. Sex and age group-depent hair mercury levels among 10 districts in Japan.

The values were highest in Chiba (4.75 μg/g for males, n=253; 2.29 μg/g for females n=232), followed by Miyagi (3.31 μg/g for male, n=561; 1.77 μg/g for female, n=624). The lowest values were found in Fukuoka (1.67

µg/g for males, n=474; 1.09 µg/g for females, n=570) and Hiroshima (2.02 µg/g for males, n=440; 1.07 µg/g for females, n=561).

The lower values were about half of those in Chiba and Miyagi. The variation seemed to depend on the total amount of daily fish/shellfish consumption and on the preference for tuna consumption. Table 2 shows the average daily intake of fish/shellfish and the rate of preference for tuna consumption. In all districts, males consumed a greater amount of fish/shellfish than females, which would partly explain the sex differences in the hair mercury levels. Tuna is a major carnivorous fish with a high methylmercury level that is often consumed in Japan. The highest rate of tuna consumption was found in Okinawa and Chiba. The highest hair mercury level in Chiba can be explained by both the high amount of fish/shellfish consumption and its high tuna consumption. Although Okinawa also showed high tuna consumption, their lower fish/shellfish consumption would tend to depress their hair mercury level. On the other hand, Fukuoka and Hiroshima with the lowest hair mercury level showed both lower fish/shellfish consumption and the lowest preference for tuna among all the districts. Therefore, not only the amount of fish/shellfish consumption but also the tuna preference rate would appear to explain the regional variations in hair mercury levels in Japan.

Table 2. Average fish consumption and tuna preference rate.

Residence	Fish consumption (g/person/day)			Frequency of tuna consumption (%) [a]
	Female	Male	Total	
Miyagi	98	95	96	69.0
Chiba	50	89	70	75.5
Nagano	56	59	57	60.5
Wakayama	53	57	55	52.5
Tottori	47	76	69	16.6
Hiroshima	43	55	48	30.6
Fukuoka	51	46	48	23.1
Kumamoto	46	54	50	38.8
Minamata	56	64	59	20.0
Okinawa	40	43	42	77.5

[a] Frequency of persons who often eat tuna

Average hair mercury levels for both sexes and age-dependent variations. Figure 4 shows the distribution of mercury concentrations in hair without permanent wave treatment.

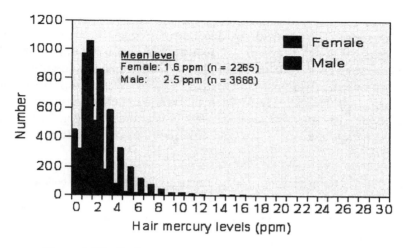

Figure 4. Distribution of hair mercury concentrations for both sexes in Japan.

We eliminated the data on hair with permanent wave treatment since thioglycolate used in the lotion for the process effectively removes some of the hair mercury. The geometric means of hair mercury concentrations were 2.40 µg/g (n=3668) for males and 1.63 µg/g (n=2265) for females.

The age group dependent distribution of hair mercury concentrations for both sexes without any permanent wave processing is shown in Figure 5.

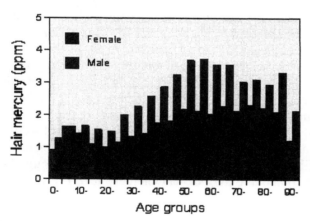

Figure 5. Age group-dependent distribution of hair mercury concentrations for both sexes in Japan.

Following a transient decline around the 20s, the level increased into the 50s in both sexes. The sex and age-dependent changes in mercury can be mainly explained by the amount of fish/shellfish consumption. Table 3 shows the average fish/shellfish consumption (g/day) among age-groups for both sexes in Japan (National Nutrition Survey by Ministry of Health in 2001). In addition, we can estimate that the average mercury concentration in fish/shellfish consumed by Japanese people is about 0.1 µg/g, from both the results of hair mercury levels and fish/shellfish consumption by age-groups and sexes.

Table 3. Average fish and shellfish consumption (g/day) in age-groups for both sexes in Japan (2001).

Age	Total	1-6	7-14	15-19	20-29	30-39	40-49	50-59	60-69	70-
Male	104.3	38.1	70.1	83.7	90.6	93.7	114.9	134	139	112
Female	84.9	38.9	59.9	76.5	72.4	72.4	85.2	109	109	93.2
Total	94.0	38.6	65.0	80.3	80.7	82.1	99.4	121	123	101
n:										
M	5,852	396	591	358	587	702	798	930	804	686
F	6,629	401	580	330	695	843	868	1,033	912	967

Soure: Date from National Nutrition Survey by Ministry of Health

Percentage of Japanese females of child-bearing age (15-49 years) exceeding recommended safe hair mercury levels . Table 4 shows the percentage of Japanese females of child-bearing age (15-49 years) above recommended safe hair mercury levels. About 0.1% of the sub-population had hair mercury exceeding 10 µg/g (a BMDL used by NAS). The Japanese Government recommended a safety exposure level for methylmercury as 3.4 µg Hg/kg bw/week (1973), corresponding to a hair mercury level of about 5 µg/g. About 1.7% of females of child-bearing age exceeded this level, and 25% of them exceeded the provisional tolerable weekly intake (PTWI) level (1.6 µg mercury/kg/week, corresponding to about 2.2 µg/g in hair mercury concentration) determined at the 61st meeting of the Joint FAO/WHO Expert Committee on Food Additives (JECFA). However, as many as 75% of females of child-bearing age exceeded the level (RfD: 0.1 µg Hg/kg bw/day corresponding to about 1 µg/g of hair mercury level) recommended by the U.S. EPA.

Table 4. Percentage of Japanese females of child-bearing age (15-49 years)
Exceeding some recommended safe hair mercury levels.

$\bigcirc \to \geq$ 1 µg/g	73.7% (RfD by EPA)
$\bigcirc \to \geq$ 2.2 µg/g	24.9% (PTWI by 61st JECFA)
$\bigcirc \to \geq$ 5 µg/g	1.7% (Japan)
$\bigcirc \to \geq$ 10 µg/g	0.1% (a BMDL used by NAS)

Hair mercury levels in some general populations of Asia and Oceania .
Such wide regional variations seemed to depend mainly on the total daily
intake of fish/shellfish and also the species of fish/shellfish consumed daily.
The inhabitants living in Lake Murray (in Papua New Guinea) area who
showed the highest hair mercury level consumed about 9 kg fish/month
(Kyle and Ghani, 1982). The Zhoushan Islands (China) which showed the
second highest hair mercury level are famous for the largest fish catches in

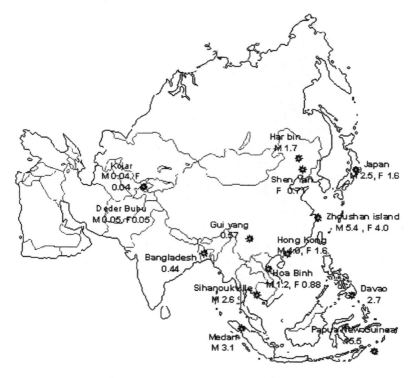

Figure 7. Hair mercury concentrations (µg/g) in some Asian and Oceania districts.

China.

The average concentration of hair mercury sampled in general populations without any particular exposure to mercury were (Fig. 7): Shen Yang, China; 0.71 μg/g for females (n=55), Zhoushan islands, China; 5.4 μg/g for males (n=28) and 4.0 μg/g for females (n=19), Gui Yang, China; 0.57 μg/g (n=28), Sihanoukville, Cambodia; 2.6 μg/g for males (n=18), Hoa Binh, Vietnam; 1.2 μg/g for males (n=384) and 0.88 μg/g for females (n=329), Deder Bubu, Kyrgyzstan; 0.05 μg/g for males (n=51) and 0.05 μg/g for females (n=40), Kojar, Kyrgyzstan; 0.04 μg/g for males (n=57) and 0.04 μg/g for females (n=99). Harbin, China; 1.7 μg/g for males (n=64). Medan, Indonesia 3.1 μg/g for males (n=55). Hong Kong; 4.0 μg/g for males (n=166) and 1.6 μg/g for females (n=35). Davao, Philippines; 2.7 μg/g (n=42), Bangladesh; 0.44 μg/g (n=219). Papua New Guinea; 15.5 μg/g (n=210), 18.0 μg/g for Lake Murray (n=114), 8.3 for Suki (n=51) and 3.2 for Rumginae (n=45).

The inhabitants eat fish/shellfish almost every day. The average hair mercury level was similar to that in Chiba, which showed the highest level in Japan. The male inhabitants in Hong Kong also showed a level similar to that of Chiba males. The inhabitants in Sihanoukville (Cambodia) and Davao (Philippines) showed a level similar to that of Japanese males. The town faces the sea, and so fish is a dietary staple. On the other hand, the inhabitants of Sheng Yang (China), Gui Yang (China) and Bangladesh who seldom eat fish/shellfish showed less than half the average Japanese hair mercury level. People living in Kojar and Deder Bubu, Kyrgyzstan, which are completely isolated from the sea, do not have many opportunities to eat fish/shellfish, and showed very minimal mercury levels.

In the natural course of events, most human exposure to MeHg is through fish/shellfish consumption. The methylmercury exposure estimated from hair mercury levels depends on the amount and species of fish/shellfish consumed daily. The developing brain in the late gestation period is known to be most vulnerable. Further, more methylmercury accumulates in the fetuses than in mothers. Therefore, efforts must be made to protect the fetuses from the risk of MeHg, especially in populations which consume a lot of fish/shellfish. For pregnant women and those who may become pregnant, Ministry of Health and Welfare, Japan recently (June 2003) issued a caution to reduce the consumption of several kinds of fish and whales that contain high mercury. If human exposure to MeHg were independent of nutrition from fish, one could aim at the zero exposure. However, the exposure is through fish, a very important source of protein and other nutrients (Clarkson and Strain, 2003; Sakamoto et al., 2004) especially for Japanese and some Asian populations. As shown in Figure 6 fish/shellfish is an important source of protein for the Japanese. Pregnant women in particular need not give up eating fish to obtain the nutritional benefits. However, they would do well to at least consume smaller fish, which contains less MeHg, thereby balancing the risks and benefits of fish consumption (Sakamoto et al., 2004).

REFERENCES

Akagi, H., Nishimura, H. Speciation of mercury in the environment. In: Suzuki, T., Imura, N., Clarkson T.W., (eds), Advances in mercury toxicology. New York, 53-76.

Amin-Zaki, L., Majeed, M.A., Greenwood, M.R., Elhassani, S.B., Clarkson, T.W., Doherry, R.A. Methylmercury poisoning in the Iraqui suckling infant: a longitudinal study over five years. *J. Appl. Exp. Toxicol.*, 1, 210-214, 1981.

Ashner, M., Clarkson, T.W. Distribution of mercury 203 in pregnant rats and their fetuses following systematic infusion with thiol-containing amino acids and glutathione during late gestation. *Teratology*, 38, 145-155, 1988.

Clarkson, T.W., Strain, J.J. Nutritional factors may modify the toxic action of methyl mercury in fish-eating populations. *J. Nutr.*, 133, 1539S-1543S, 2003.

Dickman, M.D., Leung, K.M.C. Mercury and organochlorine exposurefrom fish in Hong Kong. *Chemosphere*, 37, 991-1015, 1998.

Drash, G., Bose-O'Reilly, S., Beinhoff, C., Roider, G., Maydl, S.. The Mt. Diwata study on the Philippines 1999 – assessing mercury intoxication of the population by small scale gold mining. *Sci. Tot. Env.*, 267, 151-168, 2001.

Feng, Q., Suzuki, Y., Hisashige, A. Hair mercury levels of residents in China, Indonesia, and Japan. *Arch. Env. Health*, 53, 36-43, 1998.

Holsbeek, L., Das, H.K., Joiris, C.R. Mercury in human hair and relation to fish consumption in Bangladesh. *Sci. Tot. Env.*, 30, 181-188, 1996.

Kajiwara, Y., Yasutake, A., Adachi, T., Hirayama, K. Methylmercury transport across the placenta via neutral amino acid carrier. *Arch. Toxicol.*, 70, 310-314, 1996.

Kreshaw, T.G., Clarkson, T.W., Dhahir, P.H.. The relationship between blood levels and dose of methylmercury in man. *Arch. Env. Health*, 35, 28-36, 1980.

Kyle, J.H., Ghani, N.,. Methylmercury in human hair: A study of a Papua New Guinean population exposed to methylmercury through fish consumption. *Arch. Env. Health*, 37, 266-271, 1982.

Loi, V.D.. Mercury pollution due to gold mining activities in Vietnam. *Proceedings of NIMD Forum*, 90-92, 2003.

National Research Council.. Toxicological effects of methylmercury. Washington, DC: National Academy Press, 2000.

NIMD Mission Report. – Investigation into suspected mercury contamination at Deder Bubu, near Kojar, Osh, Kyrgyzstan, requested by UNHCR (United Nations High Commissioner for Refugees) 1997.

NIMD Mission Report. – Investigation into suspected mercury contamination at Sihanoukville, Cambodia requested by WPRO (World Health Organization Regional Office for the Western Pacific, 1999.

Sakamoto, M., Kakita, A., Wakabayashi K., et al.. Evaluation of changes in methylmercury accumulation in the developing rat brain and its effects: a study with consecutive and moderate dose exposure throughout gestation and lactation periods. *Brain Res.*, 949, 43-50, 2002a.

Sakamoto, M., Kubota, M., Matsumoto S., et al.. Declining risk of methylmercury exposure to infants during lactation. *Env. Res.*, 90, 185-189, 2002b.

Sakamoto, M., Kubota, M.. Plasma fatty acid profiles in 37 pairs of maternal and umbilical cord blood samples. *Environ. Health Prev. Med.*, 9, 67-69, 2004.

Sakamoto, M., Kubota, M., Liu, XJ, Murata, K., Nakai, K., Satoh, H.. Maternal and fetal mercury and n-3 polyunsaturated fatty acids as a risk and benefit of fish consumption to fetus. *Env. Sci. & Technol.*, 38, 3860-3863 (Web Release Date: 21-May-2004; (Article) DOI: 10.1021/es034983m) 2004.

Svensson, B-G., Schülts, A., Nilsson, A., Åkesson, I., Åkesson, B., Skerfving, S.. Fish as a source of exposure to mercury and selenium. *Sci. Tot. Env.*, 126, 61-74, 1992.

Swedish Expert Group. Mercury in fish: A toxicological-epidemiologic evaluation of risks. Nord Hygienisk Tidskrift, Suppl. 4. Stockholm, Sweden, 1971.

Takeuchi, T.. Pathology of fetal Minamata disease. The effect of methylmercury on intrauterine life of human beings. *Pediatrician*, 6, 69-87, 1997.

World Health Organization (Ed.), IPCS, Environmental Health Criteria 101. Methylmercury, World Health Organization, Geneva, 1990.

Yasuda, Y., Lia, Q., Matsuyama A., et al.. A survey of environmental pollution by mercury derived from a chemical factory drain in Gui Zhou, China. Proceedings of International Workshop on Health and Environmental Effects of Mercury, 31-46, 2002.

Yasutake, A., Matsumoto, M., Yamaguchi, M., Hachiya, N.. Current hair mercury levels in Japanese for estimation of methylmercury exposure. *J. Health Sci.*, 50, 120-125, 2004.

Chapter-18

MERCURY POLLUTION FROM ARTISANAL GOLD MINING IN BLOCK B, EL CALLAO, BOLÍVAR STATE, VENEZUELA

Marcello M. Veiga[1,2], Dario Bermudez[3,2], Heloisa Pacheco-Ferreira[4,2], Luiz Roberto Martins Pedroso[5,2], Aaron J. Gunson[1,2], Gilberto Berrios[6], Ligia Vos[7], Pablo Huidobro[8] and Monika Roeser[8]

[1] *Dept. Mining Engineering, University of British Columbia, Vancouver, Canada*
[2] *Consultants to UNIDO*
[3] *National Experimental University of Guayana, Puerto Ordaz, Venezuela*
[4] *Dept. of Preventive Medicine - Faculty of Medicine, Federal University of Rio de Janeiro, Brazil*
[5] *CETEM – Center of Mineral Technology, Rio de Janeiro, Brazil*
[6] *Hecla Mine, Venezuela*
[7] *CVG - Venezuelan Corporation of Guayana, VP Environment, Venezuela*
[8] *UNIDO – United Nations Industrial Development Organization, Vienna, Austria*

INTRODUCTION

Artisanal and small-scale gold mining (ASM) is an essential activity in many developing countries. The current number of artisanal gold miners is estimated to be between 10 and 15 million people worldwide (Veiga and Baker, 2004) with almost 30% of this contingent being women (Hinton et al, 2003). Since 1998, annual gold production from ASM has constituted 20 to 30% of the global production, ranging from 500 to 800 tonnes (UNEP, 2002; MMSD, 2002). Assuming that miners lose between 1 and 2 grams of Hg per gram of gold produced, it is estimated that annually between 650 and 1000 tonnes of Hg are released into the environment. The predominant source of ASM Hg release is China (200 to 250 tonnes/yr) followed by Indonesia, which releases 100 to 150 tonnes/yr, while Brazil, Bolivia, Colombia, Peru, Philippines, Venezuela and Zimbabwe each release from 10 to 30 tonnes/yr of Hg (Gunson and Veiga, 2004; Shoko and Veiga, 2003; Veiga, 2003; Veiga and Hinton, 2002). Mercury releases in Latin America are declining, as the most easily extractable ore has been depleted and the operating costs have increased. However, the gradual increase in the price of gold in 2003 is motivating miners to re-work abandoned ore deposits.

The southern part of Venezuela below the Orinoco River, involving the State of Bolívar, the State of Amazonas, and the Federal Territory of Delta Amacuro, is called the Guayana Region. The main mining activities are conducted in the State of Bolívar, which has an area of 240,528 km², comprising 75% of the hydroelectric potential of the country. Less than 5% of the Venezuelan population (which is 24.2 million) lives in the Guyana Region. In 1999, the labor force experienced a 1.1% decrease in number resulting in an unemployment rate of 13.2% (1,365,752 people). In 2000, 63% of the individuals making up the workforce were men. Unemployment among men reached 12.5%, 1.1% higher than 1999. In 2000, 14.4% of women did not have a job. This was 1.7% higher than in 1999 (CONAPRI, 2003).

El Callao is located in the Northeastern part of the State of Bolívar, 150 km from Ciudad Guayana (Puerto Ordaz). Mining started in 1724, when Capuchin priests explored the area.

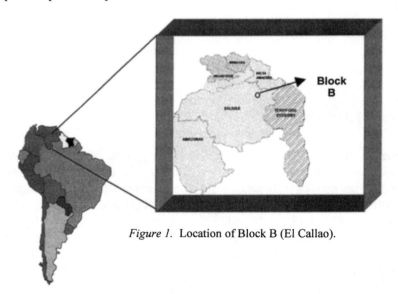

Figure 1. Location of Block B (El Callao).

The municipality of El Callao was founded in 1853 with the name of Caratal and many small gold mining companies were installed in the region. In 1970, CVG (Corporacion Venezoelana de Guayana) incorporated Minerven, a state-owned company, which currently has two cyanidation plants producing together approximately 200 to 300 kg of gold/month: the Peru Plant processes 5,200 tonnes of material/month and the Caratal Plant processes 14,000 tonnes/month. About 15% of the Peru Plant material is Hg-contaminated tailings purchased by the company from the artisanal gold miners.

CVG-Minerven owns a total area of 48,848 ha of mining concessions. The company granted 77 concessions of mining of which 59 are contracts with companies and 18 with individuals.

The main portion of CVG-Minerven mining concessions "rented" to third parties is named "Block B" (Figure 1). With an area of 1,785 ha, this site was chosen by UNIDO for this project. CVG-Minerven has also rented mining areas in Block B to organized companies (e.g. the American company Hecla Mine).

In the State of Bolívar, with a population of 1,214,486, there are about 15,000 people directly involved in ASM. This includes about 2000 "bateeros" and "suruqueros" who are those miners using pans to extract gold and diamonds from alluvial deposits and tailings, 5,000 miners using hydraulic monitors in elluvial and colluvial operations (gold and diamond), 3,000 miners working in hard rocks (quartz veins) and 5,000 miners operating in dredges and rafts in waterways all over the state.

In Venezuela, two hundred and fifty years after the beginning of (the paper says mining started in 1724) mining activities, the social and economic situation of the artisanal miners has not changed substantially. Observations of small gold miners in the State of Bolívar, Venezuela, reveal serious effects on families as well as the degradation of the community's socio-economic conditions.

In El Callao, ASM miners mostly work in the CVG-Minerven concessions but there are also some illegal miners working outside these concessions. This has been generating employment for the surrounding communities. Since the gold ore is abundant and extremely rich, people rarely consider other types of economic activity. The ASM miners, to some extent, have played the role of gold prospectors for the company. As the price of gold has been increasing since the end of 2002, the number of ASM miners has increased substantially in the region and many of them are outsiders who have never had any previous experience in mining.

Since 1995, UNIDO has provided technical assistance related to mercury pollution for government, companies and artisanal gold miners in Bolívar

State, Venezuela. The situation, as reported by UNIDO (1996), was extremely serious in El Callao with miners and millers indiscriminately using mercury to amalgamate gold.

The current work was conducted at the end of 2003 as a preliminary mission to assess the current health situation and to prepare for a more substantial project to be conducted in the future. The mission assessed the level of mercury intoxication of the miners and the surrounding population and introduced simple mineral processing techniques capable of substantially reducing mercury exposure and release in the region.

POPULATION OF BLOCK B

According to the 2001 Census, the total population of El Callao is 17,410 and there are 1,731 people living in Block B, which represents almost 10%

Table 1. Main Characteristics of the Population of Block B.

	Chile	La Fabrica	La Iguana	Monkey Town	Nuevo Mexico
Number of inhabitants	411	359	147	433	381
% Females	45.5	52.9	33.3	47.1	47.5
Main age group	Under 18 (50%)	Between 19 and 55 (51%)	Older than 56 (40%)	Between 19 and 55 (49%)	Under 18 (47%)
Average number of sons per family	2.4	2.4	1.3	2.2	2.6
% Man living with woman	46.6	72.4	42.9	67.4	71.6
Average number of people per house	3.7	3.7	2.5	3.6	3.7
% Illiterates	27.7	19.5	42.2	18.0	21.0
% Individuals with no technical education	98.1	90.0	99.3	96.1	96.1
Main education level	83% Primary	73% Primary-secondary	90% Primary	75% Primary-secondary	72% Primary-secondary
Main occupation	Unemployed - Miner	Unemployed - Miner	Unemployed-Framer	Miner - Unemployed	Unemployed - Miner
Main type of house	Zinc - Wood	Cement - Mud	Zinc -Cement	Cement - Wood	Cement - Mud
Main type of roof	Zinc	Zinc	Zinc	Zinc	Zinc
Main type of floor	Cement - earth	Cement	Earth - cement	Cement	Cement
% Houses with electricity	96.5	92.1	44.9	99.0	93.8
Main source of water	Truck	Pipe - truck	Creek - pipes	Truck	Pipe - truck
Main sanitation	Bush - toilet	Toilet	Bush	Toilet	Toilet

of the El Callao population (Hecla, 2004).

The main characteristics of the five communities located in Block B are shown in Table 1. About 47% of the population is female and 44% is younger than 18. Children under 15 years of age account for 30% of the population. Individuals between 19 and 55 years of age represent 46.5% of the Block B population. On average, the couples in the region have 2.3 children and there are 3.5 persons per family. About 1/5 of the population is illiterate and 1/4 does not have any kind of technical education. About 2/5 of the population has a primary education (6th grade), 1/4 a secondary education and just 2% has a technical or university education. About 30% of the population defines its occupation as student, 21% as unemployed, 19% as housewife and just 15% as miner. However the main activity of the population is definitely gold mining and processing. Water is available mainly through water trucks (56%) and just 1/3 of the population has water from pipes. About 70% of the population has toilets.

PROCESSING CENTERS

About 250 ASM miners in Block B excavate the ore from 30-80 m deep shafts using explosives and transport it in 50-kg polyethylene bags to the Processing Centers (locally known as "molinos") to be crushed, ground, concentrated and amalgamated. About 30% of the ore entering Block B comes from areas located outside of Block B.

Miners pay 10% of the recovered gold to the millers, known as "molineros" (Center owners). In the Block B area there are 28 operating Processing Centers, each one with 3 to 6 hammer mills making a total of 86 active hammer mills and 25 jaw crushers in the region. The Centers are basically located in 3 communities along the main Caratal-Chile road (from East to West): Nuevo México, La Fabrica and Monkey Town. There are some minor processing activities in the communities of Chile and La Iguana and 94% of the Processing Centers are located in the communities of Nuevo México and Monkey Town.

The primary ore is transported to the Processing Centers, where millers crush it to below 2 inches with jaw crushers and make a heap to feed 25 HP hammer mills. There is no concentration process, i.e. the ore is ground to −1 mm in hammer mills and passed on copper-amalgamating plates (Figure 2).

Figure 2. Hammer mill and copper-amalgamating plates.

Amalgamation Cu-plates are stationary copper sheets, usually dressed with a thin layer of mercury (usually 150g Hg/m² of plate) used to amalgamate free gold particles. Working with a slope of 10%, these 1.5 to 2 m-long plates receive pulp of auriferous ground ore (10 to 20 % solids in the pulp), and the amalgamation takes place when gold particles contact the plate surface. The efficiency of the process is low due to the short time of ore-mercury contact. About 0.3 m² of plate is required to treat 1 tonne of ore/24 h for pulps with 20% solids. Abrasion of the mercury surface releases droplets that go out with the pulp. A large majority of artisanal miners in El Callao do not use a mercury trap at the end of the plates. So, tailings from Cu-plates typically contain 60 to 80 ppm Hg.

Periodically the process is interrupted and amalgam is scraped off the plates with a sharp piece of metal. At this stage, miners are exposed to high levels of Hg vapor. Quite often the Venezuelan miners burn the amalgamation plates to "remove" fine gold trapped on the plate. The amalgam recovered from the plate is squeezed to eliminate excess mercury and burned on a tray or a shovel. Some millers have good retorts available for miners but they argue that the retorting time is too long (15 minutes) and they simply use a propane blowtorch to decompose amalgam, emitting a large amount of mercury into the atmosphere and exposing themselves to mercury vapor. This is clearly contaminating everyone directly involved in the ore processing as well as their neighbors, since the Processing Centers are very near the houses.

The ore in El Callao is usually extremely rich, with grades ranging from 12 to 20 g Au/tonne. Through interviewing several millers, it was observed that a Processing Center produces on average 100 to 200g Au/day (2.6 to 5.2 kg/month). Each hammer mill processes 1.7 to 2.5 tonnes/day or 5 to 7.5 tonnes/Processing Center/day. The daily gold production can reach as much as 1 kg/day depending on the type of ore being processed. Based on the average gold production reported by the interviewed millers, Block B production might be around 1 to 2 tonnes Au/a. In all of El Callao, the gold production can reach as much as 5 to 6 tonnes/a considering that there are 80 to 90 Processing Centers in the region. According to CVG-Minerven engineers, since the gold is very fine (size smaller than 0.074mm) artisanal miners cannot achieve the liberation size using hammer mills and in the amalgamation process just 30% of the gold is trapped. The rest is sent to the tailing ponds and later sold to the mining companies.

Gold is not melted in Block B but in the village of El Callao where there are about 25 gold shops where gold *doré* (i.e. the product of burning amalgam) is sold and consequently melted. Some of these shops are in family houses. As the gold *doré* may contain up to 10% mercury, as a result of incomplete burning, this mercury is released in the urban environment during melting.

The price of gold paid to miners is around US$10/g in Venezuelan currency (Bolívars) at the official exchange rate. As the US dollar in the black market reaches prices at least 70% higher than the official rate, it is highly probable that miners are selling gold in neighboring countries, e.g. Brazil.

After visiting the Processing Centers and discussing with local experts, an average operating cost was obtained. The mining and milling costs are quite dependent on the type of ore being processed. When grinding hard ores, the hammers are changed after grinding 1.5 to 2 tonnes of ore. This represents the major cost of the milling operation (as high as 65% of the operating costs). However, millers do not charge more for milling hard ores. Each Processing Center counts 3 to 5 "employees" who usually do not receive a salary but just live off the gold left (trapped) inside the hammer mills. At the end of the day the "employees" open the mills and clean them on the amalgamation plates to recover their earnings. Considering that most Processing Centers do not pay their employees, the operating cost of a Processing Center must be between 0.43 g Au/tonne and 0.97 g Au/tonne of ore processed. As the "molinero" (mill owner) receives 10% of the gold produced, his/her break-even point is reached when processing ores with 0.24 and 0.5 g Au/bag for soft and hard ore respectively. Many miners do not acknowledge this and they work below the break-even point.

In general, the metallic mercury used in Block B comes from Brazil and it is not legally purchased. It is observed that typically a Processing Center buys (and loses) 6 to 8 kg of Hg/month. This represents 14.4% to 32.6% of the operating costs when hard and soft ores are milled respectively. Mercury is sold in the area at a price around US$ 20-25/kg, which is 5 to 6 times higher than the international market price. Millers provide Hg for the miners who add it to the plates during operation. The ratio Hg_{lost}: $Au_{produced}$ is around 1.5 to 2. The Hg release in Block B is estimated to be between 2 and 4 tonnes/a. and can reach as much as 12 tonnes/a in all of El Callao.

Millers accumulate Hg-contaminated tailings in their ponds and sell it to CVG-Minerven and eventually to other companies applying cyanidation to extract residual gold. The companies re-grind the tailings to below 200 mesh (0.074mm) and leach the material with cyanide. CVG-Minerven plant operators do not have control of Hg in the effluents or in the gold melting room. This mercury is definitely contaminating plant operators and reducing the efficiency of the gold precipitation with zinc dust (Merrill-Crowe Process). The company does not buy tailings with less than 6g Au/tonne. They pay for tailings with 6, 12 and 20g Au/tonne, 30%, 40% and 50% of the value of pure gold respectively. As the amount of tailing produced is equal to the amount of material ground, the production of tailings in Block B must be between 44,000 and 65,000 tonnes/yr. Considering an average grade of 7 g Au/tonne (which is low for El Callao), the amount of gold going to the tailings per year in Block B is around 308 to 455 kg. This divided by 28 Processing Center owners gives 11 to 16 kg Au/per owner. As the company pays around 30% of the gold value for tailings (1 g = US$13.6 New York Market on April 7, 2004), a miller would receive at least something between US$ 44,880 and US$ 65,280 per year when selling his/her tailings. This might be the minimum received by miners since the companies do not buy tailing with grade below 6 g Au/tonne. This is clearly a better business for the Processing Center owners than the processing operation itself, where they receive 10% of the gold production. The miners and the employees of the Processing Centers are the main victims of this unfair system.

This work did not attempt to assess the environmental problems caused by mercury use in Block B. However, it is clear that the runoff water coming from the Processing Centers raises environmental concerns. The water passes through the Hg-contaminated tailings on its way to the Yuruarí River, which supplies water to the population of El Callao and nearby communities. The high level of organic matter in the region, associated with the large amount of Hg-contaminated suspended particles being carried by the water, creates conditions to oxidize and complex the metallic mercury released by miners. Soluble Hg-organic complexes may eventually be transformed into the most toxic form of mercury, methylmercury (Meech et al, 1998). However, the eventual bioaccumulation of methylmercury in the region seems to have less impact on humans than the occupational exposure to metallic mercury vapor, since just a few small fish are found in the streams and (local) fish are not a staple food.

HEALTH ASSESSMENT

The health assessment combined information of total Hg concentration in urine with medical exams to evaluate the level of impact that the pollutant caused or may cause to individuals residing in this mining hotspot.

Inhalation of Hg vapor is more significant for mining and gold shop workers directly involved in handling metallic mercury, but can also indirectly affect surrounding communities. Once in the lungs, Hg is oxidized, forming Hg (II) complexes that are soluble in many body fluids. Acute Hg poisoning, which can be fatal or can cause permanent damage to the nervous system, has resulted from inhalation of 1,200 to 8,500 $\mu g/m^3$ of Hg (Jones, 1971). Impairment of the blood-brain barrier, together with the possible inhibition by Hg of certain associated enzymes, will certainly affect the metabolism of the nervous system (Chang, 1979). Hg vapor is completely absorbed through the alveolar membrane and complexes in the blood and tissues before reacting with biologically important sites. The biological half-life of Hg absorbed as vapor in blood is about 2-4 days with 90% is excreted through urine and feces. This is followed by a second phase with a half-life of 15-30 days (WHO, 1991). The time interval between the passage of elemental Hg through the alveolar membrane and complete oxidation is long enough to produce accumulation in the central nervous system (Mitra, 1986). Mercury can irreversibly damage the nervous system. Kidneys are the most affected organs in exposures of moderate duration to considerable levels, while the brain is the dominant receptor in long-term exposure to moderate levels. Total mercury elimination through urine can take several years. Then, the Hg levels in urine would not be expected to correlate with neurological findings once exposure has stopped (Stopford, 1979).

Symptoms typically associated with high, short-term exposure to Hg vapor (1,000 to 44,000 $\mu g/m^3$), conditions which miners are subjected to when they burn amalgams in open pans, are chest pains, dyspnoea, cough, haemoptysis, impairment of pulmonary function, and interstitial pneumonitis. Long-term, low-level Hg vapor exposure has been characterized by less pronounced symptoms of fatigue, irritability, loss of memory, vivid dreams, and depression. Acute exposure has caused delirium, hallucinations and suicidal tendency as well as erethism (exaggerated emotional response), excessive shyness, insomnia, and in some cases muscular tremors. The latter symptoms are associated with long-term exposure to high levels of Hg vapor. In milder cases, erethism and tremors regress slowly over a period of years following removal from exposure pathways (WHO, 1991). A person suffering from a mild case of Hg poisoning can be unaware because the symptoms are psycho-pathological. These ambiguous symptoms may result in an incorrect diagnosis.

Since inorganic Hg poisoning affects the liver and kidneys, high Hg levels in the urine can indicate undue exposure to Hg vapor. WHO (1991) collected a large amount of evidence to conclude that a person with a urine Hg level above 100 µg/g creatinine has a high probability of developing symptoms such as tremors and erethism. For Hg levels between 30 and 100 µg/g creatinine, the incidence of certain subtle effects in psychomotor performance and impairment of the nerve conduction velocity can increase. The occurrence of several subjective symptoms such as fatigue, irritability, and loss of appetite can be observed. For Hg levels below 30-50 µg/g creatinine, mild effects can occur in sensitive individuals but it seems more difficult to observe symptoms. Drake et al. (2001) found a significant correlation between Hg in air from 0.1 to 6,315 µg/m³ and urine mercury levels from 2.5 to 912 µg/g of creatinine in gold miners from El Callao. Tsuji et al. (2003) evaluated ten studies reporting paired air and urine Hg data and obtained a strong correlation between both media at medium and high concentrations. At air concentrations below 10 µg/m³, the authors concluded that the concentration of Hg in urine was indistinguishable from background levels. The World Health Organization (1991) described a relationship between Hg in air (A) in µg/m³ and in urine of exposed workers (U) expressed as µg/g creatinine: $U = 10.2 + 1.01 A$. Thus a person exposed to about 40 µg/m³ of Hg in air should show levels of Hg in urine around 50 µg/g creatinine. This is the maximum urine Hg concentration recommended by WHO. Drasch et al. (2002) consider the Hg level in urine of 5 µg Hg/g creatinine an alert value and 20 µg Hg/g creatinine as an action level, i.e. the individual must be removed from the pollution source.

URINE ANALYSIS

The urine samples in Block B were collected in 50 mL vials and total mercury analyses were processed using a LUMEX portable atomic absorption spectrometer (RA 915+) coupled to a pyrolysis chamber (RP 91C). The equipment works according to the principle of the thermal destruction of the sample followed by the determination of the amount of elemental mercury released. A small volume of urine sample, in this case 100 μL, was obtained with a micro-pipette, introduced in a quartz crucible and then into the pyrolisis chamber (RP 91C) that operates at 800°C. The vapor released in the pyrolysis chamber then entered the atomic absorption spectrometer (RA 915+). All procedures were controlled by a laptop computer. LUMEX uses a Zeeman process (Zeeman Atomic Absorption Spectrometry using High Frequency Modulation of Light Polarisation ZAAS-HFM) that eliminates interferences and does not use a gold trap. The detection limit of the urine samples established in the Venezuelan analytical conditions was 0.2 μg Hg/L. This equipment was able to analyze 300 urine samples in 12 hours.

Creatinine analysis was performed using a Bioclin kit from the company Quibasa. Creatinine reacts with picric acid, to form a yellow-reddish chemical complex in conditions where the maximum production of the dyed complex creatinine-picrate occurs. The spectrometric analyses were conducted at wavelength of 510 nm in a Bausch & Lomb Spectronic 20 spectrophotometer.

In order to evaluate the LUMEX analytical precision, urine samples from 15 selected volunteers were collected and analyzed using LUMEX, and sent to three Venezuelan institutions: Laboratorio de Espectroscopia Molecular, Facultad de Ciencias, Universidad de los Andes (ULA), La Salle Institute, and UCV-Universidad Central de Venezuela (Caracas). The analytical method used by ULA and UCV was cold vapor atomic absorption spectrometry. La Salle used atomic absorption spectrometry with hydride generation.

The linear correlation coefficients (r)3 between Hg analyzed by LUMEX and the results from the Venezuelan laboratories were: 0.8868 (ULA), 0.9178 (UCV), and 0.9690 (La Salle). This indicates a relatively good performance of the portable analyzer. However, the largest discrepancies occurred among the Venezuelan laboratories. There are many reasons for this, such as: type of analytical equipment, analytical procedure, quality of reagents and water, cleanliness of the laboratory glassware and environment (in particular laboratory air), quality of Hg standards used in the analysis, quality control methodology, stability of the electricity source, etc.

MEDICAL EXAMS

The medical exam followed the Protocols developed by UNIDO (Veiga and Baker, 2004). Questions related to the health history of the volunteers were applied in order to exclude participants with severe diseases from the statistical evaluation (e.g. someone who has had a stroke might be excluded from the survey). Individuals were selected for a series of specific neuropsychological tests designed to detect the effects of mercury poisoning. These tests were applied by medical doctors, and were followed by local health care professionals. All volunteers involved in the medical exams signed an agreement to participate in the Health Assessment involving four questionnaires/exams:

1. Evaluation of risk of mercury exposure (personal data, occupational exposure to mercury, confounding factors to exclude candidates with other problems, diet issues including frequency and type of food),
2. General health (questions related to health conditions and subjective symptoms as described by the patient, e.g. metallic taste, salivation, fatigue, etc),
3. Clinical-neurological exams (e.g. blood pressure, signs of gingivitis, ataxia, tremors, reflexes, etc.),
4. Specific neuropsychological tests (e.g. memory, coordination, etc.).

All the results of these questionnaires/exams were compared with the mercury analysis in urine samples. All the procedures were clearly explained to the population. All questionnaires were translated into Spanish.

The clinical-neurological exams, a fundamental part of the assessment of the evolution of metallic mercury exposure, allow the observation of speech problems, walking, balance, coordination, muscular strength, tactile sensibility, autonomous features of the cranial and spinal nerves, superficial and profound reflexes, and other features.

Specific neuropsychological tests were applied to test and evaluate:

- Recent memory, using the Wechsler Memory Scale (WMS),
- Episodic memory,
- Fine motor coordination using the MOT Test (match box),
- Coordination and dexterity of the hand,
- Spatial perception,
- Fine motor and manual dexterity,
- Motor-visual coordination,

- Perception of the background figure,
- Perception of the constancy of forms,
- Visual perception (Frostig).

The questionnaires related to mercury exposure and to general health were applied by a local nurse. The clinical-neurological exams and specific neuropsychological tests were conducted by a local physician and an experienced foreign neurotoxicologist.

SELECTION OF VOLUNTEERS

The Health Assessment volunteers were preliminarily selected based on the population distribution of the different communities of Block B and based on previous work conducted by Hecla (2004) that defined the demographic distribution. About 500 possible volunteers were identified in a proportion of 40% men, 30% women and 30% children. Out of the 500, a total of 209 samples of urine (66 from women, 62 from children, 48 from millers and 33 from miners) were collected and analyzed for Hg with LUMEX and creatinine. After applying the exclusion criteria, just 105 persons were selected to perform the neurophysiological tests (questionnaire 4). Most male volunteers had been working in the Processing Centers and/or in the mines. The distinction between millers and miners is that millers work exclusively as "employees" of the Processing Centers whereas miners extract the ore, take it to the Centers and follow all concentration steps with the millers. Both workers are contaminated by mercury vapor, but millers are constantly in contact with mercury while miners spend some time in the mines. Both burn amalgam using blowtorches. All male miners and millers were older than 15 and younger than 50. Women were selected according to the proximity from their residences to the Processing Centers. Their ages ranged from 15 to 45. All children were younger than 15.

RESULTS OF URINE ANALYSIS

The overall average of total Hg concentration in the 209 samples analyzed was 104.59 µg Hg/g creatinine with standard deviation of 378.41 µg Hg/g creatinine. Classes of Hg concentrations were selected to highlight the results (Figure 3). About 61.7% of the sampled individuals had Hg levels in urine above the alert level of 5 µg/g creatinine, 38.3% of the individuals had Hg levels above the action level (20 µg/g creatinine), 20.6% above the maximum of 50 µg/g creatinine recommended by the World Health Organization (WHO, 1991a) and 15% above the critical level of 100 µg/g creatinine which is the level where neurological symptoms should be evident.

The situation with miners and millers is dramatic, as 30% and 79% of the miners and millers respectively have Hg in urine above the action level and 52% of the millers have levels above the critical level (Figure 4).

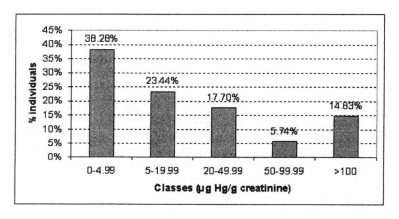

Figure 3. Classes of Hg Concentrations in Urine.

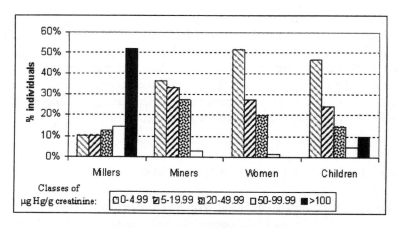

Figure 4. Distribution of Hg in Urine.

In addition, about 14.6% of millers had shown extremely high mercury concentrations in urine, ranging from 1221 to 3260 µg Hg/g creatinine.

This result allows the generalization that more than 90% of the sampled individuals working in the Processing Centers (millers) have Hg levels in urine above the alert level.

The average level of Hg in urine from the 66 women sampled in this study was 13.02 with standard deviation of 23.34 µg Hg/g creatinine. About 27% of the women had urine Hg concentrations above the alert level and 21% above the action level. The highest levels of mercury in urine from women are found in the communities of Monkey Town and Nuevo México. As already mentioned, these communities have the highest concentration of Processing Centers in Block B and houses were built very close to the mills. A number of women (21% of the group) had occasional direct contact with Hg, being sporadically involved in the amalgamation / amalgam burning process. More than 40% and 20% of the interviewees have complained of mental and physical fatigue respectively. Based on variance analysis (ANOVA) a significant difference was not found among the averages of mercury concentration in urine from women living in different communities ($p = 0.318$). No correlation was found between Hg concentrations in female urine and the distances from their houses to the Processing Centers ($r = 0.047$).

The average concentration of total Hg in urine of children from the communities around the Processing Centers is 33.30 with standard deviation of 70.80 µg Hg/g creatinine. As in the previous groups, the results show high variability as a result of differences in living and working habits of the children. About 53% of the 62 children sampled had Hg concentration in urine above the alert level and 14.5% above the action level. As also seen in Figure 6, about 10% of the sampled children had levels of mercury in urine above the critical level of 100 µg/g creatinine in which neurological symptoms of intoxication should be observed. In some cases the direct participation of children in the mining and processing activities was observed . These children work voluntarily to help their parents and relatives (Bermudez, 1999). It is also very common to see children playing (or living) in the Processing Centers. About 32% of the urine samples from children showed Hg concentrations below the detection limit of the method (0.2 µg/L). In the community of Nuevo Mexico, 84% of the urine samples were above the action level (20 µg Hg/g creatinine). In Monkey Town, 25 % of the samples had Hg concentrations above the action level, and the maximum Hg concentration was found in the urine of a 13 year-old girl (384 µg Hg/g creatinine). No correlation was found between the concentration of total mercury in urine from children and the distance between their residences and the Processing Centers (r = - 0.189).

Despite the direct contact with mercury of both groups, miners and millers, the symptoms are slightly more evident in miners. As mentioned before, the differences in working habits between miners and millers are not significant. The correlation between symptoms reported by the individuals (also known as "subjective symptoms") plus observed ("objective") symptoms, and Hg in urine of miners and millers, was stronger in the group of millers (Table 2). In the miners' group, the linear correlation coefficients were below 0.5, indicating poorer correlation between symptoms and levels of Hg in urine than in the case of millers.

In 25% of women and children, it was possible to identify objective symptoms or abnormal behavior during the clinical exams, but just 28% of the miners and millers have showed objective symptoms.

The difference in observed symptoms between the examined groups was not statistically significant, despite the higher levels of mercury in urine of

Table 2. Correlation between Symptoms and Hg in Urine (r = linear correlation coefficient).

Symptoms	r (Miners)	r (Millers)
Mental Fatigue[(*)]	0.48	0.65
Physical Fatigue[(*)]	0.35	0.60
Kidney Disorder[(*)]	0.34	0.08
Cough[(*)]	0.08	0.71
Metallic Taste[(*)]	0.02	0.42
Tremors[(o)]	0.03	0.51

[(*)] reported by the patient; [(o)] observed during medical exam.

miners and millers when compared with women and children. Regarding the statistical analysis, it was considered as a positive symptom all cases in which at least one or more objective symptoms were identified. In chronic intoxication, the individuals are exposed to lower concentrations of the pollutant for long periods of time and if the absorption exceeds excretion, the toxic substance accumulates and the effects of accumulation are perceived as subtle alterations of the neuropsychological and neurobehavioral functions. Sub-clinical alterations are more difficult to diagnose in a normal clinical evaluation and this is the most classical and usual effect of mercury on the nervous system.

SPECIFIC NEUROPSYCHOLOGICAL TESTS

A scoring process suggested in the UNIDO Protocols (Veiga and Baker, 2004) was used to express the results of the neuropsychological tests. Just three groups were evaluated: millers, miners and women. The latter were not directly involved in the amalgamation process. The scoring process ranged from 0 to 3. A score of 0 indicates good performance in the test and score 3 indicates highly poor performance; the higher the score, the greater the function deficiency.

The Wechsler Memory Scale (WMS) test evaluated recent memory. The test involved a procedure to ask the patient to repeat a list of numbers that ranges from 4 to 8 single numbers. The results showed that about 23% of miners and 10.5% of millers had scores of 2 (indicating deficiency) whereas just 9.4% of women had this score. Nearly 34.3% of miners and 26% of millers scored zero (no problem) and almost 44% of the women did not show any memory problem. The correlation of WMS with Hg-urine levels was not statistically significant to show differences between groups. However, just 23% of individuals with high mercury concentrations in urine (above 50 µg Hg/g creatinine) showed no difficulties (score 0) in completing the WMS Memory Test, whereas between 40 and 45% of individuals with Hg in urine below 20 µg Hg/g creatinine had no problem completing the series of tests. It is also possible to observe that the percentage of individuals with bad performance (score >1) reaches 76.5% for individuals with Hg in urine above 50 µg/g creatinine.

The Match Box test evaluated coordination, tremor and concentration. The test consisted in putting 20 matches on a table, half of them on each side of an open matchbox, and asking the individual to put all the matches into the box using the left and right hands alternately. The time is measured. About 15 seconds is considered a normal time for this task. Miners and millers performed better than women but the millers performed worse than miners. The Match Box Test did not show any correlation between Hg levels in urine and test performance.

The MOT Test (Finger-Tapping) evaluated spatial perception, fine motor coordination and normal dexterity. The test consisted in asking the volunteer to keep his/her elbows on the table and make as many points as possible on a piece of paper with a pencil. The number of points is counted after 10 seconds. A normal individual can easily make more than 65 points. Miners and millers had more difficulties in performing this test than women. About 72% of millers had scores of 1 and 2, indicating deficiencies in performing the test.

The Frostig test evaluated visual perception. It consists in drawing a line from one point to another across a narrow gap. The results did not show good sensibility in measuring neurological effects, as almost 100% of all tested individuals could perform the tasks without any problem. The only exception was a miller, who has evident serious chronic intoxication by metallic mercury.

In the test that estimates Episodic Memory, no women showed deficiencies, but almost 14% of miners and 11% of millers showed scores above 2. The tests that evaluate episodic memory are similar to the Mini-Mental State Examination tests. This evaluates memory, orientation, and the ability to calculate and speak. This is probably one of the most used and studied tests for quick evaluation of neurological functions.

This test has been quantified and adjusted to different ages and degrees of instruction. Its application is simple, and this test can also be applied to the illiterate (Bertolucci et al, 1994). A score of 2 or 3 in this test is a clear indication of a neurological problem. The correlation between Hg in urine and scores obtained in the Episodic Memory Test (Mini-Mental) provided a clear indication that the number of individuals who score zero (no problem) decreases when the Hg level in urine increases. In addition, the percentage of individuals with poor performance in this simple test (scores of 1, 2 and 3) increases with the level of Hg in urine (Figure 5).

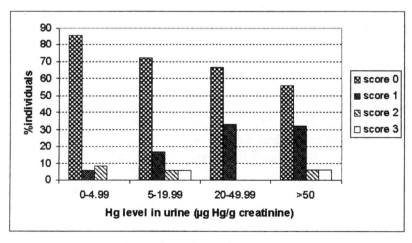

Figure 5. Relationship between Hg in Urine and Episodic Memory Test.

This is a strong indication of alteration of neurological function by mercury vapor intoxication.

About 27% of individuals who performed the specific neuropsychological tests have noticeable neurological problems detected in the clinical exams. However, no significant correlation was found between the results of the medical exams and the specific tests.

The tests of coordination and manual dexterity (drawing figures) normally evaluate patients with significant central neurological lesions. In this study a similar scoring process as the mini-mental tests was adopted, i.e. ranging from 0 (no problem) to 3 (deficient). The percentage of miners and millers with a score above 2 was, 52.2% and 34.8% respectively, which is a clear demonstration of poor performance.

DEMONSTRATION OF CLEANER TECHNOLOGIES

Extracting Gold and Mercury from Tailing

As part of the fieldwork, some simple techniques were brought to the miners' attention in order to reduce mercury emissions and exposure while decomposing amalgams. A Processing Center ("El Mago") was rented to conduct tests and demonstrate cleaner technologies of gold processing to the local artisanal miners. Five tests were conducted using tailings from the Processing Center pond. Extraction of residual gold from tailings was in fact advantageous as this material was more homogeneous than the primary ores.

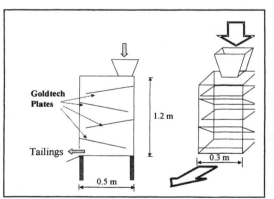 

Figure 6. Special-Amalgamating Plates (Zigzag).

In addition, this got the attention of miners and millers, as gold in tailings is usually unliberated or very fine and they are aware that they do not have methods to extract it. Ultimately, this is the gold extracted by the "company's methods" (cyanidation after extensive grinding).

Most tests were conducted using a sluice box with four special amalgamating Goldtech plates, manufactured in Brazil and locally commercialized by a company called PARECA (at US$ 200 per one 40x30 cm plate). Unlike the ordinary amalgamating Cu-plates, this special plate has a thin coating of Hg and Ag electrolytically deposited onto a copper plate. Mercury and eventually gold from tailings are captured and firmly fixed to the plate surface. Mercury losses are minimized.

When the plates are fully loaded, amalgam is removed with a plastic scraper. This kind of plate has been used successfully in Brazil to remove mercury from contaminated tailings, (Veiga et al, 1995) but they can also be used to amalgamate gravity concentrates. The support for four plates was made of wood in Venezuela. The configuration was set up to allow a cascade effect from one level to another. Another wood structure was built to hold the four Goldtech 40 x 30 cm plates, placed in a zigzag (Figure 6).

This allowed a reduction of the flow speed on each plate and rendered better results.

The four Goldtech plates were activated with vinegar before receiving Hg-contaminated tailings. Then, about 1,100 kg of tailings were fed onto the plates. Samples before and after feeding the Goldtech plates were systematically collected every 15 minutes and the whole material was dried, pulverized, homogenized and analyzed in triplicate using the LUMEX atomic absorption spectrometer (Table 3).

Table 3. Hg (mg/kg) in Samples before and after Treatment with Goldtech Plates.

subsample:	A	B	C	Ave
Before Goldtech plates	64.3	59.4	62.8	62.2
After Goldtech plates	2.98	2.93	3.80	3.24

The material before entering the Goldtech plates was contained 62.2 mg/kg of Hg and left the plates with 3.24 mg/kg Hg on average. More than 95% of the Hg was removed from the tailings. However, the gold recovery process was not so efficient. Before entering the Goldtech plates, the tailings had on average 9.53 g Au/tonne. The final tailing, after leaving the Goldtech plates, analyzed 9.05 g Au/tonne. Just 0.7 g of gold was recovered from the plates. This is a clear indication of lack of liberation of gold from the silicates.

Another test was conducted using 1,805 kg of tailings to demonstrate to miners and millers the advantages of using carpets to concentrate gold prior to amalgamation. The material fed the hammer mill and discharged on to a sluice box lined with two carpets in series. Both carpets were locally acquired: a synthetic grass and a Multiouro Tariscado. The latter is produced by the Brazilian company Sommer, and is widely used by Brazilian artisanal gold miners. This carpet costs around US$10 to 15/m² in Venezuela. The Brazilian carpet, appropriated to retain fine gold, was placed at the beginning of the flow, where the speed is slower. Visibly, the Brazilian carpet retained more gold than the synthetic grass.

The concentrates from the carpets were washed and amalgamated on the Goldtech plates. This operation was done by cleaning the carpets with water and directing the pulp of concentrates to the plates placed in the zigzag structure. The gravity concentrate was re-passed three times on the plates. After retorting, 1.3 g of gold was obtained. The final tailings had an average grade of 3.75 g Au/tonne.

Another test was set up using Cleangold sluice boxes followed by Goldtech plates. The Cleangold sluice uses polymeric magnetic sheets, with the magnetic poles aligned normally to the direction of the flow, inserted into a simple aluminum sluice box. Magnetite, a mineral usually found in gold-ore deposits, forms a corduroy-like bed on the sluice floor, which appears effective at recovering fine gold. Fine particles of steel from the hammers are also trapped and form the liner. This sluice box is available in a variety of sizes and a 2ft x 6in (60x15 cm) sluice costs US$ 75 in the USA. The main advantage of this sluice is the high concentration ratio. Gold becomes trapped in a magnetite layer and the sluice can be scraped and washed into a pan. About 628 kg of tailings were passed in a 2 ft Cleangold sluice box followed by two 8x8in (20x20 cm) sluices and the tailing was directed to the zigzag structure with four Goldtech plates. The Cleangold sluice retained visible mercury droplets and recovered 0.64 g of gold in a concentrate that analyzed 2854 mg/kg Au The gold recovered by this sluice box was extremely fine. This was demonstrated to the miners and millers. Goldtech plates recovered 0.25g Au. This configuration was capable of recovering 15.4% of the gold from the tailings in which 11% of the gold was recovered using the Cleangold sluice and 4.4% of gold in the Goldtech special amalgamating plates.

Using the Cleangold sluices, it was also possible to pan the concentrate and, using a plastic vial, gold was sucked from the concentrate, avoiding entirely the use of mercury. However, the plastic sucker recovery is probably limited to 50% of the gold as more gold remains in the concentrate than can be recovered by panning.

All results have reinforced the conclusion that lack of gold liberation is the main obstacle to increasing gold recovery in El Callao. In other words, without an appropriate grinding process, gold recovery by gravity separation will still be poor. The use of hammer mills to re-grind tailings is not appropriate. CVG-Minerven cyanidation plant uses ball mills to re-grind the tailings to at least below 200 mesh (0.074 mm). The only possibility to increase gold recovery in the Processing Centers is by using small ball mills (e.g. Ø48x60 cm) to reduce the size of the ground product and promote gold liberation. As the gold is very fine, it seems that the Cleangold sluice is an appropriate and affordable technology for the miners and millers to process primary ores via gravity concentration.

RETORTS

Miners burn amalgam in shovels and, usually, very near their noses, where they can better observe the decomposition process. Very few miners and millers believe that this is a problem and they keep burning mercury carelessly. Some millers have retorts available to be used by miners but they remove the top of the retort and use it as an open-air crucible to burn and melt gold. In many cases this operation is conducted in a confined environment.

Using the LUMEX spectrometer, mercury in the breath of miners and residents was analyzed. The normal level of mercury in the breath depends on the number of amalgam-dental fillings of each individual as well as the level of mercury in the environment. This is usually below 100 ng/m³. It is also known that just 7% of the mercury vapor dose received by an individual is released during respiration (Pogarev et al, 2002). In El Callao, miners usually have shown 10,000 ng/m³ of mercury in their breath. Children living near Processing Centers had as much as 5,000 ng/m³ of mercury in their exhaled air. Despite the rudimentary procedure adopted to analyze the breath of volunteers, this was effective to call the attention of the people working in the Centers or living around them to the high levels of mercury to which they have been exposed and accumulated in their lungs. This simple analysis has also contributed to convincing the miners and millers to watch the demonstration of different types of retorts made by UNIDO in Venezuela.

In order to introduce retorts to miners, four different types of retorts were locally manufactured and the principle of retorting was demonstrated using a

Thermex glass retort. The local retorts were fabricated in a metal shop in El Callao using crucibles of stainless steel. The crucibles were actually small stainless steel salad or sauce bowls acquired in kitchen stores. An RHYP retort made of galvanized water connection pipes was also built and tested. All retorts were demonstrated to miners and amalgam was burned using a propane torch, such as they would generally use to burn amalgams in shovels. The burning process took an average of 15 to 20 minutes.

One of the most appreciated retorts was that fabricated using a Chinese design (Gunson and Veiga, 2004). This was a more elaborate retort built on a steel table but also using a stainless steel salad bowl as crucible. A steel cover (bucket) was placed on the crucible. The table was filled with water and the amalgam burned with a blowtorch from the bottom. As the crucible wall was thin, the retorting time was short. Mercury condensed on the wall of the cover and dripped into the water. The manufacturing cost of this prototype was around US$ 80 but this could be drastically reduced. This retort took 10 to 15 minutes to eliminate most mercury from amalgam using a propane blowtorch. A serious inconvenience of this, and other retorts, is that miners can remove the cover (bucket) from the crucible while the retort is still hot. When this occurs, miners are exposed to mercury vapor. This was demonstrated by Schulz-Garban (1995) when the urine of 20 amalgamation workers using retorts were analyzed. It was noticed that 8 individuals had high mercury levels in urine because they habitually opened retorts before cooling them. This was shown to miners and millers and they were advised to cool down the cover pouring water on it and then wait some minutes before opening the system.

Analytical techniques were discussed and transferred to a chemical laboratory in Ciudad Guyana (La Salle Foundation). The methodologies for quantitative and semi-quantitative Hg analyses using a solution of dithizone with chloroform (Veiga and Fernandes, 1991) were demonstrated to the local technicians.

A two-day workshop was also conducted in the El Mago Processing Center, where the project team set up a series of practical demonstrations, and in an auditorium in El Callao where the results of the project were discussed with almost 60 stakeholders.

CONCLUSIONS AND RECOMMENDATIONS

The medical exams, the urine analyses and the specific neuropsychological tests have revealed that millers and miners show symptoms that suggest serious mercury intoxication. In just one worker it was characterized as acute mercury intoxication and his immediate removal from the polluting source for treatment was recommended to the local doctors and nurses. The specific neuropsychological tests have highlighted a deficit of the cognitive functions in some workers such as alteration of visual perception, deficiency of recent and episodic memory, as well as deficiency of spatial perception, motor coordination and manual dexterity. The cognitive deficiency showed positive correlation with the levels of total mercury in urine and this is more prominent in Hg concentrations above 6µg/Hg/g creatinine. Some of the neuropsychological tests have revealed that people not directly involved in the amalgamation work but living near the Processing Centers have been neurologically affected by mercury vapors. The alterations found in the medical exams and neuropsychological tests of people indirectly exposed to mercury vapors call for immediate action to reduce emissions and reduce exposure of innocent people to the pollutant.

Suggested are:

- The immediate removal of children from the polluting source (many children play and work in the Processing Centers) (Figure 7),
- The continuation of the monitoring of cognitive aspects of the residents of Block B, in particular children from ages 7 to 12,
- The training of local health workers to be able to monitor health conditions as well as performing the specific neuropsychological tests in residents, and
- The implementation of an interdisciplinary strategy to educate workers and residents to reduce and avoid mercury vapor exposure.

Processing Centers are usually effective in reducing widespread environmental impacts caused by artisanal gold miners, but at the same time, when the centers are not well operated, they can concentrate contamination in certain areas. The technologies used by the Processing Centers in El Callao are exposing miners, millers and the surrounding communities to high levels of mercury vapor. The technique of using hammer mills associated with amalgamating copper plates recovers, at most, 30% of the gold in the ore. As gold in El Callao seems to occur at fine grain size, a more efficient grinding process is needed. This must be associated with a gravity separation process that does not use mercury. The amalgamation of the whole ground ore using copper-plates, as extensively used in El Callao, is an ancient inefficient technology that releases mercury to the tailings and exposes operators to high levels of mercury vapor.

There are many concentration techniques that do not require mercury in the processing, such as sluice boxes using appropriate carpets, Cleangold sluices, centrifuges, etc.

The use of retorts is another critical issue in El Callao. Miners and millers do not believe in mercury pollution and keep burning amalgams in pans and shovels. In many cases this is conducted in a closed environment. Companies and the government should immediately establish an awareness campaign to introduce safer procedures for amalgam decomposition, such as retorts.

Figure 7. Definitely, there are better places to play than a Processing Center: it is just a matter of opportunity and help.

Miners and millers need to be made aware that using any retort is better than using nothing. Different types of retorts should be brought to the miners' attention.

In terms of business, the Processing Centers are clearly using a poor strategy. As they charge 10% of the gold production and the miners do not know how much gold they have mined, the millers very frequently do not produce enough gold to pay their operating costs, at 0.24 and 0.5 g Au/bag for soft ore and hard ore respectively. The method used in Zimbabwe for Custom Processing Centers (Shoko and Veiga, 2003) seems to be more adequate, since the price for processing ore is fixed based on hours of grinding. Definitely a better arrangement between miners and millers must be established in El Callao.

The business relationship between mining companies and miners/millers must be carefully revisited and improved. The companies, in particular CVG-Minerven, are acquiring gold-rich tailings from miners, potentially creating future problems for the company as the benefits become concentrated in the company and Processing Centers' owners. The relationship between companies and miners/millers can rapidly deteriorate. The amount of mercury being introduced into the companies' environments together with contaminated tailings is considerable, and this is definitely contaminating employees (especially those working in the gold melting room) and tailing ponds. As mining companies do not purchase tailings (and sometimes ore) with less than 6 g Au/tonne, it is also foreseen that millers (Processing Center owners) will soon start their own cyanidation plants. Cyanidation of Hg-contaminated tailings, as seen in many other countries, exacerbates the danger of mercury in the environment and facilitates metallic Hg solubilization and methylation (Gunson and Veiga, 2004). This is a problem that soon will come to El Callao and will need immediate attention from the authorities.

As yet, there is no educational program for miners/millers, residents and the general population to make them aware of the dangers caused by mercury. No consistent program has been implemented in the region to bring simple solutions for miners and millers such as the use of retorts to protect themselves and the surrounding population. There are reliable local equipment manufacturers with good technical capacity to develop simple types of equipment suitable for small-scale miners. These manufacturers could be trained to produce better pieces of equipment. Mercury pollution cannot be reduced if the miners/millers do not see any additional benefit in terms of gold production.

All action should take into consideration the problems associated with poverty and rudimentary living and working conditions of the people of Block B. UNIDO has been implementing Transportable Demonstration Units (TDU) in six countries in Africa, Asia and South America to bring hands-on training to miners/millers and the general public. A movable unit consists of a tent to be used as classroom and a container with small pieces of equipment to teach the miners and millers the advantages of using cleaner methods. This brings to the miners' and millers' attention a variety of technical options for gold concentration, amalgamation and retorting; it is up to them to select what is affordable, appropriate and durable according to their convenience. The unit also incorporates programs to attract miners and the public to watch skits and movies about environmental impacts and mercury pollution, highlighting local cultural aspects and incorporating concepts of environmental and health protection. The unit can also bring ideas to improve the livelihood of different mining communities such as suggesting economic diversification[N27] or value-adding techniques (e.g. handcraft, fish farming, agriculture, brick making using tailings, etc). An initiative like this is badly needed in El Callao and the collaboration of local mining companies is critical to guarantee the sustainability of this program.

Any initiative in the region must take into consideration the critical living situation of the communities. Programs cannot be solely focused on reduction of mercury pollution since this pollutant is a consequence and not a cause of the local problems. Poverty is the main cause of the misuse of mercury and intoxication of this and so many other communities around the world. Local and international institutions, government, mining companies, the private sector and NGOs should act immediately to solve the dramatic situation of the mining communities of El Callao. Some suggestions of badly needed actions are as follows:

- Generate non-mining and sustainable activities,
- Establish a latrine construction program,
- Evaluate and promote the construction of an adequate sewage system,
- Improve the supply of potable water (source of water, treatment plant and pipes),
- Establish community programs to improve/build houses far from the Processing Centers,
- Establish a non-centralized Community Health Center, so it can get more support from other regions,
- Establish labor training programs using the TDU as an example,
- Establish programs to improve the educational level of the communities, and
- Establish projects for the rehabilitation of degraded and polluted areas.

ACKNOWLEDGEMENTS

The Project team wants to express its profound gratitude to the following people and institutions: Rebeca Erebrie, VP Environment, Science and Technology, CVG and Armando John Madero, VP Mining, CVG, Franqui Patines, Benjamin Millán, Jesus Rebolledo, Walter Maciel and Joaquín Lezama from CVG Minerven; the Rivero family (Rafael, Maria Elena and Rafaelito) and Antonio "Lalo" from Molino El Mago; Dr. Salvador Penna (UDO-Universidad del Oriente), Dra Eudelis del Vale Romero Platina and Ms. Aracelis Arevalo (CVG-Minerven) as part of the medical team; Luís Perez and Luzmila Sanchez from Fundacion La Salle; Luisa de Bermúdez for filming the project; Dr. John Meech from CERM3 – The Center for Environmental Research in Minerals, Metals and Materials of the University of British Columbia, Canada, for the use of LUMEX; Metall-Technic, Munich, Germany for the donation of the Thermex retorts; José Pino and Clyde Pepppin from Hecla Mine; Jorge Paolini from the Center for Research in Management and Education for the Sustainable Development, UNEG – Universidad Experimental de Guayana; all Processing Center's owners and employees that provided information for the UNIDO team: Roberto Clark, Guillermo Herrera, Vladimir Vega, Felix Rangel and Ramon Alvorado among many others.

REFERENCES

Bermúdez, D., La Pequeña Minería en Venezuela. Report for Int. Labour Organization. IEPC-Erradicación Internacional del Trabajo Infantil, Lima, Perú. (in Spanish). 1999.

Bermúdez, D., El Ambiente y la Salud Ocupacional en la Minería en Pequeña Escala. *In*: I Encuentro Internacional de Investigación en Salud Ocupacional. Universidad de Guadalajara, Secretaría del Trabajo y Previsión Social (Delegación Jalisco) y PIENSO A.C. Guadalajara, Estado de Jalisco, México. Mayo 2003. (in Spanish). 2003.

Bertolucci, P.H.F., Brucki, S.M.D., Campacci, S.R., Juliano, Y. O Mini-Exame do Estado Mental em uma População Geral. Impacto da Escolaridade. *Arquivos de Neuropsiquiatria*, 52, 1-7 (in Portuguese). 1994.

Chang, L.W., Pathological Effects of Mercury Poisoning. In: The Biogeochemistry of Mercury in the Environment. p.519-580. Ed. J.O. Nriagu. Elsevier/North-Holland Biomedical Press, Amsterdam. 1979.

Drake, P., Rojas, M., Reh, C.M., Mueller, C.A., Jenkins, F.M. Occupational Exposure to Airborne Mercury during Gold Mining Operations near El Callao, Venezuela. *Int. Arch. Occ. Env. Health*, 74, 206-212. 2001

Drasch, G., Boese-O'Reilly, S., Maydl, S., Roider, G. Scientific Comment on the German Human Biological Monitoring Values (HBM Values) for Mercury. *Int. J. Hygiene Env. Health*, 509-512. 2002.

Gunson, A.J., Veiga, M.M. Mercury and Artisanal Gold Mines in China. *Env. Practice*, 6, 109-120. 2004

Hecla,. Estudio Socio-económico de las Comunidades del "Bloque B" de El Callao. Comp. Minera Hecla, Edo Bolívar, Venezuela. (in Spanish) 2004

Hinton, J.J., Veiga, M.M., Veiga, A.T. Clean Artisanal Mining, a Utopian Approach? *J. Cleaner Prod.*, 11, 99-115, 2003.

Jones, H.R.. Mercury Pollution Control. Noyes Data Co., New Jersey, 251p. 1971

Meech, J.A., Veiga, M.M., Tromans, D. Reactivity of Mercury from Gold Mining Activities in Darkwater Ecosystems. *Ambio* , 27, 92-98, 1998.

Mitra, S. Mercury in the Ecosystem. Trans Tech Publ., Netherlands. 327p. 1986.

MMSD – Mining, Minerals and Sustainable Development, Breaking New Ground. Int. Institute for Environment and Development and World Business Council for Sustainable Development. London, UK. 441 p. 2002.

Pogarev, S.E., Ryzhov, V., Mashyanov, N., Sholupov, S., Zharskaya, V. Direct Measurement of Mercury Content of Exhaled Air: A New Approach for Determination of the Mercury Dose Received. *Anal. Bioanal. Chem.*, 374, 1039-1044, 2002.

Schulz-Garban, K. Determination of Hg Concentration in Workers and in the Air of Several Amalgamation and Gold Processing Centers of "Bajo" Caroni, June-Novembre 1994. Master Thesis, UNEG, Ciudad Guayana. 156p. (in Spanish) 1995.

Shoko, D., Veiga, M.M. Information about the Project Site in Zimbabwe (Kadoma-Chakari region). Report to GEF/UNDP/UNIDO Global Mercury Project. 19 p., 2003.

Tsuji, J.S., Williams, P.R.D., Edwards, M.R., Allamneni, K.P., Kelsh, M.A., Paustenbach, D.J., Sheehan, P.J. Evaluation of Mercury in Urine as an Indicator of Exposure to Low Levels of Mercury Vapor. *Env. Health Perspect.*, 111, 623-630, 2003.

UNEP – United Nations Environment Programme, Global Mercury Assessment. IMOC – Inter-organizational Programme for the Sound Management of Chemicals. A cooperative agreement among UNEP, ILO, FAO, WHO, UNIDO, UNITAR and OECD.Geneva, Switzerland, 258 p, 2002.

UNIDO, Advisory Assistance on Avoidance Mercury Pollution from Artisanal Gold Mining Operations in State of Bolívar, Venezuela. Prepared by M. M. Veiga, contract SI/VEN/94/801/11-51. 147 p. 1996.

Veiga, M.M. Information about the Project Site in Indonesia (North Sulawesi and Kalimantan). Report to GEF/UNDP/UNIDO Global Mercury Project. 14 p. 2003.

Veiga, M.M. and Baker, R. Protocols for Environmental and Health Assessment of Mercury Released by Artisanal and Small-Scale Gold Miners (ASM). GEF/UNDP/UNIDO Global Mercury Project. UNIDO Project EG/GLO/01/G34. Vienna, 292 p. 2004.

Veiga, M.M., Fernandes, F.R.C., Aspectos Gerais do Projeto Poconé.*In*: Poconé, um Campo de Estudos do Impacto Ambiental do Garimpo, p.1-25. Veiga and Fernandes (eds.). Rio de Janeiro, CETEM/CNPq (in Portuguese). 1991.

Veiga, M.M., and Hinton, J.J., Abandoned Artisanal Gold Mines in the Brazilian Amazon: A Legacy of Mercury Pollution. *Nat. Res. Forum*, 15-26, 2002.

Veiga, M.M., Veiga, A.T., Franco, L.L., Bonagamba, M., Meech, J.A. An Integrated Approach to Mercury-contaminated Sites. Proc. Eco Urbs' 95, p.51-53. Rio de Janeiro, Jun. 19-23, 1995.

WHO - World Health Organization,. Environmental Health Criteria. 118. Inorganic Mercury. Geneva, Switzerland, 168p, 1991.

Chapter-19

AN ECOSYSTEM APPROACH TO DESCRIBE THE MERCURY ISSUE IN CANADA: FROM MERCURY SOURCES TO HUMAN HEALTH

Marc Lucotte[1], René Canuel[1], Sylvie Boucher de Grosbois[1,2], Marc Amyot[4], Robin Anderson[6], Paul Arp[3], Laura Atikesse[1,2], Jean Carreau[1], Laurie Chan[5], Steve Garceau[1], Donna Mergler[1,2], Charlie Ritchie[3], Martha J. Robertson[6] and Claire Vanier[1,2]

[1] COMERN, Institute of Environmental Sciences, Université du Québec à Montréal, C.P. 8888, Succursale Centre-Ville, Montréal (Québec), Canada, H3C 3P8.
[2] COMERN, CINBIOSE, Université du Québec à Montréal, C.P. 8888, Succursale Centre-Ville, Montréal (Québec), Canada, H3C 3P8.
[3] COMERN, Faculty of Forestry and Environmental Management, University of New Brunswick in Fredericton, 28 Dineen Drive, P.O.Box 44555 , Fredericton (New Brunswick), Canada E3B 6C2.
[4] COMERN, Department of Biology, Université de Montréal, C.P. 6128, Succursale Centre-Ville, Pavillon Marie-Victorin, Montréal (Québec) Canada, H3C 3J7.
[5] COMERN, Center for Indigenous People's Nutrition and Environment, McGill University, 845 Sherbrooke St. West, Montréal (Québec) Canada, H3A 2T6
[6] COMERN, Fisheries and Oceans Canada – Northwest Atlantic Fisheries Centre; St. Johns (Newfoundland) Canada, A1C 5X1

INTRODUCTION

It has been decades now since the international scientific community initially raised the issue of mercury (Hg) contamination in the global environment. The presence of Hg in ecosystems is ubiquitous, even in the absence of local/regional contamination point sources. Almost all fish consumers (occasional or frequent) are exposed to this contaminant. Governments of the industrialized countries have invested considerable financial and human resources, in order to better understand the biogeochemical behavior and cycling of Hg and its impacts on the health of populations. Indeed, our knowledge of the sources and fate of this pollutant has greatly evolved since these early reports. Numerous protocols, technical documents, epidemiological and clinical studies, detailing precise aspects of the Hg cycle have been published. However, given the complexity of environmental processes leading to the accumulation of Hg in fish tissue, and the relative importance of fish as a protein source among communities, most available literature fails to fully evaluate the level of risk to health (and/or the health benefits related to fish consumption) encountered by fish consumers in their daily lives.

This paper summarizes the learning acquired through a wide-scale integrated study of the mercury (Hg) pathways in lake environments of three distinct regions located in Eastern Canada: Lake St. Pierre (LSP), Labrador (Lab), and Abitibi (Ab). This research was accomplished by a multi-disciplinary team of researchers assembled under the auspice of the Collaborative Mercury Research Network (COMERN), a major Canadian initiative supported by numerous universities and government agencies throughout the country. The prime focus of the study was to link human exposure to Hg with particular local/regional environmental and socio-economic characteristics and settings.

Two conditions must co-occur to define a situation where higher Hg exposure can be identified for populations/sub-populations/groups:

- Frequent fish consumption;
- Mercury levels of concern in the edible fish resource.

Thus, specific scientific objectives of the study were constructed to verify the occurrence of such situations by:

1) Describing and comparing Hg pathways, from its loading in the environment, its transfer to aquatic food webs and its final accumulation in edible fish species;
2) Assessing and comparing exposure levels (and related health impacts) of populations exploiting aquatic resources, in response to tradition, economic dependency or recreation, and stating on health risks/health benefits related to fish consumption in specific regional contexts.

Description of the sites of study

Lake St. Pierre is a large fluvial lake located in the St. Lawrence (Qc) waterway, which drains the watershed of the Great Lakes (Table 1). The lake's primary catchment is impacted by heavy industrial development and agricultural practices. The lakeshores are rather densely populated and its abundant fish resources, organized in a complex food web, support an important commercial harvest.

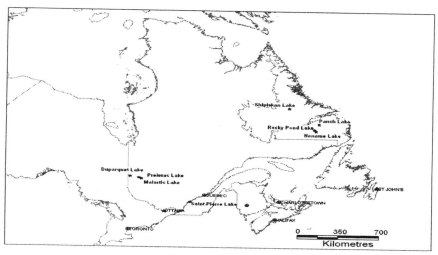

Figure 1. Localization.

Lake St. Pierre shores also exhibit an extended and unique wetland ecosystem, being recognized as such as a UNESCO Biosphere reserve. Lakes of the Abitibi and Labrador regions are located in remote areas of the Boreal forest. These lakes were selected as they are frequently fished by the local population, either for subsistence (Labrador) or recreation (Abitibi). The three lakes of Abitibi are larger than those of Labrador, subject to low to moderate agricultural activity, and the surrounding population is restricted to sparse and small villages, whereas the four Labrador lakes are remote difficult to access and exempt from immediate anthropogenic impacts. The somewhat simple fish population structure in Labrador lakes is dominated by salmonids while Lake St. Pierre and Abitibi lakes are mainly inhabited by persids.

RESULTS

Community groups were assembled in three regions to share their knowledge and to actively participate in: *(1)* elaborating and realizing the research plan; *(2)* identifying sampling sites actually used by fishers; *(3)* describing the local environmental characteristics; *(4)* sampling and measuring Hg levels in the different ecosystem compartments; *(5)* describing local fishing habits and fishing pressure; *(6)* gathering dietary, social, economic and cultural information *(7)* evaluating human exposure through indicators (hair and blood); *(8)* choosing and applying neuro-functional tests to monitor early health alterations, and *(9)* transmitting acquired knowledge within the community.

Table 1. Description of lake environments.

		Lakes under study						
	Malartic	Duparquet	Preissac	Noname	Panch	Rocky pond	Shipiskan	St. Pierre
Region	Abitbi	Abitibi	Abitibi	Labrador	Labrador	Labrador	Labrador	St. Lawrence
Longitude	78.10	79.28	78.37	59.39	59.02	59.58	62.26	72.65
Latitude	48.35	48.41	48.38	52.69	53.27	52.49	54.64	46.25
Lake area (ha)	11440	4847	8251	2743	1407	621	1721	31130
Watershed area (10^3 ha)	307.1	172.4	99.4	36.2	131.4	30.7	360.4	75050
Mean slope (%)	3.00	5.24	1.06	2.73	5.08	2.43	3.70	2.3*
Ratio lake/watershed	27	36	12	13	93	50	209	2411*
Wetland area (% watershed)	8	3	5	NA	NA	14	NA	1*
Exploitation	Sports-fishing	Sports-fishing	Sports-fishing	Subsistence	Subsistence	Subsistence	Subsistence	Sports Commercial
agriculture (% watershed)	3	1	1	0	0	0	0	27
Forestry (% watershed)	3	7	2	0	0	0	0	NA
Population density (nb/km2)	1 to 9.9	1 to 9.9	1 to 9.9	0 to 0.9	0 to 0.9	0 to 0.9	0 to 0.9	10 to 69.9
Ethnic groups	Caucasian & First Nations	Caucasian & First Nations	Caucasian & First Nations	Innu	Innu	Innu	Innu	Caucasian

* Lake St. Pierre primary catchment

A total of close to 400 participants representing sports fishers of the Lake St. Pierre area, anglers from the Abitibi region and Innu representatives of the community of Sheshashiu (Labrador) brought their invaluable contribution to the study. This study yielded numerous results. Only those relevant to the scope of the present paper are presented here.

Environmental characteristics

A wide variety of physico-chemical parameters were measured at different frequencies in the water column of the selected lakes, over a period of two years (Table 2).

Table 2. Biogeochemical Characteristics Of Lakes.

Region	Lakes under study							
	Malartic	Duparquet	Preissac	Noname	Panch	Rocky Pond	Shipiskan	St. Pierre
	Abitbi	Abitibi	Abitibi	Labrador	Labrador	Labrador	Labrador	St. Lawrence
Water color (Pt mg/L)	83.7	58.5	58.5	5.9	1.7	6.3	2.1	50.1
PO4 (umol / 1 -- P)	0.66	0.68	0.44	NA	NA	NA	NA	1.48
DOC (ppm -- C)	10.86	10.92	9.05	NA	NA	NA	NA	7.67
Dissolved Hg (ng/L)	1.19	1.67	0.75	2.8	2.2	2.7	2.2	1.10
pH	7.0	7.4	7.1	5.8	6.0	5.9	6.3	7.6
Conductivity (µS/cm)	85.99	95.45	110.00	16.52	NA	18.00	35.92	317.24
SPM (mg/L)	11.85	11.58	3.65	1.,55	0.,21	0.93	0.14	18.61
Hg SPM (ppb)	116.50	349.70	155.20	NA	NA	NA	NA	548.33
chl@	3.07	4.52	3.16	2.68	0.82	1.94	0.67	NA
Atmospheric loadings (ug Hg/m2/an)*	5			5				4.1

* Data on yearly wet deposition from the Mercury Deposition Network; closest station: Lab: Newfoundland; Ab and LSP: St-Anicet.

They include conventional environmental descriptors such as estimations of the different sources of Hg loadings and distribution in the lakes. These measurements were also intended to describe and understand the dynamics of the mercury cycle, in particular relative to the methylation potential of the lake ecosystems.

Trophic status. The available assemblage of physico-chemical data for the three regions indicate that lakes of the Labrador region are clear water oligotrophic aquatic ecosystems, compared to the colored humic waters of the Abitibi region, and the more eutrophic Lake St. Pierre. These differences are most likely attributable to the respective vegetation coverage and land use of the lake watersheds in the three regions.

Mercury loadings. Lake St. Pierre exhibits high particulate suspended material content and conductivity, attributed in part to the influence of agricultural practices and soil erosion. This particulate loading also carries a significant amount of Hg to the lake (Amyot et al., 2004). Elsewhere in Abitibi and Labrador, local industrial or agricultural Hg point sources are absent. Hg loadings to the lakes then come mainly from the slow leaching of atmospheric Hg accumulated within the watersheds (Lucotte et al., 1999). The relative proportion of anthropogenic vs. natural Hg accumulated on these soils is still debated. Nevertheless, data from the Mercury Deposition Network (NADP-MDN, 2004) suggest that the levels of atmospheric loadings in the two regions are comparable. Likewise, Hg concentrations in water do not differ significantly within the three regions.

Mercury levels in fish resources

Fish sampling was concurrently conducted in the three regions with the assistance of local human resources and knowledge. Results are presented in Table 3.

Mercury levels in fish of similar trophic level from the Abitibi and Labrador regions are fairly similar. As expected, Hg levels in predator species are higher in the three regions. However, Hg levels in top predator northern pike are about two fold lower in Lake St. Pierre than in specimens captured in the other two regions. Similarly, Hg levels in piscivorous walleye from Abitibi are significantly higher (ranging from 2 to 4 times) than those of Lake St. Pierre. The highest standardized Hg values were monitored among the lake trout populations of Labrador, while the commonly consumed yellow perch (Lake St. Pierre), Atlantic salmon and brook trout (Labrador) contain lower Hg levels.

Biomarkers of human exposure

Data of hair Hg levels and fish consumption habits are presented in Table 4. According to the conventional and established understanding of Hg trophic transfer and exposure, one had hypothesized that of the three participating groups, the Innu community, who fish for subsistence in response to traditional values, would eat more fish and clearly be more exposed to Hg than anglers occasionally harvesting the remote lakes of Abitibi. In the same sense, it could a priori be expected that the heavily industrialized and populated watershed of Lake St. Pierre, where sport fishing is part of a cultural way of life, would result in higher Hg contamination of fish, and to higher human exposure.

Table 3. Fish Data.

Region	Lakes	Species	Standard length[1] (mm)	Hg levels (ppm)	Proportion of total fish meals from the regions	Food web structure
Labrador	Data combined	Atlantic Salmon[2]	715	0.01	100 %	Salmonidae Simple
		Lake Trout	590	0.67		
		Lake Whitefish	420	0.18		
		Northern Pike	660	0.03		
		Smelt[2]	170	0.22		
		Brook Trout	440	0.05		
Abitibi	Preissac	Walleye	350	0.32	47,8%[3] Other sources : Tuna : 7,2% Freshwater fish:11,7% Marine fish : 35,3% Seafood: <1%	Persidae Simple
		Sauger	350	0.85		
		Northern Pike	545	0.45		
	Duparquet	Walleye	350	0.45		
		Sauger	350	0.43		
		Northern Pike	545	0.45		
	Malartic	Walleye	350	0.79		
		Sauger	350	0.47		
		Northern Pike	545	NA		
St. Lawrence	St. Pierre	Walleye	350	0.17	44%[3] Other sources : Tuna : 4,9% Freshwater fish:11,7% Marine fish : 24,4% Seafood:15%	Persidae Complex
		Sauger	350	0.21		
		Northern Pike	545	0.16		
		Yellow Perch	155	0.10		
		Burbot	253	0.09		

[1] Mercury concentration at standardized length corresponding to the averaged regional length of catches.
[2] Data from Bruce et al. (1979)
[3] From the 2003 dietary survey

Table 4. Biomarkers Of Human Exposure.

Labrador	118 participants			
	Fish meals	**SD**	**minimum**	**maximum**
fishmeals/year				
Salmon	29.1	57.5	1	336
Lake Trout	40.7	117.0	1	1008
Arctic Char	44.7	102.6	1	536
Northern Pike	12.6	28.3	1	120
meals/spring	(total for three-months)			
Salmon	3.2	11.9	0	84
Lake Trout	11.9	34.7	0	252
Arctic Char	9.9	41.0	0	252
Northern Pike	3.8	12.4	0	60
Smelt	7.9	15.9	0	84
Brook Trout	10.0	21.8	0	168
Mean Hg levels in first 3cm of hair (ppm)				
mean		*std*	*minimum*	*maximum*
0.39		0.39	0.2	2.47
Lake St. Pierre	*130 participants*			
	Fish meals	*SD*	*minimum*	*maximum*
fishmeals/year				
Yellow Perch	15,3	22,7	0	163
Walleye	13,0	16,4	0	112
Northern Pike	1,1	4,2	0	39
meals/spring	(total for three-months)			
Yellow Perch	3,7	9,3	0	76
Walleye	3,1	6,2	0	39
Northern Pike	0,5	3,5	0	39
Mean Hg levels in first 3cm of hair (ppm)				
mean		*std*	*minimum*	*maximum*
0.83		0,97	0,04	5,23
Abitibi	*1 46 participants*			
	Fish meals	*SD*	*minimum*	*maximum*
fishmeals/year				
Walleye	18,1	21,8	0	96
Northern Pike	5,1	10,0	0	48
meals/spring	(total for three-months)			
Walleye	6,6	12,0	0	70
Northern Pike	1,5	3,1	0	12
Mean Hg levels in first 3cm of hair (ppm)				
mean		*std*	*minimum*	*maximum*
0,78		1,47	0,01	13,47

These expectations were not met by the results of our study. Mercury levels in hair of participants of the three regions are relatively low, being well below the recognized threshold for potential early alteration of health (WHO, 1990; 2003). The average Hg levels in hair collected among Abitibi and Lake St. Pierre anglers are similar and up to two times higher than those of the Innu cohort. These data, calculated on the first three centimeters, represents recent seasonal fish consumption at the time of sampling (winter in Lake St. Pierre and spring for both Abitibi and Labrador). The observed number of seasonal/yearly fish meals is significantly higher (up to five times) in Labrador than in the two other regions where the fish consumption frequency is comparable on both temporal scales. Details on fish consumption patterns indicate that dietary habits greatly differ between regions. Fish eaten by the secluded Innu community is exclusively of local origin and people tend to prefer smelt, lake and brook trout, either on a seasonal or yearly basis. In contrast, the Lake St. Pierre and Abitibi groups consume other fish products bought in local markets (respective proportion of 44% and 48%). In the Lake St. Pierre cohort, the fish consumption from local origin is mainly composed of northern pike, walleye and yellow perch, the latter representing about half of the local fish meals. The Abitibi group also frequently enjoys northern pike and walleye species.

DISCUSSION

Numerous processes have been presented in the scientific literature to explain trends in Hg introduction in the biota of lakes and accumulation along aquatic food webs. Mercury levels in edible fish species and related human exposure are modulated by several complex factors, which interact differently according to particular environmental settings. To understand and explain the series of results presented in the last section, we developed a concept called "*ecosystems vulnerability to Hg contamination*". Factors influencing the ecosystems vulnerability can be grouped into two components, and the examination of the data set presented above will be arranged accordingly:

1. *The environmental component*, which can be divided in four key factors, each of them likely to be predominant in determining high Hg accumulation in edible fish species: (1) the sources and extent of Hg loadings to aquatic ecosystems; (2) the capability of lake ecosystems to transform the bulk loading of Hg into bioavailable MeHg; (3) the ability of food webs to assimilate MeHg and accumulate it toward the

higher trophic levels, up to predator species; (4) the overall ability of the ecosystem to cope with, or recover from changes (either an increase or a decrease) in Hg loading rates.

2. *The Human component*. Humans are an intrinsic part of ecosystems. As such, human behavior (including social, economical and traditional factors) is an important component of ecosystem functioning and must be taken into account to correctly assess the threat to the health of individuals resulting from the presence of Hg in fish. For this, three new key factors can be distinguished: (1) the extent of fish resources use, as modulated by traditions, culture, recreational activities and/or economic dependency to fish resources; (2) the ability of populations/communities to lower their levels of exposure to Hg either through changes in culture/tradition/habits, or a better management of environmental resources (fishing strategies, land use, etc), which in turn leads to lower Hg levels in edible fish species; (3) the overall sensitivity of populations/communities to the physiologic stress imposed by exposure to Hg, in light of their general health status (presence of other contaminants in food sources, diseases, alcoholism, metabolic factors, etc.).

Environmental component

Most of the descriptive data assembled here point to Lake St. Pierre as being favorable to high accumulation of Hg in the top predator species: Mercury loadings in Lake St. Pierre are undeniably higher than in the other two regions, in view of the agricultural and industrial use of the lake primary catchments. Suspended particular material from adjacent rivers certainly represents a non-negligible source of Hg to the ecosystem. Furthermore, it is widely recognized that wetlands represent privileged environmental settings for Hg methylation, considering the high biological activity occurring in these milieu. Wetland macrophytes are indirectly involved in Hg methylation, since they support numerous communities of epiphytic periphyton (Hill et al., 1995; Cleckner et al., 1998). Other preliminary results from our study indicate that periphyton assemblages of Lake St. Pierre are an efficient substrate for Hg methylation (Hamelin et al., 2004; Planas et al., 2004). Considering that Lake St. Pierre is surrounded by extended wetlands, it can be expected that the methylation efficiency of this ecosystem surpasses the one encountered in the nutrient limited lakes of Abitibi and Labrador, where the rocky primary catchments are mainly covered with black spruce and other species typical of boreal environments. Finally, more eutrophic aquatic environments were reported to be efficient methylation incubators (Bodaly et al., 1984; Hintelmann and Wilken, 1995; Lucotte et al., 1999).

Surprisingly, predator northern pikes and walleyes from Lake St. Pierre bear significantly less Hg than those of the Abitibi lakes. Putting these results in parallel with regional fish growth data, Simoneau et al. (2004) suggested that Hg dilution in fish tissue occurring in the faster growing specimens of Lake St. Pierre generates lower Hg levels for a specimen of the same size. They also suggest that these increased growth rates, most notably in Lake St. Pierre's walleye population, occur as a result of higher fishing pressure, which in turn diminishes competition for food and ecological niche (Göthberg 1983, Verta 1990, Doire et al. 2004a, b). Indeed, Lake St. Pierre's commercial fisheries are among the most prosperous within Canadian freshwater lakes. This finding could have major implications in future fishing management strategies, knowing that a proper level of the fishing intensity applied to lakes could tend to improve the quality of fish with respect to Hg and probably other contaminants, regardless of other environmental features that could influence Hg methylation and bioaccumulation.

Human component

The levels of Hg in hair have been recognized for many years as a simple and valid proxy to estimate human exposure to Hg (WHO, 1990; 2003, Schwartz, 2000). This is why most epidemiological or descriptive studies dealing with exposure to Hg fail to report Hg content of food sources and use only Hg hair levels to characterize the different exposure scenarios. The data gathered in this study allows us to go further and calculate the total seasonal Hg exposure of participants due to fish consumption using fish consumption patterns and measured Hg levels in fish species. These estimates are compared to the averaged Hg levels measured in the first three centimeters of hair locks. Then, we weighted these estimates against the outputs from simulation runs made using a simple model that assembles on a STELLA platform the published equations and constants describing Hg metabolic rates (NRC, 2000; EPA, 1997; NIEHS, 1999; Myers et al., 1997; Davidson et al., 1995; Kjellstrom et al., 1986; 1989). The graphic-interfaced model calculates either: 1) baseline Hg levels in hair according to typical fish consumption habits (frequency, specie, meal size and bodyweight) and contamination; 2) Hg levels following peak exposure to Hg via higher fish consumption or occasionally higher level of Hg in the food source. We successfully validated the model's output against another similar tool published by Carrier et al. (2001), using a variety of available data sets on Hg exposure and corresponding biomarker signals (blood and hair – Kershaw et al., 1980; Birke et al., 1972; Sherlock et al., 1984).

Table 5. Stella model - simulation runs.

Region	Number[1] fish meals local (3 months)	Number fish meals all sources (3 months)	Calculated[2] mean Hg level in fish diet (ppm)	Calculated daily exposure (ugHg/day/kg bodyweight)	Modeled Hg levels in hair using calculated exposure (ppm)	Measured Hg levels in hair (ppm) (first 3 cms)
L. St. Pierre	7,3	16,6	0,09	0,033	0,6	0,8 (0,97)[4]
Abitibi	8,1	16,9	0,27	0,100	1,6	0,8 (1,47)[4]
Labrador	46,9	46,9	0,25	0,243	4,1	0,4 (0,39)[4]
	Calculated daily exposure			Modeled Hg levels in hair		Measured Hg levels in hair
Fundy[3]	0,051 (ugHg/day/kg bodyweight)			0.8 ppm		0,3 ppm (0,33)[4]

[1] From dietary survey.
[2] Average Hg levels in all fish species consumed, according to dietary surveys (see Table 2). Measurements were performed on Hg levels in local species at standardized edible length (see Table 3), and canned tuna (average from the different types of tuna consumed: 0,2ppm). Mercury data for other fish sources from Dabeka et al., 2003.
[3] Population of Grand Manan Island, St. Andrews/St. Stephens (New-Brunswick - Canada) in 2002; number of participants: 135; Daily exposure calculated according to food consumption survey and measurements of Hg content of the different fish consumed (market and local); typical body weight: 70kg.
[4] Standard deviation on average hair values.

Simulations were made in order to calculate an averaged Hg level for the first three centimeters of hair locks that correspond to the calculated steady state daily exposure to Hg through fish consumption in the three regions. In Abitibi and Lake St. Pierre, the model responses are similar, within the standard deviation values, to the measured mean Hg contents in hair, with a closer fit observed for the Lake St. Pierre data set. However, a similar simulation made using the Labrador data set yields a modeled Hg hair signal higher by a factor of more than 10 from the level that could have been expected from actual knowledge of Hg metabolic processes.

Considering data generated by this study and other simulations made from published data on daily Hg exposure and corresponding hair Hg signals in different ethnic communities, Canuel et al. (2004) proposed that, contrarily to well received, accepted and commonly used scientific precepts, the relation between Hg dose/response - expressed as human Hg exposure through fish consumption/Hg levels in hair - can vary among certain ethnic groups in response to either dietary or metabolic factors. In a companion paper, the authors suggest from other evidence that these discrepancies might be linked to variations in MeHg binding capability with red blood cells in the gastro-intestinal tract (de Grosbois et al., 2004).

Even if future research is needed to fully decipher the processes explaining and supporting the observations and hypotheses issuing from this study, these findings could have major implications in future policies regarding the establishment of fish consumption guidelines.

CONCLUSIONS

A priori considerations on the potentiality of risk of high Hg exposure solely based on limited/partial information on whole ecosystem functioning are likely to lead to erroneous conclusions. The integrated transdisciplinary study presented here gathered researchers around a global vision of the Hg issue on a regional level, while helping identify new gaps in knowledge by forcing the integration of all hierarchical levels of ecosystem science toward the characterization of its vulnerability to Hg contamination. This framework stresses the role of humans as an important part of the ecosystem, which must not be set aside during environmental studies. The following lessons can be drawn from this exercise: (1) the Human constituent, composing part of the ecosystem vulnerability, can be one of the most important factors setting respective community exposure level to Hg; (2) While ecosystem response to the stress imposed by the presence of Hg greatly varies according to particular environmental settings, the human response to this stress can also differ greatly from one region to the other. The difficulty then arises in the identification of effective intervention strategies to minimize the risks related to Hg exposure. Our findings demonstrate the need to base such intervention on the complete description of ecosystems, including their Human components.

We are convinced that the transdisciplinary approach presented here establishes a new pattern for future efficient and conclusive studies on the Hg issue. It demonstrates also that new fish consumption advisory guides should not be established around strictly normative aspects but rather include environmental and human parameters detailing the specific vulnerability of local ecosystems to mercury.

REFERENCES

Amyot, M.,C., Beauvais, P., Constant, S., DeGrosBois, D., Gabrovska, H., Godmaire, D., Lean, L., Fisher-Rousseau, S., Garceau, E., Garcia, S., Gérome, V., Girard, R., Goulet, S., Hamelin, R., Harris, J., Hill, K., Hindle, J., Holmes, J., Laroulandie, M., Lucotte, D., Mergler, N., Milot, M., Nugent, O., Nwobo, B., Page, M., Pilote, D., Planas, L., Poissant, N., Roy, Siciliano, S., Wang, F., Zhang, H., Zhang, J. The St Lawrence River case study: linking Hg biogeochemistry, health and environmental education. Proceedings of the 7th International Conference on Mercury as a Global Pollutant, Ljubljana, Slovenia. 2004.

Birke, G., Johnels, A.G., Plantin, L.-O., Slostrand, B., Skerfving, S., Westermark T. Studies on human exposed to methylmercury through fish consumption. *Arch. Environ. Health.*, 25, 77-91, 1972.

Bodaly, R.A., Hecky, R.E., Fudge R.J.P. Increases in fish mercury levels in lakes flooded by the Churchill River diversion, northern Manitoba. *Can J. Fish Aquat. Sci.,* 41, 682-691, 1984.

Boucher de Grosbois, S., Canuel, R., Atikesse, L., Lucotte, M., Arp, P., Mergler, D., Chan, L. New evidences on methylmercury Human exposure through fish consumption and differentiation in populations' mercury uptake. *Lancet*, (submitted) 2004.

Bruce, W. J., Spencer, K. D., Arsenault E. Mercury content data for Labrador Fishes, 1977-1978. Fish. Mar. Service Data Report 142, 1989.

Canuel, R., Boucher de Grosbois, S., Atikessé, L., Lucotte, M., Arp, P., Mergler, D., Chan, L., Amyot, M., Anderson R. New evidences on Human metabolic response to Methylmercury exposure through fish consumption. *Science*, (submitted) 2004.

Carrier, G., M., Bouchard, R.C., Brunet, M., Caza, A. toxicokinetic model for predicting the tissue distribution and elimination of organic and inorganic mercury following exposure to methylmercury in animals and humans. II: Application and validation of the model in humans. *Toxicol. Appl. Pharm.*, 171, 50-60, 2001.

Cleckner, L.B., Garrison, P.G., Hurley, J.P., Olson, M.L., Krabbenhoft, D.P. Trophic transfer of methyl mercury in the northern Florida Everglades. *Biogeochemistry*, 40, 347-361, 1998.

Dabeka, R.W., McKenzie, A.D., Bradley P. Survey of total mercury in total diet food composite and an estimation of the dietary intake of mercury by adults and children from two Canadian cities, 1998-2000. *Food. Food. Add. Contam.*, 20(7) 629-638, 2003.

Davidson, P., Myers, G.J., Cox, C., Shamlaye, C., Marsh, D., Tanner, M., Berlin, M., Sloane-Reeves, J., Cernichiari, E., Choisi, O., Choi A., Clarkson, T. Longitudinal neurodevelopmental study of Sheychellois children followin g in eutero exposure to methylmercury from maternal fish ingestion: outcomes at 19 and 29 months. *Neurotoxicology*, 16, 677-688, 1995.

Doire, J., Lucotte, M., Fortin, R., Verdon, R. Influence of intensive fishing on fish diet in natural lakes from northern Québec. *Can. J. Fish Aquat. Sci.*, (submitted) *In review.* 2004-a.

Doire, J., Lucotte, M., Fortin, R., Verdon, R. Influence of intensive fishing on fish growth rate in natural lakes from northern Québec. *Can. J. Fish Aquat Sci.*, (submitted) *In review.* 2004-b.

Göthberg, A., Intensive fishing – a way to reduce the mercury level in fish. *Ambio*, 12, 259-261. 1983.

Hamelin, S., Planas, D., M., Amyot Role of epiphytes on mercury accumulation and methylation. SCL / CCFFR Conference. St John.s (NF), Canada; January 2004.

Hill, W.R., Stewart, A.J., Napolitano, G.E. Mercury speciation and bioaccumulation in lotic primary producers and primary consumers. *Can. J. Fish Aquat. Sci.*, 53, 812-819, 1996.

Hintelmann, H., Wilken R.-D. Levels of total mercury and methylmercury compounds in sediments of the polluted Elbe River: influence of seasonally and spatially varying environmental factors. *Sci. Tot. Env.*, 166 (1-3) 1-10, 1995.

Kjellstrom, T., Kennedy, P., Wallis, S., Mantell, C. Physical and mental development of children with prenatal exposure to methylmercury from fish. Stage 1: Interviews and psychological tests at age 4. Report 3080. Solna, Sweden: National Swedish Env. Protec. Board, 1986.

Kjellstrom, T., Kennedy, P., Wallis, S., Mantell, C. Physical and mental development of children with prenatal exposure to methylmercury from fish. Stage 2: Interviews and psychological tests at age 6. Report 3642. Solna, Sweden: National Swedish Environmental Protection Board, 1989.

Lucotte, M., Schetagne, R., Thérien, N., Langlois, C., Tremblay A. (Eds) Mercury in the biogeochemical cycle. Berlin: Springer, 334, 1999.

Muir, D.C.G., Banoub, J., Kwan M. Spatial trends and pathways of POPs and metals in fish, shellfish and marine mammals of northern Labrador, Nunavik and Nunavut. In: Synopsis of Research Conducted Under the 1997/98 Northern Contaminants Program, Environmental Studies No. 75. Ottawa: Indian and Northern Affairs Canada. 171-174, 1998.

Muir, D., Kwan, M., Lampe, J. Spatial trends and pathways of POPs and metals in fish, shellfish and marine mammals of northern Labrador and Nunavik. Synopsis of Research Conducted Under the 1998/99 Northern Contaminants Program, Environmental Studies, Ottawa: Indian and Northern Affairs Canada. 165-171, 1999.

Myers, G.J, Davidson, P.W., Shamlaye, C.F., Axtell, C.D., Cernichiari, E., Choisy, O., Choi, A., Cox, C., Clarkson T. W., A Effects of prenatal methylmercury exposure from a high fish diet on developmental milestones in the Sheychelles Child Development A study. *Neurotoxicology*, 18, 819-830, 1997.

NADP-MDP: National Atmospheric Deposition Program – Mercury Deposition Nework On line database available at: http://nadp.sws.uiuc.edu/mdn/ 2004.

National Institute of Environmental Health Science (NIEHS). Scientific issues relevant to assessment of health effects from exposure to methylmercury. Workshop organized by the Committee of Environmental and Natural Resources (CENR). Office of Science and Technology Policy (OSTP). The White House. Raleigh, NC, 1999.

National Research Council (NRC).Toxicological effects of methylmercury. Committee on the toxicological effects on methylmercury; Board of Environmental Studies and Toxicology Commission on Life Sciences; National Research Council, Washington DC. National Academy Press., 2000.

Planas, D., Desrosiers, M., Hamelin, S. Mercury methylation in periphyton biofilms. Proceedings of the 7th International Conference on Mercury as a Global Pollutant. Ljubljana (Slovenia) 27th June-2 July 2004.

Rice, D., Schoeny, R., Mahaffey K. Methods and rationale for derivation of a reference dose for methylmercury by the U.S. EPA. *Risk Analysis*, 23, 107-115, 2003.

Schwartz, H. Le méthylmercure au Canada : Exposition des Premières Nations et des Inuits au méthylmercure présent dans l'environnement canadien. Health Canada report; Volume 3. publication no H34-97/3-1999F; 9, 1999.

Sherlock, J., Hislop, L., Newton, D., Topping, G. Whittle, K. Elevation of mercury in human blood from controlled chronic ingestion of methylmercury in fish. *Hum. Toxicol.*, 3, 117-131, 1984.

Siciliano, S.D., Sangster, A., Daughney, C.J., Losetto, L.L., Germida, J.J., Rencz, A.N., O'Driscoll, N.J., Lean D.R.S. Are MeHg concentrations in the wetlands of Kejimkujik national park, Nova Scotia, Canada, dependent on geology? *J. Env. Quality*, 32, 2085-2094, 2003.

Simoneau, M., Garceau, S., Lucette, M. Fish growth rates control mercury concentration in walleye from Eastern Canadian lakes. *Env. Res.*, (in press) 2004.

U.S. EPA Mercury Study Report to Congress. Office of Air Quality Planning and Standards and Office of Research and Development; Research Triangle Park, NC., 1997.

Verta, M. Changes in fish mercury concentrations in an intensively fished lake. *Can J. Fish Aquat. Sci.*, 47,1888-1897, 1990.

Wagemann, R., Trebacz, E., Boila, G., Lockhart, W.L. Methylmercury and total mercury in tissues of Arctic marine mammals. *Sci. Tot. Env.*, 218, 19-31, 1998.

World Health Organization Environmental Health Criteria 101: Methylmercury. WHO document; Geneva; Switzerland, 1990.

World Health Organization Elemental Mercury and Inorganic Mercury Compounds: Human Health Aspects. Concise International Chemical Assessment; Document 50, 2003.

Chapter-20

THE GEF/UNDP/UNIDO GLOBAL MERCURY PROJECT - ENVIRONMENTAL AND HEALTH RESULTS FROM A SMALL-SCALE GOLD MINING SITE IN TANZANIA

D.[1], Appleton, G.[2], Drasch, S.[2], Böese-O'Reilly, G.[2], Roider, R.[1], Lister , H.[1], Taylor, B.[1], Smith, A.[3], Tesha and Beinhoff C.[4]

[1] *British Geological Survey, Keyworth, Nottingham, NG12 5GG, United Kingdom*
[2] *Institute of Forensic Medicine, Ludwig-Maximilians-University, Munich, Germany*
[3] *Ministry of Energy and Minerals, Box 2000, Dar es Salaam, Tanzania*
[4] *United Nations Industrial Development Organization (UNIDO), Vienna International Centre, P.O. Box 300, A 1400 Vienna, Austria*

INTRODUCTION

Artisanal gold mining is one of the major sources of mercury contamination, especially in developing countries. Whilst the gold extraction process (known as amalgamation) is a simple technology, it is potentially very harmful to the environment and can contaminate air, soil, rivers and lakes with mercury. The health of the miners and other people living within the area affected by mercury contamination may be negatively affected through inhalation of mercury vapour or contaminated dusts, direct contact with mercury, through eating fish and other food, and through the ingestion of waters and soils affected by the mercury contamination.

Environmental and health impacts resulting from the use of mercury (Hg) in the artisanal gold extraction process require concerted and coordinated global responses. The objective of the ongoing GEF/UNDP/UNIDO project *Removal of Barriers to the Introduction of Cleaner Artisanal Gold Mining and Extraction Technologies* (also referred to as the *Global Mercury Project*

(GMP)) in Brazil, Indonesia, Lao PDR, Sudan, Tanzania, and Zimbabwe is to assist these countries located in key transboundary river/lake/marine basins in assessing the extent of Hg pollution, introducing cleaner gold extraction technology, which eliminates or reduces Hg releases and developing capacity and regulatory mechanisms. The GMP Project is accompanied by the development of monitoring programmes. In order to ensure sustainability, capacity is being built to carry out continuous monitoring beyond the project three-year term. The ultimate goals of the GMP Project are:

1. to reduce Hg pollution of international waters by emissions emanating from small-scale gold mining;
2. to introduce cleaner technologies for gold extraction and to train people in their application;
3. to develop capacity and regulatory mechanisms that will enable the sector to minimize mercury pollution;
4. to introduce environmental and health monitoring programmes;
5. to build capacity of local laboratories to assess the extent and impact of Hg pollution.

The GMP Project will also aim to increase knowledge and awareness of miners, Government institutions and the public at large by explaining in detail the results of the Public Health and environmental studies conducted in the mercury "hot spot" areas.

The selection of project demonstration sites was done in accordance with the objective of alleviating the impact of Hg on international waters. Two sites were selected in Brazil (Creporizinho and Sao Chico in the Tapajos area draining into the Amazon river) and Indonesia (Galangan mine in Central Kalimantan, draining into the Java Sea and Talawan near Manado, draining into the Celebes Sea). One demonstration site was selected in each of the other four countries: Lao PDR (Luang Prabang, draining into the Mekong river), Sudan (Gugob, near Al Damzain, draining into the Blue Nile), Tanzania (Rwamagasa, draining into Lake Tanganyika) and Zimbabwe (Chakari, draining to a tributary of the Zambezi river).

In August 2003, the British Geological Survey (BGS), acting under the UK Natural Environment Research Council, signed a contract with UNIDO to carry out limited Environmental and Health surveys and assessments in the Rwamagasa artisanal gold mining area in the Republic of Tanzania (Appleton et al., 2004). The environmental assessment was executed by the BGS whilst the medical and toxicological investigations were subcontracted to the Institut für Rechtsmedizin der Ludwig-Maximilians-Universität München, Germany. The regional health authorities in Geita supported the medical investigations, whilst the environmental assessment was carried out in collaboration with staff from the Geita Mines Office and from the Kigoma and Mwanza offices of the Tanzania Fisheries Research Institute (TAFIRI).

Rwamagasa is located in Geita District, which has an area of 7,825 km², 185 villages, and a population around 712,000 (census of 2002). The number of artisanal miners in the Geita District is unknown but it is estimated to be as many as 150,000, most of whom are illegal panners. Primary artisanal workings in the Rwamagasa area are centred on quartz veins in sheared, ferruginous, chlorite mica schists. Grab samples of vein and wall rock grade 6-62 g/t Au. The only legal mining in the Rwamagasa area is carried out within the boundaries of the Primary Mining Licence held by Blue Reef Mines where approximately 150 people are involved in mining and mineral processing activities. This is the only site in the Rwamagasa area where primary ore is being mined underground. All other mineral processing activity of any significance is concentrated at the northern margin of Rwamagasa, especially on the land sloping down to the Isingile River. In this area, there are about 30 groups of historic and active tailings dumps and about ten localities where small (200 litre) ball mills are operating. The number of people actively involved, at one particular time, in ball milling, sluicing and amalgamation is probably no more than 300.

Amalgam is burned in a small charcoal fire, which releases Hg to the atmosphere. Amalgamation mainly takes place adjacent to amalgamation ponds, which are usually formed of concrete, but sometimes have only wood walls even though environmental legislation dictates that the Hg contaminated mineral concentrates and tailings should be stored in concrete lined structures.

The Blue Reef Mine is reported to produce about 1 kg Au per month whereas artisanal miners re-working tailings produce about 0.5 kg per month. On this basis, approximately 27 kg of Hg will be released to the environment from the Rwamagasa area each year. Of this, atmospheric emissions from amalgam burning will be about 14 kg from the Blue Reef mine site and 7 kg from the other amalgamation sites. About 2 to 3 kg Hg will remain in heavy mineral tailings in the amalgamation ponds, which are frequently reprocessed. It is reported that the number of miners working in the Rwamagasa area was much larger in the past, so the historical release of mercury would probably have been higher than at present.

The young and strong men, so called healthy workers, are mainly found in the bigger and more technically equipped properties. Older people, women of all ages and children mainly work in the smaller artisanal mining properties. Retorts are not used, neither is there any other protection, such as ventilation, against any kind of mercury contamination. Housing areas, food stalls and the schools are located close to the sites where amalgamation and burning of the amalgam is carried out. Mineral processing tailings containing mercury are found within the village adjacent to cultivated land or near local water wells. Mercury is usually stored in the miner's houses in small soft-drink bottles, near to where they and their families sleep. The mercury is mainly obtained from Nairobi in Kenya and the gold is either used for jewellery in Tanzania or sold to Dubai.

Hygiene standards are extremely low and are a reason for many infectious diseases such as diarrhoea, typhoid and parasitism. There is no effective waste disposal system for either mercury, sanitary or other domestic waste.

Road accidents, accidents in insecure tunnels and amalgamation plants, malaria, tuberculosis, and sexually transmitted diseases including AIDS are the dominant causes of morbidity and mortality. No special health service exists for the mining community – the nearest dispensary is about 10 km away. A local dispensary is under construction, but the construction has been stopped due to lack of money. The village lacks social welfare services and a police post for security. The nearest district hospital is in Geita, 45 km to the northeast. All non-minor illnesses have to be transferred to Geita hospital, which is adequately equipped for a district hospital.

Background information on mercury contamination associated with artisanal gold mining in Tanzania is available in a number of published reports and scientific papers (Appel et al., 2000; Asanao et al., 2000; Campbell et al., 2003a,b; Harada et al., 1999; Ikingura and Akagi, 1996; Ikingura et al., 1997; Ikingura and Akagi, 2002; Kahatano et al., 1997; Kinabo, 1996; Kinabo, 2002a,b; Kinabo and Lyimo, 2002; Kishe and Machiwa, 2003; Machiwa et al., 2003; Mutakyhwa, 2002; Semu et al., 1989; Sindayigaya, 1994; University of Dar es Salaam, 1994; van Straaten 2000a,b).

ENVIRONMENTAL ASSESSMENT FIELD PROGRAMME

The objective of the environmental assessment was to (i) identify hotspots in the project demonstration sites, (ii) conduct specified geochemical and toxicological studies and other field investigations in order to assess the extent of environmental pollution in surrounding water bodies and (iii) devise intervention measures. Although Rwamagasa is located only 37 km to the south of Lake Victoria, streams draining the Rwamagasa 'mining hotspot' actually drain SW into the Nikonga River, and then for a further 430 km via the Moyowosi swamps and the River Malagarasi before reaching Lake Tanganyika near Ilagala, about 50 km to the SSE of Kigoma. One of the major objectives of the project is to assess the impact of mercury contamination on international waters as well as in the vicinity of the 'mining hotspot', so the field programme was carried out in two areas: (a) the Rwamagasa 'mining hotspot' sub-area and (b) the Lake Tanganyika – River Malagarasi sub-area (see Figure 1). Dispersion of Hg from Rwamagasa to Lake Tanganyika is probably relatively unlikely because contaminant Hg will be adsorbed by organic material in the extensive Moyowozi and Njingwe Swamps and flooded grassland area, located from 120 km to 350 km downstream of Rwamagasa. Whereas the swamps will act as a potential biomethylation zone, they will also act as an environmental sink for Hg contamination, which is likely to inhibit migration of Hg into the lower reaches of the Malagarasi River and Lake Tanganyika. The swamp area was inaccessible within the logistical and budgetary constraints of the current project.

The environmental field programme was carried out during the dry season at which time there was little evidence that large quantities of contaminated tailings were being washed into the Isingile River. However, waste water and tailings from amalgamation 'ponds' were observed at one site to be overflowing onto an area where vegetables were being grown. If large quantities of Hg contaminated tailings are dispersed onto the seasonal swamp (*mbuga*) area adjacent to the Isingile River during the wet season, then this may lead to the significant dispersion of Hg both into the aquatic system and onto agricultural sites being used for rice, maize, and vegetable cultivation.

Previous studies in the Lake Victoria Goldfields area indicate that dispersion of Hg from tailings is relatively restricted, not least because Fe-rich laterites and seasonal swamps (*mbugas*) act as natural barriers or sinks attenuating the widespread dispersion of Hg in sediments and soils.

A field programme was carried out in September–October 2003 leading to the collection of a total of 38 water, 26 drainage sediment, 151 soil, 66 tailings, 21 vegetable and 285 fish samples. Preparation and analysis of the samples was carried out in the UK and Canada. Analytical data for duplicate field samples, replicate analyses and recovery data for Certified Reference Materials indicate a level of analytical precision and accuracy that is appropriate for this type of environmental survey. Cd, Cu, Pb and Zn were determined in drainage sediment, tailings and soil samples, in addition to Hg and As (which were specified in the ToR and BGS's proposal) on the basis that these could be useful indicators of mineralization and/or anthropogenic contamination. The range of chemical substances determined in the samples collected, and the range of media sampled should not be considered to represent a comprehensive environmental survey. In addition, the results reported here refer only to the sites sampled at the time of the survey and should not be extrapolated to infer that elevated levels of contamination are not present at other sites or elsewhere in the district or region. The results presented reflect the level of resources available for the environmental assessment.

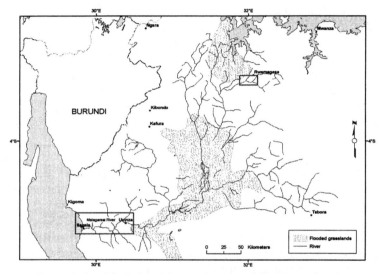

Figure 1. Location of the River Malagarasi and Rwamagasa sub-areas in northwest Tanzania.

Nature and extent of the mercury pollution in the river system

At the time of sampling, mercury in filtered drainage water samples ranged from 0.01 to 0.03 μg/L in the R. Malagarasi and from 0.01 to 0.07 μg/L in the Rwamagasa area. None of the filtered water samples exceeded any of the Tanzanian Water Quality Standards or other national and international water quality standards, or criteria for drinking water, protection of aquatic biota or the protection of human health. Arsenic in filtered water ranged from 0.1 to 2.4 μg/L and none of the samples collected exceeded any water quality standards or criteria.

Hg concentrations in the fine fraction of streams sediments from the River Malagarasi at Ilagala range up to 0.65 mg/kg, which is rather high for an area that does not appear to be unduly affected by anthropogenic contamination. Concentration of Hg in the fine fraction, together with adsorption of Hg onto Fe and organic material, may in part explain these relatively enhanced Hg concentrations, but these hypotheses need to be verified by further studies. Other possible sources include the geothermal springs at Uvinza or contamination of sediment by mercuric soap, which may be used by some people for skin lightening, although no evidence of this was observed during the field programmes.

In the Rwamagasa area, Hg in the fine fraction of drainage sediments ranges from 0.08 to 2.84 mg/kg, although Hg does not exceed the Toxic Effects Threshold (1 mg Hg/kg) of the Canadian Sediment Quality Criteria for the Protection of Aquatic Life for more than 2 km downstream from the major mineral processing centre located to the south of the Isingile River. Toxic Effects Thresholds for As, Cd, Cu, Pb and Zn are not exceeded in drainage sediments from the Rwamagasa area.

Environmental assessment of the Rwamagasa area based on sampling of mineral processing wastes and soils

There is little difference between Hg concentrations in samples taken from historic (dry) primary tailings piles (mean 5 mg/kg) and samples taken from recent sluice box tailings (mean 3 mg/kg). Hg in tailings samples from the amalgamation ponds and amalgamation pond tailings (mean 86 mg/kg) are on average about 20 times higher. The high level of Hg in the primary and sluice box tailings is the result of recycling/reprocessing of amalgamation pond tailings. An association between Cd-Cu-Hg-Zn probably reflects contamination from mercury used in amalgamation combined with metals that are possibly derived from the ball mills and galvanised roof sheets. Correlations between arsenic and iron probably reflect the influence of trace quantities of arsenopyrite and pyrite in the gold ore. Both these hypotheses need to be verified.

At the time of the survey, generally low concentrations of Hg occurred in most of the analyzed soils used for cassava, maize, and rice cultivation, as well as *mbuga* and unclassified soils located away from the urban centre of Rwamagasa and associated mineral processing areas. Higher concentrations are found in urban soils and also in *mbuga* and vegetable plot soils adjacent to the Isingile River, close to the mineral processing areas. Hg in the urban soils is probably mainly derived from air borne transport and deposition of Hg released during the burning of amalgam, although this has not been verified. High Hg appears to occur in the *mbuga* and vegetable plot soils where these are impacted by Hg-contaminated water and sediment derived from mineral processing activities located on the southern side of the Isingile River. In these soils there is a clear association between Cd-Cu-Zn, which reflects contamination from metals that are possibly derived from the ball mills and/or galvanized roof sheets. An association between As, Cu and Fe probably reflects the influence of the weathering products of arsenopyrite and pyrite found in the gold ore, although this needs to be verified.

Mercury exceeds (1) the maximum permissible concentration of Hg in agricultural soil in the UK (1 mg/kg) in 12 soil samples; (2) the Canadian Soil Quality Guideline for agricultural soils (6.6 mg/kg) in three samples; and (3) the UK soil guideline value for inorganic Hg for allotments (8 mg/kg) in two samples.

Cadmium and zinc exceed the maximum permissible concentrations for agricultural soil in the UK (3 mg Cd/kg and 200 mg Zn/kg) in only a few soil samples. Arsenic exceeds the Canadian Soil Quality Guideline for agricultural soils (12 mg/kg) in nine agricultural and urban soils.

Soil profile data demonstrate that surface contamination by mineral processing waste in some agricultural soils affects the root zone. Hoeing of the soils is likely to result in mixing of surface Hg contamination throughout the root zone, although this has not been verified.

Mercury in agricultural produce

Hg in vegetable and grains samples collected from the agricultural areas potentially impacted by mercury contamination are mainly below the detection limit of 0.004 mg/kg Hg with concentrations of 0.007 and 0.092 mg/kg Hg recorded in two yam samples and 0.035 mg/kg Hg in one rice sample. A positive correlation between Hg in agricultural crops and soil was not detected during the present survey. Hg in beans, onions and maize samples purchased at Rwamagasa market are below the detection limit (<0.004 mg Hg/kg) whilst two dehusked rice samples contain 0.011 and 0.131 mg/kg Hg. The concentrations of Hg in rice are similar to those recorded in rice grown on the highly contaminated soils of the Naboc irrigation system on the island of Mindanao in the Philippines.

Mercury in fish: biomarkers for mercury methylation and potential food sources

The main fish species used as bioindicators of mercury contamination included perch (*Lates spp*), tigerfish (*Brycinus spp*), tilapia (*Oreochromis spp*), catfish (*Clarias spp*). Fish tissue THg data indicates that the sites sampled in the immediate area of mining activities at Rwamagasa, are the worst affected (Figure 2) and should be considered environmental 'hotspots' and sites of biomethylation. Many fish tissue samples from these sites fail export market standards (0.5 mg/kg) and also exceed the WHO recommended standard for the protection of health of vulnerable groups (0.2 mg/kg). Mercury in fish collected from the Nikonga River, approximately 25 km downstream from Rwamagasa, have low Hg concentrations. This suggests that the impact of mercury contamination on aquatic biota is relatively restricted, which is confirmed by the generally low mercury concentrations in drainage sediment and *mbuga* soils at distances more than about 6 km downstream of the main mineral processing area (or 'hotspot'). However, this observation will need to be verified by more detailed studies.

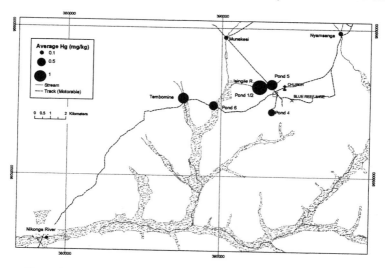

Figure 2. Average mercury concentration (mg/kg) in catfish (<u>Clarias spp.</u>), Rwamagasa area.

Fish length vs. mercury concentration plots for fish from the River Malagarasi and Malagarasi delta area of Lake Tanganyika (collected from Uvinza and Ilagala) confirm generally low mercury concentrations that are similar to levels found in similar species in Lake Victoria (Figure 3).

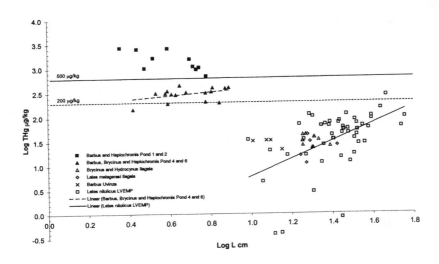

Figure 3. Hg (µg/kg) related to length (cm) in piscivorous, insectivorous and planktivorous fish from (a) the Rwamagasa area, (b) River Malagarasi – Lake Tanganyika (Ilagala, Uvinza) and (c) Lake Victoria (Lake Victoria Environmental Management Project, Machiwa et al., 2003).

All fish samples collected from the Malagarasi River area are below the WHO threshold for vulnerable groups (0.2 mg/kg). This suggests that mercury contamination from the Rwamagasa artisanal gold mining centre does not have a significant impact on fish stocks in either the lower reaches of the River Malagarasi or the international waters of Lake Tanzania.

Exposure to environmental mercury

None of the water samples collected from the river network, or associated drainage ponds exceeded the WHO or local Tanzanian guideline values of 1 µg Hg/l for drinking water. Whilst this suggests that mineral processing operations have not contaminated local surface waters and shallow groundwaters it does not indicate whether drinking water used by the local people has been contaminated with Hg or other non-related substances. More extensive monitoring of drinking water sources (which was not the focus of the current investigations) should be considered as a component of any subsequent follow up work.

The only samples of filtered water collected during the survey that contained relatively high Hg concentrations (max. 0.45 µg Hg/l) were from amalgamation ponds. This highlights the need for careful management of waste-waters from these ponds and monitoring of any nearby drinking water supplies.

The average mercury concentration recorded for samples of rice grain grown on soils potentially impacted by mercury contamination was 0.026 µg/g (dry wt.). Consequently, the amount of mercury entering the body, assuming an average consumption of 300g rice/day is 0.055 mg THg/week (equivalent to 0.46 µg MeHg/kg bw/week), which is much lower than the Provisional Tolerable Weekly Intake (PTWI) of 0.3 mg for total mercury and 1.6 µg/kg bw/week for methyl mercury (MeHg) in the diet set by the WHO and the FAO. These are likely to maximum inputs because most people in the Rwamagasa area will consume less than 300 g rice/day because they will also consume cassava and maize, which are generally grown on soils with low Hg. This observation needs to be verified by more detailed studies.

The vast majority of people in the Rwamagasa area principally eat Tilapia (*Oreochromis spp.*), Perch (*Lates spp.*) and dagaa (dagan; *Rastrineobola spp.* and equivalents) from Lake Victoria. Catfish (*Clarias spp; kamare, mumi*) is eaten by less than 10% of those people. Consumption of 250g perch, 500g tilapia and 250g of catfish each week would result in an intake of 27 µg THg/week (equivalent to 0.35 µg MeHg/kg bw/week) for residents of Ilagala-Uvinza area, 44 µg THg/week (equivalent to 0.58 µg MeHg/kg bw/week) for Rwamagasa residents consuming only fish from Lake Victoria, 56 µg THg/week (equivalent to 0.75 µg MeHg/kg bw/week) for people in the Rwamagasa background area consuming tilapia and perch from Lake Victoria and catfish from the local streams, and 259 µg THg/week (equivalent to 3.45 µg MeHg/kg bw/week) for people in the Rwamagasa area consuming tilapia and perch from Lake Victoria and catfish from mining impacted streams. Apart from the latter group, these inputs related to fish consumption are well below the WHO/FAO PTWI . It appears that only those people consuming catfish from the Isingile River, and other mining contaminated locations such as Tembomine, are likely to be at risk of exceeding the PTWI for Hg.

People consuming 300g/day of rice grown on the Hg contaminated Isingile *mbuga* soils and 1kg of fish from Lake Victoria would have a combined estimated MeHg input of 1.04 µg MeHg kg bw/week, which is two thirds of the MeHg PTWI.

Whereas it is not known whether individuals practice geophagia in the Rwamagasa area, elevated exposures to Hg could result from the occasional deliberate and habitual consumption of contaminated soils and dusts. For example, the PTWI of 0.3 mg for total Hg in the diet set by the WHO and the FAO, is equivalent to 26 µg THg/day for a 30kg child and would be exceeded by an individual practising geophagia (central estimate and worst-case) or on a case by case basis by an individual occasionally consuming soil/dust (worst-case). The practice of geophagy by pregnant females would be of particular concern in this regard given the sensitivity of the foetus to Hg.

The inadvertent ingestion of dusts and soils even those having Hg concentrations significantly above the regional background, and hence considered to be moderately contaminated, does not appear to lead to a significant excess exposure to Hg. For comparison, exposures due to inadvertent ingestion of soils and/or dusts (0.72 to 1.8 μg THg/day or 5 to 13 μg THg/week) are typically less than individual exposure via other dietary sources water, rice and fish.

However, given the uncertainties involved in estimating inadvertent dust and soil intake in the rural Rwamagasa environment, exposure via this route, in addition to more classical geophagic behaviour, should be considered when planning remedial/intervention measures. Such measures could include the marking and fencing off of waste tips and areas of enhanced contamination and improvements in hygiene (washing of hands and food preparation such as the drying of cassava and other crops directly on the ground and the use of soil as a desiccant to aid the storage of groundnuts and beans). Whilst geophagy does have an important cultural and possibly nutritional benefit, the resulting levels of potential exposure to young adults and pregnant woman are high enough to suggest that this practice should be positively discouraged within the mining districts. If it was demonstrated that geophagy is practiced in the Rwamagasa area, the importation of geophagic materials into local markets from outside the contaminated region should be encouraged and the negative effects of using local soils conveyed though local women's groups and childhood development officers.

Medical investigation methodology

The extraction of the gold with liquid mercury releases Hg, especially highly toxic Hg vapour into the local environment. The health status of 211 volunteers in Rwamagasa artisanal gold mining area and 41 non-exposed people from a nearby control area in Katoro, located 30 km distant from Rwamagasa was assessed with a standardised health assessment protocol from UNIDO (UNIDO 2003) by an expert team from the University of Munich, Germany in October/November 2003. The health assessment protocol was developed by UNIDO in collaboration with the Institut für Rechtsmedizin der Ludwig-Maximilians-Universität München, Germany and other international experts. The "Health Assessment Questionnaire" was partly translated in Swahili to be used to examine the general health condition of members of the mining community and to indicate symptoms of Hg poisoning. State of the art anamnestic, clinical, neurological, neuro-psychological and toxicological tests were used. All participants were examined to identify neurological disturbances, like ataxia, tremor and coordination problems. The data was compiled for statistical purposes and confidentiality regarding all health related issues was maintained.

Results of the medical investigation

Mercury concentrations in the bio-monitors urine, blood and hair were significantly higher statistically in the exposed population from Rwamagasa compared with the Katoro control group, but only some amalgam burners showed Hg levels above the toxicological threshold limit HBM II in urine (Figure 4), blood and hair. A speciation of Hg in hair demonstrates that mainly inorganic Hg (including Hg vapour) contributes to the high body burden of the artisanal miners. Low mercury concentrations in all biomonitors (especially blood) of volunteers not occupationally exposed to mercury in Rwamagasa indicate that there is no relevant secondary exposure of humans to Hg in this area by air, drinking water or food, especially locally caught fish.

Only a few statistically significant correlations were detected between Hg concentrations in biomonitors (urine, blood and hair) and anamnestic/clinical data for the amalgam-burners sub-group. Significant correlations included those between the anamnestic data (i) "tremor at work" with Hg in urine, blood and total Hg in hair and (ii) Hg in blood with tiredness, lack of energy, weakness, and problems with concentration and clear thinking. The only significant correlation between a classical clinical indicator and Hg in biomonitors was "heel to knee tremor" with total Hg in hair whilst significant correlations with the "Matchbox test" were found with Hg in urine and blood. Whereas on a group basis Hg in the target tissue (i.e. brain) correlates well with Hg in urine, blood and hair of people with significantly different levels of occupational or environmental exposure, the poor correlations between classical clinical indicators of mercury intoxication and Hg in bio-indicators within the group of amalgam-burners in the present study probably reflects large inter-individual differences (i.e. an individual's biomonitor Hg level may not directly indicate their target tissue (brain) Hg burden). In an individual who has suffered from chronic exposure to Hg, damage to the central nervous system may have occurred months or years before the biomonitor samples were analysed. Biomonitor data indicate an individual's recent body-burden whereas the clinical indicators probably indicate an individual's past or cumulative Hg burden. This would explain why the former occupationally exposed group shows a high median medical score whilst the group's biomonitor Hg levels are only slightly elevated. When the results of individual anamnestic, clinical and neurological tests are summed together, significant correlations exist (i) between Hg in urine and blood with the anamnestic score and (ii) between Hg in urine and the sum of all the anamnestic, clinical and neurological tests. It was shown that for the Rwamagasa amalgam-burner group, which is predominantly exposed to inorganic Hg (including Hg vapour), the Hg concentration in urine is a sound predictor for a Hg intoxication.

Typical symptoms of Hg intoxication were prevalent in the exposed group. For example, combining the medical score with the biomonitoring results made it possible to diagnose chronic Hg intoxication in 25 out of 99 amalgam burners, and in 3 out of 15 former amalgam burners (Figure 5). Table 1 shows the Hg concentrations in biomonitors for the group of intoxicated amalgam burners.

Table 1. Mercury concentrations in biomonitors of the 25 intoxicated amalgam burners (in some cases hair samples were not available).

	Hg-blood (µg/l)	Hg-Urine (µg/g crea.)	T-Hg-Hair (µg/g)	MeHg-Hair (µg/g)
N	25	25	20	18
median	8.6	13.2	4.1	0.77
maximum	33.3	36.8	48.7	5.25

Within the other population groups in Rwamagasa (i.e. people not occupationally exposed to Hg) and in the Katoro control group no cases of Hg intoxication were diagnosed. The percentage of cases diagnosed with Hg intoxication within the amalgam burners was lower in Rwamagasa than in the comparable small-scale gold mining area of Mt. Diwata in the Philippines, for example, where 85% of the amalgam burners were intoxicated (Drasch 2001).

The difference in the level of intoxication cannot be explained by a different (i.e. a safer) amalgam burning technique in Rwamagasa. Moreover, it must kept in mind, that the maximal burden (as expressed in the top Hg concentrations found in the biomonitors) was comparable to Mt. Diwata.

The impression gained during the field programme was that this difference might be explained just by a lower amount of Hg used for gold extraction in the Rwamagasa area, reflecting the lower level of gold production. This results in a lower number people exhibiting high levels of Hg intoxication.

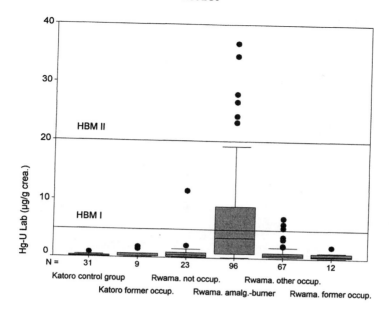

Figure 4. Total mercury concentration in urine (µg Hg/g creatinine), determined in laboratory (occup. = occupational).

Child labour in the Rwamagasa mining sites is very common from the age of 10. The children work and play with their bare hands with Hg. This is very important because Hg can cause severe damage to the developing brain.

Nursed babies of amalgam burning mothers are at special risk. Extremely high mercury concentrations were detected in two out of five breast-milk samples from nursing mothers who worked as amalgam burners. In addition to a placental transfer of Hg during pregnancy from the mother to the foetus (as has been proved in other studies) this high Hg burden of nursed babies should be a cause of great concern.

Poverty is the main reason for the poor health status of the small-scale mining communities. Struggling for survival frequently makes gold mining a necessity in order to obtain financial resources.

The daily fight for survival makes the miners put their own health and the health of their children at risk. A reduction of the release of Hg vapours from small-scale gold mining like in Tanzania into the atmosphere should not only reduce the number of Hg intoxicated people in the mining area but it will also reduce global atmospheric pollution, because a significant

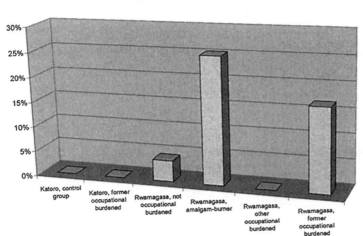

Figure 5. Frequency of the diagnosis of mercury intoxication in the different sub-groups.

proportion of mercury vapours formed by burning of amalgam in the open-air may be transported long-range distances (Lamborg, 2002).

The total (global) release of mercury vapour from artisanal gold mining is currently estimated to be up to 1,000 metric tonnes per year (MMSD, 2002), while approximately 1.900 tonnes of Hg from all other anthropogenic sources were released into the atmosphere (Pirrone, 2001).

Mercury is undoubtedly a serious health hazard in the small-scale gold mining area of Rwamagasa. Working for many years in the amalgamation process, especially amalgam burning has resulted in severe symptoms of Hg intoxication. The exposure of the whole community to Hg is reflected in raised Hg levels in the urine, and the detection of the first symptoms of brain damage such as ataxia, tremor and movement disorders. Mercury intoxication (according to the definition of UNIDO (UNIDO 2003)) was diagnosed in 25% of the amalgam burners from Rwamagasa. In addition, intoxication was also detected in some people that had formerly worked with Hg and amalgam. People from Rwamagasa who are not directly involved in amalgam burning, have a higher Hg burden than the control group, although the majority of these people are not intoxicated. The background Hg burden in the Katoro control group is the same order of magnitude as in western industrial countries.

Recommendations for monitoring water quality and biota

Monitoring is expensive and costs can be reduced if the main exposure routes are known. Hence there is a need for a more intensive study to link exposures from various pathways to Hg levels in blood prior to the development of monitoring or remediation strategies.

Monitoring in the environmental survey followed, as closely as was practicable, the internationally accepted protocols recommended by UNIDO (2003). It is recommended that water monitoring be carried out in the Rwamagasa drainage system during the wet season in order to test for Hg dispersion in solution and in the suspended sediment. The short term and medium term temporal variation in these pollution indicators should also be investigated.

Continuous monitoring equipment capable of determining Hg at low concentrations in drainage systems is, as far as the authors of this report are aware, not available commercially. So any monitoring system would be periodic rather than continuous. Quarterly monitoring will probably be adequate for the Isingile and Nikonga Rivers for a period of two years. If no significant Hg concentrations are detected during that period, and there are no significant changes in the amount of mineral processing and associated factors, then annual monitoring, following the USEPA recommendations, will probably be adequate.

The only effective option to prevent continuing Hg pollution of the Isingile River and surrounding agricultural areas is to require (a) the removal of all the existing mineral processing waste currently located close to the Isingile River and (b) the termination of all mineral processing activities in the vicinity.

Monitoring of drinking water from wells in the Rwamagasa area was not carried out during the current survey, but should be considered when designing any future water quality monitoring systems.

Monitoring of biota (fish and agricultural crops) has been carried out as part of the current study and could be carried out periodically using the UNIDO sampling protocols (UNIDO, 2003), which document procedures for the periodic monitoring of aquatic biota. Periodic monitoring of agricultural crops could also be carried out, although the results of this study indicate that little Hg is present in most of the crops. Due to time and funding constraints, the current study was able to sample only a relatively limited number of sites. For this reason it is recommended that a more comprehensive survey should be carried out, in order to verify the results presented in this study.

Recommendations for remediation and rehabilitation

The present survey did not detect any concentrations of Hg in solution that would require remediation, as they did not exceed water quality standards. Should future water quality monitoring detect concentrations that require remediation, then a number of remediation technologies may be appropriate.

From a practical point of view, there would be little justification in trying to remediate and rehabilitate the Hg contaminated bottom sediments of the Isingile River until (i) the releases of Hg contaminated mineral processing tailings from the Rwamagasa area have been terminated, and (ii) the risk of future contamination of the drainage system by progressive or catastrophic releases of Hg contaminated processing waste has been eradicated. It is, however, relatively unlikely that the tailings piles located adjacent and to the south of the Isingile are a potential source of catastrophic contamination as the waste piles are relatively small and the slopes are relatively gentle. However, both (i) the highly contaminated amalgamation pond tailings and (ii) the primary and sluice box tailings, that have been contaminated with Hg as a result of reprocessing the amalgamation pond tailings, are probably the main source of potential Hg contamination in the Rwamagasa area and dispersal of these tailings needs to be avoided. Removal of the tailings to a safe containment facility, underlain and covered with lateritic material (hydrous ferric oxides) should be considered. As far as the authors are aware, no clean-up goals for Hg have been set in Tanzania, although this needs to be verified.

The principal remediation-rehabilitation options for Hg-contaminated soils and sediments in the Isingile River – Rwamagasa area include (i) excavation of Hg-contaminated soil and disposal to an off-site secure landfill or depository, (ii) electroleaching, comprising wet extraction followed by electrolytic preparation of the leachate, an emerging and potential alternative cleanup method that is reported to offer a cheaper and more environmentally friendly alternative to thermal treatment or the acid leaching process. The cost of these potential remediation options has not been estimated.

Specific practical remediation measures cannot be recommended until a much more detailed assessment has been made of Hg concentrations in the agricultural soils, their uptake by crops and transfer into the human food chain. On the basis of evidence collected during this survey, it appears that significant amounts of Hg are not adsorbed into the grain of the agricultural plants. If this can be confirmed by more detailed site specific studies (involving further collection and analysis of soil and rice grain samples from exactly the same sites, for example) it may be possible to confirm that there is little or no potential for a direct negative impact on human health caused by the consumption of rice and other crops grown on these relatively high Hg soils. Mercury uptake by other crops (such as maize or cassava) grown on soils that are currently used for rice should also be evaluated in case such a change in agricultural practices would increase the potential exposure of the local population.

Recommendations for reduction of the release of mercury into the environment

The exposure to Hg for the miners and the community has to be drastically decreased. Proper mining techniques to reduce the accidents and Hg exposure are essential. Small-scale miners need all possible support to introduce cleaner and safer gold mining and extraction technologies.

The Local Mines Office in Geita needs to ensure that the small scale miners follow relevant mining and environmental regulations and approved practices, such as making sure that all amalgamation is carried out in cemented ponds and that all tailings from these amalgamation ponds are stored in appropriate cemented storage areas that prevent dispersal of Hg onto adjacent land and into water courses.

Exposure to Hg vapour is avoidable with the application of simple technological improvements such as retorts. Technical solutions need to go hand in hand with awareness raising campaigns.

An alliance of local, regional, governmental and intergovernmental bodies is needed to improve the social, health and environmental situation of artisanal small-scale gold miners. Cooperation between health and environmental sectors is needed on local, regional, national and intergovernmental level; for example, UNIDO and WHO in Dar es Salaam could form a nucleus of a national mercury task force.

Recommendations for reduction of mercury as a health hazard

The clinical testing and laboratory results indicate that Hg is a major health hazard in the Rwamagasa area especially for those artisanal miners who burn amalgam. A lower, but significant, level of Hg intoxication is observed in those residents of Rwamagasa who have no occupational exposure.

In order to reduce the level of risk from Hg it is suggested that:

1. Child labour with highly toxic substances must be stopped immediately. Legal restrictions on child labour need to be implemented immediately.
2. Women of childbearing age need special information campaigns on the risk of Hg to the foetus and the nursed baby and advice on how to avoid, or at least reduce, exposure.
3. Participants in the medical assessment diagnosed with Hg intoxication need medical treatment. A system is required for the diagnosis and treatment of Hg related health-problems. Capacity building, including establishing laboratory facilities to analyse Hg in human specimens is required. The financial aspect of treatment and the legal problem of importing drugs (such as chelating agents like DMPS or DMSA, required to remove Hg from the body) need to be solved. Funding of preventive campaigns and for treatment facilities is now needed.
4. Training programs for the health care providers in the district in Geita and other health centres in mining areas is required to raise awareness of Hg as a health hazard and advise people how they can reduce their own and their children's environmental and occupational exposure to Hg.
5. Clinical training of local health workers, including the use of a standardised questionnaire and examination flow scheme (MES = mercury examination score). Particular attention needs to be paid to collecting information on individual's environmental and occupational exposure as this will aid the detailed assessment of exposure routes and the design of strategies that will help to reduce Hg exposure.
6. A mobile "mercury ambulance" might ensure that small-scale miners can be reached more efficiently than from a permanent local health office. A "mercury ambulance" equipped with the necessary medical and laboratory utensils bus could be driven into the artisanal mining areas. Two or three specially trained doctors or nurses could perform the examinations, and begin to carry out treatment. The ambulance could also be used for health awareness programs (e.g. video equipment). Miners in remote areas might welcome evening entertainment and soccer videos might attract more miners to the "mercury ambulance", than other information material. Sponsors could be sought for a "mercury ambulance", which could be based on a truck or bus chassis.

Recommendations to increase awareness of the risks of mercury

(a) Assess in a different study design the possibility of Hg related birth and growth defects, increased abortion/miscarriage rates, infertility problems, learning difficulties in childhood or other neuro-psychological problems related to occupational and/or environmental mercury exposure.

(b) Assess in a more detailed study the possible transfer of Hg from the environment to, mother to child via breast-milk and related possible adverse health effects. Females at childbearing age and younger urgently require more awareness to refrain from amalgam burning, at least during pregnancy and nursing. If this is not possible, a discussion whether to provide them with milk powder and high purity drinking water together with training them to prepare hygienically appropriate formula food for their babies needs to be based on a larger data base and a different epidemiological approach.

(c) Assess the relative importance of the main potential sources of exposure for people in Rwamagasa who are not occupationally exposed to Hg (i.e. airborne Hg-vapour; ingestion of Hg-contaminated dust through hand-to-mouth contact or on unwashed or inadequately washed food; ingestion of locally grown Hg-contaminated crops; Hg-contaminated fish from local streams; deliberate occasional or habitual consumption of soil (geophagia)). This has not been evaluated adequately and requires further integrated investigations by a team of environmental, public health, medical and toxicological specialists.

Recommendations for improvement of general health

Poverty is considered to be the main reason for most of the health and environmental problems in the Rwamagasa area. At the moment it does not seem to be acceptable that children live in Rwamagasa because of inadequate sanitary standards and high exposure to Hg. The improvement of sanitary standards is needed urgently.

The relative occupational health risks of mining should be assessed in more detail (accidents, malaria, drinking water quality, sexually transmitted diseases, tuberculosis, HIV / AIDS). One option to reduce the health hazards in Rwamagasa might be a proper zoning into industrial areas, commercial areas and housing areas. The imposition of basic hygienic standards, such as proper drinking water and reduction of *Anopheles* mosquitoes would also lead to an improvement in the health of the local people.

Raising safety awareness and the introduction of appropriate mining techniques (such as better tunnel safety) will help to reduce the risk of accidents at mining sites.

The risk of sexually transmitted diseases could be reduced if campaigns for safer sex were more effective. An appropriate health service is urgently required to improve the health status of the Rwamagasa community.

REFERENCES

Appel, P.W.U. et al. An integrated approach to mineral exploration and environmental assessment in southern and eastern Africa - a pilot study in Tanzania. 2000/16, Danmarks og Grønlands Geologiske Undersøgelse, 2000.

Appleton, J.D., Taylor, H., Lister, T.R., Smith, B., Drasch, G., Boese-O'Reilly, S. Final report for an assessment of the environment and health in the Rwamagasa area, Tanzania. UNIDO Project EG/GLO/01/G34. British Geological Survey Commissioned Report CR/04/129, 2004.

Asano, S. et al., Acute inorganic mercury vapor inhalation poisoning. *Pathology International*, 50(3), 169-174. 2000.

Campbell, L. M., D. G., Dixon, et al. A review of mercury in Lake Victoria, East Africa: Implications for human and ecosystem health. *J. Toxic.and Env. Health*, Part B-Critical Reviews 6, 325-356, 2003a.

Campbell, L. M., O. Osano, et al. Mercury in fish from three rift valley lakes (Turkana, Naivasha and Baringo), Kenya, East Africa. *Env. Pollut.*, 125, 281-286, 2003b.

Drasch G., Boese-O'Reilly S., Beinhoff C., Roider G., Maydl S. The Mt. Diwata study on the Philippines 1999 - assessing mercury intoxication of the population by small scale gold mining. *Sci. Tot Env.*, 267, 151-168, 2001.

Harada, M., S. Nakachi, et al. Monitoring of mercury pollution in Tanzania: relation between head hair mercury and health. *Sci. Tot. Env.*, 227, 249-256, 1999.

Ikingura, J. R. and H. Akagi. Monitoring of fish and human exposure to mercury due to gold mining in the Lake Victoria goldfields, Tanzania. *Sci. Tot. Env.*, 191, 59-68, 1996.

Ikingura, J.R., Mutakyahwa, M.K.D. Kahatano, J.M.J., Mercury and mining in Africa with special reference to Tanzania. *Water Air Soil Pollut.*, 97, 223-232, 1997.

Ikingura, Akagi Proceedings of the International Workshop on Health and Environmental Effects of Mercury, Impacts of Mercury from Artisanal Gold Mining in Africa. University of Dar es Salaam, 19-20 November 2002, 56-65, 2002.

Kahatano et al. Study of Mercury Levels in Fish and Humans in Mwakitolyo Mine and Mwanza Town in the Lake Victoria Goldfields, Tanzania. In Landner L (ed) 1997. Small Scale Mining in African Countries. Prospects, Policy and Environmental Impacts, 159-170, 1997.

Kinabo, C., Lyimo, E., Preliminary studies of mercury contamination in various food crops at Mgusu mining village in Geita, Tanzania, Proceedings of the International Workshop on Health and Environmental Effects of Mercury - Impacts of Mercury from Artisanal Gold Mining in Africa. National Institute for Minamata Disease, Japan & University of Dar es Salaam, Tanzania, University of Dar es Salaam, Tanzania, pp. 204-214, 2002.

Kinabo, C., Mercury pollution associated with artisanal mining on Lake Victoria Goldfield, in Tanzania. In: R. Ebinghaus, G. Petersen and U. von Tuempling (Editors), Fourth international conference on mercury as a global pollutant; book of abstracts. GKSS-Forschungszentrum, Geesthacht, Federal Republic of Germany, 335, 1996.

Kinabo, C., Comparative analyses of Hg-concentration in domestic ducks (Anser sp.) from Mgusu mining village in Geita, Mwanza and Dar es Salaam, Proceedings of the International Workshop on Health and Environmental Effects of Mercury - Impacts of Mercury from Artisanal Gold Mining in Africa. National Institute for Minamata Disease, Japan & University of Dar es Salaam, Tanzania, University of Dar es Salaam, Tanzania, 215-222, 2002a.

Kinabo, C.P., Comparative analyses of mercury contents in cosmetics and soaps used in the city of Dar es Salaam, Proceedings of the International Workshop on Health and Environmental Effects of Mercury - Impacts of Mercury from Artisanal Gold Mining in Africa. National Institute for Minamata Disease, Japan & University of Dar es Salaam, Tanzania, 173-186, 2002b.

Kishe, M.A., Machiwa, J.F., Distribution of heavy metals in sediments of Mwanza Gulf of Lake Victoria, Tanzania. *Env. International,* 28, 619-625, 2003.

Lamborg, C.H., Fitzgerald, W.F., O'Donnell, J., Torgersen, T. A non-steady-state compartment model of global-scale mercury biogeochemistry with interhemispheric gradients. *Geochim. Cosmochim. Acta,* 66, 1105-1118, 2002.

Machiwa, J.F., Kishe, M.A., Mbilinyi, H.G., Mdamo, A. Mnyanza, O. Impact of gold mining in Lake Victoria Basin on mercury levels in the environment. Lake Victoria Environmental Management Project Report, 46, 2003.

MMSD, Breaking New Ground: Mining, Minerals, and Sustainable Development. International Institute for Environment and Development. Earthscan Publications Ltd, London, UK. As available at http://www.iied.org/mmsd/finalreport/index.html per September 2002.

Mutakyahwa, M.K.D., Mercury contamination due to gold mining in Tanzania, Proceedings of the International Workshop on Health and Environmental Effects of Mercury - Impacts of Mercury from Artisanal Gold Mining in Africa. National Institute for Minamata Disease, Japan & University of Dar es Salaam, Tanzania, 21-29, 2002.

Pirrone, N., Munthe, J., Barregård, L., Ehrlich, H.C., Petersen, G., Fernandez, R., Hansen, J.C., Grandjean, P., Horvat, M., Steinnes, E., Ahrens, R., Pacyna, J.M., Borowiak, A., Boffetta, P., Wichmann-Fiebig, M. Ambient Air Pollution by Mercury *(Hg)* – Position Paper. Office for Official Publications of the EC. (available on http://europa. eu.int/comm/environment/air/background.htm#mercury) 2001.

Semu, E., Guttormsen, D. Bakken, L. Microbial-Populations and Activity in 2 Soils of Tanzania as Influenced by Mercury. *Mircen J. Appl. Microbiol. Biotech.*, 5, 33-542, 1989.

Sindayigaya, E., Van Cauwenbergh, R., Robberecht, H. Deelstra, H., Copper, zinc, manganese, iron, lead, cadmium, mercury and arsenic in fish from Lake Tanganyika, Burundi. *Sci. Tot. Env.*, 144, 103-115, 1994.

UNIDO,. Protocols for Environmental and Health Assessment of Mercury Released by Artisanal and Small-Scale Gold Miners. UNIDO, Vienna, 2003.

University of Dar es Salaam,. Monitoring of mercury and other heavy metal pollution in gold mining areas around lake Victoria, Tanzania. Final report, Department of Geology, University of Dar es Salaam, Dar es Salaam, 55, 1994.

van Straaten, P., Human exposure to mercury due to small scale gold mining in northern Tanzania. *Sci. Tot. Env.*, 259, 45-53, 2000.

van Straaten, P.,. Mercury contamination associated with small-scale gold mining in Tanzania and Zimbabwe. *Sci. Tot. Env.*, 259, 105-113, 2000.

Chapter-21

DEVELOPMENT OF PROGRAMS TO MONITOR METHYL-MERCURY EXPOSURE AND ISSUE FISH CONSUMPTION ADVISORIES

Henry A. Anderson[1] and Jeffrey D. Bigler[2]

[1] MD Chief Medical Officer, Wisconsin Division of Public Health, Madison, WI,USA
[2] National Program Manager, National Fish and Wildlife Contamination Program, USEPA, Washington DC, USA

INTRODUCTION

Regular consumption of fish (1 – 2 meals per week) has been associated with substantial reduction in the risk of death from heart attacks (Daviglus et al., 1997; Albert, 2002). The health benefit attributed to a diet rich in fish and fish oils has led public health authorities to promote a balanced diet containing two meals of fish per week (AHA, 2000). Worldwide, fish and shellfish as a dietary source of protein is rapidly expanding such that fish consumption is estimated to have surpassed other animal dietary protein sources such as beef and fowl. Unlike domesticated beef and fowl that are farm produced for general population consumption, fish and shellfish are primarily harvested from the wild. Aquaculture is rapidly growing in importance but remains limited and currently only provides a small proportion, perhaps 25% of fish and shellfish consumed worldwide. The diet and geographic movement of wild fish harvested for commercial sale can not be controlled. Many sought after fish are large predators at the top of the food chain. Because of their unrestricted movement and opportunistic diet, wild fish are vulnerable to accumulating bio-persistent pollutants circulating in the environment. Methylmercury is one of those chemicals.

It is estimated that anthropogenic sources have contributed to a 2 to 5-fold increase in the global circulating pool of atmospheric mercury. Atmospheric inorganic mercury is deposited in aquatic systems where bacterial methylation converts the inorganic forms of mercury to methylmercury. Biomagnification of methylmercury results in levels in fish that are 10^4 to 10^6 times higher than levels in water (US Environmental Protection Agency, 1997)

Governmental Response to Methylmercury Fish Contamination

Fish and shellfish consumption is the predominant source of methylmercury exposure to humans. Recent epidemiological studies (NRC, 2000) have led to the conclusion that methylmercury is more toxic than previously recognized, especially in-utero exposures which are expressed as later neurodevelopment delay. Utilization of these studies in risk assessments has resulted in the reduction of acceptable human exposures (U.S. EPA, 2001; JECFA, 2003).

Market basket surveys of methyl mercury in commercial foods and surveys to establish population consumption distributions by fish and shellfish species allow exposure assessment analyses. Estimated exposures can then be compared to the target risk thresholds. Human studies of blood or hair mercury distributions in populations have confirmed the exposure assessments and have led governmental scientists to conclude that there is a slim margin of safety for a significant proportion of the population (Mahaffey et al., 2004).

Up until the recent reduction in the acceptable daily or weekly methylmercury exposure, national governments relied upon regulatory approaches to remove the most highly contaminated fish from commerce. In the United States the Food and Drug Administration "action level" has been at 1 ppm methylmercury in fish tissue for several decades. Fish found to exceed that "action level" could be removed from interstate commerce. In the European Union, most fish have a tolerance of 0.5 ppm and initially a small number of less frequently consumed fish were given a 1 ppm tolerance. Increased numbers of commercial fish species, changing catch locations, as well as increased monitoring of fish tissue mercury concentrations have created the need for the European Union (EU) to increase the number of fish species needing a tolerance of 1 ppm to remain on the market. The number of "1 ppm exemptions" has risen to 22 species.

In 2001 the USEPA RfD was reduced from 0.3 ug/kg/day to 0.1 ug/kg/day and in 2004 the EU's provisional tolerable weekly intake (PTWI) was reduced from 3.2 to 1.6 ug/kg/week. These changes required a modification of the current regulatory approach if current commercial fisheries were to remain viable while at the same time excess methylmercury exposures prevented. If individuals consumed the recommended amount of fish per week but chose higher mercury fish, exposures could significantly exceed recommended levels (Knobeloch et al., 1995). The consumer passive regulatory approach could no longer assure that a significant proportion of the general population would not exceed the new toxic thresholds.

National governments have responded to the likelihood that 5-10% or more of their population may be exceeding the hazard threshold for methylmercury with an approach that includes the use of commercial fish consumption advisories. Based upon the epidemiology studies these advisories have generally been targeted to protect the most vulnerable segment of the population, pregnant women and their developing fetus.

Each country has a different mix of subpopulations that are likely to consume more fish than the average. Such groups include commercial fisherman, coastal residents, recreational fisherman, ethnic and indigenous groups with traditional diets high in fish and individuals seeking a high fish diet for the reported heath benefits. For many countries, national consumption advisories appear to have become the principal exposure reduction strategy to protect their citizens from excess methylmercury exposure and toxicity.

Although many commercial species have worldwide markets, other fish are local. Concentrations of pollutants such as methylmercury in the fish tissue may also vary by the location where the fish reside and the size/age of the fish. This variability necessitates consumption advisories tailored to the consumption patterns and fish in commerce in each country.

There is no readily available source to locate all the national fish and shellfish consumption advisories that have been issued, to obtain the advisory development protocols, or to review evaluations of such programs. Table 1 provides a summary of characteristics of examples of such advisories. Most of these can be found on governmental web pages. Not included in the table are the positive statements all advisories include concerning the benefits of fish consumption and the recommendation to include at least 2 fish meals per week.

Eleven of the 12 countries listed in Table 1 address mercury contamination in their advisories.

Only 4 include other contaminants, specifically PCB and Dioxin. Taiwan only addresses PCB in recreational fish skin, liver and eggs. All but Taiwan address commercial fisheries. There is a commonality in the target populations. All countries address women of childbearing age, but there are some differences in how the population of women are defined. Most consistent is the direct mention of pregnant and lactating women. Fewer advisories mention infants or young children. Only three, Sweden, Taiwan and Canada include advice for the general population. All countries issue advice on how frequently fish can be consumed in terms of meals per week or month. Only Finland, Great Britain, Norway, Sweden and the United States include advice to not consume some species. The most commonly mentioned species are swordfish and shark. Nearly all advisories include these species and these are the species most commonly mentioned as "no consumption" for women of reproductive years. Countries whose populations consume whale meat typically include these on there advisories.

Recreational Fisheries

Many countries have a thriving freshwater and or marine recreational fishery that typically is carefully managed and often involves the issuing of recreational licenses and permits to fish specific waters. Such licenses and the regulation pamphlets are a convenient means to inform anglers of advisories. Some segments of the population may also rely upon these locally available fish resources for subsistence. Such groups are indigenous populations or immigrants from countries with a tradition of fish consumption. Six of the countries in Table 1 address recreational fish in their advisories. Countries are just beginning to comprehensively address fish consumption and to explicitly include sport-caught fish as part of their commercial advisory.

Wisconsin, USA: An example of a Comprehensive Fish Consumption Advisory Program

Thirty-five years ago mercury was identified in freshwater sport fish found in many states bordering the Great Lakes (Konrad 1970; Kleinert et al., 1972). From a national perspective, contamination was felt to be limited

Table 1. Worldwide National Fish Consuption Advisories.

Country	Water body Type	Year	Agency	Commercial Recreational	Pollutant	Species	Population of Concern	Meal Advice	Comments
European Union	Marine	2004	EU Commission EU Food Safety Authority (EFSA)	Commercial	Mercury	Swordfish, shark, tuna, pike	Women who might become pregnant, who are pregnant, who are breast-feeding, young children Women of childbearing age	Should not eat more than one small portion (<100g) per week of large predatory fish, such as swordfish, shark, marlin and pike. If they eat this portion they should not eat other fish during this period. Also they should not eat tuna more than twice per week. Consumers should pay attention to any more specific advice from national authorities in light of local specificities. Recommends that women of childbearing age select fish from a wide range of species, without giving undue preference to large predatory fish such as swordfish and tuna	
Denmark	Marine	2004		Commercial	Mercury PCB	Tuna, Halibut, Swordfish, Porbeagle shark, Pike Perch, Zander, Escolar, Ray	Pregnant and lactating women	Pregnant and lactating women should not eat large portions of listed fish All individuals should eat 1-2 meals of fish per week.	One can eat limited amounts of these fish. For example, tuna salad is not harmful, but one should avoid a whole tuna steak. Pregnant and lactating women can eat all other common fish without problems.
Finland	Marine Fresh Water	2004	National Nutrition Council	Commercial Recreational	Mercury PCB, Dioxin	Large Baltic herring and wild caught salmon; Pike and fish from inland waters, Pike from sea, Pike from inland waters	Children, young people, people at fertile age Pregnant women and nursing mothers	Once or twice a month Do not eat	For those who eat fish from inland waters on an almost daily basis should reduce consumption of large perches, pike perches, burbots (due to Hg) Farmed fish low in PCB, Dioxins

Table 1. cont'd.

Country	Water body Type	Year	Agency	Commercial Recreational	Pollutant	Species	Population of Concern	Meal Advice	Comments
Great Britain	Marine waters	2004	UK Food Standards Agency	Commercial	Mercury	Shark, swordfish, and marlin	Pregnant women, women of childbearing age who intend to become pregnant, infants, and children	Avoid eating	
Great Britain						Tuna	Pregnant women, women who are trying to become pregnant	Eat no more than four medium sized cans with a drained weight of 140g per can or up to two fresh tuna steak per week	
Norway	Marine waters	2003	Norway's National Veterinary Institute Scientific Panel	Commercial	Mercury	whale meat	Pregnant women and mothers who are breast feeding	Do not eat	The findings of the scientific panel must be approved by the Norwegian health authorities
Sweden	Marine and fresh waters	2004	National Food Administration	All	Mercury POP	Pike, perch, pike-perch, burbot, eel, large halibut, Baltic - herring, salmon	All	Eat no more than once a week	
Sweden	Marine and fresh waters		National Food Administration	All	Mercury POP	1. Pike, perch, pike-perch, burbot, eel, large halibut,	Women of childbearing age who intend to become pregnant, infants, and children under 16 years of age	1. Eat no more than once a week	
						2. Baltic herring or salmon		2. Eat no more than once a month	
						1. Pike, perch, pike-perch, burbot, eel, large halibut,	Pregnant women	1. Do not eat	
						2. Baltic herring or salmon		2. Eat no more than once a month	

Table 1. cont'd.

Country	Water body Type	Year	Agency	Commercial Recreational	Pollutant	Species	Population of Concern	Meal Advice	Comments
Australia and New Zealand	Marine waters	2001	Food Standards Australia and New Zealand (FSANZ)	Commercial	Mercury	shark, ray, swordfish, barramundi, gemfish, orange roughy, ling, and Southern bluefin tuna.	Pregnant women and women considering pregnancy	Limit consumption to four (150 g) meals per week	Pregnant women can eat as much other fish, including salmon, canned salmon and canned tuna as they like
	Fresh-waters			Recreational	Mercury	Freshwater fish caught in geothermal waters	Pregnant women and women considering pregnancy	Limit consumption to four (150 g) meals per week	Pregnant women can eat commercial fish as prescribed
Japan	Marine waters	2003	Health, Labor and Welfare Ministry	Commercial	Mercury	Broadbill swordfish and alfonsino	Pregnant woman	No more than two 60-80g servings per week	Fish tissue data also provided in press release
						Bottlenose dolphin	Pregnant woman	No more than one 60-80g serving every 2 months	Fish tissue data also provided in press release
Japan						Baird's beaked whale, short-finned pilot whale, sperm whale, and shark meat	Pregnant woman	No more than one 60-80g serving per week	Fish tissue data also provided in press release
Taiwan	Fresh-waters	2000	Taiwan EPA	Recreational	PCB	fresh water species	All consumers	Do not eat fish liver, skin, and eggs	There is no official fish advisory in Taiwan; just a soft recommendation
Canada	Marine waters	2002	Health Canada	Commercial	Mercury	shark, swordfish, fresh and frozen tuna	All consumers	Limit consumption of shark, swordfish, and fresh and frozen tuna to one meal per week	Note that this advisory does not apply to canned tuna
Canada	Marine waters	2002	Health Canada	Commercial	Mercury	shark, swordfish, fresh and frozen tuna	Pregnant women, women of childbearing age, young children	Limit consumption of shark, swordfish, and fresh and frozen tuna to one meal per month	Note that this advisory does not apply to canned tuna

Table 1. cont'd.

Country	Water body Type	Year	Agency	Commercial Recreational	Pollutant	Species	Population of Concern	Meal Advice	Comments
United States	Marine & Fresh-water	2004	USEPA and USFDA	Commercial	Mercury	shark, swordfish, king mackerel, tilefish	Women who may become pregnant, pregnant women, nursing mothers, young children	Do not eat	"Do not eat Shark, Swordfish, King Mackerel, or Tilefish because they contain high levels of mercury"
					Mercury	All other fish	Women who may become pregnant, pregnant women, nursing mothers, young children	"Eat up to 12 ounces (2 average meals) a week of a variety of fish and shellfish that are lower in mercury"	"Five of the most commonly eaten fish that are low in Hg are: shrimp, canned light tuna, salmon, pollock, catfish. Another commonly eaten fish, albacore ("white") tuna has more mercury than canned light tuna. When choosing your two meals of fish and shellfish, you may eat up to 6 ounces (one average meal) of albacore tuna per week.
				Recreational	Mercury	All sport caught fish species	Women who may become pregnant, pregnant women, nursing mothers, young children	"Check local advisories about the safety of fish caught by family and friends in your local lakes, rivers, and coastal areas. If no advice is available, eat up to 6 ounces (one average meal) per week of fish you catch from local waters, but don't consume any other fish during that week."

to local fresh waters and involve recreational fish species rather than commercial species. No national advisories were developed. However because of the importance of consumption of locally caught fish, individual states, including Wisconsin began active fish tissue monitoring programs and state public health authorities linked the test results with consumption frequency advice for sport caught fish. A similar program occurred in the Canadian Province of Ontario. By 2002, 48 states issued advisories for sport-fish consumers. Mercury is the most common contaminant covered by a state advisory (USEPA, 2003). In the United States, states have responsibility for recreational fisheries and the federal government regulates commercial fisheries. Although similar during the 1970s and most of the 1980s, in the 1990s many states and the US federal government began using different methods to assess chemical toxicity and translate it into advice for fish tissue levels of concern (Anderson and Liebenstein, 1989; Anderson et al., 1993). The recently revised mercury toxicity assessments led to a renewed focus on mercury contamination in fish tissue and state recognition that issuing consumption advice only for sport-caught fish while ignoring the exposure contribution from commercially consumed fish did not make sound public health practice. Over the past three to four years, some states began issuing comprehensive fish advisories that included both sport-caught and commercial fish. Currently twelve states include recommendations for commercial fish in their sport fish consumption guidelines.

There is general agreement that a comprehensive fish consumption advisory program should include: 1. Public health surveillance and reporting of mercury poisoning (health care delivery system reporting) and high exposures as measured in whole blood or hair total mercury (laboratory based surveillance); 2. Fish tissue biomonitoring; 3. Advisory development; 4. Advisory evaluation.

Public Health Surveillance

While physician and laboratory reporting is far from comprehensive, case reports are often illustrative of situations that need to be addressed and assist in raising public awareness. Wisconsin has encouraged case reporting and since 1992 has investigated 7 instances of excessive mercury exposure from fish consumption. Cases involved both commercial fish consumption (Knobeloch, 1993) and sport fish. Exposures ranged up to an estimated 100 ug per day of methymercury. With the increased awareness of and concerns for mercury, more hair and blood testing has begun. In the last 2 years there have been three case investigations.

Fish tissue monitoring

Wisconsin has 40,000 miles of rivers and 15,000 lakes. With a limited sample collection capacity and laboratory analysis budget, it was necessary to devise sampling strategies. When in the early 1970s mercury was recognized as a fish contaminant of concern (Konrad, 1970; Kleinert and Degurse, 1972) the Wisconsin Department of Natural Resources (WDNR) been systematically monitoring mercury concentrations in Wisconsin fish (Michaels and Schrank, 2003).

Initially the monitoring strategy focused on rivers receiving effluents from mercury discharging industries. Later, in the 1980s, testing from northern lakes receiving no effluents became the focus when a number of northern Wisconsin lakes had been found to have among the highest mercury concentrations found in predator fish. In the decade of the 1990s monitoring began on a statewide basis using a scheduled rotating basin approach. This strategy involved sampling sites within the major river drainage basins on a five year rotating schedule.

In 1999, a new monitoring strategy called "baseline monitoring" was devised for lakes, wadable, and non-wadable streams and rivers. Fish are being collected for contaminant analysis at a subset of baseline sites where limited or no fish contaminant data exist or where updated information is required. The goal of this strategy is to obtain a statewide distribution of fish contaminant data so the status of contaminants can be determined on a statewide basis versus the previous rotating basin or suspected source impacted sites. Fish are also collected from sites where fish consumption advisories are in place and updated data are required to maintain a 5-year return frequency.

In 2003 the WDNR described the above strategies and summarized all the fish tissue data (Michaels and Schrank, 2003). Figure 1 taken from that report graphically presents 24 years of sampling data. The table represents 12,964 samples from 1,046 locations and 810 unique waterbodies. Of the 183 known native and non-native fish in Wisconsin 54 species were sampled during this time period. These 54 species were sampled because they were targeted by anglers and sportfish consumers or were species valuable for comparisons across sites or over time.

WDNR staff primarily collected the fish using methods dependent on waterbody and species. Tissue samples were prepared using standard procedures (WDNR Field Manual). Preparation of the edible portions of fish involved thawing, measuring length, weighing, and grinding either skin-on fillets (all species except for bullhead, catfish, and sturgeon), skin-off fillets

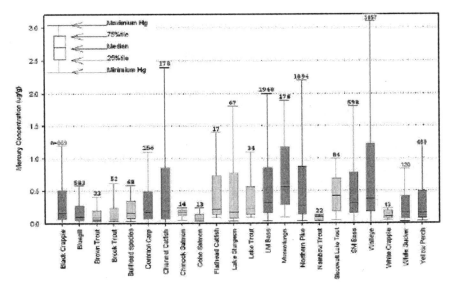

Figure 1. (from Michaels and Schrank, 2003). Mercury for 22 species (1977-2001), all waterbodies. Lighter boxplots denotes limited data (n<100) Mercury concentrations generally increase with size of fish, varies between waterbodies.

(only bullhead, catfish, and sturgeon), or cross-sections of fillets (sturgeon). Over 94% of samples were single fish samples. Figure 1 summaries the data.

The average value of all samples was 0.36 ppm, with a range of below detection to 3.1 ppm and a 75 percentile of 0.48 ppm.

Advisory development

When Wisconsin's sport fish consumption advisory protocol was initiated in the 1970s, the FDA commercial fish action level was applied to sport fish. This provided the angler a qualitative comparison to market fish rather than quantitative, risk assessment based advice. The target audience was primarily anglers and their families. However, as risk assessment procedures advanced and Wisconsin and other states gained the expertise to utilize such procedures some states felt that the advisory should be fully health based rather than utilize the FDA process that included a cost-benefit consideration. During the 1980s and 1990s Wisconsin began to advise the public on how much sport fish was "safe" to eat based upon the type of risk assessment utilized by the USEPA RfD process. Five consumption rate groupings were utilized; "unlimited" (225 11/2lb meals/year), one meal a

week (52 meals/year), one meal a month (12 meals a year), six meals a year
and "Do not eat." The advisory grew in complexity as it provided
consumption frequency advice by species and size for each specific water
body tested. By 2000 the advisory booklet included advice on 340 different
water bodies or river segments.

In 1995 the USEPA revised its RfD for methylmercury. This led to
considerable controversy and a series of external peer reviews. In 1998 the
National Academy of Sciences review panel confirmed the appropriateness
of the reduced RfD. In 1999 Wisconsin reviewed its methylmercury advisory
protocol applying the new RfD and utilized focus groups to review the
existing advisory. It was concluded that a less complex advisory was needed
and should include both sport and commercial fish, provide a simple
message by species, be consistent with neighboring states and apply to
waters not yet tested. In 2000 a new advisory was developed called the
"Statewide Safe Eating Guidelines."

Figure 2 provides the new guidelines. The guidelines and other
informational materials can be found on the Wisconsin Department of

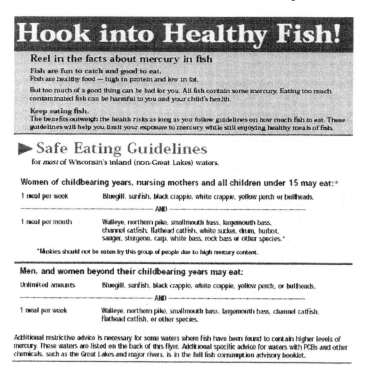

Figure 2.

Natural Resources and the Wisconsin Division of Public Health web pages (URL:http://www.dnr.state.wi.us/org/water/fhp/fish/advisories/ index.htm) - (URL: http://www.dhfs.state.wi. index.htm).

This general advice is augmented with site specific consumption advice developed where fish-monitoring data indicates that more stringent advice is necessary. Only 93 lakes and waters warranted such special advice. Special advice for PCBs is applied to 50 river reaches and lakes. The waters where more stringent advisories are in place are posted with special warnings and the specific advice for that lake.

The new advisory also includes information on both sport and commercial fish. Figure 3 shows the advisory format combining sport and commercial fish into a single advisory Figure 4 is an example of the outreach

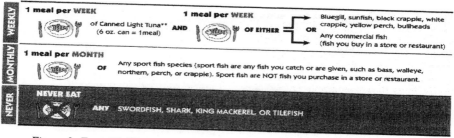

Figure 3. From "A Woman and Child's Guide to Eating Fish from Wisconsin".

materials developed to assist in educating consumers in how to identify fish that are low in methylmercury.

Wisconsin's fish consumption advisory goals remain to a) inform the public about the chemical contaminants contained in some sport-fish, b) educate consumers as to how they can minimize their exposure to contaminants, c)

remind consumers of the health benefits of fish consumption, and d) present advisory information in a manner conducive to maximal voluntary compliance. Because of potential adverse reproductive and developmental effects current advisories make specific consumption frequency recommendations for childbearing-aged women, but also provide advice for the general population. Advisories seek to help individual consumers make informed decisions regarding sport-fish consumption.

Advisory evaluation

Documenting that an advisory is effective is a significant challenge. Too frequently there are insufficient resources to determine the impact of an advisory and to track its penetration over time.

The easiest outcome to assess is awareness of the advisory and increased understanding of the chemical toxicity. However, awareness is only the first step in a successful advisory program.

Figure 4.

The goal must be to reduce exposure that can only result from behavior modification.

Wisconsin has conducted several assessments of advisory effectiveness. In 1994-1995 we surveyed adult residents of the Great Lakes Basin (Tilden et al., 1997). That survey found that over 3 million residents were consuming Great Lakes sport fish and that advisory awareness among women, minorities and low income households was nearly one half that of white males. The results of that study led to significant changes in state advisories with specific outreach to the difficult to reach.

In 2001, before a national USA advisory was being discussed, a consortium of 12 states conducted a telephone survey of 3,015 women aged 18-45.

The twelve states were selected based on their mercury sport fish advisories. Half issue state-wide advisories and half use a site-specific advisories. The states were spread throughout the United States (Figure 5). The goal was to characterize current fish consumption patterns (commercial, all sport-caught fish) and estimate the level of knowledge of mercury, advisory awareness and compliance among consumers of

12 State Mercury Awareness Survey (1999)
Women, Age 18- 45

Figure 5. 12 state Hg Awareness Survey (1999).

commercial and sport-caught fish (Anderson et al., 2004; Knobeloch et al., 2004).

We reaffirmed that fish is important in the United States' diet. Only 8% reported no fish or shellfish consumption during the previous 12 months. Included in the 92% who reported some fishmeals were 29% who reported sport-fish consumption (inter-state variability from 14–43%). Ten

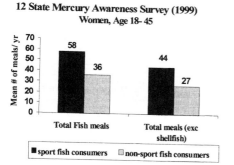

Figure 6. 12 State Mercury Awareness Survey (1999).

percent of participants reported consuming two or more fish meals per week. There was an wide range of reported consumption. The maximum reported for commercial fish consumption was 572 meals per year. The maximum for sport fish consumption was 384 meals per year. Of note was the finding that those who consumed sport fish consumed 60% more total fish and shellfish

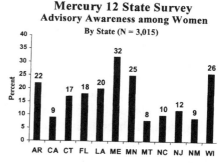

Figure 7. Mercury 12 State Survey.

than those who did not report consuming sport fish (figure 6). This finding supports the need for comprehensive advisories including both sport and commercial fish species.

Although there was considerable consumption of sport fish, awareness of specific state consumption advisories was only 20%, ranging by state from 8-32% (figure 7). Women who were older, had more than a high school education, and had a household member with a fishing license were the most informed about mercury and fish consumption advisories. Most states distributed their advisory with their sport fishing licenses so it was not surprising that households with a license holder would have greater awareness. What was encouraging is that most license holders are men and our survey was of their spouses. Previous research had shown that the men often did not share the advisory information with their wives. In this case there seems to be some improvement in communication.

We found that those aware of the advisory were more informed about the toxicity of mercury. Overall 71% of survey participants recognized that mercury harms a developing child. That rose to 87% among those aware of their state's advisory. Those aware of the advisory were also more likely to understand the characteristics of fish that predicted higher mercury contamination (figure 8).

	Aware of Advisory	Not Aware of Advisory
Harms developing child	87%	67%
Harms ability of muscles	52%	37%
Mercury not reduced by cooking	76%	47%
Higher in older fish	56%	43%
Higher in larger fish	38%	29%
Higher in fish that eat others	23%	18%
Highest in muscle/meat	8%	6%

Significant higher than among those unaware of state advisories (P<0.01)

Figure 8.

CONCLUSIONS

Most current methylmercury fish consumption advisories focus on risks to women and their infants. However a word of caution must be interjected to not overlook the potential for toxicity via another mode of action in other vulnerable populations. Studies have associated dietary methylmercury exposure with an increased risk of coronary artery disease and heart attacks in men (Salonen, 1995; Guallar, 2002). While there are also cardiovascular benefits to fish consumption, they may be negated when mercury is high. Fortunately there are fish that are low in mercury but high in beneficial fatty acids. Research on the adult cardiovascular risk warrant increased research and analysis, before advisories dismiss providing a risk message to older adults, especially men.

If fish consumption advisories are to inform and protect the public, it is important to develop and maintain a comprehensive exposure surveillance program that includes fish tissue and human biomonitoring as well as an advisory effectiveness evaluation strategy. Governments relying on advisories must continuously ask the questions, "Is the message being heard and is it being adhered to?" In most countries just beginning to issue commercial advisories, such strategies are still in the formative stages. The experience of USA states, and Sweden may help inform such efforts.

REFERENCES

AHA (American Heart Association) AHA Dietary Guidelines, revision 2000: A Statement for healthcare professionals from the nutrition committee of the American Heart Association. Circulation; 102, 2296-2311, 2000.

Albert, C.M., Campos, H., Stampfer, M.J., Ridker, P.M., Manson, J.E., Willett, W.C., Ma, J. Blood levels of long-chain n-3 fatty acids and the risk of sudden death. *N. Engl. J. Med.*, 346, 1113-1118, 2002.

Anderson, H.A., Liebenstein, L. Sport fish consumption advisories. *AJPH*, 79, 1434-1435, 1989.

Anderson, H.A., Amrhein, J.F., Shubat, P., Hesse, J. Protocol for a uniform Great Lakes sport fish consumption advisory. Great Lakes Sport Fish Advisory Task Force, Council of Great Lakes Governors, September 1993, Chicago, IL, 1993.

Anderson, H.A., Hanrahan, L.P., Smith, A., Draheim, L., Kanarek, M. Olson, J. The role of sport fish consumption advisories in mercury risk communication: The Role of Sport-fish Consumption Advisories in Mercury Risk Communication: a 1998-1999 12 State Survey of Women age 18-45, *Env. Res.*, 95, 315-324, 2004.

Daviglus, M.L., Stamler, J., Orencia, A.J., Dyer, A.R., Liu, K., Greenland, P., Walsh, M.K., Morris, D., Shekelle, R.B. Fish consumption and the 30-year risk of fatal myocardial infarction. *N. Engl. J. Med.*, 336, 1046-1053, 1997.

WI DNR Field Procedures Manual. PART B: Collection Procedure 1005.1 Fish Contaminant Monitoring Program – Field and Lab Guidelines, Wisconsin Department of Natural Resources, Madison, WI.

Guallar, E., Sanz-Gallardo, M.I., Van't Veer, P., Bode, P., Aro, A., Gomez-Aracena, J., Kark, J.D., Riemersma, R.A., Martin-Moreno, J.M., Kok, F.J. Heavy metals and myocardial infarction study group. Mercury, fish oils, and the risk of myocardial infarction. *N. Engl. J. Med.*, 347, 1747-54, 2002.

JECFA (Joint FAO/WHO Expert Committee on Food Additives). Sixty-first meeting, Rome, 10-19 June 2003. Summary and conclusions. Available from: URL: ftp://ftp.fao. org/es/esn/jecfa/jecfa61sc.pdf.

Kleinert, S., Degurse, P.E. Mercury levels in Wisconsin fish and wildlife. DNR Technical Bulletin No. 52, Madison, WI, 1972.

Knobeloch, L.M., Ziarnik, M., Anderson, H.A., Dodson, V.N. Imported Seabass as a source of Mercury exposure: a Wisconsin case study. *Env. Health Perspect.*, 103, 604-606, 1995.

Knobeloch, L.M., Anderson, H.A., Imm, P., Peters, D., Smith, A. Fish Consumption, Advisory Awareness and Hair Mercury Levels Among Women of Child-bearing Age, *Env. Res.*, 97, 220-227, 2005.

Konrad, J.G. Mercury: new-found threat. *Wisc. Cons. Bull.*, 35, 3-5, 1970.

Mahaffey, K.R., Clickner, R.P., CB Joseph. Blood organic mercury and dietary mercury intake: National Health and Nutrition Examination Survey, 1999 and 2000. *Env. Health Perspect.*, 112, 562-570, 2004.

Michaels, E Schrank, C. Summary of Wisconsin mercury data for edible portions of fish 1977-2001 December 2003, Fisheries Management and Habitat Protection, Wisconsin Department of Natural Resources, Madison, WI.

NRC (National Research Council/National Academy of Sciences), Toxicological effects of methylmercury. Committee on the Toxicological Effects of Methylmercury. Board on Environmental Studies and Toxicology, Commission on Life Sciences. 2000. National Academy Press, Washington DC.

Salonen, J.T., Seppanen, K., Nyyssonen, K., Korpela, H., Kauhanen, J., Kantola, M., Tuomilehto, J., Esterbauer, H., Tatzber, F., Salonen, R. Intake of mercury from fish, lipid peroxidation, and the risk of myocardial infarction and coronary, cardiovascular, and any death in eastern Finnish men. Circulation, Feb 1, 91, 645-55, 1995.

Tilden, J., Hanrahan, L.P., Anderson, H.A., Palit, C., Olson, J., MacKenzie, W. and the Great Lakes Consortium. Health advisories for consumers of Great Lakes sport-fish: is the message being received? *Env. Health Perspect.*, 105, 1360-1365, 1997.

USEPA (U. S. Environmental Protection Agency). Mercury Study Report to Congress, Volume III: Fate and transport of Mercury in the environment. U.S. Environmental Protection Agency, Office of Air Quality Planning and Standards and Office of Research and Development, EPA-452/R-97-005, 1997.

USEPA (U.S. Environmental Protection Agency), Office of Science and Technology, Office of Water. Water Quality Criterion for the Protection of Human Health: Methylmercury, Final. EPA-823-R-01-001, Washington, 2001.

URL: http://www.epa.gov/waterscience/criteria/methylmercury/document.html.

USEPA (U.S. Environmental Protection Agency). Update: National listing of fish and wildlife advisories. EPA: Pub. No.: EPA-823-F-03-003, Washington DC. 2003.

Chapter-22

HEALTH EFFECTS AND RISK ASSESSMENTS

Philippe Grandjean[1], Sylvaine Cordier[2], Tord Kjellström[3], Pál Weihe[4] and Esben Budtz-Jørgensen[1]

[1] *University of Southern Denmark, 5000 Odense, Denmark; and Harvard School of Public Health, Boston, MA 02115, USA;*
[2] *National Institute for Health and Medical Research (INSERM), Unit 625, Rennes, France;*
[3] *National Institute of Public Health, Stockholm, Sweden; Australian National University, Canberra, Australia; and Wellington School of Medicine and Health Sciences, Wellington, New Zealand;*
[4] *Faroese Hospital System, FR-100 Tórshavn, Faroe Islands.*

INTRODUCTION

Detailed risk assessments for methylmercury have been recently published by national and international bodies (e.g., the (U.S.) National Research Council (NRC, 2000), the U.S. Environmental Protection Agency (U.S.EPA, 2001), and the Joint FAO/WHO Expert Committee on Food Additives (JECFA, 2003). These reports concluded that the developing brain is the main target for methylmercury toxicity, and they emphasized the prospective epidemiological studies as the main basis for deriving an exposure limit. The same conclusion was reached in UNEP's global assessment (UNEP, 2002).

Evidence from poisoning outbreaks in Japan and Iraq have clearly demonstrated the severe and widespread damage that may occur to the brain when exposed to methylmercury during development. In Minamata, Japan, it was noted that the pregnant mother could appear in good health, while her child would be born with serious congenital methylmercury poisoning. However, the confirmation of methylmercury as the etiologic agent came late, and case-related exposure information was therefore difficult to obtain. After the Iraqi poisoning episode, which happened during a famine, exposure information was gleaned from segmental analysis of long hair strands from the mothers, while assuming a constant hair growth rate. Although these studies do not provide detailed dose-response relationships, they demonstrated the serious consequences of excess exposures to this neurotoxicant and documented that the developing brain is a highly sensitive target.

This paper reviews the human evidence with particular emphasis on recent epidemiological data on neurobehavioral effects, and it discusses the uncertainties involved in assessing human health risks based on observational studies. Three major prospective cohort studies have been conducted in New Zealand, the Faroe Islands, and the Seychelles. Cross-sectional studies of neurobehavioral function will also be considered, as will the recent evidence of methylmercury-associated cardiovascular disease.

New Zealand

A group of 11,000 mothers, who gave birth to children in 1978, was initially screened, and the hair-mercury concentrations were determined for the 1000 mothers, who had consumed 3 fish meals per week during pregnancy. Seventy-three mothers had a hair mercury result above 6 µg/g, thereby constituting a high-exposure group. At the first follow-up at age 4 years, 31 high-exposure children and 31 reference children with lower exposure were matched for potential confounders (i.e., mother's ethnic group, age, child's birthplace and birth date). The high-exposure group showed lower scores on the Denver Developmental Screening test (Kjellstrom et al., 1986).

A follow-up of the original cohort was carried out at age 6 years, now with three control groups with lower prenatal mercury exposure (Kjellstrom et al., 1989). During pregnancy, mothers in two of these control groups had high fish consumption and average hair mercury concentrations of 3–6 µg/g and 0–3 µg/g, respectively. Matching parameters were maternal ethnic group, age, smoking habits, residence, and sex of the child. At this time, 61 of the high exposure children were available for examinations. Lead levels in cord blood and garden soil were tested to assess potential confounding, but there was no association between lead and methylmercury exposure.

Results of the psychological performance tests correlated well. Stepwise robust multiple regression analysis showed that the full (and performance) Wechsler Intelligence Scale for Children (WISC-R), the McCarthy scales for children's abilities (perceptual and motoric) and the Test of Oral Language Development (a standardized test used in child development studies in New Zealand) were most strongly associated with the maternal hair mercury concentration (Kjellstrom et al., 1989). The proportion of the variance in test results due to hair mercury above 6 μg/g was about 2 %, which was similar to the influence of social class and home language, two of the main confounders accounted for in the analysis. Robust regression analysis reduced the impact of one extreme outlier (with maternal hair-mercury above 80 μg/g). A reanalysis of the full database of this study (Crump et al., 1998) replicated the association between high maternal mercury exposure and reduced test performance, but the statistical significance was very much influenced by the extreme outlier. However, when this subject was excluded additional associations became statistically significant.

Faroe Islands

The Faroe Islands are located in the North Atlantic between Norway, Shetland and Iceland. In this fishing community, excess exposure to methylmercury is mainly due to the traditional habit of eating meat from the pilot whale. Ingestion of whale blubber causes exposure to lipophilic contaminants, notably polychlorinated biphenyls (PCBs). The first birth cohort consisted of 1,022 children born during a 21-month period in 1986-1987 (Grandjean et al., 1997). Prenatal methylmercury exposure was determined from mercury concentrations in cord blood and maternal hair; both spanned a range of about 1000-fold. Cohort members were invited for detailed examination at school age (7 years), and a total of 917 of eligible children (90.3%) participated. The physical examination included a sensory function assessment and a functional neurological examination with emphasis on motor coordination and perceptual-motor performance. Main emphasis was placed on detailed neurophysiological and neuropsychological function tests that had been selected as sensitive indicators of abnormalities thought to be caused by methylmercury. A repeat examination was carried out at age 14 years, again with a high participation rate, with a clinical test battery similar to the one previously applied.

The main finding at the 7-year follow-up in the Faroes was that decrements in attention, language, verbal memory, and, to a lesser extent, in motor speed and visuospatial function, were associated with prenatal methylmercury exposure; the cord-blood mercury concentration was the best risk indicator (Grandjean et al., 1997). These findings were robust in the full Faroes data set in analyses controlled for age, sex and confounders, and they persisted after exclusion of high-exposure subjects. Support for these findings was seen in some of the neurological tests, but particularly in delays in brainstem auditory evoked potentials (Murata et al., 1999). Likewise, prenatal methylmercury exposure was associated with a decrease in the normal heart rate variability and a tendency of increased blood pressure (Sørensen et al., 1999).

Data on the 14-year follow-up are currently being processed, but two recent publications describe the neurophysiological outcomes that are unlikely to be affected by socioeconomic confounders (Murata et al., 2004; Grandjean, 2004). Mercury-associated delays in peak III of the brainstem auditory evoked potentials remained at age 14 years, as did the decreased heart rate variability. Both functions involve brainstem nuclei, and the results showed significant correlations that became weaker when adjusted for mercury exposure. This finding suggests that brainstem toxicity may be an important component of developmental methylmercury neurotoxicity.

Delays in brainstem auditory evoked potentials peak V (i.e., the signal elicited by the transmission of the electrical signal to the midbrain) were associated only with the current methylmercury exposure (Figure 1) (Murata

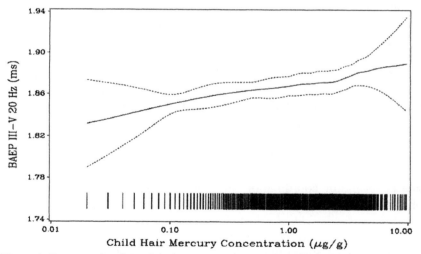

Figure 1. Latency of peak V of the brainstem auditory evoked potentials recorded in 859 Faroese children at 14 years and adjusted for sex and age (Murata et al., 2004). The association is estimated in a generalized additive model analysis, using the current hair-mercury concentration as an indicator of current exposure. The broken lines indicate the point-wise 95% confidence interval for the dose-response relationship. Each vertical line above the horizontal axis represents one observation at the exposure level indicated.

et al., 2004). The average hair-mercury concentration in the adolescents examined was about 1 µg/g (i.e., close to the current Reference Dose (RfD) established by the U.S.EPA).

This observation suggests that the vulnerability of the brain may extend into the teenage period, and that even exposures similar to the RfD may cause adverse effects.

Another prospective study (Cohort 2) included 182 singleton term births. These children were first examined by the Neurological Optimality Score (NOS) at age two weeks. Detailed information was obtained on exposures both to methylmercury and to lipophilic pollutants. The NOS showed significant decreases at higher cord-blood mercury concentrations, while PCB was not important (Steuerwald et al., 2000).

Seychelles

Two birth cohorts were formed in the Seychelles, an archipelago in the Indian Ocean, both groups involving about 800 children (i.e., about 50% of all children born during the recruitment period) (Shamlaye et al., 1995). For exposure assessment, a hair sample was obtained from the mother, in many cases six months after birth. The hair segment that represented the pregnancy period was identified from the assumption that hair grows 1.1 cm per month. The authors noted that the first pilot study was not as well-controlled as the main or longitudinal study: there were fewer covariates, medical records were not reviewed as carefully, there was less information on socio-economic status.

A subset of 217 children from the pilot cohort was evaluated at 66 months (Myers et al., 1995). Maternal hair mercury was negatively associated with four outcomes: the McCarthy General Cognitive Index and Perceptual Performance subscale; and the Preschool Language Scale Total Language and Auditory Comprehension subscale. When statistically determined outliers and observations considered to be influential were removed from the analyses, statistical significance of the association remained only for auditory comprehension.

The main Seychelles study included evaluation of the children at 6.5, 19, 29 and 66 months of age, and again at 8 years. No association with maternal hair mercury was found for most endpoints in these children (Myers et al., 1995). At 29 months there was an association between mercury exposure and decreased activity level in boys only, who also showed a possible mercury-associated delay in age for walking, but the latter was not significant when adjusted for confounders.

Table 1. Main differences between three of the prospective studies of methyl-mercury-exposed children.

Attribute	New Zealand	Faroes	Seychelles
Source of exposure	Shark and other large ocean fish	Whale, ocean fish and shellfish	Ocean fish
Exposure assessment	Maternal hair at parturition	Cord blood and maternal hair	Maternal hair ≤ 6 mo after parturition
Concomitant exposures	Lead in house paint and air	PCBs (whale blubber)	Pesticide use in tropics
Language	English (and Pacific languages)	Faroese (and Danish)	Creole (English and French)
Socioeconomic setting	Industrialized Western	Industrialized Scandinavian	Middle-income developing
Family-setting	Urban, mixed cultures	Traditional	Mainly matriarchal
Outcome tests	Omnibus	Domain-related and neurophysiological	Omnibus and domain-related
Clinical examiners	Clinical specialists	Clinical specialists	Nurse/student

The most detailed examination was then carried out at age 8 years using tests thought to be similar to those applied in New Zealand and the Faroes. In calculating possible effects of prenatal methylmercury exposure, the regression equations included adjustment for postnatal exposure. No association between deficits and maternal hair-mercury concentrations was evident (Myers et al., 2003). Despite the apparent differences between the three studies of mercury-exposed populations, they may not necessarily be in disagreement. For example, when the hair-mercury concentration is taken as the exposure biomarker in both the Faroes and the Seychelles studies, and the Boston Naming Test is used as the outcome parameter, the confounder-adjusted regression coefficient from the two studies does not differ to a statistically significant extent (Keiding et al., 2003). Further, some differences would be anticipated, because the studies used different methods for assessment of exposures and outcomes, and because the epidemiological settings are different. The New Zealand study population may be most similar to continental Europe and North America, even though about half of the high-exposure mothers were of Pacific Island descent. Their diet is high in ocean fish, and very few of them smoke, but otherwise their exposures in daily life would not be very different from other New Zealand women or, indeed, women from other Western societies.

New data will soon appear from Japan, where a prospective study has just initiated (Nakai et al., 2004), as has a cross-sectional study that includes analysis of preserved umbilical cords as a retrospective measure of the children's prenatal exposure levels (Murata et al., 2004). Further, the hair samples collected for the New Zealand cohort of 11,000 women may be used for potential follow-up studies (Kjellstrom, personal communication).

Cross-sectional studies

The neurotoxicity of methylmercury should be judged on the basis of the overall strength of the total evidence available. Several cross-sectional studies (Table 2) supplement the prospective data. As with the prospective studies, researchers have chosen populations that include representation of high-level exposures to methylmercury. However, due to the remote settings (e.g., in the Arctic or the Amazon basin), less sophisticated parameters had to be chosen, also taking into account the possible differences in culture, language, and school education. Some studies focus on the impact of methylmercury from freshwater fish, e.g., in the Amazonian region, where fish contamination is increased by pollution from gold-mining activities. In the Arctic, the traditional diet includes marine mammals and other species high in the food chain that accumulate biomagnified methylmercury. The numbers of children examined have ranged from tens to a few hundred. Selection bias, especially when studying older children, is possible; usually no information was reported about children who were unavailable for the study. Most likely such selection would result in healthier children being examined, thereby potentially obscuring exposure-related effects.

The studies also differ in regard to the exposure intervals covered. Exposure levels range from an average of about 5 μg/g maternal hair (Cree Indians in Northern Québec, and two subgroups in the Amazon basis) up to 13 μg/g in French Guiana, and even higher in Northern Greenland.

Because the developing brain is considered the main target of methylmercury toxicity, evaluation of these studies must assume that exposures measured at the time of the postnatal examination represent causative exposures at the time of the greatest vulnerability of the nervous system. Irregular exposures and the added impacts of postnatal exposure will complicate the evaluation..

Despite the differences in cultural settings and other limitations, several findings of these studies appear to be concordant (Table 2). The results tend to confirm that attention, motor coordination and speed, and visuospatial function are sensitive targets of methylmercury toxicity. The vulnerability of language and verbal memory was not evaluated.

Exposure Assessment

The purpose of an appropriate exposure assessment is to provide a correct measure of exposure in terms of the amount that has reached the toxicological target during the relevant time period. Because the validity depends on the degree to which the exposure biomarkers reflect the "true" exposure, they can be considered only proxy variables, which are always imprecise to some – usually unknown – extent. This issue is important, because exposure misclassification is likely to cause underestimation of the true effect of the exposure.

In prospective studies, samples for mercury analysis have included maternal hair, cord blood, and cord tissue. In addition, maternal dietary questionnaires have been used to obtain information on the origin and approximate magnitude of the methylmercury exposure.

Cross-sectional studies rely on surrogate measures of fetal exposure, because the children were enrolled several years after birth. In the populations studied (Table 2), the diet probably changed little over time due to the preponderant role of fish among available resources. In addition, fish contamination by mercury most likely remained fairly stable in the years following birth. Such stability was evidenced by the study in Peru, in which peak and average concentrations of mercury in maternal hair during pregnancy were very similar and in French Guiana where hair mercury levels measured five years apart in different villages were remarkably constant.

Most studies in Table 2 used maternal hair collected at time of child's examination. In two studies, the child's own exposure (blood or hair measurement) was used instead due to incomplete data on maternal exposure levels, and because a good correlation was documented between the child's and the mother's exposure levels (Grandjean et al., 1999).

This association is to be expected when older children are examined in communities where they are usually sharing the meals with their parents at home. If the child's current exposure does not even provide a correct ranking between individuals, it could introduce exposure misclassification and thus underestimate the risk. However, if the child's exposure is lower than the maternal level, an exaggeration of the risk at a certain exposure level may ensue.

A hair sample may provide a calendar of mercury exposure. Although hair growth rates are known to vary, a 9-cm hair sample obtained at parturition or shortly thereafter would be thought to represent the average mercury exposure during the whole pregnancy.

Table 2. Cross-sectional studies of neurodevelopmental effects in children exposed to methylmercury (studies with more than 100 subjects examined).

Country (reference)	Number (age range)	Main source of exposure	Exposure biomarker	Average exposure* (range)	Outcome measures	Results (ns)
Canada (McKeown-Eyssen at al, 1983)	234 children (12-30 months)	freshwater fish	maternal hair during pregnancy	m_a=6 μg/g (0-24)	Growth parameters Neurological examination	Abnormalities of tendon reflexes (boys only, no dose-response)
Peru (Marsh et al., 1995)	131 infants	marine fish	maternal hair during pregnancy	m_g=7 μg/g (1-29)	Growth parameters Developmental milestones	Ns
Madeira (Murata, et al. 1999)	153 children (7 years)	marine fish	child hair maternal hair	m_g=3.8 μg/g m_g=9.6 μg/g	Neurological examination Neuropsychological tests Brainstem auditory evoked potentials Visual evoked potentials	ns I-III (and I-V) interpeak latency correlated with maternal Hg N145 correlated with maternal Hg
Brazil (Grandjean et al., 1999)	420 children (7-12 years)	freshwater fish (gold mining area)	child hair maternal hair	m_g=11.0 μg/g m_g=11.6 μg/g	Finger-tapping Santa Ana dexterity test WISC-R Digit span Stanford-Binet Copying Recall Bead memory	ns β=-5.58 p=0.001 ns β=-3.40 p=0.003 β=-1.23 p=0.02 ns
French Guiana (Cordier et al, 2002)	248/290 children neurological ex. (6m-6 years) 206/243 children neuropsy. tests (5-12 years)	freshwater fish (gold mining area)	child hair maternal hair	m_g=10.2 μg/g m_g=12.7 μg/g	Neurological examination Finger tapping Stanford-Binet test Blocks Copying Bead memory McCarthy Digit spans forward Leg coordination	Increased tendon reflexes ns ns β=-2.98 p<0.001 ns ns β=-3.72 p=0.006

* m_a, arithmetic mean; m_g, geometric mean; ns = not significant

In New Zealand, monthly (10-mm segment) exposure levels varied by a factor of 2, and on average the peak monthly exposure was about 50% higher than the 9-month average (Kjellstrom, 1989). In the Faroes, similar variations were recorded, with coefficients of variation mostly being below 25% (Grandjean et al., 2003). Comparison of hair segments from the Faroes and the Seychelles showed similar short-term variations (Lanzirotti et al., 2002).

Mercury analyses should always be supported by detailed quality control procedures (e.g., blind comparison with other experienced laboratories and use of certified samples) as was first done in the New Zealand study (Kjellstrom et al., 1989). Until recently, the biomarker imprecision was thought to be appropriately reflected by such laboratory quality data, although the low levels of imprecision (usually about 5% or less) could not explain why associations between mercury concentrations in hair and blood often showed wide scattering. An additional consideration is that the hair mercury concentration is subject to variability, e.g., due to hair type, hair color, external contamination, and leaching due to permanent hair treatments (Grandjean et al., 2002). In international comparisons, three main types of hair structure are recognized (i.e., African, Caucasian, and Oriental), but good data for calibration with blood concentrations exist only for the latter two hair types.

When calculating an exposure level from the hair mercury concentration, an average hair-to-blood ratio of 250 has been generally used (U.S.EPA, 2001). This ratio is appropriate for Caucasian and Oriental hair (Grandjean et al., 2002), but is known to vary between individuals. The 95th percentile differs from the median by a factor between 2 and 3. It also depends on the concentration level, and it changes with age (Budtz-Jørgensen et al., 2004a). For example, in 7-year-old Caucasian children (with finer hair than adults), the ratio is about 370 (i.e., about 50% higher than the ratio of 250 attained by age 14 years).

The possible impact of irregular exposures also needs to be considered. Thus, "bolus" exposures might be more toxic than steady exposures at an average level, but the dose eventually reaching the fetal brain from a maternal seafood meal would be unlikely to represent a steep peak. Nonetheless, exposure variability is likely to introduce error in the exposure assessment, and such misclassification would lead to underestimation of the dose-response relationship. In agreement with this prediction, exclusion of subjects with variable exposures during gestation tended to increase the associations between the mercury exposure and the deficits (Grandjean et al., 2003).

The "true" exposure can be estimated if at least three exposure indicators are available. Such calculations have recently shown that the coefficient of variation for the hair-mercury imprecision is about 50% and thus maiking this biomarker about twice as imprecise as the blood concentration (Budtz-Jørgensen et al., 2004a). The greater the imprecision, the greater the impact on the regression coefficient for the exposure biomarker. At the same time, adjustment for confounders with better precision will cause additional bias toward the null hypothesis (Budtz-Jørgensen et al., 2003b).

Outcome Variables

The validity of outcome variables depends on their sensitivity to the exposure under study and the associated specificity (i.e., lack of sensitivity to the influence of other factors, including confounders). The choice of effect parameters must at the same time be feasible and appropriate for the age of the children, and for the setting of the study. Tests that depend only minimally on cooperation of the subject have the advantage of being less likely to be affected by motivation. The more advanced neuropsychological tests are only possible when a child has reached school age. However, such tests may be of uncertain validity, if they have not previously been applied in the same culture. In addition, many tests require special skills of the examiner. All of these issues need to be considered when evaluating the study findings.

Most studies employed a battery of neurobehavioral tests, some of which appeared to be more sensitive to mercury neurotoxicity than others. Simple comparison of regression coefficients may provide suggestions for the most sensitive parameter, at least within the confines of a particular study. To facilitate such comparisons, the regression coefficient may be expressed as a proportion of the standard deviation of the test result, or as a delay in mental development calculated from the regression coefficient for age.

Benchmark dose levels may also be used as a basis for comparison. Thus, the most sensitive neurological, neuropsychological, and neurophysiological effects parameters all exhibit benchmark dose levels of 5-10 μg/g hair. Despite the great variability of the study settings and the outcome variables, a substantial degree of concordance exists and that the combined evidence is quite convincing in regard to the dose-response relationship.

Neurological tests

All prospective studies and several cross-sectional studies included at least one neurological examination (e.g., McKeown - Eyssen et al., 1983;

Marsh et al., 1995; Steuerwald et al., 2000; Cordier et al., 2002). Unfortunately, the protocols of examination differ between studies: in the Canadian and French Guiana studies, evaluation of sensory functions, cranial signs, muscle tone, stretch reflexes and coordination were conducted. Both studies reported abnormal tendon reflexes in association with an increased maternal hair concentration. However, these signs were mild and isolated, and the reproducibility of the assessment in the French Guiana study was reported to be poor (Cordier et al., 2002). In the study in Peru (Marsh et al., 1995), essential details of the neurological evaluation (items, age of the children examined) were not presented, thus precluding evaluation of the results; frequencies of abnormal signs were reported to be independent of maternal mercury concentrations. In Madeira and the first Faroes cohort, the neurological examination emphasized motor coordination and perceptual motor performance (Grandjean et al., 1997; Murata et al., 1999); children who failed the most difficult of the 19 functional neurological tasks tended to have slightly higher exposures than children who performed well.

In summary, the clinical neurological tests provide limited evidence linking low-dose methylmercury exposures to detectable abnormalities. The absence of clear, positive findings most probably reflects the lack of sensitivity of this type of examination within this range of exposures. One weakness is that the performance on a clinical test is rated by the examiner, thereby introducing a potential subjective aspect. In addition, scoring is usually a simple pass-fail or pass-questionable-fail, thereby limiting the sensitivity.

Developmental tests

Among the prospective studies, the New Zealand children were examined at 4 years of age using the Denver Developmental Screen Test (DDST). It consists of four major function sectors: gross motor, fine motor, language and personal-social, where possible scores are abnormal, questionable, or normal. The prevalence of developmental delay was 52% for children in the high-exposure group and 17% for controls; the delays most frequently affected fine motor and language sectors. The Bayley Scales of Infant Development (BSID) was used in the Seychelles, but no mercury-associated deficits were reported. Developmental tests may be useful for such studies in small children, they may be less dependent on differences in culture than are tests appropriate for older children, but they may be of limited sensitivity to subtle changes.

Neuropsychological tests

While likely to be more sensitive in revealing early neurotoxic changes, neuropsychological tests require that the administration is standardized, and they may show examiner-dependence. Further, they may be sensitive to details in the test situation, such as the use of an interpreter, changes in temperature, and other aspects that may be important when a test is used for the first time in a particular culture. In New Zealand, two psychologists tested the same children with shortened version of the test battery and documented a remarkable agreement (Kjellstrom et al., 1989). In the Faroes and several other studies, each test was administered to all children by the same examiner, thus limiting the possible impact of examiner-related differences.

Traditionally, studies in this field have included standard intelligence test batteries, because of the wealth of information available on such tests and the known implications of deficient performance. The New Zealand study applied the Wechsler Intelligence Scale for Children (WISC) and McCarthy Scale of Children's Abilities at age 6. Likewise, the Seychelles examinations included the McCarthy Scales. These intelligence tests may not be the most appropriate and sensitive for methylmercury toxicity. Still, significant effects were found on WISC and McCarthy in the New Zealand study. The New Zealand study also used the "Test of Oral Language Development" (a standard test widely used in New Zealand school programs), and this test appeared to be the one most affected by methylmercury exposure.

In the Faroes, the approach taken was to emphasize tests that reflected functional domains (e.g., attention, motor speed, verbal memory). The functions chosen were those that were most likely to be affected by developmental methylmercury exposure, as judged from location of neuropathological lesions in poisoning cases and as illustrated by studies of other developmental neurotoxicants, especially lead. The Boston Naming test appeared to be the most sensitive outcome. Similar tests were included when the Seychelles cohort was examined at age 8 years. However, application of the Boston Naming test in a different culture may not be unproblematic. For example, the sequence of stimuli may not necessarily represent an increasing degree of difficulty (e.g., if pictures of an igloo or an acorn are not as familiar to children in a tropical developing country as they are to northern children from the Faroes, or the United States, where the test was developed). In addition, each stimulus was meant to have one correct answer, but since Seychellois is a mixed language, some stimuli in the Boston Naming test are known to have as many as three correct answers.

Thus, even if the same test material and instructions (after translation) are used, the validity of the tests will differ.

A variety of neuropsychological tests, including WISC subtests were used under different circumstances in the cross-sectional studies in Madeira, Greenland, Brazil and French Guiana. In the first two and in an Indian group studied in Brazil, all tests were administered with the use of an interpreter, thereby making the test results less reliable. Likewise, the Continuous Performance test used in the Seychelles was not supervised by an examiner, thereby allowing for possible untoward variability that would be less likely if the examiner was present. Results on such computer-assisted tests will also be affected by the prior computer experience of the child.

Due to the type of populations studied, the researchers attempted to avoid culture-biased and language-dependent tests, thereby precluding evaluation of important domains. The functions evaluated therefore focused on motor speed and motor coordination, visuospatial organization, attention, and short-term memory. Several tests were common to these studies and also overlap with the prospective studies (e.g., Finger Tapping, and Stanford-Binet Bead Memory). At increased exposure levels, reduced scores were evidenced on the Santa Ana dexterity test in Brazil and the McCarthy Leg coordination test in French Guyana. In these two studies, scores on the Stanford-Binet Copying test (that measures visuospatial organization) were negatively associated with mercury exposure with similar regression coefficients in the two studies. Several types of errors occurred in these tests, and the French Guiana study pointed more specifically at rotation errors among younger children (5-6 years old). Such errors would suggest possible insult in the parietal lobes of the brain resulting in developmental delay in the learning to place objects in space (Sullivan et al., 1999). Whether or not this type of test would appear sensitive in other populations living in other cultural environments needs to be established.

Neurophysiological testing

As an objective evaluation of brain dysfunction that is probably less sensitive to motivation or socioeconomic confounding, neurophysiological tests have been applied in several studies. Their applicability requires advanced instrumentation and depends on skilled examiners. An outcome that has previously been found to be sensitive to lead exposure is brainstem auditory evoked potentials. They are recorded using surface electrodes placed on the skull while the child listens to a stimulus in one ear. The transmission of the electrical signals within the brain is then recorded as peaks that represent the acoustic nerve, an intermediate connection in the pons, and the midbrain. The latency of peak III was significantly increased at higher intrauterine exposure to mercury. Parallel associations were found in 7-year-old children in the Faroes and in Madeira, and this observation was replicated in the Faroese cohort when examined at 14 years. A smaller study from Ecuador also reported delays in peak III at higher exposure levels. In addition, prolonged latencies of peak V among the 14 year-olds were linked to the current mercury exposure only (Figure 1). As a parameter primarily affected by postnatal exposure, this particular outcome seems to differ from the majority of functions sensitive to methylmercury during fetal development.

Confounding Variables

Three major reasons for confounding have been noted as to why a mercury effect might have been overestimated: (a) association of mercury intake with exposure to other neurotoxic pollutant(s); (b) other types of residual confounding; and (c) inadequate adjustment for multiple comparisons (NIEHS, 1998). The best protection against confounding problems is to select a study setting where such concerns are unlikely and, if relevant, may be adjusted for appropriately. Thus, a homogeneous society with limited differences in socioeconomic and cultural factors should be preferred. The existence of residual confounding can never be fully excluded. On the other hand, "phantom" covariates should not be invoked to explain away biologically plausible associations between methylmercury exposure and neurobehavioral deficits. In addition, most attention is usually paid to confounders that affect the outcomes in the same direction as the exposure under study, but confounders may also have the opposite effect of attenuating the apparent impact of the exposure. Thus the potential for overestimation of a toxic effect should not be raised without also paying attention to the risk of underestimation.

Standard multiple regression methods are often used for controlling for confounding effects. Even in the absence of confounding, adjustment for such established predictors as sex, age, and maternal intelligence, should be included to obtain a more precise estimate of the mercury effect. In general, the prudent approach is usually to include all covariates that may be potential confounders. However, in situations where the exposure is measured with some degree of imprecision (as is the case here), this approach may result in biased estimates. Inclusion of such covariates, which are associated with the exposure but without any explanatory power in regard to the effect, will then increase the underestimation of the effect of the exposure of interest (Budtz-

Jørgensen et al., 2003).

As a main concern in regard to confounding, socioeconomic conditions vary substantially between the study settings. Although mercury neurotoxicity was reported in almost all studies, differences within each study could be important. New Zealand and the Faroes represent relatively wealthy, industrialized populations, where socioeconomic differences are thought to be limited, but most of the other studies were carried out in developing countries, where basic sanitary problems are common, and where nutritional deficiencies may occur, both of which may be difficult to adjust for. In the Seychelles, stunting still occurs in a small percentage of children (WRI, 2003). In New Zealand, ethnic differences appeared to play a role, but the analysis was based on matching of the children in the different exposure groups for ethnicity. An additional factor of possible interest is consanguinity, which is more frequent in island populations and other isolated communities. However, to cause confounding, the degree of consanguinity would have to be associated with mercury exposure and at the same time result in neurobehavioral deficits. Both assumptions seem unlikely, but documentation from an individual study would be a major undertaking.

The family structure and home environment are documented as important determinants of childhood development. Within the populations studied, circumstances may vary. For example, only about 25% of the births in the Seychelles are nuptial, while an additional 50% are recognized by a father, but about 25% of children have no known father (MISD, 1998). Accordingly, children of the Seychelles cohort were said to be accompanied by a "care-giver", often a relative, with whom the child was living (Myers et al., 2003). The variable family structure, which may be difficult to adjust for in statistical analyses, contrasts with the more uniform circumstances of most other studies with a traditional and stable family structure. In the New Zealand study, low social class and non-English home language reduced, as anticipated, the score on some tests and more than 6 months of breastfeeding increased the score on some tests (Kjellstrom et al., 1989). These variables were accounted for in the multiple regression analysis.

Among other known developmental neurotoxicants, none is as prevalent as ethanol. Most studies reported that maternal alcohol use during pregnancy was minimal, but in some cases it may be difficult to assess because of the importance of home-brewed beverages (e.g., in the Seychelles) (Perdrix et al., 1999).

In particular the Faroese are exposed to polychlorinated biphenyls (PCBs), in this case from eating whale blubber. Detailed analyses of the Faroes data failed to show any important impact of PCB exposure on the neurotoxicity outcomes (Budtz-Jørgensen et al., 1999; Grandjean et al., 2002; Steuerwald et al., 2000). The relative importance of PCB and mercury was assessed in structural equation analysis taking into account imprecision in both variables. Inclusion of PCB exposure attenuated the mercury effect somewhat, but mercury remained statistically significant, while PCB was far from that (Budtz-Jørgensen et al., 2002). In New Zealand and the Seychelles, the ocean fish consumed is unlikely to be contaminated by PCB, and the same would be the case with freshwater fish in the Amazon Basin.

A reverse effect may occur if subjects who do not eat mercury-containing fish instead consume fruits and vegetables that contain pesticides. Such exposures are more likely in tropical developing countries. The neurotoxicity of many pesticides could then potentially cause neurodevelopmental effects in children with low-level mercury exposures, thus blurring the dose-response relationship. Although pesticide use might be a cofactor in the Seychelles, no information is available to evaluate this possibility, and this issue has apparently not been considered in the epidemiological studies. Nonetheless, the United Nations Food and Agriculture Organization (1997) helped remove large amounts of obsolete pesticides (such as malathion) from the Seychelles. Given the neurotoxic potential of these substances, and the likelihood of their use in tropical countries, exposure potentials need to be considered.

An additional consideration is natural toxins, such as cyanides present in cassava grown in the tropics (Dora, 2002). If non-fish eaters consume more cassava than those with a high fish intake, then cyanide exposure could potentially cause a confounding bias toward the null hypothesis, as has been suggested by Dora in regard to evidence from Brazil. This concern may also relate to the Seychelles, where clusters of so-called tropical myeloneuropathies have occurred as a possible effect of cyanide intoxication from cassava consumption (Roman et al., 1985).

Certain essential nutrients in fish and seafood may provide beneficial effects on brain development, thereby possibly counteracting adverse effects of the contaminants. This possibility has often been mentioned in regard to ocean fish (Myers et al., 2003). Perhaps, if ocean fish contains higher concentrations of essential n-3 fatty acids than do freshwater fish, then this difference could perhaps explain why the mercury dose-response relationship appears to be steeper in populations that rely on river fish. Although high intake of these fatty acids may cause an increase in birth weight, no effect on early neurobehavioral development was found in the Faroes (Steuerwald et al., 2000).

Selenium concentrations in ocean fish from New Zealand do not depend on fish size, while mercury concentrations increased linearly with size

(Kjellstrom, 2000). Although selenium has been considered to potentially provide protection against mercury effects, cord-blood selenium concentrations in the Faroes did not impact on mercury-associated deficits. It therefore seems likely that essential nutrients would counteract no more than limited aspects of mercury neurotoxicity.

Cardiovascular Effects

Although the developing brain is considered the critical target organ in regard to methylmercury, recent evidence has suggested that mercury from fish and seafood may promote or predispose to the development of heart disease. This evidence is yet inconclusive, but deserves attention, because it suggests that a narrow definition of subpopulations at risk, i.e., pregnant women, might leave out other vulnerable groups.

The first studies of methylmercury-associated cardiovascular disease were carried out in Finland. One important study showed that the intima-media thickness of the carotid arteries in apparent association with the degree of mercury exposure from fish (Salonen et al., 2000). A possible mechanism may be induction of lipid peroxidation. In this regard, the interesting observation was also made that, while essential fatty acids from fish may prevent cardiovascular mortality, this beneficial effect may be cancelled or overwhelmed by concomitant exposure to methylmercury (Rissanen et al., 2000). The increased risk seems to occur at hair-mercury concentrations above 2 $\mu g/g$, i.e., only twice the level corresponding to the U.S.EPA Reference Dose. More recent information tends to support these findings. A large multi-center study from Europe showed an increased risk of cardiovascular disease associated with toenail mercury concentrations (Guallar et al., 2002). However, this finding was mainly due to a particularly strong risk observed in one of the centres. In a U.S. study of health care workers, only a minimal risk was seen, but after exclusion of the dentists with high toenail-mercury concentrations likely due to amalgam exposures, the risk was similar to the one observed in the European study (Yoshizawa et al., 2002).

The possible interaction between toxic methylmercury and beneficial polyunsaturated fatty acids is of particular relevance in northern populations. Early evidence suggested that Inuits in Greenland had a low cardiovascular morbidity and mortality that was linked to their high intake of n-3 long-chain polyunsaturated fatty acids. Recent evidence indicates that this notion may be erroneous (Bjerregaard et al., 2003). The potential significance of methylmercury in this regard is therefore highly relevant.

Risk Assessment and Exposure Limits

In deriving exposure limits from epidemiological data, regulatory authorities have increasingly relied upon the use of benchmark dose estimates (Budtz-Jørgensen et al., 2001). According to usual default settings, an exposure at the benchmark dose (BMD) results in an increased frequency of a pathological outcome from 5% to 10%. The benchmark dose level (BMDL) is then the point of departure that represents the lower 95% confidence limit of the BMD. When using a linear dose scale, outliers with high exposure levels may become highly influential. The BMDLs from the New Zealand data were 17-24 µg/g, but when the child with the highest mercury level was omitted, they decreased to 7.4-10 µg/g (Crump et al., 1998).

JECFA (2003) considered the BMDL an exposure that is "without appreciable adverse effects in the offspring". This interpretation may be true under some circumstances, but in large epidemiological studies, where the confidence interval is relatively narrow, the BMDL will be closer to the BMD. For example, the results from the Faroes show that exclusion of the subjects with a maternal hair-mercury concentration above 10 µg/g (a cut-off level lower than the BMDL used by JECFA) barely altered the regression coefficients and the P-values (Grandjean et al., 1997). The BMDL is therefore not a no-adverse-effect level (NOAEL), but rather a lowest-observed-effect level (LOEL). This consideration is likely to affect the choice of uncertainty factors, especially in regard to brain function, where even small decrements may be of substantial social and economic impact.

JECFA (2003) used BMDLs based on the maternal hair-mercury concentration. In contrast to the NRC (2000), JECFA decided to exclude the New Zealand study and therefore arrived at a higher overall average BMDL. For the Faroes study, the BMDL chosen by JECFA was 12 µg/g maternal hair (i.e., an average for the linear dose-response curve for several different functions and not the most sensitive brain function, as preferred by NRC).

The problem of choosing the most sensitive function may be resolved by using a structural equation model for deriving integrated BMD and BMDL values (Budtz-Jørgensen et al., 2003b; Budtz-Jørgensen et al., 2004b). This calculation includes all exposure information, confounders, and cognitive outcomes, and also takes into regard effects of measurement uncertainty. Using this advanced statistical approach, the overall BMDL is calculated at 6 µg/g maternal hair (or 43 µg/L cord blood). Thus, by incorporating the complete data set in the assessment, the resulting hair-based BMDL is only half the size of the BMDL chosen by JECFA (2003).

This finding is in agreement with the general finding that measurement uncertainty (in the exposure or the response) leads to overestimation of the benchmark results (Budtz-Jørgensen et al., 2003a; Budtz-Jørgensen et al., 2004b). Thus, although the above calculations are based on the Faroes study only, it is likely that such refinements of the BMDL calculations using data from other studies would result in a similar, if not greater, decrease in the BMDL results.

In calculating an exposure limit from a BMDL, an uncertainty factor is usually applied to take into account sources of variation in individual susceptibility as well as insufficiencies in the data base (e.g., concerning effects on target organs other than the developing nervous system). The NRC (2000) and U.S. EPA (2001) chose a total uncertainty factor of 10. However, JECFA (2003) concluded that "the two study samples represent diverse populations", and that "no uncertainty factor is needed to account for variation in vulnerability among subgroups". This decision is also based on the assumption that the most sensitive effects are the average neurobehavioral outcomes in the two studies, on which the overall average BMDL was based. However, JECFA had included results from a study that did not identify statistically significant decrements, thus hardly representing a vulnerable population.

JECFA included only an uncertainty factor of 3.2 to account for the total human inter-individual variability for dose reconstruction (converting maternal blood concentration to a steady-state dietary intake). Although in accordance with default calculations, it omits the consideration of toxicodynamic sources of variation as well as insufficiencies in the data base. Along with an uncertainty factor of 2 for conversion of hair-mercury concentrations to intake levels, the total uncertainty factor used was 6.4.

The choice of uncertainty factors explains in part the difference in the recommended exposure limits (Table 3). Another decision is which studies to include. Most important perhaps, the adjusted BMDL (see above) will result in lower exposure limits than those arrived at in the risk assessments carried out so far.

Table 3. Calculated exposure limits for methylmercury.

	NRC (2000)	JECFA (2003)
Number of studies	One (three)	Two
Exposure biomarker	Cord blood (hair)	Hair
BMDL selected	58 µg/L cord blood	14 µg/g hair
Uncertainty factor	10	3.2 and 2
Exposure limit	0.1 µg/kg per day	1.6 µg/kg per week

Public Health Relevance

Recent discussions on methylmercury have erroneously pictured the situation as a conflict between a negative and a positive study. This misleading characteristic may be related to disagreements between regulatory agencies and has been exploited by vested interest groups. In epidemiology, the term non-positive is often used in regard to studies that were unable to detect a particular effect. Also, no matter how positive a study is, all observational studies have weaknesses, and a prudent judgment should be based on the total amount of evidence available, no on single studies, whether positive or not. The present chapter has demonstrated that the scientific evidence on methylmercury neurotoxicity is fairly consistent, and that adverse effects are likely to occur even at low-level exposures. There is no dispute about the very serious prenatal effects that occurred in Minamata at maternal hair-mercury concentrations thought to be in the range of 10–100 µg/g (Tsubaki and Irukayama, 1977; NRC, 2000; UNEP, 2002). It would seem intuitively logical that less severe effects may occur at the exposure ranges found in the more recent studies. Because of the global significance of methylmercury contamination of food, these scientific findings need to be expressed in terms that may facilitate an evaluation of their public health significance.

The Faroes study showed that each doubling in prenatal mercury exposure corresponded to a delay of one or two months in mental development at age 7 years (Grandjean et al., 1997). Because rapid development occurs at that age, such delays may be important. Also, even small shifts in a measure of central tendency may be associated with large changes in the tails of the distribution. Such developmental delays are likely to be permanent, at least in part, but the long-term implications are unknown. The experience with lead neurotoxicity suggests that such effects are likely to remain and that they may even become more apparent with time.

A shift in IQ levels was documented in the New Zealand study (Kjellstrom et al., 1989). The average WISC-R full-scale IQ for the study population (n= 237) was 93. In the group with maternal hair mercury above 6 µg/g (n=61) the average was 90. The average exposure in the latter group would be about 4-fold higher than in the study population as a whole. Another way of presenting these shifts in IQ is to estimate the increased number of subjects with very low IQ as methylmercury exposure increases. In New Zealand, an IQ below 70 (= mental retardation) was twice as common (increase from 5 to 10%) in the highest hair mercury group (> 10 µg/g) compared to the group with hair mercury below 6 µg/g (Kjellstrom, 2000). For the IQ range 71-85 the increase was from 20 to 25%, but due to the small number of children, this result was not statistically significant.

Another approach was used in the Faroes in the absence of formal IQ tests. The regression coefficients were expressed as a proportion of the standard deviation of the test results (Grandjean et al., 1999). The most sensitive outcome parameters show a decrement of about 10% of the standard deviation at each doubling of the prenatal methylmercury exposure level. Had an IQ scale been used, with a standard deviation of 15 IQ points, a doubling in the exposure could have caused a deficit of about 1.5 IQ points. These findings are in full agreement with the New Zealand data.

Each of these estimations is associated with some degree of uncertainty. Some scientific uncertainties are bound to remain, although new prospective cohort studies on methylmercury neurotoxicity are starting to provide new evidence. However, the documentation is not going to expand substantially or otherwise provide much clearer guidance for regulatory agencies. The experience with lead research (Bellinger and Needleman, 2001) has amply illustrated that apparent disagreement is likely to occur between studies carried out by different methods in different settings. In the absence of complete coherence, decisions on preventive efforts should be justified by all available evidence, taking into account its various uncertainties and inconsistencies. Potential costs and other societal consequences of policy decisions – including decisions to do nothing - also deserve fair consideration. However, these issues should be addressed separately from the discussion of toxicological and epidemiological concerns. Otherwise, the erroneous impression will be generated that disagreements on preventive measures are solely due to uncertainties in epidemiologic evidence.

CONCLUSIONS

As previously documented (e.g. with lead pollution), a full coherence among all available evidence should not be anticipated. Decisions on preventive efforts should therefore be justified by the scientific database at large. Still, the current evidence on adverse health effects of methylmercury reveals a substantial consistency, also in regard to low-level dietary exposures.

Calculated exposure limits for methylmercury published by national and international bodies differ only little, although the approaches were not the same. However, none of the reports took into account the impact of measurement imprecision, and benchmark results used by the committees are therefore biased toward higher values. The NRC and the U.S. EPA used a larger total uncertainty factor that may have compensated for this problem. While the focus was on protecting the fetus against adverse effects, recent evidence on postnatal neurotoxicity suggests that the protection also include adolescence. Emerging evidence that cardiovascular disease risks in adults may be associated with methylmercury exposure suggests that this protection could prudently be extended to the population at large.

ACKNOWLEDGMENTS

This work was supported by the U.S. National Institute of Environmental Health Sciences (ES09797 and ES11681) and the Danish Medical Research Council. The contents of this paper are solely the responsibility of the authors and do not represent the official views of the NIEHS, NIH or any other funding agency.

REFERENCES

Bjerregaard, P., Kue, Young, T., Hegele, R.A. Low incidence of cardiovascular disease among the Inuit-what is the evidence? *Atherosclerosis*, 166, 351-7, 2003.

Budtz-Jørgensen, E., Keiding, N., Grandjean, P. White R.F., Weihe P. Methylmercury neurotoxicity independent of PCB exposure [letter]. *Environ. Health Perspect.*, 107A, 236-237, 1999.

Budtz-Jørgensen, E., Keiding, N., Grandjean, P. Benchmark dose calculation from epidemiological data. *Biometrics*, 57, 698-706, 2001.

Budtz-Jørgensen, E., Keiding, N., Grandjean, P., Weihe, P. Estimation of health effects of prenatal methylmercury exposure using structural equation models. *Environ. Health*, 1, 2, 2002

Budtz-Jørgensen, E., Keiding, N., Grandjean, P. Application of structural equation models for evaluating epidemiological data and for calculation of the benchmark dose. In: Proceedings of the ISI International Conference on Environmental Statistics and Health; 2003 July 16-18; Santiago de Compostela, Spain; p. 183-194, 2003a.

Budtz-Jørgensen, E., Keiding, N., Grandjean, P., Weihe, P., White R.F. Consequences of exposure measurement error for confounder identification in environmental epidemiology. *Stat. Med.*, 22, 3089-3100, 2003b.

Budtz-Jørgensen, E., Grandjean, P., Jørgensen, P.J., Weihe, P., Keiding, N. Association between mercury concentrations in blood and hair in methylmercury-exposed subjects at different ages. *Environ. Res.*, (in press) 2004a.

Budtz-Jørgensen, E., Keiding, N., Grandjean, P. Effects of exposure imprecision on estimation of the benchmark dose. *Risk Anal.*, (in press) 2004b.

Cordier, S., Garel, M., Mandereau, L., Morcel, H., Doineau, P., Gosme-Seguret, S., Josse, D., White, R., Amiel-Tison, C. Neurodevelopmental investigations among methylmercury-exposed children in French Guiana. *Environ. Res.*, 89, 1-11, 2002.

Counter, S.A., Buchanan, L.H., Laurell, G., Ortega, F. Blood mercury and auditory neuro-sensory responses in children and adults in the Nambija gold mining area of Ecuador. *Neurotoxicology*, 19, 185-196, 1998.

Crump, K.S., Kjellstrom, T., Shipp, A.M., Silvers, A., Stewart, A. Influence of prenatal mercury exposure upon scholastic and psychological test performance: benchmark analysis of a New Zealand cohort. *Risk Anal.*, 18, 701-713, 1998.

Davidson, P.W., Myers, G.J., Cox, C., Axtell, C., Shamlaye, C., Sloane-Reeves, J., et al. Effects of prenatal and postnatal methylmercury exposure from fish consumption on neurodevelopment: outcomes at 66 months of age in the Seychelles Child Development Study. *JAMA*, 280, 701-707, 1998.

Dorea, J.G. Fish are central in the diet of Amazonian riparians: should we worry about their mercury concentrations? *Environ. Res.*, 92, 232-44, 2003.

FAO (United Nations Food and Agriculture Organization). Dangerous Pesticide Stocks Removed from Zambia and the Seychelles - Large Stocks Continue to Threaten Health and Environment, FAO Says. Press release 97/31. http://www.fao.org/WAICENT/FAOINFO/AGRICULT/AGP/AGPP/Pesticid/Disposal/PR97-31.htm

Grandjean, P., Weihe, P., White, R.F., Debes, F., Araki, S., Murata, K. Sørensen, N., Dahl, D., Yokoyama, K., Jørgensen, P.J. Cognitive deficit in 7-year-old children with prenatal exposure to methylmercury. *Neurotoxicol. Teratol.*, 19, 417-428, 1997.

Grandjean, P., Weihe, P., Burse, V.W., Needham, L.L., Storr-Hansen, E., Heinzow, B. et al. Neurobehavioral deficits associated with PCB in 7-year-old children prenatally exposed to seafood neurotoxicants. *Neurotoxicol. Teratol.*, 23, 305-317, 2001.

Grandjean, P., White, R.F., Weihe, P., Jørgensen, P.J. Neurotoxic risk caused by stable and variable exposure to methylmercury from seafood. *Ambul. Pediatr.*, 3, 18-23, 2003.

Grandjean, P., White, R., Nielsen, A., Cleary, D., deOliveira-Santos, E. Methylmercury neurotoxicity in Amazonian children downstream from gold mining. *Environ. Health Perspect.*, 107, 587-591, 1999.

Guallar, E., Sanz-Gallardo, M.I., van't Veer, P., Bode, P., Aro, A., Gomez, Aracena, J., Kark, J.D., Riemersma, R.A., Martin, Moreno, J.M., Kok, F.J. Mercury, fish oils, and the risk of myocardial infarction. *N. Engl. J. Med.*, 347, 1747-54, 2002.

JECFA (Joint FAO/WHO Expert Committee on Food Additives). Sixty-first meeting, Rome, 10-19 June 2003. Summary and conclusions. Available from: URL: ftp://ftp.fao.org/es/esn/jecfa/jecfa61sc.pdf.

Keiding N, Budtz-Jørgensen E, Grandjean P. Prenatal methylmercury exposure in the Seychelles [letter]. *Lancet*, 362, 664-665, 2003.

Kjellstrom, T. Methylmercury exposure and intellectual development in vulnerable groups in New Zealand. In: Proceedings of the US-Japan workshop, Nov. 2000. Minamata, Japan, National Institute for Minamata Disease; 2000.

Kjellström, T., Kennedy, P., Wallis, S., Stewart, A., Friberg, L., Lind, B., et al. Physical and mental development of children with prenatal exposure to mercury from fish. Stage 2, interviews and psychological tests at age 6. (Report 3642) Stockholm, National Swedish Environmental Protection Board; 1989.

Kjellström, T., Kennedy, P., Wallis, S., Mantell, C. Physical and mental development of children with prenatal exposure to mercury from fish. Stage 1: Preliminary tests at age 4. (Report 3080) Stockholm, National Swedish Environmental Protection Board; 1986.

Lanzirotti, A., Jones, K.W., Clarkson, T.W., Grandjean, P. Human health risks from methyl mercury in fish. In: Science Highlights - National Synchroton Light Source Activity Report. Upton, NY: Brookhaven National Laboratory;. 97-99, 2002.

Marsh, D.O., Turner, M.D., Smith, J.C., Perez, V.M.H., Allen, P., Richdale, N. Fetal MeHg study in a Peruvian fish eating population. *Neurotoxicology*, 16, 717-726, 1995.

McKeown-Eyssen, G., Ruedy, J., Neims, A. Methylmercury exposure in northern Quebec. II. Neurologic findings in children. *Am. J. Epidemiol.*, 118, 470-479, 1983.

Murata, K., Weihe, P., Renzoni, A., Debes, F., Vasconcelos, R., Zino, F., Araki, S., Jorgensen, P., White, R., Grandjean, P. Delayed evoked potentials in children exposed to methylmercury from seafood. *Neurotoxicol Teratol.*, 21, 343-348, 1999.

Murata, K., Weihe, P., Budtz-Jørgensen, E., Jørgensen, P.J., Grandjean, P. Delayed brainstem auditory evoked potential latencies in 14-year-old children exposed to methylmercury. *J. Pediatr.*, 144, 177-183, 2004.

Murata, K., Sakamoto, M., Nakai, K., Weihe, P., Dakeishi, M., Iwata, T., et al. Effects of prenatal methylmercury exposure on neurodevelopment in Japanese children. In: Proceedings of the National Institute of Minamata Disease Conference; Niigata, Japan; 2004 (in press).

Myers, G.J., Davidson, P.W., Cox, C., Shamlaye, C.F., Tanner, M.A., Marsh, D.O., et al. Summary of the Seychelles child development study on the relationship of fetal methylmercury exposure to neurodevelopment. *Neurotoxicology*, 16, 711-716, 1995.

Myers, G.J., Davidson, P.W., Cox, C., Shamlaye, C.F., Palumbo, D., Cernichiari, E., et al.

Prenatal methylmercury exposure from ocean fish consumption in the Seychelles child development study. *Lancet*, 361, 1686-1692, 2003.

MISD (Management & Information Systems Division), Statistics & Database Administration Section (SDAS) of the Ministry of Information Technology & Communication of the Republic of Seychelles. Statistical Bulletin Quarterly: Population and Vital Statistics 2002. Victoria, Mahe, 2003: table 3 (Registered Live Births By Year, Month of Registration, Sex and Status, 1998-2002).

Nakai, K., Suzuki, K., Oka, T., Murata, K., Sakamoto, M., Okamura, K., et al. The Tohoku Study of Child Development: A Cohort Study of the Effects of Perinatal Exposures to Methylmercury and Environmentally Persistent Organic Pollutants on Neurobehavioral Development in Japanese Children. *Tohoku J. Exp. Med.*, 202, 227-237, 2004.

Needleman, H.L., Bellinger, D. Studies of lead exposure and the developing central nervous system: a reply to Kaufman. *Arch. Clin. Neuropsychol.*, 16, 359-374, 2001.

NIEHS. Workshop organized by Committee on Environmental and Natural Resources (CENR), Office of Science and Technology Policy (OSTP), The White House: Scientific Issues Relevant to Assessment of Health Effects from Exposure to Methylmercury, November 18-20, 1998. Available from: URL: http://ntp-server.niehs.nih.gov/main_pages/PUBS/MethMercWkshpRpt.html.

NRC (National Research Council). Toxicological Effects of Methylmercury. Washington: National Academy Press; 2000.

Perdrix, J., Bovetmm P., Larue, D., Yersin, B., Burnand, B., Paccaud, F. Patterns of alcohol consumption in the Seychelles Islands (Indian Ocean). *Alcohol*, 34, 773-785, 1999.

Rissanen, T., Voutilainen, S., Nyyssonen, K., Lakka, T.A., Salonen, J.T. Fish oil-derived fatty acids, docosahexaenoic acid and docosapentaenoic acid, and the risk of acute coronary events: the Kuopio ischaemic heart disease risk factor study. *Circulation*, 102, 2677-9, 2000.

Roman, G.C., Spencer, P.S., Schoenberg, B.S. Tropical myeloneuropathies: the hidden endemias. *Neurology*, 35, 1158-70, 1985.

Salonen, J.T., Seppanen, K., Lakka, T.A., Salonen, R., Kaplan, G.A. Mercury accumulation and accelerated progression of carotid atherosclerosis: a population-based prospective 4-year follow-up study in men in eastern Finland. *Atherosclerosis*, 148, 265-73, 2000.

Shamlaye, C.F., Marsh, D.O., Myers, G.J., Cox, C., Davidson, P.W., Choisy, O., et al. The Seychelles child development study on neurodevelopmental outcomes in children following in utero exposure to methylmercury from a maternal fish diet: background and demographics. *Neurotoxicology*, 16, 597-612, 1995.

Sullivan, K. Neurodevelopmental aspects of methylmercury exposure: neuropsychological consequences and cultural issues. PhD. Thesis in Behavioral Neuroscience, Boston University School of Medicine; 1999.

Tsubaki T, Irukayama K. Minamata disease: methylmercury poisoning in Minamata and Niigata, Japan. Amsterdam, Elsevier Scientific Publ. Co., 1977.

UNEP (United Nations Environment Programme). Global Mercury Assessment. Geneva, Switzerland; 2002.

U.S. EPA (Environmental Protection Agency), Office of Science and Technology, Office of Water. Water Quality Criterion for the Protection of Human Health: Methylmercury, Final. EPA-823-R-01-001, 2001. Washington. URL: http://www.epa.gov/waterscience/criteria/methylmercury/document.html.

Weihe, P., Hansen, J.C., Murata, K., Debes, F., Jorgensen, P.J., Steuerwald, U., et al.

Neurobehavioral performance of Inuit children with increased prenatal exposure to methylmercury. *Int. J. Circumpolar Health*, 61, 41-49, 2002.

WRI (World Resources Institute) Earth Trends: The Environmental Information Portal. Variable: Children's Health: Stunting in children under 5. 2003. URL: http://earthtrends.wri.org/text/POP/variables/387.htm.

Yoshizawa, K., Rimm, E.B., Morris, J.S., Spate, V.L., Hsieh, C.C., Spiegelman, D., Stampfer, M.J., Willett, W.C. Mercury and the risk of coronary heart disease in men. *N. Engl. J. Med.*, 347, 1755-60, 2002.

PART-V:
REGIONAL CASE STUDIES

Chapter-23

DYNAMIC PROCESSES OF ATMOSPHERIC MERCURY IN THE MEDITERRANEAN REGION

Nicola Pirrone, Francesca Sprovieri, Ian M. Hedgecock, Giuseppe A.Trunfio and Sergio Cinnirella

CNR-Institute for Atmospheric Pollution, Division of Rende, 87036 Rende, Italy

INTRODUCTION

During the 80's and part of 90's most research on atmospheric mercury focused on sources and deposition to terrestrial watersheds and freshwater lakes. The recent advances in automated techniques for the accurate measurement of gaseous oxidised mercury species, at the concentrations typically encountered in the atmosphere, has had a major impact on the study of mercury cycling in the environment. The use of these instruments at contaminated sites and in remote locations has produced not only more, but also more reliable data with greater time resolution in the last 3-4 years than in all the time atmospheric mercury has been studied (e.g., Keeler et al., 1995; Pirrone et al., 1995; Munthe et al., 2001; Wangberg et al., 2001; Lindberg et al., 2002a; Landis et al., 2002; Sprovieri et al., 2002; Ebinghaus et al., 2002; Pirrone et al., 2003a; Sprovieri et al., 2003).

Mercury is released into the European environment from a multitude of natural and anthropogenic sources. Natural sources (including re-emission of previously deposited mercury) account for about 40% of the total mercury released annually ($\sim$ 550 tonnes) to the European atmosphere (Pirrone et al., 1996; Pacyna et al., 2001; Pirrone et al., 2001b); therefore the understanding of the relative contribution of natural vs. anthropogenic sources is of fundamental importance in shaping strategies for reducing the impact of this highly toxic element on the European environment and population. In recent years mercury pollution issues have been investigated in the context of several integrated research projects, funded by the European Commission as part of the 4th, 5th and 6th Framework Programmes, and by national funding agencies in most Mediterranean countries.

The first coordinated atmospheric mercury monitoring network established in Europe in the framework of MAMCS and MOE projects (Pirrone et al., 2003a; Munthe et al., 2003) showed an absence of spatial gradients in ambient concentrations of total gaseous mercury (TGM) at background sites over Europe (Wangberg et al., 2001), whereas ambient concentrations of oxidised and particle bound mercury in southern Europe may be much higher (up to a factor of 2) than that observed at North European sites. It has been suggested that chemical and physical processes in the Marine Boundary Layer (MBL) could be at least partly responsible for this phenomenon (Pirrone et al., 2000; Hedgecock and Pirrone, 2001; Hedgecock et al., 2003; Pirrone et al., 2003a).

In understanding the mercury cycling in the MBL a better assessment of the lifetime of elemental mercury with changing meteorological conditions and location is of fundamental importance for a correct evaluation of transport and deposition patterns over a region. In a recent paper Hedgecock and Pirrone (2004) proposed a much shorter lifetime for Hg^0. The shortest lifetime would occur when air temperatures are low and sunlight and deliquescent aerosol particles are plentiful. The reason for this, briefly, is that at lower temperatures sea salt aerosol acidification is more rapid and therefore so is halogen activation. The halogens, particularly bromine released from the aerosols photolyse and modelling studies suggest that it is the reaction between $Hg^0_{(g)}$ and Br which is the most important in determining the Hg^0 lifetime. Thus the modelled lifetime for clear-sky conditions was estimated to be actually shorter at mid-latitudes and high latitudes than near the equator, and for a given latitude and time of year, cooler temperatures enhance the rate of Hg oxidation (Hedgecock and Pirrone, 2004). Under typical summer conditions (for a given latitude) of temperature and cloudiness, the lifetime of $Hg^0(g)$ in the MBL is calculated to be around 10 days at all latitudes between the equator and 60° N. This is much shorter than the generally accepted atmospheric residence time for $Hg^0(g)$ of a year or more. Given the relatively stable background concentrations of $Hg^0(g)$ which have been measured, continual replenishment of $Hg^0(g)$ must take place, suggesting a *"multi-hop"* mechanism for the distribution of Hg, rather than solely aeolian transport with little or no chemical transformation between source and receptor.

The following sections of this chapter present an overview of natural and anthropogenic sources of atmospheric mercury in the Mediterranean basin, the available data on ambient levels of atmospheric speciated mercury at coastal and open sea locations, and the integrated modelling system developed to evaluate spatial and temporal distributions of ambient concentrations, dry and wet deposition fluxes and re-emission rates with changing meteorological conditions.

METEOROLOGICAL PATTERNS

The Mediterranean Sea is surrounded by high peninsulas and important mountain barriers. The most important are the Alps and the Balkan Peninsula to the north, the Iberian Peninsula to the west, the Atlas Mountains to the southwest, and the Asia Minor to the northeast. The gaps between these major mountainous regions act as channels for air mass transport from/to the Mediterranean. This kind of transport is considered very important for cyclogenetic activities and the air quality in the Mediterranean region. These topographic features along with the significant variation of the physiographic characteristics are partially responsible for the development of various-scale atmospheric circulations. These locally produced atmospheric circulations are quite strong, especially during the warm period of the year (Kallos et al., 1998 and ref. herein).

The Mediterranean climate has some distinctive characteristics. It cannot be characterized as either maritime or continental. The cold season (end of October - beginning of March) is the rainy period. The warm season (June - September) is the dry period with almost no rain. The rest of the year consists of the transient seasons (spring and autumn) where the winter and summer-type weather patterns are interchanging.

Cyclogenic activity in the Mediterranean is a result of positive vorticity advection and invasion of cold air over relatively warm waters. These cold air masses originate from Western, Central or Eastern Europe and Scandinavia. Usually, the cold outbreak is associated either with cyclogenesis or with the rejuvenation of dissipating lows moving into the area. Anticyclonic circulation during winter is associated with a cold core anticyclone laying over Central Europe and/or the Mediterranean. This high-pressure system has a relatively long duration (two to four weeks) and is associated with weak northerly flow in the Mediterranean.

The summertime Mediterranean basin is directly under the descending branch of the Hadley circulation, driven by deep convection in the Inter-Tropical Convergence Zone. Owing to cloud-free conditions and high solar radiation intensity, the region is particularly sensitive to photochemical air pollution (Millan et al, 1997; Kouvarakis et al, 2000). Indeed, concentrations of important trace gases and aerosols are typically 2 to 20 times higher over the Mediterranean than in the hemispheric background troposphere, i.e. over the North Pacific Ocean in summer (Lelieveld et al., 2002). In summer, a strong east-west surface pressure difference (>20 hPa) is generated between the Azores high and Asian monsoon low-pressure regimes. This quasi-permanent weather systems cause northerly flow in the lower troposphere over the Mediterranean, feeding into the trade winds further south. The flow is strongest and most persistent in August. In the free troposphere, on the other hand, westerly winds prevails. In the upper troposphere, over the eastern Mediterranean, the flow is additionally influenced by the extended anticyclone centred over the Tibetan Plateau.

During the transient seasons, the weather changes between summer and winter type. This change occurs rapidly during spring and more slowly during autumn. Because of these complicated flow patterns in the Mediterranean Region air pollutants released from various sources located in the surrounding areas can be transported over long distances, in a complex fashion.

EMISSIONS TO THE ATMOSPHERE

A number of attempts have been made to estimate the regional emissions of mercury from natural and anthropogenic sources. Recent studies have shown that natural sources in the Mediterranean region are significant, if not the most important, source of mercury released annually to the atmosphere.

Natural Sources

Natural sources include volcanoes, evaporation from soil and water surfaces, degradation of minerals and forest fires. Mercury in small, but varying concentrations can be found virtually in all geological media. Elemental and some forms of oxidized mercury are permanently coming to the atmosphere due to their volatility. High temperature in the Earth mantle results in high mercury mobility and mercury continuously diffuses to the surface. In the zones of deep geological fractures these processes go on more intensively. In the Mediterranean region are located the so-called mercury geochemical belts where mercury concentrations in the upper layer appreciably exceed their average values. In some parts of mercury belts the intensive accumulation of mercury resulted in the formation of (extractable) deposits (Jonasson and Boyle, 1971; Bailey et al., 1973). Regions with high concentrations in surface rocks are characterized by high mercury emissions to the atmosphere. A first assessment of mercury emissions from two major natural sources in the Mediterranean basin which includes the contribution from volcanoes and seawater, has been considered comparable (nearly 110 t yr-1) with that estimated for major anthropogenic sources (nearly 105 t yr^{-1}) in the region (Pirrone et al., 2001a). Recent measurements of mercury distributions in the atmosphere (Pirrone et al., 2003a, Sprovieri et al., 2003), in surface and deep seawaters (e.g., Ferrara et al., 2001; Horvat et al., 2003; Gårdfeldt et al., 2003b) and calderas and volcanoes (Bacci, 1989; Ferrara et al., 1998) allowed to revise the previous estimates published by Pirrone et al. (2001a).

Emissions from Volcanoes. Table-1 reports available measurements for major Mediterranean volcanoes and calderas. Mercury emissions from the three major volcanoes and abandoned mercury mines in the region, although considered an important source of mercury on a local scale, do not represent an important source in the regional atmospheric budget. Emissions from Etna, Vulcano and Stromboli were measured by Ferrara *et al.* (2000a) and Edner et al. (1994) using LIDAR (LIght Detection And Ranging) and manual (gold traps followed by the Gardis mercury analyser) techniques. The emission of mercury from these three volcanoes was estimated using an Hg/SO$_2$ ratio of $1.5 *10^{-7}$ that led to an annual mercury emission estimate from these three volcanoes in the range of 0.6-1.3 tons yr^{-1} (Ferrara et al., 2000a) which is well below one percent of that derived for other point and diffuse sources in the region. Differences between degassing and explosion activity should be remarked and an Hg/SO$_2$ ratio of $0.3-2 *10^{-4}$ is adopted in the latter case (Pyle and Mather, 2003). Anomalous levels of mercury have been observed near cinnabar deposits. A well known site characterised by intensive geothermal activities is the Monte Amiata, located in Tuscany, Italy where high levels of mercury emissions and ambient concentrations have been observed (Ferrara et al., 1998; Bacci, 2000). Mercury is emitted from volcanoes primarily as gaseous Hg^0 with a concentration around 1-10 mg m^{-3} in the flue gases from which an emission rate of 300-400 g h^{-1} is estimated (Bacci, 2000).

Also mercury emissions from calderas may represent an important natural source of mercury. The Phlegrean fields (Pozzuoli, Italy) have been monitored with a LIDAR system and fluxes of Hg associated as Hg-S complexes were in the range of 0.9 to 4.5 g day^{-1} (Ferrara et al., 1998).

Concentrations in the condensed water fluxes reported by Ferrara et al. (1998) lie in the range 2-690 ng m^{-3}.

Table 1. Mercury emission rates of volcanoes and calderas in the Mediterranean region.

Country	Name	Type	Last eruption	Degassing with no eruption (10^{-5} t day^{-1})	Degassing during eruption (10^{-5} t day^{-1})	Reference
Greece	Santorini	shield	1950			SNMH, 2004
Greece	Nisyros	strato	1888			SNMH, 2004
Italy	Etna	shield	2003	16.9-147	5753	Ferrara et al., 2000a
Italy	Stromboli	strato	2003	2-21		Ferrara et al., 2000a
Italy	Vesuvio	complex	1944			SNMH, 2004
Italy	Vulcano	strato	1890	0.4-1.5		Ferrara et al., 2000a
Italy	Campi flegrei	caldera	1538	0.09-0.4		Ferrara et al., 1994
Italy	Ischia	complex	1302			SNMH, 2004
Italy	Lipari	strato	729			SNMH, 2004
Italy	Mar Sicilia	submarine	1911			SNMH, 2004
Italy	Pantelleria	shield	1891			SNMH, 2004
Turkey	Nemrut Dagi	strato	1441			SNMH, 2004

Emissions from Seawater. Several studies suggest that the evasion of elemental mercury from surface water is primarily driven by (1) the concentration of mercury in the surface water, (2) solar irradiation which is responsible for the photo-reduction of oxidised mercury available in the top-water microlayer, and (3) the temperature of the top-water microlayer and air above the surface water (air-water interface). Mercury emission rates from surface water have been measured in different areas of Europe and in different seasons using the floating chamber technique (Schroeder et al., 1989; Xiao et al., 1991; Cossa et al., 1996; Ferrara et al., 2000b) and/or derived from DGM measurements (Gårdeldt et al., 2003b; Horvat et al., 2003). The evasion of mercury from lake surfaces (see Table-2) is generally higher than that observed over the sea. Average emission rates in the North Sea were in the range 1.6 to 2.5 ng m^{-2} h^{-1} (Cossa et al., 1996 and ref. therein), whereas higher values (5.8 ng m^{-2} h^{-1}) have been observed in the Scheldt outer estuary (Belgium) and over lakes in Sweden (up to 20.5 ng m^{-2} h^{-1}). In the Mediterranean Sea mercury emission rates were in the range 0.7 to 10.1 ng m^{-2} h^{-1} in unpolluted areas (dissolved mercury concentration in the top water microlayer_was 6.3 ng L^{-1}) and between 2.4 and 11.25 ng m^{-2} h^{-1} in polluted coastal areas (dissolved mercury concentration in the top water microlayer was 15 ng L^{-1}) close to a chlor-alkali plant (Ferrara et al., 2000b). In open sea, mercury emission rates were much lower (1.16-2.5 ng m^{-2} h^{-1}) and less variable between day and night, though the dissolved mercury concentration in the top water microlayer (6.0 ng L^{-1}) was very similar to that observed in unpolluted coastal areas.

Recent studies carried out in the Mediterranean Sea aboard the Research Vessel Urania have provided for the first time spatial and temporal distribution of mercury evasions from the Mediterranean seawater (Pirrone et al., 2003a). A two-layer air-water exchange model was applied to estimate the flux of mercury (elemental) released from the seawater with changing ambient concentrations in air, dissolved gaseous mercury (DGM) concentrations and temperature of the top seawater layer and ambient air at the interface. The estimates of mercury fluxes from the top seawater were based on continuous atmospheric and water measurements of total and elemental gaseous mercury (in air) and DGM in the top layer of seawater (Pirrone et al., 2003a; Horvat et al., 2003; Gårdeldt et al., 2003b). In order to evaluate the spatial distribution of mercury evasional fluxes from the Mediterranean seawater, DGM concentrations measured in the top water microlayer and deep seawater along 6000 km cruise path aboard the R.V. Urania (Horvat et al., 2003; Gårdeldt et al., 2003b) have been used as input, along with other ancillary data, to the integrated mercury modelling system, discussed later in this chapter, to estimate spatial and temporal distribution of Hg$^0_{(g)}$ evasional fluxes from the Mediterranean sea.

Emissions from Soil and Vegetation. Mercury emissions from top soils and vegetation is significantly influenced by historical atmospheric deposition, type of vegetation and soil and meteorological conditions. Mercury exchange fluxes at the soil-air interface are measured with dynamic flux chamber or micrometeorological methods (Lindberg, 2002b).

Mercury fluxes from unaltered or background sites in North America have been observed in the range of -3.7 to 9.3 ng m^{-1} hr^{-1}, while in altered sites the flux ranges from -15.4 to 3334 ng m^{-1} hr^{-1} (Nacht and Gustin, 2004), unfortunately no high quality data are available for Mediterranean soil and vegetation. The total mercury concentration observed in conifer sap flow is around 12.3-13.5 ng l^{-1} (Bishop et al., 1998) while the uptake of ground vegetation shows highest accumulation values in roots (82-88%) followed by rhizome (8-17%) and leaf (0.03-4%), highlighting the barrier function for the transport of inorganic Hg (Cavallini et al., 1999; Patra and Sharma, 2000; Schwesig and Krebs, 2003). Foliar Hg flux up to 20 ng m^{-2} h^{-1} have been

Table 2. Mercury emission rates (ng m^{-2} h^{-1}) from surface waters of the Mediterranean Sea. (revised after Pirrone et al. 2003a).

Country	Method	Period		Emission Rate (ng m^{-2}h^{-1})	References
North-Western Mediterranean	N.A.	N.A.		1.16	Cossa et al., 1996 and reference herein
Northern-Tyrrenian Sea (Poll. Coastal zone)	FC	Summer	Day	11.25	Ferrara et al., 2000b
			Night	2.4	
Northern-Tyrrenian Sea (Off-shore)	FC	Summer	Day	2.5	Ferrara et al., 2000b
			Night	1.16	
Northern-Tyrrenian Sea (Unpoll. Coastal zone)	FC	Summer	Day	up to 10.1	Ferrara et al., 2000b
			Night	1	
		Winter	Day	0.7-2.0	
Eastern Mediterranean	FC	Summer		7.9	Gårdeldt et al., 2003b
Strait of Sicily	FC	Summer	Day	2.3	Gårdeldt et al., 2003b
			Night	40.5	
Tyrrenhian Sea	FC	Summer		4.2	Gårdeldt et al., 2003b
Western Mediterranean Sea	FC	Summer		2.5	Gårdeldt et al., 2003b
Tyrrenian Sea (Unpoll. Coastal zone)	FC	Summer		2.8	Ferrara et al., 2000b
Tyrrenian Sea (Off-shore)	FC	Summer		1.8	Ferrara et al., 2000b
Tyrrenian Sea near shore sites round Sardinia	FC	Summer		3.8	Gårdeldt et al., 2003b
Mediterranean Sea (Poll. Coastal zone)	FC	Summer		6.8	Ferrara et al., 2000b
Mediterranean Sea (Unpoll. Coastal zone)	FC	Summer		2.16	Ferrara et al., 2000b
Mediterranean Sea (offshore)	FC	Summer		1.83	Ferrara et al., 2000b

observed and it is a function of soil and air Hg concentrations and depends on light conditions (Ericksen and Gustin, 2004).

Emissions from Biomass Burning. Mercury emissions from biomass burning is not often well considered in regional emissions estimates, however, in very dry regions, such as the south of Mediterranean and several regions of Africa it may represent an important contribution to the global atmospheric mercury budget (Veiga et al., 1994; Carvalho et al., 1998; Friedli et al., 2001; Friedli et al., 2003). Mercury in vegetation originate from several sources, including uptake from the atmosphere, foliar accumulation of depositions, uptake from roots (Rea et al., 2002), but also the proximity to natural or anthropogenic source may change Hg content in vegetation (Lodenius, 1998; Carballeira, and Fernandez, 2002; Lodenius et al., 2003). Atmospheric gaseous mercury is taken up and released by the stomata activity (Browne, 1978 in Rea, 2002; Mosbaek, 1988 in Rea, 2002), and is thus closely linked with the transpiration of plants. All three major forms of atmospheric mercury (Hg^0, Hg^{II} species, $Hg_{(p)}$) contribute to mercury accumulation in vegetation (Ericksen et al., 2003; Ericksen and Gustin, 2004). Dissolved Hg is transported through the xylem sap from the roots to the foliage but it contributes with a small amount. Indeed, Hg in plant roots relate to soil Hg concentration and most remains in the root zone (Bishop et al., 1998, Ericksen and Gustin 2004). Different contents of mercury may be found in different tissues, predominantly leaves, bark and root, in addition forest typologies as well as contaminated sites proximity may change Hg concentration (Table-3). Moreover, the foliar age and the species physiology contribute to the final Hg amount in vegetation (Rea et al., 1996, Schwesig and Krebs, 2003), while the phenological cycle of the plant affects its retention because of deciduous plants transfer accumulated mercury to soil through litterfall. The initial vegetation conditions are determinant for mercury emissions from burning biomass but also the typology of fire constrains Hg content in smokes.

In particular, crown fires contribute smaller that litter fires because of the highest concentration reaching up to 71 ng g^{-1}. Mercury emitted from burning biomass reaches an elevated percentage (97-99%) leading to a complete transfer of Hg from vegetation to atmosphere.

Under the above conditions emission factors for wildfires amount up to 112±17 µg kg^{-1} (Friedli et al., 2001).

Table 3. Mercury content (ng g^{-1}) in live vegetation.

Forest Typology	Soil Hg	Hg foliar concentration (ng g^{-1})	References
Deciduous	Unimpacted	52.3	Lindberg, 1996
Deciduous	Unimpacted	3.3-28.8	Lindberg, 1996
Deciduous	Low	0.24	Eriksen et al., 2003
Deciduous	High	22-138	Eriksen et al., 2003
Deciduous	Contaminated	29-38.3	Friedli et al., 2001
Coniferous	Unimpacted	20-65.5	Rasmussen, 1994
Coniferous	Low	13.9-30.1	Friedli et al., 2001
Coniferous	High	58.7	Friedli et al., 2001
Vegetation	Contaminated	10^5-10^6	Fisher et al, 1995
Litterfall		39.7-52.9	Amirbahman et al., 2004

The global emission should then be accounted considering the forest biomas s distribution and consistency of fires. In Table-4 are reported trends of burnt surface for European Countries since 1990. Data of biomass are reported in the Forest Resource Assessment (FAO, 2000) while burnt surface is derived from the Global Forest Fire Assessment 1990-2000 (FAO, 2001), integrat ed with European Forest Fires Information System (Barbosa et al., 2003). Despite data on burnt surface are incomplete and not congruent, and biomass estimation revels the limits of a mean value, emission trends for EU-25 is estimated between 0.9 and 3.6 tonnes per year (Figure-1).

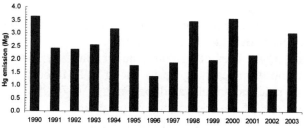

Figure 1. Annual mercury emission from forest fires in EU-25 Countries including Norway and Switzerland.

Anthropogenic Sources

Emissions from Industrial Sources. Mediterranean mercury emissions (2000 estimates) from fossil fuels combustion in electric power plants and industrial facilities, cement production and incineration of municipal solid wastes represent the major emission sources of mercury released annually, nearly 82% of the regional total (Pirrone et al., 2001a). Cement manufacturing represents the major single anthropogenic source in the region (~25% of the total), whereas fossil fuels combustion and incineration of municipal solid waste account for about 30% and 27% (see above, both are greater than 25 %) of the regional total, respectively. Seventy percent of the emission from coal combustion is from electric utility boilers, whereas the remaining fraction is from coal combustion for other industrial uses. Primary and secondary smelters, iron-steel production and a number of minor sources in the miscellaneous category represent less then 15% of the regional total. On a country-by-country basis the emission sources pattern may be different from that observed in the region as a whole (Pirrone et al., 2001a).

The incineration of solid wastes is the leading anthropogenic source of mercury in France (51% of the national total), due to the high annual rate (over 30% in 1995) of solid wastes disposed of through incineration facilities, whereas cement production represents the major single source in Italy (36.4%) and Morocco (53.1%).

Coal combustion in electric power plants and industrial facilities is the dominant source of mercury released annually in Turkey (44.6%) and in the former Yugoslavia (46.1%), while oil combustion is an important source in France, Italy and Spain. The regional distribution in the Mediterranean region for the 2000 suggests that about 57% of the regional total is due to emissions from the European Union (EU-15) countries. France is the leading Hg emitter with 21.4% of the regional total followed by Turkey (15.2%), Italy (10.8%), Spain (8.6%), former Yugoslavia (7.4%), Morocco (6.5%), Bulgaria (6.4%), Egypt (5.8%), Syria (3.4%), Libya (2.8%), Tunisia (2.6%), Greece (2.5%), Algeria (1.9%), Jordan (1,5%), Lebanon (1.4%), Israel (0.8%), Cyprus (0.8%) and Albania (0.2%) (Pirrone et al., 2001).

Table 4. Annual mercury emissions (kg) from forest fires in European Countries.

Country	Biomass (t ha⁻¹)	1990	1991	1992	1993	1994	1995	1996	1997	1998	1999	2000	2001	2002	2003	2004
Austria	250	5.6	1.5	3.7	3.1	1.6	0.9	0.8	0.6	2.6	0.2	1.2	-	-	-	-
Belgium	101	0.2	0.6	0.2	1.2	0.6	0.7	16.4	3.2	0.3	0.0	0.0	0.0	-	-	-
Cyprus	21	3.4	0.3	0.0	0.2	0.4	0.2	0.3	0.4	1.3	0.0	18.9	11.4	5.2	5.5	-
Czech Republic	125	10.9	1.1	17.9	16.1	11.3	5.6	28.6	48.7	15.8	4.7	5.3	8.6	-	-	-
Denmark	58	0.9	0.9	1.8	0.1	0.0	0.0	0.4	0.1	0.2	0.0	0.0	0.0	-	-	-
Estonia	85	8.7	0.6	17.0	6.2	4.3	1.5	5.0	10.6	0.5	10.5	6.5	5.8	-	-	-
Finland	49	2.4	1.2	5.9	3.2	8.6	3.5	4.5	8.1	1.1	5.8	5.0	0.9	3.1	7.5	-
France	92	748.3	104.4	171.1	172.0	257.6	186.9	117.4	222.4	198.7	163.5	210.8	210.9	64.9	633.8	
Germany	134	14.2	13.8	73.7	22.4	16.7	8.9	20.7	9.0	6.0	6.2	8.7	1.8	1.8	19.7	-
Greece	25	108.1	36.5	199.9	151.3	162.1	76.2	70.9	143.3	315.8	53.3	467.6	51.4	112.7	9.5	-
Hungary	112	16.9	16.9	16.9	16.9	16.9	16.9	16.9	16.9	16.9	0.0	0.0	16.9	-	-	-
Ireland	25	2.4	1.9	1.6	1.5	0.8	1.3	1.3	1.3	1.3	1.3	1.3	1.3	-	-	-
Italy	74	1574.3	819.2	860.7	1617.6	1129.9	405.2	480.6	921.9	1289.2	747.0	1163.5	633.4	167.6	365.2	138.8
Latvia	93	24.6	32.2	87.1	5.3	3.6	5.6	9.7	6.3	2.2	16.1	14.0	10.7	-	-	-
Lithuania	99	4.6	0.6	9.6	3.4	3.3	3.6	4.3	1.5	1.0	2.4	2.4	1.2	-	-	-
Netherlands	107	2.7	5.0	2.1	1.2	3.3	2.8	1.9	2.7	2.5	2.5	2.5	2.5	-	-	-
Norway	49	0.5	2.9	7.5	1.2	1.3	0.6	2.8	4.8	1.6	4.7	9.4	5.3	-	-	-
Poland	94	52.9	22.2	350.9	38.7	26.3	18.3	55.9	22.9	614.7	432.7	516.1	521.2	-	-	-
Portugal	33	479.9	674.5	218.3	184.7	285.8	626.9	328.5	112.9	585.3	261.0	589.9	413.3	433.5	1559.1	127.2
Slovakia	142	3.1	3.7	3.7	8.2	1.5	1.4	3.5	0.6	0.5	8.9	14.4	4.9	9.5	24.9	-
Slovenia	178	0.7	10.5	7.8	34.7	17.2	3.9	4.9	9.8	25.0	8.6	5.3	13.0	-	-	-
Spain	24	548.5	657.8	281.1	240.1	1176.4	385.7	160.8	264.8	359.2	221.0	502.7	248.3	67.7	401.1	123.8
Sweden	63	-	-	41.0	7.1	17.5	2.8	15.4	45.1	21.1	21.1	21.1	8.8	-	-	-
Switzerland	165	20.4	2.7	1.0	0.8	5.4	8.1	4.3	35.7	4.6	0.4	1.2	2.1	-	-	-
United Kingdom	76	3.9	1.0	1.7	1.3	8.8	4.6	5.0	2.8	0.5	1.5	2.3	1.9	-	-	-
Sum		13990	9760	10218	10751	8289	5086	16315	9093	37733	8491	17519	11044	937	3085	390

THE SPECIATION OF ATMOSPHERIC MERCURY AT COASTAL SITES AND OVER THE OPEN SEA

In order to assess the relative contributions of different patterns affecting the cycle of mercury in the Mediterranean region, in the last six years a significant effort has been addressed to evaluate the levels of mercury species at coastal sites and at open sea locations.

Figure 2. Atmospheric mercury measurement sites in coastal areas and cruise paths of three oceanographic campaigns carried out in 2000, 2003 and 2004.

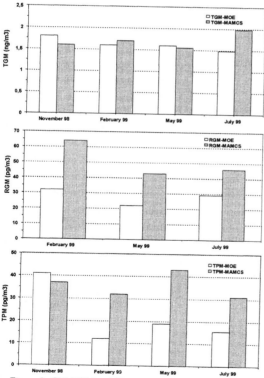

Figure 3. GM, RGM and TPM concentrations at costal sites of Mediterranean region and North Europe (in Wangberg et al., 2001).

framework of several national and European funded projects several intensive field campaigns have been carried out at coastal sites at in open sea, the aim was to determine the ambient concentration of atmospheric mercury species along with other ancillary atmospheric parameters (i.e., trace elements, NO_x, SO_2, O_3, surface and deep seawater) (Figure 2) (Pirrone et al., 2003b). Measurements carried out at coastal sites in the framework of MAMCS during the 1998-1999 period showed for the first time that ambient concentrations of oxidised forms of mercury and particulate mercury were generally higher than that observed at background sites in North Europe (Figure 3) (Wangberg et al., 2001; Munthe et al., 2003; Pirrone et al., 2003b; Sprovieri et al., 2003). These findings suggested to pay more attention to all those mechanisms, not yet considered at that time, that could be responsible for the oxidation of elemental mercury in the MBL which were believed to be driven primarily by atmospheric conditions (fog or clear sky, summer or winter periods), temporal variations of other atmospheric pollutants (e.g., O_3, Br and Cl compounds, H_2O_2) and re-emissions (volatilisation) of $Hg^0_{(g)}$ from surface seawater.

In the summer of 2000 an oceanographic campaign took place (the first MED-OCENAOR project, the cruise path is shown in Figure 1, an ongoing series of projects aboard the R. V. Urania). The aim was (and is) to perform integrated atmospheric and water measurements in order to make a first assessment of spatial distributions

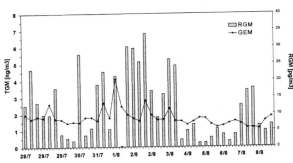

Figure 4. TGM and RGM concentrations observed along 6000 km cruise path over the Mediterranean Sea (in Sprovieri et al., 2003).

of mercury compounds in ambient air, emissions rates (possibly exchange fluxes) from seawater, chemical and physical characterisations of mercury in dissolved and particulate phase in surface and deep seawater as well as in sediment samples (see papers by Horvat et al., 2003, Ferrara et al., 2003 , Gårdeldt et al., 2003; Sprovieri et al., 2003. For the first time (Figure 4) RGM was observed at open sea locations far from the influence of continental sources with ambient concentrations in the range of 5 to 32 pg m$^-$^{3} (well above the detection limit of the measurement method used); similar behaviour was observed during the following cruises performed in the summer of 2003 and winter 2004 (Figure 2). Similar observations made at a coastal sites during MAMCS (1998-1999) and MERCYMS (2003-2004) campaigns where significant RGM levels were seen when air masses originated from open sea sectors.

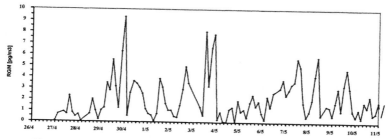

Figure 5. RGM concentrations observed at a coastal site, in Fuscaldo,
Italy in Jan-Feb 2004 (Pirrone, 2004: unpublished data).

Figure 5 shows recent measurements of RGM performed in January-February 2004 at a coastal site in the South of Italy (Pirrone, unpublished data). Ambient concentrations of RGM followed a diurnal pattern with highest values at midday and lowest in the night, which are in good agreement with that obtained during previous intensive campaigns carried out at the same location and at open sea (Pirrone et al., 2003a; Sprovieri et al., 2003; Wangberg et al., 2001).

36 hour-backward trajectories (Figure-6) allow the origin of air masses crossing the monitoring site to be determined. Considering the recent new estimate of lifetime of $Hg^0_{(g)}$ in the MBL by Hedgecock and Pirrone (2004), the good agreement between modelled and measured RGM concentrations in the MBL obtained by Hedgecock and Pirrone (2001) and Hedgecock et al. (2003), and the

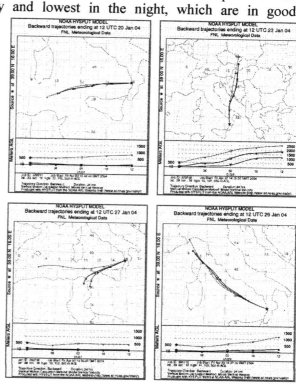

Figure 6. 36 hour backward trajectories
for the Fuscaldo site, Italy.

incidence of higher RGM concentrations with air masses that have spent a longer time over the sea goes a long way to confirming the earlier hypothesis that the *in-situ* oxidation of $Hg^0_{(g)}$ may represent a source of reactive mercury present in the MBL. The RGM produced is much more efficiently deposited to surface water by wet scavenging and precipitation events than $Hg^0_{(g)}$. All these findings and improvements in understanding the chemical and physical processed of mercury in the MBL have allowed the development of an Integrated Mercury Modelling System which is presented in the following section of this chapter.

REGIONAL ATMOSPHERIC MERCURY MODELLING

The Integrated Modelling System

The integrated mercury model development has been performed within the atmospheric meteorological/dispersion model RAMS. The resulting framework permits the incorporation of almost any type of source (point or diffuse source), gas and aqueous phase chemistry, altitude and cloud cover dependent photolysis rate calculation, wet and dry deposition and air-water exchange processes. RAMS has a number of features which make it extremely useful for air quality studies. It has numerous options for the boundary conditions, two-way interactive grid nesting capabilities, terrain following coordinate surfaces and non-hydrostatic time-split time differencing, a detailed cloud microphysics parameterization, various turbulence parameterization schemes, radiative transfer parameterizations (short and long wave) through clear and cloudy atmospheres, a detailed surface-layer parameterization (soil, vegetation type, lakes and seas, etc). The RAMS code is currently released under GPL (General Public Licence) by ATMET (Atmospheric, Meteorological and Environmental Technologies). In particular, the representation of cloud and precipitation microphysics in RAMS includes the treatment of each water species (cloud water, rain, pristine ice, snow, aggregates, graupel, hail) as a generalized Gamma distribution (see Pielke et al.,1992; Walko et al., 1995 and Meyers, 1997). The scheme allows hail to contain liquid water and contains the description of homogeneous and heterogeneous ice nucleation, and ice size change by means of vapour deposition and sublimation.

The surface heterogeneities connected to vegetation cover and land use are assimilated and described in detail in RAMS by means of the LEAF-2

(Land Ecosystem Atmosphere Feedback version 2) model (Walko et al., 2000). This model represents the storage and vertical exchange of water and energy in multiple soil layers, including the effects of freezing and thawing soil, temporary surface water or snow-cover, vegetation, and canopy air. The surface domain meshes are further subdivided into patches, each with a different vegetation or land surface type, soil textural class, and wetness index to represent natural sub-grid variability in surface characteristics. Each patch contains separate prognosed values of energy and moisture in soil, surface water, vegetation, and canopy air, and calculates exchange with the overlying atmosphere weighted by the fractional area of each patch. The LEAF model assimilates standard land use datasets to define the prevailing land cover in each grid mesh and possibly the patches, then parameterizes the vegetation effects by means of biophysical quantities.

MPI (Message Passing Interface) based parallel processing is implemented in RAMS by the method of domain decomposition. Dynamic load balancing is available where computational load differs between subdomains of a grid, or if computer nodes differ in computing capacity and, even with the extra load from the mercury processes modules, RAMS results in very good parallel efficiency.

It is worth noting that two relevant reasons for using RAMS are its full microphysical parameterization for wet processes, which greatly influence the wet mercury chemistry and deposition, and the detailed parameterization of surface processes which aids proper descriptions of the mercury air-surface exchange processes.

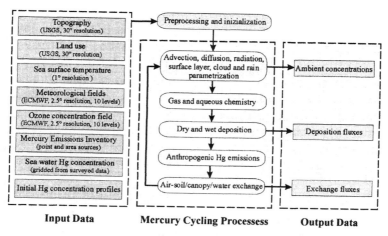

Input Data   **Mercury Cycling Processess**   **Output Data**

Figure 7. Flow diagram highlighting major components of the integrated mercury modeling system used for the Mediterranean region.

Model Components: RAMS allows a desired number of prognostic scalar fields to be added to the model simulation, which are then automatically advected and diffused forward in time. In the present version of the model four additional scalars have been added, $Hg^0_{(g)}$, $Hg^{II}_{(g)}$, Hg^P and $Hg^{II}_{(aq)}$. $Hg^{II}_{(aq)}$ represents the sum of oxidised Hg species in the aqueous phase, and in the absence of liquid water, for instance, after the evaporation of non-precipitating cloud droplets it is transported as if it were Hg^P. However, unlike Hg^P, it is considered soluble so that should it once again be in the presence of liquid water it is assumed to be scavenged and returns to the aqueous phase. Hg^P will in time be replaced by Hg^P and Hg_{ads} where Hg^P will represent Hg irreversibly bound to particulate matter, and Hg_{ads} Hg which is reversibly adsorbed (Pirrone et al., 2000). In order to model accurately the dynamics of gaseous and particulate mercury, a number of *ad-hoc* modules were developed and coupled to RAMS. These modules account for major processes (Figure 7) that affect the fate of mercury in the atmosphere:

- natural and anthropogenic mercury emissions;
- dry and wet deposition;
- chemical transformation;
- air-water exchange;
- air-soil exchange;
- canopy emissions.

Model Input/Output. Standard model inputs are the USGS global land use data at 30'', or about 1 km resolution (based on 1-km Advanced Very High Resolution Radiometer data spanning April 1992 through March 1993); the global USGS topography dataset at 30'', or about 1km resolution; the global monthly climatological sea surface temperature data at 1 degree resolution (about 100 km). The European Centre for Medium-Range Weather Forecasting (ECMWF) meteorological data have been used. The data is available at 6-hourly intervals, with a spatial resolution of 2.5°×2.5° latitude and longitude, and 10 pressure levels (100–1000 hPa). Horizontal winds and temperature were nudged with the lateral nudging method implemented in RAMS. The database for anthropogenic mercury emissions in Europe employed in the model calculations has been compiled for 1995 (Pacyna et al., 2001) and 2000 (Pacyna et al., 2003). The Hg emission inventory provides annually averaged emissions from point and area sources. The emission inventory also provides information on Hg speciation (i.e. the proportions of Hg^0, Hg^{II} and Hg^P, Hg associated with particulate material, depending on emission source category).

Given the importance of Hg oxidation by OH, the production of OH needs to be well described as does the O_3 concentration. Therefore the initialisation of RAMS has been adapted to include 6 hourly O_3 concentration fields from ECMWF. In order to improve air-water exchange modelling, the gridded sea water Hg concentration in the Mediterranean Sea (surveyed during the MED-OCEANOR 2000 cruise) has been included in the model.

Model output includes the gas phase concentration fields of Hg^0, Hg^{II} and Hg^P (total particulate Hg, soluble and insoluble), dry deposition velocities, emissions and deposition fluxes, and the concentration of Hg^{II} in rain water.

Chemical Processes in the Atmosphere

The Hg atmospheric chemistry model currently used in this initial version of the integrated modelling system has been designed specifically for the purpose and includes both gaseous and aqueous phase chemistry relevant to Hg and its oxidation and deposition. Oxidation is fundamental to Hg deposition, both dry and wet, as RGM species are much less volatile and much more soluble than Hg^0. Given the importance of Hg oxidation by OH, the production of OH needs to be well described and therefore so does the O_3 concentration. This is particularly true in the case of the Mediterranean where modelling studies suggest that it is oxidation by OH rather than Br - modelling studies suggest that Br rather than Br_2, Cl_2 or Cl is the major oxidant in the remote MBL – which accounts for the major part of the conversion of $Hg^0_{(g)}$ to $Hg^{II}_{(g)}$ even in the MBL (Hedgecock and Pirrone, 2004; Hedgecock et al., 2004). One of the major reasons for this is the high O_3 concentrations experienced all over the Mediterranean Basin, especially in the summer (see Lelieveld et al., 2002). Ozone fields from the European Centre for Medium range Weather Forecasting (ECMWF) have been included in the input to RAMS, and an on line photolysis rate constant calculation procedure is used. The program fast-j (Wild et al., 2000) has been linked to the meteorological model in order to provide photolysis rate constants which are a function of latitude, longitude, date, time of day, altitude and column optical depth. In order to reduce the calculation time required by the model, operator splitting between the meteorological model and the chemistry model has been employed. RAMS typically uses 90s time steps for the grid resolution chosen for these simulations, but the chemistry model is called every 15 minutes. The chemistry model is run for each cell, whilst fast-j is called for each column, the altitude dependent photolysis rate constants being stored in a temporary array. $Hg^0_{(g)}$, $Hg^{II}_{(g)}$, $Hg^{II}_{(aq)}$ and Hg^P are all transported by the model, other chemical species are not transported but their concentrations stored 'locally' for each cell after each chemistry time step. $Hg^{II}_{(aq)}$ represents the sum of oxidised Hg species in the aqueous phase, which in the absence of liquid water, for instance, after the evaporation of non-precipitating cloud droplets is transported as if it were Hg^P, thus is capable of being dry deposited.

$Hg^{II}_{(aq)}$, unlike Hg^P, which is emitted and considered irreversibly bound to the particulate matter, is considered soluble so that should it once again be in the presence of liquid water it is assumed to be scavenged and returns to the aqueous phase. $Hg^{II}_{(aq)}$ which partitions to soot is treated as $Hg^{II}_{(aq)}$. The non-Hg chemistry currently used in the preliminary simulations includes detailed HO_x and SO_x chemistry in both the gas and aqueous phases, in order to obtain reasonable values for the OH concentration and the cloud droplet pH. The inclusion of the chemistry and mass transfer model has been incorporated into the overall modelling framework in such a way that the chemistry model may be expanded or reduced, or updated with new reaction coefficients separately from the rest of the model. The database is then recompiled with the integrator and the resulting files are then recompiled together with RAMS, thus avoiding any necessity to manually alter the RAMS code to include new chemical species or reactions. This was considered to be of great importance as the model was conceived with the intention of testing numerous and various chemical modelling schemes.

In the near future the model will be tested including detailed NO_x emissions and chemistry. The inclusion of a VOC emissions database is also foreseen; chemistry models based on both the CBM-IV and SAPRC 99 mechanisms have already been prepared. The coupling of a parametrised sea salt aerosol production model to the integrated model will allow the inclusion of the detailed MBL photochemical model described in Chapter-13. The complexity of this model will necessarily have a major influence on calculation times, however the full MBL model will only be applicable over part of the modelling domain, and investigations are under way to refine the integrated model in order to use the most appropriate chemistry models in a given situation, using for example different chemical mechanisms for the MBL, urban/industrial boundary layer and remote continental boundary and possibly a separate chemistry mechanism again for the free troposphere. Such an approach would give the best possibility of ensuring accurate chemistry modelling whilst making the most efficient use of computing resources. This methodology has already been tested in the prototype integrated modelling system where two chemical mechanisms have been used, one for 'dry' situations and one for 'wet', the difference being the inclusion of aqueous phase chemistry if the model cell had a liquid water content above a threshold value. The method functions without problems, there is a slight time saving but the difference between the mechanisms is small in the sense that even using the 'wet' mechanism in a 'dry' cell means that the initial concentrations of the aqueous phase species is zero and the rate constants of aqueous phase reactions and mass transfer are all zero, thus the complexity of the problem is the same only the initialisation takes longer. However the feasibility of the approach has been demonstrated. The present version of the chemistry model contains over 100 gas and aqueous phase reactions and mass transfer between the gas and aqueous phase of 10 species. The Hg chemistry included in the model is summarised below:

Gas phase oxidation: The reactions of Hg with O_3, H_2O_2, OH, HCl and Cl_2 are included, the reaction with Br will be included in the MBL chemistry mechanism.

Aqueous phase oxidation: The oxidation reactions included are those with O_3, HOCl / OCl-, and OH.

Aqueous phase reduction: The reduction of Hg^{II} compounds to Hg^0 by reaction with HO_2 / O_2^-, are included as are the reduction of $HgSO_3$ and $(HgSO_3)_2^{2-}$. However the HO_2 / O_2^- reaction may be removed, (Gårdeldt and Jonsson, 2003).

Gas – Aqueous Phase Equilibrium: The mass transfer between the gas phase and cloud droplets of Hg^0, HgO, and $HgCl_2$ are considered.

Aqueous Equilibria: The equilibria between Hg^{2+}, and the OH^-, Cl^-, and SO_3^{2-}, are included. For Cl^- complexes up to $HgCl_4^{2-}$ are considered. Br^- complexes will be included for the MBL mechanism.

Aerosol – Liquid Equilibria: $soot_{(air)}$ – $soot_{(aq)}$ is accounted for according to the method used by Petersen *et al.*, (1995).

Aqueous – Soot Equilibria: The partitioning of Hg compounds between the aqueous phase and the solid phase within droplets ($Hg^{II}_{(soot)}$ – $Hg^{II}_{(aq)}$), is based on Petersen *et al.*, (1995).

Where the temperature dependence of a reaction or equilibrium constant is known it has been included in the chemical mechanism database and the value is recalculated at each time step, as are the photolysis rate constants. The rate and equilibrium constants of the Hg chemistry included in the model can be found in Table-I of Chapter-13. It should be mentioned that although the chemical mechanism includes the equilibria between soot in the aqueous and gas phases, and that between Hg^{II} in the aqueous phase and Hg^{II} associated with soot scavenged by the cloud droplets, neither of these is actually activated in these preliminary simulations, which have been performed without including the soot emission inventory.

Dry Deposition

Dry deposition of $Hg^0{}_{(g)}$. The low solubility and relatively high volatility of $Hg^0{}_{(g)}$, mean that dry deposition of elemental mercury is unlikely to be a significant pathway for removal of atmospheric mercury, although approximately 95% or more of atmospheric mercury is elemental mercury (Lindberg et al., 1992; Fitzgerald, 1995). Therefore as in a number of other modelling studies (Petersen et al., 1995; Pai et al., 1997, Bullock and Brehme, 2002) the $Hg^0{}_{(g)}$ dry deposition is assumed to be zero in the model. However, emission fluxes of Hg^0 from the ground are calculated as a net flux, that is emission minus deposition, see Air-Soil Exchange of $Hg^0{}_{(g)}$, below.

Dry Deposition of $Hg^{II}{}_{(g)}$. A surface and meteorological variable dependent deposition velocity is used to calculate the $Hg^{II}{}_{(g)}$ deposition flux. Because in RAMS, to represent natural sub-grid variability in surface characteristics, the surface grid cells are divided into sub-grid patches, each with a different vegetation or land surface type, the overall deposition velocity is computed as

$$\overline{v}_d = \sum_{i=1}^{np} v_i \, f_i$$

where *np* is the number of patches (which is specified during RAMS pre-processing phase) and f_i is the fractional index coverage of *i-th* patch (with $\sum_{i=1}^{np} f_i = 1$). The patch deposition velocity v_i is calculated using the resistance model (Wesely and Hicks, 1977) as:

$$v_i = (r_a + r_d + r_c)^{-1}$$

where r_a is the atmospheric resistance through the surface layer, r_d is the deposition layer resistance and r_c is the canopy or surface resistance.

The atmospheric resistance, which represents bulk transport by turbulent diffusion through the lowest 10 meters of the atmosphere, is obtained according to Wesely and Hicks (1977) as:

$$r_a = \frac{1}{k\,u_*}\left[\ln(z_s/z_0) - \varphi_H\right]$$

where z_s is the reference height (= 10 m), z_0 is the surface roughness length, k is the von Karman constant (≈ 0.4), u_* is the friction velocity, φ_H is a stability correction term which accounts for the effects of buoyancy on the eddy diffusivity of $Hg^{II}_{(g)}$. Note that both u_* and z_0 are directly provided by RAMS for each land patch.

The deposition layer resistance, which represents molecular diffusion through the lowest thin layer of air, is parameterized for $Hg^{II}_{(g)}$ in terms of the Schmidt number by the equation developed by Wesely and Hicks (1977):

$$r_d = \left(\frac{2}{k\,u_*}\right)\left(\frac{Sc}{Pr}\right)^{2/3}$$

where Pr is Prandtl number of air (≈ 0.72), $Sc = v/D_A$ is the Schmidt number, v is the kinematic viscosity of air ($\approx 0.15 \times 10^{-4}\,m^2 s^{-1}$) and D_A is the molecular diffusion coefficient of $Hg^{II}_{(g)}$ in air. Canopy or surface resistance over land is computed according to the Wesely's formulation (1989):

$$r_c = \left[(r_{st} + r_m)^{-1} + r_{uc}^{-1} + (r_{dc} + r_{lc})^{-1} + (r_{ac} + r_{gs})^{-1}\right]^{-1}$$

where $r_{st} + r_m$ is the leaf stomata and mesophyllic resistance, r_{uc} is the upper canopy resistance, r_{dc} is related to the gas phase transfer by buoyant convection in canopies, r_{lc} is the lower canopy resistance, r_{ac} is a resistance that depends on canopy height and density, and r_{gs} is the ground surface resistance. All the latter contributions are computed according to the Wesely's formulae as modified by Walmsley and Wesely (1996). Many of these resistances are season and landuse dependent and some are adjusted using solar radiation, moisture stress and surface wetness provided by the specific RAMS submodels. Some parameterizations have been developed by Wesely for SO_2 and O_3 and are scaled for $Hg^{II}_{(g)}$ based on the molecular diffusivity, Henry's law constant and chemical reactivity toward oxidation. As in Pai et al., (1997), for $Hg^{II}_{(g)}$ deposition properties we have assumed similarity with HNO_3 (whose properties are reported in Wesely, (1989)) because $HgCl_2$ and HNO_3 have a similar solubility.

Over water, the surface resistance r_c is assumed to be zero because of the high Henry's Law constant of $Hg^{II}_{(g)}$.

Dry deposition of $Hg_{(p)}$. Surface dry deposition velocity of particulate mercury $Hg_{(p)}$ is largely dependent on particle size. As in Pai et al. (1997) the particle size distribution is assumed to be log-normal with a geometric mass mean diameter of 0.3 µm and a geometric standard deviation of 1.5 µm. The deposition velocity is determined by dividing the distribution into a fixed number of size intervals, calculating the velocity v_i for each interval and aggregating them in a weighted mean. The resistance approach has been adopted for v_i calculation. Particulate matter does not interact with vegetation as gases do, and in particular, particles are usually assumed to stick to the vegetation surface (e.g. Voldner et al., 1986) and therefore $r_c \approx 0$. Thus using the parameterization of Slinn and Slinn (1980) and Pleim et al., (1984) we assume:

$$v_i = \frac{1}{r_a + r_b + r_a r_d v_g} + v_g$$

where $v_g \approx 0$ is the gravitational settling velocity, proportional to the square of particle diameter and which, as pointed out by Pai et al., (1997), is negligible for $Hg_{(p)}$; aerodynamic resistance r_a is identical to the value used for $Hg^{II}_{(g)}$ dry deposition, while the resistance to diffusion through the quasi-laminar sub-layer r_b depends on aerosol Brownian diffusion and inertial impaction. In particular we parameterize (e.g., Pleim et al., 1984) the deposition layer resistance in terms of the Schmidt number $Sc = v / D_p$, where v is the air viscosity and D_p is the Brownian diffusivity of the current class size of $Hg_{(p)}$ in air (neglecting, with respect to Pleim's parameterization, the small term related to the Stokes number):

$$r_d = Sc^{2/3} / u_*$$

The Brownian diffusivity is given by (Shimada *et al.*, 1993) $D_p = k\, T\, C_c / (3\pi\, \mu\, d_p)$, where k is the Boltzmann constant, T and μ are the air temperature and viscosity, respectively; C_c is the Cunningham correction factor (a function of the mean free path of air molecules and particle diameter d_p).

Wet Deposition

Wet deposition is an important removal process for both RGM and particulate mercury while, because of its low solubility, direct wet removal of $Hg^0_{(g)}$ is negligible, if compared with the first two contribution. Obviously the aqueous phase oxidation of $Hg^0_{(g)}$ is important here, and so therefore is the rate at which droplets take up $Hg^0_{(g)}$. This process is included as two reactions in the chemistry model describing uptake and out-gassing of $Hg^0_{(g)}$ and thereby representing the equilibrium between the gas and aqueous phases.

A relevant fraction of particulate mercury is removed by both in-cloud scavenging (rainout) and below-cloud scavenging (washout). The rate of particle wet removal depends upon ambient concentration, cloud type, rainfall rate and particle size distribution, but usually the $Hg_{(p)}$ wet deposition is determined by a synthetic scavenging efficiency coefficient from the ambient concentration.

Wet deposition of the insoluble fraction of particulate Hg is modelled assuming that the local depletion rate is proportional to the concentration C (mercury mass per unit volume of air), introducing a wet scavenging rate Λ, dependent on the precipitation intensity: $\partial C / \partial t = -\Lambda\, C$. The scavenging rate is used to reflect the propensity of mercury to be removed by the current precipitation, including all possible below-cloud and in-cloud processes. In a RAMS time step, assuming the scavenging coefficient to be constant over the time step, wet removal changes the cell concentration C, according to $\Delta C = C(1 - e^{-\Lambda \Delta t})$, where Δt is the time step size. Then, the wet deposition rate at the ground is simply the cumulative sum of the depletion occurring at all levels of the atmospheric column in which rain occurs. The scavenging rate is obtained by (Junge and Gustafson, 1957) $\Lambda = \varepsilon\, P_r /(\Delta z\, L_w)$, where ε denotes the scavenging efficiency, P_r is the precipitation rate, Δz is the cell height and L_w is the volume fraction of liquid water in clouds (defined as the ratio of the liquid water content of the cell and the water density). Both P_γ and L_w are provided for each time-step during the simulation by the RAMS microphysics submodel (Walko et al., 1995; Meyers et al., 1997).

The $Hg^{II}_{(aq)}$ cloud water concentration is directly obtained from the chemistry module and uses $\varepsilon = 1$ because $Hg^{II}_{(aq)}$ is already present in precipitating cloud water. For $Hg_{(p)}$ a value of 1 is currently assumed for scavenging efficiency as in Pai et al., (1997).

Canopy Emissions $Hg^0_{(g)}$

As pointed out by many researchers, (e.g. Lindberg et al., 1992; Hanson et al., 1995; Linderberg, 1996) in vegetated areas the Hg^0 dissolved in soil water is transported to leaves and emitted to the atmosphere via the transpiration stream. As in Xu et al. (1999) the elementary mercury plant emissions flux F_c (ng m^{-2} s^{-1}) is determined, assuming a constant Hg^0 concentration in the evapotranspiration stream (Linderberg, 1996), as $F_c = E_c C_s$ where E_c (m^3 H$_2$O m^{-2} s^{-1}) is the evapotranspiration rate and C_s (ng m^3-H2O) is the $Hg^0_{(g)}$ concentration in the soil water. The evapotranspiration rate E_c which in Xu et al. (1999) is determined by the Penman-Monteith equation, is here directly provided by the detailed LEAF-2 RAMS submodule (Walko et al., 2000). The latter is able to take into account correctly the type of vegetation in each cell and its minimum stomatal resistance. It also takes into account stomatal closure caused by excessively warm temperatures or cold (freezing) temperatures, lack of solar radiative flux (which is separately parameterized in RAMS), lack of water in the soil layers comprising the plant root zone and the canopy water vapour mixing ratio at the leaf surface which effects evaporation from the stomata. C_s is currently assumed to be 100 ng l^{-1} (Xu et al., 1999).

Air-Soil Exchange of $Hg^0_{(g)}$

$Hg^0_{(g)}$ natural soil emission is mainly dependent on soil temperature, soil moisture and solar radiation (Kim et al., 1995; Carpi and Lindberg, 1998). Here, as in Xu et al., (1999), the elementary mercury net flux from soil F_s (ng m^{-2} h^{-1}) is parameterized as a function of soil temperature T_s (°C) using the relation found by Carpi and Lindberg (1998) as

$$\log F_s = 0.057 \ T_s - 1.7$$

The latter relation empirically accounts for the net exchange rate (re-emission minus deposition) and it is worth noting that, differently from Xu et al., (1999), because the $Hg^0_{(g)}$ dry deposition is not explicitly simulated, the original regression constants can be used. During the simulation the soil state is checked and the $Hg^0_{(g)}$ emissions are disallowed in presence of frozen soil (Xu et al., 1999).

Air-Water Exchange of $Hg^0_{(g)}$

In modelling $Hg^0_{(g)}$ air-water exchange the film theory has been used for mass transfer calculations at the interface (Xu et al., 1999, 2000; Poissant et al., 2000; Mason et al., 2001; Lin and Tao 2003). The elementary mercury evasional flux F_w (ng m^{-2} s^{-1}) is driven by the fugacity difference between the overlying air and surface water:

$$F_w = K(C_w - \frac{C_g}{\bar{H}})$$

where K (m s^{-1}) is the overall mass transfer coefficient, C_w and C_g are respectively the Hg^0 concentration (ng m^{-3}) in water and air and $\bar{H}$ is the dimensionless inverse Henry's Law constant of Hg. The low solubility of $Hg^0_{(g)}$ means that most of the resistance to gas exchange lies in the water film and K can be expressed by just the liquid-phase transfer coefficient K_w. As in Lin and Tao (2003) for the latter mass transfer coefficient we adopt the approach by Poissant et al. (2000) in which K_w (cm h^{-1}) is correlated with the mass transfer of CO_2 across the air-water interface:

$$K_w = (0.45 \cdot u_{10}^{1.64})\left[Sc_w^{Hg^0} / Sc_w^{CO_2}\right]^{-0.5}$$

where Sc is the Schmidt number in water and u_{10} is the wind speed (m s^{-1}) at 10 m which is derived from RAMS. The Schmidt number of CO_2 was calculated using the temperature-corrected dependency (Poissant et al., 2000):

$$Sc_w^{CO_2} = 0.11\ T^2 - 6.16\ T + 644.7$$

where T is in $°C$. The Schmidt number of $Hg^0_{(g)}$ was calculated using its definition $Sc = \nu / D$, where for the kinematic viscosity of water (cm^2 s^{-1}) and $Hg^0_{(g)}$ diffusivity in water (cm^2 s^{-1}) the following temperature ($°C$) – dependent parameterisations of Thibodeaux (1996) and Kim and Fitzgerald (1986), respectively, are used:

$$\nu = 0.017\ e^{-0.025\ T}$$
$$D = 6.0 \times 10^{-7} T + 10^{-5}$$

The dimensionless inverse Henry's Law constant is calculated following Sanemasa's expression (1975) as a function of temperature (K):

$$\log \bar{H} = -1078/T - \log(T) + 5.592$$

While C_g is simulated by the model, a mean value of 0.04 ng l^1 is assumed for C_w (as in Xu et al., 1999).

Modelling Applications

The Figure 8 show the balance between deposition (dry deposition of $Hg^{II}{}_{(g)}$ and primary $Hg_{(p)}$ and $Hg_{(p)}$ resulting from the evaporation of cloud droplets containing Hg^{II} compounds, and wet deposition) and Hg emission (evasion of Hg^0 from the top water micro-layer) for the Mediterranean Sea. The deposition is certainly underestimated as the deposition due to secondary $Hg_{(p)}$ (that is Hg^0 and Hg^{II} adsorbed on particulates during transport) are not yet included. However, even with secondary $Hg_{(p)}$ deposition included, emissions from the sea surface will still far outweigh the input from atmospheric deposition. The Mediterranean Sea is therefore a net emitter of Hg to the atmosphere, in contrast to the generally held concept that oceans are the eventual sink for atmospheric Hg. However the Mediterranean is not an ocean and has quite distinctive regional geological, hydrological and meteorological characteristics. Whether the Hg in Mediterranean sea water is a result of local geology and geological activity, or due to Hg transported to the sea by rivers is not yet clear, but is an active area of field and modelling research (http://www.cs.iia.cnr.it/MERCYMS/project.htm).

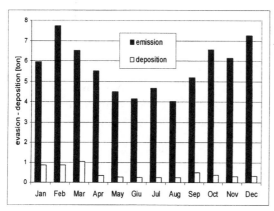

Figure 8. Modelled Mediterranean Hg air-water exchange balance for the year 1999.

	Emission [ton]	Deposition [ton]
Jan	5.95	0.87
Feb	7.74	0.87
Mar	6.51	1.04
Apr	5.50	0.35
May	4.50	0.27
Jun	4.16	0.24
Jul	4.66	0.25
Aug	4.02	0.24
Sep	5.19	0.50
Oct	6.56	0.38
Nov	6.15	0.29
Dec	7.27	0.32
TOT	68.21	5.61

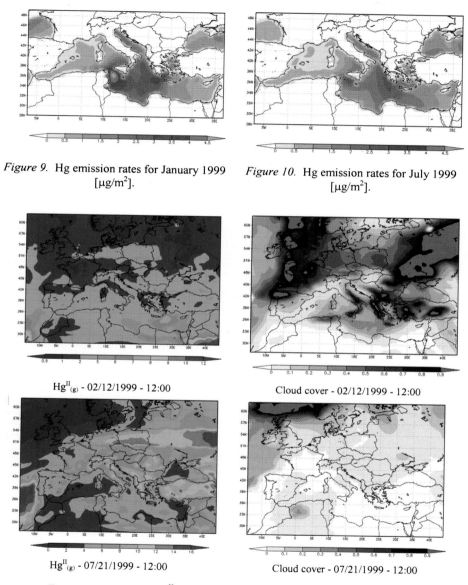

Figure 9. Hg emission rates for January 1999 [μg/m²].

Figure 10. Hg emission rates for July 1999 [μg/m²].

Hg$^{II}_{(g)}$ - 02/12/1999 - 12:00

Cloud cover - 02/12/1999 - 12:00

Hg$^{II}_{(g)}$ - 07/21/1999 - 12:00

Cloud cover - 07/21/1999 - 12:00

Figure 11. Modelled Hg$^{II}_{(g)}$ concentration [pg/m³] and cloud cover [frac].

The Figures 9 and 10 show the emission rates for January and July 1999. The rates depend on local sea water concentrations, the velocity of the wind (raised to the power of 1.64), and the water surface temperature.

The summer months in the Mediterranean are often dominated by anticyclonic weather with low wind speeds, the winter is often windier, however more detailed analysis of the results are required before hard and fast conclusions may be drawn about the seasonality of Hg emissions.

The influence of cloud cover on the photolysis of O_3 and thus the production of the OH radical partially explains the lowest model layer Hg^{II} concentration patterns in the Figure 11. OH production also requires water vapour and the higher Hg^{II} values are seen near the edge of the cloudy regions. The Figure 12 gives an idea of the dry and wet deposition patterns obtained from the integrated model. Wet deposition obviously depends on rainfall, and the low values over N. Africa are easily explained. The low values seen at higher latitudes are because the precipitation was falling as snow. The model does not consider interactions between snow, hail, ice and mercury compounds. Dry deposition is lower over areas which have experienced rainfall because the $Hg_{(p)}$ are $Hg^{II}_{(g)}$ are efficiently scavenged by rain.

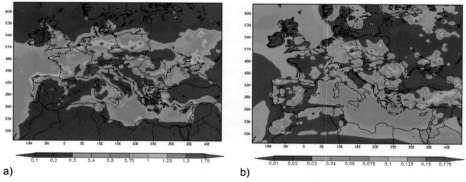

a)   b)

Figure 12. Modelled accumulated wet (a) and dry (b) Hg depositions in $\mu g/m^2$ for January and February 1999.

It is worth noting that from this point of view that the regularity of the rainfall rather than the intensity is the major factor in reducing dry deposition fluxes. The British Isles and north-western France provide a good example of this fact.

The effect of the emissions in the boundary layer is clear from the Figure 13 and their effect decreases with height through the mixed layer, after which the concentration of Hg remains constant throughout the free troposphere. The higher concentration point is in northern Europe and the lower on the Mediterranean coast. Note that the concentration units are ng kg^{-1}(air) rather than the more common, ng m^{-3}(air), using ng m^{-3}(air) would result in an apparent decrease in Hg concentration with altitude because of the pressure difference between the bottom and top model layers.

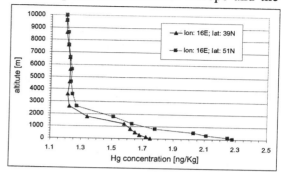

Figure 13. Modelled vertical profile of total Hg [ng kg^{-1} of air].

CONCLUSIONS

The points listed below summarise the further work which is required to be able to characterise the relationship between emissions and spatial patterns of ambient concentrations and deposition fluxes of mercury and its species with greater accuracy:

♦ there is a strong need to promote measurement programs to assess the level of mercury and its compounds (Hg^0, Hg^{II} and particulate Hg) on a European scale and at major urban, industrial and remote sites;

♦ in order to reduce the uncertainty associated with ambient measurements and assure data comparability at European level, there is a strong need to develop standard methods for assessing the Hg^0, Hg^{II} and particulate Hg concentrations in ambient air;

♦ an improved mercury emission inventory is needed for major anthropogenic sources, possibly on a 0.5 x 0.5 degree spatially resolved grid including North Africa and the Middle East regions along with speciation at the stack;

♦ our investigations carried out in the last decade have highlighted the importance played by the sea salt aerosol and sea spray formation in the cycling of mercury and its compounds in the MBL and thus on its deposition to marine waters, therefore besides improving Hg speciation measurement techniques, selected atmospheric measurements aimed to characterise the atmosphere composition and its variations in space and time should be encouraged and should be part of future Hg research programmes;

♦ the improvement of mesoscale and regional scale models are very much related on the future progress of kinetic studies to assess the interaction of gas phase mercury and halogen containing radicals;

♦ a number of preliminary modeling studies performed in recent years have highlighted the need of accounting for the time-dependent vertical profile of Hg^0 concentrations at the model inflow boundaries. Meanwhile advanced hemispherical/global models would contribute to improve modelling capability by providing a better assessment of the boundary conditions on a regional scale;

♦ the changes occurring in chemical speciation of Hg compounds in the top-water micro-layer, deep sea water and sediments may have substantial effect on the exchange of gaseous mercury at the air-water interface and thus on its cycling between the lower atmosphere and the ocean, therefore integrated atmospheric and oceanographic studies should be encouraged;

♦ one of the major source of uncertainty in mesoscale and regional scale mercury modeling is the lack of knowledge of the mechanisms controlling the exchange fluxes of gaseous mercury at the air-water, air-soil and air-vegetation interfaces with changing meteorological conditions, geophysical parameters and the occurrence of biotic and abiotic processes in the top-water micro-layer that may affect the exchange of gaseous mercury between surface water and lower atmosphere.

ACKNOWLEDGEMENTS

The authors wish to acknowledge the financial contributions received from the European Commission in the framework of several European projects including MAMCS (ENV4-CT97-0593) and MERCYMS (EVK3-2002-00070), the Italian Ministry of Research and the Italian National Research Council in the framework of MED-OCEANOR 2000, 2003 and 2004 projects. Special thanks to all students and technicians for their valuable contribution provided during the field campaigns.

REFERENCES

Amirbahman, A., Ruck, P.L., Fernandez, I.J., Haines, T.A., Kahl, J.S. The Effect of Fire on Mercury Cycling in the Soils of Forested Watersheds: Acadia National Park, Maine, U.S.A. *Water, Air, Soil Pollut.*, 152, 315-331, 2004.

Bacci, E. Mercury in the Mediterranean. *Mar. Pollut. Bull.*, 20, 59-63, 1989.

Bacci E., Gaggi C., Lanzillotti E., Ferrozzi S., L. Valli L. Geothermal power plants at Mt. Amiata (Tuscany-Italy): mercury and hydrogen sulphide deposition revealed by vegetation. *Chemosphere*, 40, 907-911.

Bailey E.H., Clark A.L., and Smith R.M. Mercury. *US Geol. Surv. Prof. Pap.*, 821, 410-414. 1973.

Barbosa P., San-Miguel-Ayanz J., Camia A., Gimeno M., Libertà G., Schmuck G. Assessment of fire damages in the EU Mediterranean Countries during the 2003 Forest Fire Campaign. Special Report, EC-JRC, Ispra, 2003.

Bishop, K.H.; Lee, Y.H.; Munthe, J.; Dambrine, E. Xylem sap as a pathway for total mercury and methyl mercury transport from soil to tree canopy in a boreal forest. *Biogeochemistry*, 40, 101-113, 1998.

Browne, C.L. and Fang, S.C. Uptake of Mercury Vapor by Wheat, *Plant Physiology* 61, 430-433, 1978.

Bullock, O.R.Jr., Brehme, K.A., Description and evaluation of atmospheric mercury simulation using the CMAQ model. *Atmos. Environ.*, 36, 2135-2146, 2002.

Carballeira, A., Fernandez, J.A. Bioconcentration of metals in the moss Scleropodium purum in the area surrounding a power plant. *Chemosphere*, 47, 1041-1048, 2002.

Carpi, A., Lindberg, S.E., Application of a Teflon dynamic flux chamber for quantifyng soil mercury flux: tests and results over background soil. *Atmos. Environ.*, 32, 873-882, 1998.

Carvalho, J.A.; Higuchi, N.; Araujo, T.; Santos, J.C., Combustion completeness in a rain forest clearing experiment in Manaus, Brazil. *J. Geophys. Res.*, 103 (D11), 13195-13200, 1998.

Cavallini, A., Natali, L. Durante, M. Maserti, B.E. Mercury uptake, distribution and DNA affinity in durum wheat (Triticum durum Desf.) plants, *Sci. Tot. Env.*, 243/244, 119, 1999.

Cossa, D., Coquery, M., Gobeil, C., and Martin, J.M. Mercury fluxes at the ocean margins. In Global and regional mercury cycles: sources, fluxes and mass balances, pp. 229-247. Ed. by W. Baeyens et al. Kluwer Academic Publishers, 1996.

Ebinghaus, R., Kock, H.H., Temme, Ch., Einax, J.W., Löwe, A.G., Richter, A., Burrows, J.P.; Schroeder, W.H. Antarctic springtime depletion of atmospheric mercury, *Env. Sci. Technol.*, 36, 1238-1244, 2002.

Edner, H., P. Ragnarson, S. Svanberg, E. Wallinder, R. Ferrara, R. Cioni, B. Raco, and G. Taddeucci, Total Fluxes of Sulfur Dioxide from the Italian Volcanoes Etna, Stromboli, and Volcano Measured by Differential Absorption Lidar and Passive Differential Optical Absorption Spectroscopy, *J. Geophys. Res.*, 99, 18827-18838, 1994.

Ericksen, J.A., Gustin, M.S., Schorran, D.E., Johnson, D.W., Lindberg, S.E., Coleman, J.S. Accumulation of atmospheric mercury in forest foliage. *Atmos. Environ.*, 37, 1613-1622, 2003.

Ericksen, J.A., Gustin, M.S. Foliar exchange of mercury as a function of soil and air mercury concentrations. *Sci. Tot. Environ.*, 324, 271-279, 2004.

FAO, Forest Resources Assessment 2000 - Global synthesis. FAO Forestry Paper, 140. Rome, 2000.

FAO. Global Forest Fire Assessment 1990-2000. Forest Resources Assessment Working Paper - 55, Rome, 2001.

Ferrara R., Maserti B.E., De Liso A., Cioni R., Raco B., Taddeucci G., Edner H., Ragnarson P., Svanberg S., and Wallinder E. Atmospheric mercury emission at Solfatara volcano (Pozzuoli, Phlegraean Fields - Italy). *Chemosphere*, 29, 1421-1428, 1994.

Ferrara, R., Mazzolai B., Edner H., Svanberg S., Wallinder E. Atmospheric mercury sources in the Mt. Amiata area, Italy. *Sci. Tot. Env.*, 213, 13-23, 1998.

Ferrara, R., Mazzolai, B., Lanzillotta, E., Nucaro, E., Pirrone, N. Volcanoes as Emission Sources of Atmospheric Mercury in the Mediterranean Basin. *Sci. Tot. Env.*, 259, 115-121, 2000a.

Ferrara, R., Mazzolai, B., Lanzillotta, E., Nucaro, E., Pirrone, N. Temporal trends in gaseous mercury evasion from the Mediterranean Seawaters. *Sci. Tot. Env.*, 259, 183-190, 2000b.

Ferrara, R., Lanzillotta, E., Ceccarini, C. Dissolved gaseous mercury concentration and mercury evasional flux from seawater in front of a chlor-alkali plant. *Environ. Techn.*, 971-978, 2001.

Ferrara, R.; Ceccarini, C.; Lanzillotta, E.; Gårdfeldt, K.; Sommar, J.; Horvat, M.; Logar, M.; Fajon, V.; Kotnik, J. Profiles of dissolved gaseous mercury concentration in the Mediterranean seawater, *Atmos. Environ.*, 37, S85 -S92, 2003.

Fischer, R.G., Rapsomanikis, S., Andrea,e M.O. Bioaccumulation of Methylmercury and Transformation of Inorganic Mercury by Macrofungi. *Env. Sci. Technol.*, 29: 993-999, 1995.

Fitzgerald, W.F. Is mercury increasing in the atmosphere ? The need for an atmospheric mercury network. *Water, Air Soil Pollut.*, 80, 245-254, 1995.

Friedli, H.R., Radke, L.F., Lu, J.Y., Mercury in Smoke from Biomass Fires. *Geophys. Res. Lett.*, 28, 3223- 3226, 2001.

Friedli, H.R., Radke, L.F., Lu, J.Y., Banic, C.M., Leaitch, W.R., MacPherson, J.I. Mercury emissions from burning of biomass from temperate North American forests: laboratory and airborne measurements. *Atmos. Environ.*, 37, 253-267, 2003.

Gårdfeldt, K., Jonsson, M., Is bimolecular reduction of divalent mercury complexes possible in aqueous systems of environmental importance? *J. Phys. Chem. A.*, 107, 4478-4482, 2003a.

Gårdeldt, K., Sommar, J., Ferrara, R., Ceccarini, C., Lanzillotta, E., Munthe, J., Wangberg, I., Lindqvist, O., Pirrone N., Sprovieri, F., Pesenti, E. Release of Mercury to the Atmosphere from Atlantic Coastal Water and the Mediterranean Coastal and Open Water. *Atmos. Environ.*, 37/S1, 73-84, 2003b.

Hanson, P.J., Lindberg, S.E., Tabberer, T.A., Owens, J.G., Kim, K.-H., Foliar exchange of mercury vapor: evidence for a compensation point. *Water, Air, Soil Pollu.,t* 80, 373-382, 1995.

Hedgecock, I., Pirrone, N. Mercury and Photochemistry in the Marine Boundary Layer– Modeling Studies for in-situ Production of Reactive Gas Phase Mercury. *Atmos. Environ.*, 35, 3055-3062, 2001.

Hedgecock, I., Pirrone, N., Sprovieri, F., Pesenti, E. Reactive Gaseous Mercury in the Marine Boundary Layer: Modeling and Experimental Evidence of its Formation in the Mediterranean. *Atmos. Environ.*, 37 (S1), 41-50, 2003.

Hedgecock, I. M., and Pirrone, N. Chasing Quicksilver: Modeling the Atmospheric Lifetime of $Hg^0_{(g)}$ in the Marine Boundary Layer at Various Latitudes. *Env. Sci. Technol.*, 38, 69 – 76, 2004.

Horvat, M., Kotnik, J., Fajon, V., Logar, M., Zvonaric, T., Pirrone, N. Speciation of Mercury in Surface and Deep Seawater in the Mediterranean Sea. *Atmos. Environ.*, 37/S1, 93-108, 2003.

Jonasson I. R. Boyle R. W. Geochemistry of mercury. Spatial Symposium on Mercury in Man's Environment, Environment Canada, Ottawa, Canada, 5-21. 1971

Junge, C. E., Gustafson, P. E.: On the distribution of sea salt over the United States and its removal by precipitation. *Tellus*, 9, 164–173, 1957.

Kallos, G., Kotroni V., Lagouvardos K., Papadopoulos A., "On the long range transport of air pollutants from Europe to Africa", *Geophys. Res. Lett.*, 25, 5 619-622, 1998.

Keeler, G.J., Glinsorn, G., Pirrone, N. Particulate mercury in the atmosphere: its significance, transport, transformation and sources. *Water, Air Soil Pollut.*, 56, 553-564, 1995.

Kim, J.P. and W.F. Fitzgerald. Sea-air partitioning of mercury in the equatorial Pacific Ocean. *Science*, 231, 1131-1133, 1986.

Kim, K.-H., Lindberg, S.E., Meyers, T.P. Micrometeorological measurements of mercury vapor fluxes over background forest soils in eastern Tennessee. *Atmos. Environ.*, 29, 267-282, 1995.

Kouvarakis G., Tsigaridis, K., Kanakidou, M., and Mihalopoulos, N. Temporal variations of surface regional background ozone over Crete Island in southeast Mediterranean, *J. Geophys. Res.*, 105, 4399-4407, 2000.

Landis, M.S., Stevens, R.K., Schaedlich, F., Prestbo, E., Development and characterization of an annular denuder methodology for the measurement of divalent inorganic reactive gaseous mercury in ambient air. *Env. Sci. Technol.* 36, 3000-3009, 2002.

Lelieveld, J., Berresheim, H., Borrmann, S., Crutzen, P. J., Dentener, F. J., Fischer, H., Feichter, J., Flatau, P. J., Heland, J., Holzinger, R., Korrmann, R., Lawrence, M.G., Levin, Z., Markowicz, K. M., Mihalopoulos, N., Minikin, A., Ramanathan, V., De Reus, M., Roelofs, G. J., Scheeren, H. A., Sciare, J., Schlager, H., Schultz, M., Siegmund, P., Steil, B., Stephanou, E.G., Stier, P., Traub, M., Warneke, C., Williams, J., Ziereis, H. Global air pollution crossroads over the Mediterranean. *Science*, 298, 794-799, 2002.

Lin, X., Tao, Y. A numerical modelling study on regional mercury budget for eastern North America. *Atmos. Chem. Phys.*, 3, 535-548, 2003.

Lindberg, S.E., Meyers, T.P., Taylor Jr., G.E., Turner, R.R., Schroeder, W.H. Atmosphere-surface exchange of mercury in a forest: results of modeling and gradient approaches. *J. Geophys. Res.*, 97, 2519-2528, 1992.

Lindberg, S.E.,. Forests and the global biogeochemical cycle of mercury: the impotence of understanding air/vegetation exchange processes. In: Baeyens, W., *et al.* (Ed.), Global and Regional Mercury Cycles: Sources, Fluxes, and Mass Balances. Kluwer Academic, Boston, MA, 359-380, 1996.

Lindberg, S.; Brooks, S.; Lin, C-J.; Scott, K.; Landis, M. S.; Stevens, R. K.; Goodsite, M.; Richter, A. The dynamic oxidation of gaseous mercury in the arctic atmosphere at polar sunrise. *Environ. Sci. Technol.*, 36, 1245-1256, 2002a.

Lindberg, S.E., Zhang, H., Vette, A.F., Gustin, M.S., Barnett, M.O., Kuiken, T. Dynamic flux chamber measurement of gaseous mercury emission fluxes over soils. Part 2: Effect of flushing flow rate and verification of a two-resistance exchange interface simulation model. *Atmos. Environ.*, 36, 847-859, 2002b.

Lodenius, M. Dry and wet deposition of mercury near a chlor-alkali plant. *Sci. Tot. Env.*, 213, 53-56, 1998.

Lodenius, M., Tulisalo, E., Soltanpour-Gargari, A. Exchange of mercury between atmosphere and vegetation under contaminated conditions. *Sci. Tot. Envi.*, 304, 169-174, 2003.

Mason, R.P., Lawson, N.M., Sheu, G.R. Mercury in the Atlantic Ocean: factors controlling air-sea exchange of mercury and its distribution in the upper waters. *Deep-Sea Research II*, 48, 2829-2853, 2001.

Meyers, M.P., Walko, R.L., Harrington, J.Y., Cotton, W.R. New RAMS cloud microphysics parameterization. Part II: The two-moment scheme. *Atmos Res.*, 45, 3-39, 1997.

Millan, M., Salvador R., Mantilla E., Kallos G. Photo-oxidant dynamics in the Mediterranean basin in summer - Results from European research projects. *J. Geophys. Res.*, 102, 8811-8823, 1997.

Mosbaek, H., Tjell, J.C., Sevel, T. Plant Uptake of Airborne Mercury in Background Areas. *Chemosphere*, 17, 1227-1236, 1988.

Munthe, J., Wangberg, I., Pirrone, N., Iverfeld, A., Ferrara, R., Ebinghaus, R., Feng., R., Gerdfelt, K., Keeler, G.J., Lanzillotta, E., Lindberg, S.E., Lu, J., Mamane, Y., Prestbo, E., Schmolke, S., Schroder, W.H., Sommar, J., Sprovieri, F., Stevens, R.K., Stratton, W., Tuncel, G., Urba, A. Intercomparison of Methods for Sampling and Analysis of Atmospheric Mercury Species. *Atmos. Environ.*, 35, 3007-3017, 2001.

Munthe, J., Wangberg, I., Iverfeld, A., Lindqvist O., Strömberg, D.,, Sommar, J., Gårdfeldt, K., Petersen, G., Ebinghaus, R., Prestbo, E., Larjava, K., Siemens, K.V. Distribution of atmospheric mercury species in Northern Europe: final results from the MOE project. *Atmos. Environ.*, 37 (S1), 9-20, 2003.

Nacht, D.M.,, Gustin, M.S. Mercury emissions from background and altered geologic units throughout Nevada. *Water, Air Soil Pollut.*, 151, 179-193, 2004.

Nriagu J. and Becker C. Volcanic emissions of mercury to the atmosphere: global and regional inventories. *Sci. Tot. Env.*, 304, 3–12, 2003.

Pacyna, E., Pacyna, J.M., Pirrone, N. Atmospheric Mercury Emissions in Europe from Anthropogenic Sources. *Atmos. Environ.*, 35, 2987-2996, 2001.

Pacyna, J.M., Pacyna, E.G., Steenhuisen, F., Wilson, S. Mapping 1995 global anthropogenic emissions of mercury. *Atmos. Environ.*, 37 (S1) 109-117, 2003.

Pai, P., Karamchandani, P., Seigneur, C. Simulation of the regional atmospheric transport and fate of mercury using a comprehensive Eulerian model. *Atmos. Environ.*, 31, 2717-2732, 1997.

Patra, M. Sharma, A. Mercury Toxicity In Plants. *Botanical Review*, 66, 379-422, 2000.

Petersen, G., Iverfeldt. A., Munthe, J. Atmospheric mercury species over central and northern Europe. Model calculations and comparison with observations from the Nordic Air and Precipitation Network for 1987 and 1988. *Atmos. Environ.*, 29, 47-67, 1995.

Pielke, R.A., Cotton, W.R., Walko, R.L., Tremback, C.J., Lyons, W.A., Grasso, L.D., Nicholls, M.E., Moran, M.D., Wesley, D.A., Lee, T.J., Copeland, J.H. A comprehensive meteorological modeling system - RAMS. *Meteor. Atmos. Physics*, 49, 69-91, 1992.

Pirrone, N., Glinsorn, G., Keeler, G.J. Ambient Levels and Dry Deposition Fluxes of Mercury to Lakes Huron, Erie and St. Clair. *Water, Air Soil Pollut.*, 80, 179-188, 1995.

Pirrone, N., Keeler, G.J., Nriagu, J.O. Regional Differences in Worldwide Emissions of Mercury to the Atmosphere. *Atmos. Environ.*, 30, 2981-2987, 1996.

Pirrone, N., Hedgecock, I., Forlano, L. The Role of the Ambient Aerosol in the Atmospheric Processing of Semi-Volatile Contaminants: A Parameterised Numerical Model (GASPAR). *J. Geophys. Res.*, 105, D8, 9773-9790. 2000.

Pirrone, N., Costa, P., Pacyna, J.M., Ferrara, R. Atmospheric Mercury Emissions from Anthropogenic and Natural Sources in the Mediterranean Region. *Atmos. Environ.*, 35, 2997-3006. 2001a.

Pirrone, N., Pacyna, J.M., Barth, H. Atmospheric Mercury Research in Europe. *Special Issue of Atmos. Environ.*, Vol. 35 (17) Elsevier Science, Amsterdam, Netherlands, 2001b.

Pirrone, N., Ferrara, R., Hedgecock, I.M., Kallos. G., Mamane, Y., Munthe, J., Pacyna, J. M., Pytharoulis, I., Sprovieri, F., Voudouri, A., Wangberg, I. Dynamic Processes of Atmospheric Mercury Over the Mediterranean Region. *Atmos. Environ.*, 37 (S1), 21-40, 2003a.

Pirrone, N., Pacyna, J.M., Munthe, J., Barth, H. Dynamic Processes of Mercury and Other Atmospheric Contaminants in the Marine Boundary Layer of European Seas. *Special Issue of Atmos. Environ.*, 37 (S1), 2977-3074, 2003b.

Pleim, J., Venkatram A., Yamartino, R. ADOM/TADAP model development program. Volume 4. The dry deposition module. Ontario Ministry of the Environment, Rexdale, Ontario. 1984.

Poissant, L., Amyot, M., Pilote, M., Lean, D., Mercury Water-Air Exchange over the Upper St. Lawrence River and Lake Ontario. *Env. Sci. Technol.*, 34, 3069-3078, 2000.

Pyle, D.M., Mather, T.A. The importance of volcanic emissions for the global atmospheric mercury cycle, *Atmos. Environ.*, 37, 5115-5124, 2003

Rea, A.W., Keeler, G.J., Scherbatskoy, T., The deposition of mercury in throughfall and litterfall in the Lake Champlain watershed: a short-term study. *Atmos. Environ.*, 30, 3257-3263, 1996.

Rea, A.W.; Lindberg, S.E.; Scherbatskoy, T., Mercury accumulation in foliage over time in two northern mixed-hardwood forests. *Water, Soil Air Pollut.*, 133, 49-67, 2002.

Rasmussen P.E. Mercury in Vegetation of the Precambrian Shield. In: Mercury Pollution: Integration and Synthesis (Watras C.J. and Huckabee J.W. Eds.) Lewis Publishers, Palo Alto, CA, U.S.A.: 417-425, 1994.

Sanemasa, I. The solubility of elemental mercury vapour in water. *Bul.l Chem. Soc. Jpn.*, 48, 1795–98, 1975.

Schroeder, W.H., Munthe, J., Lindqvist, O. Cycling of Mercury between Water, Air, and Soil Compartments of the Environment. *Water, Air, Soil Pollut.*, 48, 337-347, 1989.

Schwesig D., Krebs O. The role of ground vegetation in the uptake of mercury and methylmercury in a forest ecosystem. *Plant and Soil*, 253, 445-455, 2003.

Shannon, J.D., Voldner, E.C., Modeling Atmospheric Concentrations of Mercury and Deposition to the Great Lakes. *Atmos. Environ.*, 29, 1649-1661, 1995.

Shimada, M. Okuyama, K., Asai, M. Deposition of Submicron Aerosol Particles in Turbolent and Transitional Flow. *AIChE. J.*, 39, 17, 1993.

Slinn, S.A., Slinn, W.G.N. Predictions for particle deposition and natural waters. *Atmos. Environ.*, 14, 1013-1016, 1980.

Sprovieri, F.; Pirrone, N.; Hedgecock, I. M.; Landis, M. S.; Stevens, R. K. Intensive atmospheric mercury measurements at Terra Nova Bay in Antarctica during November and December 2000. *J. Geophys. Res.*, 107, 4722, 2002.

Sprovieri, F., Pirrone, N., Gårdeldt, K., Sommar, J. Atmospheric Mercury Speciation in the Marine Boundary Layer along 6000 km Cruise path over the Mediterranean Sea. *Atmos. Environ.*, 37 (S1), 63-72, 2003.

SNMH-Smithsonian National Museum of Natural Hystory Global Volcanism Program.http://www.volcano.si.edu/index.cfm .2004.

Thibodeaux, L.J. Environmental Chemodynamics. 2nd ed. Wiley and Sons, New York. 1996

Veiga, M.M.; Meech, J.A. Onante, N. Mercury pollution from deforestation. *Nature*, 368, 816-817, 1994.

Voldner, E.C., Barrie, L.A., Sirois, A. A literature review of dry deposition of oxides of sulphur and nitrogen with emphasis on long range transport modelling in North America. *Atmos. Environ.*, 20, 2101-2123, 1986.

Walko, R. L., Cotton, W.R., Meyers, M.P., Harrington, J.Y., New RAMS cloud microphysics parameterization. Part I: The single-moment scheme. *Atmos. Res.*, 38, 29-62, 1995.

Walko, R.L., et al. Coupled atmosphere-biophysics-hydrology models for environmental modeling. *J. Applied Meteor.*, 39, 931-944, 2000.

Walmsley, J.L., Wesely, M.L. Modification of coded parametrization of surface resistances to gaseous dry deposition. *Atmos. Environ.*, 30, 1181-1188, 1996.

Wängberg, I., Munthe, J., Pirrone, N., Iverfeldt, Å., Bahlman, E., Costa, P., Ebinghaus, R., Feng, X., Ferrara, R., Gårdfeldt, K., Kock, H., Lanzillotta, E., Mamane, Y., Mas, F., Melamed, E., Osnat, Y., Prestbo E., Sommar, J., Schmolke, S., Spain, G., Sprovieri, F., Tuncel, G. Atmospheric Mercury Distributions in North Europe and in the Mediterranean Region. *Atmos. Environ.*, 35, 3019-3025, 2001.

Wesely, M.L. Parametrization of surface resistances to gaseous dry deposition in regional-scale numerical models. *Atmos. Environ.*, 23, 1293-1304, 1989.

Wesely, M.L.; Hicks, B.B., "Some factors that affect the deposition of sulfur dioxide and similar gases on vegetation" *J. Air Pollut. Control Assoc.*,, 1110-1116, 1977.

Wild, O., X. Zhu, and M.J. Prather, Fast-J: Accurate simulation of in- and below-cloud photolysis in tropospheric chemical models, *J. Atmos. Chem.*, 37, 245-282, 2000.

Xiao, Z. F., Munthe, J., Schroeder, W. H. & Lindqvist, O. Vertical fluxes of volatile mercury over forest soil and lake surfaces in Sweden, *Tellus*, 43B, 267-279, 1991.

Xu, X., Yang, X., Miller, D.R., Helble, J.J., Carley. R.J. Formulation of bi-directional air-surface exchange of elemental mercury. *Atmos. Environ.*, 33, 4345-4355, 1999.

Xu, X., Yang, X., Miller, D.R., Helble, J.J., Carley. R.J. A regional scale modeling study of atmospheric transport and transformation of mercury. I. Model development and evaluation. *Atmos. Environ.*, 34, 4933-4944, 2000.

Chapter-24

SPATIAL AND TEMPORAL VARIABILITY OF ATMOSPHERIC MERCURY IN NORTH-WESTERN AND CENTRAL EUROPE - OBSERVATIONS ON DIFFERENT TIME SCALES

Ralf Ebinghaus[1], Hans. H. Kock[1], John Munthe[2] and Ingvar Wängberg[2]

[1] GKSS Research Centre Geesthacht GmbH, Institute for Coastal Research, 21502 Geesthacht, Germany
[2] IVL Swedish Environmental Research Institute, P.O.Box 47086, S-402 58 Göteborg, Sweden

INTRODUCTION

This work summarizes field studies that have been carried out to investigate the chemistry and environmental behavior of atmospheric mercury on different spatial and temporal scales in North-western and Central Europe. The spatial scales cover local and regional dimensions. Temporal scales include short-term variations of atmospheric mercury concentrations within less than one hour, variations that typically occur within time steps of a few hours to several days and long-term observations that cover almost three decades.

The knowledge of the worldwide trend of atmospheric mercury concentrations during the last few decades is valuable for at least two reasons. The trend may reveal the impact of the control measures (OECD, 1994; US EPA, 1997) on the global cycle of atmospheric mercury. The response of mercury concentrations to the control measures may also provide information about the poorly defined ratio of anthropogenic to natural emissions (Schroeder and Munthe, 1998; Ebinghaus et al., 1999a). Slemr et al. (2003) have attempted to reconstruct the worldwide trend of total gaseous mercury (TGM) concentrations from long term measurements of known documented quality (Ebinghaus et al., 1999b) on 6 sites in the northern hemisphere (NH), 2 sites in the southern hemisphere (SH), and 8 ship cruises over the Atlantic Ocean made intermittently since 1977.

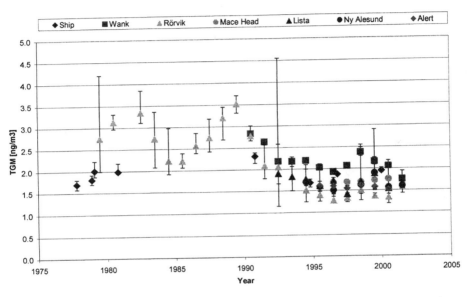

Figure 1. Trends of TGM concentrations in a) the northern hemisphere. STP concentrations are given in ng m^{-3} with standard T of 273.2 K and standard P of 1013 mbar.

The data presented in Figure 1 shows a generally good agreement between ship TGM measurements over the northern Atlantic Ocean and land-based measurements at Mace Head (Ireland), Lista, and Ny Ålesund (both Norway). The measurements at Mace Head also agree excellently with those made at Alert (Canada). TGM median values at the summit of the Wank mountain in southern Germany tend to be higher, most likely due to emissions in western and central Europe. Measurements at Rörvik in Sweden tend to provide the lowest values in the NH but are still in reasonable agreement with measurements at Alert and Lista (see also discussion in Munthe et al., 2003). All data, when taken together, suggest that the TGM concentrations in the NH had been increasing since the first measurements in 1977 to a maximum in the 1980s (most likely in the 2nd half (Slemr et al., 2003), then decreased to a minimum in 1996 and have remained constant since that time at a level of about 1.7 ng m^{-3}. Slemr et al. (2003) claim that the observed temporal profile is primarily the result of the temporal change of mercury emissions. The temporal trend is qualitatively consistent with changes in global anthropogenic emissions (e.g. Pacyna and Pacyna, 2002; Pirrone et al., 1996) emphasising that the analysis of emission changes may help to explain observed long-term trends. However, these estimates do not reflect the measured decrease from 1990 to 1996 (Slemr et al., 2003).

EMISSIONS AND THEIR IMPORTANCE FOR THE LOCAL-SCALE VARIABILITY OF ATMOSPHERIC MERCURY

Mercury is emitted into the atmosphere from a number of natural and anthropogenic sources that may directly bias measurement results generated at single locations. The exchange of mercury at any interface is driven by a concentration gradient. However, emission fluxes are based on complex processes and not only a simple phase transfer of Hg^0 (Lindberg et al., 1998). The speciation of mercury emitted into the atmosphere is of fundamental importance for its atmospheric fate. Hg^0 will add to the regional and hemispherical background whereas total particulate-phase mercury (TPM) and reactive gaseous mercury (RGM) will deposit on much smaller scales.

The discussion about the relative significances of natural and anthropogenic emissions is complicated by processes that we classify here as indirect anthropogenic re-emissions. This term is chosen to describe secondary re-emission of mercury from anthropogenic sources following partial initial deposition of primary anthropogenic mercury emissions. Unfortunately, these processes usually take place in compartments that also exhibit a natural mercury surface-exchange like waters and soils, thereby making it impossible to distinguish between the two parallel ongoing natural and anthropogenically induced processes and to separate them quantitatively. Most natural surfaces have been impacted by atmospheric mercury deposition of anthropogenic origin and hundreds of sites exist worldwide where local sources have contributed to the contamination. The total contribution of these diffuse sources to local and regional mercury budgets is unknown but could be significant.

Anthropogenic emissions

For entire Europe the annual anthropogenic emissions of mercury has been estimated to be about 726 tons, originating from 928 sources by base year 1988 (Axenfeldt et al., 1991). The 1995 anthropogenic emissions were estimated to about 342 tons, a decrease of 45% compared to these emissions in 1990 (Pacyna, et.al., 2001). In 1995 the European emissions of anthropogenic mercury contributed about 13% to the global emissions of this element from anthropogenic sources. The dominating source categories in Europe are fossil fuel combustion and chlor-alkali plants. Waste incineration and non-ferrous metal smelting contribute less than 10 % to the European total. Emissions from the former German Democratic Republic (GDR), including the different species of elemental mercury, divalent inorganic mercury and particulate-phase mercury, accounted for more than 40% of the European total. The contribution of the GDR originated from a few relatively small but in most cases highly industrialized areas. According to Helwig and Neske (1990), extremely high amounts were emitted in the region Halle/Leipzig/Bitterfeld, due to both burning of lignite coal in power plants with flue gas desulfurisation equipment and high losses of mercury from the chlor alkali factories.

An important emitter of air- pollutants in eastern Germany was the former Chemische Werke Buna (hereafter referred to as the BSL Werk Schkopau), located near Halle/Saale (Krüger et al., 1999). As a consequence of the reunification of Germany, about 50 production installations which were considered to be ecologically harmful were closed at the BSL Werk Schkopau in the period between 1990 and 1992. Among them were three chlor-alkali production facilities. In these facilities, mercury was used as an electrode for chlorine and sodium hydroxide production and as a catalyst for acetaldehyde production.

An example how to handle the difficult question of mercury emissions from a (partly) inactivated production plant was given for the industrial BSL complex. It could be shown by a combination of field measurements with numerical modelling, emissions of mercury from inactive plants could be estimated with an uncertainty of a factor of two. Emission estimates of the inactivated industrial complex on the order of 2 to 4 kg per day are reflecting the importance of more or less diffuse emissions of mercury from formerly active production facilities (Ebinghaus and Krüger, 1996). An attempt has been made to derive a mercury mass-balance on a local scale around the BSL complex. It could be estimate that 30% of the emitted mercury is deposited in the local vicinity of the plant, whereas 70% are available for long-range transport at least on a regional scale.

In the recently completed EMECAP project (Mazzolai et al., 2004), measurements of airborne mercury species around chlor-alkali plants in Sweden, Italy and Poland were made. In Sweden, the measurements were performed at the EKA Chemicals chlor alkali plant in Bohus, located 50 km north east of Gothenburg. Total Gaseous Mercury (TGM) was measured using manual Au-traps but also using Tekran Model 2537A gas-phase mercury vapour analysers. In addition an automatic unit for sequential sampling of GEM, RGM and PM was employed. The unit was connected to a Tekran 2357A analyser. More details on methods and the equipment used for mercury measurements is found in Sommar 2004 and Wängberg 2003.

Dispersion modelling using the TAPM model (Hurley et al., 2001) was used to integrate measurement results over a larger area and longer time. Emission estimates were provided both from the company's own measurements and LIDAR measurements performed during the sampling campaigns (Wängberg et al., 2001; Grönlund et al., 2004). Concentration ranges and mean concentrations obtained at two sites in the vicinity of the plant (Figure 2) are shown in Table 1. The average annual distribution of GEM and RGM in the Bohus area as obtained from TAPM model calculations is shown in Figure 3 and Figure 4 respectively. A mercury emission value of 74 kg Hg per year was used as input parameter in the model. This value corresponds to the average mercury flux from the cell house measured with the LIDAR system, Gronlund 2004.

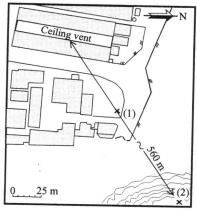

Figure 2. The north end of the Eka Chemicals chlor-alkali plant in Bohus. The numbers denote Site 1 and Site 2 were measurements were performed.

The influence from the mercury cell chlor-alkali plant (MCCA) is obtained if subtracting 1.6 ng m^{-3}, which is the mercury background concentration in this area of Sweden. The average mercury concentration in residential areas of Bohus is 1-3.5 ng m^{-3} higher than the background.

Table 1. Mercury concentrations at site 1 and 2.

Hg species at site 1	Conc. Range (ng m⁻³)	Mean Conc.(ng m⁻³)
GEM	1.5 - 540	55
PM	< 0.01 - 0.176	0.028
RGM	< 0.04 – 1.07	0.17
Hg species at site 2		
GEM	1.4 - 40	3.5
PM	> 0.01	> 0.01
RGM	< 0.04 – 0.11	0.007

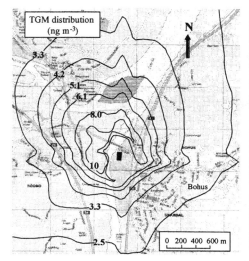

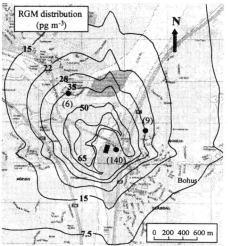

Figure 3. Yearly average distribution of TGM calculated using the TAPM model. The values presented correspond to the mercury emitted plus 1.6 ng/m³, which is the background concentration of mercury in this area. The position of the cell house is marked by a filled rectangle.

Figure 4. Yearly average distribution of RGM calculated using the TAPM model. The concentrations are due to RGM emission from the cell house, i.e. background concentrations are not included. Numbers in brackets correspond to annual mercury wet deposition in □g m⁻². Sites where wet deposition were measurement are denoted by red dots.

Most of the mercury emitted from the cell-house (50-100 kg per year), is quickly dispersed and transported far from Bohus. Hence, the impact on the local area is rather small. The present measurements show that RGM influences the deposition of mercury in the area. During rain, all RGM is likely to be deposited within an area of about 1 km radius around the MCCA plant. However, since it is not raining all the time only a small fraction (0.03 kg) of the RGM emitted is being wet deposited and the total contribution to the area of 1 km proximity to the plant is only enough to double the deposition in comparison to the ambient mercury deposition.

It can be concluded that the local impact on mercury emission to air from the MCCA plant in Bohus is basically restricted to an area of a few hundred meters around the cell house. Most of the mercury emitted is removed from the area by dispersion and thus contributes to the global atmospheric budget of mercury.

Re-emissions from rivers and estuaries

It is well established that rivers transport mercury as a result of anthropogenic discharges and/or natural surface run-off, and both transport and deposition behavior in the river and especially in the estuaries with respect to discharges into the oceans have been studied. One case study (Bahlmann, 1997) deals with the occurrence of volatile mercury species along the transect of the highly contaminated Elbe river in Germany. This investigation shows that volatile mercury compounds (defined as dissolved gaseous mercury (DGM)) in the waterbody can be detected along the course of the river. It is shown that the highest concentrations of "free" volatile mercury compounds are found in the estuarine region. The levels of DGM found in this study are illustrated in Table 2, which compares DGM levels in several waters.

Table 2. Range of DGM Concentrations Measured in European Rivers, Marine Systems and Lakes.

Location	DGM Concentrations (pg L^{-1})	Reference
Elbe River	47 - 152	Bahlmann, 1997
Elbe River Estuary	54 - 122	Coquery, 1995
German Bight (North Sea)	18 - 284	Bahlmann, 1997
German Bight (North Sea)	17 - 87	Reich, 1995
North Sea	< 20 - 90	Coquery, 1995
Baltic Sea	14 - 22	Schmolke et al., 1997
Scheldt Estuary	190 - 500	Baeyens et al., 1991
Scheldt Estuary	45 - 80	Baeyens and Leermakers, 1996
Seine Estuary	<10 - 90	Coquery, 1995
Lake Borrsjön	0.5 - 9	Gardfeldt, 2001
Swedish West coast	40 - 100	Gardfeldt, 2001
Baltic Sea	average 17.5	Wängberg 2001
Mediterranean Sea	18 - 40	Ferrara 2003
Mediterranean Sea	12 - 87	Gårdfeldt et al., 2003
near Island of Capraia	10 - 20	Lanzillotta 2002

588 CHAPTER-24: MERCURY IN EUROPE

The Elbe seems to represent one extreme in that respect for more than 99% of the transported mercury has been shown to be particulate (Wilken and Hintelmann, 1991), while other contaminated rivers and streams seem to exhibit quite different behavior (Turner and Lindberg, 1978). This process is quite likely since the suspended particles in the waterbody of the Elbe are rich in both mercury and bacteria which have been shown to be resistant to the high mercury levels. These bacteria were shown to demethylate MeHg-compounds as a detoxification mechanism and might also be able to reduce divalent mercury compounds (Ebinghaus and Wilken, 1993). On the base of the determined concentrations and volatilization potentials, it has been calculated that the Elbe river could emit between 100 and 500 kg of mercury to the atmosphere annually (Bahlmann, 1997). On a global base, it has been estimated that only 10% of the rivers' annual mercury loads originate from direct anthropogenic sources, while the remainder results from natural and indirect anthropogenic emissions (Cossa et al., 1996); however, on a regional base, there may be vast differences in those proportions.

Re-emissions from Lakes

Nriagu estimated that an important fraction (10-50%) of dissolved mercury in lakes is in the elemental form (Nriagu, 1994). However, more recent estimates in surface waters place this fraction closer to 5-10% (Amyot et al., 1995; Fitzgerald and Mason, 1996; Schmolke et al., 1997). Flux measurements in Sweden indicate that the emitted mercury is also in the elemental form (Lindberg et al., 1996). Several studies have reported measurements of DGM in lakes, some of which have been used to model evasion. Earlier flux chamber measurements (Xiao et al., 1991; Schroeder et al., 1992) over five Swedish lakes found fluxes between 3 and 20 ng m^{-2} h^{-1} from the lakes to the overlying atmosphere. At one occasion, a net deposition was observed. It was also found that fluxes during daytime were larger than at night, indicating that sunlight, biological activity and temperature might play an important role in the mercury volatilization processes. The authors were also able to demonstrate pronounced seasonal differences in the mercury volatilization rate with fluxes in May and June being much larger than in November. During June, 1994 the first Modified Bowen Ratio measurements of mercury vapor fluxes over a boreal forest lake were made at Lake Gårdsjön, Sweden (Lindberg et al., 1996). Using highly accurate methods with multiple replicate samplers, the authors measured concentration gradients of mercury vapor, CO_2, and H_2O over the lake surface. Mercury was found to be readily emitted from the lake surface, and there was no evidence of Hg(0) dry deposition to the lake surface. Emission rates over the lake averaged 8.5 ng m^{-2} h^{-1}, and appeared to be weakly influenced by water temperature and solar radiation. Overall, the fluxes ranged from ~2 to 18 ng m^{-2} h^{-1}. Overall, the surface water of the lake appears to be a more active zone for mercury exchange than the surrounding soils based on two independent studies (Xiao et. al., 1991; Lindberg et. al., 1998). In the Lake Gårdsjön study, it was concluded that re-emissions occur from the lake surface during most of the year but some dry deposition may occur during winter.

Re-emissions from Wetland Areas

The Elbe floodplains have an estimated area of 1,100 km^2 and total deposited mercury amount of 1,500 t (Wallschläger 1996). Average mercury concentrations in the soils range from 1 to 10 µg g^{-1} and most of the mercury is bound by high molecular weight organic matter. In this study, flux chamber measurements were employed to estimate mercury volatilization, but they were also compared to the modified Bowen-ratio method (MBR) method by measuring the concentration gradient of atmospheric mercury in the soil atmosphere boundary layer. Additionally, mercury volatilization was estimated from soil air concentration data assuming laminar diffusion. Results for mercury emissions from soils containing 10 µg g^{-1} range from 28 to 260 ng m^{-2} h^{-1}. However, it was noted that the individual applied techniques did not agree very well due to their inherent methodological differences, and, that there is a need for intercomparison studies (Wallschläger et al., 2002). The overall annual emission of mercury from the Elbe floodplains to the atmosphere was estimated to be on the order of 1 t y^{-1}, thereby representing a significant, but not dominant mercury source on the regional scale. Observations of increased Hg volatilization from contaminated floodplains during rain events led to the suggestion of a two step mechanism comprising the same reactions: a small initial displacement of soil air containing Hg(0) and DMM (Wallschläger et al., 1995) followed by formation of volatile mercury compounds in the liquid phase as a result of reduction of Hg(II) and dismutation of MeHg$^+$ (Wallschläger et al., 2000). These studies also calculated that direct reduction of wet deposited Hg(II) is not likely to be a major source of mercury volatilization.

Studies over background forest soils in Sweden (Xiao et al., 1991) found fluxes of lower magnitudes. It was noted that net mercury emissions occur during summer at a rate of 0.3 ± 0.4 ng m^{-2} h^{-1}, while net deposition was observed in winter at a rate of 0.9 ± 0.4 ng m^{-2} h^{-1}.

Re-emissions from Marine Systems

The distribution, transport and fate of mercury at the earth's surface is critically dependent on the biogeochemical cycling and atmospheric exchange of mercury in the marine environment since, by its shear size. The GEM content in air over the southern Baltic Sea indicated that, in general, during the summer conditions, the sea-to-air transport of gaseous mercury dominated, while during the winter season, a tendency of gaseous mercury to deposit into the water has been found (Marks, 2001). Four campaingns were performed over Swedish coastal seawater surface during the summer and winter of 1997 and the summer of 1998 (Gardfeldt, 2001). Mercury evasion was found in the interval between -2.7 to +8.8 ng m^{-2} h^{-1} with an average evasion of + 0.6 ng m^{-2} h^{-1} . GEM measurements performed at two Baltic Sea coastal stations (Peninsula Hel, Poland and Preila, Lithuania) during summer 1997 were correlated with meteorological parameters, directional distribution and diurnal variability (Urba, 2000). Analysis of the data implies that the Baltic sea, in particular its southern part and the Gulf of Gdansk, are the main gaseous mercury source for the region during summer month. Atlantic seawater measurements were performed during September 1999 at the Mace Head Atmospheric Research Station situated on the Irish west coast (Gardfeldt, 2003). The predicted average mercury evasion from the coastal Atlantic water was 2.7 ng m^{-2} h^{-1} implying that the concentration of GEM in the Atlantic air is enhanced by mercury evasion from the sea. Another important factor is that all these flux calculations are based on calm sea and moderate wind conditions (under which also all of the sampling campaigns were probably performed). It has been modeled, though, that fluxes may increase by a factor of up to 25 during storms and rough sea conditions (Baeyens et al., 1991), so this estimate may be systematically low. A number of studies have measured volatile mercury specieswell above saturation levels in Baltic and Atlantic seawaters primarily as DGM (Gardfeldt, 2001 ; Marks, 2002).

Measurements of DGM were performed during two cruises in the Baltic Sea during the BASYS project (Wängberg et al., 1999). The concentrations of total gaseous mercury (TGM) in air over the Southern Baltic Sea and dissolved gaseous mercury (DGM) in the surface seawater were measured during summer and winter. The summer expedition was performed on July 02-15, 1997, and the winter expedition on March 02-15, 1998. Average TGM and DGM values obtained were 1.70 and 17.6 ng m^{-3} in the summer and 1.39 and 17.4 ng m^{-3} in the winter, respectively. Based on the TGM and DGM data, surface water saturation and air-water fluxes were calculated. The results indicate that the seawater was supersaturated with gaseous mercury during both seasons, with the highest values occurring in the summer. Flux estimates were made using the thin film gas-exchange model. The average Hg fluxes obtained for the summer and winter measurements were 38 and 20 ng m^{-2} day^{-1}, respectively. The annual mercury flux from this area was estimated by a combination of the TGM and DGM data with monthly average water temperatures and wind velocities, resulting in an annual flux of 9.5 µg m^{-2} y^{-1}. This flux is of the same order of magnitude as the average wet deposition input of mercury in this area. This indicates that re-emissions from the water surface need to be considered when making mass-balance estimates of mercury in the Baltic Sea as well as modelling calculations of long-range transboundary transport of mercury in Northern Europe. The data presented here is published in Wängberg et al., 2001.

SYNOPTIC VIEW OF THE REGIONAL GEM DISTRIBUTION

Extremely valuable information has been generated during the Nordic Network for atmospheric mercury in the late 80's. In the Nordic study, manual methods for the analysis of atmospheric mercury with a time resolution of several days was applied. A north to south increasing gradient of approximately 15% in the annual average GEM concentration was established (Iverfeldt, 1991a). This effect was attributed to an increasing impact of the major atmospheric mercury source areas in eastern Germany.

In the following section the results from two field experiments carried out for two weeks in summer 1995 (June/July) and three weeks in winter 1997 (March/April) at four European sites along a 800 km line between Berlin and Stockholm will be described (Schmolke, 1999).

The aim of the South-to-North Transect Experiment was on the one hand to get new information about the regional horizontal distribution and transboundary transport of mercury in middle-northern Europe. On the other hand it was an effort to improve the knowledge about the short time variability of GEM and the influence of the main meteorological and atmospheric chemical factors on its diurnal pattern.

Neuglobsow is the most southern sampling site, about 150 km north of the heavily industrialized region between Halle and Leipzig. Considering the area of the former GDR as important source area for atmospheric mercury (Berdowskie et al., 1997).

Zingst, the second German sampling site is located adjacent to the southern shore line of the Baltic Sea.

The Rörvik site is located on the west coast of Sweden approximately 40 km south of Gothenburg and 1km east of the shore.

The most northerly sampling site, Aspvreten, an EMEP monitoring station as well, is located south east of Stockholm.

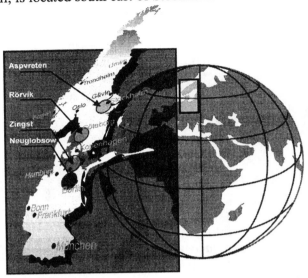

Figure 5. Location of the study sites during the two transect experiments.

The sampling locations are shown in Figure 5. At all four sites GEM was measured using Tekran Gas Phase Mercury Analyzers (Model 2537A). During both, the summer and the winter experiment a decreasing variability in the GEM concentration from south to north was evident.

During both experiments a south to north decreasing GEM gradient was found. A GEM difference between the mean concentrations observed at the southernmost site Neuglobsow and the most northerly site Aspvreten of 0.60 1995 and 0.38 1997 ng m^{-3} was calculated. To avoid the influence of single peak events on the mean concentration levels, the more robust median GEM levels were also compared. During winter 1997 a north to south increasing median GEM concentration gradient of approximately 20% was found. During the summer 1995 experiment the gradient was less pronounced but with 14 % also significant. Comparing the median concentrations between the 1995 and 1997 experiments the systematically elevated levels at all sites during the winter experiment become obvious. This finding is in accordance to model results which predicts elevated concentrations during the late winter/ early spring (Petersen et al., 1995).

The data sets shown here are currently used as experimental reference data for an international model intercomparison exercise under the UN-ECE

Table 3. Summary of 1-hour Average Concentrations Measured during Transect
1995 and 1997 and Related Basic Statistics.

Site	Valid N (n)	Mean (ng m⁻³)	Minimum (ng m⁻³)	Maximum (ng m⁻³)	Std.Dev (ng m⁻³)
Aspvreten	408	1.75	1.38	4.81	0.21
Rörvik	427	1.94	1.53	2.87	0.18
Zingst	427	2.09	1.62	3.79	0.30
Neuglobsow	425	2.13	1.49	4.03	0.34
1995 June-July					
Aspvreten	222	1.51	1.10	1.79	0.11
Rörvik	232	1.54	1.20	1.91	0.11
Zingst	233	1.83	1.45	3.79	0.26
Neuglobsow	250	2.11	1.42	4.66	0.50

Convention on Long-Range Transport of Air Pollutants (CLTRAP) (WMO/EMEP/UNEP, 1999). Comparing these results with measurements carried out under the umbrella of the Nordic Network for Atmospheric Mercury during the years 1985 to 1989, and recent data from the years 1990 to 1992, two trends are supported. On one hand, the trend of declining GEM background concentration during recent years continues. During the years 1985-1989 and 1990-1992 (Iverfeldt et al., 1995) the median concentrations of GEM, observed at the Swedish west coast, were 2.8, and 2.6 ng m⁻³ respectively. On the other hand, the absence of elevated GEM peak events at the Swedish sampling sites during the time of this study corresponds well with investigations carried out in the years 1991 and 1992.

Aircraft Measurements of Atmospheric Mercury over Northern and Eastern Europe

Since almost all our knowledge on the regional distribution of atmospheric mercury is derived from ground-based measurements at single locations for different time periods, little information is available on the vertical and horizontal distribution in the troposphere. Based on the generally accepted view that elemental mercury with an atmospheric residence time of about 1 year is by far the dominating component of total atmospheric mercury (Lindqvist and Rodhe, 1985; Slemr et al., 1985; Schroeder and Jackson, 1987), rather even vertical and horizontal distribution of atmospheric mercury in the troposphere of a hemisphere is expected. The very few attempts to measure mercury concentrations onboard airborne platforms, however, have provided conflicting results so far. Slemr et al. (1985) measured GEM concentration above central Europe at altitudes varying from 6,000 to 12,000 m in a fairly small range from 1.2 to 3.1 ng m^{-3} without any pronounced vertical gradient. This is roughly in agreement with the suppositions outlined above. In contrast to this data set, Ionov et al. (1976), Kvietkus et al. (1985), Brosset (1987) and Kvietkus (1995) report the occurrence of pronounced gradients with decreasing GEM concentrations with increasing altitude. Ionov et al. (1976) reported a decrease in mercury concentrations with increasing altitude which was similar to the concentration decrease of radon decay daughters and from this they estimated an atmospheric residence time of mercury to be about 10 days. Measurements of Brosset (1987) were made above sea west of Göteborg where no local sources are to be expected. They cover altitudes up to 3,000 m and the mercury concentrations decreased roughly proportional to the pressure decrease with altitude. Kvietkus et al. (1985) and Kvietkus (1995) reported measurements over different areas of the former Soviet Union. Mercury concentrations varied strongly depending on the location but generally decreased with increasing altitude of up to 3,500 m. A more detailed analysis of the vertical profile over the eastern Lithuania in June 1988 (Kvietkus, 1995) revealed the almost exact proportionality of measured mercury concentrations to the pressure at the sampling altitudes. A possible dependency of the detector response on ambient pressure is not discussed in either of these works. However, a recent study by Ebinghaus and Slemr (2000) has shown, that the response of the most commonly used Atomic Fluorescence Spectroscopy detectors is significantly dependent on the ambient pressure.

They reported mercury measurements onboard an aircraft during a level flight from Oberpfaffenhofen, southwest of Munich, to Halle (400 km distance) and back, made on June 13, 1996. GEM concentrations measured during the horizontal cruises at constant altitudes of 900 and 2,500 m above sea level (a.s.l.) are described. Consequently, an air mass of an entire volume of about 8000 km^3 has been investigated during the flight by sampling over a time period of less than 6 hours. At an altitude of 900 m a.s.l., GEM concentrations showed a slight gradient with decreasing concentration to the north. The average GEM concentration was 1.77 ± 0.1 ng m^{-3} (n=17). According to the radiosonde vertical soundings this entire flight leg was within the mixing layer in a humid air mass. During the return level flight at 2,500 m a.s.l., free tropospheric extremely dry air mass was encountered with an average GEM concentration of 1.64 ± 0.1 ng m^{-3} (n=22). Within the mixing layer the horizontal distribution of atmospheric mercury is very homogeneous over a distance of 400 km with slightly decreasing concentrations to the north. This could be explained by an incomplete exchange of an air mass with low GEM concentrations replacing from the north the air mass with higher GEM concentrations. Major conclusions of our aircraft measurements are: a) atmospheric mercury is evenly distributed within an air mass over long distances, b) concentrations may change with the change of air masses, and c) slight differences between atmospheric mercury concentrations in the mixing layer and the free troposphere may be due to different air masses rather than a vertical gradient.

TEMPORAL TRENDS AND SEASONAL VARIABILITY OF GEM

One of the major questions connected with the present mercury research worldwide is that of trends:

a. have the sources of mercury, its long range transport and deposition increased substantially in comparison with pre-industrial times ?

b. how do they reflect the control measures adopted to control and reduce anthropogenic mercury emissions ?

Almost all information on historical trends in atmospheric mercury concentrations and subsequent deposition has been derived from analysis of dated soils, sediments and peat cores (e.g. Lockhart, 1995; Coggins, 2000; Handong, 2003). These data suggest that the present mercury deposition is 2 to 5 times higher than the pre-industrial ones. Inventories of natural and anthropogenic mercury sources suggest that the present anthropogenic mercury sources are in the same order of magnitude as the natural ones (Nriagu and Pacyna, 1988). Although these inventories are influenced by large uncertainties, they imply an increase in mercury deposition by about a factor of 3 compared with pre-industrial times, which is in fairly good agreement with experimental data derived from sediment, soil and peat analyses (Slemr, 1996).

Because of an atmospheric residence time of about 1 year (Slemr et al., 1985; Lindqvist and Rodhe, 1985), long term monitoring of atmospheric mercury concentrations may provide direct evidence of temporal trends.

Continuous measurements of GEM concentrations with 15 minutes time resolution have been carried out between September 1995 and December 2003 at the atmospheric research station at Mace Head, western Irish coast line (Ebinghaus, 2002). The Mace Head atmospheric research station is ideally placed to study western inflow boundary conditions of atmospheric trace gases travelling from the Atlantic Ocean into northwestern Europe.

Between September 1995 and December 2003 no trend in the atmospheric mercury concentrations could be seen at this location. The annual average concentration levels at Mace Head derived from the entire measurement data between 1995 and 2003 remain fairly constant at 1.74 ng m^{-3}. Comparison with 4 short-term monitoring data sets of two Swedish sites that measured GEM in 1998/99 with similar instrumentation reveals that the atmospheric mercury concentration data obtained at Mace Head are on an average about 0.2 to 0.3 ng m^{-3} higher than those measured at the continental sites. Higher concentrations at Mace Head may partly be explained by emissions of mercury from the ocean surface (Gardfeldt, 2003). However, it should be noted that in general the concentration levels measured at Mace Head are on an average lower than those found at continental European sites that are influenced by anthropogenic mercury emissions.

Measurements of GEM at Harwell, a rural site in southern central England using a similar instrumentation between June 1995 and April 1996 revealed an average concentration of 1.68 ng m^{-3} (Lee et al., 1998).

Between 1990 and 1996 Slemr and Scheel (1998) observed a decreasing trend of 7% in yearly GEM mean concentrations at the Wank Mountain in Southern Germany reducing from 2.97 ng m^{-3} in 1990 to 1.82 ng m^{-3} in 1996. Iverfeldt (1995) observed a decreasing trend in GEM measurements made in Sweden from a mean GEM concentration value of 3.2 ng m^{-3} for the period of 1985 to 1989 to a mean concentration of 2.7 ng m^{-3} for the period 1990 to 1992. These cited studies report measurements taken during earlier time periods compared to those at Mace Head started in September 1995. Significant reductions in the total anthropogenic mercury emissions into the atmosphere have been calculated for the time after 1990 in the course of the political changes in eastern Europe. Emissions were reduced from 726 tons annually before 1990 (Axenfeldt et. al., 1991) to 342 tons per year in 1995 (Pacyna et. al., 2001). Further emission reduction to about 200 tons year^{-1} has been calculated since then however, the percentage is relatively small compared with previous decreases. This estimates are in good agreement with the GEM data set obtained at Mace Head for the years between 1995 and 2001.

Between 1995 and 2003 average GEM-concentrations are remaining fairly stable at a concentration value of 1.75 ng Hg m^{-3}. However, on an annual basis (averaged monthly means) the winter months show higher concentrations compared with the summer months. Monthly averages for the individual months have been used to evaluate seasonal variations in the

GEM background levels. Lowest GEM concentrations have been observed in summer (April to September), with approx. 1.6 ng m^{-3}, whereas the average concentrations during wintertime (October to March) are around 1.9 ng m^{-3}.

Summer minimum GEM concentrations have also been observed by Slemr (1996) at the Wank Summit and by Brosset (1982) in Sweden. GEM measurements at the Wank Mountain in Southern Germany increased from a minimum during December and January to maximum concentration in February, March and April. From April onwards, concentrations decreased towards a minimum. The peak to peak amplitude of seasonal variation observed at the Wank summit was 0.75 ng m^{-3} which corresponded to 30% of the average GEM concentration observed at the site. A spring maximum is also consistent with GEM measurements made by Brosset, (1982) between October 1979 and September 1980, in Sweden. However a second maximum in September and October was also observed in this data. Brosset, (1987) observed another seasonal variation in measurements again in Sweden between July 1983 and June 1984 in which GEM concentrations were at a maximum between October and December.

These two types of seasonal variation are explained by Slemr (1996). The seasonal variation, with summer minimums observed at the Wank Mountain, and in Sweden (Brosset, 1982), is characteristic of the majority of trace gases of which almost all are removed from the atmosphere by oxidation processes (Warneck, 1988). The major oxidation species in the troposphere is the OH radical which has a pronounced seasonal cycle at middle and higher latitudes (Warneck 1988 ; Logan et. al, 1985).

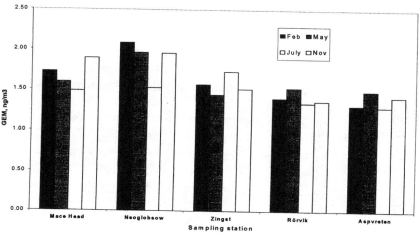

Figure 6. Seasonal variation of GEM during MOE campaigns.

Higher OH concentrations in summer lead to faster removal by oxidation and to a summer minimum in pollutant concentrations. Polluted air masses move in regions with low photochemical activity (Beine, 1997).

A seasonal cycle in washout of mercury has also been observed in many studies (Jensen and Iverfeldt, 1994). Maximum washout occurs in summer, and may result from greater oxidation of to Hg^0 to more soluble species.

Five 2-week campaigns of measurements of atmospheric mercury species were performed at 5 sampling sites in the MOE project (Munthe et al., 2002) in 1998 and 1999. Main features of the results have been presented in Wängberg et al. (2003; 2003a) and Munthe et al. (2003). Four of the measurement sites were located in a south-to-north transect. In Figure 6 the GEM results has been sorted after seasons. Each bar represents the median value observed during one 2-week campaign. The seasonal variation is not very clear. GEM measured in July is the lowest at all sites except Zingst. GEM from the November campaign is the highest at Mace Head and Neuglobsow whereas the data from the May campaign are the highest for the Swedish Stations Rörvik and Aspvreten. Elevated concentrations during colder seasons would be expected in areas where coal is used for heating and electricity production and could explain the Neuglobsow results.

SYNOPTIC VIEW OF MERCURY-SPECIES DISTRIBUTION ON LOCAL AND REGIONAL SCALES

In accordance with recommendations of EMEP/WMO meetings MSC-East has initiated an intercomparison study for mercury transport and chemistry models. Experimental input data have been generated between July 6 to 27, 2000 at GKSS Research Centre Geesthacht, Germany.

Table 4. Summary of average concentrations of airborne mercury species.

Species	Average concentration	Minimum concentration	Maximum concentration	Number of samples
Elemental Hg (GEM)	1.59 ng m^{-3}	1.0 ng m^{-3}	2.4 ng m^{-3}	Ca. 5000
RGM; Denuder	4.0 pg m^{-3}	0.3 pg m^{-3}	90 pg m^{-3}	122
RGM;Mist Chamber	9.0 pg m^{-3}	below detection limit	20 pg m^{-3}	25
TPM	40 pg m^{-3}	below detection limit	275 pg m^{-3}	27

Table 4.2. Summary of average concentrations of mercury species in precipitation.

Species	Average concentration	Minimum concentration	Maximum concentration	Number of samples
Total mercury	7.2 ng L^{-1}	5.0 ng L^{-1}	11.0 ng L^{-1}	7
Reactive mercury	2.2 ng L^{-1}	1.3 ng L^{-1}	3.5 ng L^{-1}	4

Table 4.3. Meteorological conditions.

Parameter	Average value	Minimum value	Maximum value	Number of samples
Air temperature	15.4° C	9° C	27.5° C	
Rel. humidity	77 %	31 %	96 %	
Precipitation rate	3.5 mm / 48 hrs	0.9 mm / 48 hrs	5.9 mm / 48 hrs	8

The study area is a rural site located in a wooded area. Metropolitan Hamburg is located approximately 30 km west of GKSS.

On the basis on the information summarized in the previous tables, the following initial concentration values were adopted by EMEP :

$$\text{GEM: } 1.7 \text{ ng m}^{-3}; \quad \text{RGM: } 5 \text{ pg m}^{-3}; \quad \text{TPM: } 40 \text{ pg m}^{-3}$$

Because of the application in numerical simulation models for Europe, no measurement uncertainty is given, since model simulations can not handle this important analytical parameter appropriately (Ryaboshapko, 2002).

The results of the field measurement activities in the MOE project are summarised in Figure 7 (Munthe et al., 2002). The TGM data show an expected pattern with relatively uniform concentrations and slightly higher concentrations at the Neuglobsow stations and gradually decreasing in a south to north gradient. However, the TGM concentrations at Mace Head are unexpectedly high and average concentrations are higher than all the 3 German and Swedish stations. For RGM, the variability is considerably higher and no clear conclusions can be drawn from data avareged over longer time periods. It can be noted that the RGM measured at Mace Head is similar to that measured at other stations indicating that formation in the marine boundary layer may occur.

TPM is the only species which behaves as expected based on the location of the main source areas i.e. with highest concentrations at the stations closest to te source areas and lowest at the background station Mace Head. Based on these results and detailed evaluations of individual campaign data, it has been concluded that TPM is a good tracer of anthropogenic point source emissions (Wängberg et al., 2003). The MeHg(g) data are highly variable and no significant differences related to distance from source areas can be detected. The results from Mace Head are elevated in comparison to the other stations and also exhibits a higher variability suggesting a possible oceanic source of MeHg(g). Detailed discussions of these results can be found in Wängberg et al. (2003) and Munthe et al.(2003).

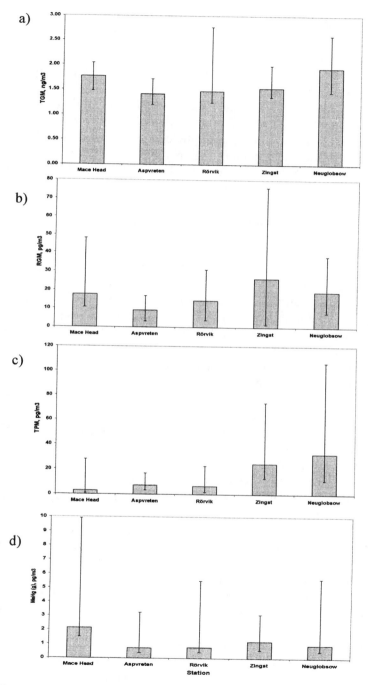

Figure 7. Geographical variation of Hg species at the 5 MOE measurement sites.
a) TGM; b) RGM; c) TPM; d) MeHg(g).

SUMMARY AND CONCLUSIONS

Local emissions contribute to the regional or even hemispherical mercury burden. The potential for long-range transport is strongly dependent on the mercury species emitted.

Observational data on the spatial and temporal distribution of atmospheric mercury are very limited. Available data sets are mainly based on ground-based measurements at single locations for different time periods. Within this work measurement campaigns have been carried out to investigate the (quasi-)real time distribution of atmospheric mercury in the horizontal and in the vertical dimension. Just as significant as differences in regional variability of atmospheric mercury concentrations is the seasonal variability with generally higher concentrations in winter. This observation is consistent with our long-term data set obtained at Mace Head and with independent model results as well.

Aircraft measurements revealed data from two horizontal flight legs at different altitudes (below and above the boundary layer) over a distance of 400 km, and from spiral ascents and descents between 400 and 4000 m. The diameter of the spirals was about 5 km. Consequently, an air mass of an entire volume of about 8000 km^3 has been investigated during the flight by sampling over a time period of less than 6 hours.

Numerical models generally use a constant vertical mixing ratio of atmospheric mercury as an initial parameter however, reliable experimental data on the vertical distribution to support or to validate this important boundary condition hardly existed and gave conflicting results.

It can be concluded that further TGM measurements onboard aircrafts are needed. In the meantime aircraft studies on atmospheric mercury have been increasingly carried out in North America.

The Mace Head atmospheric research station is ideally placed to study western inflow boundary conditions of atmospheric trace gases traveling from the Atlantic Ocean into northwestern Europe. Between September 1995 and December 2003 no trend in the atmospheric mercury concentrations could be seen at this location. The annual average concentration levels at Mace Head derived from the entire measurement data between 1995 and 2003 remain fairly constant at 1.74 ng m^{-3}. Comparison with 4 short-term monitoring data sets of two Swedish sites that measured TGM in 1998/99 with similar instrumentation reveals that the atmospheric mercury concentration data obtained at Mace Head are on an average higher than those measured at the continental sites. We observed that the Mace Head data are about 0.2 to 0.3 ng m^{-3} higher than those of the two Scandinavian background stations. Transport from North America across the Atlantic may well occur, but such large scale processes should not only affect the Mace Head station, but also the Swedish sites close to Stockholm and Gothenburg, respectively. No local anthropogenic emission source exists near Mace Head. Higher concentrations at Mace Head may partly be explained by emissions of mercury from the ocean surface or by oxidative removal processes of TGM while the air masses are traveling from west to east. However, it should be noted that in general the concentration levels measured at Mace Head are on an average lower than those found at continental European sites that are influenced by anthropogenic mercury emissions.

A south to north concentration gradient of atmospheric mercury between highly industrialized Eastern Germany and Central Sweden is well documented and could be attributed to regional anthropogenic emissions especially from the area around Halle/Leipzig/Bitterfeld in Eastern Germany. The reason for the observed west-to-east decreasing gradient from Mace Head to remote sites in Scandinavia is not yet clear.

Another interesting aspect is the comparison of the Mace Head data with those obtained at Alert (82.5°N; 62.3°W) in the Canadian Arctic. Both data sets start in 1995 (January and September respectively) and are ongoing. The long-term average at both sites is around 1.7 ng m^{-3} over the years, both sites do not show a trend of concentrations over time. However, at Alert the phenomenon of the so-called "Mercury Depletion Events (MDEs)" can be observed each year after polar sunrise. By a complex sequence of photolytically mediated chemical and physical processes (which are not yet completely understood) atmospheric mercury concentrations drop below the background value several times for several hours up to days. Recently, MDEs have also been detected at Barrow, Alaska (71° 19'N;156° 37'W) the Eastern part of Hudson Bay (Kuujjuarapik, 55°30'N; 77°73'W) and at a coastal site in the Antarctic (Neumayer, 70°39'S, 8°15'W) as described in this work.

Although Mace Head is only 2 degrees latitude south of the Canadian station Kuujjuarapik, MDEs have never been observed in our data set between 1995 and 2001. We propose that the total absence of sea ice at this coastal location may be a possible explanation for the significantly different environmental chemistry of atmospheric mercury at these two sites.

Monitoring of atmospheric mercury at Mace Head is planned to continue until at least October 2004.

The atmospheric cycling of mercury is driven by the volatility of the elemental form, oxidation and adsorption processes and the subsequent deposition of ionic species to the underlying surface. Once deposited, oxidized mercury species can be converted into the elemental form and are available for evasion fluxes again. The role of vegetation in the overall picture seems to be ambiguous and needs further investigation, since all processes are apparently very specific to both the site and plant species.

To obtain a more complete and accurate picture on the global biogeochemical cycling of mercury including the relative importance of natural sources these pathways and their terrestrial counterparts definitely have to be investigated much more thoroughly before the large discrepancies in global mercury budgets can be resolved. Actual flux measurements could be a major tool to help to answer these questions in the future and are presently increasing in quantity and quality worldwide.

In general it can be concluded that mercury concentrations measured at any specified location at any specified location are influenced by meteorological, photochemical and biosphere/atmosphere exchange processes on variable spatial and temporal scales.

REFERENCES

Amyot, M., Mierle, G., Lean, D.R.S., McQueen, D.J. Volatilization of Hg from lakes mediated by solar radiation, Proc. 1995 Canadian Mercury Network Workshop, York University, Toronto, Ontario, Canada, 1195.

Axenfeldt, F., Münch, J., Pacyna, J. Europäische Test-Emissionsdatenbasis von Quecksilberkomponenten für Modellrechnungen. In: Petersen, G. (Ed.) Belastung von Nord-und Ostsee durch ökologisch gefährliche Stoffe am Beispiel atmosphärischer Quecksilberverbindungen. Abschlussbericht des Forschungsvorhabens 10402726 des Umweltforschungsplanes des Bundesministers für Umwelt, Naturschutz und Reaktorsicherheit. Im Auftrag des Umweltbundesamtes, Berlin, External Report GKSS 92/E/111, GKSS Research Centre Geesthacht, Max-Planck-Str., D-21502 Geesthacht, Germany, 1991.

Baeyens, W. Leermakers, M. Particulate, dissolved and methylmercury budgets for the Scheldt estuary (Belgium and The Netherlands), In: Baeyens W., Ebinghaus, R. and Vasiliev, O. (eds): Global and Regional Mercury Cycles: Sources, Fluxes and Mass Balances. NATO-ASI-Series, 2. Environment, 21, Kluwer Academic Publishers, Dordrecht, The Netherlands, 285-301, 1996.

Baeyens, W., Leermakers, M., Dedeurwaerder, H. Lansens, P. Modelization of the mercury fluxes at the air-sea interface, *Water, Air, Soil Pollut.*, 56, 731-744, 1991

Bahlmann, E. Untersuchungen einer GC/AFS Kopplung zur Bestimmung flüchtiger Quecksilberspezies in der Elbe, Magisterarbeit Universität Lüneburg.

Beine, H.J. Measurements of CO in the high Arctic. AGU Fall meeting, 1997.

Berdowski J.J.M., Baas J.,Bloos J.-P.J., Visschedijk A.J.H., Zandveld P.Y.J. The European Emission Inventory of Heavy Metals and Persistent Organic Pollutants. TNO , Report UBA-FB, UFOPLAN. Ref,No. 104.02 672/03, Apeldoorn, 23, 1997.

Bloom, N.S., Watras, C.J. Observations of methylmercury in precipitation, *Sci. Tot. Env.*, 87/88, 199-207, 1988.

Brosset, C. Total airborne mercury and its possible origin, *Water, Air Soil Pollut.*, 17, 37-50, 1982.

Brosset, C. The behavior of mercury in the physical environment. *Water Air Soil Pollut.*, 34, 145-166, 1987.

Coggins, A.M. Long term measurements of total gaseous mercury (TGM) at Mace Head and atmospheric deposition of heavy metals in ombrogenous peats in the West of Ireland, PhD Thesis, National University of Ireland, Galway, Ireland, 2000.

Coquery, M. Mercury speciation in surface waters of the North Sea, *Neth. J. Sea Res.*, 34, 245-259, 1995.

Cossa, D. and Gobeil, C. Speciation and mass balance of mercury in the lower St. Lawrence estuary and Saguenay Fjord (Canada), In Ebinghaus, R., Petersen, G., Tümpling, U. v. (eds.),: Fourth International Conference on Mercury as a Global Pollutant, Book of Abstracts, Hamburg August, 4 - 8, 1996, p. 458 (GKSS Forschungszentrum Geesthacht GmbH, Max-Planck-Str., D-21502 Geesthacht, Germany), 1996.

Ebinghaus, R. and Wilken, R.D. Transformations of mercury species in the presence of Elbe river bacteria, *Appl. Organomet. Chem.*, 7, 127-135, 1993.

Ebinghaus, R., Kock, H.H., Jennings, S.G., McCartin, P., Orren, M.J. Measurements of Atmospheric Mercury Concentrations in Northwestern and Central Europe --- Comparison of Experimental Data and Model Results, *Atmos. Environ.*, 29, 22, 3333-3344, 1995.

Ebinghaus, R. Krüger, O. Emission and Local Deposition Estimates of Atmospheric Mercury in North Western and Central Europe, In: W. Baeyens, R.Ebinghaus, and O.Vasiliev (eds.): Global and Regional Mercury Cycles: Sources, Fluxes and Mass Balances. NATO-ASI-Series, 2. Environment - 21, Kluwer Academic Publishers, Dordrecht, The Netherlands, 135-159, 1996.

Ebinghaus, R., Tripathi, R.M., Wallschläger, D. Lindberg, S.E. Natural and anthropogenic mercury sources and their impact on the air-surface exchange of mercury on regional and global scales, in Mercury Contaminated Sites – Characterization, Risk Assessment and Remediation, edited by R. Ebinghaus et al., pp. 1-50, Springer-Verlag, Berlin, 1999a.

Ebinghaus, R., et al. International field intercomparison measurements of atmospheric mercury species at Mace Head, Ireland, Atmos. Environ., 33, 3063-3073, 1999b.

Ebinghaus, R. Slemr, F. Aircraft measurements of atmospheric mercury over southern and eastern Germany, Atmos. Environ., 34 , 895-903, 2000.

Ebinghaus, R., Kock, H.H., Coggins, A.M., Spain, T.G., Jennings, S.G., Temme, Ch. Long-term measurements of atmospheric mercury at Mace Head, Irish west coast between 1995 and 2001, Atmos. Environ., 36, 5267-5276, 2002.

Ferrara, R., Ceccarini, C., Lanzillotta, E., Gardfeldt, K., Sommar, J., Horvat, M., Logar, M., Fajon, V., Kotnik, J. Profiles of dissolved gaseous mercury concentration in the Mediterranean seawater, Atmos. Environ., 37, S85-S92, 2003.

Ferrara, R., Maserti, B.E., Petrosino, A., Bargagli, R. Mercury levels in rain and air and the subsequent wash-out mechanism in a central Italian region, Atmos. Environ., 20, 125-128, 1988.

Fitzgerald, W.F. and Mason, R.P. The global mercury cycle: oceanic and anthropogenic aspects, in: Baeyens, W., Ebinghaus, R., Vasiliev, O. (eds) Global and Regional Mercury Cycles: Sources, fluxes and mass balances, NATO ASI Series 2. Environment, 21, Kluwer, Dordrecht, The Netherlands, 85-108, 1996.

Gardfeldt, K., Feng, X., Sommar, J., Lindqvist, O. Total gaseous mercury exchange between air and water at river and sea surfaces in Swedish coastal regions, Atmos. Environ., 35, 3027-3038, 2001.

Gardfeldt, K., Sommar, J., Ferrara, R., Ceccarini, C., Lanzillotta, E., Munthe, J., Wängberg, I., Lindqvist, O., Pirrone, N., Sprovieri, F., Pesenti, E., Strömberg, D. Evasion of mercury from coastal and open waters of the Atlantic Ocean and the Mediterranean Sea, Atmos. Environ, 37, S73– S84, 2003.

Handong Yang, Neil L.R. Distribution of mercury in six lake sediment cores across the UK, Sci. Tot. Env., 304, 391-404, 2003.

Hanson, P.J., Lindberg, S.E., Tabberer, T.A., Owens, J.G., Kim, K.H. Foliar exchange of mercury vapor: evidence for a compensation point, Water, Air Soil Pollut, 80, 373-382, 1995.

Helwig, A., Neske, P. Zur komplexen Erfassung von Quecksilber in der Luft. In: Aktuelle Aufgaben der Messtechnik in der Luftreinhaltung, VDI Berichte 838, 457-466, VDI Verlag Düsseldorf, Germany, 1990.

Hultberg, H., Munthe, J., Iverfeldt, Å. Mechanisms of deposition of methylmercury and merury to coniferous forests, Water, Air Soil Pollut., 80, 363-371, 1995.

Hurley P., Blockley A., Rayner K. Verification of a prognostic meteorological and air pollution model for year-long predictions in the Kwinana region of Western Australia Atmos. Environ., 35, 1871-1880, 2001.

Ionov, V.A., Nazarov, I.M. Fursov, V.Z. Mercury transport in the atmosphere, Reports Acad. Science USSR 228, 456-459, 1976.

Iverfeldt Å. Mercury in forest canopy throughfall water and its relation to atmospheric deposition. *Water Air Soil Pollut.*, 56, 553-564, 1991.

Iverfeldt Å. Occurrence and turnover of atmospheric mercury over the Nordic countries . *Water Air Soil Pollut.*, 56, 251-265, 1991a.

Iverfeldt, A., Munthe, J., Brosset, C., Pacyna, J. Long-term changes in concentration and deposition of atmospheric mercury over Scandinavia, *Water, Air Soil Pollut.*, 80, 227-233, 1995.

Jensen, A., Iverfeldt, A. Atmospheric bulk deposition of mercury to the southern Baltic sea area, In: Watras, C.J. and Huckabee, J.W. (eds) Mercury Pollution: Integration and synthesis, Lewis Publishers, Ann Arbor, Michigan, 221-229, 1994.

Krüger, O., Ebinghaus, R., Kock, H.H., Richter-Politz, I., Geilhufe, Ch. Inverse modelling of gaseous mercury emissions at the contaminated industrial site BSL Werk Schkopau. In: Mercury Contaminated Sites - Characterization, Risk Assessment and Remediation, eds. R. Ebinghaus, R.R., 1999.

Kvietkus, K. Investigation of the gaseous and particulate mercury concentrations along horizontal and vertical profiles in the lower troposphere. In *Proceedings of the 10th World Clean Air Congress,* eds. P. Anttila, J. Kämäri, M. Tolvanen, Espoo, Finland, May 28 – June 2, 284, 1995.

Kvietkus, K., Shakalis, I. Rosenberg, G. Results of measurements of mercury concentrations in the atmosphere on horizontal and vertical profiles, *Phys. Atmos.,* 10, 69-72, 1985.

Lanzilotta, E., Ceccarini, C., Ferrara, R. Photo-induced formation of dissolved gaseous mercuri in coastal and offshore seawater of the Mediterranean basin, *Sci. Tot. Env.,* 300, 179-187, 2002.

Lee, D.S., Dollard, G.J., and Pepler, S. Gas phase mercury in the atmosphere of the United Kingdom, *Atmos. Environ.,* 32, 855-864, 1998.

Lindberg, S.E., Jackson, D.R., Huckabee, J.W., Janzen., S.A., Levin, M.J., Lund, J.R. Atmospheric emission and plant uptake of mercury from agricultural soils near the Almadén mercury mine, *J. Environ Qual.,* 8, 572-578, 1979.

Lindberg, S.E., Meyers, T.P., Taylor, G.E., Turner, R.R., Schroeder, W.H. Atmospheric /surface exchange of mercury in a forest: results of modeling and gradient approaches, *J. Geophys Res.,* 97, 2519-2528, 1992.

Lindberg, S.E., Hanson, P.J., Meyers, T.P., Kim K-Y. Micrometeorological studies of air/surface exchange of mercury over forest vegetation and a reassessment of continental biogenic mercury emissions. *Atmos. Environ.,* 32, 895-908, 1998.

Lindberg, S.E., Meyers, T.P., Munthe, J. Evasion of mercury vapor from the surface of a recently limed acid forest lake in Sweden, *Water, Air Soil Pollut.,* 85, 2265-2270, 1996.

Lindberg, S.E., Stratton, W.J. Atmospheric mercury speciation: Concentrations and behaviour of reactive gaseous mercury in ambient air, *Env. Sci. Technol.,* 32, 49-57, 1998.

Lindqvist, O., Rodhe, H. Atmospheric mercury – a review, *Tellus,* 37B, 136-159, 1985.

Lockhart, W.L., Wilkinson, P., Billeck, B.N., Hunt, R.V., Wagemann, R., Brunskill, G.J. *Water, Air Soil Pollut.,* 80, 603-610, 1995.

Lodenius, M. Dry and wet deposition of mercury near a chlor-alkali plant, *Sci. Total Environ.,* 213, 53-56, 1998.

Logan, J.A. Tropospheric ozone: seasonal behaviour, trends, and anthropogenic influence, *J. Geophys. Res.,* 86, 7210-7254, 1985.

Marks, R., Be dowska, M. Air-sea exchange of mercury vapour over the Gulf of Gdansk and southern Baltic Sea, *J. Mar. Sys.,* 27, 315-324, 2001.

Marks, R. Preliminary investigation of mercury saturation in the Baltic Sea winter surface water, *Sci. Tot. Environ.*, 229, 227-236, 2002.

Munthe, J. The aqueous oxidation of elemental mercury by ozone, *Atmos. Environ.*, 26A, 1461-1468, 1992.

Munthe, J., Wängberg, I., Pirrone, N., Iverfeldt, Å., Ferrara, R., Ebinghaus, R., Feng, X., Gårdfeldt, K., Keeler, G., Lanzillotta, E., Lindberg, S.E., Lu, J., Mamane, Y., Prestbo, E., Schmolke, S., Schroeder, W.H., Sommar, J., Sprovieri, F., Stevens, R.K., Stratton, W., Tuncel, G., Urba, A. Intercomparison of methods for sampling and analysis of atmospheric mercury species. *Atmos. Environ.*, 35, 3007-3017, 2001.

Munthe, J., Wängberg, I., Iverfeldt, Å. Lindqvist, O., Strömberg, D., Sommar, D., Gårdfeldt, K., Petersen, G., Ebinghaus, R, Prestbo, Larjava, K. Siemens, V. Distribution of atmospheric mercury species in Northern Europe: Final Results from the MOE Project. *Atmos. Environ.*, 37, S9-S20, 2003.

Nriagu, J. Mechanistic steps in the photoreduction of mercury in natural waters, *Sci. Tot. Env.*, 154, 1-8, 1994.

Nriagu, J.O. and Pacyna, J.M. Quantitative assessment of worldwide contamination of air, water and soils by trace metals. *Nature*, 33, 134-139, 1988.

OECD, Mercury, Co-operative Risk Reduction Activities for Certain Dangerous Chemicals, Environment Directorate, 1994.

Pacyna, E.G., Pacyna, J.M., Pirrone, N. European emissions of atmospheric mercury from antropogenic sources in 1995, *Atmos. Environ.*, 35, 2987-2996, 2001.

Pai, P., Karamandchandani, P., Seigneur, Ch. Simulation of the regional atmospheric transport and fate of mercury using a comprehensive Eulerian model, *Atmos. Environ.*, 31, 2717-2732, 1997.

Petersen, G., Iverfeldt, Å., Munthe, J. Atmospheric mercury species over central and northern Europe. Model calculations and comparison with observations from the Nordic air and precipitation network for 1987 and 1988, *Atmos. Environ.*, 29, 47-67, 1995.

Petersen, G., Munthe J., Bloxam, R. Numerical modeling of regional transport, chemical transformations and deposition fluxes of airborne mercury species, in: Baeyens, W., Ebinghaus, R., Vasiliev, O. (eds) Global and Regional Mercury Cycles: Sources, fluxes and mass balances, NATO ASI Series 2. Environment, Vol. 21, Kluwer, Dordrecht, The Netherlands, 191-218, 1996.

Pirrone, N., Keeler, G.J. and Nriagu, J.O. Regional Differences in Worldwide Emissions of Mercury to the Atmosphere. *Atmos. Environ.*, 30, 2981-2987, 1996.

Reich, S. Untersuchungen zur Speziesanalytik von Quecksilber in Meerwasser, Diplomarbeit, Fachhochschule Hamburg. 1995

Ryaboshapko, A., Bullock, R., Ebinghaus, R., Ilyin, I., Lohman, K., Munthe, J., Petersen, G., Seigneur, C., Wängberg, I. Comparison of Mercury Chemistry Models, *Atmos. Environ.*, 36, 3881-3898, 2002.

Schmolke, S.R., Wängberg, I., Schager, P., Kock, H.H., Otten, S., Ebinghaus, R., Iverfeldt, Å. Estimates of the air/sea exchange of mercury derived from the Lagrangian experiment, summer 1997, 1st Ann. Meet. EU MAST III Project BASYS, Sept 29-Oct 1, Warnemünde, Germany, 1997.

Schmolke, S.R., Schroeder, W.H., Kock, H.H., Schneeberger, D., Munthe, J., Ebinghaus, R. Simultaneous measurements of total gaseous mercury at four sites on a 800 km transect: spatial distribution and short-time variability of total gaseous mercury over central Europe, *Atmos. Environ.*, 33, 1725-1733, 1999.

Schroeder, W.H. Jackson, R.A. Environmental measurements with an atmospheric mercury monitor having speciation capabilities, *Chemosphere*, 16, 183 –199, 1987.

Schroeder, W.H., Lindqvist, O., Munthe, J. and Xiao, Z.F. Volatilization of mercury from lake surfaces, *Sci. Tot. Env.*, 125, 47-66, 1992.

Schroeder, W.H. Munthe, J. Atmospheric mercury – an overview, *Atmos. Environ.*, 32, 809-822, 1998.

Slemr, F., Schuster, G., Seiler, W. Distribution, speciation and budget of atmospheric mercury, *J. Atmos. Chem.*, 3, 407-434, 1985.

Slemr, F. Trends in atmospheric mercury concentrations over the Atlantic Ocean and the Wank summit and the resulting constraints on the budget of atmospheric mercury, in: Baeyens, W., Ebinghaus, R., Vasiliev, O. (eds) Global and Regional Mercury Cycles: Sources, fluxes and mass balances, NATO ASI Series 2. Environment, 21, Kluwer, Dordrecht, The Netherlands, 33-84, 1996.

Slemr, F., Robertson, P. Brunke, E. Monitoring of atmospheric mercury at Cape Point and Wank. In: Proceedings of the 4th International Conference on Mercury as a Global Pollutant, eds. R. Ebinghaus, G. Petersen and U. v. Tümpling, August 4 - 8, 1996, Hamburg, 127, 1996.

Slemr, F., Scheel, H.E. Trends in atmospheric mercury concentrations at the summit of the Wank mountain, southern Germany, *Atmos. Environ.* 32, 845-853, 1998.

Slemr, F., Brunke, E., Ebinghaus, R., Temme, Ch., Munthe, J., Wängberg, I., Schroeder, W., Steffen, A., Berg, T. Worldwide trend of atmospheric mercury since 1977, *Geophys. Res. Lett.*, 30, 10, 1516, doi:10.1029/2003GL016954, 2003.

Turner, D. Lacerda, O. Vasiliev, W. Salomons, Springer Environmental Science, Springer Verlag, Heidelberg, 377-392.

Turner, R. R., Lindberg., S. E. Behavior and transport of mercury in a river-reservoir system downstream of an inactive chloralkali plant. *Env. Sci. Technol.*, 12, 918-923, 1978.

Urba, A., Kvietkus, K., Marks, R. Gas-phase mercury in the atmosphere over the southern Baltic Sea coast, *Sci. Tot. Env.*, 259, 203-210, 2000.

US EPA, Mercury Study Report to Congress, US EPA-425/R-97-0004, 1997.

Wallschläger D., Wilken, R.-D. The Elber River: A special example for a European river contaminated heavily with mercury, in: Baeyens, W., Ebinghaus, R., Vasiliev, O. (eds) Global and Regional Mercury Cycles: Sources, fluxes and mass balances, NATO ASI Series 2. Environment, 21, Kluwer, Dordrecht, The Netherlands, 317-328, 1996.

Wallschläger, D., Hintelmann, H., Evans, R.D. Wilken, R.D. Volatilization of dimethylmercury and elemental mercuryfrom river Elbe floodplain soils, *Water, Air, Soil Pollut.*, 80, 1325-1329, 1995.

Wallschläger, D., Kock, H.H., Schroeder, W.H., Lindberg, S.E., Ebinghaus, R. Wilken, R.D. Mechnism and significance of mercury volatilization from contaminated floodplains of the German river Elbe, *Atmos. Environ.*, 34, 3745-3755, 2000.

Wallschläger, D., Kock, H.H., Schroeder, W.H., Lindberg, S.E., Ebinghaus, R. and Wilken, R.D. Estimating gaseous mercury emissions from contaminated floodplain soils to the atmosphere with simple field measurement techniques, *Water, Air Soil Pollut.*, 135, 39-54, 2002.

Warneck, P. Chemistry of the natural atmosphere, Academic Publ., New York. 1988

Wilken, R.D. and Hintelmann, H.: Mercury and methylmercury in sediments and suspended particles from the river Elbe, North Germany, *Water, Air, Soil Pollut.*, 56, 427-437, 1991.

WMO/EMEP /UNEP: Workshop on modelling of Atmospheric Transport and Deposition of Persistent Organic Pollutants and Heavy Metals, Geneva, Switzerland, 16 to 19 November 1999

Wängberg, I., Schmolke, St., Schager, P., Munthe, J., Ebinghaus, R., Iverfeldt, A. Estimates of air-sea exchange of mercury in the Baltic Sea, *Atmos. Environ.*, 35, 5477-5484, 2001.

Wängberg I, Munthe J., Pirrone N., Iverfeldt Å., Bahlman E., Costa P., EbinghausR., Feng X., Ferrara R., Gårdfeldt K., Kock H., Lanzillotta E., Mamane Y., Mas F., Melamed E., Nucaro E. Osnat Y., Prestbo E., Sommar J., Spain G., Sprovieri F., Tuncel G. Atmospheric Mercury Distribution In Northern Europe and In the Mediterranean Region. *Atmos. Environ.*, 35 3019-3025, 2001a.

Wängberg I, Munthe J., Ebinghaus R., Gårdfeldt K., Sommar J. Distribution of TPM in Northern Europe. *Sci. Tot. Environ.* 304, 53-59, 2003.

Wängberg, I., Edner, H., Ferrara, R., Lanzillotta, Munthe, J., Sommar, J., Sjöholm, M., Svanberg, S. and Weibring, P.. Atmospheric mercury near a chlor-alkali plant in Sweden. *Sci. Tot. Env.*, 304, 29-41, 2003a.

Xiao, Z., Munthe, J., Schroeder, W.H. and Lindqvist O.: Vertical fluxes of mercury over forest soil and lake surfaces in Sweden. *Tellus*, 43B, 267-279, 1991.

Chapter-25

ATMOSPHERIC MERCURY: A DECADE OF OBSERVATIONS IN THE GREAT LAKES

Gerald J. Keeler and Timothy J. Dvonch

University of Michigan Air Quality Laboratory, Ann Arbor, MI 48109, USA

INTRODUCTION

The Great Lakes contain approximately 20 percent of the world's surface freshwater and are one of the world's most precious resources. While the Great Lakes are cleaner today than they have been since the 1950s, many factors continue to threaten their health. Contaminant levels remain a concern, especially those that are bioaccumulative, and fish consumption advisories continue to be issued for all of the Great Lakes States and for the Province of Ontario. Environmental threats to the Great Lakes ecosystems extend far outside the region and for many these threats are global in nature.

Mercury (Hg) is one of these persistent, bioaccumulative toxic pollutants of concern. Once mercury is released into the environment it can be converted to an extremely persistent, bioaccumulative organic form, methylmercury. Methylmercury can then build up in organisms high within the food chain, such as fish, posing a risk to wildlife and humans that consume these fish. Mercury continues to be targeted as a pollutant of concern for source identification, reduction and/or elimination through a variety of state, federal and international efforts. Recently, the Great Lakes Governors identified reducing the input of toxic substances to the lakes and reducing human health impacts as major priorities for restoration efforts in the Great Lakes. The atmosphere has been determined to be the most significant source of Hg to Michigan's inland lakes and for some of the

Great Lakes (Fitzgerald et al., 1991; Landis et al., 2002a). Mercury monitoring was identified by the Great Lakes Commission in 2003 as one of the most urgent priority among the air toxic programs in the Great Lakes.

On a global basis, it is estimated that between 50 to 75% of total atmospheric Hg emissions are of anthropogenic origin (Pirrone et al., 1996). Natural emissions are typically assumed to be elemental gaseous Hg^0 (Pacyna and Pacyna, 2002), however, a lack of measurement data make this assumption highly uncertain. Anthropogenic emissions are primarily Hg^0, divalent reactive gaseous mercury (RGM), and particulate Hg (Hg(p)). The dominant form of Hg in the atmosphere is Hg^0. Because it's relatively insoluble and deposits very inefficiently, the mean residence time for Hg^0 in the atmosphere is estimated to be approximately one year (Schroeder and Munthe, 1998) allowing for global redistribution. However, this lifetime was recently challenged due to new insights on the atmospheric chemistry of Hg, and these studies suggest the lifetime of Hg will likely be much shorter. RGM directly emitted to the atmosphere is expected to deposit efficiently on a local or regional scale near major sources largely because of its solubility, as is the case for Hg_p. Atmospheric deposition at any particular location can, therefore, be a complex combination of local, regional, and global emissions and transport/ transformation processes (EPMAP, 1994).

Major anthropogenic Hg sources in the Great Lakes Region and preliminary estimates of their annual emissions into the atmosphere have been reported (Pirrone et al., 1996; USEPA, 1994). Sources include: fossil fuel utility boilers, municipal and hospital waste incinerators, iron and steel production, coke production, lime production, hazardous waste recycling facilities, and secondary copper, petroleum refining, and mobile sources. However, the sources of Hg are numerous and many are not well characterized. It appears that an accurate emissions inventory that includes speciated anthropogenic as well as natural Hg sources is still years away.

Research aimed at understanding atmospheric mercury concentrations, chemistry, and deposition in the Great Lakes Region has been carried out at the University of Michigan Air Quality Laboratory (UMAQL) since 1990. Early studies focused on the relative importance of urban/source areas like Detroit and the Chicago/Gary area on loadings to the Great Lakes. These large-scale collaborative efforts included the Lake Michigan Mass Balance Study (LMMBS) and the Atmospheric Exchange Over Lakes and Oceans Study (AEOLOS). These early investigations required the development and refinement of methods for the measurement and analysis of samples collected in networks. The Great Lakes Atmospheric Mercury Assessment Project (GLAMAP) was such a network and provided the first comprehensive regional atmospheric mercury measurements in the Great Lakes Region. This international study showed the importance of a regional approach to understanding mercury sources and transport. The research to develop and refine measurement methods for automated wet and dry deposition which began with these early studies continues to this day. In the mid to late 1990s it became clear that methods were needed for speciated gaseous mercury and for the accurate determination of mercury associated with particulates. These efforts as well as intensive studies carried out in the Great Lakes to investigate the atmospheric processes that control the transport and fate of mercury are discussed and the major conclusions from our efforts are presented.

Here we report on studies performed in the Great Lakes Region over the past decade and the insights gained from this work into the sources,

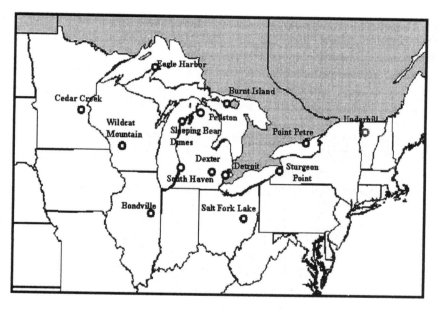

Figure 1. Great Lakes Atmospheric Mercury Monitoring Sites.

transport, chemistry, and deposition of atmospheric mercury. A summary of the major findings from this work are presented and implications discussed.

METHODS

Sampling Sites

The atmospheric measurements discussed in this chapter were collected at multiple sites in the Great Lakes Region as part of a number of mercury studies with different objectives and goals. The characteristics of each of the monitoring sites established and utilized have varied from urban, suburban, rural, lakeshore, upwind, downwind, etc. The sites are shown in Figure 1 and complete site descriptions are provided in the references cited in the text.

Ambient Mercury –Development of USEPA IO-5

Sampling and analysis of vapor and particulate phase Hg in the atmosphere was not routinely performed in the early 1990's. Techniques employed at this time included the use of Au-coated sand traps for the collection of vapor phase Hg followed by Cold Vapor Atomic Fluorescence Spectroscopy (CVAFS) for analysis (Fitzgerald and Gill, 1979). The high pressure drop induced by the Au-coated sand led to the development of Au-coated bead traps which were shown to be ideal for ambient sampling of total vapor phase Hg. Particulate-phase Hg sampling and analysis were also not well developed at that time, and were prone to many sampling biases and artifacts. The importance of well-tested methods that could be employed in a network led to the development of new and refined techniques (Keeler et al., 1995). The new vapor and particulate phase sampling and analysis procedures were included in the USEPA Compendium of Methods for the Determination of Inorganic Compounds in Ambient Air (IO-5) that was published in 1999 (USEPA, 1999).

Automated Speciation Systems

Semi-continuous sampling of speciated Hg in the atmosphere has been carried out using the Tekran (Toronto, Canada) mercury speciation units as described by Landis et al., 2004. Measurements of Hg^0, Hg_p, and RGM were performed using an integrated Tekran®, Inc. 2537A, 1130, 1135-P automated mercury measurement system. This system measured the three Hg species in a semi-continuous fashion, for twelve sample cycles performed over a 24-hour period. A complete single operation cycle consisted of a 1-hr sampling period and a 1-hr desorption period. During the sampling period, ambient air drawn at a rate of 10 L min⁻¹ first passed through a KCl-coated denuder housed in the Model 1130. The air then flowed into the Regenerable Particulate Filter (RPF) assembly located within the Model 1135-P. RGM in the air stream was collected by the KCl-coated denuder followed by the capture of Hg_p (<2.5µm) by a quartz filter disk inside the RPF assembly. Hg^0 which remained in the air stream was then directed into 2537A at a flow rate of 1 L min⁻¹. The Model 2537A was equipped with a pair of matched gold cartridges which allow alternate sampling and desorption. The adsorbent trap thermally desorbs and detects Hg^0 using Cold Vapor Atomic Florescence Spectrometry (CVAFS). During the sampling hour, Hg^0 was measured and detected at a 5-min resolution while the RGM and Hg_p were collected. During desorption, zero air generated by the Model 1130 replaced the ambient air stream into the 2537A at a flow rate of 7 L min⁻¹. The previously sampled Hg_p and RGM were thermally desorbed and converted to elemental mercury at 800 °C and at 500 °C respectively. The Model 2537A sequentially detects the amount of Hg_p and RGM.

Freshly coated denuders were replaced on a weekly basis. Internal calibrations were performed on a daily basis using an internal permeation tube. Calibration of the permeation tube was carried out prior to each intensive campaign. The detection limit for RGM and Hg_p was 4 pg m⁻³, which was calculated as three times the standard deviation of the blank.

The automated mercury methods were compared against the manual methods developed by the UM for the USEPA IO-5. Table 1 shows a comparison of the total gaseous Hg concentration collected in both urban and rural locations using the manual and automated systems.

Table 1. Comparison of Vapor Phase Hg Measurement Techniques: Au-Coated Glass Beads vs. Tekran 2537 Continuous Hg Vapor Analyzer.

Location	Dates	N	Mean Conc. (ng/m³)	Mean % Diff.
Chicago, IL	Jan. 15 - 22, '95	14	4.2	4.4
Davie, FL	Aug. 27 - Sep. 7, '95	21	2.6	5.7
Dexter, MI	Oct. 16 - 30, '95	22	1.6	6.1

The mean absolute percent difference calculated between the daily Au-coated bead method employed in USEPA IO-5 and the 24-hour average of the 5-minute concentrations collected using the Tekran 2537 ranged from 4.4-6.1%. The difference in the two methods is within the analytical and sampling uncertainties of the methods used. The Tekran 1130/1135 automated sampling systems for RGM and Hg_p were also compared against manual methods and the results are reported in Landis et al (2002b) and Lynam and Keeler (2004).

Event Precipitation System

The importance of collecting wet deposition on an event basis for receptor modelling and meteorological analysis has been established (Burke et al., 1995; Hoyer et al., 1995; Dvonch et al., 1999; Landis and Keeler, 1997). The first event-based network in the Great Lakes basin was operated at four sites in Michigan from 1992-1994 (Hoyer et al., 1995). Mercury wet deposition was collected using a MIC-B (MIC Co., Thornhill Ontario) following the design of Merle et al., 1990. This collector was a wet-only automated design which collected the precipitation into a large Teflon-coated funnel which was attached to a 10-L borosilicate glass bottle (Keeler et al., 1994). The desire for a more discrete sampling system that could collect event samples into individual sample bottles without pouring off from a larger collection bottle led to the UM Modified MIC-B sampler described by Landis and Keeler 1997.

The UM sampler allows for the automated collection of up to four different samples using independent sampling trains. The Hg sampling train is composed of glass and Teflon and samples were collected into acid-cleaned Teflon sampling bottles. A separate sampling train was used for trace elements and a third was used for major ions. Precipitation samples were analyzed for total Hg using (CVAFS) and a suite of trace elements (Fe, As, Cd, Pb, V, Cr, Ni, Mn, Zn, etc.) using a Finnigan MAT "Element" magnetic sector (high-resolution) ICP-MS. Analyses of major ions were performed using ion chromatography (Dionex 600) and provides excellent MDLs for the major ions (sulfate MDL ~0.04 mg L^{-1}). All sample handling, processing, and analysis methods employed ultra-clean techniques (Landis and Keeler, 1997; Hoyer et al., 1995).

Dry Deposition

Atmospheric dry deposition remains one of the most difficult measurements to perform for Hg as well other pollutants. Techniques used to estimate the dry deposition flux include micrometeorological approaches, throughfall collection, ambient measurements used in inferential models, and direct measurements using surrogate surfaces. Throughfall measurements, foliar washing, and surrogate surface techniques were employed in the Great Lakes and were described by Rea et al. (2000; 2001; 2002).

Two types of surrogate surfaces were used to directly measure the dry deposition flux of atmospheric mercury. The first was a novel water surface sampler developed concurrently with researchers developing a water surface sampler for nitrogen and sulfur compounds, under a cooperative agreement with the USEPA (Yi et al. 2002). The UM Mercury Water Surface Sampler (MWSS) sampler allowed the collection of both reactive gaseous Hg and particulate Hg forms. The second type of surrogate sampler, described below, used a low-Hg blank, greased surface for the collection of particulate bound Hg. A brief description of the techniques are provided.

Mercury Water Surface Sampler (MWSS)

Total (particulate + gas phase) Hg fluxes were measured using an aerodynamically smooth, Frisbee shaped, recirculating water collection surface during the Atmospheric Exchange Over Lakes and Oceans Study (AEOLOS). The MWSS used in these studies contained a rigorously acid-cleaned glass insert that fit into the center of the aerodynamic surface that was filled with MQ water before each sampling period. The water surface developed by the UMAQL and used in the Lake Michigan studies was patterned after the symmetric low speed airfoil used at Carnegie Melon University (Wu et al., 1992). The airfoil has a sharp leading edge to minimize airflow disruptions caused by collector geometry. The airfoil was constructed of plastic and acid cleaned before the start of the intensive sampling campaign. The expense and difficulties associated with properly cleaning the sampler before each sampling period led to the development of a simpler water surface sampler described below.

The simpler MWSS used in subsequent studies was similar to the recirculating water surface sampler described above except a static water surface was employed. Eliminating the recirculating sampling stream prevented the loss of particulate Hg in the sampler and simplified the cleaning procedures. The static sampler contained a rigorously acid-cleaned Teflon insert that fit into the center of the aerodynamic surface that was filled with MQ water before each sampling period. The water surface plate (39.4 cm diameter, 0.65 cm depth) was made of Teflon, and was replaced with a new acid cleaned plate before each sample. The Teflon plate was placed inside the center area of the aerodynamic surface holder so that the water surface is level with the top of the Frisbee. All fittings downstream of the collection surface were made of Teflon or plastic to minimize adsorption. The dry deposition flux (ng m^{-2} h^{-1}) for Hg collected with the water surface sampler represents the combined RGM and particulate-bound Hg mass in the solution divided by the duration of the sampling period (nominally 24-h). The dry deposition samplers were covered at the onset of any precipitation and uncovered immediately following the rain.

Dry Deposition Greased Surface

The particulate phase dry deposition flux was measured using a smooth greased plate with a sharp leading edge (<10°) mounted on a wind vane or using a symmetric low speed airfoil with a greased center section as described in Wu et al. (1992). The greased plate type of deposition system was used successfully to directly measure particulate dry deposition (Holsen et al. 1992; Sofuoglu et al. 1998; Gilemeister, 2001). The dimensions of the each greased strip used was 5.7 x 1.8 cm with five plates and 20 strips with a total collection area of 205.2 cm^2 exposed. All sample components of the surrogate surfaces were acid cleaned following the procedure defined in Landis and Keeler (1997). The dry deposition flux (ng m^{-2} h^{-1}) for Hg collected with the greased surfaces was determined by measurement of the particulate-bound Hg mass on the surface divided by the area of each greased surface and the duration of the sampling period (nominally 24-h). Experiments were performed to show that Hg^0 did not significantly contribute to the mass of Hg collected by the water and the greased surrogate surfaces. These experiments were performed in the laboratory using the entire surrogate surface samplers under controlled conditions at ambient concentrations similar or greater than typical urban levels (5.5 – 8.7 ng m^{-3}).

MERCURY LEVELS IN THE GREAT LAKES

Ambient Mercury Measurements

The Lake Michigan Urban Air Toxics Study (LMUATS) provided new insight on the levels and behavior of atmospheric mercury and other hazardous air pollutants in the southern Lake Michigan Basin (Keeler, 1994; Holsen et al., 1992; Pirrone and Keeler, 1993; Pirrone et al., 1995). Total mercury measurements were performed simultaneously at three locations during the summer of 1991 as part of the month-long intensive study. The project was one of the first designed to observe the behavior of many different classes of compounds as they were advected from the urban/industrial source regions across Lake Michigan.

The LMUATS revealed that ambient mercury concentrations, both vapor and particulate phase, were significantly elevated in the Chicago urban/industrial area relative to the levels measured concurrently in surrounding areas. The levels of atmospheric mercury varied greatly from day to day at the urban Chicago location. In addition, the total vapor phase mercury concentrations varied diurnally with the highest concentrations observed during the daytime.

Measurements of particulate mercury provided new data on the levels, particle size, and form of this critical pollutant. The concentrations of particle phase mercury during LMUATS were significantly greater than those observed previously at rural sites in the Great Lakes Region, as much as 50 times greater. Particulate mercury was measured on coarse particles >2.5 μm in size as well as on fine particles <2.5 μm. Furthermore, coarse particle Hg was measured in both urban and rural locations, and that the chemical form and reactivity of the particulate Hg varied depending upon the source and meteorological conditions. Since the 1991 study on Lake Michigan, over-water measurements of mercury have been performed in the southern Lake Michigan Basin with levels exceeding those measured during LMUATS (particulate Hg concentrations >1 ng/m^3) (Keeler et al., 1994), and confirmed that particulate Hg was associated with both fine and coarse particulate matter. These findings suggested that dry deposition estimates for mercury had likely underestimated the mass loading of this toxic compound to both terrestrial and aquatic systems. The LMUATS and subsequent over-water cruise Hg data provided the impetus for the later Lake Michigan studies that followed, and made it clear that the factors controlling the transport and deposition of mercury from the atmosphere were not well-understood.

Some of the first long-term atmospheric Hg data in the region was collected as part of a Great Lakes Protection Fund sponsored project that commenced in 1992 (Hoyer et al., 1995). Daily total vapor and particulate phase Hg samples and event wet deposition (discussed later), were collected at three rural sites in Michigan (Pellston, South Haven and Dexter) over a two-year period. Extensive development studies provided sampling and analytical methods for routine atmospheric Hg determinations, and were designed to be robust under very harsh conditions experienced at sampling sites located in three different climatic regions within the Great Lakes Basin.

Regional and local–scale spatial gradients were reported for both the atmospheric concentrations and wet deposition of Hg. Meteorological analysis indicated that the elevated levels of Hg observed in the atmosphere were associated with transport from the urban/industrial area in Detroit as well as with transport from the Chicago/Gary corridor (Hoyer, 1995; Keeler and Hoyer, 1997). The findings demonstrated that source-receptor relationships for atmospheric Hg could be determined, and that short-duration ($\leq$ daily) ambient sampling and event-precipitation sampling were critical for this determination.

The Great Lakes Atmospheric Mercury Assessment Project (GLAMAP) extended the Hg measurements performed in Michigan to a region-wide network of ambient sites in the Great Lakes Region aimed to determine the influence of the large anthropogenic source areas on mercury levels. The GLAMAP commenced in 1994 to address the dearth of atmospheric mercury data for most of the Great Lakes region, and provided a unique database for investigating source-receptor relationships for atmospheric mercury. Measurements performed included gas- and particle-phase mercury, as well as particulate trace elements, from 11 rural monitoring locations across the entire region (Burke, 1998). A total of more than 1,300 sets of 24-hour measurements were collected from the 11 sites over a two-year period from December 1994 through December 1996. Atmospheric mercury concentrations measured during GLAMAP were typical of rural locations, with daily mean concentrations ranging from 1.0 to 3.5 ng m^{-3} for gas-phase mercury and from 1 to 100 pg m^{-3} for particle-phase mercury.

Statistically significant spatial and seasonal differences were observed for both gas- and particle-phase mercury measured across the Great Lakes region. Sub-regions were identified within the region where the GLAMAP sites had similar trends in atmospheric mercury levels. These trends are discussed in terms of their spatial and temporal trends.

Spatial and temporal trends for atmospheric mercury

Atmospheric mercury concentrations measured during GLAMAP were statistically different across the Great Lakes region. Average gas-phase mercury concentrations for the two-year study period differed by as much as 25% between GLAMAP sites (1.63 - 2.03 ng m^{-3}), while average particle-phase mercury levels differed by nearly a factor of three (8.7 - 24.5 pg m^{-3}). These differences were greater than previously reported spatial gradients for atmospheric mercury across smaller geographic scales (Keeler and Hoyer,

1997; Olmez et al., 1996; Keeler et al., 1995; Iverfeldt and Lindquist, 1986). Concentrations of both gas- and particle-phase mercury were consistently higher at the sites in the *east* and *south* sub-regions compared to the sites in the *north* and *west* sub-regions. This spatial trend reflected the proximity of the sites to anthropogenic source areas for atmospheric mercury within the Great Lakes region.

Although concentrations for both gas- and particle-phase mercury were not statistically different between the two sampling years at any of the GLAMAP sites, seasonal differences were statistically significant. Additionally, the observed seasonal trends differed for the two forms of atmospheric mercury. Seasonally averaged gas-phase mercury concentrations were typically highest for the spring seasons and lowest for the autumn seasons during the study. This seasonal trend was consistent across most of the GLAMAP sites, indicating that regional-scale (or larger scale) processes were important for gas-phase mercury. In addition, the magnitude of these seasonal differences was significantly greater at the sites in the east and south parts of the Great Lakes region. Seasonally averaged particle-phase mercury concentrations were significantly higher for the winter season during GLAMAP, but only at the sites in the east and south. Particle-phase mercury concentrations were not statistically different between other seasons at these sites, or between all seasons at the sites to the north and west in the region.

Meteorological factors clearly played a significant role in the seasonal trends for both gas- and particle-phase mercury. For example, specific synoptic-scale meteorological conditions were consistently associated with both above and below average concentrations of particle-phase mercury at the sites in the eastern and southern portions of the Great Lakes region. Periods with elevated atmospheric pressure (≥ 1020 mbar) across the region during the winter and autumn months with lower mixed-layer heights (≤ 800 m) were associated with above average particle-phase mercury concentrations (30-50 pg m^{-3}). The highest concentrations of particle-phase mercury were observed during wintertime high pressure conditions with air mass transport from known anthropogenic source areas. Spatial differences in the seasonal behavior indicated that source influences also contributed to these trends.

Spatial differences in particle-phase mercury concentrations across the Great Lakes region were influenced by proximity to known anthropogenic source areas within the region, since synoptic conditions conducive to air mass transport from these areas were associated with the highest particle-phase mercury concentrations. Distance from the major source areas for the region likely influenced the lower range of concentrations at the sites in the north sub-region compared to the other GLAMAP sites.

Synoptic-scale meteorological features also influenced gas-phase mercury levels in the region but the significance of these relationships was not as strong as observed for particle-phase mercury. Periods with lower atmospheric pressure (daily mean sea-level pressure $\leq$ 1015 mbar) during the spring and summer seasons were associated with above average gas-phase mercury concentrations ($\geq$ 2.0 ng m^{-3}) at the sites in the east and south sub-regions. Precipitation ahead of the frontal boundary typically associated with low pressure systems also occurred with above average concentrations at these sites. Below average gas-phase mercury concentrations (1.4-1.6 ng m^{-3}) occurred during the autumn season with strong pressure gradients between high and low pressure systems, and fast transport across the region (daily mean wind speeds $\geq$ 6 m sec^{-1}). The highest concentrations of gas-phase mercury were observed with low pressure conditions and air mass transport from known source areas. Thus, it was shown that source-receptor relationships for ambient Hg were strongly influenced by the distance from anthropogenic source regions and atmospheric transport that was controlled by synoptic-scale meteorology.

ATMOSPHERIC DEPOSITION PROCESSES

Wet Deposition

From 1992-1994 the UMAQL operated the first Hg event precipitation network in the Great Lakes, which included 3 sites: Dexter, South Haven, and Pellston, MI (see Figure 1). This two-year record of data clearly indicated a strong gradient in the wet deposition of Hg from the elevated levels in the south to the lower levels observed at the Pellston site (Hoyer et al, 1995). This study also found that air mass transport from source regions in summer often led to highly elevated Hg concentrations in precipitation, whereas in winter, a similar air mass trajectory resulted in extremely low levels of Hg in precipitation (~1.5 ng L^{-1}) if the precipitation was snow. This was explained based on the difference in formation of rain droplets versus ice crystals, their different growth processes in-cloud with a potential influence on difference in below-cloud processes during deposition. The cloud microphysical processes, together with the atmospheric Hg speciation, were thought to be responsible for the strong seasonal variations that were observed in the event Hg concentrations and deposition.

The Lake Michigan Mass Balance Study (LMMBS) was performed from July 1994 through October 1995 at five sites (Bondville, IL, Chicago, IL, Kenosha, WI, South Haven, MI, and Sleeping Bear Dunes, MI). Similar to the earlier Michigan network, elevated concentrations of Hg in precipitation were observed at the southern lake sites when compared to the northern site at Sleeping Bear Dunes (Landis et al. 2002a). These observed gradients for Hg in wet deposition were also similar to the gradients observed in ambient gas and particle phase Hg from the GLAMAP project, which were largely the result of anthropogenic point source emissions in the southern Great Lakes region (Landis and Keeler, 2002a). The annual wet deposition of Hg to Lake Michigan averaged over the entire lake was 10.6 μg m^{-2}, or 895 kg to the lake. There was significant spatial and temporal variability in the mercury wet deposition flux over Lake Michigan. The summertime flux of Hg was much larger than the wintertime flux due to the higher concentrations of Hg in rain than in snow and the greater precipitation amounts observed in the summer.

The Atmospheric Exchange Over Lakes and Oceans Study (AEOLOS) conducted concurrently to the LMMBS added an over-water measurement component using the EPA research vessel *Lake Guardian* to the LMMBS network of land-based sites (Landis and Keeler, 2002). A meteorological cluster analysis was conducted which found the Chicago/Gary urban area had a significant impact on atmospheric Hg concentrations across the entire Lake Michigan Basin, and estimated that this area contributed almost 20% of the total deposition to Lake Michigan, and 14% to the wet deposition.

The total deposition due to the sources in the urban/industrial area would be even greater had RGM deposition been considered in that analysis, but due to the lack of a reliable method these measurements were not performed.

The importance of urban sources on deposition of Hg in the urban areas was also investigated in Detroit MI in 1996. Hg event precipitation was collected as part of a study to investigate the atmospheric contributions of Hg in urban runoff (Gildemeister et al., 2004). Mercury wet deposition measured in Detroit over the nine-month period was three times that measured at the Eagle Harbor site for the same period. At the conclusion of this study it was unclear how representative these findings were and whether this trend would continue after changes in Hg emissions.

Recognizing that long-term precipitation records are essential for establishing trends and understanding the impacts of changes in Hg emissions, a decade of event precipitation sampling has been conducted at three sites in Michigan (Dexter, Pellston, and Eagle Harbor, see site locations in Figure 1).

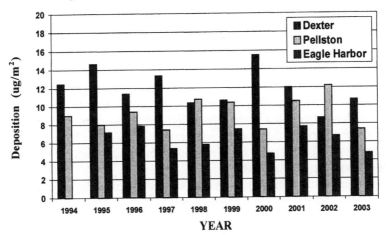

Figure 2. Annual Wet Deposition measured at Dexter, Pellston, and Eagle Harbor, Michigan 1994-2003.

Figure 2 displays the annual Hg wet deposition measured at these sites for the period 1994-2003. Over the 10-year deposition record, a clear decreasing gradient from south to north was observed. While the year-to-year variability in the deposition was on average 18% at each site, the 10-year total wet deposition sum at Dexter was 1.6 times the deposition collected at Pellston and 2.1 times that measured at the Eagle Harbor site. With the exception of the 2002 Hg deposition for Pellston (the maximum annual deposition over the 10 year record) the south to north decreasing gradient in deposition was observed each year. Futhermore, there was not an obvious trend in the deposition rates at the three sites over the decade of measurements. While there have been recent attempts to control Hg emissions within the region and nationally over the past decade, this data illustrates the consistent long-term impact that anthropogenic sources in the southern part of the Great Lakes region have had on Hg deposition across the Great Lakes Basin.

To date, only a limited number of studies have been performed simultaneously in urban areas and in downwind areas impacted by the sources. Studies in both Chicago and south Florida have found as much as 2/3 of the Hg wet deposition to be of local anthropogenic origin (Landis and

Keeler, 2002a; Dvonch et al., 1999). In light of this, the UMAQL has established a new urban Hg wet deposition network, which adds three urban sites, Detroit, Grand Rapids and Flint, to the long-term data collection at the three rural sites of Dexter, Pellston, and Eagle Harbor, in order to assess the long-term influence of urban sources relative to background regional sources through the central region of the Great Lakes. In addition, a new comprehensive monitoring site was established in Stuebenville, OH to specifically assess the impacts of coal-combustion emissions in the southern Great Lakes Region relative to other regional sources contributing to the Hg wet deposition at this site. Two additional years of data collection at the seven sites will allow for a full quantitative source apportionment and assessment of the role of urban anthropogenic sources on Hg wet deposition across the central Great Lakes Region.

Dry Deposition

During periods without precipitation, Hg can be removed from the atmosphere by particle deposition and by gas exchange between the air, water, and earth's surface. The importance of dry deposition as a source of Hg to the Great Lakes and inland aquatic environments was the focus of several studies (Pirrone et al., 1993, 1995a,b; Rea et al., 2001, 2002; Landis et al., 2002a; Vette et al., 2002; Gildmeister et al., 2004). These studies have shown the importance of Hg speciation on the deposition to both urban and remote areas of the region.

Mercury dry deposition flux measurements were performed using surrogate surfaces techniques in Chicago, IL as part of the AEOLOS and LMMBS.

The dry deposition flux (ng m^{-2} h^{-1}) for Hg collected with the water surface sampler represents the combined RGM and particulate-bound Hg mass in the solution, while the greased surface gives the particulate Hg flux only, thus providing a method to determine the RGM flux. The daily dry deposition fluxes measured in July 1994 are shown in Figure 3. The difference between the deposition measured with the two surrogate surfaces revealed that 52% of the dry deposition measured was due to particulate Hg and the remainder was due to RGM deposition.

As part of a whole-ecosystem Hg cycling study Rea et al. (2002) measured Hg in the foliage of deciduous trees in Pellston, MI over the course of the growing season and found that total foliar Hg accumulation was substantially less than vapor phase Hg^0 deposition estimated following Lindberg et al. (1992). It was determined that Hg$_p$ and RGM dry deposition were rapidly washed off foliar surfaces, and therefore foliar accumulation of

Hg most likely represents vapor phase Hg^0 assimilation Rea et al. (2001). In controlled pot and chamber studies with aspen, Ericksen et al. (2003) determined that all foliar accumulation of Hg was due to vapor uptake, regardless of soil Hg concentration.

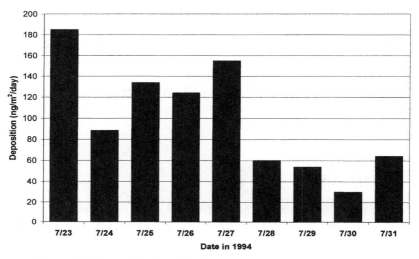

Figure 3. Mercury dry deposition flux measured in Chicago, Illinois using the UM aerodynamic mercury water surface sampler in 1994.

The rate of Hg accumulation in foliage was linear with no significant difference between accumulation rates measured by Rea et al. (2002) in two different forests, with significantly different meteorological conditions and modeled vapor phase Hg^0 deposition velocities at each site. Miller et al. (2004) suggested that the lack of difference in foliar accumulation rates for the two sites studied by Rea et al. (2002) which indicated that foliar Hg accumulation was limited by biological processes mediating sequestration of the Hg. Since the annual transfer of Hg from foliage to forest floor via leaf fall represented the net vapor phase Hg^0 deposition (Rea et al., 2002; Ericksen et al., 2003), Miller et al (2004) developed an empirical method to estimate the accumulation of Hg in foliage of the study area, with the Hg content of deciduous foliage found to be a linear function of growing season length. To date, Hg^0 deposition and accumulation has not been adequately treated in Hg transport and deposition models and represents a significant source of uncertainty in our impact assessments.

In 1996, wet and dry deposition samples were collected at three sites in the City of Detroit to investigate the atmospheric contributions of Hg to urban runoff (Gildemeister, 2001; Gildemeister et al., 2004). The monthly dry particulate deposition flux for the April-October period was similar to the monthly wet deposition flux (10.2 $\mu g\ m^{-2}$ vs 14.8 $\mu g\ m^{-2}$, respectively) at the Livernois site in Detroit. It is anticipated that the total dry deposition flux due to both particulate and gaseous Hg would have been greater than the wet deposition flux, based upon the flux measurements performed in Chicago which suggested that the about 52% of the dry deposition was due to Hgp.

While it was evident that urban sources were impacting Hg deposition to downwind lakes and ecosystems, studies performed to date were limited by the lack of RGM measurements. The RGM data is essential for estimating the dry deposition of Hg and for identifying the source or sources of the Hg deposited.

ATMOSPHERIC CHEMISTRY AND SPECIATION

Hg has been measured in the atmosphere in both gas and particle phases. While greater than 95% of global gaseous Hg is in the elemental state (Hg^0), it is the divalent gaseous form of Hg (Hg^{2+}), as well as particulate Hg, which are highly water soluble (Fitzgerald et al., 1991). It is this water solubility which makes Hg^{2+} species the critical components in understanding Hg removal processes and deposition rates from the atmosphere (Lindberg et al., 1992). The fact that the gaseous forms of Hg interact in a complex way with particulate matter suggest that gas-particle partitioning of Hg also controlled the deposition from the atmosphere. A large fraction (as much as 95%) of the Hg emitted by various source types was in a water soluble, reactive gaseous form (Prestbo and Bloom 1995; Dvonch et al., 1999; Lindberg and Stratton, 1998). While much progress has been made in identifying and quantifying Hg emission sources, few field-based studies have attempted to identify the mechanisms and processes critical to enable predictive modelling of Hg transport, transformation, and deposition. These include the characterization of speciated Hg in emissions, ambient air, and ultimately deposition.

Beginning in 1998 the UMAQL, in collaboration with the USEPA-NERL and Florida DEP, performed a series of field-based evaluations of a prototype ambient Hg speciation instrument designed by Tekran Inc. (Model 1130P). In subsequent years, revisions of this prototype instrument were employed in several UMAQL studies of ambient mercury speciation in both urban and remote locations within and outside the Great Lakes region.

Local Source Impacts in Urban Areas

Measurements of speciated gaseous Hg were made in Detroit, MI during each summer from 2000-2002. The sampling site was located in close proximity (within 4 km) to a large heavy industrial source complex (Dvonch et al 2004), which included coal combustion, oil refineries, coke ovens, iron/steel mills, and sewage sludge incineration. Strong local source impacts were observed at the site with the maximum hourly RGM value reached 208 pg/m^3 and the Hg^0 exceeded 14 ng/m^3 on July 17, 2001. An analysis of the surface meteorological data collected on-site indicated that winds were from the SW during this period, the direction of the nearby industrial source complex. These results provide evidence that Hg^{2+} may remain in a divalent form downwind from the source. The maximum values observed in 2001 were quite similar to those measured in Detroit in 2000 and 2002 (Lynam and Keeler, 2002; Lynam and Keeler, 2004). The maximum RGM values in Detroit were also similar those previously measured by the UMAQL in Baltimore, MD in 1998, when levels reached 211 pg/m^3 after plume impaction at the measurement site by a nearby municipal waste incinerator (Dvonch et al., 2004). Elevated RGM may be expected immediately downwind of waste incinerators since previous in-stack measurements have shown that 75-95% of the Hg is emitted as RGM (Dvonch et al., 1999).

Production of RGM in Ambient Air

Speciated measurements of gaseous Hg were made on the University of Michigan's North Campus in Ann Arbor, MI during the summer of 1999 (Dvonch et al., 2004). A clear diurnal pattern was observed in the RGM concentrations similar to that observed in Detroit. This pattern was particularly pronounced on certain days, such as those shown in June and July, 1999. The highest levels of RGM occurred during the daytime, after solar noon, as seen in Figure 4. A clear positive relationship between RGM and O_3 was also observed on these summer days, as RGM maximums exceeded 140 $pg \ m^{-3}$ on both June 22 and June 23, 1999. Overall for the 16-week sampling period at Ann Arbor, Dvonch et al. (2004) determined the diurnal patterns observed in RGM were found to significantly co-vary with ambient O_3 ($r = 0.50$, $n = 916$, $\alpha = 0.01$). Since O_3 is a photochemically produced secondary pollutant that serves as an indicator of increased photochemistry and increased oxidant production, the positive relationship observed between with RGM points to the real-time production of RGM as a result of photochemical oxidants.

An analysis of concurrent Hg^0 concentrations provided additional evidence for the photochemical production of RGM. Overall for the 16-week sampling period at Ann Arbor, a significant negative correlation was found

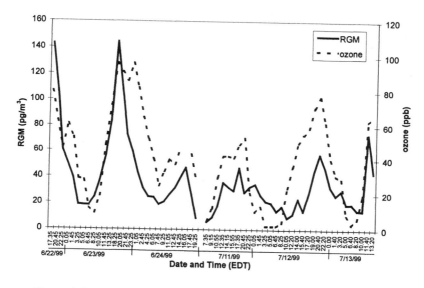

Figure 4. Reactive Gaseous Mercury and Ozone Measured at Ann Arbor, MI (June-July, 1999).

between Hg^0 and O_3 (r = -0.18, n = 526, α = 0.01) (Dvonch et al., 2004). This relationship was particularly pronounced during periods of enhanced O_3, such as September 1-5, 1999 (r = -0.77, n = 58, α = 0.01). A significant negative correlation between Hg^0 and RGM was also found during this period (r = -0.35, n = 66, α = 0.01), as illustrated in Figure 5 by the sharp decrease in Hg^0 together with the increase in afternoon RGM and O_3.

The strong diurnal patterns observed provide additional evidence to suggest that RGM is produced via photochemical reactions. As part of their analysis Dvonch et al. (2004) also calculated daily air mass back-trajectories from the site, which suggested that the diurnal RGM maximums observed at Ann Arbor were not due to local source impacts, but instead were a result of RGM production during transport of the air mass. It was also noted that while the increases in RGM represented roughly only 10% of the decreases in Hg^0 during periods of elevated O_3, a mass balance of the two species should not be expected given the high solubility of RGM relative to Hg^0 and the expected deposition during air mass transport.

While a small amount of field-based data have been published to date to assess atmospheric Hg oxidation in northern temperate climates, recent measurements along the west coast of Washington State identified that large and frequent Hg^0 losses occurred during summertime periods with increased O$_3$ (Weiss-Penzias et al., 2003), and measurements along the west coast of Ireland suggest BrO$^-$ as responsible for Hg^0 oxidation (Munthe et al., 2003). The loss of Hg^0 and production of RGM on days with increased

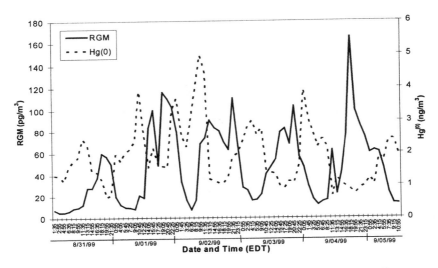

Figure 5. Speciated Gaseous Mercury Measured at Ann Arbor, MI (August-September 1999)

photochemistry observed in Michigan differ from the above investigations in that these measurements are far removed from influences of the marine environment. Because of this, reactive halogen species which evolved from sea salt would not be expected to be responsible for the observed Hg^0 oxidation. Species such as the OH radical are more likely to be important in temperate climates far removed from marine influences, and require further study in future investigations.

Long-Term Continuous Measurements of Speciated Ambient Hg in Detroit

The intensive speciated Hg data collected in urban areas during 1998-2002 made clear the need for long-term measurements. In September of 2002, the UMAQL in conjunction with the Michigan DEQ established a long-term speciated ambient Hg monitoring site in Detroit, MI. Measurements of Hg^0, RGM, and particulate Hg utilizing the Tekran 2537A/1130/1135 Mercury Speciation System were performed. Data are presented here for the first full year of data collection (thru September 2003). Mean levels (± standard deviation) of Hg^0, RGM, and particulate Hg were 2.4 ± 1.4 ng m^{-3}, 16.5 ± 28.9 pg m^{-3}, and 22 ± 30 pg m^{-3}, respectively. The UMAQL has also subsequently established long-term speciated ambient Hg monitoring sites at a rural site in Dexter, MI as well as at a regionally impacted site in Stuebenville, OH to quantify the source impacts and levels of speciated ambient Hg across the Great Lakes region.

Summary and Future Work

Our understanding of the environmental cycle of Hg has drastically improved over the past decade. The importance of urban/industrial areas on the levels and deposition of Hg in the urban areas has been documented. Dry deposition in urban areas likely exceeds wet deposition and the importance of Hg bound to large particles was revealed. Direct emissions of reactive gaseous Hg (RGM) and its role in the total loading of this contaminant were shown to be significant. Deposition of Hg0 and its subsequent accumulation in plant materials results in significant fluxes of Hg to vegetated ecosystems. Elevated RGM was observed during periods of enhanced photochemical activity with high ozone, warm temperatures, and high solar insolation, which indicated that RGM was produced in the atmosphere during atmospheric transport. Changes in the form of both vapor and particulate phase Hg in response to regional changes in atmospheric chemistry suggest that more research is needed to understand the chemical reactions controlling the deposition of this persistent bioaccumulative pollutant.

The Great Lakes region has received considerable attention with respect to the levels and effects of bioaccumulative persistent pollutants (PBTs) such as Hg. Some of the highest concentrations of Hg in precipitation, vapor, and particulate forms have been observed in the region which is consistent with our understanding of the emissions density in the major urban areas in the region. Significant south to north gradients in the levels and deposition of Hg were observed and air mass transport from known source areas could explain the majority of the variability in the Hg deposition recorded.

ACKNOWLEDGEMENTS

The extensive research performed by the UMAQL over the past decade would not have been possible without the support of grants from the US EPA Great Lakes National Program Office, US EPA National Exposure Research Laboratory, USEPA Great Waters Program, and the Great Lakes Protection Fund. The authors would like to thank the dedicated students and staff of the UMAQL for their contributions to our understanding of atmospheric mercury in the Great Lakes Basin including: Marion Hoyer, Lisa Cleckner, Carl Lamborg, Ganda Glinsorn, Steve Mischler, Nicola Pirrone, Alan Vette, Ann Rea, Matt Landis, Janet Burke, Joe Graney, Mary Lynam, Elizabeth Malcolm, Amy Gildemeister, Jim Barres, Julie Peterson, Minghao Zhou, Bian Liu, Emily White-Christainsen, Ali Kamal, Frank Marsik, Khalid Al-Wali and Molly Lyons.

REFERENCES

Burke, J. An Investigation of source-receptor relationships for atmospheric mercury in the Great Lakes Region using receptor modeling techniques. Ph.D. Thesis, University of Michigan, 1998.

Burke, J., Hoyer, M., Keeler, G., Scherbatskoy, T. Wet deposition of mercury and ambient mercury concentrations at a site in the Lake Champlain basin. *Water, Air Soil Pollut.*, 80, 353-362, 1995.

Dvonch, J.T., Graney, J.R., Marsik, F.J., Keeler, G.J., Stevens, R.K. Use of elemental tracers to source apportion mercury in south Florida precipitation. *Env. Sci. Technol.*, 33, 4522-4527, 1999.

Dvonch, J.T., Keeler, G.J., Lynam, M.M., Marsik, F.J., Barres, J.A. The production of reactive gaseous mercury (RGM) in ambient air and its removal by precipitation. *J. Environ. Monit.*, (submitted) 2004.

Ericksen, J., Gustin, M.S., Schorran, D., Johnson, D., Lindberg, S., Coleman J. Accumulation of atmospheric mercury in forest foliage. *Atmos. Envir.*, 37, 1613-1622, 2003.

Expert Panel on Mercury Atmospheric Processes (EPMAP). Mercury Atmospheric Processes: A Synthesis Report. EPRI/TR-104214, 1994.

Fitzgerald, W., Gill G. Sub-nanogram determination of mercury by two-stage gold amalgamation and vapor phase detection applied to atmospheric analysis. *Anal. Chem.* 15, 1714, 1979.

Fitzgerald, W.F., Mason, R.P., Vandal, G.M. Atmospheric cycling and air-water exchange of mercury over mid-continental lacustrine regions. *Water, Air Soil Pollut.*, 56, 745, 1991.

Gildemeister, A.E Urban Atmospheric Mercury: The Impact of Local Sources on Deposition and Ambient Concentration in Detroit, Michigan. Ph.D. Thesis, University of Michigan, 2001.

Gildemeister, A.E, Keeler, G.J.and Graney, J.R. Source proximity reflected in spatial and temporal variability in particle and vapor phase Hg concentrations in Detroit, MI. *Atmos. Environ.*, 39, 353-358, 2005.

Holsen, T.M., Noll, K.E., Fang, G.C., Lee, W.J., Lin, J.M. and Keeler, G.J. Dry Deposition and particle size distributions measured during Lake Michigan Urban Air Toxics study. *Env. Sci. Technol.*, 27, 1141-1150, 1992.

Hoyer, M.E., Burke, J.B., Keeler, G.J. Atmospheric sources, transport and deposition of mercury in Michigan: two years of event precipitation. *Water, Air Soil Pollut.*, 80, 199-208, 1995.

Iverfeldt, A., Lindqvist, O. Atmospheric oxidation of elemental mercury by ozone in the aqueous phase. *Atmos. Environ.*, 20, 1567-1573, 1986;.

Keeler, G.J. Atmospheric Monitoring for the Lake Michigan Mass Balance and the Lake Michigan and Lake Superior Loading Studies: Quality Assurance Project Plan, submitted to the U.S. EPA Great Lakes National Program Office, Chicago, IL: 1994.

Keeler, G., Glinsorn, G., Pirrone, N. Particulate mercury in the atmosphere: its significance, transport, transformation and sources. *Water Air Soil Pollut.*, 80, 159-168, 1995.

Keeler, G. J., Hoyer, M. "Recent measurements of atmospheric mercury in the Great Lakes region." In Atmospheric Deposition of Contaminants to the Great Lakes and Coastal Waters. J.E. Baker, ed. Pensacola, FL: SETAC Press, 1997.

Keeler, G.J., Hoyer, M.E., Lamborg, C.H. Measurements of atmospheric mercury in the Great Lakes basin. Mercury Pollution: Integration and Synthesis. Lewis Publishers, Boca Raton, FL, 231-241, 1994.

Landis, M.S. Keeler, G.J. Critical evaluation of a modified automatic wet-only precipitation collector for mercury and trace element determinations. *Env. Sci. Technol.*, 29, 2123-2132, 1997.

Landis, M.S., Keeler, G.J. Atmospheric mercury deposition to Lake Michigan during the Lake Michigan Mass Balance Study. *Env. Sci. Technol.*, 36, 4518-4524, 2002.

Landis, M.S., Stevens, R.K., Schaedlich, F., Prestbo, E. Development and characterization of an annular denuder methodology for the measurement of divalent inorganic reactive gaseous mercury in ambient air. *Env. Sci. Technol.*, 36, 3000-3009, 2002.

Landis, M.S., Vette, A., Keeler, G.J. Atmospheric mercury in the Lake Michigan Basin: influence of the Chicago/Gary urban area. *Env. Sci. Technol.*, 36, 4508-4517, 2002.

Landis, M.S. Keeler, G.J., Al-Wali, K.I., Stevens, R.K. Divalent inorganic reactive gaseous mercury emissions from a mercury cell chlor-alkali plant and its impact on near-field atmospheric dry deposition. *Atmos. Environ.*, 38, 613-622, 2004.

Lindberg, S.E., Meyers, T.P., Taylor, G.E., Turner, R.R., Schroeder, W.H. Atmosphere-surface exchange of mercury in a forest: Results of modeling and gradient approaches. *J. Geophys. Res.*, 97, 2519-2528, 1992.

Lindberg, S.E., Stratton, W.J. Atmospheric mercury speciation: concentrations and behavior of reactive gaseous mercury in ambient air. *Env. Sci. Technol.*, 32, 49-57, 1998.

Lynam, M.M., Keeler, G.J. Comparison of methods for particulate phase mercury: sampling and analysis. *Anal. Bioanal. Chem.*, 374, 1009-1014, 2002.

Lynam, M.M., Keeler, G.J. Automated speciated mercury measurements in Detroit, Michigan: implications for deposition to the surrounding watershed. *Atmos. Environ.*; (submitted) 2004.

Mierle, G. Aqueous inputs of mercury to precambrian shield lakes in Ontario. *Environ. Tox. and Chem.*, 9, 843-851, 1990.

Miller, E.K., Van Arsdale A., Keeler, G.J., Chalmers, A., Poissant, L. Kammen, N. Estimation and Mapping of Wet and Dry Mercury Deposition Across Northeastern North America. *Ecotoxicology*, (in Press) 2004.

Munthe, J., Wangberg, I., Iverfeldt, A., Lindqvist, O., Stromberg, D., Sommar, J., Gardfeldt, K., Petersen, G., Ebinghaus, R., Prestbo, E., Larjava, K., Siemens, V. Distribution of atmospheric mercury species in Northern Europe: final results from the MOE project. *Atmos. Environ.*, 37 (S1), 9-20, 2003.

Olmez, I. M.R. Ames, G. Gullu, J. Che, J.K. Gone Upstate New York trace metals program Vol. I. "Mercury" MIT Report No. MITNRL-064, 1996.

Pacyna, E.G., Pacyna, J.M. Global emission of mercury from anthropogenic sources in 1995. *Water Air Soil Pollut.*, 137, 149-165, 2002.

Pirrone, N., Keeler, G.J. Deposition of trace metals in urban and rural areas in the Lake Michigan Basin. *Water Sci. & Technol.*, 28, 261-271, 1993.

Pirrone, N., Keeler, G.J., Holsen, T.M. Dry Deposition of semivolatile organic compounds to Lake Michigan. *Env. Sci. Technol.*, 29, 2123-2132, 1995.

Pirrone, N., Keeler, G.J. and Holsen, T.M. Dry Deposition of trace elements over Lake Michigan: A hybrid-receptor deposition modeling approach. *Env. Sci. Technol.*, 29, 2112-2122, 1995.

Pirrone, N., Keeler, G.J., Nriagu, J.O. Regional differences in worldwide emissions of mercury to the atmosphere. *Atmos. Environ.*, 30, 2981-2987, 1996.

Prestbo, E.M., Bloom, N.S. Mercury speciation adsorption method (MESA) method for combustion flue gas: methodology, artifacts, intercomparison, and atmospheric implications. *Water, Air Soil Pollut.*, 80, 145-158, 1995.

Rea, A.W., Lindberg, S.E., Keeler, G.J. Assessment of dry deposition and foliar leaching of mercury and selected trace elements based on washed foliar and surrogate surfaces. *Env. Sci. Technol.*, 34, 2418-2425, 2000.

Rea, A.W., Lindberg, S.E., Keeler, G.J. Dry deposition and foliar leaching of mercury and selected trace elements in deciduous forest throughfall. *Atmos. Environ.*, 35, 3453-3462, 2001.

Rea, A.W., Lindberg, S.E., Scherbatskoy, T. Keeler, G.J. Mercury accumulation over time in two northern mixed-hardwood forests. *Water, Air Soil Pollut.*, 133, 49-67, 2002.

Schroeder, W. H., Munthe. J. Atmospheric Mercury - An Overview. *Atmos. Environ.*, 32:809-822, 1998.

Sofuoglu, S.C., Paode, R.D., Sivadechathep, J., Noll K.E., Holsen, T.M., Keeler, G.J. Dry Deposition Fluxes and Atmospheric Size Distributions of Mass, Al, and Mg Measured in Southern Lake Michigan during AEOLOS. *Aerosol Sci. Technol.*, 29, 4, 281-293, 1998.

U.S. EPA Office of Air Quality Planning and Standards, Deposition of Air of Pollutants to the Great Waters. First Report to Congress. Research Triangle Park, NC: GEPA-453/R-93-055, 1994.

U.S. EPA, Compendium of Methods for the Determination of Inorganic Compounds in Ambient Air, IO-5, Sampling and Analysis for Vapor and Particle Phase Mercury Utilizing Cold Vapor Atomic Fluorescence Spectrometry. U.S. EPA National Risk Management Research Laboratory, Cincinnati, OH: EPA-625/R-96-010a, 1999.

Vette, A.F., Landis, M.S., Keeler, G.J. Deposition and emission of gaseous mercury to and from Lake Michigan during the Lake Michigan Mass Balance Study. *Env. Sci. Technol.*, 36, 4525-4532, 2002.

Weiss-Penzias, P., Jaffe, D.A., McClintick, A., Prestbo, E.M., Landis, M.S. Gaseous elemental mercury in the marine boundary layer: evidence for rapid removal in anthropogenic pollution. *Env. Sci. Technol.*, 37, 3755-3763, 2003.

Yee Lin Wu, Davidson, Cliff I., Lindberg, S., Armistead, E. Russell, G. Resuspension of particulate chemical species at forested sites. *Env. Sci. Technol.*, 26, 2428-2435, 1992.

Yi, S.M., Holsen, T.M., Noll, K.E. Development of a novel water surface sampler for dry deposition. *Env. Sci. Technol.*, 36, 2815-2821, 2002.

Chapter-26

RECENT TRENDS IN MERCURY EMISSIONS, DEPOSITION, AND BIOTA IN THE FLORIDA EVERGLADES: A MONITORING AND MODELLING ANALYSIS

Thomas D. Atkeson[1], Curtis D. Pollman[2] and Donald M. Axelrad[1]

[1] *Florida Department of Environmental Protection, MS 6540, 2600 Blair Stone Road Tallahassee, FL 32399-2400 USA*
[2] *Tetra Tech, Inc., Suite 301, 401 W. University Ave., Gainesville, Florida 32601-5280, USA*

INTRODUCTION

With the discovery in 1989 of widespread, severe contamination of biota by mercury in the Florida Everglades, the Florida Department of Environmental Protection and its many collaborators mounted a sustained program of monitoring, modelling and research to plumb the causes of this problem and propose solutions. This analysis describes the 15-year trend records and reconstructions developed to put the research findings in historical context, whereas the process-level research findings are reported elsewhere. At the time these studies began neither the causes of this problem were apparent nor was there any expectation among the scientists that any would live to see the fruits of their labors – *viz.,* a significant decline in mercury in Everglades biota.

However, and most gratifyingly, beginning mid-decade of the 1990's and continuing into the new millennium, progressive, statistically significant declines in mercury concentrations have been observed in both largemouth bass and great egret nestlings at a number of sites located throughout the Everglades (Pollman et al., 2002; Frederick et al., 2001). It was not until later that retrospective studies determined that marked declines in local emissions and deposition of mercury (RMB, 2002; Husar and Husar, 2002) from the major point emissions sources in southern Florida antedated the declines in emissions and deposition.

Atmospheric sources of wet and dry deposition are now accepted as the major sources of mercury to the Everglades (Stober et al., 2001), and because local emissions have been inferred to be the predominant source of mercury deposited in south Florida rainfall, including the Everglades (Dvonch et al., 1999), the question arises whether the observed declines in biota mercury concentrations can be related to declines in local emissions.

This chapter reviews the existing data on mercury emissions, deposition, and biota trends in south Florida in order to address this question. Much of this discussion is based on work previously published by Pollman et al. (2002) and Pollman and Porcella (2003), but extends that work by including more recently available, longer time series for biota concentrations, as well as incorporating new analyses on wet deposition trends of mercury and some exploratory model hindcasting to examine the relationship between emissions and deposition, and aquatic biota response.

This analysis integrates information from numerous studies to evaluate the causal relationship between mercury emissions, deposition and biotic response in the Florida Everglades. We began first with an examination of the recent trends in both mercury emissions and, as a surrogate to test the robustness of the emissions trends, mercury use. The mercury emissions and usage trends data were compiled by RMB Consulting & Research (RMB; 2002) and Husar and Husar (2002), respectively, for the major sources of mercury emissions in south Florida including municipal waste combustion (MWC) and medical waste incineration (MWI) facilities. These results indicate that large reductions in emissions (approximately 90% relative to peak emissions), occurring ca. 1991.

Second, we statistically analyzed wet deposition fluxes for mercury from November 1993 through December 2002 for samples collected in Everglades National Park to determine whether these trends can be related to changes in the atmospheric signal, or are related to changes in rainfall patterns. Although the monitoring period for wet deposition began well after the largest fraction of the reductions in local emissions had occurred, the wet deposition signal still showed a significant decline (ca. 25%; $p = 0.0413$) that agrees reasonably well with the emissions declines during the same period.

Third, we compare the biota trends to examine whether the timing and magnitude of changes in largemouth bass mercury observed in the Everglades are consistent with predicted changes produced by the changing deposition trajectory. These results suggest that the biota changes are indeed consistent with the estimated declines in local emissions and deposition, although additional analyses to test other hypotheses should be conducted before more definitive conclusions are reached.

Trends in Mercury Emissions

Two fundamentally different types of analyses have been conducted to reconstruct recent trends of mercury emissions in south Florida. The first was a direct approach where a historical emissions inventory was compiled for the period 1980 to 2000 for southeastern Florida (i.e. Broward, Dade and Palm Beach Counties (RMB, 2002). Emissions were estimated from plant operational data and emission factors typical for the source under consideration. These three counties were selected as the region containing sources most likely to be important local contributors to mercury deposition in the Everglades and south Florida. The second approach was an inferential or indirect approach, where the trend in local emissions was inferred by reconstructing a mass balance on the flows of mercury ascribed to various use categories or major economic sectors (Husar and Husar, 2002). This latter analysis first focused on mercury use on a national scale, beginning in 1850 and continuing to 2000, then reduced the scale of analysis to the state level for Florida, and finally concluded with a regional analysis for Broward, Dade, and Palm Beach counties in southeast Florida for the period 1950 through 2000.

The emissions estimates compiled by RMB (2002) indicated very large changes occurred between 1980-2000 as a function of the major combustion sources in south Florida (power generation, sugar industry, incineration of municipal and medical wastes; Figure). Total emissions were quite low between 1980 and 1982, and then increased in 1983 by 3.5 times above 1982 levels as both municipal waste combustors (MWC) and medical waste incinerators (MWI) came on line. Local emissions continued to increase through the 1980's until 1991, when a peak emission flux of nearly 3,100 kg/yr of total mercury was estimated. Throughout the peak emission period of 1983-1991, the predominant local mercury emissions source was medical waste incinerators (MWI's, 47 to 76% of the total); when combined with municipal waste combustors (MWC's), these two sources comprised 92 to 96% of the total. The contribution of power generation was never above 0.4%, while sugar processing accounted for 4 to 8% of the estimated emissions.

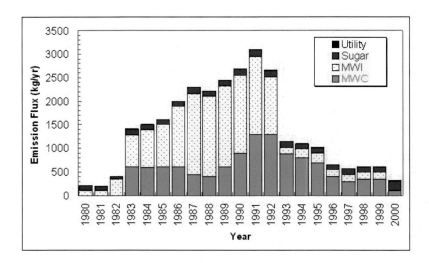

Figure 9. Annual atmospheric mercury emissions in south Florida, 1980 – 2000, estimated by
RMB (2002) as a function of major combustion source category. Sources include power
generation facilities (Utility), municipal waste combustors (MWC), medical waste incinerators
(MWI), and sugar refineries (Sugar). Note: Utility emissions are too
small to be discernable in this graphic.

As more stringent regulatory requirements took effect in 1994, many
MWI's ceased operations and medical waste was sent offsite for autoclaving
and landfilling rather than incinerated. As a result, local emissions declined
sharply through 1993 (65% compared to 1991 levels), followed by a slower
and nearly monotonic rate of decline through 2000. The total estimated
decline in local emissions between 1991 and 2000 is 2,846 kg/yr, which
equates to a total reduction of nearly 93%.

Figure 2 shows the results from the materials flows analysis conducted by
Husar and Husar (2002) for Broward, Dade, and Palm Beach counties. Use
categories that contributed most greatly to the flow of mercury through south
Florida included electrical (e.g., batteries, lighting, and switches), laboratory
use, and control (measuring and control instruments) categories. Although
coal is the largest source (45%; 65 Mg/yr) of mercury emissions for the US
(total 144 Mg/yr), little coal combustion occurs in south Florida and only oil
and product-related emissions occur.

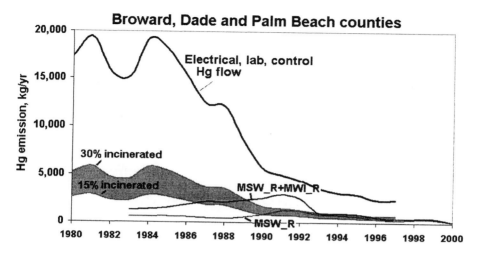

Figure 2. Waste incineration emissions for Dade, Broward, and Palm Beach counties, Florida, inferred from analysis of mercury usage, 1980 to 1997. Upper line shows annual total mercury usage based on different usages. Emission fluxes are based on 30% and 15% incineration rates (complete mobilization of combusted fraction). Plot also shows emissions for MSW and combined MSW and MWI sources estimated by RMB (2002). From Pollman et al. (2002).

Trends in Mercury Deposition

An essentially continuous record of wet deposition fluxes and concentrations of mercury are available from November 1993 through December 2002 for samples collected from the Beard Research Center in Everglades National Park as part of the Florida Atmospheric Mercury Study (FAMS, 1993-1996, Pollman et al., 1995) and the Mercury Deposition Network (MDN; MDN, 1996-2002; http://nadp.sws.uiuc.edu/mdn). The FAMS data consist of integrated monthly wet deposition measurements (Guentzel et al., 2002), while the MDN data consist of integrated weekly samples. During 1996, monitoring from both studies overlapped for the entire year, and comparison of monthly results demonstrated excellent agreement between the two programs (Pollman and Porcella, 2002). As a result, we combined the two studies to form a period of record of eight full years.
Smoothed time series were constructed for mercury deposition, rainfall depth, and volume weighted mean (VWM) mercury concentrations in wet deposition using 12-month running averages derived from the integrated FAMS-MDN data set. As illustrated in Figure 3, rainfall depth and deposition flux are very closely related – this is, of course, because deposition fluxes are the product of weekly volume-weighted mean

concentration and rainfall depth – and it is difficult to discern without further analysis whether any declines in wet deposition fluxes have occurred unrelated to changes in precipitation. Changes in VWM mercury concentrations are a less ambiguous indicator of whether changes in the atmospheric mercury signal have occurred, although precipitation depth does exert some influence on wet deposition concentrations through washout, particularly when the sample integration period is short. Plotting the running average annual VWM as a function of time indicates that VWM mercury concentrations have declined by 25% since late 1993 (in Figure 4).

An alternative analytical approach using analysis of variance (ANOVA; SAS, 1995) was used to eliminate possible confounding effects of both rainfall depth and seasonal dynamics on wet deposition concentrations.

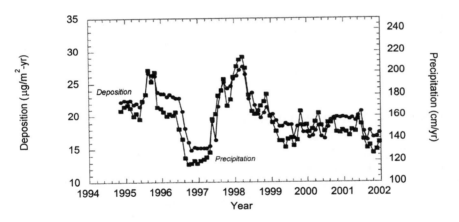

Figure 3. Annual precipitation depth and wet deposition fluxes of mercury measured at Beard Research Station in Everglades National Park, 1993-2002. Data are plotted on a monthly basis as the 12-month running total flux or depth. Data are from the FAMS study (Guentzel et al., 2002) and the MDN network. Squares represent precipitation; circles represent deposition.

Guentzel et al. (2002) demonstrated that very strong seasonal dynamics consistently underlie wet deposition mercury concentrations in Florida within any given year; as a result, a seasonal dummy variable (D_{month}) based on a sinusoidal transformation on the month of year the sample was collected was created and input to the model.

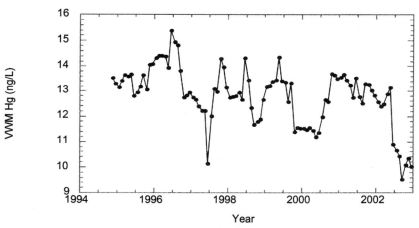

Figure 4. Annual volume weighted mean (VWM) mercury concentration in wet deposition at Beard Research Station, Everglades National Park. Plotted on a monthly basis is the 12-month running average VWM concentration. Data are from the FAMS study (Guentzel et al., 2002) and the MDN network.

The dummy variable had the following form:

$$D_{month} = A \cdot \sin(\frac{M^* \cdot \pi}{12}) + B$$

where A and B are fitted using non linear least squares regression (SAS, 1995) and are equal to 8.8827 and 6.6954, respectively, and M^* is the number of the month (*viz.*, 1 through 12), adjusted using a one month offset so that predicted and observed peak values occurred during the same month. Residuals from the ANOVA model for VWM mercury plotted as a function of time are shown in Figure 5 and demonstrate that a statistically significant decline ($p = 0.0413$) in VWM mercury concentrations occurred over the period of record. Between 1994 and 2002, the analysis indicates that VWM

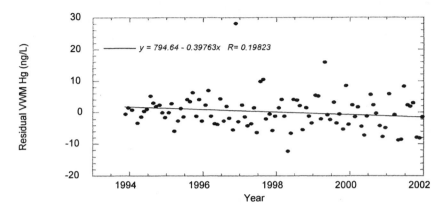

Figure 5. Plot of monthly residuals of ANOVA model of mercury deposition as a function of time. Slope of regression line is significant at $p = 0.0413$.

mercury concentrations declined by approximately 3 ng/L due to factors other than seasonal dynamics and rainfall depth.

The declines in measured VWM concentrations are considerably smaller than the overall decline in local emissions estimated to have occurred since the late 1980's and early 1990's (Figure 1 and Figure 2). However, most of the decline in emissions occurred prior to late 1993 when monitoring of mercury concentrations in wet deposition first began. Indeed, the relatively

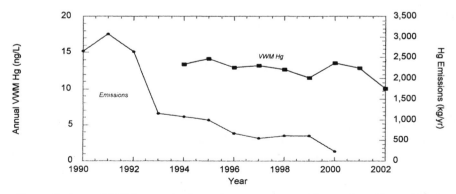

Figure 6. Annual VWM concentrations of mercury measured in wet deposition at Beard Research Center, Everglades National Park, and estimated mercury emissions from Dade, Broward, and Palm Beach counties. Emissions estimates from RMB (2002).

modest change in VWM concentrations agrees reasonably well with the emissions declines after 1993 (Figure 6).

Trends of Mercury in Everglades Biota

Two data sets are available to examine recent trends in mercury concentrations in biota in the Everglades: (1) the data of Lange et al. (2003), who have collected and analyzed largemouth bass for tissue concentrations of mercury from sites throughout Florida since 1988; and (2) the data of Frederick et al. (2001), who examined mercury concentrations in the feathers of great egret chicks, also throughout Florida and including seven sites in south Florida since 1994. Pollman et al. (2002) analyzed the significance of biota temporal trends using the Mann-Kendall Slope Test-of-Sign. This method is a non-parametric test for zero slope that calculates the slope for each possible pairwise combination of observations in the data set, and then ascribes a value of 1, 0, or −1 to the result based on the whether the slope is positive, zero, or negative.

Largemouth bass mercury concentrations for 12 sites across Florida (including 9 sites in the Everglades) were analyzed for trend significance. The period of record analyzed extended from as early as 1988 to as late as 2003. The data were stratified according to age class since different age classes in any given year reflect different exposure histories. Of a possible 120 categories (*i.e.*, 10 age classes x 12 sites), 66 had sufficient data to test for sign significance (Table 1). The results were split relatively evenly between a significant decline at the 95% confidence level (29 site-cohort combinations) and no trend (34 site-cohort combinations). Significant declines were observed across the state, suggesting a regional effect (e.g., atmospheric deposition), with the most consistent declines across cohorts observed for the two Everglades canal sites, L-67A and L-35B (Figure 7).

The three sites in WCA-3A near site 3A-15 (located near the so-called "hot spot" of high fish tissue mercury concentrations in WCA-3A) also showed some cohorts with significant declines, although nearly as many site-cohort combinations also showed no change (Figure 8). Only three site-cohort combinations showed a significant increasing trend, and these all were observed at the U3 site in WCA-2A. This increase likely reflects a highly localized effect both in time and space, such as peat burning and oxidation that occurred in the Everglades following the intense drought and dry-down in May and June 1999 (Pollman et al., 2002). This period of peat oxidation induced a series of short-term but substantial changes in mercury biogeochemistry, including large scale increases in mosquitofish mercury concentrations at site U3, while the response at 3A-15, which remained wet during this period, was more muted (Krabbenhoft and Fink, 2001).

Table 1. Summary of Mann-Kendall Slope Test-of-Sign for trends in mercury concentrations in largemouth bass. Test results are given for individual sites and age cohorts. (-) indicates significant declining trend; (0) indicates no significant trend; and (+) indicates significant increasing trend. Site-cohort combinations with insufficient data are left blank. All results reported at the 95% significance level.

Location\Age Class	0	1	2	3	4	5	6	7	8	9
Northern Florida										
Fowlers Bluff		0	0	-	0	0	-	0	0	
Central Florida										
Lake Tohopekaliga		0	-	0	-	0	0			0
East Lake Tohopekaliga	-	-	-	-	-	-	0			
Everglades										
Miami Canal and L-67A		-	-	-	-	-	0	-		
L-35B Canal		0	-	-	-	-	0			
Indian Camp Creek-Rogers		0	0		0	0				
Marsh-15	-	-	0	0	0					
Marsh-GH	0	-	0	-						
Marsh-OM		-	-							
Marsh-U3	+	+	+	0	0					
Big Lostmans Creek	0	0	0	0	0					
North Prong	0	-	0	-	-	0				

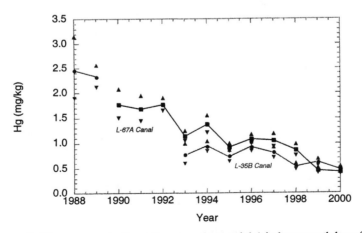

Figure 7. Tissue concentrations of mercury (wet weight) in largemouth bass from the L-67A and L-35B canals in the Florida Everglades. Filled circles and squares show the geometric mean for a respective site each year; filled triangles show ± one standard error of the mean.

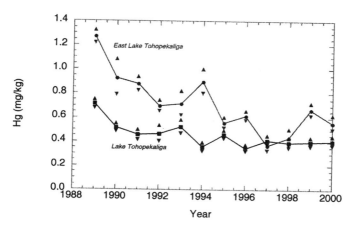

Figure 8. Tissue concentrations of mercury (wet weight) in largemouth bass in East Lake Tohopekaliga and Lake Tohopekaliga located in central Florida. Points show the geometric mean for each year; filled triangles show ± one standard error of the mean.

Great egret chick feather data from all seven sites studied by Frederick et al. (2001) also were tested for trend significance. When Pollman et al. (2002) conducted their trend significance analysis, the time frame spanned by the great egret study extended from 1994 to 2001. Additional data have since been collected, and the full period of record now extends to 2003 (Figure 9).

Four sites (Alley, Hidden, JW1, and L67) showed significant downward trends through 2001 based on both the Mann-Kendall test and Sen's median slope analysis. The data from 2002 and 2003 further substantiate the overall robustness of the downward trend. Consistent with the largemouth bass results from the same region, results from great egret colonies located in the mid-Everglades indicate over an 80% decrease in mercury concentrations over the period of 1994-2003.

Model Hindcasting

The Everglades Mercury Cycling Model (E-MCM) was used to predict changes in age 3 largemouth bass mercury concentrations in response to assumed changes in atmospheric loadings of mercury to site 3A-15, a site which has long been considered a mercury 'hot spot' in the central Everglades, and which has also experienced recent declines in LMB mercury concentrations. Originally adapted to the Everglades using the Mercury

Cycling Model (MCM; Hudson et al., 1994) as the model framework, E-MCM in its present incarnation has been calibrated and applied to at least six Everglades sites (Harris et al., 2003), including site 3A-15. A simplified trajectory of changing deposition rates from 1900 through 2000 was developed with several assumptions or constraints imposed:

1. Based on mercury accumulation rates measured in soil cores in WCA-2A (Rood et al., 1995), an increase in modern deposition rates of 8.7-fold (1985 to 1991) over "pre-industrial" (ca. 1900) was assumed. Rood et al. measured an average accumulation rate of 8 $\mu g/m^2 \cdot yr^{-1}$ for ca. 1900, which yields a modern peak deposition flux of 69.6 $\mu g/m^2 \cdot yr^{-1}$ for 1985-1991.

2. We assume that, superimposed upon the long-term background deposition of 8 $\mu g/m^2$-yr inferred from Rood et al., there has been a deposition signal derived from anthropogenic sources (local and larger geographic scale) that tracks the 1970-2000 mercury trend in the municipal solid waste (MSW) inventory compiled by Kearney and Franklin Associates (1991). This inventory shows that mercury in MSW peaked between 1985 and 1990, with a comparatively sharp decline through 1995, followed by relatively stable inventory quantities. As a first order analysis, we also assumed that anthropogenic emissions and associated deposition fluxes increased linearly from 1900 through 1991, with 1991 corresponding to the

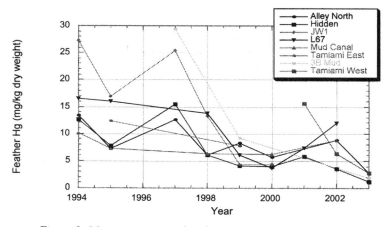

Figure 9. Mercury concentrations in great egret nestlings at various colony locations in the Florida Everglades, 1994 – 2003. Discontinuities in the period of record reflect

year of peak local emissions.

3. After emissions and deposition reached peak levels in 1991, we assumed that deposition declined linearly until 1996, with total mercury deposition reduced to 35 μg m^{-2} yr^{-1}. Following 1996, we assume deposition fluxes declined another 25% consistent with declines in VWM concentrations in wet deposition observed in south Florida (Figure 4). Figure 10 shows the mercury deposition trajectory that resulted from these assumptions.

The mercury deposition trajectory was then used as the input forcing function to reconstruct a predicted time series of biotic (largemouth bass) response in south Florida using the E-MCM model. E-MCM was initially run at pre-1900 deposition conditions until steady-state was achieved in all the model compartments (water, sediments, and biota).

The model then was perturbed by imposing the reconstructed deposition time series, and the predicted biota response compared to the observed trends for ca. 1990-2000.

Testing of the response time of the E-MCM model to perturbations in mercury loadings has shown that the recovery period predicted by the model is sensitive to the size of the pool of Hg(II) in the sediments available for methylation. Mesocosm experiments currently underway in the Everglades

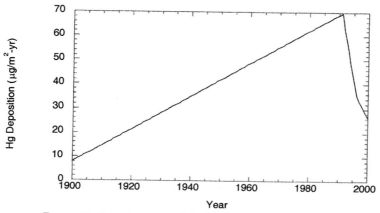

Figure 10. Total (wet + dry deposition) mercury deposition trajectory used in E-MCM model hindcast.

indicate that mercury methylation rates and transfer to the aquatic food chain respond very rapidly in response to new inputs of Hg(II) (D. Krabbenhoft, pers. comm.). These experiments are being conducted using isotopic tracers to elucidate the magnitude and timing of changes in mercury cycling to changes in mercury inputs. Similar results are emerging from the *Mercury Experiment To Assess Atmospheric Loading in Canada and the United States* (METAALICUS; R. Harris, pers. comm.), which also employs isotopic tracers. E-MCM predicts that the primary pathway for introducing mercury into the food chain at site 3A-15 is via methylation in the sediments and through the benthic food web. Thus, the magnitude of the predicted response is governed by the residence time of bioavailable mercury in the sediments, which in turn is governed at least in part by the mixed depth of actively exchanging surficial sediments. In light of the recent isotopic tracer experimental results, the traditionally assumed exchange depth of 3 cm often employed in the model is likely to prove to be a large overestimate of the size and residence time of the Hg(II) pool available for methylation. As a result we conducted the hindcast simulation assuming a sediment exchange depth of 1 cm.

The resultant hindcast indicates that the assumed declines in atmospheric

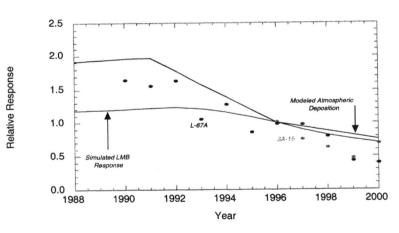

Figure 11. E-MCM simulation hindcast of changes in mercury concentrations in largemouth bass age 1 cohorts at 3A-15 in response to assumed changes in atmospheric deposition. Analysis assumes that the depth of surficial sediments actively exchanging Hg(II) is 1 cm. Shown are normalized (relative to 1996) changes in atmospheric deposition inputs, observed concentrations in largemouth bass (filled circles) for both 3A-15 and the L-67A canal located approximately 20 km east of 3A-15.

inputs can explain nearly 60% of the observed decline in observed trends in largemouth mercury concentrations (a predicted 48% decline vs. an observed 75% decline;Figure). Likewise, the predicted timing of the biota decline is concordant with the observed results. The fraction of the observed decline unexplained by the model hindcast may reflect several factors. First, the reconstructed deposition trajectory may reflect an underestimation of the effects local sources historically exerted on mercury deposition rates in the Everglades. For example, the simulation assumes that the decline in deposition between 1991 and 2000 was 62%, compared to a biota decline approximating 75%. Given the essential linearity of the longterm response of the model (Harris et al., 2001), the hindcast at most would predict a 62% decline. Second, we likely have underestimated the rapidity of the sediment response by assuming an exchange depth of 1cm rather a shallower depth interval. Reducing the exchange depth would bring the slope of the predicted response closer to the slope of the deposition trajectory, which in turn is nearly coincident with the slope of the observed biotic response. Third, the larger observed response may also reflect changes in other environmental variables during the same period – in particular declining surface water sulfate concentrations observed at the site which can influence rates of methyl mercury production (Pollman et al., 2004). The effects of these other variables need to be examined more carefully before more definitive conclusions on causality can be reached.

DISCUSSION AND CONCLUSIONS

Local atmospheric emission rates of mercury in south Florida appear to have declined by over 90% since peak levels occurring in the late 1980's to early 1990's. This estimate is supported by two independent approaches to estimating emissions. Whether these changes in emissions have had a corresponding effect on local deposition rates of mercury is in part a function of the chemical speciation of the emissions. There are two major types of gas phase mercury species present in emissions from combustion sources: elemental mercury or Hg(0), and reactive gaseous mercury (RGM) or Hg(II).

Speciation of emissions is critical because it influences greatly how far emitted mercury likely will be transported. Hg(0) reacts in and is deposited from the atmosphere only very slowly, and has a characteristic residence time in the troposphere on the order of 6 months. RGM, on the other hand, is highly reactive and is scavenged rapidly from the lower troposphere by either wet deposition or by dry deposition processes.

If, for example, there had been a decline in Hg(0) emissions from south Florida, but RGM emissions remained constant, we would expect little or no change in biota mercury concentrations in the Everglades as a result. On the other hand, if local RGM emissions had declined, but Hg(0) emissions remained constant, one would expect to see more of a biotic response. By not considering speciation, we risk misinterpreting the true significance of the relationship between local emissions and biotic response. This would be particularly true if Hg(0) emissions greatly predominate. Unfortunately only limited data are available on the speciation of mercury emissions as a function of source, including speciation measurements conducted by Dvonch et al. (1999) from a municipal waste incinerator (8 measurements), a medical waste incinerator (3 measurements) and a cement kiln (3 measurements) in Dade and Broward counties. The fraction of RGM emitted ranged from 25% of the total (cement kiln) to nearly 95% for the medical waste incinerator. The fraction of RGM emitted by the municipal waste incinerator averaged ca. 75%. Since the local emissions inventory for Dade and Broward counties in 1995-96 was dominated by municipal waste and medical waste incineration (ca. 86% of total emissions), it appears likely that RGM emissions were predominant, at least for 1995-96. If these speciation results are similar for historical emission patterns (and there is no reason to expect that Hg(0) emissions were more important), then our approach of examining total emissions and linking the trends to local biota response appears reasonable.

Coupled with changes in local emission rates is evidence that mercury concentrations in wet deposition (annual VWM) in south Florida have declined by about 25% since late 1993. Statistical analysis indicates that the trends are significant, and are due to factors other than seasonal dynamics and changes in precipitation rates. Although the declines in measured VWM concentrations are considerably smaller than the overall decline in local emissions, most of the decline in emissions occurred prior to late 1993 when monitoring of mercury concentrations in wet deposition first began. Indeed, the relatively modest change in VWM concentrations agrees reasonably well with the emissions declines after 1993.

Statistically significant declines in mercury concentrations in both largemouth bass and great egret chicks have been observed for a number of sites in the Everglades. Declines for both species are on the order of 75 to 80% over approximately the past decade. Model hindcasting using the E-MCM model calibrated for site 3A-15 indicates that changes in atmospheric deposition inferred from sediment core analyses may account for most if not all of the recent changes in largemouth bass mercury concentrations, both in terms of timing and magnitude of change, although the effects of concomitant shifts in other environmental variables (*viz.*, surface water sulfate concentrations) need to be elucidated further. These results are predicated on rapid rates of turnover of the pool of Hg(II) that is readily bioavailable in surficial sediments for methylation, and are consistent with recent isotopic tracer experiments indicating that mercury cycling in aquatic systems responds very rapidly to recent inputs.

Further research, analyses and modelling will explore the uncertainties in the model with the aim of better constraining the processes that govern interactions at the sediment-water interface and the timing of this interaction. Additional improvements will be pursued to refine the E-MCM to improve it as a tool to support environmental policies such as total maximum daily load analyses throughout Florida.

REFERENCES

Atkeson, T.D. Pollman, C.D. Trends of mercury in Florida's environment: 1989-2001. Presented at the 23rd ACS National Meeting, April 7-11; Orlando, FL. 2002.

Dvonch, J.T., J.R. Graney, G.J. Keeler and R.K. Stevens. Utilization of elemental tracers to source apportion mercury in South Florida Precipitation. *Env. Sci. Technol.*, 33, 4522-4527, 1999.

Frederick, P. C., Spalding, M.G., Dusek, R. Wading birds as bioindicators of mercury contamination in Florida: annual and geographic variation. *Env. Toxicol. Chem.*, 21, 262-264, 2001.

Guentzel, J.L., Landing, W.M., Gill, G.A., Pollman, C.D. Processes Influencing Rainfall Deposition of Mercury in Florida. *Env. Sci. Technol.*, 35, 863-873, 2001.

Harris, R., Pollman, C., Hutchinson, D. Beals, D. Florida Pilot Mercury . Total Maximum Daily Load (TMDL) Study: Application of the Everglades Mercury Cycling Model (E-MCM) to Site WCA 3A-15. Report submitted to Florida Department of Environmental Protection. Tetra Tech, Inc., Lafayette, CA. 2001.

Harris, R., Beals, D., Hutchinson, D., Pollman, C. Mercury Cycling and Bioaccumulation in Everglades Marshes: Phase III Report. Final Draft. Report submitted to Florida Department of Environmental Protection. Tetra Tech, Inc., Lafayette, CA. 2003.

Hudson, R.J.M., Gherini, S.A., Watras, C.J., Porcella, D.B. Modeling the biogeochemical cycle of mercury in lakes: The Mercury Cycling Model (MCM) and its application to to the MTL study lakes. In C.J. Watras and J. W. Huckabee [eds.], Mercury Pollution – Integration and Synthesis. CRC Press Inc. Lewis Publishers, 1994.

Husar, J.D. Husar, R.B. Trends of anthropogenic mercury mass flows and emissions in Florida. Final Report, Lantern Corporation, Clayton, MO. 74 pp, Typescript, 2002.

Kearney and Franklin, Assoc. Characterization of Products Containing Mercury in Municipal Solid Waste in the United States, 1970-2000 by A. T. Kearney, Inc. and Franklin Associates, Inc. to US EPA, 4/92 NTIS# PB92-162 569. 1991.

Krabbenhoft, D.P. Fink, L.E. The effect of dry down and natural fires on mercury methylation in the Florida Everglades. Appendix 7-8 in Everglades Consolidated Report, South Florida Water Management District, West Palm Beach, FL http://www.sfwmd.gov/org/ema/everglades/previous.html, 2001.

Lange, T., Richard, D., Sargent, B., Lundy, E. Fish Tissue Monitoring Report. Florida Fish and Wildlife Conservation Commission, Environmental Research Institute, Eustis, Florida. 14, 2003.

NADP, National Atmospheric Deposition Program, Mercury Deposition Network. National Atmospheric Deposition Program (NRSP-3)/Mercury Deposition Network. (2002). NADP Program Office, Illinois State Water Survey, 2204 Griffith Drive, Champaign, IL 61820). http://nadp.sws.uiuc.edu/mdn/ 2002.

Pollman, C., Gill, G., Landing, W., Guentzel, J., Bare, D., Porcella, D., Zillioux, E., Atkeson, T. Overview of the Florida Atmospheric Mercury Study. *Water, Air & Soil Pollut.*, 80, 285-290, 1995.

Pollman, C.D., G.A. Gill, W.M. Landing, J.L. Guentzel, D.A. Bare, D. Porcella, E. Zillioux, Atkeson, T. Overview of the Florida Atmospheric Mercury Study (FAMS). *Water, Air, Soil Pollut.*, 80, 285-290, 1995.

Pollman, C.D. Porcella, D.B. Trends in mercury emissions and concentrations in biota in south Florida: A critical Analysis. Final Report to the Florida Electric Power Coordinating Group. 19 June, 2002. 46 pp, Typescript. 2002.

Pollman, C.D., Porcella, D.B. Assessment of trends in mercury-related data sets. Air Waste Management Association, Florida Section, Annual Meeting. 15-17 September 2002, Jupiter Beach, FL. 2002.

Pollman, C.D. Porcella, D.B. Assessment of trends in mercury-related data sets. J. Phys. IV France 107, 1083-1090, 2003.

Pollman, C.D., Harris, R., Porcella, D.B., Hutchinson, D. Have Changes in Atmospheric Deposition Caused Concomitant Changes in Largemouth Bass Concentrations in the Florida Everglades – A Model Hindcast Analysis. 7th International Conference on Mercury as a Global Pollutant, 27 June – 2 July, 2004, Ljubljana, Slovenia, 2004.

RMB Consulting & Research, Inc. Atmospheric mercury emissions from major point sources – Broward, Dade, and Palm Beach Counties, 1980-2000. Draft Final Report, RMB Consulting & Research, Inc., Raleigh, NC., 2002.

Rood, B.E., Gottgens, J.F., Delfino, J.J., Earle, C.D., Crisman, T.L. Mercury accumulation trends in Florida Everglades and Savannas Marsh flooded soils. Water, Air & Soil Poll.; 80:981-990. 1995.

Stober, Q.J., Thornton, K., Jones, R., Richards, J., Ivey, C., Welch, R., Madden, M., Trexler, J., Gaiser, E., Scheidt, D., Rathbun, S. South Florida Ecosystem Assessment: Phase I/II – Everglades Stressor Interactions: Hydropatterns, Eutrophication, Habitat Alteration, and Mercury Contamination (Summary). EPA 904-R-01-002. USEPA Region 4 Science & Ecosystem Support Division, Water Management Division, and Office of Research and Development. 2001.

Chapter-27

MERCURY POLLUTION IN CHINA – AN OVERVIEW

Xinbin Feng

State Key Laboratory of Environmental Geochemistry, Institute of Geochemistry, Chinese Academy of Sciences, Guiyang 550002, P.R. China

MERCURY POLLUTION IN AQUATIC SYSTEM

Mercury pollution in Songhua River, Northeastern China

Songhua River, situated in Northeastern China, is one of the seven largest rivers in China and was seriously contaminated with mercury since the Acetic Acid Plant of Jilin Chemical Company went into operation in 1958. The plant, which was the largest producer of acetaldehyde in 1960s in China, utilized mercury sulfate as a catalyzer to manufacture acetaldehyde. The process is basically the same as that used once by the Chisso Company in Minamata, Japan, which eventually caused painful Minamata disease. Wastewater from the plant containing both inorganic and methylmercury was directly discharged into the Songhua River. The old technique was completely substituted by a new one without utilization of mercury in 1982. From 1958 to 1982, the plant discharged in total113.2 t total mercury and 5.4 t methyl mercury to the Songhua River, which constituted 69.8% and 99.3% of total anthropogenic Hg-tot and methyl mercury input to this river, respectively. In addition to this contamination source, some small plants which also used mercury, such as a chloroethylene plant from Jilin Chemical Company, Jilin Dye Plant, Changchun Meteorological Instrument Plant, a few gold mining companies using the amalgamation method to extract gold,

and a few chlor- alkaline plants, are distributed inside the drainage area of
the Songhua River and also discharged mercury into the river.

A few, large research projects related to integrated control and
countermeasure of methylmercury pollution of the Songhua River were
carried out between 1980s and 1994. The results from these research projects
were published in two books (Liu, 1994; Liu et al., 1997) and in a number of
papers in Chinese journals (Zhang et al., 1994; Wang and Qi, 1984; Guo,
1990; Zhang et al., 1993; Hou et al., 1994; Yu et al., 1994; Zhai et al., 1991;
Yu et al., 1994; Lin, 1995; Zhang and Li, 1994; li et al., 2001; Wang et al.,
1986; Zhang et al., 1985; Wang et al., 1985).

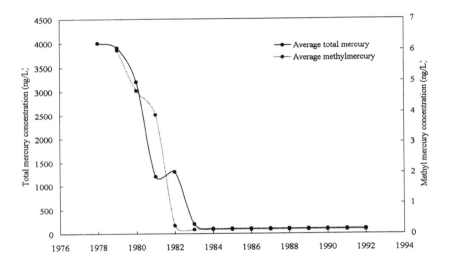

Figure 1. Total and methyl mercury distributions in river water from Jilin City to
Shanzakuo section from 1978 to 1992 (adopted from Yu et al., 1994).

After the largest mercury emission source, the acetic acid plant of Jilin
Chemical Company, completely terminated discharge of mercury to the
Songhua River in 1982, total mercury and methylmercury concentrations in
river water decreased significantly as shown in Figure 1 (Yu et al., 1994).

Total mercury concentrations in surface sediment from different
sampling sites of the Songhua River in 1973 and 1986 were compared, as
shown in Figure 2.

Even though the total mercury concentrations in surface sediment
declined significantly after the major mercury discharge source to the river
was completely stopped, mercury concentrations in sediment at certain

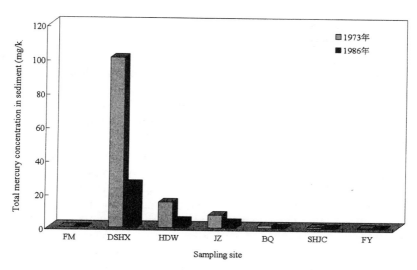

Figure 2. Comparison of mercury concentrations in sediments between samples collected in 1973 and those in 1986. FM: Fengman; DSHX: Dongshihaoxian; HDW: Hadawan; JZ: Jiuzhan; BQ: Baiqi; SHJC: Songhuajiang Chun; FY: Fuyu (Adopted from Zhang et al., 1994).

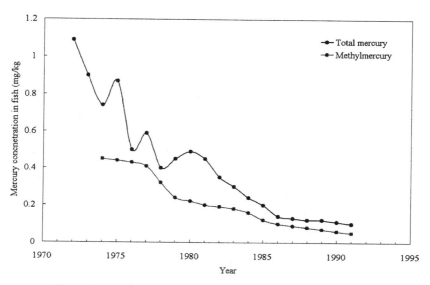

Figure 3. Temporal changes of mercury concentration in fishes from 1972 to 1991 (from Yu et al., 1994).

highly contaminated sites were still elevated compared to the background mercury concentration in sediment of the Songhua River, which is 0.14 mg/kg (Zhang et al., 1994). Both average total mercury and methyl mercury concentrations in fish from the Songhua River were monitored from 1974 to 1991 (Yu et al., 1994).

It is also shown that mercury concentrations in fish decreased gradually after mercury discharge to the river was completely stopped in 1982, as shown in Figure 3.

The health impacts of methylmercury pollution in the Songhua River to local fishermen and inhabitants were intensively investigated (Lin, 1995; Zhang et al., 1993; Hou et al., 1994; Zhai et al., 1991; Li et al., 2001). Lin (1994) systemically reviewed the research progress on the health impact to the local fishermen from methylmercury pollution in the Songhua River from 1970s to late 1980s. The average mercury concentrations in hair of fishermen reached 13.5 mg/kg in 1970s, and some fishermen showed the symptoms of Minamata disease, including concentric constriction of the visual field, loss of sensation in hands and feet, hearing impairment, and ataxia. A survey carried out in 1997 showed that the average total mercury concentration in hair of fishermen dropped to 1.8 mg/kg (Li et al., 2001), and it is demonstrated that the threat of health impact to local fishermen from methylmercury pollution in Songhua River dramatically declined since the major pollution source stopped discharging mercury to the river.

Mercury pollution to Jiyun River, Eastern China

The Jiyun River, located in Tianjin, eastern China, was seriously contaminated with mercury since the 1960s when a Chloralkali plant started to operate at the middle reaches of the river. Before 1977, mercury contaminated wastewater from the plant had been directly discharged into the river. After 1977, a mercury control project went into operation and mercury concentration in wastewater decreased tremendously (Zhang et al., 1981). Many studies (Zhang et al., 1981; Lin et al., 1983, 1984; Lin and Kang, 1984; Kang, 1986; Peng et al., 1983; Yang et al., 1982) were carried out during late 1970s and early 1980s to elucidate mercury distribution and speciation in water and sediment, and the mercury methylation rate in the sediment of this river. It was revealed (Zhang et al., 1981) that methylmercury concentrations in water and sediment varied from below detection limit to 64 ng/L and from below detection limit to 0.35 mg/kg, respectively, and that total mercury concentrations in river water and sediment ranged from 20 to 24,000 ng/L, from 0.03 to 845 mg/kg, respectively. Unfortunately, no studies on mercury concentration in fish from Jiyun River have been reported.

Mercury pollution to other aquatic systems

The Guizhou Organic Chemical Factory (GOCF), a state-owned company, has been producing acetic acid from acetaldehyde synthesized by addition reaction of acetylene and water using mercury as a catalyst. The production of acetic acid in the plant continued from 1971 to 2000 in the eastern urban section of Qingzhen City, Guizhou Province, Southwestern China. The total loss of mercury into the environment was 134.6 t for the 30 year period. The wastewater from the plant was directly discharged into the Dongmenqiao River, which is the water source for paddy rice fields of about 120 ha farmland. Yasuda et al. (2004) investigated mercury distribution in the farmland, and the average total mercury concentration is 15.73±42.98 mg/kg with the maximum concentration of 321.38 mg/kg. GOCF is located about 10 km upstream of Baihua Reservior, which was dammed in 1966. Mercury contaminated wastewater finally reached the reservoir through the Dongmenqiao River. Total mercury concentrations in the sediments of the reservoir ranged from 5 to 40 mg/kg, and total mercury concentrations in water varied from 25.5 to 70.0 ng/l (Yan et al., 2003 and Yan et al., unpublished data). Since the bedrocks of the drainage areas of the reservoir are limestone and dolomite, it is an alkaline reservoir and the pH of the water is around 8. Though the reservoir is seriously contaminated with mercury, total and methylmercury concentrations in fish are usually very low and less than 0.15 mg/kg (wet weight). A research project is on-going in our research group to elucidate the biogeochemical cycling of mercury in the reservoir (e.g. Feng et al., 2004b).

An et al. (1982) reported mercury pollution from the Yanguoxia Chloralkali Plant in Yongjing County, Ganshu Province, Northwestern China. The total mercury concentrations in the air of the workshop varied from 23 to 111 $\mu g\ m^{-3}$. Total mercury concentrations in hair of workers from the chloralkali plant ranged from 6.1 to 202.8 mg/kg and averaged out to 45.4 mg/kg, which were significantly elevated compared to the average total mercury concentration of 0.25 mg/kg in hair from people at a reference site. The plant discharged approximately 4.8 t mercury to the Yellow River each year during late 1970s, and total mercury concentrations in river water of the Yellow River within 14 km down reach of the pollution source varied from 0.04 to 3480$\mu g\ m^{-3}$. Mercury concentrations in wheat grain cultivated around the chloralkali plant varied from 0.02 to 0.11 mg/kg and averaged at 0.068

mg/kg, which already exceeded the national advisory limit of mercury in food, which is 0.02 mg/kg. No further report on mercury pollution status in this area was found from the open literature since 1982. Zhang (1993) reported fish mercury concentrations from the Yellow river in the section of Ningxia Province, which is downstream of Ganshu Province. Total mercury concentrations in fish varied from 0.09 to 0.90 mg/kg in 1982 and 1983, and ranged from 0.45 to 0.98 mg/kg in 1988. Thirty-two percent of the fish in theYellow River in Ningxia section had mercury concentrations exceeding the national advisory limit for fish, which is 0.3 mg/kg in 1982, while 100% of fish exceeded the limit in 1988. Though other mercury pollution sources could not be ruled out, the Yanguoxia Chlor alkali Plant may be responsible for the elevation of mercury in fish in the Ningxia section of the Yellow River.

Table 1. The statistical data of mercury production, imports and exports, and total consumption from 1980 to 1995.

Year	Total national production (ton)	Total import (ton)	Total export (ton)	Total consumption (ton)
1950-79	31432		17156	14276
1980	894		985	
1981	826		946	
1982	912		476	
1983	895		300	320
1985	800			
1990	930			
1991	768	200	63	905
1992	580	538	52	1066
1993	515	344	53	806
1994	466	909	48	1327
1995	779	799	16	1562

ANTHROPOGENIC MERCURY EMISSION INVENTORY IN CHINA

China is one of the largest mercury production and consumption countries, as shown in Table 1. Before 1991, China was a large mercury exporter, but after that became a large mercury importer because the resources in most Chinese mercury mines were exhausted, on the one hand,

and on the other hand, the national demand for mercury has been increasing since the early 1990s when artisanal gold mining activities started.

Asia, especially China, has been regarded as the largest atmospheric mercury emission source from a global perspective (Pirrone et al., 1996, Pacyna and Pacyna, 2002). There is, however, tremendous uncertainty on mercury emission inventory estimates for Asia, especially for China because of lack of direct measurement data to establish reliable emission factors for different anthropogenic sources. The efforts to establish an accurate anthropogenic mercury emission inventory in China are underway (Feng and Hong, 1997; Feng, 1997; Qi, 1997; Feng and Hong, 1999; Wang et al., 2000; Feng et al., 2002). In this section, an estimation of anthropogenic mercury emissions in China in 1995 is made, depending on recent domestic measurements and literature data.

Table 2. Emission factors and total mercury emissions from different anthropogenic sources.

Source category	Emission factor	Total Hg emission (tonnes)
1. Coal combustion	0.12 (g/T)	145
2. Non-ferrous metal production		27
-Primary Cu	10 (g/T)	6
-Primary Pb	3 (g/T)	2
-Primary Zn	20 (g/T)	19
3. Pig iron and steel production	0.04 (g/T)	8
4. Cement	0.1 (g/T)	48
5. Gold extraction		107
-Large scale	0.68 (g/g)	21
-Artisinal	15 (g/g)	86
6. Hg mining	45 (g/kg)	35
7. Chlor-alkali production	18 (g/T NaOH)	2
8. Battery, electrical light, thermal meter (do you mean thermometer?)	5% mercury used	20
9. Others		10
Total		**402**

Emission factors of different sources

The major anthropogenic mercury emission sources are categorized as: coal combustion; copper production; lead production; zinc production; pig iron and steel production; cement production; gold extraction; mercury mining; chlor-alkali production; battery, electrical light, thermometer production; and other sources. Emission factors from different sources are listed in Table 2. The emission factor from coal combustion is synthesized from the literature (Feng, 1997; Qi, 1997; Feng et al., 2002; Wang et al., 2000; Zheng et al., 2004). There is large uncertainty on the estimation of the mercury emission factor from coal combustion in China. Wang et al. (2000) reported an emission factor of 0.177 g/T based on the assumption that average mercury concentration in Chinese coal is 0.22 mg/kg. An intensive investigation of mercury concentrations in Chinese coals was carried out in collaboration between the Institute of Geochemistry, the Chinese Academy of Sciences and the US Geological Survey, and it was revealed that the average mercury concentration in Chinese coal is about 0.15 mg/kg (personal communication with Prof. Zheng Baoshan, 2004), which is much lower than the value used by Wang et al. (2000). Therefore we recalculated the mercury emission factor, which is 0.12 g/T, as listed in Table 2.

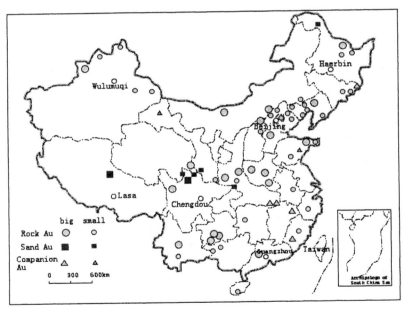

Figure 4. Distribution of gold mines in China.

Gold mining is also one of the most important atmospheric mercury emission sources in China. Gold mines are distributed in most provinces as shown in Figure 4. It is estimated that about 1/3 of annual gold production in

China was produced using the mercury amalgamation technique during the 1980s and 1990s. In the last decade, many surveys were conducted to investigate health impacts to gold miners from different provinces (Wang et al., 1999; Xi et al., 1996; Wang, 2001; Duan et al., 2001; Zhong and Lou, 2000; Guan et al., 2002; Zhang et al., 1999), but only a few studies regarding mercury pollution to the local environment were carried out (Lin et al., 1997; Dai et al., 2003). There are basically two different techniques regarding gold extraction using mercury in China, namely large scale mining and small scale, or artisanal mining. For large scale mining activities, mercury is well recovered, and the emission factor was estimated to be 0.68 g Hg/ g gold produced (Qi, 1997). However, mercury is poorly recovered in the artisanal mining process, and the mercury emission factor is estimated to be 15 g Hg/ g gold produced (Qi, 1997). The annual production of gold from artisanal mining generally constituted about 16% of gold produced using amalgamation techniques during the 1980s and 1990s. From 2000, artisanal gold mining activities were officially banned, but a few illegal artisanal mining workshops are still operating in some remote areas.

Battery production is an important mercury consumption industry, and Table 3 lists the annual estimated mercury consumption in battery production from China in 1992-1999. It is rather difficult to estimate the mercury emission rate from battery, light and thermometer production, and Qi (1997) estimated that, in general, 5% of the used mercury will be emitted to the air. Mercury emission factors from mercury mining and chloralkaline plants were adopted from Qi (1997), and the emission factors from other sources, such as non-ferrous metal production, pig iron and steel production, and cement production are adopted from Pacyna and Pacyna (2002).

Table 3. Estimated mercury consumption (ton) in battery production in China in 1992-1999.

Battery type	1992	1993	1994	1995	1996	1997	1998	1999
Ordinary Zn-Mn	24.2	32.4	26.1	37.3	21.2	23.2	21.3	12.7
Alkaline Zn-Mn	10.3	18.7	24.4	55.2	76.0	137.6	199.1	325.8
Hg Oxide	150.2	300.5	344.3	369.3	237.9	494.5	570.0	463.2
Total	184.7	351.6	394.8	461.8	355.1	655.3	790.4	801.7

Estimates of total mercury emissions in 1995

The 1995 Chinese emissions of total mercury from anthropogenic sources are presented in Table 2. About 402 metric tons of total mercury were emitted in 1995. Coal combustion (including power plant and domestic uses) is the largest source, which accounted for 36.1% of total anthropogenic emissions. Meanwhile, our mercury group is investigating mercury speciation in flue gases in coal combustion using the Ontario Hydro method, which will result in a more precise mercury emission factor from coal combustion in China. All those efforts will definitely improve the precision of estimating mercury emissions from coal combustion in China in the near future.

Gold extraction using mercury may also be an important atmospheric mercury emission source, especially artisanal gold mining activity (Lin et al., 1997). Fortunately the artisanal gold mining activities were completely banned by the government recently, which will significantly reduce mercury emissions from gold extraction from now on. In addition, cement and non-ferrous metal production are still important atmospheric mercury emission sources.

Waste incineration was not employed in China to dispose of waste in 1995, though some metropolitan areas, such as Shanghai, Beijing and Guangzhou, are now using this technology to dispose of municipal waste. The total amount of waste incinerated is, however, still limited in China. Thus, mercury emission from waste incineration, one of the major atmospheric mercury emission sources in western countries, was ignored in our estimation. So far, landfills have been the major way to dispose of municipal waste in China. Though studies showed that landfills are important atmospheric mercury emission source (Lindberg and Price, 1999), we do not yet have enough information to estimate the emission rate from this source category. A research project is, however, on-going in our group to study mercury emission flux from landfills in China, and hopefully the knowledge gap will be bridged soon.

Uses of mercury in chloralkali production, in battery production, and in production of measuring instruments and electrical lights were not banned but decreased in China until 2002, which is in contrast with Western countries, where uses of mercury in these industries have already ceased. However, emissions of mercury from these sources were quite low compared to emissions from other sources.

MERCURY CONCENTRATIONS IN AMBIENT AIR OF URBAN AREAS IN CHINA

Using an acidified, sulfurized cotton trap to pre-concentrate mercury coupled with CVAFS detection method, Yu (1985) for the first time measured total gaseous mercury (TGM) concentrations in ambient air at some tourist attraction areas in Beijing. TGM concentration varied from 1 to 87 ng m^{-3}, and at most sampling sites, the average TGM concentration exceeded 20 ng m^{-3}. Quality assurance (QA) and quality control (QC) information of the method applied is, unfortunately, not available from his work. Using an automated mercury vapor analyzer (Tekran 2537A), Liu et al. (2000) monitored TGM concentrations at one industrial, two urban, three suburban, and two rural sampling locations in January, February and September of 1998 in Beijing, China. In the three suburban sampling stations, mean TGM concentrations during the winter sampling period were 8.6, 10.7, and 6.2 ng m^{-3}, respectively. In the two urban sampling locations, mean TGM concentrations during winter and summer sampling periods were 24.7, 8.3, 10, and 12.7 ng m^{-3}, respectively. In the suburban-industrial and the two rural sampling locations, mean mercury concentrations ranged from 3.1-5.3 ng m^{-3} in winter to 4.1-7.7 ng m^{-3} in summer. It is clear that TGM concentrations in ambient air of Beijing are elevated compared to the global background values that are believed to be within the range from 1.0 to 1.5 ng m^{-3}. Wang et al. (1996) investigated preliminarily TGM concentrations in ambient air in metropolitan Chongqing and its suburb and TGM concentrations varied from 9.2 to 101.5 ng m^{-3} with an average of 34.4 ngm^{3}.

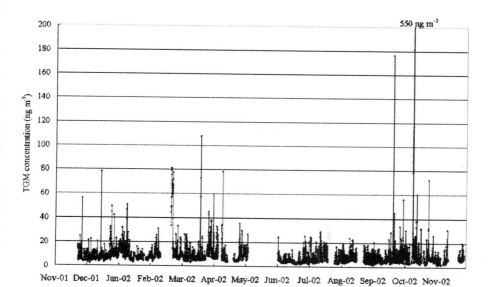

Figure 5. Hourly averaged total gaseous mercury (TGM) concentration in the air of Guiyang, China from November 23,2001 to November 30, 2002
(From Feng et al., 2004).

TGM concentrations in ambient air were occasionally monitored at an urban site in Guiyang between 1996 and 2000 (Feng et al., 2002; 2003). The sampling site where the Institute of Geochemistry, Chinese Academy of Sciences is located is a dense residential area, and a number of industries are located southwest of the sampling site within 20 km.

The average TGM concentration in the ambient air was 11 ± 4 ng m^{-3} in December 1996, and was 13 ± 9 ng m^{-3} in October 1999 (Feng et al., 2002). Four measurement campaigns were carried out to monitor TGM at this site in the following periods: April 19-30, 2000; February 26- March 14, 2001; June 26- July 20, 2001 and October 9 – November 22, 2001, respectively. High temporal resolved data were obtained by using automated mercury analyzers Gardis 1A and Tekran 2537A.

The geometric means of TGM from different seasons were 8.56, 7.45, 5.20 and 8.33 ng m-3 in spring 2000, winter 2001, summer 2001 and fall 2001, respectively (Feng et al., 2003).

A more intensive TGM measurement campaign was carried out at this sampling site from 23 November 2001 to 30 November 2002 using a high temporal resolution 5 min mercury vapor analyzer (Tekran 2537A) (Feng et al., 2004a). A total of 77,541 individual data points were collected and the hourly averaged TGM concentrations over the sampling period are depicted in Figure 5.

TGM concentrations follow lognormal frequency distribution pattern and the mean TGM concentration is 8.40 ng m^{-3} on the basis of one year observation. The elevation of TGM in Guiyang is attributed to coal combustion from both industrial and domestic uses. A seasonal distribution pattern of TGM, with a descending order of winter, spring, fall and summer, was observed. The highest TGM concentration in winter is attributed to household heating using coal.

Only a few data are published on reactive gaseous mercury (RGM) concentrations in ambient air in China. Using the KCl coated tubular denuder technique (Feng et al., 2000), Feng et al. (2002) carried out a short-term measurement campaign in Guiyang in October 1999. The average RGM concentration was 453.8 pg m^{-3}. Shang et al. (2003) also conducted a short-term measurement campaign at the same sampling site in March 2002 using the same technique and the average RGM concentration was 37.5 pg m^{-3}. Coal combustion in Guiyang was demonstrated to be RGM emission source (Tang et al., 2003), and these limited data showed that RGM concentrations in ambient air of Guiyang were significantly higher than the background RGM concentrations < 10 pg m^{-3} measured in Northern America and Northern Europe (Lindberg et al., 1998; Sommar et al., 1999). The significant discrepancy of RGM concentrations between the two campaigns is mainly attributed to the weather conditions. It is dry season in October and it is relatively wet season in March in Guiyang. Since RGM consists of water soluble mercury species such as $HgCl_2$ and $Hg(NO_3)_2$, the lifetime of RGM in ambient air is strongly dependent on the relative humidity. Therefore, the dry season favors the retention of RGM in the air, and RGM concentration could be very high since there are RGM emission sources in the city.

Fang et al. (2001) studied total particulate mercury (TPM) concentrations in ambient air at five sampling sites representing the tourism district, an ordinary industrial district, a scattering heating residential district, a special industrial district and a reference area in Changchun City, Northeastern China from July 1999 to January 2000. The daily average TPM concentrations ranged from 22 to 1984 pg m^{-3}. A significant correlation was obtained between the TPM concentrations and the total suspended particulate concentrations. It is demonstrated that coal combustion and wind-blown soil material are the two main sources of particulate mercury in Changchun City. Wang et al. (2002) reported TPM measurement data at three sampling sites in Beijing, and TPM concentrations ranged from 360 to 440 pg m^{-3}. Duan and Yang 1995 reported TPM concentrations in ambient air at six sampling locations in Lanzhou City, Northwestern China. TPM concentrations varied from 100 to 1000 pg m^{-3}, and a clearly seasonal TPM distribution pattern was obtained showing that TPM concentrations during winter and summer are usually higher than during spring and fall. Again, the authors claimed that coal combustion emissions and wind blown dust were the main sources of particulate mercury in Lanzhou City. The limited reported data showed that TPM concentrations in urban ambient air in China are much higher than that of the background value of 1 to 86 pg m^{-3} (Keeler et al., 1995).

MERCURY POLLUTION TO SOIL COMPARTMENT

A survey conducted by the China National Environmental Monitoring Center 1990 showed that the national background mercury concentration in soil is 0.038 mg/kg. Zhang and Zhu 1994 reported that the average mercury concentration in soil from Tibet, which is the less impacted by human activities, is 0.022 mg/kg. Figure 6 shows the distribution of mercury in soil over the whole of China (China National Environmental Monitoring Center, 1994).

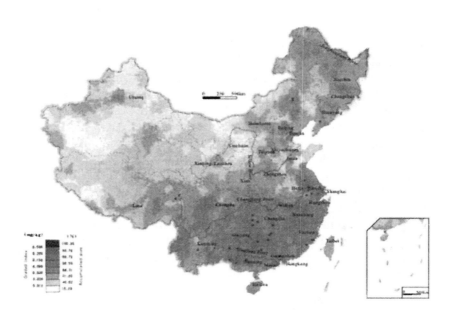

Figure 6. Distribution of mercury in surface soil of China (From China National Environmental Monitoring Centre, 1994).

We can see that mercury concentrations in soil are not evenly distributed over China, and in certain areas, such as southwestern and southern China, mercury concentrations are quite elevated. These elevations are mainly related to the geological background since the Global Circum- Pacific Mercury Belt crosses these areas. Of course, mercury emissions from human activities also caused the elevation of mercury concentrations in urban soils.

Tao et al. (1993) investigated mercury concentrations in soil from Shenzhen area, South China. The average mercury concentration in soil is 0.068 mg/kg, which is significantly higher than the national background value of < 0.01 mg/kg. The utilization of fertilizers containing high mercury during the 1970s is believed to be the main cause of the elevation of mercury in soil (Tao et al., 1993). Of course, the deposition of atmospheric mercury emitted from industrial activities could not be totally ruled out. Guo et al. (1996) reported that average mercury concentration in soil from Taiyuan City is 0.110 mg/kg, and Wang et al. (2003) attributed the elevation of mercury in soil to industrial emissions.

MERCURY POLLUTION IN GUIZHOU

Guizhou is located in Southwestern China with an area of 170,000 km², accounting for about 1.8% of the total area of China. It is known as the "mercury capital" of China because more than 60 % of total national mercury resources were discovered in this province. Guizhou is located in the Global Circum-Pacific Mercury Belt, and at least 13 large and super large-scale mercury mines have already been discovered in the province. As shown in Figure 5, the background mercury concentrations in soil from this area are very elevated. Meanwhile, Guizhou is one of the largest coal producing provinces and is also one of the areas where acid deposition occurs frequently due to coal combustion emissions. Mercury concentrations in coal from Guizhou are significantly elevated compared to the average mercury concentration in Chinese coal due to low-temperature thermal fluid activities in the Mesozoic and Cenozoic Eras in this area (Feng et al., 2002). Mining of mercury and other metals, coal combustion, and other human activities significantly polluted the environment with mercury. Tremendous national and international attention has been given to mercury pollution in the province (Horvat et al., 2003; Xiao et al., 1998; Tan et al., 2000; Yasuda et al., 2004; Feng et al., 2002, 2003, 2004a, 2004b, 2004c).

A study conducted by Xiao et al. (1998) showed that total mercury deposition flux in Fanjing Mountain Nature Reserve (FMNR) area in the Northeast of Guizhou, which is one of over 3000 nature reserves established in the world, was calculated to be 115 μg km⁻² y⁻¹. The deposition flux is very elevated compared to the estimated total deposition rate over Southwestern Sweden, which is about 40 μg km⁻² y⁻¹, implying that mercury pollution in the province is even affecting the biogeochemical process in FMNR (Xiao et al., 1998). Tan et al. 2000 monitored total mercury deposition fluxes at 12 sampling sites in the province, and the total deposition fluxes ranged from 336 to 2340 μg km⁻² y⁻¹. TGM concentrations in ambient air measured at these sites varied from 2.7 to 12.2 ng m⁻³ (Tan et al., 2000). Horvat et al. (2003) assessed the level of contamination with mercury in two geographical areas of Guizhou province. Mercury pollution in the areas concerned originate from mercury mining and ore processing in the area of Wanshan, while in Qingzhen, mercury pollution originates from GOCF, as stated in section 1.3. The results of this study confirmed high contamination of Hg in soil sediments and rice in the mercury mining area in Wanshan, and high levels of mercury in soil and rice were also found in the vicinity of GOCF. Mercury contamination in Wanshan is geographically more widespread due to deposition and scavenging of Hg from contaminated air and deposition on land. In Qingzhen, Hg contamination of soil is very high near the chemical plant, but the levels reach background concentrations at a distance of several km. Even though the major source of Hg in both areas is inorganic mercury, it is observed that active transformation of inorganic Hg to organic Hg species (MeHg) takes place in water, sediments and soils. The concentration of Hg in rice grains can reach up to 569 μg/kg of total Hg, of which 145 μg/kg was in MeHg form. The percentage of Hg as MeHg varied from 5 to 83%. It was concluded in this study that the population mostly at risk is located in the vicinity of smelting facilities, mining activities and close to the waste disposal sites in a wide area of Wanshan.

Feng et al. (2003b) investigated the status of mercury pollution in the groundwater systems in the Wanshan area and found that total mercury concentrations in river waters ranged from 29.7 to 585.8 ng/L. The mercury mining wastes and processing residues are the mercury contamination sources to surface water systems. Feng et al. (2004c) estimated mercury emissions from artisanal zinc smelting using indigenous methods in Hezhang, Guizhou, and mercury emission factors were estimated to be 154.7 and 78.5 g Hg t^{-1} of Zn produced from sulfide ore and oxide ore, respectively. These emission factors are much higher than the literature value used to estimate mercury emissions from zinc smelting in developing countries, which is 25 g Hg t^{-1} of Zn produced. Annual mercury emission rates from zinc smelting in Hezhang from 1989 to 2001 are listed in Table 4. The local surface water, soil and crops were also contaminated with mercury due to zinc smelting activities.

CONCLUSIONS

Mercury pollution to aquatic systems from acetic acid plants and chloralkaline plants, both of which use mercury is an environmental concern in China, and some of the plants are still in operation. The remediation of mercury polluted aquatic systems is an imperative task, and the biogeochemical cycling of this metal and its health impacts to local inhabitants need to be scrutinized as well.

Total anthropogenic mercury emissions in 1995 in China was estimated to be 402 tons. Coal combustion, gold mining and cement production are the most important mercury emission sources.

Table 4. Annual zinc production and mercury emission from artisanal zinc smelting in Hezhang.

Year	Zn production (t)	Hg emission (Kg)
1989	7610.5	1119
1990	11289	1660
1991	11639	1712
1992	11989	1763
1993	17403	2560
1994	23453	3449
1995	26731	3932
1996	23038	3388
1997	22700	3339
1998	31100	4574
1999	45200	6648
2000	48098	7074
2001	32700	4809

Mercury emission from coal combustion is projected to be increasing due to the increasing need of energy with the rapid economic increase since 1995. On the other hand, mercury emission from gold mining has been decreasing since the late 1990s because artisanal gold mining, which is one of the largest mercury emission sources, was totally banned then. Cement production is, and will still be, an important mercury emission source in China. Of course, there are large uncertainties on the estimation of mercury emissions from anthropogenic sources in China because the emission factors from most of sources were obtained only by indirect methods, such as using literature values obtained from other countries. More studies are needed to conduct intensive field measurement of not only total mercury emissions, but also mercury speciation from different mercury emission sources to precisely estimate mercury emissions from anthropogenic sources in China.

From limited data, we can conclude that mercury concentrations in urban air are very elevated. More atmospheric mercury measurement data, especially data from remote background areas in China, are needed to correctly understand the global cycle of mercury in the troposphere. Soils from urban areas are usually contaminated with mercury deposited after being emitted to the air from industrial processes, which poses a threat to terrestrial ecosystems. Mercury pollution in Guizhou is a regional, environmental issue. The biogeochemical cycle of mercury at the regional scale there, and its environmental and health risks, need to be studied.

ACKNOWLEDGEMENT

The author would like to thank Dr. Marilyn Engle, Dr. Bob Stevens and Dr. Jozef Pacyna for their valuable comments on the previous version of this manuscript.

REFERENCES

An, X., Xu, Y., Deng, R., Miao, X. Investigation of mercury pollution in Yanguoxia. *Environ. Res.*, 4, 74-83 (in Chinese) 1982.

Dastoor, A. P., Larocque, Y. Global circulation of atmospheric mercury: a modeling study. *Atmos. Environ.*, 38, 147-161, 2004.

Duan, S., Liu, X., Li, H., Yuan, Z. A survey of health impact from mercury of miner in a gold mining workshop. Henen *J. Prevention Medicine*, 12, 358-359 (in Chinese) 2001.

Duan, S., Yang, H. Pollution status of airborne particulate mercury in the air of Lanzhou. *Environ. Monitor. Manag. Technol.*, 7, 19-20 (in Chinese) 1995.

Fang, F., Wang, Q., Liu, R., Ma, Z., Hao, Q. Atmospheric particulate mercury in Changchun City, China. *Atmos. Environ.*, 35, 4265-4272, 2001.

Feng, X. Analytical techniques of mercury in environmental samples and distribution and modes of occurrence of mercury in coals from Longtan formation formed in Permian period in Guizhou. Ph.D. Thesis, Institute of Geochemistry, Chinese Academy of Sciences, Guiyang, China (in Chinese). 1997.

Feng, X. Hong, Y. Modes of occurrence of mercury in coals from Guizhou, People's Republic of China. *Fuel*, 78, 1181-1188, 1999.

Feng, X., Sommar, J., Lindqvist, O., Hong, Y. Occurrence, emissions and deposition of mercury during coal combustion in the province Guizhou, China. *Water, Air Soil Pollut.*, 139, 311-324, 2002.

Feng, X., Tang, S., Shang, L., Yan, H., Sommar, J., Lindqvist, O. Total gaseous mercury in the atmosphere of Guiyang, PR China. *Sci. Tot. Env.*, 304, 61-72, 2003a.

Feng, X., Qiu, G., Wang, S., Shang, L. Distribution and speciation of mercury in surface waters in mercury mining areas in Wanshan, Southwestern China. *J. Physique IV*, 107, 455-458, 2003b.

Feng, X., Shang, L., Wang, S., Tang, S., Zheng, W. Temporal variation of total gaseous mercury in the air of Guiyang, China. *J. Geophys. Res.* 109, D03303, doi:10.1029/2003JD004159, 2004a.

Feng, X., Yan, H., Wang, S., Qiu, G., Shang, L., Dai, Q., Hou, Y. Seasonal variation of gaseous mercury exchange rate between air and water surface over Baihua Reservoir, Guizhou, China. *Atmos. Environ.*, 38, 4721-4732, 2004b.

Feng, X., Li, G., Qiu, G. A preliminary study on mercury contaminations from artisinal zinc smelting using indigenous method in Hezhang county, Guizhou, China. Part 1 mercury emissions from zinc smelting and its influences on surface waters. *Atmos. Environ.*, 38, 6223-6230, 2004c.

Guan, M., Xue, G., Li, X., Fu, X. Epidemiological survey of mercury pollution in a gold mine. Chinese *J. Public Health Engin.*, 1, 146-148 (in Chinese), 2002.

Guo, L. Transportation of mercury in water of Songhua River. *Environ. Pollut. Control*, 12, 6-8 (in Chinese), 1990.

Guo, C., Wang, Y., Ren, Y., Shi, J. The distribution characteristics of mercury content in the surface soil in Taiyuan. *Journal of Shanxi University*, 19, 339-344 (in Chinese). 1996.

Horvat, M., Nolde, N., Fajon, V., Jereb, V., Logar, M., Lojen, S., Jacimovic, R., Falnoga, I., Qu, L., Faganeli, J., Drobne, D. Total mercury, methylmercury and selenium in mercury polluted areas in the province Guizhou, China. *Sci. Tot. Env.*, 304, 231-256, 2003.

Hou, T., Gu, G., Zhao, J., Guo, X., Xu, J., Bao, L. Clinical survey and analysis of fishermen's health effected by MeHg pollution in Songhua River. *Environ. Sci.,* 11, 147-149 (in Chinese). 1994.

Kang, M. The distribution pattern of methylmercury in sediments of Ji Yun River. *Huan Jin Ke Xue Chong Kan,* 7, 38-41 (in Chinese). 1986.

Keeler, G, J., Glinson, G., Pirrone, N. Particulate mercury in the atmosphere: its significance transport transformation and sources. *Water, Air Soil Pollut.,* 80, 159-168, 1995.

Li Y., Zhang, H., Po, P., Shi, Y., Shen, H., Yuan, S. Study on accumulative levels of mercury among fishing population in the area of second Songhua River. *J. Labour Medicine,* 18, 142-144 (in Chinese) 2001.

Lin, Y., Kang, M., Liu, J. Speciation of mercury in sediment of Jiyun River. *Environ. Chem.,* 2, 10-19 (in Chinese) 1983.

Lin, Y., Kang M. A study on hydro- transportation of mercury in sediment of down reaches of Jiyun river. *Environ. Sci.,* 5, 25-29 (in Chinese) 1984.

Lin Y., Kang, M., Liu, J. Transformation and transportation of mercury in sediment of down reaches of Jiyun River. *Environ. Chem.,* 5, 9-14 (in Chinese). 1984.

Lin, X. Review of methylmercury pollution in the second Songhua river to the health of fishing man in the past 20 years. *J. Environ. Health,* 12, 238-240 (in Chinese) 1995.

Lin, Y., Guo, M., Gan, W. Mercury pollution from small gold mines in China. *Water, Air Soil Pollut.,* 97, 233-239, 1997.

Lindberg, S. E. Stratton, W. J. Atmospheric mercury speciation: concentrations and behavior of reactive gaseous mercury in ambient air. *Env. Sci. Technol.,* 32, 49-57, 1998.

Lindberg, S. E., Price, J. Measurements of airborne emission of mercury from municipal landfill operations: a short-term study in Florida. *J.A.W.M.A.,* 49, 174-185, 1999.

Liu, S. L., Nadim, F., Perkins, C., Carley, R. J., Hoag, G. E., Lin, Y. H., Chen, L. T. Atmospheric mercury survey in Beijing, China. *Chemosphere,* 48, 97-107, 2000.

Liu, Y. Report on integrated control and countermeasure of methylmercury pollution of Songhua River, 1-367 (in Chinese) 1994.

Liu, Y., Wang, N., Zhai, P. Controls and standards on methylmercury pollution of Songhua River in China. Science Publisher, Beijing, 1-385(in Chinese). 1997.

National Environmental Monitoring Center, Soil environmental background value of elements in the people's republic of China. *China Environ. Sci. Press,* Beijing, 1-87 (in Chinese) 1990.

National Environmental Monitoring Center. The atalas of soil environmental background value in the people's republic of China. *China Environ. Sci. Press,* Beijing, 1-195 (in Chinese) 1994.

Pacyna E. G, Pacyna J. M. Global emission of mercury from anthropogenic sources in 1995. *Water, Air Soil Pollut.,* 137, 149-165, 2002.

Pacyna, J. M., Pacyna, E. G., Steenhuisen, F., Wilson, S. Mapping 1995 global anthropogenic emissions of mercury. *Atmos. Environ.,* 37, S109-S117, 2003.

Peng, A., Wang, W., Sun, J. The impacts of humid acids on transportation of mercury in Ji Yun River. *Environ. Chem.,* 2, 33-38 (in Chinese) 1983.

Pirrone, N., Keeler, G.J., Nriagu, J.O. Regional differences in worldwide emissions of mercury to the atmosphere. *Atmos. Environ.,* 30, 2981-2987, 1996.

Qi, X. Development and application of an information administration system on mercury. Master Thesis, Research Center for Eco-Environmental Sciences, Chinese Academy of Sciences, Beijing, China (in Chinese). 1997.

Shang, L., Feng, X., Zheng, W., Yan, H. Preliminary study of the distribution of gaseous mercury species in the air of Guiyang City, China. *J. de Physique IV,* 107, 1219-1222, 2003.

Sommar, J., Feng, X., Gardfeldt, K., Lindqvist, O. Measurement of fractionated gaseous mercury concentrations over northern and central Europe, 1995-1999. *J. Environ. Monitoring,* 1, 435-439, 1999.

Tan, H., He, J., Liang, L., Lazoff, S., Sommar, J., Xiao, Z., Lindqvist, O. Atmospheric mercury deposition in Guizhou, China. *Sci. Tot. Env.,* 259, 223-230, 2000.

Tao, S., Deng, B., Chen, W. Content distribution pattern and pollution of mercury in soil from Shenzhen area. *China Environ. Sci,* 13, 35-38 (in Chinese) 1993.

Wang, S., Qi, S. Preliminary evaluation of methylmercury pollution in water of Songhua River. Huajinkexueqingbao, (9), 34-40 (in Chinese) 1984.

Wang, Q., Wang, N., Wang, S. Speciation of mercury in sediment of the second Songhua River. *Environ. Pollut. Control,* 4, 11-15 (in Chinese) 1985.

Wang, S., Wang, N., Wang, Q. Evaluation standard of methylmercury pollution in aquatic system and evaluation of methylmercury pollution in the second Songhua River. *Environ. Sci.,* 7, 73-76 (in Chinese) 1986.

Wang, D., Li, X., Wu, C. Preliminary investigation on the atmospheric mercury in Chingqing. *Chongqing Environ. Sci.,* 18, 58-61 (in Chinese) 1996.

Wang, W., Shui, C., Du, W. An investigation of health impact of mercury to miner in a gold ming workshop. *China Public Hygiene,* 15, 432 (in Chinese) 1999.

Wang, Q., Shen, W., Ma, Z. Estimation of mercury emission from coal combustion in China. *Env. Sci. Technol.,* 34, 2711-2713, 2000.

Wang, S. An investigation of health impact of mercury to miner in a gold ming workshop. *Occupation and Health,* 17, 20 (in Chinese) 2001.

Wang, W., Liu, H., Yang, S., Peng, A. Distribution of mercury on the aerosol in the atmosphere of Beijing. *J. of Shanghai Jiaotong University* 36(1), 134-137 (in Chinese) 2002.

Wang Y., Xing X., Guo C. Mercury pollution to soil from Taiyuan and its causes. *Chinese J. Ecology,* 22, 40-42 (in Chinese) 2003.

Xi, S., Chen, M., Weng, Z,m Ma, Q., Zhu, M., Xu, H. An investigation of mercury poisoning in a gold mining workshop. *Labor Medicine,* 13, 92-93 (in Chinese) 1996.

Xiao, Z., Sommar, J., Lindqvist, O., Tan H., He J. Atmospheric mercury depositon on Fanjing Mountain Nature Reserve, Guizhou, China. *Chemosphere,* 36(10), 2191-2200, 1998.

Yan, H., Feng, X., Tang, S., Wang, S., Dai, Q., Hou, Y. The concentration and distribution of different mercury species in the water columns of Baihua reservoir. *Journal de Physique IV France,* 107, 1385-1388, 2003.

Yang, H., Jia S., Zhang, H., Wang, B. Microbial methylation of mercury in sediment of Ji Yun River at Hangu area. ACTA ECOLOGICA SINICA, 2, 211-215 (in Chinese). 1982.

Yu, G. Total gaseous mercury concentrations in ambient air at some tourist attraction areas in Beijing. Environmental Chemistry 4, 74-75 (in Chinese). 1985.

Yu, C., Liang, D., Chao, J., Hao, Y., Zhang, Y., Zhang, J. The status of mercury pollution of fish from Songhua River. *Environ. Sc.,* 15, 35-38 (in Chinese) 1994.

Yu, C., Wang, W., Liang, D., Chao, Z., He Y., Zhong, H. A study on the pollution characteristics of total mercury and methylmercury in Songhua River. *J. Changchun University of Earth Sciences,* 24(1), 102-108 (in Chinese) 1994.

Zhai, P., Liu, A., Zhen, G., Ma B. Investigation on the hair methylmercury content of the residents along the Songhua river. *China Environ. Sci.,* 11, 75-78 (in Chinese) 1991.

Zhang, S., Tang, Y., Yang, W., Rao, L., Feng, F., Qu, C. Characteristics of mercury polluted chemical geography of Ji Yun River. ACTA SCIENTIEA CIRCUMSTANTIAE, 1(4), 349-362 (in Chinese) 1981.

Zhang, L., Wu, J., Zhang, Y., Ma, X. Speciation of mercury in sediment from Hadawan to Shaokuo of the second Songhua River. *Environ. Chem.*, 4, 40-46 (in Chinese) 1985.

Zhang, H., Wang, X., Gao, X., Wang, X., Qiu, B., Lin, X. A survey of the health status of inhabitants along Songhua River after control of mercury pollution. *Chinese J. of Public Health,* 9, 349-350 (in Chinese). 1993.

Zhang, P., Mercury pollution to Yellow River in Ningxia section and its impacts on fish. *Arid Environmental Monitoring,* 7, 234-237 (in Chinese) 1993.

Zhang, Q., Du, N., Li, Z., Wang, W. Studies of mercury pollution of water and deposit in second Songhua River (Jilin-Fuyu reach). *J. Environ. Health,* 11(5), 193-195 (in Chinese) 1994.

Zhang, X., Zhu, Y. Mercury concentrations and geographical distribution in soil from Tibet. *Environ. Sci.,* 15(4), 27-30 (in Chinese) 1994.

Zhang, B., Li, Y. Causes and control of methylmercury pollution of the second Songhua River. *Hydraulic Geology and Engineering Geology,* (2), 49-51 (in Chinese) 1994.

Zhang, C., Zhang, X., Chang, Z. Mercury pollution of gold extraction from gold mercury compound in Henen province Xiaoqinling gold mineral belt. *J. Geolog. Hazards Environ. Preservation,* 10(2), 18-23 (in Chinese) 1999.

Zheng, B. Personal communication. 2004.

Zhong, G., Lou, J. A survey of occupational risk of mercury in gold mine. *Journal of Sichuan Continuing Education College of MS,* 19(1), 61-62 (in Chinese) 2000.

MERCURY POLLUTION IN THE ARCTIC AND ANTARCTIC REGIONS

Francesca Sprovieri and Nicola Pirrone

CNR-Institute for Atmospheric Pollution, Division of Rende, 87036 Rende, Italy

INTRODUCTION

There is increasing evidence that the Arctic region, once considered a remote environment, is greatly impacted by global pollution. While there are virtually no anthropogenic pollution sources within the Arctic itself, it is, in fact, impacted by emissions from many of the world's largest industrial regions, resulting in surprisingly high pollution levels, especially during winter and spring, when "Arctic Haze"(the first indication of global pollution impact) forms. The occurrence of contaminants in the Arctic is influenced by the nature and rate of emissions from sources, the processes by which these compounds are transported to the Arctic, removal processes, and the direct atmospheric exchange with snow-pack, lake ice, sea ice, and the surface ocean. To assess mercury airborne transport and deposition in the Northern Hemisphere long-term measurements performed at stations of the EMEP monitoring network (Berg and Hjellbrekke, 1998) and the North American NADP/MDN monitoring network (*NADP/MDN*, 2002) were used in addition to the data of short-term measurements performed during episodic measurement campaigns. Numerous measurements have shown that the global background concentration of mercury in surface air ranges from 1 to 2 ng m^{-3}, though local mercury concentrations in industrial regions can be several times higher. The highest concentrations (above 2.2 ng m^{-3}) are typical of Europe, South-eastern Asia and the eastern part of North America, which are also the main emission regions and, therefore, characterized by the highest depositions. However, even in the remote parts of the Atlantic and Pacific oceans, as well as in the Arctic, mercury in surface air does not fall below 1.4 ng m^{-3}. Trajectory cluster analysis and a PSCF model (Potential Source Contribution Function) have been used by Cheng and Schroeder (2001) to investigate the relationship between sources and receptors for Hg concentrations observed at Alert, in the Canadian high Arctic, during 1995 and, therefore, to identify major anthropogenic Hg emission source regions in the northern hemisphere that contribute significantly to the Hg^0 concentrations measured at Alert throughout 1995. The analysis of 10-day back trajectories have shown that atmospheric transport of Hg to Alert is dominated by air masses from Eurasia which cross the Arctic Ocean. . In particular, long-range transport of Hg^0 only occurred in the cold seasons (Autumn and Winter), while summertime flows tend to circulate in the Arctic Ocean and do not travel very far (Cheng and Schroeder, 2001). Atmospheric transport, therefore, is shorter in the summer than in the winter season showing the seasonal variability of the flow patterns. The contribution of specific regions of the Northern Hemisphere to the mercury pollution of the Arctic environment is shown in Figure 1. The most significant contribution is made by Asian (33%) and European (22%) sources (Ryboshapko, 2003).

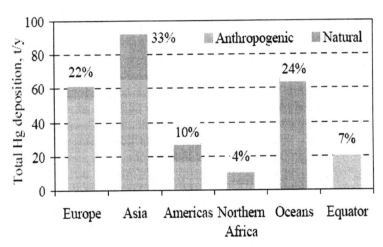

Figure 1 Global mercury impact to the Arctic
Source: Ryboshapko, 2003

The total mercury deposition to the Arctic is assessed at about 270 t/y. Of this amount, about 50% is contributed by anthropogenic emission sources. Several studies suggest that there has been a recent increase in Hg levels in Arctic biota, lake sediments, arctic food supplies (Lockhart et al., 1998) as well as in native people of circumpolar countries (AMAP, 1997; AMAP, 1998), despite a 20-yr decrease in global atmospheric Hg emissions since 1989 (Pirrone, et al., 1996; Pacyna, et al., 2001). If global emissions clearly decreased, another explanation for the marked Hg deposition increase in the Arctic must be sought.

CLIMATE AND METEOROLOGY

The Antarctic continent is in many ways quite different from its northern counterpart. Antarctica is a high, ice-covered land mass surrounded by the southern extensions of the Atlantic, Pacific, and Indian Oceans (Figure 2).

On the other hand, the vast region of the Arctic consists of the ice-covered Arctic Ocean surrounded by many islands and the continental mass lands of northern North America, northern Europe and northern Asia (Figure 3).

As a result of these differences, the climates of the two regions are very different. The overwhelming characteristic of polar regions, in terms of both intensity and duration, is the cold. The long winter night ensures very low temperatures. The polar regions

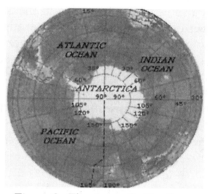

Figure 2. The Antarctic Continent.

thus, show large seasonal variations in incoming solar radiation, from none during the winter months to 24 hours of sunlight at mid-summer receiving, on an annual basis, less solar radiation than other parts of the world. However, the radiation levels vary greatly depending on the season and in their respective mid-summers the daily totals are greater than at other places on earth.

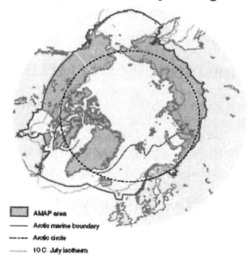

AMAP area
Arctic marine boundary
Arctic circle
10 C July isotherm

Figure 3. The Arctic Region as defined by Temperature, the Arctic marine boundary AMAP and assessment area.

Source: AMAP 1998.

Large part of the solar radiation that reaches the Earth is reflected by extensive cloud, snow and ice cover. The high albedo of polar regions, from the persistent snow and ice and the large loss of long-wave radiation due to the exceptional clarity and dryness of the atmosphere, is a key factor in the surface energy budget and ensures a net loss of radiation in all or most of the months of the year. This radiation imbalance produces low temperatures and results in a redistribution of heat from southern latitudes via air and ocean currents (Varjo and Tietze 1987). The loss is, in fact, compensated through transport of sensible and latent heat from lower latitudes, usually within cyclones, and by heat carried within ocean currents. Because of the lack of transport of warm current to the Southern Ocean and the pressure of strong westerlies, which blocks heat supply over the Antarctic, the Antarctic is colder than the Arctic. Summer temperatures in most of the Antarctic continent remain well below freezing. In the Arctic, however, rapid and strong snowmelt produces a large influx of fresh water to the rivers and Arctic Ocean in the spring and summer and supports a burst of life during a brief and intense summer. Important circulation systems of the world's oceans are driven by sinking cold water at the periphery of polar regions. On the basis of temperature, the Arctic is defined as the area north of the 10°C July isotherm, i.e., north of the region which has a mean July isotherm, i.e., north of the region which has a mean July temperature of 10°C (Figure-3) (Linell and Tedrow 1981, Woo and Gregor 1992). Based on oceanographic characteristics, the marine boundary of the Arctic is situated along the convergence of cool, less saline surface waters from the Arctic Ocean and warmer, saltier waters from oceans to the south. The frequency and the position of the persistent high-low-pressure systems not only play an

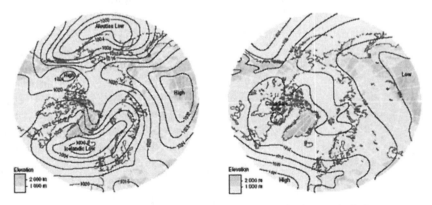

Figure 4. Mean atmospheric sea-level pressure (mb) in the Arctic during the winter and spring seasonal.

Source: AMAP, 1998.

important role in the existence of the regional and local climates in the polar regions, but also link the Arctic and Antarctic climatic system to the world climatic system influencing also the atmospheric air mass circulation which deliver contaminants to the poles. During the winter and spring seasons, the lower tropospheric circulation of the northern polar region is dominated by high pressure over the continents and low pressure over the northern Pacific and Atlantic Oceans (Figure 4). In particular, the intense Siberian high pressure cell tends to force air on its western side northward into the Arctic. The high pressure ridge over North America generally drives air out of the Arctic southward, resulting in the transport of air over Siberia into the High Arctic, and south over North America.

Consequently, contaminants which are introduced into the atmosphere in Siberia or Eastern Europe (in either vapour form or bound to small particles which have a low deposition velocity) are frequently transported into the Arctic region. Moreover, anticyclones are characterized by relatively low wind speed, and thus stagnant conditions. Near the surface, the relative lack of cloud cover and low incident solar radiation during Arctic winter can produce extended periods of surface radiation inversions. These factors reduce the effectiveness of vertical mixing and removal at the surface, resulting in the accumulation of contaminants in the lower Arctic atmosphere. Some air is also exchanged with the south when low pressure vortices along the Arctic front mix warm southern air with cold northern air in a large-scale turbulent eddy. In summer, the continental high pressure cells disappear, and the oceanic low pressure cells weaken, particularly in the north Pacific. Northward transport from mid-latitudes decreases accordingly. In contrast to the winter period, in the summer, south-to-north transport from Eurasia is, therefore, much weaker as the Siberian high dissipates. The mean air circulation pattern in the lower part of the atmosphere yields to a more circular clockwise flow around the North Polar Region. Marked variation in cloud cover and precipitation accompanies this seasonal variation.

ATMOSPHERIC MERCURY DEPLETION EVENTS (AMDES)

A surprising discovery that provided a great impetus for Arctic atmospheric chemistry research in several nations interested in preventing pollution of polar regions, was the observing of an unusual phenomenon called Atmospheric Mercury Depletion Events (AMDEs) in the atmospheric boundary layer of the Arctic and sub-arctic regions. Figure 5 shows several sampling Arctic and sub-arctic locations where intensive mercury measurement campaigns were performed in order to study the spatial extent of the AMDEs and above all, to understand the atmospheric mechanism behind these springtime phenomena. During the 3-month period following polar sunrise, it has been identified for the first time at Alert in the Canadian High Arctic (Schroeder et al. 1998; 2003), an atmospheric mechanism by which gaseous elemental mercury (Hg^0) may be converted to reactive and water-soluble forms (Reactive Gaseous Mercury (RGM) and/or Particulate Mercury (Hg-p)) that deposit quickly thus increasing the mercury fluxes and deposition

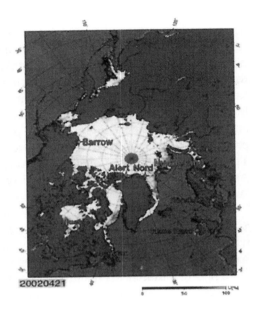

Figure 5. Mercury sampling locations in the Arctic and sub-Arctic region.

Source: Berg et al., 2003.

processes in the fragile ecosystems of the Arctic and/or Antarctica.

Springtime AMDEs first noted in the Arctic and observed each spring, have also been observed in Antarctica (Ebinghaus et al., 2002). The recent increase in Hg accumulation in the Arctic is, therefore, a direct consequence of the AMDEs. Figure 6 illustrates the net effect of AMDEs on the annual deposition of mercury in the Arctic. The contribution of AMDEs (occurring over several weeks) to the annual mercury deposition may be as high as 50% in coastal Arctic regions. AMDEs occur at the same time as tropospheric ozone depletion events suggesting

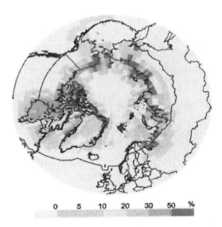

Figure 6. The contribution of MDEs to annual mercury deposition in the Arctic.

Source: AMAP, 1998.

that both species were removed by similar unknown homogeneous and/or heterogeneous chemical reactions. The AMDEs are in fact, strongly correlated with ground-level ozone depletion events (Schroeder et al., 1998) which had been discovered 10 years earlier (Barrie et al., 1988). These events cannot be explained by changing meteorology alone but occur when the atmosphere is in a chemically "perturbed" condition. The results of a number of studies suggest that catalytic destruction involving the $BrO_{(g)}$ radical is responsible for the O_3 depletion. Ozone depletions are caused by photochemically-initiated autocatalytic reactions involving reactive halogen species (such as Br and BrO) derived from sea-salt aerosols in areas of active open water (opened series of open leads and polynyas where the wave activity and sea-salt aerosol generation is very high). Hence, it is very likely that mercury depletion events, having temporal ground level concentrations similar to those of ozone, occur in polar areas through similar mechanisms. Given, thus, the correlation between O_3 and $Hg^0_{(g)}$, and the simultaneous increase in Hg associated with particulate matter (Schroeder et al., 1998; Lu et al., 2001) and RGM (Lindberg et al., 2001) it is thought that BrO, or another Br containing radical formed in the period around the polar sunrise, is responsible for the sudden increase in the oxidation rate of $Hg^0_{(g)}$ and the formation of less volatile and more soluble $Hg^{(II)}$ compounds which are deposited or condense onto pre-existing particulate matter (Lu et al., 2001; Lindberg et al., 2001; Berg et al., 2003a,b; Steffen and Schroeder, 2002; Poissant and Pilote, 2001, Ebinghaus et al., 2002). Although, the kinetic data on the reactions of mercury with atomic halogens is actually scarce, the interaction with halogen radicals seems to be the only plausible process accounting for the fast mercury depletion in the Arctic and/or Antarctic troposphere. The systematic seasonality of AMDEs leads to the supposition that the depletions of O_3 and Hg^0 require both sunlight conditions and a frozen aerosol or snow surface. The frozen surfaces on the Arctic Ocean are, in fact, enriched by concentrations of halide ions (Br-, Cl-), sunlight conditions probably trigger the release of halogen containing compounds to the atmosphere. Other reactive halogen species present in the Arctic could be thermodynamically able to oxidize Hg^0 to form RGM in the gaseous phase, including Cl_2, Br_2, BrCl, HOCl and HOBr. However, molecular Cl_2, Br_2, and BrCl are not likely to produce in-situ RGM formation observed in the Arctic during springtime since they can be rapidly photolyzed by solar radiations (Fan, S.-M. Et al., 1992). HOCl and HOBr are more resistant under sunlight conditions but they do not exhibit a strong diurnal cycle in the remote marine boundary layer (Mozurkewich, 1995), therefore, they may not account for the reactive mercury species production during polar sunrise. Two RGM species have been considered as products: mercury oxide (HgO) and mercury halides ($HgBr_2/HgCl_2$). Based on published studies of reactive halogens in the Arctic (Foster, et al., 2001), $HgBr_2$ should be favored considering also that the KCl-Annular denuders may preferentially collect $HgBr_2/HgCl_2$ over HgO (Landis et al., 2003). During AMDEs elevated RGM (and/or Hg-p) also coincided with periods characterized by increased levels of column BrO. One potential indicator of the overall spatial extent of these events is illustrated by the monthly GOME maps of BrO distribution. These and related maps (Lu et al., 2001) clearly confirm that MDEs and associated RGM (Hg-p) production should be concentrated in coastal zones and might not be expected in other locations (e.g., continental areas).

Photolyzable bromine which builds up under the ice, escapes through constantly changing patterns of open leads and polynyas. These dynamic open water areas are also sources of sea-salt aerosols, water vapour, and heat from the comparatively warm ocean waters. All these products are concentrated in the near surface air due to the lack of vertical convection, (caused by limited solar input, the high-albedo of the snow/ice surfaces, and a positive temperature inversion). The bromine source regions are, thus, concentrated in the dynamic areas of annual sea ice, and the advection of Br compounds to inland and ice-shelf regions is controlled by prevailing winds and is effectively influenced by topography. Oxidation of Hg^0 and enhanced deposition of RGM (Hg-p) would not be expected in areas without advection of Br compounds. Under these conditions, several arctic experiments have shown that RGM reached high levels exceeding those measured near industrial point sources (Sheu et al., 2001). Garbarino et al. (2002) shown that mercury concentrations in snow over sea ice were highest in the predominately downwind direction of the open water leads and polynyas surrounding Point Barrow (e.g., to the west), an area that often shows enhanced BrO. When these surface emissions of photolyzable bromine encounter air-mass containing Hg^0 emissions from southern latitudes under sunlight conditions, mercury depletion/deposition events will occur.

The springtime mercury deposition rates in the Arctic could therefore be related to a function of the spatial coverage of annual sea ice, the air-mass transport of mercury emissions to this region, and local air-mass circulation. These phenomena are, in turn, controlled by average spring and summertime temperatures. The atmospheric oxidation of elemental mercury (Hg^0) to Hg(II) after polar sunrise is also evident from the increase in concentrations of mercury observed in surface snow from the polar night to the Arctic spring (Lu et al., 2001; Lindberg et al., 2002). The Arctic environment may, therefore, act as a global sink for atmospheric mercury. More recently, with the improvement of RGM and Hg-p sampling techniques, investigations of the atmospheric mercury chemistry have been performed at different maritime circumpolar stations to examine and to better understand the mercury depletion mechanism through integrated experiments involving mercury concentrations also in the snow/ice surfaces.

SAMPLING LOCATION AND SPATIAL EXTEND OF THE MERCURY DEPLETION EVENTS IN THE ARCTIC REGION

Several field experiments have been performed at different Arctic locations (see Table 1) in order to better understand the chemical processes that may act to enhance the capture of Hg from the global atmosphere and its deleterious impact on Arctic ecosystems.

High-temporal-resolution Hg measurements performed at Alert, Nunavut, Canada (82.5N, 62.3W), showed for the first time in 1995, an unsuspected atmospheric process during the 3-month period following polar sunrise (mid-March to mid-June at Alert); atmospheric mercury concentrations showed extraordinary fluctuations from background concentrations to undetectable levels (<0.1ng m^{-3}) (Schroeder et al., 1998).

Table 1. Atmospheric mercury species measurements performed at different Arctic locations from 1995 to 2003.

Measurement Sites	Coordinates	Period	Measurement type	Techniques	References
Alert (Canada)	82°5'N 62°3'W	1995-2002	Hg^0; TPM	Tekran 2537A; AE-TPM Traps; CRPU-(Cold Regions Pyrolysis Unit)	Schoroeder et al., 1998; Lu et al., 2001; Steffen et al., 2002
Barrow (Alaska)	71°19'N 156°37'W	1999-2003	Hg^0; RGM; Hg-p	Tekran 2537A; Tekran 1130 and KCl-Coated Annular Denuders; Tekran 1135;	Lindberg et al., 2002
Nord (Greenland)	81°30'N 16°40' W	1998-2002	Hg^0	Tekran 2537A;	Skov et al., 2004
Ny-Alesund (Svalbard Islands)	78°54' N 11°53' E	2000-2003	TGM; Hg^0; RGM; Hg-p; TPM	Tekran 2537A; Tekran 1130 and KCl-Coated Annular Denuders; Tekran 1135; AE-TPM Traps ;	Berg et al., 2001;2003; Sommar et al., 2004; Sprovieri et al., submitted
Amderma (Russia)	69°43'N 61°37'E	2000-2001	Hg^0	Tekran 2537A;	AMAP, 2002

Three periods were clearly distinguished due to different Hg behaviour observed during the year: (i) fall/winter period, with background Hg^0 concentrations (1.5 – 1.6 ng m^{-3}); (ii) springtime, with highly variable Hg^0 levels often to well below the detection limit and (iii) summertime, with higher Hg^0 concentrations probably due to the warm season in which temperatures and/or sunlight conditions induced volatile Hg^0 emission and/or re-emission from water and land surfaces to the atmosphere (Schroeder et al., 1998).

Moreover, Hg^0 depleted as O_3 showing a very strong correlation with ground level ozone concentrations. The discovery of the AMDEs, have enhanced research interest in this region thus several and intensive measurement expeditions were performed from the 1995 to 2001 (Steffen and Schroeder, 2002). In their Alert study, Steffen and Schroeder (2002) found that Hg concentrations associated with particles increased at the same time of the Hg^0 concentrations decreased. This was the first indication that atmospheric chemical reactions were converting Hg^0 to inorganic Hg(II) species, less volatile than the Hg^0 and more likely to associate to particles. The Spring 1998 campaigns confirmed the strong anti-correlation between Hg^0 and Hg-p (Lu et al., 2001) observed at Alert.

Intensive measurement campaigns performed at Pt.

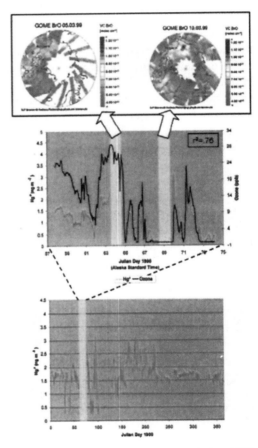

Figure 7. Hg° concentrations at Barrow during 1999. *Source:* Lindberg et al. 2002.

Barrow, Alaska (Lindberg et al., 2001; Lindberg et al., 2002) from 1999 to 2003 showed for the first time, that during depletion events, significant concentrations of RGM were reported. All of the data were collected at the NOAA-Climate monitoring and Diagnostic Laboratory (CMDL) in Barrow, AK. The sampling site is located at Point Barrow and is surrounded by water to the north, east, and west. Barrow is geographically the northern-most point in Alaska, located at 71°19'N, 156°37'W about 1600 km south of Alert. The 1999 data (Figure 7) provide the first confirmation of AMDEs at this more southerly Arctic site with the RGM production during AMDEs at significant levels when Hg^0 is being depleted. Although others have suggested that the depleted Hg^0 at Alert accumulates in the aerosol-phase Hg (Schroeder et al., 1998), Barrow data clearly indicate an important change in gaseous speciation during AMDEs, producing levels of RGM unprecedented at remote and rural sites (Lindberg et al., 1998; Ebinghaus et al., 1999; Munthe et al., 2001). The depletion events observed begin within a few days of polar sunrise (late January) and persist until snowmelt (early June), suggesting a role of both sunlight and frozen surfaces.

Gaseous and aerosol Br also exhibit strong seasonal cycles at Barrow and, like RGM, peak annually between January and June. During this period, aerosol Br increases nearly 20-fold over typical concentrations and can exceed 100 ng m^{-3}. Hypotheses for the sources of this Br include aerosol enrichment by bubble bursting from the sea-surface microlayer, and/or other aerosol-related reactions.

The most probable mechanism involves heterogeneous reactions at the interface of hygroscopic sea-salt aerosol many of which are initiated in the surface microlayer of snowflakes or the snowpack (Barrie et al., 1997). Particulate Hg (Hg-p) samples collected at Barrow (Landis et al., 2001) indicate that RGM is clearly the primary species being formed during this AMDE period and that the two species behave differently. For example, during a period with 3-5 hours of darkness, particulate Hg and RGM were anti-correlated, with Hg-p peaking just prior to sunrise when RGM was at its daily minimum but decreasing rapidly upon sunrise (Landis et al., 2001).

The authors suggest that different reaction pathways or reactants may be responsible for creating Hg-p as compared to RGM and that the Hg-p produced at night is photosensitive. One candidate reaction would involve aerosol-bound BrCl that would readily oxidize any sorbed Hg^0 but that is rapidly decomposed under sunlight (Fan et al., 1992). However, upon the advent of 24-h were now positively correlated (Landis et al., 2001). The authors speculate that the Hg-p detected after 24-h sun reflects RGM sorbed onto the existing aerosol. These observations may help explain why the air at Alert appears to be characterized by a larger Hg-p/RGM ratio than at Barrow (Lu et al., 2001). The surface reactivity of airborne RGM suggests that it would readily partition to the aerosol phase upon formation. Hence, Hg-p/RGM ratios may be useful indicators of the age (time since oxidation of

Hg^0), and hence transport distance, of depleted air masses. Barrow data suggested that at least some RGM is being formed in situ at ground level, while Hg^0 in the air sampled at Alert may have undergone significant depletion/oxidation events over the sea ice prior to being sampled at the Alert station.

Atmospheric mercury depletion events have also been studied at Station North, Northeast Greenland, 81°36'N, 16°40'W, during the Arctic Spring (Skov et al., 2004). GEM and O3 were measured starting from 1998 and 1999, respectively, until August 2002. Weekly average concentrations of atmospheric bromine were also determined from samples collected on particle filters. Figure-8 shows the results of ozone and Hg^0 measurements together with concentrations of fBr. Ozone and Hg^0 were rather stable from September/October until the end of February/beginning of March. Then, a highly perturbed period appeared where both ozone and Hg^0 were depleted to zero from, respectively, about 40 ppbv and 1.5 ng m^{-3}. At the same time fBr increases and reaches a maximum of about 10 ng m^{-3}. Ozone and Hg^0 have been observed to deplete simultaneously and to be highly correlated during atmospheric AMDEs (Schroeder et al., 1998). After the depletion period, some very high concentrations of Hg^0 appeared with values above 2 ng m^{-3} in 2000, up to 1.9 ng/m3 in 2001 and at a maximum of 5.7 ng m^{-3} in 2002. High values after AMDEs are also observed at Alert (Schroeder et al., 1998), Barrow (Lindberg et al., 2002), and Svalbard (Berg et al., 2001), and they are attributed to to reemission of mercury to the atmosphere. In the spring 2001-2002 Canadian and Russian scientists collaboratively measured Hg^0 concentrations for the first time in the Russian high Arctic at Amderma (69°43'N, 61°37'E) and were observed AMDEs in the period from the end of March until the middle. These results confirm that the springtime depletion of mercury after polar sunrise is a mechanism endemic to Polar Regions and that Hg^0 is chemically converted to more reactive mercury species during the AMDEs. Atmospheric mercury measurements were carried out on routine basis at Ny-Alesund (78.9°N, 11.9°E) (Spitzbergen, Svalbard Islands) from February 2000 to the end of May 2003. Additionally, intensive measurement campaigns were performed in the presence of 24 hours daylight after polar sunrise, from about the middle of April to middle of May during the years 2000-2003 (Berg et al., 2003; Sommar et al., 2004; Sprovieri et al., submitted). Ny-Ålesund is situated at the southern shore of Kongsfjorden on the West Coast of Spitzbergen in the Norwegian high Arctic at about 15 km from the northernmost Barents Sea. An offshoot of the waning Gulf Current (West Spitzbergen current) causes ice-free conditions along the west coast of Spitzbergen during the entire year. Measurements were performed at the Global Atmospheric Watch monitoring observatory operated by the Norwegian Institute for Air Research (NILU) at Zeppelin

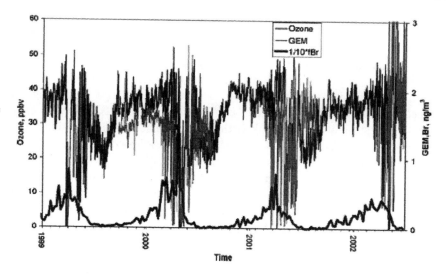

Figure 8. Hourly O3 mixing ratios and weekly concentrations of fBr measured from 1999 to 2002 at Station North, Greenland. GEM was measured each Spring from 2000 to 2002.

Source: Skov et al. 2004.

Mountain (474-m a.s.l). The theory of RGM and PM formation during depletion of elemental Hg in the Arctic, has been strengthened by the results from the intensive measurement campaigns performed each spring from 2000 to 2003. Hg^0concentrations at Zeppelin were within the range <0.1 - 3 ng m^{-3}. The annual averages for 2000, 2001 and 2002 were 1.47, 1.56 and 1.59 ng m^{-3}, respectively.

Episodic depletions of Hg^0, closely resembling ozone depletions in Arctic surface level air, were frequently observed during an intensive atmospheric mercury campaign performed at Ny-Alesund, (Svalbard Islands, Spitzbergen) from April 18th to May 12th 2003, during the arctic springtime (Sprovieri et al., submitted). Hg^0 concentrations dropped from around 1.5ng m^{-3} to undetectable levels (< 0.1 g m^{-3}) and were strongly correlated with surface O_3 depletion. GEM, RGM and Hg-p concentrations were measured using a method extensively tested by Landis et al. (2002), which can provide automated measurements of speciated atmospheric mercury concentration levels. GEM, RGM and Hg-p concentrations were simultaneously quantified by an automatic integrated system (a Tekran automated unit, Model 1130 and a Tekran 1135 particulate unit, coupled to an automated gas-phase mercury analyser, Tekran 2537A). Three GEM depletion events were clearly observed during the springtime campaign at Zeppelin Station (474-m a.s.l.). Regarding the first and major depletion event, depicted from 21th to 22th of April, Hg^0 levels dropped very fast within 24 hours and the concentrations changed from about 1.7 ng m^{-3} to below the detection limit. This depletion, instead of other two minor events, is characterized by several dips and spikes in the Hg^0 trend with concentrations always below 0.9 ng m^{-3} which could be interpreted as minor depletion episodes. During the following minor depletion episodes Hg^0 dropped below 0.25 and 0.5 ng m^{-3} respectively.During the ODEs, Hg^0 exhibits a strong correlation with O$_3$ (r^2 = 0.8) suggesting a possible link with the chemical reactions which destroy tropospheric ozone, as first suggested by Schroeder et al. (1998). Dramatically increased levels of Hg-p and RGM were measured as Hg^0 concentrations dropped to undetectable levels at Ny-Alesund, as well as at several Arctic sites. The RGM data show that during springtime AMDEs, RGM appears at significant levels only when Hg^0 is being depleted, clearly indicating an important change in *gaseous* speciation during AMDEs. RGM concentration-time series were in the range between 2.5 pg m^{-3} to 228.1 pg m^{-3} and Hg-p concentration-time series were in the range between 0.26 pg m^{-3} to 98.6 pg m^{-3}, peaking just during the first and major depletion event. The data obtained at Ny-Alesund in particular, indicate that RGM is clearly the primary species formed during the AMDE period and that behaved differently from Hg-p.

Particulate mercury samples collected at Barrow (Lindberg et al., 2002) also showed RGM as the primary species during the AMDE period with a different behaviour of both mercury species. Lindberg et al. (2002), however, observed that Hg-p peaks during not 24-hour sunlight but just prior to sunrise. Hg-p data at Ny-Alesund were obtained during 24-hour sunlight conditions suggesting that different reaction pathways or reactants may be, probably, responsible for creating Hg-p as compared to RGM during the first and major AMDE. In addition, the general low Hg-p/RGM ratio observed during 24-hr sunlight conditions at Barrow led to speculate that the Hg-p detected after 24-h sun could reflect RGM sorbed onto the existing aerosol. BrO distribution from GOME satellite data and back-trajectories (by HYSPLIT Model) have been compared with mercury species concentrations obtained at Ny-Alesund during the 2003 experiment. The results show evidence, in a preliminary analysis, that the most probable mechanism, driving the Hg^0 depletion and the simultaneous RGM production, is not solely action of atmospheric halogen chemistry at Ny-Alesund but also that an important role is played by the transport of air masses already depleted from mercury prior to arrive at the sampling site at Zeppelin.

THE MERCURY DEPLETION EVENTS IN THE ANTARCTIC REGION

The first extended baseline data for the concentration and speciation of atmospheric mercury (TGM and dimethylmercury, DMM) in Antarctica were reported by De Mora et al. (1993). Annual average TGM values were 0.52, 0.60 and 0.52 over three consecutive years, and DMM values were less than 10% of the TGM values, and often below the detection limit. The seawater and atmospheric concentration of DMM have also been measured in the Antarctic Ocean by Pongratz and Heumann (1999) who found the

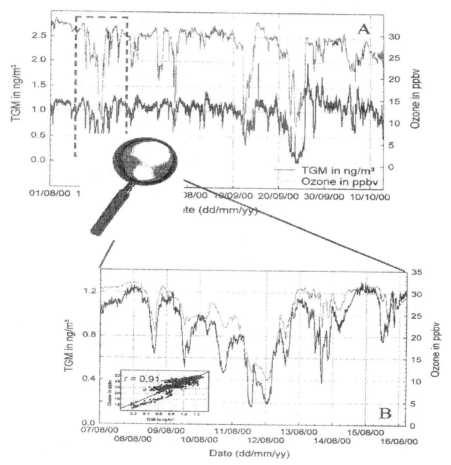

Figure 9. O₃ and TGM concentrations during the MDEs observed at Neumayer, Antartica from August to October 2000.

Source: Ebinghaus et al. 2002.

mean atmospheric concentration to be 6 pg m^{-3}. However, Hg speciation and the variation of Hg species in the Antarctic region are currently not known.

The measurements reported by Ebinghaus et al. (2002) comprise the first annual time series of ground-level TGM (Total Gaseous Mercury) concentrations in the Antarctic to investigate the occurrence of possible AMDEs in south polar regions. The study also provides high-resolution data that can be compared with existing data sets of AMDEs in the Arctic revealing similarities and differences in the temporal and quantitative sequence of AMDEs after polar sunrise. The TGM series measured in Antarctica at the German Research Station at Neumayer (70°39'S, 8°15'W) by Ebinghaus et al. (2002) showed several Hg depletion events during Antarctic springtime (between August and November) 2000 (Figure 9), with minimum daily average concentrations of about 0.1 ng m^{-3}; TGM and O3 were strongly positively correlated as seen in the Arctic boundary layer after polar sunrise. At Neumayer (relative to the beginning of spring in both hemispheres) the ozone depletion events are less frequent, and are shorter (Lehrer, 1999). Friess (2001) documented that the numerous strong and sudden enhancements of BrO detected during August and September 1999 and 2000 at Neumayer are caused by BrO located in the lower troposphere, released by well-known autocatalytic processes on acidified sea-salt surfaces.

The TGM data can be characterized by three different time periods: (i) Between January and February 2000 and December 2000-February 2001, in which TGM concentrations were highly variable. During this time period, TGM and ozone concentrations are frequently negatively correlated; (ii) the period between March and July 2000, which shows TGM concentrations at constant level of background (about 1.146 ng m^{-3}) while ozone concentrations constantly increased. (iii) the period between August and November 2000, in which several simultaneous depletion events of surface-level TGM and ozone were detected with minimum daily average TGM concentrations of about 0.1 ng m^{-3}. Ebinghaus et al. (2002) also found that these Hg depletion events coincided with enhanced column densities of BrO over the sea ice around the Antarctic continent after polar sunrise (September) as shown from measurements performed using the satellite borne GOME instrument. Air masses at ground level coming from the sea ice surface, accompanied by BrO enhancements, could be a necessary prerequisite for the MDEs at Neumayer. Most of the sea ice where enhanced BrO concentrations are found is located north of Neumayer station at lower latitudes. Therefore, the sea ice is a possible place where the photochemical reaction of ozone and Br atoms and/or the following reaction of BrO radicals and Hg^0 can take place in the Antarctic during the springtime.

After an intensive measurement campaign carried out at Terra Nova Bay, between November 1999 and January 2000, during which opposite trends between TGM concentrations and the quantity of Hg associated with particulate matter was observed (Sprovieri and Pirrone, 2000), the desirability of more detailed measurements of atmospheric Hg species in Antarctica was clear, in order to understand Hg cycling in polar environment. Atmospheric mercury measurements were performed at the Italian Antarctic Station, in Terra Nova Bay (74°41'S, 164°70'E) from the middle of November 2000 to the middle of January 2001. Terra Nova Bay is a small gulf inlet of the Ross Sea on the western coast of the Antarctic continent; the atmospheric measurement site is at Icaro Camp situated at 3.5 km from the main Italian Antarctic Station (IAS), on a hilltop 90 m a.s.l. The results obtained provide simultaneous measurements of both Hg^0 and RGM performed in Antarctica. Hg^0 concentrations were in the range of 0.29 ng m^{-3} to 2.3 ng m^{-3} with a mean value of 0.9±0.3 ng m^{-3}. The average Hg^0 concentration was substantially lower than values obtained elsewhere. In the Northern Hemisphere mercury fluxes to the atmosphere derived from emission estimates for major natural and anthropogenic sources and ambient concentration measurements are higher than those evaluated for the Southern Hemisphere. This adds weight to the widely accepted hypothesis that the background Hg concentrations between the two hemispheres are different. Figure-10 shows Hg^0 and O_3 ambient concentrations for the measurement period. The data is comparable to that of Ebinghaus et al. (2002) for the periods January and February 2000, and December 2000 to January 2001, which were characterized by variable TGM concentrations (Ebinghaus et al., 2002) measured TGM as there were no denuders on their instrument to remove RGM) and during which TGM and O3 were often negatively correlated.

RGM concentrations during the measurement period were surprisingly high and comparable with those at sites directly influenced by significant anthropogenic Hg sources (in the range of 10.53 pg m^{-3} to 334.2 pg m^{-3} with a mean value of 116.3 ± 77.8 pg m^{-3}). Recent studies performed in the Arctic (Lindberg et al., 2001) also report very high RGM concentrations between polar dawn and snowmelt, suggesting that there are specific mechanisms and/or characteristics of polar environments that at certain times, and apparently in the presence of surface snow are extremely favourable to the production of RGM. Comparable RGM results have been reported by Temme et al. (2003) during an intensive measurement campaign performed at the German Research Station (Neumayer) during Antarctic summertime. They found RGM values ranged between 5 pg m^{-3} and maximum levels of more than 300 pg m^{-3} from December 2000 to February 2001. The high concentrations of RGM measured at Terra Nova Bay are somewhat surprising above all because the period in which the measurements were performed was not one in which simultaneous ozone and mercury depletion events occurred, when it has been shown that RGM production increases significantly (Lindberg et al., 2002). The only possible source of RGM is, in these particular conditions, the gas phase oxidation of Hg^0.

According with Temme et al. (2003), the very high RGM concentrations at Terra Nova Bay as well as Neumayer should be influenced by local production of oxidized gaseous mercury species over the Antarctic continent or shelf ice during polar summer. Recent investigations in the Arctic, have

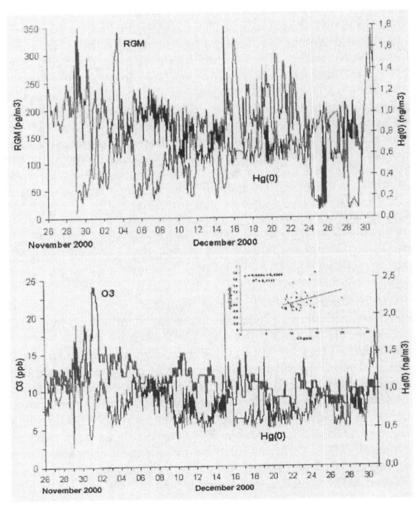

Figure 10. Two-hourly mean GEM and RGM concentrations at Terra Nova Bay, Antarctica from November to December 2000.

Source: Sprovieri et al. (2002).

highlighted that there are gas phase oxidation processes occurring which are as unknown, probably involving bromine-containing radicals. However, although in the Arctic the highest RGM concentrations were found during MDEs elevated concentrations were found at Barrow until snowmelt. Snowmelt is much more limited even at coastal Antarctic sites than it is at Arctic sites such as Barrow, which suggests that the snow-pack is directly involved in maintaining high RGM concentrations. Further studies are necessary to explain the reaction mechanism and the kinetics of the MDEs and the RGM production identified during measurement campaigns in the Antarctic during springtime depletion period (Ebinghaus et al., 2002; Temme et al., 2003) and for the no depletion period (Sprovieri et al., 2002). It is also important to combine the results observed with trajectory calculations in combination with sea ice maps in order to investigate the origin of the depleted air masses and the actual places where the chemical reactions involving ozone, reactive bromine species, and elemental mercury take places.

MERCURY DEPOSITION TO THE SNOW AND ICE SURFACES IN THE POLAR REGIONS AND AIR-SNOW-ICE EXCHANGE PROCESS

The mechanism of the conversion of Hg^0 into RGM (and/or Hg-p) is not well understood neither are the roles of the snow pack and the host of chemicals existing therein. For this reason various conversion and air–snow interaction studies were undertaken during measurement campaigns from the 1997-1998. The first observations of the seasonal variation of Hg content in snow on the Arctic Ocean are reported by Lu et al. (2001). Fresh snow samples were taken in the spring snow-pack of the Eastern Canadian Arctic, Hudson Bay, and Greenland. In the Canadian Island Archipelago, and at coastal sites on Hudson Bay, concentrations were generally much higher (25 - 160 ng L^{-1}) and enhanced Hg levels in snow were clearly reflected in the 4-fold increase from the dark winter months (Nov. 1997 to Jan. 1998: 7.8 ng L−1) to the sunlit spring months (Feb. to May, 1998: 34 ng L−1). These snow data are consistent with observed tropospheric O_3 and Hg^0 depletion events, and satellite observations of atmospheric BrO concentrations in the northern hemisphere (Richter et al., 1998). These observations constitute direct evidence of a link between sunlight-assisted Hg^0 oxidation, greatly enhanced atmospheric Hg(II) wet and/or dry deposition, and elevated mercury concentrations in Arctic snow and snow-pack during springtime. Similar results were obtained in subsequent years and, during springtime

2000 at Alert, a 20-fold increase in Hg concentration in the snow samples from before an AMDE to after an AMDE was observed (Steffen et al., 2002). Using, the technique of the CRPU- Cold Regions Pyrolysis Unit, it was found that during AMDEs, when Hg^0 concentrations decrease, a significant amount of mercury (in forms not identify by the instrument) remain in the air. Hence it was deduced that the remaining component not recovered throughout pyrolysis, was removed from the air, presumably having been deposited to the frozen surface (Steffen et al., 2002). This finding substantiates the argument that the converted Hg^0 during AMDEs is deposited onto the snow. In support of this hypothesis, closer to the surface at Alert, vertical concentration gradients of mercury were measured in the interstitial air within the snow-pack and in the air above the snow at 190 and 120 cm height. Profiles were measured during both the dark and light periods and before and during AMDEs. These profile data showed no significant difference in Hg^0 concentrations between near surface and those at 1.9 and 1.2m heights during dark and 24-hours daylight conditions before an AMDE; in contrast, while a depletion event occurred, the Hg^0 concentration in the air column above 10 cm remained invariant with height but close to the surface a noticeable increase in the Hg^0 concentration was observed. Concurrent ozone profiles exhibited a sharp decrease at the surface level. These data suggested that the snow pack was a source of Hg^0 but a sink for ozone. Hg^0 behaves in a manner that is different from ozone as the two substances approach and interact with the frozen surface. The nature of this interaction is complex and is not yet fully understood at this time. Other profile studies between the interstitial air and air above the snow pack show a positive gradient suggesting that Hg^0 is emitted from the snow pack. The interstitial air in the snow pack was, in fact, alternately measured and, in particular, these measurements showed that the concentration of Hg^0 in the interstitial air in the snow pack was almost continually higher than in ambient air, especially when an AMDE occurred. This shows that Hg^0 was almost continually released from the snow pack indicating that the snow was a source of mercury. During and after the AMDEs, pulses of Hg^0 were observed to be emitted from the snow pack. It is thought that some of the oxidised species of mercury that are previously deposited during the AMDEs, under sunlight conditions, are reduced back to Hg^0 through chemical photo-reduction processes and are then released from the snow pack to the air above the snow surface. Flux measurements obtained during the following years at several arctic locations (Alert, Ny-Alesund) suggested a slightly increased flux of Hg^0 following an AMDE, confirming the complex atmospheric mercury chemistry in the Arctic. During Barrow expeditions (2000-2001) Lindberg et al. (2002) observed an increase of Hg^0 from <1 to >90 ng/L over this period. The 1998 data from a ship frozen in the Beaufort Sea 550 km north of Barrow have shown similar trends in surface snow (Welch et al., 1999). During snowmelt at Barrow under the 24-h daylight mercury exhibited surprisingly dynamic behaviour: MDEs ended abruptly, airborne Hg^0 spiked and airborne RGM decreased to detection limit (1 pg m^{-3}), while total Hg concentrations in snow decreased drastically, by 92%. Since a possible source of this airborne Hg^0 spike probably derived from the interaction between the snow pack and the air immediately above it leading to Hg^0 evasion from the snow-pack (Schroeder et al., 1998). Hg2+(aq) is readily photo-reduced to Hg^0 and evaded from surface waters (Lindberg et al., 2000) The flux of snow-pack Hg to air and water is thus clearly influenced by the melting process. To investigate the behaviour of Hg in snow Lalonde et al. (2002) studied Hg^0 and total Hg concentrations at different snow-pack depths above a frozen lake (Sainte-Foy, QC Canada). Results indicate that deposited Hg is highly labile in snow-pack samples, decreasing, on average, by 60% within 24 h after deposition in the first snow stratum while total Hg levels at depth were lower than near the surface, and remain constant over time. It was hypothesized that Hg depletion in snow could be caused by a rapid snow-air Hg exchange resulting from Hg(II) photo-induced reduction to volatile Hg^0. Hence, since once deposited, Hg could be rapidly reduced and re-emitted and the massive Hg deposition events observed in springtime in polar regions may probably have less impact than previously anticipated in research studies. Ferrari et al. (2004) studied Hg concentrations pattern in the interstitial air of snow during night and day conditions at Station Nord, Greenland, during March, 2002. The sampling period was shortly after polar sunrise and before any ozone and mercury depletion events. Results obtained indicated mercury oxidation and reduction processes in the top layers of the snow-pack. Recently it was proposed that homogeneous and/ or heterogeneous processes between Hg^0 and Br /BrO radicals were leading to the deposition of oxidised Hg from the interstitial air of snow onto the snow grains (Ferrari et al., 2004). The snow-pack can also be a source of Hg^0 to the atmosphere through Hg(II) reduction by HO_2 and/or photo-dissociation of some Hg(II) complexes (Dommergue et al., 2003a). The snow samples were taken at different depths. Hg^0 concentrations at 60 and 100 cm depth do not exhibit the same pattern as those at 20 and 40 cm. No night and day profile was observed for these deeper depths indicating that the production was affecting mainly the top layers of the snow-pack. Levels of Hg^0 in the atmosphere were relatively constant at Station Nord in Greenland shortly after polar sunrise, whereas the concentrations in interstitial air of snow varied much more with a depletion during the day and a production at night. The observed production must then be driven by internal chemical processes, and not by change in atmospheric conditions. Depletion of Hg^0 in the interstitial air of snow is probably the result of homogeneous and/or heterogeneous chemistry with halogenated species (especially Br and BrO) leading to the formation of Hg(II) and its adsorption on the snow surfaces. The Artic snow pack is known to produce active bromine and chlorine species in the interaction process between sea-salts, ozone and acid species in the snow (Fan and Jacob, 1992; Tang and Mc Connel, 1996). The production of reactive species is even possible during low irradiation periods. Moreover, reaction between Hg^0 and Br radicals is fast (Ariya et al., 2002). Thus, similarly to the depletions of atmospheric Hg^0 and ozone in polar regions during springtime, Hg^0 depletion in the interstitial air of the snow could be the result of fast oxidation processes involving halogenated radicals. Therefore, during the day, under sunlight conditions with solar irradiation strong enough to produce Br-radicals, mercury, in its oxidized form Hg(II), accumulate in the snow-pack and the oxidation processes in the snow-pack is, therefore, more active than reduction. After the sunset, with solar radiation weaker, oxidation is less active and allows the reduction to be predominant. Hg^0 production occurring in the snow-pack during the night requires the presence of a potential reductant of Hg(II). However, research studies recently performed (Dommergue et al., 2003a) have shown a production of Hg^0 in the air of the snow in Kuujjuarapik (Canada) during the day, directly correlated with solar radiation suggesting that Hg^0 production is driven by a photo-chemically induced reduction and proposing, as a potential reductant, hydroperoxyl radicals (HO_2) (Lin and Pehkonen, 1999). The major source of HO_2 is assumed to be the photolysis of formaldehyde (HCHO) both in the troposphere and in the snow-pack interstitial air (Sumner and Shepson, 1999). The mechanism by which Hg^0 is emitted from the snow during night time is at present unknown, however this reduction process could be explained if a night time source of HO_2 exists in the air of the snow pack. Hg fluxes to the atmosphere calculated at Nord-Greenland (0.06-0.4 ng m^{-2} h^{-1}) were weaker compared to those observed by Schroeder (2003) at different Arctic sites (1-8 ng m^{-2} h^{-1}). However, Steffen et al. (2002) have shown that without any mercury depletion event, gaseous mercury concentration was homogeneous from the snow surface to a height of a few meters above the snow. Enhanced concentrations of Hg in surface snow are clearly evident throughout the AMDE period also at Ny-Alesund. Results from snow samples performed during several spring campaigns have shown Hg concentrations variable from a background of about 2 ng L^{-1} during the polar night to about 40 ng L^{-1} in the Arctic spring. After Hg^0 events, Hg concentrations increased from background values to 100 ng L^{-1}. The ratio between deposition and reemission is an important parameter that determines the impact of mercury depletion in the Arctic environment. More research is needed to improve our understanding of this exceptional aspect of environmental mercury cycling.

CONCLUSIONS

The substantial different geographical distribution of landmasses around both poles influences the Hg^0 annual mean observed in the Arctic (1.59 ng m^{-3} to 1.60 ng m^{-3}) and Antarctica (1.04 ng m^{-3}). The Antarctic region is remote from human activities on the other continents being far from landmasses; in contrast, the Arctic region is surrounded by northern North America, northern Europe and northern Asia, therefore, it undergoes to the anthropogenic influences came from the continents themselves. However, several emission Hg data sets have shown that during the last 20-years there is been in the Northern Hemisphere a decrease in global atmospheric Hg emission of about 30% despite the recent increase in Hg levels observed in Arctic biota, lake sediments, arctic food as well as in native people of circumpolar countries as recently carried out by different studies. A potential explanation for the opposite trends observed concerning the Hg increase in polar biota and the Hg decrease in global emissions, could be the AMDEs, probably a relatively recent atmospheric mechanism by which Hg^0 (Hg^0) may be converted to reactive and water-soluble forms (RGM and/or Hg-p) that deposit quickly thus increasing the mercury fluxes and deposition processes in the fragile ecosystems of the Arctic and/or Antarctica. Enhanced concentrations of Hg in surface snow are clearly evident during the AMDEs that lead, therefore, to enhanced Hg inputs into polar ecosystems (about 300 Tonnes in the Arctic and from 50 to 100 tonnes in the Antarctica). The observations seen in the polar regions, thus constitute direct evidence of a link between sunlight assisted Hg^0 oxidation, greatly enhanced atmospheric Hg(II) wet and/or dry deposition, and elevated Hg concentrations in the polar snow-pack in spring. Several studies on the Hg fate carried out in the polar regions suggested that a significant fraction of deposited Hg is bioavailable (up to 40%), about 25% can be re-emitted to the atmosphere and the fraction in meltwater is more than 50%.

It has been thought, in fact, that the AMDEs are, probably, a recent phenomena due to the climate global changes in the polar regions and, in general, to the global warming of the planet over the last 30 – 40 years. This last leading to a decreasing trend in multi-year ice coverage, earlier timing of snowmelt, increasing ocean temperature and increasing atmospheric circulation can impact the dynamics of the AMDEs. These climate changes have, in fact, increased atmospheric transport of photo-oxidants and production of reactive halogens (Br/Cl) in the polar regions enhancing Hg oxidation reactions. The decreasing, in addition, in total column ozone amounts over the polar areas and the subsequent increasing of the incident solar UV-B which influence the production of reactive halogen species could lead to increase of the Hg^0 oxidation processes and Hg accumulation in polar ecosystems.

ACKNOWLEDGEMENTS

The authors wish to acknowledge the contribution received from European Commission, USEPA, Italian CNR, Italian Ministry of Research, and Italian Polar Research Programmes (PNRA). The authors wish also acknowledge that the ozone data at the Zeppelin Station (Svalbards Islands) during the 2003 Arctic Mercury Study were kindly provided by Torunn Berg at NILU.

REFERENCES

AMAP Arctic pollution issues: State of the arctic environment", Arctic Monitoring & Assessment Program, Oslo, 103, 1997.

AMAP Assessment report: Arctic pollution issues. Arctic Monitoring and Assessment Programme (AMAP), Oslo, Norway, 859, 1998.

Ariya, P.A., Khalizov, A., Gidas, A. Reactions of gaseous mercury with atomic and molecular halogens: kinetics, product studies, and atmospheric implications. *J. Phys. Chemistry A*, 106, 7310–7320, 2002.

Barrie, L.A., Platt, U. Arctic tropospheric chemistry: an overview. *Tellus*, 49 (B), 450-454, 1997.

Barrie, L. A., Bottenheim, J. W., Schnell, R. C., Crutzen, P. J., Rasmussen, R. A. Ozone Destruction and Photochemical-Reactions at Polar Sunrise in the Lower Arctic Atmosphere, 15 *Nature*, 334, 138-141, 1988.

Berg., T. Hjellbrekke A.G. Heavy Metals and POPs within the ECE region. EMEP/CCC Report 7/98, Norwegian Institute for Air Research, Kjeller, Norway, 1998.

Berg, T., Bartnicki, J., Munthe, J., Lattila, H., Hrehoruk, J., and Mazur, A. Atmospheric mercury species in the European Arctic: measurements and modelling, *Atmos. Environ.*, 35, 2569–2582, 2001.

Berg, T., Aspmo, K. Atmospheric Mercury at Zeppelin Station. Report: 889/2003; TA-1994/2003, Norwegian Institute for Air Research (NILU), 2003.

Berg, T., Sekkesæter, S., Steinnes, E., Valdal, A. Wibetoe, G. Arctic springtime depletion of mercury in the European Arctic as observed at Svalbard. *Sci. Tot. Env.*, 304, 43-51, 2003a.

Berg, T., Sommar, J., Wängberg, I., Gårdfeldt, K., Munthe, J. Schroeder, W.H. Arctic mercury depletion events at two elevations as observed at the Zeppelin Station andDirigibile Italia , Ny-Ålesund, spring 2002. *J. Phys. IV*, 107, 151-154, 2003b.

Cheng and Schroeder Transport patterns and potential sources of total gaseous mercury measured in Canadian high Arctic in 1995. *Atmos.Environ.*, 35, 1141-1154, 2001.

De Mora, S.J., Patterson, J. E., Bibby, aD. M. Baseline atmospheric mercury studies at Ross Island, Antarctica, *Antarctic Science*, 5, 323– 326, 1993.

Dommergue, A., Ferrari, C.P., Poissant, L., Gauchard, P.A., Boutron, C.F. Diurnal cycles of gaseous mercury within the snowpack at Kuujjuarapik/Whapmagoostui, Quebec, Canada. *Environ. Sci. Technol.*, 37, doi :10.1021/es026242b, 2003a.

Ebinghaus, R., Kock, H.H., Temme, C., Einax, J.W., Lowe, A. G., Richter, A., Burrows, J. P., Schroeder, W.H. Antarctic springtime depletion of atmospheric mercury, *Env. Sci. Technol.*, 36, 1238–1244, 2002.

Ebinghaus, R., Jennings, S.G., Schroeder, W.H., Berg, T., Donaghy, T., Guntzel, J., Kenny, C., Kock, H.H., Kvietkus, K., Landing, W., Munthe, J., Prestbo, E., Schneeberger, D., Slemr, F., Sommar, J., Urba, A., Wallschläger, D., Xiao, Z. International field intercomparison measurements of atmospheric mercury species at Mace Head, Ireland, *Atmos. Environ.*, 33, 3063-3073, 1999.

Fan, S.M., Jacob, D.J. Surface ozone depletion in Arctic spring sustained by bromine reactions on aerosols. *Nature*, 359, 522–524, 1992.

Ferrari, C.P., Dommergue, A., Boutron, C.F., Skov, H., Goodsite, M., Jensen, B. Night time production of elemental mercury in interstitial air of snow at Station Nord, Greenland. *Atmos.Environ.*, 38, 2727-2735, 2004.

Foster, K.L., Plastridge, R.A., Bottenheim, J.W., Shepson, P.B., Finlayson-Pitts, B.J., and Spicer, C.W. The role of Br_2 and BrCl in surface ozone destruction at polar sunrise. *Science*, 291, 471-474, 2001.

Friess, U. Spectroscopic Measurements of Atmospheric Trace Gases at Neumayer-Station, Antarctica. Ph.D. Thesis, University of Heidelberg, Heidelberg, Germany, 2001.

Garbarino, J. R., Snyder-Conn, E., Leiker, T. J., Hoffman, G. L. Contaminants in arctic snow collected over northwest Alaskan sea ice, *Water, Air Soil Pollut.*, 139, 183-214, 2002.

Lalonde, J.D., Poulain, A.J., Amyot, M. The Role of Mercury Redox Reactions in Snow-to-Air Mercury Transfer. *Env. Sci. Technol.*, 36, 174–178, 2002.

Landis, M. S.; Stevens, R. K.; McConville, G.; Brooks, S. R. Presented at the 6th International Conference on Mercury as a Global Pollutant, Minamata, Japan, October 2001.

Landis, M.S., Stevens, R.K., Schaedlich, F., Prestbo, E.M. Development and Characterization of an annular denuder methodology for the measurement of divalent inorganic RGM in ambient air, *Env. Sci. Technol.*, 36, 3000-3009, 2002.

Landis, M. S. and Stevens, R. K. Comment on "Measurements of Atmospheric Mercury Species at a Coastal Site in the Antarctic and over the South Atlantic Ocean during Polar Summer". *Env. Sci. Technol.*, 37, 3239–3240, 2003.

Lehrer, E. Polar Tropospheric Ozone Loss. Ph.D. Thesis, University of Heidelberg, Heidelberg, Germany, 1999.

Lin, C.-J., Pehkonen, S.O. The chemistry of atmospheric mercury: a review. *Atmos. Environ.*, 33, 2067-2079, 1999.

Lindberg, S.E., and Stratton, W.J. Atmospheric mercury speciation: Concentrations and behaviour of RGM in ambient air, *Env. Sci. Technol.*, 32, 49-57, 1998.

Lindberg, S. E., Vette, A., Miles, C., Schaedlich, F. *Biogeochemistry*, 48, 237, 2000.

Lindberg, S.E., Brooks, S., Lin, C.-J., Scott, K., Meyers, T., Chambers, L., Landis, M., Stevens, R. Formation of RGM in the arctic: evidence of oxidation of Hg-II compounds after arctic sunrise. *Water, Air Soil Pollut.*, 1, 295-302, 2001.

Lindberg, S. E., Brooks, S., Lin, C. J., Scott, K. J., Landis, M. S., Stevens, R. K. Goodsite, M., and Richter, A. Dynamic oxidation of gaseous mercury in the Arctic troposphere at polar sunrise, *Env. Sci. Technol.*, 36, 1245–1256, 2002.

Linell K.A., Tedrow J.F.C. Soil and permafrost surveys in the Arctic. Clarendon Press, Oxford, 279, 1981.

Lockhart, W.L., Wilkinson, P., Billeck, B.N., Danell, R.A., Hunt, R.V., Brunskill, G.J., St.Louis, J.V., Fluxes of mercury to lake sediments in central and northern Canada inferred from dated sediment cores, *Biogeochemistry*, 40, 163-173, 1998.

Lu, J. Y., Schroeder, W. H., Barrie, L. A., Steffen, A., Welch, H. E., Martin, K., Lockhart, L., Hunt, R. V., Boila, G., Richter, A. Magnification of atmospheric mercury deposition to 15 polar regions in springtime: the link to tropospheric ozone depletion chemistry, *Geophys. Res. Lett.*, 28, 3219-3222, 2001.

Lu, J. Y., Schroeder, W. H., Keeler, G. Field intercomparison studies for evaluation and validation of the AESminiSamplR™ technique for sampling and analysis of total particulate mercury in the atmosphere, *Sci. Tot. Env.*, 304, 115–125, 2003.

Mozurkewich, M. *J. Geophys. Res.*, 100, 14, 199, 1995.

NADP/MDN National Atmospheric Deposition Program (NRSP-3)/Mercury Deposition Network, 2002.

Pacyna, E.G., Pacyna, J.M., Pirrone, N. European emissions of atmospheric mercury from antropogenic sources in 1995, *Atmos Environ.*, 35, 2987-2996, 2001.

Pirrone, N., Keeler, G.J., Nriagu, J. Regional differences in worldwide emissions of mercury to the atmosphere. *Atmos. Environ.*, 17, 2981-2987, 1996.

Poissant, L., Pilote, M Atmospheric mercury and ozone depletion events observed at low latitude along the Hudson Bay in northern Quebec (Kuujjuarapik: 55°N). In: Sixth International Conference on Mercury as a Global Pollutant, Minamata, Japan, October 15–19. Book of Abstracts (AT-23) 2001.

Pongratz, R., Heumann, K.G. Production of methylated mercury, lead, and cadmium by marine bacteria as a significant natural source for atmospheric heavy metals in polar regions, *Chemosphere*, 39, 89– 102, 1999.

Richter, A., Wittrock, F., Eisinger, M., Burrows, J.P., GOME observations of tropospheric BrO in northern hemispheric spring and summer 1997, *Geophys. Res. Lett.*, 25, 2683-2686, 1998.

Schroeder, W.H., Anlauf, K.G., Barrie, L.A., Lu, J.Y., Steffen, A., Schneeberger, D.R., Berg, T. Arctic springtime depletion of mercury, *Nature*, 394, 331-332,1998.

Schroeder, W.H., Steffen, A., Scott, K., Bender, T., Prestbo, E., Ebinghaus, R., Lu, J.Y., Lindberg, S. E. Summary report: first international Arctic atmospheric mercury research workshop, *Atmos. Environ.*, 37, 2551–2555, 2003.

Sheu, G.R. Mason, R.P. An examination of methods for the measurements of RGM in the atmosphere. *Env. Sci. Technol.*, 35, 1209-1216, 2001.

Skov, H., Christensen, J.H., Goodsite, M.E., Heidam, N.Z., Jensen, B., Wahlin, P., Geernaert, G. Fate of Elemental Mercury in the Arctic during Atmospheric Mercury Depletion Episodes and the Load of Atmospheric Mercury to the Arctic. *Env. Sci. Technol.*, 38, 2373–2382, 2004.

Sommar, J., Wängberg, I., Berg, T., Gårdfeldt, K., Munthe, J., Richter, A., Schroeder, W.H., Urba, A., Wittrock, F. Circumpolar transport and air-surface exchange of atmospheric mercury at Ny-Ålesund (79°N), Svalbard, spring 2002. *Atmos. Chem. Phys. Discuss.*, 4, 1727-1771, 2004.

Sprovieri, F., Pirrone, N., Hedgecock, I. M., Landis, M. S., Stevens, R. K. Intensive atmospheric mercury measurements at Terra Nova Bay in Antarctica during November and December 2000. *J. Geophys. Res.*, 107 (D23), 4722, 2002.

Sprovieri, F., Pirrone N. A preliminary assessment of mercury levels in the Antarctic and Arctic troposphere, *J. Aerosol. Sci.*, 31, 757–758, 2000.

Steffen, A. Schroeder, W.H. Atmospheric mercury in the high Arctic from 1995 to 2002. Second AMAP International Symposium on Environmental Pollution of the Arctic, Rovaniemi, Finland, October 1–4, 2002.

Steffen, A., Schroeder, W., Bottenheim, J., Narayan, J., Fuentes, J.D. Atmospheric mercury concentrations: measurements and profiles near snow and ice surfaces in the Canadian Arctic during Alert 2000, *Atmos. Environ.*, 36, 2653–2661,2002.

Sumner, A.L., Shepson, P.B. Snowpack production of formaldehyde and its effect on the Arctic troposphere. *Nature*, 398, 230-233, 1999.

Tang, T., McConnell, J.C. Autocatalytic release of bromine from Arctic snowpack during polar sunrise. *Geophys. Res. Lett.*, 23, 2633-2636, 1996.

Temme, C., Einax, J.W., Ebinghaus, R., Schroeder, W.H. Measurements of atmospheric 10 mercury species at a coastal site in the Antarctic and over the south Atlantic Ocean during polar summer, *Env. Sci. Technol.*, 37, 22–31, 2003.

Varjo U. Tietze W. Norden. Man and Environment. Gebr. Borntraeger, Berlin, Stuttgart, 1987.

Welch, H. K.; Martin, K.; Lockhart, W. L.; Hunt, R. V.; Boila, G. In Synopsis of Research Conducted under the 1997/98 Northern Contaminants Program; Jensen, J., Ed.; Department of Indian Affairs and Northern Development: Ottawa, 93, 1999.

Woo M.K., Gregor D.J. Arctic Environment: Past, present and future. McMaster University, Department of Geography, Hamilton, 164, 1992.

Chapter-29

EMEP REGIONAL/HEMISPHERIC MERCURY MODELLING: ACHIVEMENTS AND PROBLEMS

Alexey Ryaboshapko, Sergey Dutchak, Alexey Gusev, Ilia Ilyin and Oleg Travnikov

EMEP Meteorological Synthesizing Center "East", Arhitektor Vlasov Str., 51, Moscow 117393, Russia

INTRODUCTION

Investigation and control of transboundary air pollution in Europe has a relatively deep history. Considerable practical and scientific progress in this direction has been achieved in the framework of the Convention on Long-Range Transboundary Air Pollution. A number of international binding instruments (Protocols to the Convention) on reduction of air pollution were developed and entered into force during the last 25 years. Scientific support for the evaluation of long-range air pollution, as well as development and implementation of the Protocols is provided by the Cooperative Programme for Monitoring and Evaluation of the Long-range Transmission of Air Pollutants in Europe (EMEP).

EMEP was established in 1977 under the United Nations Economic Commission for Europe. After the Convention entered into force in 1983 EMEP became an operational programme of the Convention. Routine activity of EMEP is based on joint efforts of the participating countries and 4 international centres of EMEP. One of these centres – the Meteorological Synthesizing Centre - East (MSC-E) is responsible for the development and application of atmospheric transport deposition models for the assessment of air pollution by heavy metals and persistent organic pollutants.

Heavy metals in line with acid compounds are included in the priority list of substances considered under the Convention. A Protocol on Heavy Metals was signed by 36 Parties to the Convention in 1998. The Protocol is aimed at control of heavy metal emissions into the atmosphere to reduce their transboundary transport and to prevent adverse effects on human health and the environment. In accordance with the Protocol EMEP is responsible for use of appropriate models and measurements for providing to European countries calculations of transboundary fluxes and depositions of lead, cadmium and mercury.

This paper is focused on the assessment of the long-range transboundary transport of mercury. Mercury is widely recognized as a global pollutant. To evaluate mercury pollution of Europe mercury emissions all over the globe and intercontinental transport should be taken into account. To meet these requirements MSC-E is developing mathematical models for the evaluation of mercury atmospheric transport on regional (Europe) and hemispherical (Northern Hemisphere) scales.

EMEP REGIONAL MODEL

The EMEP regional model considers basic processes governing the transport and deposition of mercury - advection, diffusion, dry and wet removal and chemical transformations. This is an Eulerian three-dimensional atmospheric transport model. The model operates within the so

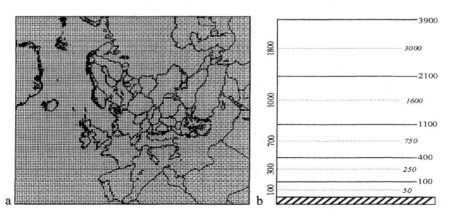

Figure 1. Horizontal (a) and vertical (b) structure of EMEP regional model.

called EMEP domain (Posh et al., 1997). This region includes the European continent, the northern part of Africa, part of the Middle East, the North Atlantic and part of the Arctic (see Figure 1). The EMEP grid consists of 135x111 grid cells with spatial resolution 50 km at 60°N. latitude.

As seen from Figure 1, the model domain consists of five non-uniform layers along the vertical. The top of the model is at a height of about 4 km. Therefore, the model domain covers the entire atmospheric boundary layer and part of the middle troposphere. Depths of the layers are 100, 300, 700, 1000 and 1800 m (from bottom to top). The advection scheme is conservative, stable and positively defined (Pekar, 1996). The model description of the vertical turbulent diffusion is based on the first order closure approach.

The model deals with three physical-chemical mercury forms: gaseous elemental mercury (GEM), reactive (oxidised) gaseous mercury (RGM) and total particulate mercury (TPM). They possess very different characteristics, which determine very different lifetimes of each form in the atmosphere. Detailed description of the parameterisation of all modelled processes can be found in MSC-E technical reports (Ilyin et al., 2001; 2002; www.msceast.org).

Scavenging of all mercury forms encompasses wet removal and uptake by the underlying surface. Wet removal of TPM and RGM is described using a washout ratio approach. It is accepted that particles containing mercury behave like sulfate particles and the equilibrium washout ratio is equal to 5×10^5 (Petersen et al., 1998; Iversen et al., 1989). Washout of RGM by the liquid phase is prescribed by the equilibrium washout ratio characteristic of nitric acid: 1.4×10^6 (Petersen et al., 1995; Jonsen and Berge, 1995). The reason for this is the similar solubilities of these two species in water. Besides, RGM and GEM can be dissolved in the aqueous phase of clouds.

Dry deposition of TPM is described in the framework of an electric resistance analogy. Mercury containing particles are in the submicron size range (Milford and Davidson, 1985; Keeler et al., 1995), hence the effect of the gravity can be ignored. Dry deposition efficiency is differentiated with regard to the land-cover category of the underlying surface and depends basically on properties of the underlying surface and atmospheric stability (Ruijgrok et al., 1997; Wesely and Hicks, 2000).

Dry uptake of GEM depends on a number of parameters. On the basis of literature data (Lindberg et al., 1992; Xu et al., 1999; Petersen et al., 2001; Seigneur et al., 2001) it is assumed that at negative temperatures uptake does not occur. Uptake also does not occur by water and vegetation-free surfaces.

For the surface covered with forest the velocity is accepted to be equal to 0.03 cm/s at 20°C and higher. For other types of vegetation the maximum value is 0.01 cm/s. The uptake velocity decreases linearly to zero as temperature decrease to 0°C.

When describing the dry uptake of gaseous oxidised mercury a similarity of dry uptake velocity to that of nitric acid is assumed (Petersen et al., 1998). This assumption comes from their similar solubilities in water. Keeping in mind the obvious lack of knowledge on this, the dry uptake velocity for RGM assumed in the model is 0.5 cm/s for all seasons and types of underlying surfaces.

Parameterisation of chemical processes includes both aqueous-phase and gaseous-phase reactions and equilibria. It is based on the chemical scheme suggested by Petersen et al. 1998. However, the scheme has been simplified - only key reactions are used in the model. They are gas-phase oxidation of GEM by ozone, dissolution of GEM and RGM in cloud droplets, aqueous phase oxidation of GEM by ozone with further sorption of the reaction products on insoluble particles within droplets, and mercury reduction to the elemental state through decomposition of mercury-sulfite complexes. All products of gaseous-phase oxidation are treated as aerosol particles. It is accepted (Brosset and Lord, 1991; Iverfeldt, 1991; Lamborg et al., 1995) that half of TPM being captured by cloud or rainwater droplets can be dissolved. After drop evaporation an aerosol particle is formed containing in its composition all earlier dissolved and insoluble mercury compounds.

An important distinction of the scheme from analogous ones typically used in atmospheric mercury models (Petersen et al, 1998; Bullock and Brehme, 2002; Shia et al., 1999) is in usage of temperature dependencies of reaction rates and equilibrium constants. For Henry's law constants the following equations are used (Sander, 1997; Ilyin et al., 2002):

For GEM: $$K_{Hg^0} = 0.00984 \cdot T \cdot \exp\left(2800 \cdot \left(\frac{1}{T} - 0.003356\right)\right),$$

for RGM: $$K_{HgCl_2} = 1.054 \cdot 10^5 \cdot T \cdot \exp\left(5590 \cdot \left(\frac{1}{T} - 0.003356\right)\right),$$

for O₃: $$K_{O_3} = 0.000951 \cdot T \cdot \exp\left(2325 \cdot \left(\frac{1}{T} - 0.003356\right)\right).$$

The dependence of the rate of gas-phase GEM oxidation by ozone on temperature is described by the following equation delivered from the data published by Hall [1995]:

$$k = 2.1 \cdot 10^{-18} \cdot \exp\left(-\frac{1246}{T}\right) \quad cm^3/molec/s$$

Since temperature in the troposphere can vary within the range of 100 degrees, the accepted dependencies can significantly change rates of mercury chemical transformations and its removal from the atmosphere.

EMEP HEMISPHERIC MODEL

The EMEP hemispheric model has been developed in order to evaluate the atmospheric transport of mercury over the Northern Hemisphere. This is a three-dimensional chemical transport model of Eulerian type. The detailed description of the model can be found in MSC-E technical reports (Travnikov and Ryaboshapko, 2002; www.msceast.org).

The model computation domain covers the whole Northern Hemisphere with a spatial resolution of 2.5° both in zonal and meridional directions. The surface grid structure of the model domain is shown in Figure 2. To avoid a singularity at the pole point, peculiar to the spherical co-ordinates, the grid has a special circular mesh of radius 1.25° including the North Pole. In the vertical direction the model domain consists of eight irregular levels of terrain-following sigma-pressure co-ordinates defined as a ratio of local atmospheric pressure to the ground surface pressure (Jacobson, 1999). The vertical grid structure of the model is presented in Figure 2.

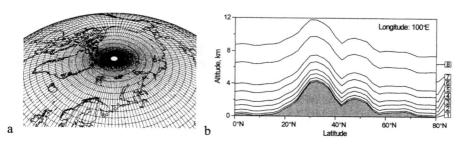

Figure 2. Horizontal (a) and vertical (b) structure of EMEP hemispheric model.

Advection is treated using the Bott flux-form advection scheme (Bott, 1992). This scheme is mass conservative, positive-definite, monotone, and is characterised by comparatively low artificial diffusion. In order to reduce the time-splitting error in strong deformational flows the scheme has been modified according to (Easter, 1993). The vertical movements are solved using the original Bott scheme generalised for a grid with variable step. Non-linear diffusion has been approximated by the second-order implicit numerical scheme in order to avoid restrictions of the time step caused by possible sharp gradients of species mixing ratio.

The modelling domain has two borders - upper layer and the Equator. At the upper boundary a uniform distribution of GEM concentration of 0.185 pptv (corresponding to about 1.5 ng/m^3 at 1 atm and 20°C) is prescribed. Within the equatorial zone some gradient of gaseous mercury was observed between the Northern and Southern Hemispheres (Slemr, 1996). In the model the gradient of GEM is set to 0.05 ng/m^3/degree at the equatorial boundary. Since the atmospheric residence times of the other mercury species are considerably shorter their input through the boundaries is neglected.

The parameterisation of mercury scavenging processes in the hemispheric model does not differ from that used in the regional model. The hemispheric model takes into account the same chemical transformations of mercury as the regional model described above. However, GEM oxidation by chlorine is introduced into the chemical scheme because in the oceanic atmosphere this reaction can give a noticeable effect.

EMISSIONS AND OTHER INPUT PARAMETERS

The Convention envisages that all participating countries should evaluate their national emissions using the same inventory methodology. Currently national data on total mercury emissions (for at least one year for the period of 1990-2000) were submitted by 34 countries. For the other countries, which have not reported national emission data, expert estimates are applied (Berdowski et al., 1997; Pacyna and Pacyna, 2002). Mercury emission data for 2000 used for the assessment of pollution level in Europe are demonstrated in Table 1 (Ilyn and Travnikov, 2003). During the last decade mercury emissions into the atmosphere in most European countries were reduced. Thus, the total European mercury emission decreased from 463 t/yr in 1990 (Berdowski et al., 1997) to 201 t/yr in 2000 (Table 1). Accuracy of the emission data is quite uncertain. It is believed that the expert estimates can be within ±30% (Pacyna and Pacyna, 2002).

Table 1. Mercury anthropogenic emissions* (AE) to the atmosphere in the EMEP region in 2000, t/y.

Country	AE	Country	AE	Country	AE
Albania	0.5	Georgia	0.5	Portugal	4.85
Armenia	0.001	Germany	29	Moldova	0.18
Austria	1.15	Greece	13	Romania	6.55
Azerbaijan	0.6	Hungary	4.21	Russia	10
Belarus	0.36	Iceland	0.05	Serbia & Mont.	3.3
Belgium	2.88	Ireland	1.95	Slovakia	4.37
Bosnia & Herz.	0.2	Italy	13.2	Slovenia	0.58
Bulgaria	4.19	Kazakhstan	0.1	Spain	23.4
Croatia	0.41	Latvia	0.21	Sweden	0.81
Cyprus	0.30	Lithuania	0.25	Switzerland	2.63
Czech Rep.	3.84	Luxembourg	0.27	Macedonia	0.05
Denmark	1.96	Monaco	0.08	Turkey	4.30
Estonia	0.55	Netherlands	0.58	Ukraine	9.03
Finland	0.6	Norway	1.00	UK	8.79
France	15	Poland	25.6	Total	201

* Values obtained by extrapolation of official data and on the base of expert estimates are in italic.

For modelling purposes the emissions need to be spatially distributed over the domain. Some European countries assess the distribution of their national emissions in accordance with the EMEP grid (50x50 km resolution). For the others the total national emissions were distributed in accordance with (Berdowski et al., 1997).

During recent years the emission density has changed very significantly. In 1990 the emission density in "hot spots" reached 3200 g/km²/yr. A very detailed emission inventory of the three mercury forms for each European country was implemented recently by Pacyna et al. (2003). The emissions were spatially distributed in accordance with the EMEP grid taking into account the locations of main point sources. In the vertical direction three emission layers were distinguished: <50, 50-150, and >150 m. It is possible to see (Figure 3b) that from 1990 to 2000 emissions declined in most European countries.

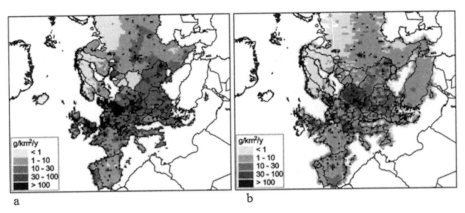

Figure 3. Spatial distribution of mercury anthropogenic emissions in Europe: (a) for 1990 (Berdowski et al., 1997) and (b) for 2000 (b) Pacyna et al. (2003).

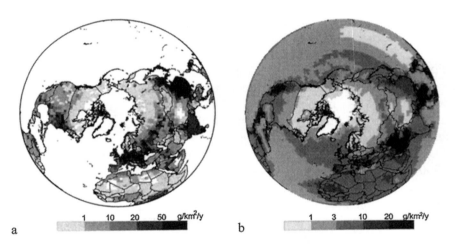

Figure 4. Hemispheric distribution of anthropogenic (a) and natural (b) mercury emissions.

To simulate mercury atmospheric transport on the hemispheric level a global emission inventory prepared by Pacyna and Pacyna (2001) is used. The inventory includes the data for three mercury forms. The spatial resolution of the emission field is 1x1 degree. In accordance with these estimates the mercury anthropogenic emission in the Northern Hemisphere totalled 1900 tons in 1995. Its spatial distribution is shown in Figure 4a.

It is well-known that mercury enters the atmosphere from different natural emission sources. For modelling purposes this emission should be assessed and spatially distributed. Lamborg et al. (2002) suggested that global natural emissions reach about 1800 t/y (1000 over land and 800 over the ocean). In order to obtain a spatial distribution of natural emission fluxes the total emission value was scattered throughout the globe depending on the mercury content in soils and the surface temperature. It was assumed that mercury evasion from the ocean was proportional to the primary organic carbon production (Travnikov and Ryaboshapko, 2002).

The obtained distribution of natural mercury emissions in the Northern Hemisphere is shown in Figure 4b. The highest emission values are typical of the so-called geochemical mercuriferous belts with increased content of mercury in soils. The emission flux from seawater is lowest in the Middle Pacific and highest in internal seas and coastal waters at low latitudes. The total natural emission of mercury in the Northern Hemisphere constitutes about 1220 t/y.

In addition to the emission data the models require some other input information like meteorological data, characteristics of underlying surface, concentrations of different reactants involved into mercury atmospheric chemistry and so on. Meteorological data used in the calculations on the regional level are based on the Re-analysis project data (Kalnay et al., 1996). These data are prepared by National Centers for Environmental Predictions together with National Center of the Atmospheric Research (NCEP/NCAR). Meteorological data necessary for the hemispheric model are based on NCEP/NCAR Re-analysis data and processed by the low atmosphere diagnostics system (SDA) developed in co-operation with Hydro-meteorological Centre of Russia. The system provides 6-hour weather prediction data along with estimates of the atmospheric boundary layer parameters and covers the Northern Hemisphere.

To take into account information on land cover of the Earth surface the regional model uses data on fifteen types of underlying surface based on (Posh et al., 2001). In hemispheric model 25-category land cover data set from NCAR Mesoscale Modelling System (MM5) is used (Guo and Chen, 1994). Since the model formulation does not require highly detailed specification, the original 25-category data were reduced to five general categories (urban, forests, grassland, bare land, and glaciers) and redistributed over the model grid.

To describe chemical transformations one has to know spatial and temporal distribution of the reactants concentration (such as ozone and sulfur dioxide) in the atmosphere. The regional model uses the calculated fields of main reactants provided by EMEP Meteorological Synthesizing Centre - West. Global monthly mean data on ozone and SO_2 concentrations in the atmosphere were kindly presented by Dr. Malcolm Ko (Wang et al., 1998; Chin et al., 1996). Besides, for the aqueous-phase chemistry cloud water was characterised by pH value equal to 4.5 and chloride ion concentration in cloud water equal to $7 \cdot 10^{-5}$ M (Acker et al., 1998). Following Seigneur et al. (2001) air concentration of molecular chlorine in the lowest model layer over the ocean is assumed to be 100 ppt at night-time, 10 ppt during the day and zero concentration over land.

POLLUTION LEVELS IN EUROPE

In accordance with the EMEP work programme MSC-E carries out a modelling assessment of mercury transboundary pollution within Europe. The main objective of the work is to evaluate mercury concentration levels in air and in precipitation. Besides, the modelling approach gives a possibility to calculate dry and wet deposition of mercury over Europe and transboundary transport between countries.

The atmospheric mercury budget for the EMEP region can be described by the following items: emissions, depositions, inflow and outflow fluxes. The main items of the budget of mercury emitted in Europe for the year 2000 are shown in Table 2. As seen from the table GEM enters mainly into the global mercury cycle. Since other mercury forms - TPM and RGM have lifetimes from hours to days, most of these species are deposited within the region. In general Europe is a net source of mercury for the global atmosphere (its emission exceed deposition). It should be kept in mind that a huge mass of atmospheric mercury (in comparison with annual European emission) enters and leaves the EMEP reservoir via lateral boundaries. This is conditioned mostly by the long-living form of elemental mercury. However, TPM can also be generated in the atmosphere due to chemical transformations. This can explain the fact that deposition of TPM plus its transport outside the EMEP region is higher than TPM emission.

Levels of mercury concentration in air are rather smooth due to its long lifetime in the atmosphere and due to the significant contribution of globally distributed mercury (Figure 5a). The highest concentration values exceed the global background only by a factor of 2-3. Maximum values of GEM concentrations in 2000 were obtained in Greece (6.3 ng/m^3), Slovakia (4.2 ng/m^3), Poland (4.1 ng/m^3) and the eastern part of Germany (4.0 ng/m^3). Regions with relatively low air concentrations are in the north of Scandinavia, where computed concentrations lay within 1.7 - 1.9 ng/m^3.

Table 2. Atmospheric budget of mercury emitted in Europe for EMEP region in 2000, t/y.

Budget item	Mercury forms			
	GEM	**RGM**	**TPM**	**ΣHg**
Total emission	369	59	31	459
Natural and re-emission	258	0	0	258
Direct anthropogenic	111	59	31	201
Total depositions *	4	57	29	90
Output from EMEP domain	360	2	7	369

* taking into account chemical transformations within the atmosphere: 5 tonnes of GEM are
oxidized and deposited as TPM.

In contrast to concentrations in air mercury depositions reveal high gradients from "hot spots" in Central Europe to the periphery of the continent (Figure 5b). The deposition intensity in different parts of Europe can differ by more than an order of magnitude. Total depositions of mercury are mostly formed by depositions of oxidised mercury forms - TPM and RGM. Since a considerable fraction of these forms has basically anthropogenic origin, the deposition maxima are usually strongly associated with the anthropogenic sources. High deposition levels are characteristic of Central and Southern Europe. The highest values of average mercury deposition per country – about 40 g/km^2/y are in Slovakia, Poland and Belgium. This is caused both by high national emissions and transboundary transport from neighbouring countries. It should be kept in mind that for individual grid cells the deposition values could exceed country average ones by an order of magnitude (the maximum value was 250 g/km^2/y).

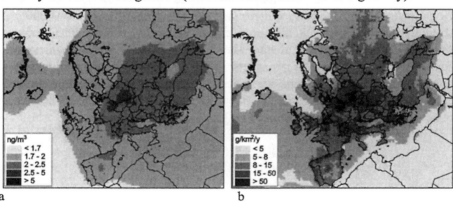

a   b

Figure 5. Annual mean concentrations of GEM (a) and total deposition (b) within EMEP in 2000.

In most part of the European region the levels of mercury deposition fluxes are between 10-20 g/km^2/y. Relatively low depositions are seen in the European North. In the central part of Scandinavia and in Northern Russia deposition fluxes as a rule do not exceed 10 g/km^2/y.

TRANSBOUNDARY POLLUTION

Assessment of transboundary depositions for each European country due to long-range atmospheric transport is the main task of EMEP model calculations. The results of the calculations are presented as a matrix of country-to-country depositions (Ilyin and Travnikov, 2003). In Table 3 a simplified version of the mercury deposition matrix is presented. Here for each EMEP country the two major sources countries of transboundary pollution are shown for each receiving country. In addition to this, contributions from totality of natural emission, secondary anthropogenic re-emission and remote (non-European) anthropogenic sources (NSR sources) are given. It is important to stress that this fraction does not indicate the pure natural contribution to depositions, but in fact it is a combination of natural inputs, inputs from remote anthropogenic sources and inputs due to previous anthropogenic pollution (re-emission).

Analysis of the table demonstrates that transboundary pollution can be very important for most European countries. For example, two neighbouring countries - France and Germany - contribute 40% of total deposition to Belgium. In some countries the main contribution is given by national sources. The highest absolute input of transboundary transport to mercury pollution (above 1 t/y) is characteristic of countries with large territories such as Russia, Poland, France, etc. It is typical for all countries that a considerable share of mercury deposition is caused by NSR sources, located all over the globe.

POLLUTION BUDGETS FOR INDIVIDUAL COUNTRIES

EMEP should provide each member-country with a detailed analysis of transboundary pollution.

Table 3. Mercury depositions on countries-receptors and contributions of different sources into the depositions (a fragment of total country-to-country matrix).

Country - receptor	Total deposition, (tonnes)	Contribution to the deposition from different sources, %				
		Main countries – sources[*]		Other EMEP countries[*]	Own sources[*]	NSR sources
Austria	1.35	Italy 8	Germany 4	17	15	56
Belgium	1.05	France 38	Germany 4	8	30	20
Bulgaria	1.88	Romania 9	Greece 7	8	36	40
Czech Rep.	1.97	Germany 18	Poland 14	11	31	26
Denmark	0.65	Germany 16	Poland 3	10	40	31
Finland	2.41	Poland 4	Germany 3	8	3	82
France	8.43	Spain 7	Switzerl. 2	5	43	43
Germany	10.48	France 4	Switzerl. 2	10	61	23
Greece	3.04	Bulgaria 3	Romania 1	3	69	24
Hungary	1.87	Slovakia 14	Romania 3	12	42	29
Italy	4.82	France 3	Spain 2	2	52	41
Netherlands	0.69	France 20	Belgium 13	21	16	30
Norway	2.60	Germany 3	Poland 3	5	6	83
Poland	11.99	Germany 10	Czech R. 4	7	61	18
Romania	3.69	Hungary 4	Poland 3	13	41	39
Russia	26.92	Ukraine 3	Poland 3	5	13	76
Slovakia	1.61	Hungary 11	Poland 6	10	49	24
Slovenia	0.43	Italy 12	Austria 2	10	24	52
Spain	6.65	Portugal 3	France 1	1	55	40
Sweden	2.88	Germany 7	Poland 7	12	1	73
Switzerland	1.09	France 10	Italy 7	3	47	33
Ukraine	7.92	Poland 7	Romania 3	12	32	46
UK	3.43	France 2	Ireland 2	3	56	37

* Only anthropogenic emissions.

Examples of two countries - Austria and Poland are considered below. The first one is a typical country-receiver of mercury pollution while the second one is a country-source (see Table 3). Information on pollution of any other European countries is available on the Internet: www.msceast.org/countries/. This information is intended to help national experts in developing abatement strategies concerning mercury emissions. Indeed, even very significant reduction of national emission can give no effect in a given country if the pollution levels are determined mainly by transboundary pollution.

The pie diagrams in Figure 6 present mercury depositions to Austria and Poland caused by national and external sources in 2000. In the case of Austria the main contribution to the deposition is made by neighbouring countries and NSR. Own sources give only 15% of the total value. The opposite situation is seen in Poland. Here national emission sources dominate. Nevertheless, the contribution of NSR is significant.

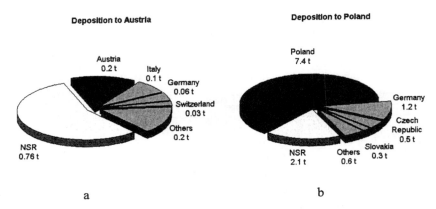

Figure 6. Mercury depositions over Austria (a) and Poland (b) from different emission sources in 2000.

The contribution of transboundary transport is non-uniformly distributed over a country. To develop national abatement strategies it is important to know the spatial distribution of transboundary pollution within a given country. Figure 7 illustrates the patterns of transboundary contributions from anthropogenic sources over the territories of the considered countries. Their regions neighbouring the countries with powerful emission sources are mostly impacted by external anthropogenic sources. In some regions of Austria the external contribution can reach 50%. In Poland noticeable contribution of transboundary mercury pollution (up to 85%) can be found in western parts of the country.

LONG-TERM POLLUTION TRENDS

According to the modelling results, emission reductions have resulted in the decrease of heavy metal depositions over the major part of the European territory. On the whole, in the period from 1990 to 2000, mercury deposition in Europe decreased 1.5 times. During the period of 1990–2000 anthropogenic mercury emissions in Europe reduced more than 2 times – from 420 to 201 t/y.

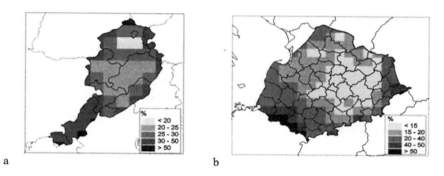

a   b

Figure 7. Contribution of external anthropogenic sources to mercury depositions to Austria
(a) and Poland (b) in 2000, % (calculated in 50*50 km grid).

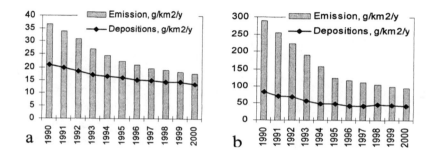

a   b

Figure 8. Trends of mercury emissions and calculated depositions during the period
of 1990-2000: in Europe as a whole (a) and in Belgium (b).

The smaller decrease of deposition in comparison with anthropogenic
emissions is conditioned by the contribution of natural sources and re-
emission, as well as by growing global emissions. This difference in the
trends for Europe as a whole is demonstrated by Figure 8a.

Some countries reduced their emissions significantly. However, the
effects of the reduction on mercury deposition can be different. In Belgium,
for example, the national emission dropped 3.5 times (Figure 8b). However,
this did not lead to the same decrease in deposition, which reduced less than
twice. Such lack of correspondence can be explained by the fact that
mercury deposition in Belgium is determined primarily by European
transboundary pollution.

ROLE OF MERCURY SPECIATION

From the viewpoint of developing abatement strategies, both on national and pan-European levels, it is very important to know which individual mercury forms are mostly responsible for elevated levels of deposition. Figure 9a demonstrates the mercury deposition field caused by emissions of only elemental mercury. The field is even and the deposition values are not high. This means that the reduction of emissions of this mercury form could not lead to a considerable decline of the deposition in Europe. If only oxidised mercury emissions are considered (Figure 9b) one can see that just these forms are primarily responsible for the elevated mercury depositions. Hence, emission reduction of oxidised mercury forms is more important to decrease atmospheric loads in the most polluted areas of Europe. At the same time reduction of GEM emissions is more important in a global context.

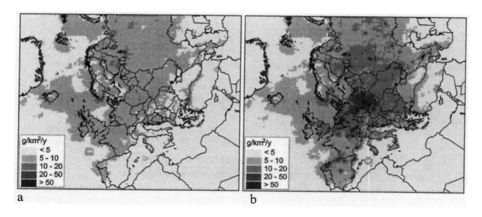

Figure 9. Mercury depositions caused by elemental mercury emission only (a) and oxidized mercury emission only (b).

HEMISPHERIC TRANSPORT AND DEPOSITION

As shown above the role of globally distributed anthropogenic sources can be significant for mercury deposition levels in different regions of the Northern Hemisphere. The results obtained by the hemispheric model demonstrate that the contribution of intercontinental transport to mercury deposition over Europe is about 40% of the total value. About half of mercury deposition to such a remote region as the Arctic is due to long-range atmospheric transport from anthropogenic emission sources. Asian emissions nowadays play the most important role on the global level.

Figure 10a displays the calculated distribution pattern of mercury concentrations in the surface air for the Northern Hemisphere. On the global level it possible to distinguish some "hot spots". The highest concentrations are typical of Europe and South-eastern Asia. As a result of mixing processes, levelling of mercury concentrations in the troposphere takes place, and the global mercury background is established. Even in the remote parts of the Atlantic and Pacific oceans, as well as in the Arctic, mercury concentration in the surface air does not fall below 1.4 ng/m^3. In accordance with the calculations the background mercury concentrations in air masses coming to Europe from the Atlantic are about 1.6 ng/m^3 for GEM, 10 pg/m^3 for TPM and 0.3 pg/m^3 for RGM. The background values calculated by the hemispheric model are used as boundary conditions for regional calculations.

The distribution of mercury depositions in the Northern Hemisphere is shown in Figure 10b. The highest deposition values are characteristic of regions of high anthropogenic emissions (Eastern Asia, Europe, North America). Depositions in different parts of the hemisphere depend significantly on intercontinental atmospheric transport. For example, the deposition levels in Europe from global non-European sources amount to about 10 g/km^2/yr. This value is comparable with those from European sources.

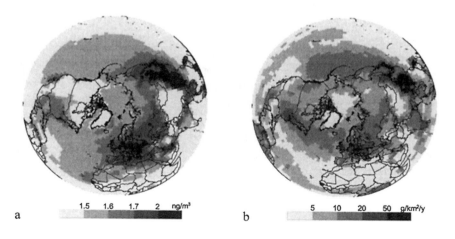

Figure 10. Hemispheric distribution of mean annual concentration of total gaseous mercury, ng/m^3 (a) and total depositions, g/km^2/yr (b).

Relative contributions of own and external sources to mercury deposition in Europe are presented in Table 4. The main contribution is given by European anthropogenic sources. However, the contribution of external sources is comparable and makes up about 40%. The most significant non-European input is made by Asian sources (15%) and mercury evasion from the ocean surface (12%). American sources contribute about 5%. It should be noted that the anthropogenic component of mercury deposition to Europe considerably exceeds the natural one and amounts to 75% of the total.

Table 4. Contributions of different regions to the total annual mercury deposition over the European region, %.

Source	Europe	Asia	Americas	North Africa	World Ocean	SH*
Anthropogenic	59	11	3	1	-	4
Natural	2	4	2	1.5	12	

* SH – Southern Hemisphere

MODEL VALIDATION

To confirm the quality and reliability of the modelling results the modelled data were compared with monitoring data obtained mainly by the EMEP monitoring network. Unfortunately, only few EMEP monitoring stations measure mercury on a routine basis, and practically all of them are located in North-western Europe. Locations of EMEP monitoring stations are shown in Figure 11a. The co-ordinates and description of the stations can be found in EMEP technical reports (Ilyin et al., 2002; www.nilu. no/projects/ccc). The comparison was carried out for annual mean mercury concentrations in air and in precipitation. To verify the hemispheric model long-term measurements performed in Europe and in North America (NADP/MDN, 2002) are used. In addition, data from short-term measurements performed during episodic measurements over the Atlantic and in Eastern Asia are considered. The locations of monitoring stations and sites of episodic measurements are shown in Figure 11b.

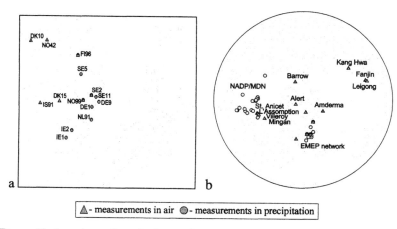

A - measurements in air   ● - measurements in precipitation

Figure 11. Locations of monitoring stations and sites of episodic measurements.

As shown in Figure 12a, the measured and calculated GEM concentrations vary within narrow limits. The agreement between average measured and calculated GEM concentrations is within 10-15%, however, the difference for individual stations can exceed 25%. Generally, the model somewhat overestimates measured values. For almost all stations measuring mercury content in precipitation the measured and calculated deposition fluxes agree within a factor of 2 (Figure 12b). On the whole mercury wet deposition fluxes measured at stations in 2001 were slightly overestimated by the model. An appreciable overestimation is noted for German station DE9. The reason for this can be connected with uncertainties in spatial distribution of the anthropogenic emissions of different mercury forms.

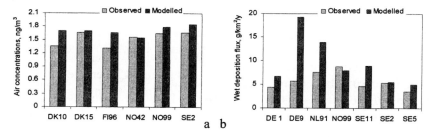

Figure 12. The comparison of measured and calculated values: (a) for GEM concentrations; (b) for wet deposition.

A 2-week measuring campaign focused on TPM and RGM (Schmolke et al., 1999; Munthe et al., 2003; Wängberg et al., 2003) gave a unique opportunity to verify the model for these mercury forms which are not measured on a routine basis. The measurements were performed simultaneously at two German, two Swedish and one Irish monitoring stations. The results of the comparison for TPM and RGM measured at the most polluted German station are presented in Figure 13. In the case of RGM the model generally overpredicts the measurements by a factor of about 2. For TPM the agreement is much better – the difference makes up less than 25%, and the correlation factor is high (0.72). It should be kept in mind that the measurements of TPM and especially RGM are very uncertain. Hence, the mentioned disagreement can be partly explained by this fact.

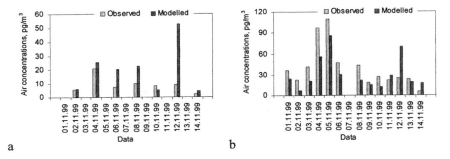

a   b

Figure 13. Modelled RGM (a) and TPM (b) concentrations by EMEP regional model against observations at German monitoring station Neuglobsow.

For the hemispheric model the results of the comparison of calculated and measured GEM concentrations are given in Figure 14a. As seen from these data, the model predicts air concentrations of mercury in background regions (~1.5 ng/m^3) rather accurately. Some underestimation of measured values takes place in the regions with an increased concentration of mercury (South-eastern Asia). In general, the difference between measured and predicted values does not exceed 30%.

The difference between predicted and measured values of the annual wet depositions of mercury is displayed in Figure 14b. The accuracy of model prediction in this case is somewhat lower, because deposition fluxes highly depend on the precipitation amount - the model input parameter with a considerable degree of uncertainty. However, in general, the ratio between measured and predicted values is close to unity, and the maximum difference between them does not exceed a factor of two.

A very useful approach to model validation is the comparison of the EMEP operational model with other scientific models used by national experts. Nine different models have been included in this intercomparison study: ADOM (Germany), CMAQ, HYSPLIT, AER (USA), GRAHM (Canada), EMAP (Bulgaria), MCM (Sweden), DEHM (Denmark) and EMEP operational model. Their descriptions can be found in (Ryaboshapko et al., 2002; Ryaboshapko et al., 2003).

At the first stage of the study only schemes of chemical transformations were compared. The results demonstrated that all the models predicted increases in mercury concentration in cloud water during the first hours of modelling experiment. The range of the predicted maximum concentrations was from 80 to 150 ng/L.

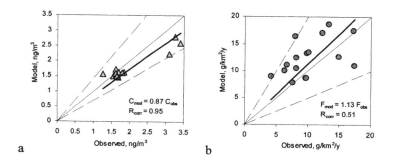

Figure 14. Comparison of measured and calculated by the hemispheric model: (a) GEM concentrations; (b) wet depositions. Dashed lines: (a) – ±30%, (b) – factor 2.

For the second stage of the comparison the data from the short-term measuring campaign mentioned above was used. Seven models calculated EGM, TPM and RGM on sample-by-sample basis for four monitoring stations in Germany, Sweden and Ireland. The scattering of the calculated values for TPM and especially for RGM in the individual samples was very high. However, the mean values for all stations and for all samples were in rather good agreement (Ryaboshapko et al., 2003). Figure 15 presents comparison results for "models vs. observations" and "models vs. models". In the case of elemental mercury all the models are in good agreement both with the observations and between each other. The data for RGM and TPM are characterised by much larger scattering. In the case of RGM the difference between the lowest and the highest modelling values reaches an order of magnitude. For all parameters of the comparison the EMEP operational model demonstrated acceptable agreement both with the observations and with the results of the other scientific models. This gives some confidence that the model can be used for the purposes of the

Convention and can provide the participating countries with information of acceptable reliability.

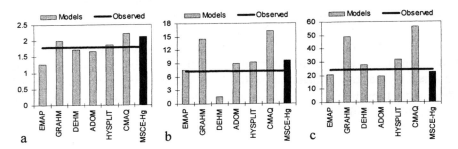

Figure 15. Comparison of different models between each other and with measurements for GEM (a), TPM (b) and RGM (c).

FURTHER DIRECTIONS OF MODEL DEVELOPMENT

It is possible to foresee two very important directions of further development of EMEP mercury models. The first one is connected with the necessity to assess mercury accumulation in environmental compartments and its secondary emission or re-emission to the atmosphere. The second direction is consideration in the modelling scheme of a newly discovered phenomenon - Arctic mercury depletion.

During a period of active usage of mercury in human activity more than one million tonnes was extracted from the lithosphere, and at least half of that came to the atmosphere (Travnikov and Ryaboshapko, 2002). A great amount of mercury was emitted in the process of coal combustion. Fitzgerald and Mason (1996) believe that 95% of previously emitted mercury has being accumulated in soil over the globe. The enhanced content of mercury in soils should inevitably lead to its re-emission to the atmosphere.

Consideration of mercury re-emission processes is very important for operational modelling of mercury transport in the atmosphere. One of the possible ways to describe this process is application of a dynamic multi-component model describing the mercury cycle in the environment during the entire period of pronounced anthropogenic impact (about 500 years Hylander & Meili, 2003). Due to very long period of supposed calculations and the contemporary level of knowledge on mercury behaviour in different environmental compartments the model cannot be very detailed. In this content a box-modelling approach seems to be the most acceptable for estimation of mercury accumulation in the environment and re-emission. On the other hand, it should have enough spatial resolution to provide input information for operational atmospheric transport models at regional and global levels. The concept of the box-modelling approach was evolved in works by Jonasson and Boyle (1971), Ribeyre et al. (1991), Hudson et al. (1995), Jackson (1997), and Lamborg et al. (2002).

The first attempt to assess mercury re-emission from European soils was done using the EMEP regional mercury model in conjunction with a simple box model (Ryaboshapko and Ilyin, 2000). Accumulated depositions during the last century were calculated by the regional model. The box model considered European soils as a single reservoir with two output fluxes – re-emission and hydrological leaching. Mercury lifetime in the box according to re-emission was assumed to be 400 yr, and according to the leaching – 950 yr. Under accepted assumptions the model predicted that by the end of 20th century the total re-emission in Europe could make up 50 t/yr. The value seems to be not very high but one should keep in mind that it is the total value for the whole European territory. In some heavily polluted areas (Eastern Germany, for example) the re-emission can nowadays exceed the current direct anthropogenic emission.

To provide more accurate re-emission assessment a modeller should possess quantitative information on the mercury cycle in soils, because the retention time of mercury in soils is the most crucial parameter. Besides, mercury deposition and accumulation should be calculated on the global scale. Finally, the fate of mercury in the environment compartments should be considered at least during last 500 years (Hylander & Meili, 2003).

Modelling assessment of the role of mercury depletion events (MDE) in the Arctic is a very challenging problem. Measurements show that MDE can provide very significant fluxes of mercury from the atmosphere into vulnerable Arctic ecosystems. To what extent this phenomenon is connected with anthropogenic influence on the mercury cycle is still an open question. Unfortunately, the mechanism of MDE is not fully understood.

Two attempts to model MDE were made recently (Christensen, 2001; Ilyin and Travnikov, 2003). EMEP calculations show that additional deposition of mercury due to MDE can be significant – about 50 t/yr or 20% of the total deposition. Figure 16 demonstrates the pattern of annual mercury deposition in the Arctic and the effect caused by MDE. One can see that the effect is the most pronounced along the shoreline of the Arctic Ocean. Here MDE can contribute more then 50% of the total deposition, and just here the life in the Arctic is the most active.

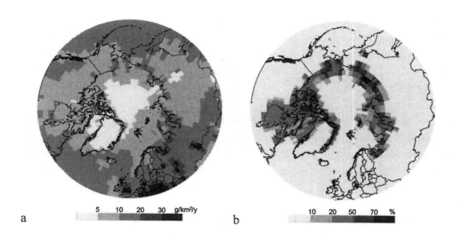

Figure 16. Spatial distribution of annual mercury depositions in the Arctic (a)
and the contribution of MDE to the total deposition (b).

The main problem for a modeller to parameterise MDE is in the fact that the real "trigger mechanism" of the phenomenon is unknown. Christensen 2001 used a certain sun zenith angle as the trigger to start MDE. In the EMEP hemispheric model (Ilyin and Travnikov, 2003) prescribed temperature changes were applied. In both cases the approaches are phenomenological. They give a possibility to tie modelled MDE with known geophysical parameters. However, they do not allow one to predict a moment of the real onset of MDE. Probably, the beginning of the depletion is connected with explosive emission into the atmosphere of some bromine species from open water when leads appear very quickly during the spring drift of Arctic ice cover.

REFERENCES

Acker, K., Möller, D., Wieprecht, W., Kalass, D., and Auel, R. Investigations of ground-based clouds at the Mt. Brocken. *Fresenius J. Anal. Chem.*, 361, 59-64, 1998.

Berdowski, J.J.M., Baas, J., Bloos, J.P.J., Visschedijk, A.J.H., Zandveld, P.Y.J. The European Emission Inventory of Heavy Metals and Persistent Organic Pollutants for 1990. TNO Institute of Environmental Sciences, Energy Research and Process Innovation, 1997; UBA-FB report 104 02 672/03, Apeldoorn, 239, 1997.

Bott, A. Monotone flix limitation in the area-preserving flux-form advection algorithm. *Month. Weath. Rev.*, 120, 2592-2602, 1992.

Brosset, C., Lord E. Mercury in precipitation and ambient air - A new scenario. *Water, Air Soil Pollut.*, 56, 493-506, 1991.

Bullock Jr. O.R., Brehme K.A. Atmospheric mercury simulation using the CMAQ model: formulation, description and analysis of wet deposition results. *Atmos. Environ.*, 36, 2135-2146, 2002.

Chin, M., Jacob D.J., Gardner, G.M., Foreman-Fowler M.S., Spiro, P.A., Savoie, D.L. A global three-dimensional model of tropospheric sulfate. *J. Geophys. Res.*, 101 (D13), 18,667-18690, 1996.

Christensen, J. Modelling of Mercury with the Danish Eulerian Hemispheric Model. The International Workshop on Trends and Effects of Heavy Metals in the Arctic, 18-22 June 2001, McLean, Virginia, USA, 2001

Easter, R.C. Two modified versions of Bott's positive-definite numerical advection scheme. *Month. Weath. Rev.*, 121, 297-304, 1993.

Fitzgerald, W.F., Mason, R.P. "The global mercury cycle: oceanic and anthropogenic aspects." In Global and Regional Mercury Cycles: Sources, Flaxes and Mass Balances." W.Baeyens, R.Ebinghaus and O.Vasiliev, eds. Dordrecht: NATO ASI Series, 2. Environment, Vol. 21., Kluwer Academic Publ., 1996; 85-108.

Guo, Y.-R., Chen, S. Terrain and land use for the fifth-generation Penn State/NCAR mesoscale modeling system (MM5): Program TERRAIN. NCAR Technical Note NCAR/TN-397+IA, National Center for Atmospheric Research, Boulder, Colorado, USA, 1994.

Hall, B. The gas phase oxidation of mercury by ozone. *Water, Air Soil Pollut.*, 80, 301-315, 1995.

Hudson, R.J.M., Gherini, S.A., Fitzgerald, W.F., Porcella, D.B. Anthropogenic influences on the global mercury cycle: a model-based analysis. *Water, Air Soil Pollut.*, 80, 265-272, 1995.

Hylander, L.D., Meili, M. 500 years of mercury production: global annual inventory by region until 2000 and associated emissions. *Sci. Tot. Env.*, 304, 13-27, 2002.

Ilyin, I., Ryaboshapko, A., Afinigenova, O., Berg, T., Hjellbrekke, A.-G. Evaluation of Transboundary Transport of Heavy Metals in 1999. Trend Analysis. EMEP MSC-E/CCC Technical Report 3/2001, 129 p., 2001.

Ilyin, I., Ryaboshapko, A., Afinigenova, O., Berg, T., Hjellbrekke, A.-G., Lee, D. Lead, Cadmium and Mecury Transboundary Pollution in 2000. EMEP MSC-E/CCC Technical Report 5/2002, 131 p., 2002.

Ilyn, I., Travnikov, O. Heavy Metals: Transboundary Pollution of the Environment. EMEP MSC-E Technical Report 5/2003, 96 p., 2003.

Iverfeldt, Å. Occurence and turnover of atmospheric mercury over the Nordic countries. *Water, Air Soil Pollut.*, 56, 252-265, 1991.

Iversen, T., Saltbones, J., Sandnes, H., Eliassen, A., Hov, O. Airborn tronsboundary transport of sulphur ans nitrogen over Europe - model descriptions and calculations. EMEP MSC-W Report 2/89, 1989; 92 p.

Jackson, T.A. Long-range atmospheric transport of mercury to ecosystems, and the importance of anthropogenic emissions - a critical review and evaluation of the published evidence. *Environ. Rev.*, 5, 99-120, 1997.

Jacobson, M.Z. Fundamentals of Atmospheric Modeling. Cambridge University Press, 656 p., 1999.

Jonasson I. R., Boyle R. W. Geochemistry of mercury. Spatial Symposium on Mercury in Man's Environment, Environment Canada, 15-16 February 1971, Ottawa, Canada, 1971.

Jonsen, J.E., Berge, E. Some preliminary results on transport and deposition of nitrogen components by use of Multilayer Eulerian Model. EMEP Meteorological Synthesizing Centre - West, Report 4/95, 25 p., 1995.

Kalnay, E., Kanamitsu, M., Kistle, R., Collins, W., Deaven, D., Gandin, L., Iredell, M., Saha S., White, G., Woollen, J., Zhu, Y., Leetmaa, A., Reynolds, R., Chelliah, M., Ebisuzaki, W., Higgins, W., Janowiak, J., Mo, K.C., Ropelewski, C., Wang, J., Jenne, R., Dennis, J. The NCEP/NCAR 40-Year Reanalysis Project. *Bull. Am. Met. Soc.*, 77, 437-471, 1996.

Keeler, G., Glinsorn, G., Pirrone N. Particulate mercury in the atmosphere: its significance, transport, transformation and sources. *Water, Air and Soil Pollut.*, 80, 159-168, 1995.

Lamborg, C.H., Fitzgerald, W.F., Vandal, G.M., Rolfhus, K.R. Atmospheric mercury in northern Wisconsin: sources and species. *Water, Air and Soil Pollut.*, 80, 189-198, 1995.

Lamborg, C.H., Fitzgerald, W.F., O'Donnell, J., Torgersen, T. A non-steady-state compartmental model of global-scale mercury biogeochemistry with interhemispheric atmospheric gradients. *Geochim. Cosmochim. Acta*, 66, 1105-1118, 2002.

Lindberg, S.E., Meyers, T.P., Taylor, G.E., Turner, R.R., Schroeder, W.H. Atmospheric-surface exhange of mercury in a forest: results of modeling and gradient approaches. *J. Geophys. Res.*, 97, 2519-2528, 1992.

Milford, J.B., Davidson, C.I. The sizes of particulate trace elements in the atmosphere – a review. *J. Air Pollut. Control Assoc.*, 35, 1249-1260, 1985.

Munthe, J., Wängberg, I., Iverfeldt, Å., Lindqvist, O., Strömberg, D., Sommar, J., Gårdfeldt, K., Petersen, G., Ebinghaus, R., Prestbo, E., Larjava, K., Siemens, V. Distribution of atmospheric mercury species in Northern Europe: Final Results from the MOE Project. *Atmos. Environ.*, 304, S9-S20, 2003.

NADP/MDN. National Atmospheric Deposition Program (NRSP-3) / Mercury Deposition Network 2002. (Online Version: http://nadp.sws.uiuc.edu/mdn/)

Pacyna, E., Pacyna, J. Global emission of mercury from anthropogenic sources in 1995. *Water, Air and Soil Pollut.*, 137, 149-165, 2002.

Pacyna, J., Pacyna, E., Steenhuisen, F., Wilson, S. Global Mercury emissions. International Workshop "Long Range Transport", 16-17 September 2003, Ann Arbor, Michigan, USA. 2003.

Pekar, M. Regional models LPMOD and ASIMD. Algorithms, parameterization and results of application to Pb and Cd in Europe scale for 1990. 1996; MSC-E/EMEP, Technical Report 9/96.

Petersen, G., Munthe, J., Pleijel, K., Bloxam, R., Kumar, A.V. A comprehensive Eulerian modeling framework for airborne mercury species: Development and testing of the tropospheric chemistry module (TCM). *Atmos. Environ.*, 32, 829-843, 1998.

Petersen, G., Bloxam, R., Wong, S., Munthe, J., Krüger, O., Schmolke, S., Kumar, A.V.. A comprehensive Eulerian modelling framework for airborne mercury species: model development and applications in Europe. *Atmos. Environ.*, 35, 3063-3074, 2001.

Posh, M., Hettelingh, J.-P., de Smet, P.A.M., Downing, R.J. Calculations and mapping of critical thresholds in Europe. Status report 1997. RIVM, National Institute of Public Health and the Environment. P.O. Box 1, 3720 BA Bilthoven, The Netherlands, 1997.

Ribeyre, F., Boudou, A., Maury-Brachet, R. Multicompartment ecotoxicological models to study mercury bioaccumulation and transfer in freshwater system. *Water, Air and Soil Pollut.*, 56, 641-652, 1991.

Ruijgrok, W., Davidson, C.I., Nicholson, K.W. Dry deposition of particles. Implications and recommendations for mapping of deposition over Europe. *Tellus*, 47B, 587 – 601, 1995.

Ryaboshapko, A., Ilyin I. Mercury re-emission to the atmosphere in Europe. In: Proceedings of EUROTRAC Symposium 2000 Transport and Chemical Transformation in the Troposphere. P.M.Midgley, M.J.Reuther, M.Williams eds. Springer-Verlag Berlin Heidelberg (CD-annex), 2001.

Ryaboshapko, A., Bullock, R., Ebinghaus, R., Ilyin, I., Lohman, K., Munthe, J., Petersen, G., Seigneur, C., Wängberg, I. Comparison of mercury chemistry models. *Atmos. Environ.*, 36, 3881- 898, 2002.

Ryaboshapko, A., Artz, R., Bullock, R, Christensen, J., Cohen, M., Dastoor, A., Davignon, D., Draxler, R., Ebinghaus, R., Ilyin, I., Munthe, J., Petersen, G., Syrakov, D. Intercomparison study of numerical models for long-range atmospheric transport of mercury. Stage II. Comparison of modeling results with observations obtained during short-term measuring campaigns. MSC-E Technical Report 1/2003, 2003.

Sander, R. Henry's law constants available on the Web. EUROTRAC Newsletter 1997; No.18, 24-25

Schmolke, S.R., Schroeder, W.H., Munthe, J., Kock, H.H., Schneeberger, D., Ebinghaus, R. Simultaneous Measurements of Total Gaseous Mercury at Four Sites on a 800 km Transect: Spatial Distribution and Short Time Variability of Total Gaseous Mercury over Central Europe. *Atmos. Environ.*, 33, 1725-1734, 1999.

Seigneur, C., Karamchandani, P., Lohman, K., Vijayaraghavan, K., Shia, R.-L. Multiscale modeling of the atmospheric fate and transport of mercury. *J. Geophys. Res.*, 106 (D21), 27795-27809, 2001.

Shia, R.-L., Seigneur, Ch., Pai, P., Ko, M., Sze, N.D. Global simulation of atmospheric mercury concentrations and deposition fluxes. *J. Geophys. Res.*, 104 (D19), 23747-23760, 1999.

Slemr, F. Trends in atmospheric mercury concentrations over the Atlantic ocean and at the Wank summit, and the resulting constraints on the budget of atmospheric mercury. In Global and Regional Mercury Cycles: Sources, Flaxes and Mass Balances. W.Baeyens, R.Ebinghaus and O.Vasiliev eds. NATO ASI Series, 2. Environment, Vol. 21. Kluwer Academic Publ., Dordrecht, 33-84, 1996.

Travnikov, O., Ryaboshapko, A. Modelling of Mercury Hemispheric Transport and Depositions. EMEP MSC-E Technical Report 6/2002, 2002, 67 p.

Wang, Y., Logan, J.A., Jacob, D.J. Global simulation of tropospheric O3-NOx-hydrocarbon chemistry, 2, Model evaluation and global ozone budget. *J. Geophys. Res.*, 103 (D9), 10727-10755, 1998.

Wesley, M.L., Hicks ,B.B. A review of the current status of knowledge on dry deposition. *Atmos. Environ.*, 34, 2261-2282, 2000.

Wängberg, I, Munthe, J., Ebinghaus, R., Gårdfeldt, K., Sommar, J. Distribution of TPM in Northern Europe. *Sci. Tot. Env.*, 304, 53-59, 2003.

Xu, X., Yang X., Miller, D.R., Helble, J.J., Carley, R.J. Formulation of bi-directional atmosphere-surface exchanges of elemental mercury. *Atmos. Environ.*, 33, 4345-4355, 1999.

ACRONYMS

ADA: *American Dental Association*
AED: *Atomic Emission Detector*
AEOLOS: *Atmospheric Exchange Over Lakes and Oceans Study*
AMAP: *Arctic Monitoring and Assessment Program*
AMCOTS: *Atmospheric Mercury Chemistry Over The Sea*
AMDES: *Atmospheric Mercury Depletion Events*
ASM: *Artisanal and Small-scale gold Mining*
ASV: *Anodic Stripping Voltammetry*
ATSDR: *Agency for Hazardous Substances and Disease Registry*
BAT: *Best Available Techniques*
BGS: *British Geological Survey*
BMD: *Benchmark Dose*
BMDL: *Benchmark Dose Level*
BSID: *Bayley Scales of Infant Development*
CBL: *Chesapeake Biological Laboratory*
CBS: *Complete Basis Set*
CCSD(T): Coupled Cluster Calculation
CDC: *Center for Desease Control and Prevention*
CI: *Chemical Ionization*
CLTRAP: *Convention on Long-Range Transport of Air Pollutants*
CMAQ: *Community Multi-scale Air Quality*
COMERN: *Canadian Collaborative Mercury Research Network*
COMTRADE: *International commercial trade statistics (UN Department of Economic and Social Affairs – Statistics Division)*
CRMs: *Certified Reference Materials*
CV AAS: *Cold Vapour Atomic Absorption Spectroscopy*
CV AFS: *Cold-Vapor Atomic Fluorescence Spectrometry*
DDST: *Denver Developmental Screen Test*
DFT: *Density Functional Theory*
DGM: *Dissolved Gaseous Mercury*
DMDCS: *Deactivating Glass Surfaces with Dimethyldichlorosilane*
DMM: *Dimethyl Mercury*
ECD: *Electron Capture Detector*
ECMWF: *European Centre for Medium range Weather Forecasting*
EFSA: *European Food Safety Authority*
EI: *Electron Impact*
ELV: *End-of-Life Vehicles*
E-MCM: *Everglades Mercury Cycling Model*

EMEP: *Co-operative programme for monitoring and evaluation of long-range transmissions of air pollutants in Europe*
EPER: *European Pollutant Emission Register*
ERA: *European Research Area*
ES: *Electron Spray*
EU: *European Union*
EXAFS: *Extended X-ray Absorption Fine Structure*
FCI: *Full Configuration Interaction*
FGD: *Flue Gas Desulfurisation*
GC-MSD: *Gas Chromatograph with Mass Selective Detection*
GEF: *Global Environmental Facility*
GEM: *Gaseous Elementary Mercury*
GEMDEs: *Gaseous Elemental Mercury Depletion Events*
GLAMAP: *Great Lakes Atmospheric Mercury Assessment Project*
GLC: *Gas-Liquid Chromatography*
GMP: *Global Mercury Project*
HPLC: *High Performance Liquid Chromatography*
HRICP-MS: *High Resolution Inductively Coupled Plasma Mass Spectrometry*
ICP-MS: *Inductively Coupled Mass Spectrometry*
ICP-MS: *Inductively Coupled Plasma Mass Spectrometer*
ICZM: *Integrated Coastal Zone Management*
IDA: *Isotope Dilution Analysis*
IPPC: *Integrated Pollution Prevention and Control*
JECFA: *Joint FAO/WHO Expert Committee on Food Additives*
LEAF-2: *Land Ecosystem Atmosphere Feedback version 2*
LIF: *Laser Induced Fluorescence*
LMMBS: *Lake Michigan Mass Balance Study*
LMUATS: *Lake Michigan Urban Air Toxics Study*
LOEL: *Lowest-Observed-Effect Level*
LRTAP: *Long-Range Transboundary Air Pollution*
LWC: *Liquid Water Content*
MALDI-TOF-MS: *Matrix-Assisted Laser Desorption Ionization Time of Flight Mass Spectrometry*
MAMCS: *Mediterranean Atmospheric Mercury Cycle System*
MB/WW: *Mass Burn/Waterwall*
MBL: *Marine Boundary Layer*
MBR: *Bowen-Ratio Method*
MCCAP: *Mercury Cell Chlor-Alkali Plant*
MCM: *Mercury Cycling Model*
MDEs: *Mercury Depletion Events*
MDN: *Mercury Deposition Network*

MERCYMS: *An Integrated Approach to Assess the Mercury Cycle into the Mediterranean Basin*

METAALICUS: *Mercury Experiment To Assess Atmospheric Loading in Canada and the United States*

MMSD: *Mining, Minerals and Sustainable Development*

MOE: *Mercury Over Europe*

MSC-East: *Meteorological Synthesizing Center – East*

MWC: *Municipal Waste Combustion*

MWI: *Medical Waste Incineration*

MWSS: *Mercury Water Surface Sampler*

NAA: *Neutron Activation Analyses*

NCAR: *National Center for Atmospheric Research*

NCEP: *National Centers for Environmental Predictions*

NHANES: *National Health and Examination Surveys*

NIMD: *National Institute for Minamata Disease*

NOAEL: *No-Adverse-Effect Level*

NOS: *Neurological Optimality Score*

ODEs: *Ordinary Differential Equations*

OECD: *Organisation for Economic Co-operation and Development*

PAH: *Polycyclic Aromatic Hydrocarbons*

PBL: *Planetary Boundary Layer*

PBTs: *Bioaccumulative Persistent Pollutants*

PCBs: *Polychlorinated Biphenyls*

PPs: *Pseudopotentials*

PSCF: model *Potential Source Contribution Function*

PTWI: *Provisional Tolerable Weekly Intake*

QA: *Quality Assurance*

QC: *Quality Control*

QTBA: *Quantitative Transport Bias Analysis*

RAMS: *Regional Atmospheric Modelling System*

RBC: *Red Blood Cells*

RECPs: *Relativistic Effective Core Potentials*

RfD: *Reference Dose*

RGM: *Reactive Gaseous Mercury*

RoHS: *Restriction of Hazardous Substances*

RPF: *Regenerable Particulate Filter*

RRKM: theory *Rice-Ramsberger-Kassel-Marcus theory*

SCOEL: *Scientific Committee on Occupational Exposure Limits*

SDA: *Atmosphere Diagnostics System*

SIM: *Scanning or Single Ion Monitoring*

TAFIRI: *Tanzania Fisheries Research Institute*

TAM: *Total Atmospheric Mercury*

TEOM: *Tapered Element Oscillating Microbalance*
TGM: *Total Gaseous Mercury*
TKE: *Turbulent Kinetic Energy*
TLC: *Thin Layer Chromatography*
TPM: *Total Particulate Mercury*
TST: *Transition State Theory*
UMAQL: *University of Michigan Air Quality Laboratory*
UN-ECE: *United Nations Economic Commission for Europe*
UNEP: *United Nations Environmental Programme*
UNESCO: *United Nations Educational Scientific and Cultural Organization*
UNIDO: *United Nations Industrial Development Organization*
US EPA: *United States Environmental Protection Agency*
US FDA: *United States Food and Drug Administration*
US NIEHS: *United States National Institute of Environmental Health Sciences*
VCM: *Vinyl Chloride Monomer*
VTST: *Variational Transition State Theory*
VWM: *Volume Weighted Mean*
WEEE: *Waste Electrical and Electronic Equipment*
WHO: *World Health Organization*
WISC-R: *Wechsler Intelligence Scale for Children*
WMS: *Wechsler Memory Scale*
XAS *X-ray: Absorption Spectroscopy*
ZAAS-HFM: *Zeeman Atomic Absorption Spectrometry using High Frequency Modulation*

SUBJECT INDEX